AF358834

Cellular and Molecular Mechanisms in Oxidative Stress-Related Diseases 2.0

Cellular and Molecular Mechanisms in Oxidative Stress-Related Diseases 2.0

Editors

Rossana Morabito
Alessia Remigante

Basel • Beijing • Wuhan • Barcelona • Belgrade • Novi Sad • Cluj • Manchester

Editors
Rossana Morabito
Department of Chemical,
Biological, Pharmaceutical
and Environmental Sciences
University of Messina
Messina
Italy

Alessia Remigante
Department of Chemical,
Biological, Pharmaceutical
and Environmental Sciences
University of Messina
Messina
Italy

Editorial Office
MDPI
St. Alban-Anlage 66
4052 Basel, Switzerland

This is a reprint of articles from the Special Issue published online in the open access journal *International Journal of Molecular Sciences* (ISSN 1422-0067) (available at: www.mdpi.com/journal/ijms/special_issues/Q57DW938K9).

For citation purposes, cite each article independently as indicated on the article page online and as indicated below:

Lastname, A.A.; Lastname, B.B. Article Title. *Journal Name* **Year**, *Volume Number*, Page Range.

ISBN 978-3-7258-1374-2 (Hbk)
ISBN 978-3-7258-1373-5 (PDF)
doi.org/10.3390/books978-3-7258-1373-5

Contents

International Journal of
Molecular Sciences

Editorial

Cellular and Molecular Mechanisms in Oxidative Stress-Related Diseases 2.0/3.0

Alessia Remigante * and Rossana Morabito

Department of Chemical, Biological, Pharmaceutical and Environmental Sciences, University of Messina, 98166 Messina, Italy; rmorabito@unime.it
* Correspondence: aremigante@unime.it

Oxidative stress is frequently described as the balance between the production of reactive species (including oxygen and nitrogen) in biological systems and the ability of the latter to defend itself through the sophisticated antioxidant machinery. At physiological levels, some oxidants, in controlled amounts, possess important signaling functions within the cell [1,2]. Specifically, cells can generate reactive species with the function of second messengers, using them for intracellular signaling and for stimulating the redox-sensitive signaling pathways to modify the cellular content of the cytoprotective regulatory proteins [3]. In fact, the redox state of the cell is normally regulated by a complex endogenous antioxidant system, composed of proteins with enzymatic activity and non-enzymatic proteins able to quickly neutralize, or ensure a low production of, reactive species [4]. Nevertheless, when oxidants are produced in excess, or when the antioxidant defenses that regulate them are ineffective, this balance can be perturbed, thus resulting in an oxidative condition. Oxidative products are highly reactive, and can directly or indirectly modulate the functions of many enzymes and transcription factors through complex signaling cascades. In particular, some of the pathways are preferentially linked to enhanced survival, while others are more frequently associated with cell death, and constitute important avenues for therapeutic interventions aimed at limiting oxidative damage or, alternatively, attenuating its consequences [5–11]. Furthermore, the magnitude and exposure of the insult, as well as the cell type involved, are key elements in defining which pathways are activated, as well as the final cell outcome. This Special Issue has been conceived to collect and contribute to the dissemination of novel findings unraveling the impact of oxidative stress on cells, their subcellular components, and biological macromolecules. Here, we offer an overview of the content of this Special Issue, which collects 13 original articles and seven reviews.

1. Original Articles

(1) "Polymorphisms in Genes Encoding VDR, CALCR and Antioxidant Enzymes as Predictors of Bone Tissue Condition in Young, Healthy Men" by Jowko, E. and collaborators [12]. The aim of the study was to assess significant predictors of bone mineral content (BMC) and bone mineral density (BMD) in a group of young, healthy men at the time of reaching peak bone mass. Regression analyses showed that age, BMI, and practicing combat and team sports at a competitive level (trained vs. untrained group; TR vs. CON, respectively) were positive predictors of BMD/BMC values at various skeletal sites. In addition, genetic polymorphisms were among the predictors. In the whole population studied, at almost all measured skeletal sites, the SOD2 AG genotype proved to be a negative predictor of BMC, while the VDR FokI GG genotype was a negative predictor of BMD. In contrast, the CALCR AG genotype was a positive predictor of arm BMD. ANOVA analyses showed that, regarding the SOD2 polymorphism, the TR group was responsible for the significant intergenotypic differences in BMC that were observed in the whole study population (i.e., lower BMC values of leg, trunk, and whole body were observed in AG TR compared to AA TR). On the other hand, a higher BMC at L1–L4 was observed in the SOD2 GG

genotype of the TR group compared to in the same genotype of the CON group. For the FokI polymorphism, the BMD at L1–L4 was higher in AG TR than in AG CON. In turn, the CALCR AA genotype in the TR group had higher arm BMD compared to the same genotype in the CON group. In conclusion, SOD2, VDR FokI, and CALCR polymorphisms seem to affect the association of BMC/BMD values with training status. In general, at least within the VDR FokI and CALCR polymorphisms, less favorable genotypes in terms of BMD (i.e., FokI AG and CALCR AA) appear to be associated with a greater BMD response to sports training. This suggests that, in healthy men during the period of bone mass formation, sports training (combat and team sports) may attenuate the negative impact of genetic factors on bone tissue condition, possibly reducing the risk of osteoporosis in later age.

(2) "Usefulness of Urinary Biomarkers for Assessing Bladder Condition and Histopathology in Patients with Interstitial Cystitis/Bladder Pain Syndrome" by Jiang, Y. H. and collaborators [13]. This study investigated the usefulness of urinary biomarkers for assessing the bladder condition and histopathology in patients with interstitial cystitis/bladder pain syndrome (IC/BPS). We retrospectively enrolled 315 patients (267 women and 48 men) diagnosed with IC/BPS and 30 controls. Data on clinical and urodynamic characteristics (visual analog scale (VAS) score and bladder capacity) and cystoscopic hydrodistention findings (Hunner's lesion, glomerulation grade, and maximal bladder capacity (MBC)) were recorded. Urine samples were utilized to assay inflammatory, neurogenic, and oxidative stress biomarkers, including interleukin (IL)-8, C-X-C motif chemokine ligand 10 (CXCL10), monocyte chemoattractant protein-1 (MCP-1), brain-derived neurotrophic factor (BDNF), eotaxin, IL-6, macrophage inflammatory protein 1 beta (MIP-1β), regulated on activation, normal T cell expressed and secreted (RANTES), tumor necrosis factor-alpha (TNF-α), prostaglandin E2 (PGE2), 8-hydroxy-2′-deoxyguanosine (8-OHdG), 8-isoproatane, and total antioxidant capacity. Further, specific histopathological findings were identified via bladder biopsies. The associations between urinary biomarker levels, bladder conditions, and histopathological findings were evaluated. The results reveal that patients with IC/BPS had significantly higher urinary MCP-1, eotaxin, TNF-α, PGE2, 8-OHdG, and 8-isoprostane levels than the controls. Patients with Hunner's IC (HIC) had significantly higher IL-8, CXCL10, BDNF, eotaxin, IL-6, MIP-1β, and RANTES levels than those with non-Hunner's IC (NHIC). Patients with NHIC who had an MBC of $\leq$760 mL had significantly high urinary CXCL10, MCP-1, eotaxin, IL-6, MIP-1β, RANTES, PGE2, and 8-isoprostane levels and total antioxidant capacity. Patients with NHIC who had a higher glomerulation grade had significantly high urinary MCP-1, IL-6, RANTES, 8-OHdG, and 8-isoprostane levels. A significant association was observed between urinary biomarkers and glomerulation grade, MBC, VAS score, and bladder sensation. However, bladder-specific histopathological findings were not well correlated with urinary biomarker levels. The urinary biomarker levels can be useful for identifying HIC and different NHIC subtypes. Higher urinary inflammatory and oxidative stress biomarker levels are associated with IC/BPS. Most urinary biomarkers are not correlated with specific bladder histopathological findings; nevertheless, they are more important in the assessment of bladder condition than in bladder histopathology.

(3) "Endoplasmic Reticulum Stress Promotes the Expression of TNF-α in THP-1 Cells by Mechanisms Involving ROS/CHOP/HIF-1α and MAPK/NF-κB Pathways" by Akhter, N. and collaborators [14]. Obesity and metabolic syndrome involve chronic low-grade inflammation, called metabolic inflammation, as well as metabolic derangements from increased endotoxin and free fatty acids. It is debated whether the endoplasmic reticulum (ER) stress in monocytic cells can contribute to amplify metabolic inflammation and, if so, by which mechanism(s). To test this, metabolic stress was induced in THP-1 cells and primary human monocytes by treatments with lipopolysaccharide (LPS), palmitic acid (PA), or oleic acid (OA), in the presence or absence of

the ER stressor thapsigargin (TG). Gene expression of tumor necrosis factor (*TNF*)-α and markers of ER/oxidative stress were determined by qRT-PCR, TNF-α protein by ELISA, reactive oxygen species (ROS) by DCFH-DA assay, hypoxia-inducible factor 1-alpha (HIF-1α), p38, extracellular signal-regulated kinase (ERK)-1,2, and nuclear factor kappa B (NF-κB) phosphorylation by immunoblotting, and insulin sensitivity by glucose-uptake assay. Regarding clinical analyses, adipose TNF-α was assessed using qRT-PCR/IHC and plasma TNF-α, high-sensitivity C-reactive protein (hs-CRP), malondialdehyde (MDA), and oxidized low-density lipoprotein (OX-LDL) via ELISA. We found that the cooperative interaction between metabolic and ER stresses promoted TNF-α, ROS, CCAAT-enhancer-binding protein homologous protein (*CHOP*), activating transcription factor 6 (*ATF6*), superoxide dismutase 2 (*SOD2*), and nuclear factor erythroid 2-related factor 2 (*NRF2*) expression ($p \leq 0.0183$). However, glucose uptake was not impaired. TNF-α amplification was dependent on HIF-1α stabilization and p38 MAPK/p65 NF-κB phosphorylation, while the MAPK/NF-κB pathway inhibitors and antioxidants/ROS scavengers, such as curcumin, allopurinol, and apocynin, attenuated the TNF-α production ($p \leq 0.05$). Individuals with obesity displayed increased adipose TNF-α gene/protein expression, as well as elevated plasma levels of TNF-α, CRP, MDA, and OX-LDL ($p \leq 0.05$). Our findings support a metabolic–ER stress cooperativity model, favoring inflammation by triggering TNF-α production via the ROS/CHOP/HIF-1α and MAPK/NF-κB dependent mechanisms. This study also highlights the therapeutic potential of antioxidants in inflammatory conditions involving metabolic/ER stresses.

(4) "Impact of Truncated Oxidized Phosphatidylcholines on Phospholipase A2 Activity in Mono- and Polyunsaturated Biomimetic Vesicles" by Yordanova, V. and collaborators [15]. The interplay between inflammatory and redox processes is a ubiquitous and critical phenomenon in cell biology that involves numerous biological factors. Among them, secretory phospholipases A2 (sPLA2), which catalyze the hydrolysis of the sn-2 ester bond of phospholipids, are key players. They can interact or be modulated by the presence of truncated oxidized phosphatidylcholines (OxPCs) produced under oxidative stress from phosphatidylcholine (PC) species. The present study examined this important, but rarely considered, sPLA2 modulation induced by changes in the biophysical properties of PC vesicles comprising various OxPC ratios in mono- or poly-unsaturated PCs. Being the most physiologically active OxPCs, 1-palmitoyl-2-(5′-oxo-valeroyl)-sn-glycero-3-phosphocholine (POVPC) and 1-palmitoyl-2-glutaryl-sn-glycero-3-phosphocholine (PGPC) have been selected for this study. Using fluorescence spectroscopy methods, we compared the effect of OxPCs on the lipid order and sPLA2 activity in large unilamellar vesicles (LUVs) made of the heteroacid PC, either monounsaturated (1-palmitoyl-2-oleoyl-sn-glycero-3-phosphocholine (POPC)) or polyunsaturated (1-palmitoyl-2-docosahexaenoyl-sn-glycero-3-phosphocholine (PDPC)), at a physiological temperature. The effect of OxPCs on vesicle size was also assessed in both the mono- and polyunsaturated PC matrices. The results showed that OxPCs decrease the membrane lipid order of POPC and PDPC mixtures with PGPC, inducing a much larger decrease in comparison with POVPC, indicative that the difference takes place at the glycerol level. Compared to POPC, PDPC was able to inhibit sPLA2 activity, showing a protective effect of PDPC against enzyme hydrolysis. Furthermore, sPLA2 activity on its PC substrates was modulated by the OxPC membrane content. POVPC down-regulated sPLA2 activity, suggesting anti-inflammatory properties of this truncated oxidized lipid. Interestingly, PGPC had a dual and opposite effect, either inhibiting or enhancing sPLA2 activity, depending on the protocol of lipid mixing. This difference may result from the chemical properties of the shortened sn-2-acyl chain residues (aldehyde group for POVPC, and carboxyl for PGPC), being, respectively, zwitterionic or anionic under hydration at physiological conditions.

(5) "Electrophoretic Determination of L-Carnosine in Health Supplements Using an Integrated Lab-on-a-Chip Platform with Contactless Conductivity Detection" by Pukleš, I. and collaborators [16]. The health supplement industry is one of the fastest growing industries in the world, but there is a lack of suitable analytical methods for the determination of active compounds in health supplements, such as peptides. The present work describes an implementation of contactless conductivity detection on microchip technology as a new strategy for the electrophoretic determination of L-carnosine in complex health supplement formulations without pre-concentration and derivatization steps. The best results were obtained in the case of +1.00 kV applied for 20 s for injection and +2.75 kV applied for 260 s for the separation step. Under the selected conditions, a linear detector response of 5×10^{-6} to 5×10^{-5} M was achieved. L-carnosine retention time was 61 s. The excellent reproducibility of both migration time and detector response confirmed the high precision of the method. The applicability of the method was demonstrated by the determination of L-carnosine in three different samples of health supplements. The recoveries ranged from 91 to 105%. Subsequent analyses of the samples by CE-UV-VIS and HPLC-DAD confirmed the accuracy of the obtained results.

(6) "Evaluation of Oxidative Stress and Metabolic Profile in a Preclinical Kidney Transplantation Model According to Different Preservation Modalities" by Mrakic-Sposta and collaborators [17]. This study addresses a joint nuclear magnetic resonance (NMR) and electron paramagnetic resonance (EPR) spectroscopy approach to providing a platform for the dynamic assessment of kidney viability and metabolism. On porcine kidney models, ROS production, oxidative damage kinetics, and metabolic changes occurring both during the period between organ retrieval and im- plantation and after kidney graft were examined. The ^{1}H-NMR metabolic profile-valine, alanine, acetate, trimetylamine-N-oxide, glutathione, lactate, and the EPR oxidative stress resulting from ischemia/reperfusion injury after preservation (8 h) by static cold storage (SCS) and ex vivo machine perfusion (HMP) methods were monitored. The functional recovery after transplantation (14 days) was evaluated by serum creatinine (SCr), oxidative stress (ROS), and damage (thiobarbituric-acid reactive substances and protein carbonyl enzymatic) assessments. At 8 h of preservation storage, a significantly ($p < 0.0001$) higher ROS production was measured in the SCS group, as compared to the HMP group. Significantly higher concentration data ($p < 0.05$–0.0001) in HMP vs. SCS for all the monitored metabolites were found as well. The HMP group showed a better function recovery. The comparison of the areas under the SCr curves (AUCs) returned a significantly smaller (-12.5%) AUC in the HMP vs. SCS. The EPR-ROS concentrations ($\mu mol \cdot g^{-1}$) from bioptic kidney tissue samples were significantly lower in HMP vs. SCS. The same result was found for the NMR monitored metabolites: lactate: -59.76%, alanine: -43.17%; valine: -58.56%; and TMAO: -77.96%. No changes were observed in either group under light microscopy. In conclusion, a better and more rapid normalization of oxidative stress and functional recovery after transplantation were observed by HMP utilization.

(7) "Protective Effect of Resveratrol in an Experimental Model of Salicylate-Induced Tinnitus" by Song, A. and collaborators [18]. To date, the effect of resveratrol on tinnitus has not been reported. The attenuative effects of resveratrol (RSV) on a salicylate-induced tinnitus model were evaluated by in vitro and in vivo experiments. The gene expression of the activity-regulated cytoskeleton-associated protein (*ARC*), tumor necrosis factor-alpha (TNF-α), and NMDA receptor subunit 2B (*NR2B*) in SH-SY5Y cells was examined using qPCR. Phosphorylated cAMP response element-binding protein (p-CREB), apoptosis markers, and reactive oxygen species (ROS) were evaluated by in vitro experiments. The in vivo experiment evaluated the gap-prepulse inhibition of the acoustic startle reflex (GPIAS) and auditory brainstem response (ABR) level. The *NR2B* expression in the auditory cortex (AC) was determined by immunohistochemistry. RSV significantly reduced the salicylate-induced expression

of *NR2B*, *ARC*, and TNF-α in neuronal cells; the GPIAS and ABR thresholds altered by salicylate in rats were recovered close to their normal range. RSV also reduced the salicylate-induced NR2B overexpression of the AC. These results confirmed that resveratrol exerted an attenuative effect on salicylate-induced tinnitus, and may have therapeutic potential.

(8) "Molecular Mechanisms of Oxidative Stress Relief by CAPE in ARPE−19 Cells" by Ren, C. and collaborators [19]. Caffeic acid phenylethyl ester (CAPE) is an antioxidative agent originally derived from propolis. Oxidative stress is a significant pathogenic factor in most retinal diseases. Our previous study revealed that CAPE suppresses mitochondrial ROS production in ARPE−19 cells by regulating UCP2. The present study explores the ability of CAPE to provide longer-term protection to RPE cells and the underlying signal pathways involved. ARPE−19 cells were given CAPE pretreatment, followed by t-BHP stimulation. We used in situ live cell staining with CellROX and MitoSOX to measure ROS accumulation; an Annexin V-FITC/PI assay to evaluate cell apoptosis; ZO−1 immunostaining to observe the tight junction integrity in the cells; RNA-seq to analyze changes in gene expression; q-PCR to validate the RNA-seq data; and a Western blot to examine MAPK signal pathway activation. CAPE significantly reduced both cellular and mitochondria ROS overproduction, restored the loss of ZO−1 expression, and inhibited apoptosis induced by t-BHP stimulation. We also demonstrated that CAPE reverses the overexpression of immediate early genes (IEGs) and the activation of the p38-MAPK/CREB signal pathway. Either genetic or chemical deletion of UCP2 largely abolished the protective effects of CAPE. CAPE restrained ROS generation and preserved the tight junction structure of ARPE−19 cells against oxidative-stress-induced apoptosis. These effects were mediated via UCP2 regulation of p38/MAPK-CREB-IEGs pathway.

(9) "The Critical Assessment of Oxidative Stress Parameters as Potential Biomarkers of Carbon Monoxide Poisoning" by Hydzik, P. and collaborators [20]. In conventional clinical toxicology practice, the blood level of carboxyhemoglobin is a biomarker of carbon monoxide (CO) poisoning, but does not correspond to the complete clinical picture and the severity of the poisoning. Taking into account articles suggesting the relationship between oxidative stress parameters and CO poisoning, it seems reasonable to consider this topic more broadly, including experimental biochemical data (oxidative stress parameters) and patients poisoned with CO. This article aimed to critically assess oxidative-stress-related parameters as potential biomarkers to evaluate the severity of CO poisoning, and their possible role in the decision to treat. The critically set parameters were antioxidative, including catalase, 2,2-diphenyl-1-picryl-hydrazyl, glutathione, thiol, and carbonyl groups. The preliminary studies involved patients (n = 82) admitted to the Toxicology Clinical Department of the University Hospital of Jagiellonian University Medical College (Kraków, Poland) during 2015–2020. The poisoning was diagnosed based on medical history, clinical symptoms, and carboxyhemoglobin blood level. Blood samples for carboxyhemoglobin and antioxidative parameters were collected immediately after admission to the emergency department. To evaluate the severity of the poisoning, the Pach scale was applied. The final analysis included a significant decrease in catalase activity and a reduction in glutathione level in all poisoned patients based on the severity of the Pach scale (I°–III°) compared to the control group. It follows from the experimental data that the poisoned patients had a significant increase in level due to thiol groups and the 2,2-diphenyl-1-picryl-hydrazyl radical, with no significant differences according to the severity of poisoning. The catalase-to-glutathione and thiol-to-glutathione ratios showed the most important differences between the poisoned patients and the control group, with a significant increase in the poisoned group. The ratios did not differentiate the severity of the poisoning. The carbonyl level was highest in the control group compared to the poisoned group, but was not statistically significant. Our

critical assessment shows that using oxidative-stress-related parameters to evaluate the severity of CO poisoning, the outcome, and treatment options is challenging.

(10) "Microplastics Exacerbate Cadmium-Induced Kidney Injury by Enhancing Oxidative Stress, Autophagy, Apoptosis, and Fibrosis" by Zou, H. and collaborators [21]. Cadmium (Cd) is a potential pathogenic factor in the urinary system that is associated with various kidney diseases. Microplastics (MPs), comprising of plastic particles less than 5 mm in diameter, are a major carrier of contaminants. We applied 10 mg/L particle 5 μm MPs and 50 mg/L CdCl2 in water for a three month in vivo assay to assess the damaging effects of MPs and Cd exposure on the kidneys. In vivo tests showed that MPs exacerbated Cd-induced kidney injury. In addition, the involvement of oxidative stress, autophagy, apoptosis, and fibrosis in the damaging effects of MPs and Cd on mouse kidneys were investigated. The results showed that MPs aggravated Cd-induced kidney injury by enhancing oxidative stress, autophagy, apoptosis, and fibrosis. These findings provide new insights into the toxic effects of MPs on mouse kidneys.

(11) "Toxicity of Metal Ions Released from a Fixed Orthodontic Appliance to Gastrointestinal Tract Cell Lines" by Durgo, K. and collaborators [22]. The mechanisms of toxicity and cellular responses to metal ions present in the environment are still a very current area of research. In this work, which is a continuation of the study of the toxicity of metal ions released by fixed orthodontic appliances, eluates of archwires, brackets, ligatures, and bands are used to test the prooxidant effect, cytotoxicity, and genotoxicity on cell lines of the gastrointestinal tract. Eluates obtained after three immersion periods (3, 7, and 14 days) and with known amounts and types of metal ions were used. Four cell lines—CAL 27 (human tongue), Hep-G2 (liver), AGS (stomach), and CaCo-2 (colon)—were treated with each type of eluate at four concentrations ($0.1\times$, $0.5\times$, $1.0\times$, and $2.0\times$) for 24 h. Most eluates had toxic effects on CAL 27 cells over the entire concentration range, regardless of exposure time, while CaCo-2 proved to be the most resistant. In AGS and Hep-G2 cells, all samples tested induced free radical formation, with the highest concentration ($2\times$) causing a decrease in free radicals formed, compared to the lowest concentrations. Eluates containing Cr, Mn, and Al showed a slight pro-oxidant effect on DNA (on plasmid φX-174 RF I) and slight genotoxicity (comet assay), but these effects are not so great that the human body could not "resist" them. Statistical analyses of the data on chemical composition, cytotoxicity, ROS, genotoxicity, and prooxidative DNA damage shows the influence of metal ions present in some eluates on the toxicity obtained. Fe and Ni are responsible for the production of ROS, while Mn and Cr have a great influence on hydroxyl radicals, which cause single-strand breaks in supercoiled plasmid DNA, in addition to the production of ROS. On the other hand, Fe, Cr, Mn, and Al are responsible for the cytotoxic effect of the studied eluates. The obtained results confirm that this type of research is useful and brings us closer to more accurate in vivo conditions.

(12) "Lead Exposure Causes Spinal Curvature during Embryonic Development in Zebrafish" by Li, X. and collaborators [23]. Lead (Pb^{2+}) is an important raw material for modern industrial production, which enters the aquatic environment in several ways and cause serious harm to aquatic ecosystems. Lead ions (Pb^{2+}) are highly toxic, and can accumulate continuously in organisms. In addition to causing biological deaths, it can also cause neurological damage in vertebrates. Our experiment found that Pb^{2+} caused decreased survival, delayed hatching, decreased the frequency of voluntary movements at 24 hpf, increased the heart rate at 48 hpf, and increased the malformation rate in zebrafish embryos. Among them, the morphology of spinal malformations varied, with 0.4 mg/L Pb^{2+} causing a dorsal bending of the spine of 72 hpf zebrafish and a ventral bending in 120 hpf zebrafish. It was determined that spinal malformations were mainly caused by Pb-induced endoplasmic reticulum stress and apoptosis. The genetic changes in somatic segment development, which disrupted developmental polarity as well as osteogenesis, resulted in uneven my-

otomal development. In contrast, calcium ions can rescue the series of responses induced by lead exposure and reduce the occurrence of spinal curvature. This article proposes new findings for lead pollution toxicity in zebrafish.

(13) "Anti-*Candida albicans* Effects and Mechanisms of Theasaponin E1 and Assamsaponin A" by Chen, Y. and collaborators [24]. *Candida albicans* is an opportunistic human fungal pathogen, and its drug resistance is becoming a serious problem. *Camellia sinensis* seed saponins showed inhibitory effects on resistant *Candida albicans* strains, but the active components and mechanisms are unclear. In this study, the effects and mechanisms of two *Camellia sinensis* seed saponin monomers, theasaponin E1 (TE1) and assamsaponin A (ASA), on a resistant *Candida albicans* strain (ATCC 10231) were explored. The minimum inhibitory concentration and minimum fungicidal concentration of TE1 and ASA were equivalent. The time–kill curves showed that the fungicidal efficiency of ASA was higher than that of TE1. TE1 and ASA significantly increased the cell membrane permeability and disrupted the cell membrane integrity of *C. albicans* cells, probably by interacting with membrane-bound sterols. Moreover, TE1 and ASA induced the accumulation of intracellular ROSs and decreased the mitochondrial membrane potential. Transcriptome and qRT-PCR analyses revealed that the differentially expressed genes were concentrated in the cell wall, plasma membrane, glycolysis, and ergosterol synthesis pathways. In conclusion, the antifungal mechanisms of TE1 and ASA included the interference with the biosynthesis of ergosterol in fungal cell membranes, damage to the mitochondria, and the regulation of energy metabolism and lipid metabolism. Tea seed saponins have the potential to be novel anti-*Candida albicans* agents.

2. Reviews

(1) "Significance of Singlet Oxygen Molecule in Pathologies" by Murotomi, K. and collaborators [25]. Reactive oxygen species, including singlet oxygen, play an important role in the onset and progression of disease, as well as in aging. Singlet oxygen can be formed non-enzymatically by chemical, photochemical, and electron transfer reactions, or as a byproduct of endogenous enzymatic reactions in phagocytosis during inflammation. The imbalance of antioxidant enzymes and antioxidant networks with the generation of singlet oxygen increases oxidative stress, resulting in the undesirable oxidation and modification of biomolecules, such as proteins, DNA, and lipids. This review describes the molecular mechanisms of singlet oxygen production in vivo and methods for the evaluation of damage induced by singlet oxygen. The involvement of singlet oxygen in the pathogenesis of skin and eye diseases is also discussed from the biomolecular perspective. We also present our findings on lipid oxidation products derived from singlet oxygen-mediated oxidation in glaucoma, early diabetes patients, and a mouse model of bronchial asthma. Even in these diseases, oxidation products, due to singlet oxygen, have not been measured clinically. This review discusses their potential as biomarkers for diagnosis. Recent developments in singlet oxygen scavengers, such as carotenoids, which can be utilized to prevent the onset and progression of disease, are also described.

(2) "Manipulation of Oxidative Stress Responses by Non-Thermal Plasma to Treat Herpes Simplex Virus Type 1 Infection and Disease" by Sutter, J. and collaborators [26]. Herpes simplex virus type 1 (HSV-1) is a contagious pathogen with a large global footprint, due to its ability to cause lifelong infection in patients. Current antiviral therapies are effective in limiting viral replication in the epithelial cells to alleviate clinical symptoms, but ineffective in eliminating latent viral reservoirs in neurons. Much of HSV-1 pathogenesis is dependent on its ability to manipulate oxidative stress responses to craft a cellular environment that favors HSV-1 replication. However, to maintain redox homeostasis and to promote antiviral immune responses, the infected cell can upregulate reactive oxygen and nitrogen species (RONS) while having a tight control on antioxidant concentrations to prevent cellular damage. Non-thermal

plasma (NTP), which we propose as a potential therapy alternative directed against HSV-1 infection, is a means to deliver RONS that affect redox homeostasis in the infected cell. This review emphasizes how NTP can be an effective therapy for HSV-1 infections through the direct antiviral activity of RONS and via immunomodulatory changes in the infected cells that will stimulate anti-HSV-1 adaptive immune responses. Overall, NTP applications can control HSV-1 replication and address the challenges of latency by decreasing the size of the viral reservoir in the nervous system.

(3) "Recent Developments in the Understanding of Immunity, Pathogenesis and Management of COVID-19" by Yegiazaryan, A. and collaborators [27]. Coronaviruses represent a diverse family of enveloped positive-sense single stranded RNA viruses. COVID-19, caused by Severe Acute Respiratory Syndrome Coronavirus-2, is a highly contagious respiratory disease transmissible mainly via close contact and respiratory droplets, which can result in severe, life-threatening respiratory pathologies. It is understood that glutathione, a naturally occurring antioxidant known for its role in immune response and cellular detoxification, is the target of various proinflammatory cytokines and transcription factors resulting in the infection, replication, and production of reactive oxygen species. This leads to more severe symptoms of COVID-19 and increased susceptibility to other illnesses, such as tuberculosis. The emergence of vaccines against COVID-19, the usage of monoclonal antibodies as treatments for infection, and the implementation of pharmaceutical drugs, have been effective methods for preventing and treating symptoms. However, with the mutating nature of the virus, other treatment modalities have been studied. With its role in antiviral defense and immune response, glutathione has been heavily explored in regard to COVID-19. Glutathione has demonstrated protective effects on inflammation and downregulation of reactive oxygen species, thereby resulting in less severe symptoms of COVID-19 infection and warranting the discussion of glutathione as a treatment mechanism.

(4) "The Yin and Yang Effect of the Apelinergic System in Oxidative Stress" by Fibbi, B. and collaborators [28]. Apelin is an endogenous ligand for the G protein-coupled receptor APJ and has multiple biological activities in human tissues and organs, including the heart, blood vessels, adipose tissue, central nervous system, lungs, kidneys, and liver. This article reviews the crucial role of apelin in regulating oxidative-stress-related processes by promoting prooxidant or antioxidant mechanisms. Following the binding of APJ to different active apelin isoforms and the interaction with several G proteins according to cell types, the apelin/APJ system is able to modulate different intracellular signaling pathways and biological functions, such as vascular tone, platelet aggregation and leukocytes adhesion, myocardial activity, ischemia/reperfusion injury, insulin resistance, inflammation, and cell proliferation and invasion. As a consequence of these multifaceted properties, the role of the apelinergic axis in the pathogenesis of degenerative and proliferative conditions (e.g., Alzheimer's and Parkinson's diseases, osteoporosis, and cancer) is currently investigated. In this view, the dual effect of the apelin/APJ system in the regulation of oxidative stress needs to be more extensively clarified, in order to identify new potential strategies and tools able to selectively modulate this axis according to the tissue-specific profile.

(5) "Molecular Genetics of Abnormal Redox Homeostasis in Type 2 Diabetes Mellitus" by Azarova, I. and collaborators [29]. Numerous studies have shown that oxidative stress resulting from an imbalance between the production of free radicals and their neutralization by antioxidant enzymes is one of the major pathological disorders underlying the development and progression of type 2 diabetes (T2D). This review summarizes the current state-of-the-art advances in understanding the role of abnormal redox homeostasis in the molecular mechanisms of T2D, and provides comprehensive information on the characteristics and biological functions of antioxidant and oxidative enzymes, as well as discussing genetic studies conducted so far in order to investigate the contribution of polymorphisms in genes encoding redox state-regulating enzymes to the disease pathogenesis.

(6) "Molecular Pathology, Oxidative Stress, and Biomarkers in Obstructive Sleep Apnea" by Meliante, P. G. and collaborators [30]. Obstructive sleep apnea syndrome (OSAS) is characterized by intermittent hypoxia (IH) during sleep due to recurrent upper airway obstruction. The derived oxidative stress (OS) leads to complications that do not only concern the sleep–wake rhythm, but also systemic dysfunctions. The aim of this narrative literature review is to investigate molecular alterations, diagnostic markers, and potential medical therapies for OSAS. We analyzed the literature and synthesized the evidence collected. IH increases oxygen free radicals (ROS) and reduces antioxidant capacities. OS and metabolic alterations lead OSAS patients to undergo endothelial dysfunction, osteoporosis, systemic inflammation, increased cardiovascular risk, pulmonary remodeling, and neurological alterations. We treated molecular alterations known to date as useful for understanding the pathogenetic mechanisms and for their potential application as diagnostic markers. The most promising pharmacological therapies are those based on N-acetylcysteine (NAC), vitamin C, leptin, dronabinol, or atomoxetine + oxybutynin, but all require further experimentation. CPAP remains the approved therapy capable of reversing most of the known molecular alterations; future drugs may be useful in treating the remaining dysfunctions.

(7) "The Potential of Cylindromatosis (CYLD) as a Therapeutic Target in Oxidative Stress-Associated Pathologies: A Comprehensive Evaluation" by Zhenzhou, H. and collaborators [31]. Oxidative stress (OS) arises as a consequence of an imbalance between the formation of reactive oxygen species (ROS) and the capacity of antioxidant defense mechanisms to neutralize them. Excessive ROS production can lead to the damage of critical biomolecules, such as lipids, proteins, and DNA, ultimately contributing to the onset and progression of a multitude of diseases, including atherosclerosis, chronic obstructive pulmonary disease, Alzheimer's disease, and cancer. Cylindromatosis (CYLD), initially identified as a gene linked to familial cylindromatosis, has a well-established and increasingly well-characterized function in tumor inhibition and anti-inflammatory processes. Nevertheless, burgeoning evidence suggests that CYLD, as a conserved deubiquitination enzyme, also plays a pivotal role in various key signaling pathways, and is implicated in the pathogenesis of numerous diseases driven by oxidative stress. In this review, we systematically examine the current research on the function and pathogenesis of CYLD in diseases instigated by oxidative stress. Therapeutic interventions targeting CYLD may hold significant promise for the treatment and management of oxidative stress-induced human diseases.

In conclusion, the publications collected are well representative of the variety of consequences due increased oxidative stress and highlight the multitude of the experimental models and approaches currently used to unravel the potential harmful effects. We hope that this editorial effort will contribute to fueling the burgeoning field of cells and tissues responses to oxidant stress, an area we believe to be relevant not only to human physiopathology, but also medical investigations at large. Thus, the precise understanding of oxidative-stress-related disease pathophysiology could favor the identification of novel therapeutic targets, as well as antioxidant strategies.

Author Contributions: A.R. and R.M. writing—original draft preparation. All authors have read and agreed to the published version of the manuscript.

Funding: This research received no external funding.

Conflicts of Interest: The authors declare that the research was conducted in the absence of any commercial or financial relationships that could be construed as a potential conflict of interest.

References

1. Akki, R.; Siracusa, R.; Morabito, R.; Remigante, A.; Campolo, M.; Errami, M.; La Spada, G.; Cuzzocrea, S.; Marino, A. Neuronal-like differentiated SH-SY5Y cells adaptation to a mild and transient $H_{(2)}O_{(2)}$ -induced oxidative stress. *Cell Biochem. Funct.* **2018**, *36*, 56–64. [CrossRef] [PubMed]
2. Akki, R.; Siracusa, R.; Cordaro, M.; Remigante, A.; Morabito, R.; Errami, M.; Marino, A. Adaptation to oxidative stress at cellular and tissue level. *Arch. Physiol. Biochem.* **2019**, *128*, 521–531. [CrossRef] [PubMed]
3. Remigante, A.; Morabito, R. Cellular and Molecular Mechanisms in Oxidative Stress-Related Diseases. *Int. J. Mol. Sci.* **2022**, *23*, 8017. [CrossRef] [PubMed]
4. Remigante, A.; Spinelli, S.; Straface, E.; Gambardella, L.; Caruso, D.; Falliti, G.; Dossena, S.; Marino, A.; Morabito, R. Antioxidant Activity of Quercetin in a $H_{(2)}O_{(2)}$-Induced Oxidative Stress Model in Red Blood Cells: Functional Role of Band 3 Protein. *Int. J. Mol. Sci.* **2022**, *23*, 10991. [CrossRef] [PubMed]
5. Spinelli, S.; Straface, E.; Gambardella, L.; Caruso, D.; Falliti, G.; Remigante, A.; Marino, A.; Morabito, R. Aging Injury Impairs Structural Properties and Cell Signaling in Human Red Blood Cells; Acai Berry Is a Keystone. *Antioxidants* **2023**, *12*, 848. [CrossRef]
6. Ferrera, L.; Barbieri, R.; Picco, C.; Zuccolini, P.; Remigante, A.; Bertelli, S.; Fumagalli, M.R.; Zifarelli, G.; La Porta, C.A.M.; Gavazzo, P.; et al. TRPM2 Oxidation Activates Two Distinct Potassium Channels in Melanoma Cells through Intracellular Calcium Increase. *Int. J. Mol. Sci.* **2021**, *22*, 8359. [CrossRef]
7. Remigante, A.; Spinelli, S.; Pusch, M.; Sarikas, A.; Morabito, R.; Marino, A.; Dossena, S. Role of SLC4 and SLC26 solute carriers during oxidative stress. *Acta Physiol.* **2022**, *235*, e13796. [CrossRef]
8. Perrone, P.; Spinelli, S.; Mantegna, G.; Notariale, R.; Straface, E.; Caruso, D.; Falliti, G.; Marino, A.; Manna, C.; Remigante, A.; et al. Mercury Chloride Affects Band 3 Protein-Mediated Anionic Transport in Red Blood Cells: Role of Oxidative Stress and Protective Effect of Olive Oil Polyphenols. *Cells* **2023**, *12*, 424. [CrossRef]
9. Remigante, A.; Spinelli, S.; Basile, N.; Caruso, D.; Falliti, G.; Dossena, S.; Marino, A.; Morabito, R. Oxidation Stress as a Mechanism of Aging in Human Erythrocytes: Protective Effect of Quercetin. *Int. J. Mol. Sci.* **2022**, *23*, 7781. [CrossRef]
10. Remigante, A.; Spinelli, S.; Marino, A.; Pusch, M.; Morabito, R.; Dossena, S. Oxidative Stress and Immune Response in Melanoma: Ion Channels as Targets of Therapy. *Int. J. Mol. Sci.* **2023**, *24*, 887. [CrossRef]
11. Remigante, A.; Spinelli, S.; Straface, E.; Gambardella, L.; Russo, M.; Cafeo, G.; Caruso, D.; Falliti, G.; Dugo, P.; Dossena, S.; et al. Mechanisms underlying the anti-aging activity of bergamot (*Citrus bergamia*) extract in human red blood cells. *Front. Physiol.* **2023**, *14*, 1225552. [CrossRef]
12. Jowko, E.; Dlugolecka, B.; Cieslinski, I.; Kotowska, J. Polymorphisms in Genes Encoding VDR, CALCR and Antioxidant Enzymes as Predictors of Bone Tissue Condition in Young, Healthy Men. *Int. J. Mol. Sci.* **2023**, *24*, 3373. [CrossRef] [PubMed]
13. Jiang, Y.H.; Jhang, J.F.; Hsu, Y.H.; Kuo, H.C. Usefulness of Urinary Biomarkers for Assessing Bladder Condition and Histopathology in Patients with Interstitial Cystitis/Bladder Pain Syndrome. *Int. J. Mol. Sci.* **2022**, *23*, 12044. [CrossRef] [PubMed]
14. Akhter, N.; Wilson, A.; Arefanian, H.; Thomas, R.; Kochumon, S.; Al-Rashed, F.; Abu-Farha, M.; Al-Madhoun, A.; Al-Mulla, F.; Ahmad, R. Endoplasmic Reticulum Stress Promotes the Expression of TNF-α in THP-1 Cells by Mechanisms Involving ROS/CHOP/HIF-1α and MAPK/NF-κB Pathways. *Int. J. Mol. Sci.* **2023**, *24*, 15186. [CrossRef] [PubMed]
15. Yordanova, V.; Hazarosova, R.; Vitkova, V.; Momchilova, A.; Robev, B.; Nikolova, B.; Krastev, P.; Nuss, P.; Angelova, M.I.; Staneva, G. Impact of Truncated Oxidized Phosphatidylcholines on Phospholipase A_2 Activity in Mono-and Polyunsaturated Biomimetic Vesicles. *Int. J. Mol. Sci.* **2023**, *24*, 11166. [CrossRef] [PubMed]
16. Pukleš, I.; Páger, C.; Sakač, N.; Šarkanj, B.; Matasović, B.; Samardžić, M.; Budetić, M.; Marković, D.; Jozanović, M. Electrophoretic Determination of L-Carnosine in Health Supplements Using an Integrated Lab-on-a-Chip Platform with Contactless Conductivity Detection. *Int. J. Mol. Sci.* **2023**, *24*, 14705. [CrossRef]
17. Simona, M.-S.; Alessandra, V.; Emanuela, C.; Elena, T.; Michela, M.; Fulvia, G.; Vincenzo, S.; Ilaria, B.; Federica, M.; Eloisa, A. Evaluation of Oxidative Stress and Metabolic Profile in a Preclinical Kidney Transplantation Model According to Different Preservation Modalities. *Int. J. Mol. Sci.* **2023**, *24*, 1029.
18. Song, A.; Cho, G.-W.; Moon, C.; Park, I.; Jang, C.H. Protective effect of resveratrol in an experimental model of salicylate-induced tinnitus. *Int. J. Mol. Sci.* **2022**, *23*, 14183. [CrossRef]
19. Ren, C.; Zhou, P.; Zhang, M.; Yu, Z.; Zhang, X.; Tombran-Tink, J.; Barnstable, C.J.; Li, X. Molecular Mechanisms of Oxidative Stress Relief by CAPE in ARPE− 19 Cells. *Int. J. Mol. Sci.* **2023**, *24*, 3565. [CrossRef]
20. Hydzik, P.; Francik, R.; Francik, S.; Gomółka, E.; Eker, E.D.; Krośniak, M.; Noga, M.; Jurowski, K. The critical assessment of oxidative stress parameters as potential biomarkers of carbon monoxide poisoning. *Int. J. Mol. Sci.* **2023**, *24*, 10784. [CrossRef]
21. Zou, H.; Chen, Y.; Qu, H.; Sun, J.; Wang, T.; Ma, Y.; Yuan, Y.; Bian, J.; Liu, Z. Microplastics Exacerbate Cadmium-Induced Kidney Injury by Enhancing Oxidative Stress, Autophagy, Apoptosis, and Fibrosis. *Int. J. Mol. Sci.* **2022**, *23*, 14411. [CrossRef] [PubMed]
22. Durgo, K.; Oresic, S.; Rincic Mlinaric, M.; Fiket, Z.; Juresic, G.C. Toxicity of Metal Ions Released from a Fixed Orthodontic Appliance to Gastrointestinal Tract Cell Lines. *Int. J. Mol. Sci.* **2023**, *24*, 9940. [CrossRef] [PubMed]
23. Li, X.; Chen, C.; He, M.; Yu, L.; Liu, R.; Ma, C.; Zhang, Y.; Jia, J.; Li, B.; Li, L. Lead Exposure Causes Spinal Curvature during Embryonic Development in Zebrafish. *Int. J. Mol. Sci.* **2022**, *23*, 9571. [CrossRef] [PubMed]
24. Chen, Y.; Gao, Y.; Yuan, M.; Zheng, Z.; Yin, J. Anti-*Candida albicans* Effects and Mechanisms of Theasaponin E1 and Assamsaponin A. *Int. J. Mol. Sci.* **2023**, *24*, 9350. [CrossRef] [PubMed]

25. Murotomi, K.; Umeno, A.; Shichiri, M.; Tanito, M.; Yoshida, Y. Significance of Singlet Oxygen Molecule in Pathologies. *Int. J. Mol. Sci.* **2023**, *24*, 2739. [CrossRef]
26. Sutter, J.; Bruggeman, P.J.; Wigdahl, B.; Krebs, F.C.; Miller, V. Manipulation of Oxidative Stress Responses by Non-Thermal Plasma to Treat Herpes Simplex Virus Type 1 Infection and Disease. *Int. J. Mol. Sci.* **2023**, *24*, 4673. [CrossRef]
27. Yegiazaryan, A.; Abnousian, A.; Alexander, L.J.; Badaoui, A.; Flaig, B.; Sheren, N.; Aghazarian, A.; Alsaigh, D.; Amin, A.; Mundra, A.; et al. Recent Developments in the Understanding of Immunity, Pathogenesis and Management of COVID-19. *Int. J. Mol. Sci.* **2022**, *23*, 9297. [CrossRef]
28. Fibbi, B.; Marroncini, G.; Naldi, L.; Peri, A. The Yin and Yang Effect of the Apelinergic System in Oxidative Stress. *Int. J. Mol. Sci.* **2023**, *24*, 4745. [CrossRef]
29. Azarova, I.; Polonikov, A.; Klyosova, E. Molecular Genetics of Abnormal Redox Homeostasis in Type 2 Diabetes Mellitus. *Int. J. Mol. Sci.* **2023**, *24*, 4738. [CrossRef]
30. Meliante, P.G.; Zoccali, F.; Cascone, F.; Di Stefano, V.; Greco, A.; de Vincentiis, M.; Petrella, C.; Fiore, M.; Minni, A.; Barbato, C. Molecular Pathology, Oxidative Stress, and Biomarkers in Obstructive Sleep Apnea. *Int. J. Mol. Sci.* **2023**, *24*, 5478. [CrossRef]
31. Huang, Z.; Tan, Y. The Potential of Cylindromatosis (CYLD) as a Therapeutic Target in Oxidative Stress-Associated Pathologies: A Comprehensive Evaluation. *Int. J. Mol. Sci.* **2023**, *24*, 8368. [CrossRef]

International Journal of
Molecular Sciences

Article

The Critical Assessment of Oxidative Stress Parameters as Potential Biomarkers of Carbon Monoxide Poisoning

Piotr Hydzik [1], Renata Francik [2,3], Sławomir Francik [4], Ewa Gomółka [5], Ebru Derici Eker [6], Mirosław Krośniak [7], Maciej Noga [8] and Kamil Jurowski [8,9,*]

[1] Toxicology Clinical Department, University Hospital, Jagiellonian University Medical College, 31-008 Kraków, Poland

[2] Institute of Health, State Higher Vocational School, 33-320 Nowy Sącz, Poland

[3] Department of Bioorganic Chemistry, Faculty of Pharmacy, Jagiellonian University Medical College, 31-008 Krakow, Poland

[4] Department of Mechanical Engineering and Agrophysics, Faculty of Production and Power Engineering, University of Agriculture in Krakow, 31-103 Krakow, Poland

[5] Toxicological Information and Laboratory Analysis Laboratory University Hospital, Jagiellonian University Medical College, Jakubowskiego 2, 30-688 Kraków, Poland

[6] Faculty of Pharmacy, Mersin University, 33343 Mersin, Turkey

[7] Department of Food Chemistry and Nutrition, Faculty of Pharmacy, Jagiellonian University Medical College, 31-008 Krakow, Poland

[8] Department of Regulatory and Forensic Toxicology, Institute of Medical Expertises, 91-205 Łódź, Poland

[9] The Laboratory of Innovative Research and Analyzes, Institute of Medical Studies, Medical College, Rzeszów University, 35-310 Rzeszow, Poland

* Correspondence: toksykologia@ur.edu.pl

Abstract: In conventional clinical toxicology practice, the blood level of carboxyhemoglobin is a biomarker of carbon monoxide (CO) poisoning but does not correspond to the complete clinical picture and the severity of the poisoning. Taking into account articles suggesting the relationship between oxidative stress parameters and CO poisoning, it seems reasonable to consider this topic more broadly, including experimental biochemical data (oxidative stress parameters) and patients poisoned with CO. This article aimed to critically assess oxidative-stress-related parameters as potential biomarkers to evaluate the severity of CO poisoning and their possible role in the decision to treat. The critically set parameters were antioxidative, including catalase, 2,2-diphenyl-1-picryl-hydrazyl, glutathione, thiol and carbonyl groups. Our preliminary studies involved patients (n = 82) admitted to the Toxicology Clinical Department of the University Hospital of Jagiellonian University Medical College (Kraków, Poland) during 2015–2020. The poisoning was diagnosed based on medical history, clinical symptoms, and carboxyhemoglobin blood level. Blood samples for carboxyhemoglobin and antioxidative parameters were collected immediately after admission to the emergency department. To evaluate the severity of the poisoning, the Pach scale was applied. The final analysis included a significant decrease in catalase activity and a reduction in glutathione level in all poisoned patients based on the severity of the Pach scale: I°–III° compared to the control group. It follows from the experimental data that the poisoned patients had a significant increase in level due to thiol groups and the 2,2-diphenyl-1-picryl-hydrazyl radical, with no significant differences according to the severity of poisoning. The catalase-to-glutathione and thiol-to-glutathione ratios showed the most important differences between the poisoned patients and the control group, with a significant increase in the poisoned group. The ratios did not differentiate the severity of the poisoning. The carbonyl level was highest in the control group compared to the poisoned group but was not statistically significant. Our critical assessment shows that using oxidative-stress-related parameters to evaluate the severity of CO poisoning, the outcome, and treatment options is challenging.

Keywords: poisoning; carbon monoxide; oxidative stress; treatment options

check for
updates

Citation: Hydzik, P.; Francik, R.; Francik, S.; Gomółka, E.; Eker, E.D.; Krośniak, M.; Noga, M.; Jurowski, K. The Critical Assessment of Oxidative Stress Parameters as Potential Biomarkers of Carbon Monoxide Poisoning. *Int. J. Mol. Sci.* **2023**, *24*, 10784. https://doi.org/10.3390/ijms241310784

Academic Editors: Rossana Morabito and Alessia Remigante

Received: 20 March 2023
Revised: 28 April 2023
Accepted: 27 June 2023
Published: 28 June 2023

1. Introduction

In Poland, approximately 5800 hospital admissions are reported yearly due to carbon monoxide poisoning [1]. Carbon monoxide (CO) is a colorless and odorless poisonous gas produced by the incomplete oxidation of fossil fuels and carbonaceous organic compounds (e.g., coal, natural gas, wood, and kerosene) [2]. The most frequently reported cases of CO intoxication are unintentional poisonings, i.e., accidental. Most deaths occur due to malfunctioning or obstructed exhaust systems, weather changes, and poor ventilation, restricting outdoor airflow into a building (especially bathing in a non-ventilated bathroom and staying in the car with the engine running in a closed garage).

It is well known that CO reversibly binds to hemoglobin and shifts the oxyhemoglobin dissociation curve to the left. In this way, it decreases the oxygen-carrying capacity of the blood and interferes with the release of oxygen at the peripheral tissue level [3]. These two main mechanisms of action underlie the potentially toxic effects of low-level CO exposure. However, the principal cause of CO-induced toxicity at low exposure levels is believed to be increased carboxyhemoglobin (COHb) formation. Clinical observation shows blood COHb levels do not correlate well with clinical symptoms and patient status [4].

Furthermore, it is known that the sole effect of CO is not only to block oxygen transfer in the blood but also to bind to other extravascular proteins such as myoglobin and non-Hb hemoproteins such as cytochrome c oxidase and cytochrome P-450. Adaptable CO (which is involved in cellular adaptation to oxidative stress and vascular dysfunction, leading to the maintenance of cellular and vascular homeostasis) regulates mitochondrial function to generate reactive oxygen species (ROS), which are responsible for controlling cellular redox states and adaptive responses to oxidative stress [5]. Numerous experiments demonstrate that CO is involved in the adaptation of cells to oxidative stress leading to the maintenance of cellular homeostasis [6]. At physiological state, approximately 1 to 3% of the oxygen consumed is incompletely reduced to anion superoxide, quickly transformed into hydrogen peroxide by superoxide dismutase (SOD) located in the mitochondrial matrix. A possible model for CO action is the generation of mitochondrial ROS based on partially and/or reversely inhibiting cytochrome c oxidase (complex IV), leading to electron accumulation at the complex III levels, which facilitates anion superoxide generation [6]. On the other hand, at low levels of CO, it is possible to improve mitochondrial respiration [7]; It can be speculated that CO induces mitochondrial ROS generation due to accelerated oxidative phosphorylation.

Cytochrome prefers oxygen to CO by a factor of 9:1, which may explain the disparity between the blood level of COHb and the clinical effects of poisoning and some beneficial effects of hyperbaric oxygen therapy (HBO). The biological half-life of carboxycytochrome c oxidase is not known. However, it may be an essential factor in the genesis of late CO poisoning sequelae and provides a rational basis for determining the duration of oxygen therapy. Other mechanisms of CO-induced toxicity have been hypothesized and evaluated, such as hydroxyl radical production and lipid peroxidation, in animals. However, as of yet, none have been demonstrated to operate at relatively low CO exposure levels [7].

The inhibition of energy coupling and aerobic metabolism violates the steady-state equilibrium between pro-oxidants and antioxidants. An imbalance in favor of pro-oxidants potentially leads to damage called oxidative stress [8]. CO-mediated ROS production initiates intracellular signal events, which regulate the expression of adaptive genes implicated in oxidative stress and function as a signaling molecule to promote vascular functions, including angiogenesis and mitochondrial biogenesis [6]. Animal studies indicate that low blood COHb levels can cause perivascular oxidative changes by releasing free radical nitric oxide from the vascular endothelium and platelet [9]. This also suggests the role of oxidative stress and hypoperfusion in CO toxicity. Organs most sensitive to CO toxicity are those with high blood flow and oxygen requirements, such as the brain, heart, and skeletal muscles. The most observed symptoms of acute CO poisoning are weakness, fatigue, malaise, headache, drowsiness, confusion, syncope, seizure, nausea, vomiting, vertigo, palpitation, and chest pain.

Prehospital care is generally limited to an interruption of CO exposure with 100% oxygen therapy with a non-rebreather mask. If the patient is in a deep comatose state and unstable, artificial ventilation intubation is necessary. Oxygen therapy reduces the biological half-time of COHb and increases oxygen concentration in the blood. This improves oxygen delivery to tissues and cells and reduces hypoxia. Most of the time, oxygen therapy is effective with the complete resolution of symptoms. In 10–30% of CO intoxication cases, delayed neuropsychiatric syndrome (DNS) could present with the most common manifestations such as cognitive and behavioral impairment, memory loss, movement disorders with parkinsonian features, seizures, visual impairment, depression, hallucinations, and urine incontinence [10–14]. The exact pathogenesis of DNS remains unknown. Increasing evidence indicates that the brain damage caused by CO intoxication results from mitochondrial oxidative stress in the central nervous system and the oxidation of proteins (the oxidation of thiol groups and the formation of carbonyl derivatives). These reactions lead to protein damage leading to abnormal immune response with white matter demyelination and leukoencephalopathy and the appearance of the delayed effects seen in DNS.

Generally, the severity of the poisoning is assessed after the end of the acute phase and the evaluation of organ damage caused by poisoning. In Poland, the Pach scale is applied for such an assessment [15,16]—Table 1. The Pach scale of CO poisoning (developed initially by Janusz Pach) consists of factors that influence and determine the severity of the poisoning. Each factor is classified from 0 to 3 points, and the total points indicate the severity of CO poisoning ranging from I° to III°.

Table 1. Pach scale—table for calculation of CO poisoning severity in Poland.

Points	0	1	2	3
Age (years)	<29	30–39	40–49	>50
CO exposition time (min)	<30	31–60	61–120	>120
Neurological injury degree	I° - No consciousness disturbances and other neurological changes in physical examination.	II° - Consciousness disturbances (somnolence and agitation); - Hyperreflexia; - Positive Babinski reflex; - Tonic–clonic convulsions; - Increased muscular tone.	III° - Unconsciousness.	IV° - Loss of consciousness; - Hyperreflexia; - Positive Babinski reflex; - Tonic–clonic convulsions; - Decreased muscle tone; - Bradyreflexia.
COHb serum level (%)	0	<15	15–30	>30
Lactate serum level (mmol/L)	1.0–1.78	1.8–3.6	3.7–5.4	>5.5

The severity of CO poisoning: I° light 1–4 points, II° medium 5–8 points, III° sever > 9 points

Basic laboratory blood tests include evaluating biomarkers such as COHb and lactate blood levels, troponin level, creatine kinase (CPK), and aminotransaminase activity of aminotransaminases (e.g., AST and ASPAT) in the blood.

The multidirectional mechanism of CO toxicity releases ROS and reactive nitrogen species (RNS), significantly disrupting redox homeostasis. Several assays are available to measure oxidative stress. The basic parameters used to estimate the antioxidant status are glutathione (GSH), carbonyl groups (=CO), sulfhydryl (thiol) groups (–SH), 1,1-diphenyl-2-picryl hydrazyl radical (DPPH), catalase (CAT) and glutathione peroxidase (GPX activity). These parameters are not used routinely in clinical practice but could be suitable potential biomarkers to assess the severity of CO poisoning and predict poisoning complications. To date, no strategy is available to evaluate antioxidative parameters in the blood that could help select therapy and minimize oxidative stress in CO poisoning patients.

This work aimed to evaluate parameters related to oxidative stress, such as the activity and levels of selected antioxidative parameters (CAT, DPPH, GSH, –SH, =CO, CAT/GSH ratio and SG/GSH ratio) as potentially valuable biomarkers to assess the severity of CO poisoning from monoxide and their possible role in deciding treatment options.

2. Results and Discussion

Serum antioxidant parameters were evaluated for patients (n = 82) divided into four groups, i.e., the C control group, S-Pach scale = I°, II°, III°. The hypothesis of the influence of exposure time (factor) on the severity of poisoning (value of the Pach scale) was also evaluated. A one-way ANOVA test was performed for three groups (levels of factor: the Pach scale = I°, II°, III°), where the time of exposure to CO (T-CO) was adopted as a dependent variable based on an interview with an intoxicated patient.

The ANOVA results (Table 2) show a statistically significant effect of the independent variable—the Pach scale (S-Pach)—on seven dependent variables: CAT, DPPH, SH, GSH, CAT/GSH, SH/GSH, and T-CO. No statistically significant effect of S-Pach on =CO derivates was found. The results of antioxidant parameters for different groups are presented in Table 3. Data are presented as means from independent measurements ± standard deviation (SD).

Table 2. One-way ANOVA results for catalase (CAT), DPPH, sulfhydryl (SH), reduced glutathione (GSH), protein carbonyl group (=CO derivates), quotient CAT and GSH (CAT/GSH), quotient SH and GSH (SH/GSH), and time of exposure to carbon monoxide (T-CO).

	Source of Variation	SS [1]	df [1]	MS [1]	F [1]	p [1]	Significant
CAT	S-Pach [2]	530530	3	176843	50.194	<0.001	Yes
DPPH	S-Pach	8352	3	2784	15.597	<0.001	Yes
SH	S-Pach	169.1	3	56.4	10.338	<0.001	Yes
GSH	S-Pach	132439	3	44146	52.231	<0.001	Yes
=CO derivates	S-Pach	2.051	3	0.684	2.030	0.117	No
CAT/GSH	S-Pach	701.6	3	233.9	15.929	<0.001	Yes
SH/GSH	S-Pach	269.3	3	89.8	404.347	<0.001	Yes
T-CO	S-Pach	40945	2	20473	10.794	<0.001	Yes

[1] SS—the sum of squares between groups, df—the number of degrees of freedom, MS—the mean sum of squares between groups, F—the test statistic value, p—probability. [2] S-Pach—Pach scale.

Table 3. Results of antioxidant parameters for different groups (value of S-Pach): catalase (CAT), DPPH, sulfhydryl (SH), reduced glutathione (GSH), protein carbonyl group (=CO derivates), quotient CAT and GSH (CAT/GSH), quotient SH and GSH (SH/GSH), and time of exposure to carbon monoxide (T-CO). Data are presented as means from independent measurements ± standard deviation (SD).

	Control	I°—Pach	II°—Pach	III°—Pach
CAT (U/mg protein)	196 ± 82.2	35.0 ± 18.3	27.3 ± 9.6	37.8 ± 22.4
DPPH (%)	5.9 ± 6.5	23.9 ± 16.6	30.1 ± 20.5	25.6 ± 11.4
SH (mmol/mg protein)	11.4 ± 2.4	13.8 ± 2.6	14.9 ± 1.6	13.5 ± 1.1
GSH (μmol/mg protein)	84.8 ± 41.9	3.77 ± 0.2	3.73 ± 0.16	3.75 ± 0.14
=CO derivates (nmol/mg protein)	1.333 ± 0.57	1.025 ± 0.58	1.005 ± 0.58	1.001 ± 0.66
CAT/GSH ratio	3.12 ± 2.96	9.34 ± 4.88	7.35 ± 2.62	10.1 ± 6.08
SH/GSH ratio	0.158 ± 0.086	3.678 ± 0.692	4.022 ± 0.523	3.842 ± 0.711
T-CO (min)	-	19.8 ± 7.0	42 ± 37	130 ± 125.7

Post hoc Tukey's tests were performed to identify homogeneous groups for the dependent variables, for which ANOVA allowed rejection of the null hypothesis. In Figures 1–4, homogeneous groups are marked with the same letters. According to Tukey's test, different letters indicate significant differences between the groups ($p < 0.05$).

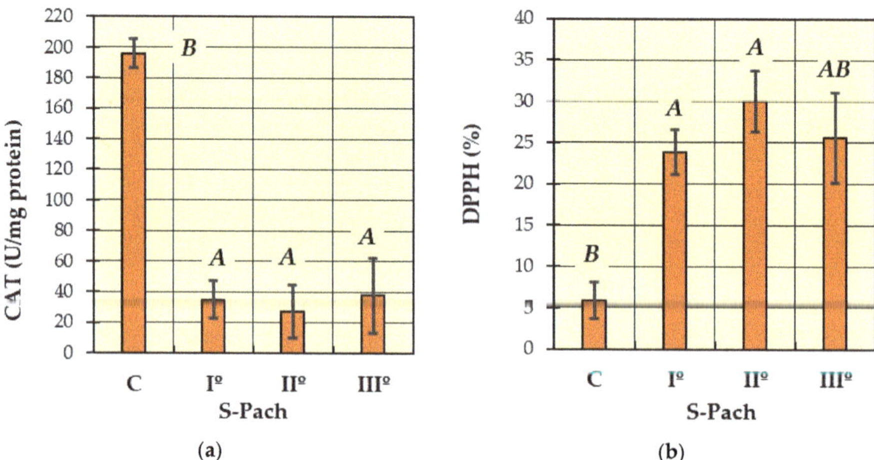

Figure 1. The values of measured antioxidative parameters in the serum of patients with CO poisoning (S-Pach = I°, II°, III°) compared to the control group (C): (**a**) activity of catalase—CAT; (**b**) inhibition of the DPPH radical. The data are presented as means from independent measurements ± standard error. Bars with a different letter indicate significant differences according to HDS Tukey's test ($p < 0.05$ was accepted as statistically significant). Homogeneous groups are marked with the same letters.

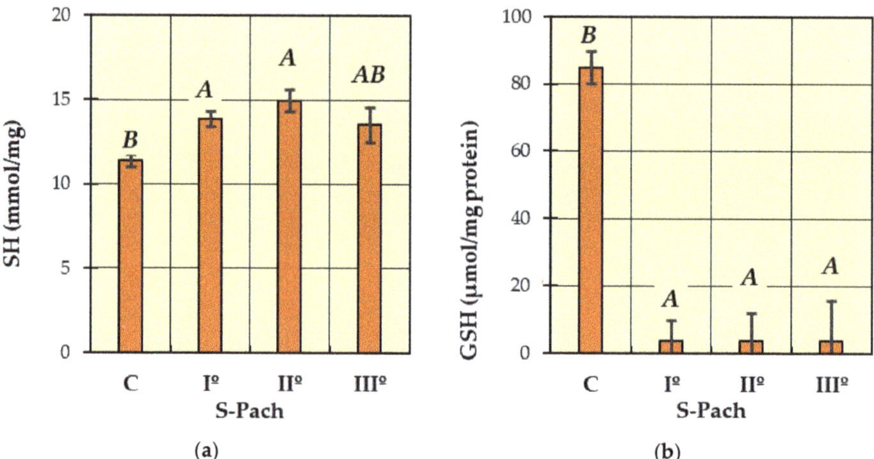

Figure 2. The concentrations of measured antioxidative parameters in the serum of patients with CO poisoning (S-Pach = I°, II°, III°) compared to the control group (C): (**a**) SH; (**b**) GSH. The data are presented as means from independent measurements ± standard error. Bars with a different letter indicate significant differences according to HDS Tukey's test ($p < 0.05$ was accepted as statistically significant). Homogeneous groups are marked with the same letters.

In Figure 1a (catalase activity—CAT), two homogeneous groups were identified: A (the patients with carbon monoxide poisoning—S-Pach = I°, II°, III°) and B (the patients without carbon monoxide poisoning—control group C). The analysis of catalase activity in the blood serum showed a statistically significant decrease in enzyme activity in all patients with carbon monoxide poisoning (S-Pach = I°, II°, III°) compared to the control group (the patients without carbon monoxide poisoning). The average values of catalase enzyme activity in people with carbon monoxide poisoning were about six times lower compared to the control group.

In Figure 1b (DPPH), two homogeneous groups were identified: A (the patients with carbon monoxide poisoning—S-Pach = I°, II°, III°) and B (S-Pach = C, III°). Compared to the control group, a statistically significant increase in the inhibition of the DPPH radical was observed in the patients with the CO poisoning group (S-Pach = I°, II°). There were no significant differences between the level of DPPH and the severity of the poisoning (S-Pach = I°, II°, III°).

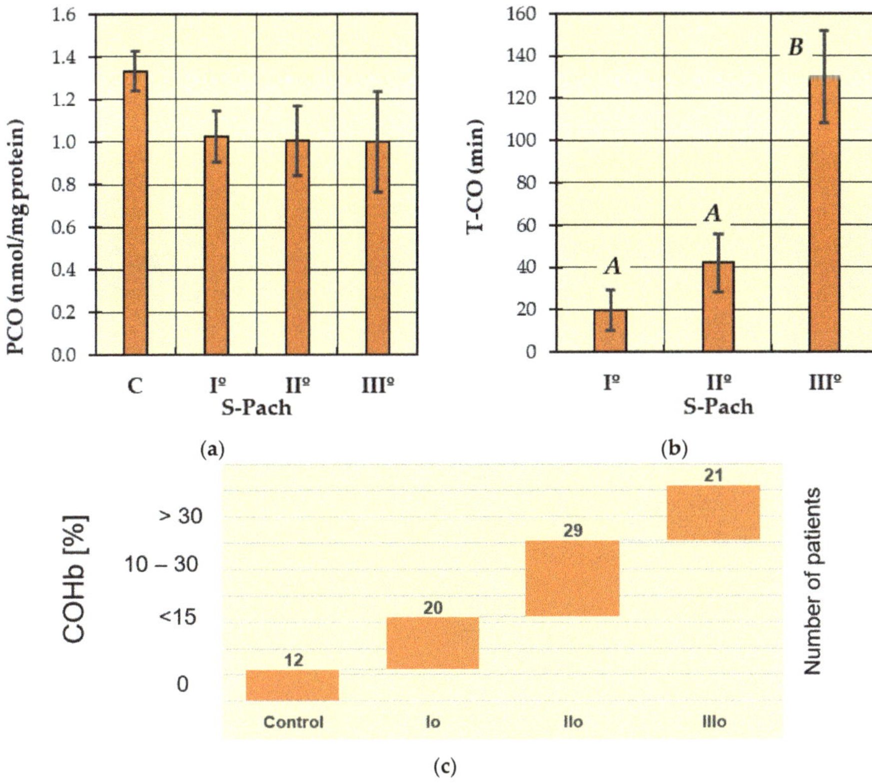

Figure 3. The concentrations of measured antioxidative parameters in the serum of patients with CO poisoning (S-Pach = I°, II°, III°) compared to the control group (C): (**a**) Carbonyl derivates (=CO derivates); (**b**) exposure time of CO. The data are presented as means from independent measurements ± standard error. (**c**) The range of COHb levels for the different Pach scores. Bars with a different letter indicate significant differences according to HDS Tukey's test ($p < 0.05$ was accepted as statistically significant). Homogeneous groups are marked with the same letters.

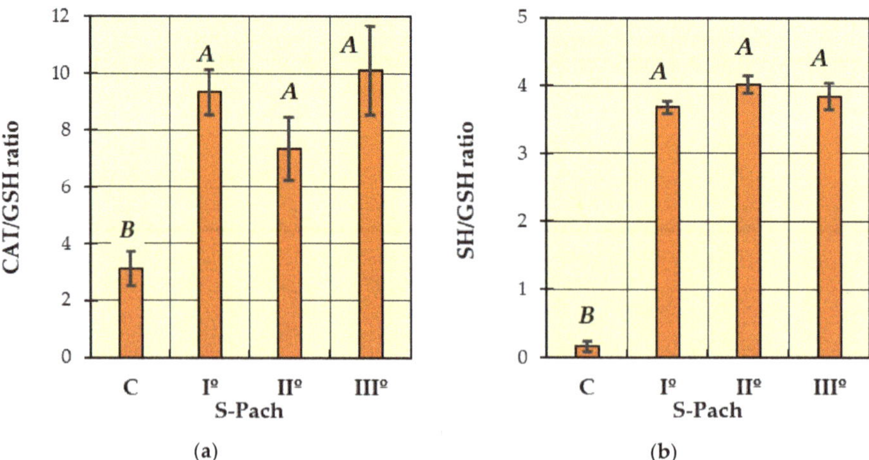

Figure 4. Comparison of patients with CO poisoning (S-Pach = I°, II°, III°) and the control group (C): (**a**) CAT/GSH ratio; (**b**) SH/GSH ratio. The data are presented as means from independent measurements ± standard error. Bars with a different letter indicate significant differences according to HDS Tukey's test ($p < 0.05$ was accepted as statistically significant). Homogeneous groups are marked with the same letters.

In Figure 2a (SH), two homogeneous groups were identified: A (the patients with carbon monoxide poisoning—S-Pach = I°, II°, III°) and B (S-Pach = C, III°). There were no significant differences between the level of SH and the severity of the poisoning (S-Pach = I°, II°, III°).

In Figure 2b (GSH), two homogeneous groups were identified: A (the patients with carbon monoxide poisoning—S-Pach = I°, II°, III°) and B (the patients without carbon monoxide poisoning—control group C). The level of reduced glutathione was significantly elevated in the control group compared to all patients with carbon monoxide poisoning groups. In all degrees of the Pach scale, the level of reduced glutathione in blood serum was close to zero, while in the control group, it fluctuated between 80 and 90 µmol/mg protein.

The level of carbonyl groups (Figure 3a) was highest in the control group compared to the patients with carbon monoxide poisoning, but the differences are not statistically significant.

In Figure 3b (T-CO), two homogeneous groups were identified: A (S-Pach = I°, II°) and B (S-Pach = I°, II°). Based on data from the literature and clinical experience, we statistically confirmed that the severity of poisoning increased with increasing duration of exposure to carbon monoxide.

In Figure 4a (CAT/GSH ratio) and Figure 4b (SH/GSH ratio), two homogeneous groups were identified: A (the patients with carbon monoxide poisoning—S-Pach = I°, II°, III°) and B (the patients without carbon monoxide poisoning—control group C). Compared to the control group, a statistically significant increase in the CAT/GSH ratio was observed in the patients with carbon monoxide poisoning (Figure 4a). The SH/GSH ratio showed more remarkable differences (Figure 4b)—the values were more than 15 times higher in the patients with CO poisoning compared to the control group. There was no statistically significant difference between the CAT/GSH ratio, SH/GSH ratio and the degree of severity of poisoning patients (S-Pach = I°, II°, III°).

Blood COHb level is well recognized as a biomarker for determining CO poisoning exposure and its effects. Although this is an essential parameter in routine toxicological diagnostics practice, more is needed to provide a complete picture of the biochemical disturbances at the cellular and tissue levels in the case of acute CO poisoning. At this time, it is difficult to explain all clinical problems related to the course of CO poisoning, treatment, and outcome [17,18]. Despite that, the pathophysiology and clinical findings of acute CO poisoning have been extensively reviewed, and many factors contribute to the severity of the poisoning and its outcome. There is still more scientific evidence on the relationship between CO poisoning and parameters related to oxidative stress. CO poisoning is a dynamic pathophysiological process, starting from the hypoxic phase and then reoxygenation with postischemic reperfusion injury. At the tissue level, mitochondrial activity requires oxygen for aerobic ATP synthesis for cellular activity. Aerobic metabolism needs the oxygen cascade, which means the gradient from the atmosphere at sea level (PaO_2 = 159 mmHg) to the cell cytoplasm (PO_2 = 4–22 mmHg). Many factors in this cascade affect final mitochondrial PO_2, including alveolar gas exchange, oxygen transport in the blood, tissue perfusion, tissue type, and local metabolic activity [19]. Under the pathological conditions of CO poisoning, any change in any of the steps in this cascade can result in hypoxia at the mitochondrial level and the formation of oxidative stress.

Many reports have described that when PO_2 is high in tissues, the generation of ROS, such as hydrogen peroxide and superoxide, increases, causing oxidative stress and free radical damage [19–22]. It has also been emphasized that ROS cause reperfusion injury during the reoxygenation phase after exposure to CO by decreasing the brain catalase activity that demonstrates hydrogen peroxide production. There was also a decrease in the ratio of reduced/oxidized glutathione, which was quickly corrected by hyperbaric oxygen as opposed to normobaric oxygen, under which it continued to decrease during the first 45 min [23]. Another possible mechanism of ROS generation is the inhibition of xanthine oxidase by allopurinol, which reduces the degree of lipid peroxidation [24]. Most clinical practice guidelines in uncomplicated CO intoxication recommend 100% oxygen therapy until normalization of COHb blood level and 50% oxygen in the next 6 h, and 30% oxygen until 24 h after interruption of CO exposure. Oxygen toxicity is more likely when a non-rebreathing mask (oxygen mask with reservoir) reaches concentrations higher than 60% [19,22,25]. It should be emphasized that in our study, patients received 100% oxygen therapy for an average of 30 to 60 min at the time of blood sampling. Such iatrogenic hyperoxia induces the formation of free radicals, which could be additive to the free radicals generated in CO poisoning. There is a lack of data on the direct or indirect impact of a very high concentration of free oxygen on antioxidative blood parameters. Many complex antioxidant mechanisms offer protection against peroxidation, involving many molecular pathways and molecules. Antioxidant molecules are not independent, and variable biochemical integration can be observed between different antioxidants. In our study, to evaluate oxidative stress and antioxidative defense (scavenger) systems, we used in vivo analyses that included the detection of blood biomarkers such as free radicals (DPPH), sulfhydryl (thiol) and carbonyl groups as a marker of protein damage, non-enzymatic antioxidants (reduced glutathione), and enzymatic antioxidants (catalase). Considering the limitations concerning the sensitivity, specificity, and timing of ROS and antioxidative biomarker analysis, we propose two main theses for our study.

(1) In the case of acute CO poisoning, ROS production with free radicals, increased consumption and/or reduced stores of antioxidants, and/or decreased activity of antioxidants [14,26–29].

(2) During intense treatment of CO poisoning with oxygen therapy at very high concentrations of free oxygen, it is suspected that the antioxidant systems are eventually overwhelmed, and the rate of cell damage exceeds the capacity of the systems that prevent or repair it.

The results showed that the antioxidant parameters in CO-poisoned patients changed. Two primary changes were observed: a reduction in CAT activity and a decreased level of reduced GSH. These parameters are more closely related to changes in cellular antioxidant status. SOD eliminates superoxide radicals ($O, O_2\bullet$) by converting them into oxygen and hydrogen peroxide (H_2O_2). CAT and glutathione peroxidase (GPx) are direct antioxidant enzymes that decompose hydrogen peroxide into oxygen and water [30,31]. The pro-oxidant activity of H_2O_2 is due to its reduction by one electron in hydroxyl radical (OH).

The formation and reactions of (OH) are associated with chemical, non-enzymatic reactions, which are beyond any cellular control or antioxidant defense mechanism. These processes are known as Haber-Weiss and Fenton reactions [32].

CAT decomposes hydrogen peroxide at high rates but with low affinity. It is most useful when H_2O_2 production or accumulation peaks are presented. The lack of catalase (acatalasemia) increases oxidative stress and induces human pathologies [20,33]. According to the Pach scale, the mean reduction in CAT observed in CO-poisoned patients during the first hour was six times lower than in the controls and was not correlated with the severity of CO poisoning. There are few data on the influence of CO and oxygen toxicity on the activity of antioxidant enzymes. For this reason, comparing our results with those of the other authors' reports is a challenge. CAT has been suggested to play its role when the glutathione peroxidase (Gpx) pathway reaches saturation with the substrate (H_2O_2). Gpx slowly decomposes H_2O_2 but with higher affinity, so it is most beneficial to decompose the small amounts of hydrogen peroxide produced under physiological conditions inside cells. Gpx needs a cofactor like reduced GSH or another sulfhydryl(thiols) to function. In the present study, we did not measure the activity of Gpx, but we observed a decrease in the level of reduced glutathione (GSH), which Gpx probably used to decompose H_2O_2. The increased peaks of H_2O_2 with Gpx saturation and CAT saturation could explain the low CAT activity observed in our study. Glutathione also acts as a cofactor for glutathione S-transferases (GSTs), enzymes of phase II metabolism. GSTs conjugate glutathione with other electrophilic compounds and lead to the depletion of reduced glutathione (GSH) [34]. The increased ratios of CAT/GSH and SH/GSH obtained in our study confirm the lack of reduced GSH in the CO-poisoned group. Both calculated ratios significantly differentiate CO-poisoned patients from the control group but did not show statistical significance in determining the stage of CO poisoning.

Another statistically significant change was observed due to the sulfhydryl (thiol) groups (–SH), an increased level in all CO-poisoned patients, irrespective of the severity of the poisoning. In fact, in the case of oxidative stress, the oxidation of cellular and extracellular proteins sulfhydryl (thiol) groups occurs directly due to ROS. Oxidation of the sulfhydryl (thiol) group alters the tertiary structure of many structural and functional proteins, causing their inactivation and aggregation. It may also lead to impairment of calcium homeostasis with calcium accumulation in the cytosol and increased activation of calcium-dependent proteases and phospholipases [24]. A high SH/GSH ratio may also indicate a significant rise in the sulfhydryl (thiol) group. Many studies show that the assessment of the content of sulfhydryl (thiol) groups serves as a better indicator of oxidative stress than measuring the total oxidative state [35–37]. On the other hand, some data show that acute oxygen toxicity during CO treatment is mainly due to the oxidation and polymerization of the -SH groups of enzymes that lead to their inactivation.

No statistically significant changes were observed in the carbonyl groups, reflecting protein oxidations. The difference was insignificant due to the relatively small number of patients with CO poisoning. Including more significant CO-poisoned patients in the study would increase the statistical significance of this parameter.

Next, increased (at statistical significance) inhibition of the DPPH radical was observed in the CO-poisoned group. According to the Pach scale, this increase in DPPH inhibition was independent of the severity of CO poisoning. This observation is difficult to interpret because DPPH (1,1-diphenyl-2-picryl hydrazyl), a free radical, can react with antioxidant and oxidative molecules. According to the Pach scale, a positive correlation was also found between the time of exposure and the severity of CO poisoning; despite significant variability, the trend remains clear.

In carbon monoxide patients, the lungs, hearts, and brain are often affected and cause coma and death in severe poisoning, resulting in immediate and delayed neuronal damage in some brain regions that cannot be easily explained by tissue poisoning. One possibility is that cell injuries during and after CO poisoning are related to brain reactions to oxygen species. Exposure to CO can cause a variety of perivascular processes, including oxidative stress that activates N-methyl-D-aspartate (NMDA) and nitrate oxide synthase (nNOS) in neurons. These events are essential for the progression of CO-mediated neuropathology. It follows from the experimental data that the poisoned patients had a significant increase in level due to thiol groups and the 2,2-diphenyl-1-picryl-hydrazyl radical, with no significant differences according to the severity of poisoning. The catalase-to-glutathione and thiol-to-glutathione ratios showed the most important differences between the poisoned patients and the control group, with a significant increase in the poisoned group. The ratios did not differentiate the severity of the poisoning. The carbonyl level was highest in the control group compared to the poisoned group but was not statistically significant. Our critical assessment is that it is challenging to use oxidative-stress-related parameters to evaluate the severity of CO poisoning, the outcome and treatment options. This is the first critical observation compared to the widespread belief in the literature that such relationships occur. However, careful statistical analysis revealed that the problem is more complex, and the discussion cannot be led without additional research. Thus, it should be noted that further studies are required, focusing on molecular mechanisms.

3. Materials and Methods

3.1. Materials

Our studies involved patients ($n = 82$, age range: 23–43 years, 51.2% females, 48.8% males) admitted to the Clinical Department of Toxicology of Jagiellonian University Medical College (Kraków, Poland) during 2015–2020. For the control group, we chose 12 healthy volunteers (31–63 years old). The control group comprised six women (mean age = 46 years) and six men (mean age = 47 years). The control group consisted only of healthy volunteers who were non-smokers with no digestive problems (e.g., malnutrition, obesity). The poisoned patients involved ($n = 70$) were divided into three groups according to the S-Pach scale: I°, II°, III°. The study group consisted of patients with confirmed CO poisoning who were included in our study without any exclusions, especially without other exposures to other xenobiotics. In both groups of patients, the antioxidant parameters in the plasma were measured. The characteristics of the research participants, taking into account their age, sex, and the group to which they were assigned, are presented in Supplementary Materials (Table S1). Poisoning was diagnosed based on medical history, clinical symptoms, and COHb blood level. Blood samples to determine COHb were taken immediately after admission to the emergency department, and COHb analysis was performed using point-of-care (POCT) and blood gas analyzers. The Department of Toxicology is located in the main city center of Krakow. According to emergency medicine principles, an ambulance should reach the accident site (a patient with CO poisoning) in 7 min. Within 15 min, it should contact the nearest emergency department. Thus, our patients were admitted to the emergency department up to 30 min after the termination of CO exposure. Patients received oxygen during transport to the hospital. Whole blood was collected in anticoagulant-treated tubes (EDTA treated). To obtain plasma, cells were removed by centrifugation in a refrigerated centrifuge for 10 min at $1500\times g$. Platelets were then depleted from the plasma sample by centrifugation for 15 min at $2000\times g$. The plasma was then immediately transferred to a clean polypropylene tube after centrifugation. During handling, the plasma samples were kept at -2 to 8 °C. Acute CO poisoning was diagnosed based on medical history, particularly the source and time of exposure to CO, clinical symptoms, and toxicological laboratory parameters. Exposure time to CO was calculated based on data obtained during the medical history of patients and family members. In addition, the information contained in the rescue operation report prepared by paramedics was also based on information. The Pach scale (see Table 1) was used to assess the severity of the poisoning. The flow chart shows the research participants' division into the control group and the CO poisoning patients' groups according to the S-Pach scale (Figure 5).

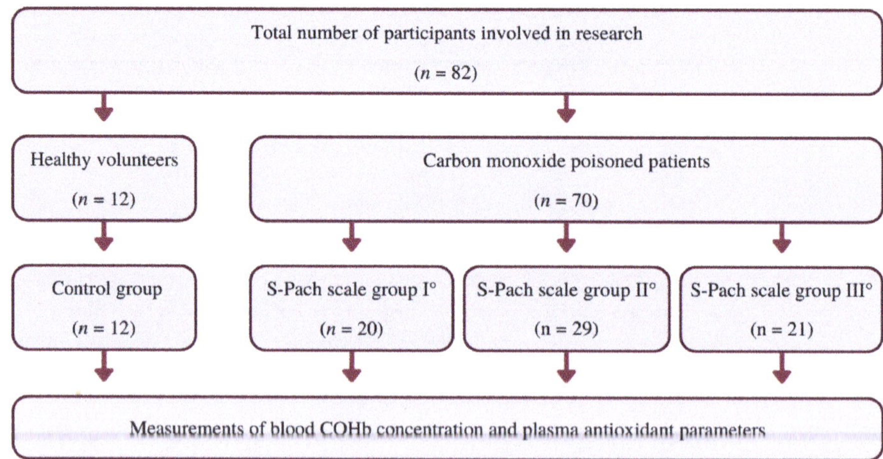

Figure 5. Flow chart showing the division of research participants into the control group and also the groups of CO-poisoned patients according to the S-Pach scale.

3.2. Measurement of Oxidative-Stress-Related Parameters

3.2.1. Measurement of Catalase (CAT)

Catalase (CAT; EC 1.11.1.6) activity in plasma samples was measured following the kinetic method described by Aebi [38]. Changes in absorbance from the beginning to the end of the reaction were measured at a wavelength of 240 nm. The enzymatic activity was calculated as U/mL of plasma. In this case, one unit of CAT activity was defined as the enzyme that decomposes 1 mol of H_2O_2 per minute. Catalase activity was expressed in U/mg protein.

3.2.2. Measurement of 2,2-Diphenyl-1-picryl-hydrazyl (DPPH)

Plasma antioxidant properties were measured using the DPPH test based on the method reported by Molyneux and modified by Annegowda [39,40]. In general, DPPH, in its stable radical form, absorbs at 517 nm, but its absorption decreases upon reduction by an antioxidant in a sample. A 0.6 mM solution of 2,2-diphenyl-1-picryl-hydrazyl (DPPH) was prepared in methanol. All plasma samples were deproteinated with 10% TCA (trichloroacetic acid) and centrifuged (Micro 200 R) at 4 °C and $2500 \times g$. The percentage of DPPH in the supernatant obtained was measured at 30 min. All samples were measured in triplicate. The absorption of the control sample (a DPPH methanol solution) was measured at the beginning and end of the analysis. The percentage of inhibition of DPPH was calculated using the following formula. DPPH % = [(ADPPH − A30 min)/ADPPH] × 100, where A30 min is the mean absorbance of the sample, and ADPPH is the absorbance of the control sample.

3.2.3. Sulfhydryl Groups (-SH)

The total sulfhydryl content was determined using Ellman's method with some modifications [41]. The absorbance was measured at 412 nm, and the SH content was calculated using a molar extinction coefficient of $13,600 \ M^{-1} \ cm^{-1}$. The -SH group content was expressed in mmol/mg protein.

3.2.4. Measurement of Reduced Glutathione (GSH)

Reduced glutathione levels were determined in blood plasma after deproteination with trichloroacetic acid (TCA). The free -SH groups were determined using the Ellman method in the supernatant obtained, and the GSH content was expressed in the GSH μmol/mg protein [41].

3.2.5. Measurement of the Carbonyl Group (=CO Derivates)

The protein carbonyl group contents were measured using the method described by Levine et al. [42]. The absorbance was measured at 370 nm. The carbonyl group content (=CO derivates) was expressed in nmol/mg protein.

3.3. Statistical Analysis

The antioxidative parameters of serum (catalase—CAT, DPPH, total sulfhydryl contents—SH, reduced glutathione levels—GSH, protein carbonyl groups—=CO derivates) were evaluated for 82 patients divided into four groups: C control group, Pach scale I°, Pach scale II°, and Pach scale III°. Using Dixon calculations, the outliers' results were removed (a total of eight measurements) [43]. The statistical hypothesis that the impact of carbon monoxide poisoning severity on antioxidative parameters is significant was verified using an analysis of variance (ANOVA) [44–51]. Preliminary analysis (two-factor ANOVA) showed no statistically significant differences in dependent variables between men and women. For this reason, one-way ANOVA was performed to verify the null hypothesis. ANOVA was conducted for each of the following dependent variables: CAT, DPPH, SH, GSH, =CO derivates, quotient CAT and GSH (CAT/GSH), quotient SH and GSH (SH/GSH), and time of exposure to carbon monoxide (T-CO). The Pach scale (S-Pach) was the intra-group factor. The S-Pach was analyzed on four levels: control—C (the patients without carbon monoxide poisoning), I°, II°, and III° (the patients with carbon monoxide poisoning). If the null hypothesis is rejected based on the ANOVA results (no significant differences between the groups), post hoc tests (multiple comparison tests) must be performed. To determine homogeneous groups, the honest significant difference (HSD) of Tukey's test was used [44,45,51]. Data were analyzed using the STATISTICA v.14.0 software (Dell Inc. (Tulsa, OK, USA, 2016)), Dell Statistica (data analysis software system), version 13 (software.dell.com). Values of $p < 0.05$ were considered statistically significant.

4. Conclusions

In conclusion, the results emphasized the need to observe the redox status of CO-poisoned patients in the acute phase of intoxication and after the discontinuation of therapy. Differences in antioxidant status are difficult to explain due to the diversity of patients in the study group. Therefore, after critical evaluation, it can be concluded that it is challenging to use oxidative-stress-related parameters to evaluate the severity of CO poisoning, outcome, and treatment options. Compared to the literature [28], a correlation is observed only with a few parameters. To our knowledge, this is the first critical observation to date compared to the widespread belief in the literature that such relationships occur. Many variables affect the course and severity of CO poisoning, such as exposure time, age, comorbidity, personal sensitivity, COHb level, and oxygen therapy. More detailed observations on antioxidant status could only be collected in experiments conducted on animal models under more strictly controlled conditions. Recognition of these relationships allows for a proper approach to CO poisoning with oxygen therapy and antioxidant use. However, careful statistical analyses reveal that the problem is more complex and that discussions cannot be conducted without additional research. Therefore, further research is necessary, focusing on the molecular mechanism.

Supplementary Materials: The supporting information can be downloaded at: https://www.mdpi.com/article/10.3390/ijms241310784/s1.

Author Contributions: Conceptualization, P.H., R.F., S.F., E.G., E.D.E., K.J., M.K. and M.N.; methodology, R.F., S.F., E.G. and E.D.E.; formal analysis, S.F. and K.J.; investigation, P.H., R.F., S.F., E.G., E.D.E., K.J. and M.K.; data curation, P.H., R.F., S.F., E.G., E.D.E., K.J. and M.K.; writing—original draft preparation, P.H., R.F., S.F., E.G., E.D.E., K.J., M.K. and M.N.; writing—review and editing, S.F., K.J. and M.N.; supervision, K.J. All authors have read and agreed to the published version of the manuscript.

Funding: Funding for this work was provided by Jagiellonian University-Medical College (grant K/ZDS/002430). The authors declare no competing financial interest. Ebru Derici Eker's (Faculty of Pharmacy, Mersin University, Mersin, Turkey) participation was financed by the students program Erasmus in Department of Food Chemistry and Nutrition, Faculty of Pharmacy, Jagiellonian University Medical College, Medyczna 9 str., 30-688 Krakow, Poland.

Institutional Review Board Statement: The study was conducted in accordance with the Declaration of Helsinki and approved by the Research Ethics Commission of the Jagiellonian University, Kraków; Approval Nr. Not applicable.

Informed Consent Statement: Informed consent was obtained from all the study participants. Patient information has been anonymized.

Data Availability Statement: The datasets generated during and/or analyzed during the current study are available from Kamil Jurowski (toksykologia@ur.edu.pl) upon reasonable request.

Conflicts of Interest: The authors declare no conflict of interest.

References

1. Świderska, A.; Sein Anand, J. Wybrane zagadnienia dotyczące ostrych zatruć ksenobiotykami w Polsce w 2010 roku. *Przegl. Lek.* **2012**, *69*, 409–414.
2. Nelson, L.; Goldfrank, L.R. (Eds.) *Goldfrank's Toxicologic Emergencies*, 9th ed.; McGraw-Hill Medical: New York, NY, USA, 2011; ISBN 978-0-07-160593-9.
3. Kao, L.W.; Nañagas, K.A. Toxicity Associated with Carbon Monoxide. *Clin. Lab. Med.* **2006**, *26*, 99–125. [CrossRef]
4. Myers, R.A.; Snyder, S.K.; Emhoff, T.A. Subacute Sequelae of Carbon Monoxide Poisoning. *Ann. Emerg. Med.* **1985**, *14*, 1163–1167. [CrossRef]
5. Choi, Y.K.; Por, E.D.; Kwon, Y.-G.; Kim, Y.-M. Regulation of ROS Production and Vascular Function by Carbon Monoxide. *Oxidative Med. Cell. Longev.* **2012**, *2012*, 1–17. [CrossRef] [PubMed]
6. Queiroga, C.S.F.; Almeida, A.S.; Vieira, H.L.A. Carbon Monoxide Targeting Mitochondria. *Biochem. Res. Int.* **2012**, *2012*, 749845. [CrossRef] [PubMed]
7. Wilson, J.L.; Bouillaud, F.; Almeida, A.S.; Vieira, H.L.; Ouidja, M.O.; Dubois-Randé, J.-L.; Foresti, R.; Motterlini, R. Carbon Monoxide Reverses the Metabolic Adaptation of Microglia Cells to an Inflammatory Stimulus. *Free Radic. Biol. Med.* **2017**, *104*, 311–323. [CrossRef] [PubMed]
8. Chance, B.; Erecinska, M.; Wagner, M. Mitochondrial Responses to Carbon Monoxide Toxicity. *Ann. N. Y. Acad. Sci.* **1970**, *174*, 193–204. [CrossRef]
9. Thom, S.R.; Ohnishi, S.T.; Ischiropoulos, H. Nitric Oxide Released by Platelets Inhibits Neutrophil B2 Integrin Function Following Acute Carbon Monoxide Poisoning. *Toxicol. Appl. Pharmacol.* **1994**, *128*, 105–110. [CrossRef] [PubMed]
10. Wright, J. Chronic and Occult Carbon Monoxide Poisoning: We Don't Know What We're Missing. *Emerg. Med. J.* **2002**, *19*, 386–390. [CrossRef]
11. Min, S.K. A Brain Syndrome Associated with Delayed Neuropsychiatric Sequelae Following Acute Carbon Monoxide Intoxication. *Acta Psychiatr. Scand.* **1986**, *73*, 80–86. [CrossRef]
12. Mathieu, D.; Nolf, M.; Durocher, A.; Saulnier, F.; Frimat, P.; Furon, D.; Wattel, F. Acute Carbon Monoxide Poisoning Risk of Late Sequelae and Treatment by Hyperbaric Oxygen. *J. Toxicol. Clin. Toxicol.* **1985**, *23*, 315–324. [CrossRef]
13. Neubauer, R.A.; Neubauer, V.; Nu, A.K.C.; Maxfield, W.S. Treatment of Late Neurologic Sequelae of Carbon Monoxide Poisoning with Hyperbaric Oxygenation: A Case Series. *J. Am. Physicians Surg.* **2006**, *11*, 56–59.
14. Mannaioni, P.F.; Vannacci, A.; Masini, E. Carbon Monoxide: The Bad and the Good Side of the Coin, from Neuronal Death to Anti-Inflammatory Activity. *Inflamm. Res.* **2006**, *55*, 261–273. [CrossRef] [PubMed]
15. Persson, H.E.; Sjöberg, G.K.; Haines, J.A.; de Garbino, J.P. Poisoning Severity Score. Grading of Acute Poisoning. *J. Toxicol. Clin. Toxicol.* **1998**, *36*, 205–213. [CrossRef] [PubMed]
16. Pach, J.; Persson, H.; Sancewicz-Pach, K.; Groszek, B. Comparison between the poisoning severity score and specific grading scales used at the Department of Clinical Toxicology in Krakow. *Przegl. Lek.* **1999**, *56*, 401–408. [PubMed]
17. Thom, S.R.; Keim, L.W. Carbon Monoxide Poisoning: A Review Epidemiology, Pathophysiology, Clinical Findings, and Treatment Options Including Hyperbaric Oxygen Therapy. *J. Toxicol. Clin. Toxicol.* **1989**, *27*, 141–156. [CrossRef]
18. Gorman, D.; Drewry, A.; Huang, Y.L.; Sames, C. The Clinical Toxicology of Carbon Monoxide. *Toxicology* **2003**, *187*, 25–38. [CrossRef]
19. O'Driscoll, B.R.; Howard, L.S.; Davison, A.G. British Thoracic Society BTS Guideline for Emergency Oxygen Use in Adult Patients. *Thorax* **2008**, *63* (Suppl. S6), vi1-68. [CrossRef]
20. Fridovich, I. Oxygen Toxicity: A Radical Explanation. *J. Exp. Biol.* **1998**, *201*, 1203–1209. [CrossRef]
21. Piantadosi, C.A. Carbon Monoxide, Reactive Oxygen Signaling, and Oxidative Stress. *Free Radic. Biol. Med.* **2008**, *45*, 562–569. [CrossRef]
22. Bryan, C.L.; Jenkinson, S.G. Oxygen Toxicity. *Clin. Chest Med.* **1988**, *9*, 141–152. [CrossRef] [PubMed]

23. Brown, S.D.; Piantadosi, C.A. Recovery of Energy Metabolism in Rat Brain after Carbon Monoxide Hypoxia. *J. Clin. Invest.* **1992**, *89*, 666–672. [CrossRef] [PubMed]
24. Skrzydlewska, E.; Farbiszewski, R.; Gacko, M. Wpływ oksydacyjnej modyfikacji białek na równowagę proteazy-antyproteazy i proteolizę komórkową. *Postępy Hig. Med. Doświadczalnej* **1997**, *51*, 443–456.
25. Kallstrom, T.J. American Association for Respiratory Care (AARC) Clinical Practice Guideline: Oxygen Therapy for Adults in the Acute Care Facility—2002 Revision & Update. *Respir. Care* **2002**, *47*, 717–720.
26. Abilés, J.; de la Cruz, A.P.; Castaño, J.; Rodríguez-Elvira, M.; Aguayo, E.; Moreno-Torres, R.; Llopis, J.; Aranda, P.; Argüelles, S.; Ayala, A.; et al. Oxidative Stress Is Increased in Critically Ill Patients According to Antioxidant Vitamins Intake, Independent of Severity: A Cohort Study. *Crit. Care* **2006**, *10*, R146. [CrossRef]
27. Thérond, P.; Bonnefont-Rousselot, D.; Davit-Spraul, A.; Conti, M.; Legrand, A. Biomarkers of Oxidative Stress: An Analytical Approach. *Curr. Opin. Clin. Nutr. Metab. Care* **2000**, *3*, 373–384. [CrossRef]
28. Kavakli, H.S.; Erel, O.; Delice, O.; Gormez, G.; Isikoglu, S.; Tanriverdi, F. Oxidative Stress Increases in Carbon Monoxide Poisoning Patients. *Hum. Exp. Toxicol.* **2011**, *30*, 160–164. [CrossRef]
29. Angelova, P.R.; Myers, I.; Abramov, A.Y. Carbon Monoxide Neurotoxicity Is Triggered by Oxidative Stress Induced by ROS Production from Three Distinct Cellular Sources. *Redox Biol.* **2023**, *60*, 102598. [CrossRef]
30. Li, S.; Yan, T.; Yang, J.Q.; Oberley, T.D.; Oberley, L.W. The Role of Cellular Glutathione Peroxidase Redox Regulation in the Suppression of Tumor Cell Growth by Manganese Superoxide Dismutase. *Cancer Res.* **2000**, *60*, 3927–3939.
31. Pamplona, R.; Costantini, D. Molecular and Structural Antioxidant Defenses against Oxidative Stress in Animals. *Am. J. Physiol. Regul. Integr. Comp. Physiol.* **2011**, *301*, R843–R863. [CrossRef]
32. Halliwell, B.; Gutteridge, J.M.C. *Free Radicals in Biology and Medicine*; Oxford University Press. Oxford, UK, 2015, ISBN 978-0-19-871747-8.
33. Góth, L. Lipid and Carbohydrate Metabolism in Acatalasemia. *Clin. Chem.* **2000**, *46*, 564–566. [CrossRef]
34. L'Ecuyer, T.; Allebban, Z.; Thomas, R.; Vander Heide, R. Glutathione S-Transferase Overexpression Protects against Anthracycline-Induced H9C2 Cell Death. *Am. J. Physiol. Heart Circ. Physiol.* **2004**, *286*, H2057–H2064. [CrossRef] [PubMed]
35. Balcerczyk, A.; Bartosz, G. Thiols Are Main Determinants of Total Antioxidant Capacity of Cellular Homogenates. *Free. Radic. Res.* **2003**, *37*, 537–541. [CrossRef] [PubMed]
36. de Zwart, L.L.; Meerman, J.H.; Commandeur, J.N.; Vermeulen, N.P. Biomarkers of Free Radical Damage Applications in Experimental Animals and in Humans. *Free Radic. Biol. Med.* **1999**, *26*, 202–226. [CrossRef] [PubMed]
37. Kadiiska, M.B.; Gladen, B.C.; Baird, D.D.; Germolec, D.; Graham, L.B.; Parker, C.E.; Nyska, A.; Wachsman, J.T.; Ames, B.N.; Basu, S.; et al. Biomarkers of Oxidative Stress Study II: Are Oxidation Products of Lipids, Proteins, and DNA Markers of CCl4 Poisoning? *Free Radic. Biol. Med.* **2005**, *38*, 698–710. [CrossRef] [PubMed]
38. Aebi, H. Catalase in Vitro. In *Methods in Enzymology*; Elsevier: Amsterdam, The Netherlands, 1984; Volume 105, pp. 121–126, ISBN 978-0-12-182005-3.
39. Annegowda, H.; Anwar, L.; Mordi, M.; Ramanathan, S.; Mansor, S. Influence of Sonication on the Phenolic Content and Antioxidant Activity of *Terminalia Catappa* L. Leaves. *Pharmacogn. Res.* **2010**, *2*, 368. [CrossRef]
40. Molyneux, P. The Use of the Stable Free Radical Diphenylpicrylhydrazyl (DPPH) for Estimating Antioxidant Activity. *Songklanakarin J. Sci. Technol.* **2004**, *26*, 211–219.
41. Ellman, G.L. Tissue Sulfhydryl Groups. *Arch. Biochem. Biophys.* **1959**, *82*, 70–77. [CrossRef]
42. Levine, R.L.; Williams, J.A.; Stadtman, E.P.; Shacter, E. Carbonyl Assays for Determination of Oxidatively Modified Proteins. In *Methods in Enzymology*; Elsevier: Amsterdam, The Netherlands, 1994; Volume 233, pp. 346–357, ISBN 978-0-12-182134-0.
43. Verma, S.P.; Quiroz-Ruiz, A. Critical Values for Six Dixon Tests for Outliers in Normal Samples up to Sizes 100, and Applications in Science and Engineering. *Rev. Mex. Cienc. Geol.* **2006**, *23*, 133–161.
44. Francik, R.; Kryczyk-Kozioł, J.; Krośniak, M.; Francik, S.; Hebda, T.; Pedryc, N.; Knapczyk, A.; Berköz, M.; Ślipek, Z. The Influence of Organic Vanadium Complexes on an Antioxidant Profile in Adipose Tissue in Wistar Rats. *Materials* **2022**, *15*, 1952. [CrossRef]
45. Francik, S.; Knapik, P.; Łapczyńska-Kordon, B.; Francik, R.; Ślipek, Z. Values of Selected Strength Parameters of Miscanthus × Giganteus Stalk Depending on Water Content and Internode Number. *Materials* **2022**, *15*, 1480. [CrossRef] [PubMed]
46. Jówko, E.; Długołęcka, B.; Cieśliński, I.; Kotowska, J. Polymorphisms in Genes Encoding VDR, CALCR and Antioxidant Enzymes as Predictors of Bone Tissue Condition in Young, Healthy Men. *Int. J. Mol. Sci.* **2023**, *24*, 3373. [CrossRef] [PubMed]
47. Remigante, A.; Morabito, R.; Spinelli, S.; Trichilo, V.; Loddo, S.; Sarikas, A.; Dossena, S.; Marino, A. D-Galactose Decreases Anion Exchange Capability through Band 3 Protein in Human Erythrocytes. *Antioxidants* **2020**, *9*, 689. [CrossRef]
48. Remigante, A.; Spinelli, S.; Basile, N.; Caruso, D.; Falliti, G.; Dossena, S.; Marino, A.; Morabito, R. Oxidation Stress as a Mechanism of Aging in Human Erythrocytes: Protective Effect of Quercetin. *Int. J. Mol. Sci.* **2022**, *23*, 7781. [CrossRef]
49. Remigante, A.; Spinelli, S.; Straface, E.; Gambardella, L.; Caruso, D.; Falliti, G.; Dossena, S.; Marino, A.; Morabito, R. Acai (*Euterpe Oleracea*) Extract Protects Human Erythrocytes from Age-Related Oxidative Stress. *Cells* **2022**, *11*, 2391. [CrossRef] [PubMed]

50. Remigante, A.; Spinelli, S.; Straface, E.; Gambardella, L.; Caruso, D.; Falliti, G.; Dossena, S.; Marino, A.; Morabito, R. Antioxidant Activity of Quercetin in a H2O2-Induced Oxidative Stress Model in Red Blood Cells: Functional Role of Band 3 Protein. *Int. J. Mol. Sci.* **2022**, *23*, 991. [CrossRef] [PubMed]

51. Zawiślak, A.; Francik, R.; Francik, S.; Knapczyk, A. Impact of Drying Conditions on Antioxidant Activity of Red Clover (*Trifolium Pratense*), Sweet Violet (*Viola Odorata*) and Elderberry Flowers (*Sambucus Nigra*). *Materials* **2022**, *15*, 3317. [CrossRef]

International Journal of
Molecular Sciences

Article

Toxicity of Metal Ions Released from a Fixed Orthodontic Appliance to Gastrointestinal Tract Cell Lines

Ksenija Durgo [1], Sunčana Orešić [1], Marijana Rinčić Mlinarić [2], Željka Fiket [3] and Gordana Čanadi Jurešić [4,*]

[1] Department of Biochemical Engineering, Faculty of Food Technology and Biotechnology, Universtiy of Zagreb, Pierrotijeva 6, 10000 Zagreb, Croatia; kdurgo@pbf.hr (K.D.); suncanaoresic@gmail.com (S.O.)

[2] Private Orthodontic Practice, Katarine Zrinske 1b, 23000 Zadar, Croatia; orto.rincic@gmail.com

[3] Division for Marine and Environmental Research, Ruđer Bošković Institute, Bijenička 54, 10000 Zagreb, Croatia

[4] Department of Medical Chemistry, Biochemistry and Clinical Chemistry, Faculty of Medicine, Universtiy of Rijeka, B. Branchetta 20, 51000 Rijeka, Croatia

* Correspondence: gordanacj@uniri.hr

Abstract: The mechanism of toxicity and cellular response to metal ions present in the environment is still a very current area of research. In this work, which is a continuation of the study of the toxicity of metal ions released by fixed orthodontic appliances, eluates of archwires, brackets, ligatures, and bands are used to test the prooxidant effect, cytotoxicity, and genotoxicity on cell lines of the gastrointestinal tract. Eluates obtained after three immersion periods (3, 7, and 14 days) and with known amounts and types of metal ions were used. Four cell lines—CAL 27 (human tongue), Hep-G2 (liver), AGS (stomach) and CaCo-2 (colon)—were treated with each type of eluate at four concentrations ($0.1\times$, $0.5\times$, $1.0\times$, and $2.0\times$) for 24 h. Most eluates had toxic effects on CAL 27 cells over the entire concentration range regardless of exposure time, while CaCo-2 proved to be the most resistant. In AGS and Hep-G2 cells, all samples tested induced free radical formation, with the highest concentration ($2\times$) causing a decrease in free radicals formed compared to the lowest concentrations. Eluates containing Cr, Mn, and Al showed a slight pro-oxidant effect on DNA (on plasmid φX-174 RF I) and slight genotoxicity (comet assay), but these effects are not so great that the human body could not "resist" them. Statistical analysis of data on chemical composition, cytotoxicity, ROS, genotoxicity, and prooxidative DNA damage shows the influence of metal ions present in some eluates on the toxicity obtained. Fe and Ni are responsible for the production of ROS, while Mn and Cr have a great influence on hydroxyl radicals, which cause single-strand breaks in supercoiled plasmid DNA in addition to the production of ROS. On the other hand, Fe, Cr, Mn, and Al are responsible for the cytotoxic effect of the studied eluates. The obtained results confirm that this type of research is useful and brings us closer to more accurate in vivo conditions.

Keywords: metals; orthodontic appliance; prooxidant effect; cytotoxicity; genotoxicity; gastrointestinal tract cell lines

Citation: Durgo, K.; Orešić, S.; Rinčić Mlinarić, M.; Fiket, Ž.; Jurešić, G.Č. Toxicity of Metal Ions Released from a Fixed Orthodontic Appliance to Gastrointestinal Tract Cell Lines. *Int. J. Mol. Sci.* **2023**, *24*, 9940. https://doi.org/10.3390/ijms24129940

Academic Editors: Rossana Morabito and Alessia Remigante

Received: 10 May 2023
Revised: 6 June 2023
Accepted: 7 June 2023
Published: 9 June 2023

1. Introduction

Although there are still numerous studies and tests on the biocompatibility (corrosion behavior and toxicity) of orthodontic materials, the results obtained are not uniform and no clear conclusions can be drawn because of the increasing variability of the materials, their composition, and the manufacturing processes [1,2]. Therefore, it is still useful to test each type of orthodontic material/alloy individually, make different combinations of devices from different materials, immerse and elute them in different media, and test the toxicity of the different/specific combinations by using in vitro models.

Main base metal orthodontic alloys commonly used for orthodontic appliances are stainless steel (SS), cobalt—chromium, nickel—titanium (NiTi), and beta titanium [3].

Fixed orthodontic appliances, the most common and effective form of orthodontic treatment, consist of archwires, bands, ligatures, and brackets that are attached to the teeth with a special adhesive, usually acrylic resin [4]. In this study, we decided to combine NiTi archwires and SS brackets, bands and ligatures as components of a fixed orthodontic appliance and use them in the cytotoxicity study. NiTi alloy is used in orthodontics mainly for wires. Its unique properties—superelasticity and shape memory effect—are limited to a narrow composition range close to its equiatomic composition. All stainless steel parts are made of austenitic steel, which is graded differently according to specific requirements of every component and has good corrosion resistance under many different environmental conditions [3].

The oral cavity is a very dynamic environment, subject to constant changes in composition, temperature, pH, and saliva quantity [3,4]. Changes in pH, as we see in patients with poor oral hygiene and plaque accumulation, can affect the stability of the used materials by accelerating corrosion processes [5,6]. Glossitis, metallic taste in the mouth, dry lips, inflammatory redness, and hypertrophy of the gums are commonly observed in patients and are the consequences of the release of these metals from dental alloys into saliva. Even if the device itself does not cause serious damage to the oral cavity, it may make it more susceptible to damage and toxic substances later in life [1,7,8]. Although studies on the cytotoxicity of these metals and orthodontic appliances have been conducted frequently and most show that cytotoxicity is negligible, it cannot be ruled out that low concentrations of metals are sufficient to cause biological changes in the oral mucosa, especially because they are in the form of mixtures [1,9,10]. The mucosal tissue of the oral cavity can absorb these low levels of metal ions over a long period of time, leading to changes in the genome of the cells of the oral cavity. In addition, metals are not biodegradable and can have irreversible toxic effects due to their accumulation in tissues. In addition, cobalt and nickel have been shown to be carcinogenic to mucous membranes [1]. These metals can accumulate locally and be found in body fluids and tissues unrelated to the oral cavity. In general, exposure to metal ions causes various pathological effects such as inflammatory responses, alterations in the oxidative stress response, increased lipid peroxidation, and changes in DNA repair mechanisms [1].

Nowadays, fibroblast cell lines (periodontal ligament, gingival, or dermal [9–14]), gastrointestinal tract cell lines [15,16], and sometimes yeast [17,18] are most commonly used for testing. Research has focused mainly on NiTi or stainless (arch)wires [2,13,19,20], brackets [2,12,14], and bands [2,9,10,16,21].

The aim of this study is to determine the cytotoxic, prooxidant, and genotoxic effects of metal ion eluates prepared separately from each part of a fixed orthodontic appliance (archwires, brackets, ligatures, and bands). The eluates were prepared in artificial saliva after three different exposure times (3, 7, and 14 days) and determined qualitatively and quantitatively by ICP-MS. Thus, the metal ion composition was known for each experimental eluate used. Cell lines from the oral cavity, stomach, liver, and colon, selected to mimic the passage of saliva through the gastrointestinal tract, were used as the biological test system. Possible correlations between ion release profiles and observed prooxidant activity, cytotoxicity, and genotoxicity were investigated. The experimental data are used to determine whether the selected fixed orthodontic appliance is safe to use and what side effects it may cause.

2. Results

This work is a continuation of the study on the toxicity of metal ions released by fixed orthodontic appliances, and for a better understanding it is necessary to mention some of the main conclusions of the previous study [22]. It was found that metal ions were released from all parts of dental alloys studied. The results from ICP-MS indicate that the bands released the highest amounts of metal ions in artificial saliva samples. Bands immersed in artificial saliva for 7 days (B7D) released more than 6 mg/L, while bands immersed for 3 (B3D) or 14 days (B14D) released 3.7 and 3 mg/L, respectively, of all ions detected. According to the manufacturer's specifications, the band should contain 17–19% Cr, 8–10.5% Ni, and the rest Fe. Mn, although not specified, was also detected in all the samples tested. Although Ni accounts for only about 10% of the bands, the content of this ion in artificial saliva samples is very high—highest in sample B14D— accounting for one-third of all detected ions. In sample B7D, which has the highest content of all detected ions, iron accounts for one-third of all detected ions, but Fe and Ni together account for more than 3.5 mg/L.

The content of released nickel ions in the samples of the archwires is slightly higher than the value published by Rinčić-Mlinarić et al. [15] under similar experimental conditions after an exposure time of 14 days (38.61 µg/L in this work versus 15.1–30.4 µg/L in the cited work). In archwires eluates, the content of Ni was highest after 7 days of elution (271 µg/L approx., is consistent with the results of Kuhta et al. [23] and Hwang et al. [24]), while the content of all certain metals was highest after 14 days of elution (1261 µg/L, approx.). According to the results presented in Petković Didović et al. [22], immersion of NiTi archwires in artificial saliva did not show pronounced morphological changes (SEM and EDX). Highly polished NiTi wires did not appear to contain an oxide layer in the as-received condition, but the gradual decrease in eluted Ti concentrations indicated that a thin protective oxide layer (composed mainly of TiO_2) developed during the time of immersion. The immersion in artificial saliva did result in the development of an adherent layer on the surface of the SS bands and ligatures, but based on the obtained results was not of the same type. The quantity of released ions correlates with the formation of layer.

The bands had the highest risk in terms of release of (heavy) metals. During the 7-day elution period, significant amounts of chromium, iron, and nickel, together with aluminum and lead, and copper, zinc, and cobalt were released in saliva.

2.1. Determination of Cytotoxicity

Increased concentrations of metal mixtures may pose a serious threat to cellular stability and function of cellular macromolecules. In this work, the potential cytotoxic effect of eluted saliva samples was investigated using the neutral red method, which measures the ability of viable cells to take up the supravital dye neutral red and bind it in lysosomes. Tongue cells were the most sensitive cells; all saliva samples (eluates) tested were toxic to this cell line at all concentration ranges tested. In addition, a most intriguing and interesting response of CAL27 cells to metal ion treatment was noted. For most samples, viability was lowest at the lowest dose tested ($0.1\times$), with a tendency to increase at a dose of $2\times$. One of the possible explanations for this phenomenon is that Fe depletion leads to G(1)/S arrest and apoptosis, so the presence of increased iron may stimulate cell proliferation [25]. Alberto et al. [26] have shown that manganese can also stimulate cell proliferation. Thus, it is possible that these metals, once present in the mixture, have effects on cell division processes and stimulate cell proliferation at very low concentrations (the dose was $0.1\times$). Considering the cytotoxic effect of the eluates on other cell lines, only Hep 27 cells and a decrease in viability after treatment with eluates of B14D stand out. No cytotoxic effect on stomach and colon cell lines was detected (Figure 1).

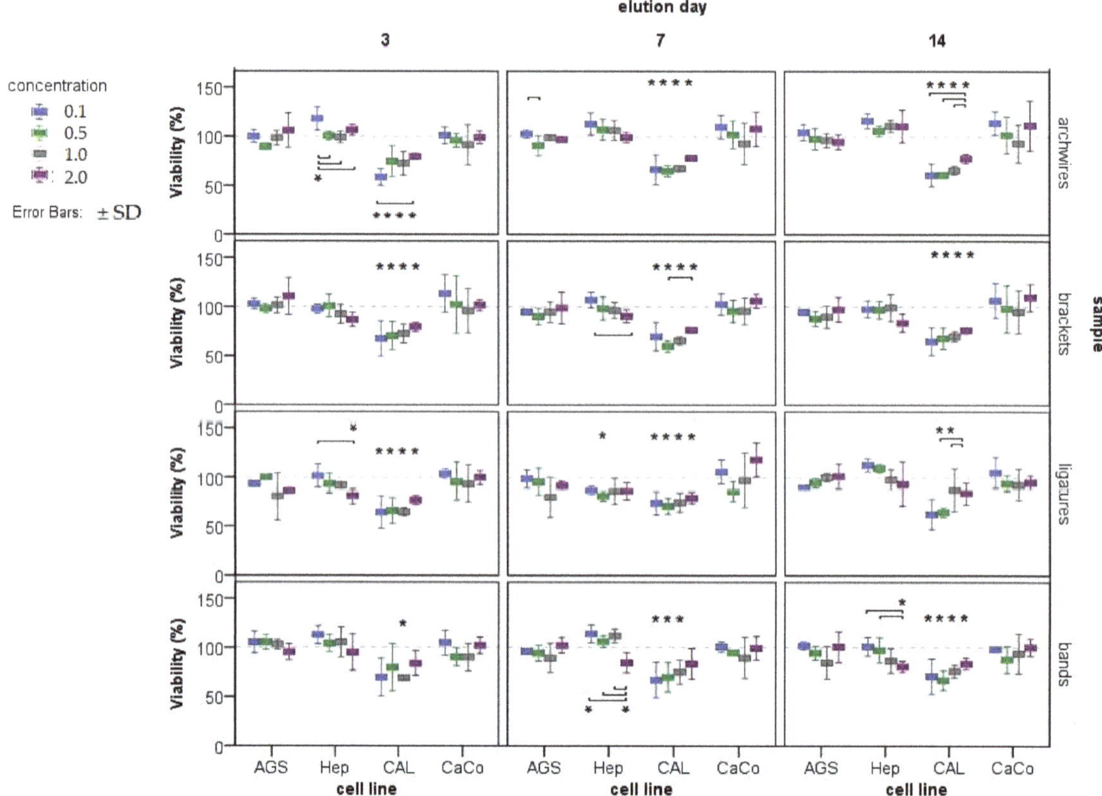

Figure 1. Estimation of cytotoxicity (expressed as cell viability in %) in four cell lines (AGS, Hep-G2, CAL 27, and CaCo-2) after exposure to eluates of metal ions released from various parts of fixed orthodontic appliances (archwires, brackets, ligatures, and bands) at three different immersion periods (3, 7, and 14 days) and four different concentrations of each eluate (0.1×; 0.5×, 1.0×, and 2.0×, represented by different colored boxes). Results are given as mean ± SD. The control group (treated with artificial saliva only) is set at 100% and presented as a dashed line. * indicates a concentration at which viability (%) is significantly different from 100% (control sample), each * is located directly above the data to which it refers, and refers only to that information and not to others in that group; more * means that more concentrations tested are significantly different from 100%; lines/connections are associated concentrations at which cell viability differs significantly (Student–Newman–Keuls post hoc test, $p < 0.05$).

2.2. Determination of Free Radicals

The presence of free radicals tested in four gastrointestinal track cancer cell lines after exposure to eluates of metal ions released from different parts of fixed orthodontic appliance (archwires, brackets, ligatures, bands, and all these parts combined) at four different concentrations (0.1×; 0.5×, 1.0×, and 2.0×) during the exposure period (3, 7, and 14 days) is presented in Figure 2.

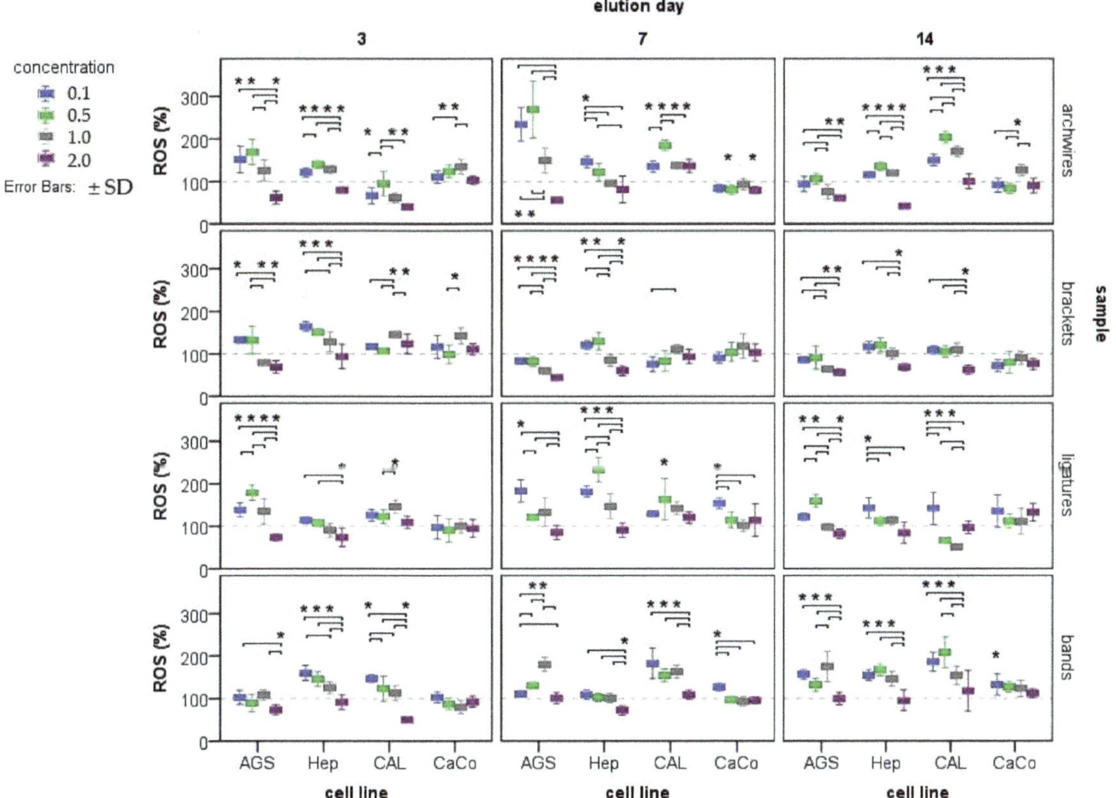

Figure 2. The presence of free radicals was tested in four cell lines (AGS, Hep-G2, CAL 27, and CaCo-2) after exposure to eluates of metal ions released from various orthodontic appliances (archwires, brackets, ligatures, and bands) at three different immersion periods (3, 7, and 14 days) and four different concentrations of each eluate ($0.1\times$; $0.5\times$, $1.0\times$, and $2.0\times$ represented by different colored boxes). Results are expressed as a percentage of sample fluorescence relative to the fluorescence of the control (artificial saliva). The control group (treated with artificial saliva only) is set at 100% and presented as a dashed line. Results are given as mean $\pm$ SD. * indicates a concentration at which ROS content is significantly different from 100%; each * is located just above the data to which it refers, and refers only to that information and not to others in that group; more * means that more concentrations tested are significantly different from 100%; lines/connections are associated concentrations at which produced ROS differ significantly (Student–Newman–Keuls post hoc test, $p < 0.05$).

There is a visible linear or arch-shaped trend of decreased presence of free radicals in the AGS, HEP-G2, and CAL27 cell lines, as the eluate tested concentrations increased. Most often in low tested concentration the effect is prooxidative, while in the $2\times$ concentration the effect is antioxidative. Surprisingly, the highest ROS production was noticed in AGS A7D, $0.5\times$ concentration.

2.3. Testing for DNA Damage

In this work, the plasmid DNA cleavage assay using supercoiled φX-174 RF I plasmid DNA was used to determine whether saliva increases oxidative DNA damage after exposure to UV light. UV production of oxygen radicals leads to oxidative damage and single and double strand breaks.

Figure 3 shows the results of the influence of the studied saliva samples on DNA stability. All samples showed a slight prooxidant effect on DNA (lower ratio of supercoiled/linear form of the plasmid compared to the negative control, but higher ratio compared to the positive control).

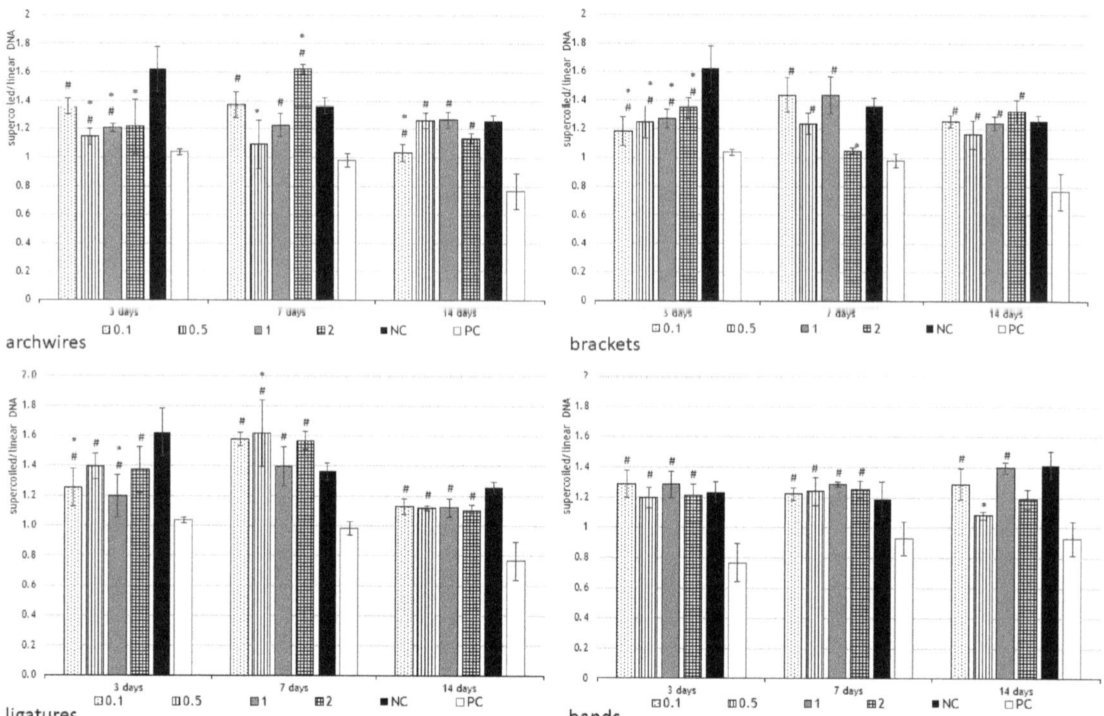

Figure 3. Testing for DNA damage. Eluates prepared from various parts of fixed orthodontic appliance (archwires, brackets, ligatures, and bands) after three immersion periods (3, 7, and 14 days) at four different concentrations (0.1×; 0.5×, 1.0×, and 2.0×) were used to test DNA stability. The plasmid φX-174 RF I was used for this experiment. The plasmid in the form of supercoiled circular DNA converts to a linear form in the presence of metal ions and H_2O_2, producing hydroxyl radicals. After electrophoresis, two bands of different intensities are formed in the gel; the upper band represents the coiled (damaged, converted to linear form) DNA, and the lower band represents the supercoiled DNA. The ratio between the lower and upper bands and comparison to the negative control indicates the potential for the dental alloy eluate to cause DNA damage. Negative control (NC)—plasmid + hydrogen peroxide; positive control (PC)—plasmid + hydrogen peroxide + UV irradiation for 15 min. Results are expressed as mean ± SD. * indicates significantly different compared to the negative control; # indicates significantly different compared to the positive control. Mann–Whitney U test, $p < 0.05$.

2.4. Determination of Genotoxicity

Not all samples are eligible for the genotoxicity test of the comet assay, but only those that exhibit some prooxidant activity but also result in cell survival greater than 80%. Based on these parameters, specific samples with specific concentrations were selected and tested only on the appropriate cell lines (Figure 4). Ligatures 14 days in concentration 1.0× on the CAL 27 cells; then, bands 3 and 7 days, in all samples 0.5× concentration, on the AGS; and finally bands 7 days in 0.5× concentration, on the CaCo2 cell line showed genotoxic effect (having greater values than control).

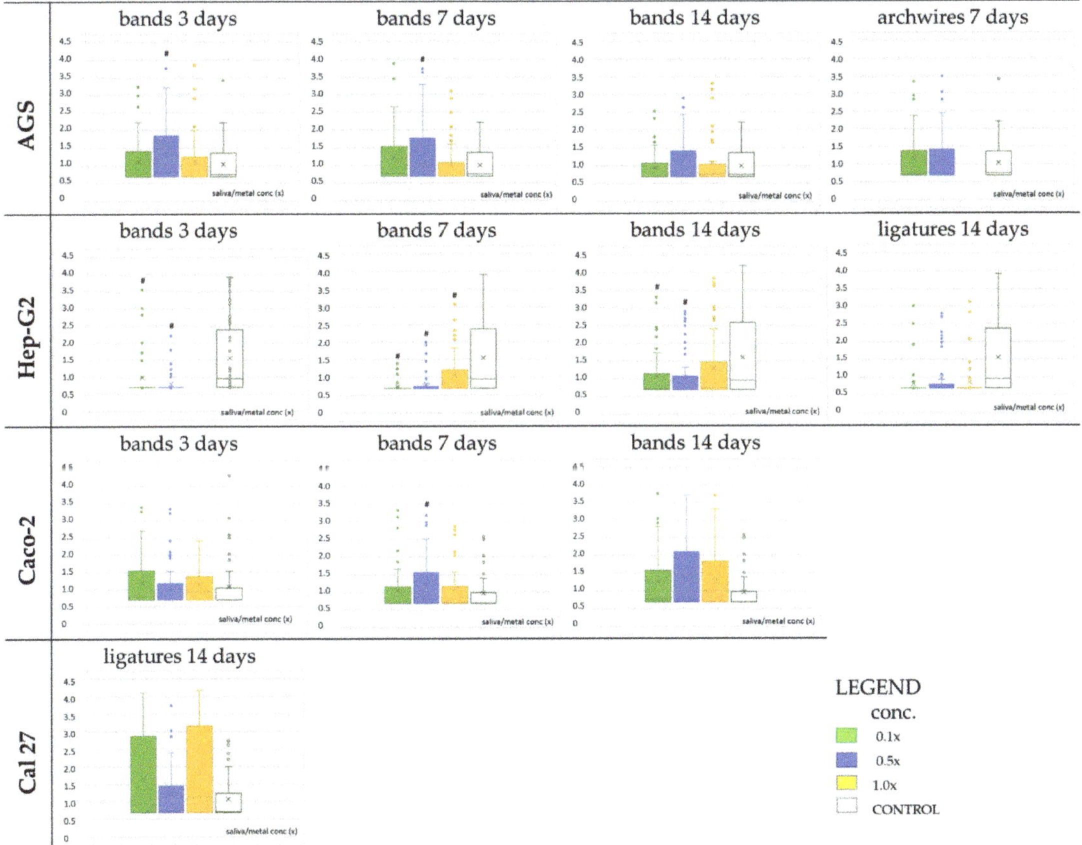

Figure 4. Genotoxicity (comet assay: tail moment) in cultures of selected cell lines induced by eluates obtained from bands immersed for 3, 7, and 14 days, ligatures 14 days, and archwires 7 days, and three different concentrations of each eluate ($0.1\times$; $0.5\times$, and $1.0\times$ represented by different colored boxes) # indicates significantly different compared to the control (LSD test, $p < 0.05$).

Figure 4 shows the results of the genotoxic effect.

These samples have increased concentration of iron, nickel, and chromium compared to the control: in L14D, a certain amount of eluted Al contributes to genotoxicity [27]; in B7D, Fe (>2.3 mg/L) indicates ferroptosis [28]. Synergistic effects of metals can also lead to increased genotoxicity of the samples [29].

Although a significant change compared with the control sample is noticed, none of the samples tested on Hep-G2 cells caused an increase in tail intensity, but instead a decrease or diminished value. In all these experiments, tail intensity was very high in the control samples. The most sensitive cells were CAL 27, on which most eluates had a toxic effect throughout the concentration range, regardless of exposure time. The consequences of the cytotoxic effect were also evident in the prooxidation effect, which could not be detected in these cells due to the increased toxicity.

2.5. Principal Component Analysis of the Potential Correlation

Principal component analysis (PCA) was used to determine the potential correlation between the parts of orthodontic appliance (archwires, brackets, ligatures, and bands in three time periods were used as variables) and the analyses performed (cytotoxicity and the presence of free radicals in four different cell lines, the four most highly expressed metals and aluminum eluted from the orthodontic appliances, and the DNA damage caused by eluted metals at four different concentrations were used as cases). Figure 5 shows the values of the principal components and their contribution to the total variance.

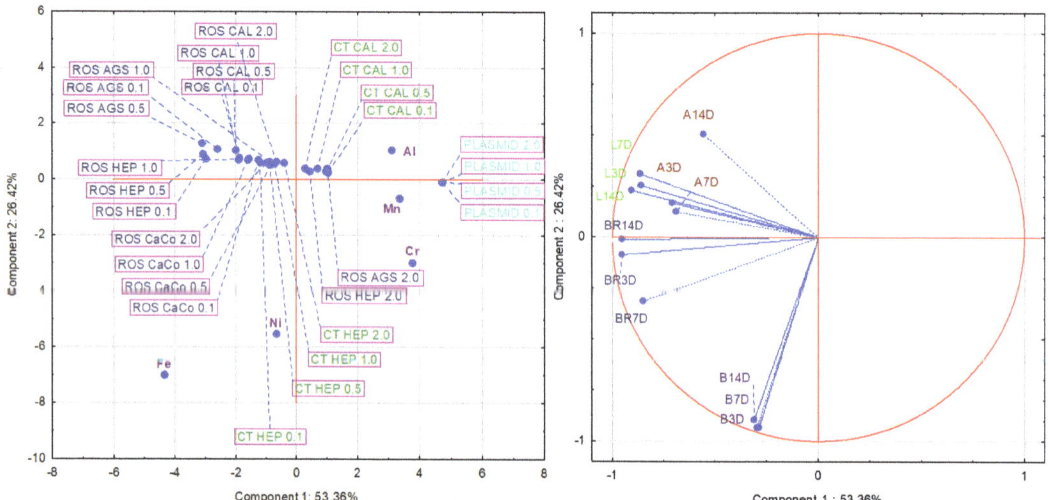

Figure 5. Principal component analysis (PCA) applied to orthodontic appliances (archwires (A), bands (B), ligatures (L), and brackets (BR) for 3, 7, and 14 days), five released metals (Fe, Cr, Ni, Mn, and Al), cytotoxicity (CT) in two different cell lines (CAL 27 and Hep 2 cells); and four different dilutions (0.1, 0.5, 1.0, and 2.0), the presence of free radicals (ROS) in all four cell lines, and all tested dilutions (0.1, 0.5, 1.0, and 2.0) and DNA damage (plasmid) in four dilutions (0.1, 0.5, 1.0, and 2.0) in the experimental group. The planes correlate components 1 and 2 (with eigenvalue 6.40 vs. 3.17). The left side of the figure projects the cases, while the right side shows the variables.

The number of components used in the PCA was determined using the Cattell scree test. On this basis, three principal components (expressed with eigenvalue > 1) were used for further analysis, and their values along with their contribution to the total variance are shown in Table 1. The first component explains 53.36% of the total variance, the second 26.42%, and the third 9.57%.

The eigenvectors of the correlation matrix (presented in Table 2) were used to interpret the principal component analysis. Component 1 is defined mainly by the brackets 3D and 14D and the ligatures 3D, 7D, and 14D. Component 2 is defined by the archwires 14D and the bands 3D, 7D, and 14D. Component 3 is defined by only three variables: archwires 3D and 7D and brackets 7D. It is worth noting that all variables defining component 1 are on the left side of the factor plane and have a negative value. Combining the results of both factor planes shows that the variables defining component 2 are related to the iron and nickel content of these samples, while that of the cases correlate genotoxicity with the content of manganese, chromium, and aluminum.

Table 1. Eigenvalues of the principal component analysis (PCA) with total variance analysis for orthodontic appliances (archwires, bands, ligatures, and brackets, for 3, 7, and 14 days), five released metals (Fe, Cr, Ni, Mn, and Al), cytotoxicity in four different cell lines (CAL cells—dilutions $0.1\times$, $0.5\times$, $1.0\times$, and $2.0\times$; Hep 2 cells—dilutions $1.0\times$ and $2.0\times$; AGS cells—dilutions $1.0\times$ and $2.0\times$; and CaCo2 cells—$1.0\times$), the presence of free radicals in four cell lines (CAL cells—dilutions $1.0\times$ and $2.0\times$; Hep 2 cells—dilutions $1.0\times$ and $2.0\times$; AGS cells—dilutions $1.0\times$ and $2.0\times$; and CaCo2 cells—dilutions: $0.1\times$, $1.0\times$, and $2.0\times$), and DNA damage in four dilutions ($0.1\times$, $0.5\times$, $1.0\times$, and $2.0\times$) in the experimental group. The number of components is five.

Component	Eigenvalue	Total Variance (%)	Cumulative Eigenvalue	Cumulative (%)
1	6.40	53.36	6.40	53.36
2	3.17	26.42	9.57	29.78
3	1.17	9.57	10.74	89.57
4	0.54	4.54	11.29	94.08
5	0.24	2.04	11.53	96.12

Table 2. Eigenvector of correlation matrix for orthodontic appliances (archwires, brackets, ligatures, and bands for 3, 7, and 14 days), five released metals (Fe, Cr, Ni, Mn, and Al), cytotoxicity in four different cell lines (CAL cells—dilutions $0.1\times$, $0.5\times$, $1.0\times$, and $2.0\times$; Hep 2 cells—dilutions $1.0\times$ and $2.0\times$; AGS cells—dilutions $1.0\times$ and $2.0\times$; and CaCo2 cells—$1.0\times$), the presence of free radicals in four cell lines (CAL cells—dilutions $1.0\times$ and $2.0\times$; Hep 2 cells—dilutions $1.0\times$ and $2.0\times$; AGS cells—dilutions $1.0\times$ and $2.0\times$; and CaCo2 cells—dilutions $0.1\times$, $1.0\times$, and $2.0\times$), and DNA damage in four dilutions ($0.1\times$, $0.5\times$, $1.0\times$, and $2.0\times$) in the experimental group.

Variable	Component 1	Component 2	Component 3
archwires 3 days	−0.71	0.17	0.59
archwires 7 days	−0.69	0.13	0.67
archwires 14 days	−0.56	0.51	0.09
brackets 3 days	−0.95	−0.09	−0.23
brackets 7 days	−0.85	−0.31	−0.36
brackets 14 days	−0.95	−0.01	−0.15
ligature 3 days	−0.91	0.23	−0.10
ligature 7 days	−0.87	0.31	−0.12
ligature 14 days	−0.86	0.26	−0.17
bands 3 days	−0.29	−0.93	0.12
bands 7 days	−0.30	−0.94	−0.11
bands 14 days	−0.31	−0.90	0.28

3. Discussion

By definition, dental alloys are a mixture of two or more metals with more or less affinity for migration in the environment. The strength and concentration of migration of the metals depends largely on the composition of the dental alloy and the physicochemical properties of the environment. The grain structure of the alloy also predisposes to corrosion and often plays a different role in different dental alloys [2,19,30]. Since there is no "ideal" dental alloy from which no metals migrate, the question is which metals migrate and what is the migration rate from the alloys to the environment? Finally, the most important question is how the constant contact of the buccal cells and the epithelial cells lining the digestive system with the metal mixture affects the stability of the cellular macromolecules.

The tendency of a metal to corrode is defined by its electrode potential, so metals with higher negative potential (such as nickel, iron, chromium, manganese) are more likely to corrode. Noble metals with positive potential, such as gold or platinum, are less reactive. The most negative potential of abovementioned metals is chromium, but it is often added into dental alloys since it reacts in the way it forms oxide that will, in further reactions, protect dental alloy, and prevent flow of metals from alloys [30]. The explanation for the unpredictable release of heavy metals from the dental alloys and bands obtained in this study, which does not depend on the days of incubation in saliva, lies in the mechanism mentioned above, which includes the manufacturing process and the type of alloy, the progress of galvanic corrosion, and the formation of a passive protective or adherent layer over time [22].

Cytotoxicity is only preliminary information about the potential for a chemical to disrupt the normal functioning of the cell and, ultimately, to cause cell death, so it is very important to determine the possible mechanisms responsible for this effect. The results of in vitro analyses may be indicative of the effects observed in vivo, although the presence of a cytotoxic effect in vitro does not mean that the material is toxic for use in vivo. This is due to the conditions of laboratory testing, which differ significantly from the complex oral environment [1,16]. Nevertheless, the absence of a cytotoxic effect is a guarantee of a good clinical response [3]. The SS bands released significant amounts of nickel, iron, and chromium. According to the data from several studies [22,31,32], stainless steel is less desirable as a biomaterial for the production of fixed orthodontic appliances because it releases the largest amount of nickel and chromium and is the least biocompatible material. The total amount of metals released from archwires, brackets, and ligatures was significantly lower than for bands, in this study. Similar findings were obtained by Wendl et al. [2]. The highest concentration of nickel measured was 1497 µg/mL (B7D incubation), iron 2385 µg/mL (B7D), chromium 723 µg/mL (B7D), and manganese 177 µg/mL (B14D). Toxicity data for the most abundant metals suggest that these concentrations are below those that cause toxic effects [29]. Wen et al. [33] tested various concentrations of nickel on seven cell lines and showed that concentrations below 1200 mg/mL were not toxic. None of the samples tested in this work exceeded this concentration, so it can be concluded that nickel is not responsible for the toxic effects of saliva. Costa et al. [32] found that the highest nickel concentration was detected in the extracts of an AISI SS bracket after an exposure time of 42 days (4.46 ± 0.68 g/mL) and in the low-nickel extracts of the SS bracket after 63 days (0.07 ± 0.01 g/mL). Maximum manganese release was also observed from the AISI 304 SS bracket extracts after 42 days (0.90 ± 0.05 g/mL) and from the low-nickel SS bracket extracts after 21 days (0.11 ± 0.02 g/mL), indicating that metal elution is highly dependent on the material used. The researchers found varying degrees of cytotoxicity of SS and low-nickel SS bracket extracts in L929 cell cultures. Neither the extracts from the SS bracket nor from the low-nickel SS bracket significantly altered cell viability, but a slight decrease in metabolic activity was observed with the 42-day SS bracket extract. Extracts from the low-nickel SS bracket did not result in significant changes in cell metabolism, and the metabolic activity of cells was not altered by any of the 21-day extracts [32]. Kanaji et al. [34] demonstrated that chromium has no toxic effect on cells at concentrations less than 0.5 mM, equivalent to 25.998 µg/mL, but metal ions released from metal-on-metal bearing surfaces have potentially cytotoxic effects on MLO-Y4 osteocytes in vitro. Wu et al. [35] showed that exposure to 10 µM potassium dichromate ($K_2Cr_2O_7$) significantly decreased the viability of HK-2 cells after 24 and 48 h of incubation and induced the intracellular formation of ROS. They also demonstrated that the expression levels of markers that activate the apoptotic pathway, including cleaved caspase-3 and poly(ADP-ribose) polymerase, were significantly upregulated after $K_2Cr_2O_7$ treatment of HK-2 cells. The threshold for acute oral toxicity of chromium (III) was 1900–3300 mg/kg [35]. The saliva obtained from the elution of the bands after 7 days with the highest chromium concentration was toxic to the tongue cells, so one of the reasons is exceeding the 700 µg/mL concentration. Concerning toxicity, it should be kept in mind that an important point is the synergistic and additive effect in toxicity when two or more metals are present in the same mixture. Rinčić Mlinarić et al. [15] demonstrated that Ni and Ti ions alone do not have a major cytotoxic effect, but their combination does, indicating their synergistic effect. Although for most of the samples studied in this work the content of individual metals (with the exception of the bands) did not exceed the toxic concentration, they proved to be toxic to the epithelial cells of the tongue. One of the explanations for this is the synergistic effect of the different ions present in these samples. On the other side, considering that the CAL cells are the cancer cells whose viability decreases (from 60% to 80%) when treated with released metal ions, the question arises whether released ions from orthodontic appliances can be considered as anticarcinogenic, which puts the effect of released metal ions into a completely different light.

Nickel, iron, and chromium, the most common metals found in saliva after 3, 7, and 14 days of elution, catalyze free radical formation through a Fenton-like reaction, whereas cadmium and other redox-inactive metals can trigger the formation of ROS through an indirect mechanism. ROS include both free radicals and non-free radical oxygen-containing molecules, such as hydrogen peroxide (H_2O_2), superoxide (O_2^-), singlet oxygen ($1/2O_2$), and the hydroxyl radical (HO). On the other hand, free metal ions can enhance oxidative stress because they can amplify the redox cycle while the metal ion can accept or donate a single electron. This process catalyzes reactions in which reactive radicals can be generated and reactive oxygen species can be formed [29]. In this work, all the tested samples induced the formation of ROS (Figure 2). In some cases, the highest concentration ($2\times$) resulted in a decrease in free radicals compared to the control. In the case of epithelial cells of the tongue, one of the reasons could be the cytotoxic effect of the tested samples. In the case of hepatocarcinoma cells and gastric cell lines, the reason could be that when the cells are exposed to a certain concentration of a metal mixture, they begin to develop an antioxidant system that degrades the already-formed reactive species. It is known that the endogenous defense system includes enzymatic and non-enzymatic antioxidants such as superoxide dismutase, glutathione peroxidase, catalase, peroxiredoxins, glutathione (GSH), thioredoxin, uric acid, and a system for repairing oxidative damage to molecules [29]. Thus, different mechanisms are triggered at lower and higher concentrations of released metal ions.

The most resistant cell line was the adenocarcinoma cell line of the colon. CaCo-2 cells have been an excellent model system for many different tests—as a monolayer, it mimics the human intestinal epithelium. Although CaCo-2 cells express many different proteins (for transport and efflux) and enzymes (phase II conjugation) to model a variety of transcellular pathways and metabolic transformation of test compounds, they lack expression of cytochrome P450 isozymes (particularly CYP3A4) [36].

The samples containing a high concentration of nickel, but also cobalt, zinc, and chromium, exhibited the highest prooxidant activity, which increased with the duration of incubation in saliva, as did the metal concentration. In addition, all samples that had significantly higher concentrations of chromium and nickel compared to the control, especially those with exposure times of 7 and 14 days, showed the potential to cause increased ROS production. Oxidative DNA damage is one of the most common consequences of exposure to external environmental factors or endogenous genotoxic substances. These reactions are associated with the action of ROS such as hydroxyl radicals, superoxides, peroxides, or simple oxygen. In the combined UV/H_2O_2 process, UV radiation activates H_2O_2, leading to the formation of the OH radicals [37,38]. The effectiveness of the UV/H_2O_2 process depends on several conditions that affect its ability to degrade organic molecules. The conversion of the supercoiled form of plasmid DNA to an open-circular form is used as an indicator of DNA damage [38]. The structure of supercoiled DNA is a good model for studying DNA damage because it is very sensitive to environmental conditions, because even a single-strand break (which occurs when the plasmid is exposed to a hydroxyl radical) in a supercoiled DNA causes the super helical structure to loosen and the plasmid to assume a loose form [39,40]. Double-strand breaks open the plasmid and convert its structure to a linear form. When there are many double-strand breaks, the plasmid is fragmented into linear duplexes that are shorter compared to the original length [37,38]. All these changes can be analyzed and quantified by gel electrophoresis and gel analysis software (Gel Analyzer 19.1). Figure 3 shows the results of the influence of the studied saliva samples on DNA stability; all samples showed a slight prooxidant effect on DNA (lower ratio of supercoiled/linear form of the plasmid compared to the negative control, but higher ratio compared to the positive control). Concerning the metals present in saliva, the results obtained from this work, indicating a slight prooxidative effect of almost all tested eluates, were expected since previous research has also confirmed the role of the prooxidant activity of metals in the development of DNA damage, using the M13 mutation assay and DNA sequencing methods and the plasmid linearization method [41,42]. Thorough studies on DNA damage mediated by some first-row transition metals (iron, cobalt, nickel, copper, zinc, chromium, manganese) under oxidative stress conditions are presented in the work of Angelé Martínez et al. [42]. Higher concentrations of cobalt, chromium, and nickel were also found in the appliances used in this study, which could be related to the prooxidant effect obtained. In addition, the plasmid–aluminum relationship is also confirmed.

The metals that enter saliva during the initial period after orthodontic appliance insertion can cause DNA damage, but to a relatively small extent because the proportions of the supercoiled plasmid form are still closer to the negative control than to the positive control. Genotoxicity is often associated with oxidative stress and free radical formation [41]. Loyola-Rodriguez et al. [14] demonstrated that all types of orthodontic brackets, regardless of their components, cause DNA damage when their eluates are exposed to human gingival fibroblasts using the comet assay. The results of the comet test performed in this work, expressed as tail length, are shown in Figure 4. Only samples showing some prooxidant activity and having a cell survival rate higher than 80% are eligible for the genotoxicity test of the comet assay. Genotoxicity was mainly found in the samples with increased concentration of iron, nickel, and chromium compared to the control. The eluate incubated with all parts for 3 days has nine times more aluminum than the control, which also contributes to genotoxicity [27]. Synergistic effects of metals can also lead to increased genotoxicity of samples [29,31]. Samples with bands have up to 100 times higher concentration of nickel and chromium, so significant genotoxicity of these samples is expected [41,42], but the results did not show significant genotoxicity in all samples. It is possible that the DNA repair mechanism was activated in the cells, so the effects of the metal on DNA are not so visible [43]. The samples obtained after eluting the bands and all parts combined for 3 and 7 days, respectively, showed an increase in genotoxicity levels. Samples obtained after 14 days of elution showed stagnation or a slight decrease in tail intensity (toxicity is detected only for CAL27 cells). This is consistent with the results that Petković Didović et al. [22] obtained since the concentration of certain metals such as nickel and chromium, which have a genotoxic effect, decreases on the 14th day compared to the 7th day of elution. This result is in accordance with the previously conducted studies, which, as mentioned above, also showed that genotoxicity increases with the duration of metal exposure, but also that it decreases again after long-term elution of dental alloys [43–45].

Unusual results in Hep-G2 cells (very high value of tail intensity in control samples and diminished values in treated samples), was noticed, probably due to activation of DNA repair mechanisms in these cells. Liver cells rich in proteins that have affinity for binding various metals are able to reduce the concentration of metals in free form that could have a genotoxic effect [43–45]. These results could be encouraging because the genotoxicity of metals from orthodontic appliances is not so great that the human body cannot "defend" itself against them.

From the results of the principal component analysis (PCA), it can be concluded that Fe and Ni are responsible for the production of ROS, while Mn and Cr have a major impact on hydroxyl radicals causing single-strand breaks in supercoiled plasmid DNA, in addition to the production of ROS. On the other hand, Fe, Cr, Mn, and Al are responsible for the cytotoxic effect of the eluates studied. There is a significant correspondence between the generation of ROS in the cells and the single strand breaks in the plasmid. Cr, Mn, and Al are highly responsible for genotoxicity as these cause linearization of the plasmid at all concentrations tested. This result is in agreement with the data found in the literature on the toxicity of Cr (VI), which is mainly present in stainless products [31,41,46].

The only limitation in the study that the authors were able to identify is the fact that the experiments were performed on cell lines that are different from normal cells. However, a negative control was included in all experiments, and all results obtained are presented in comparison with the control. All observed changes are the result of exposure to the eluates and the different metal concentrations in the eluates. Finally, the use of cell lines in toxicological studies is widely accepted as a biological test system, and the results can be well interpreted by appropriate statistical analysis.

4. Materials and Methods

4.1. Materials

4.1.1. Cell Lines

Epithelial lingual (CAL 27), hepatic (HepG2), colon (CaCo2), and stomach carcinoma (AGS) cell lines were provided by the European Collection of Authenticated Cell Cultures (ECACC). The cell lines were used to determine the cytotoxic, prooxidant, and genotoxic effects of dental alloy eluates. Cells were grown as monolayer cultures in growth media recommended by the cell line supplier and supplemented with 10% fetal bovine serum (GIBCO, New York, NY, USA), 4500 mg/L glucose, and 1% penicillin/streptomycin. The supercoiled plasmid φX 174 RF I (Promega, Madison, WI, USA) was used to determine the prooxidative effect of the dental alloy eluates on DNA.

4.1.2. Orthodontic Appliances

Orthodontic archwires used in the study were nickel and titanium alloy (rematitan® LITE ideal arches, φ 0.43 × 0.64 mm/17 × 25, Dentaurum, Baden-Wurttemberg, Germany), while bands (Dentaform, Zahn 36, size 23/Roth 22, Dentaurum), brackets (equilibrium® 2, φ 0.56 × 0, 76 mm/22 × 30, Roth 22, Dentaurum), and ligatures (remanium®, short, soft, φ 0.25 mm/10, Dentaurum) were made of stainless steel. Detailed alloy composition is given in our previous work [22].

4.2. Methods

4.2.1. Preparation of the Test Samples

All tested samples including the control were prepared in artificial saliva, according to the Tani–Zucchi receipt pH 4.8 (composition: 1.5 g/L KCl; 1.5 g/L NaHCO$_3$; 0.5 g/L NaH$_2$PO$_4$ × H$_2$O; 0.5 g/L KSCN; 0.9 g/L lactic acid). This pH value (4.8) corresponds to the pH of one- and two-day-old dental biofilm and serves to simulate patients with poor oral hygiene [5]. A low pH associated with a high lactic acid content also tends to be associated with a high caries rate [6].

Eluates in artificial saliva were prepared according to the current ISO standard (ISO 10993-5:2009) [47], briefly, in this way: one orthodontic archwire, ten brackets, ten ligatures, and two bands were immersed individually in 20 mL of artificial saliva and autoclaved at 121 °C for 15 min (CertoClav, Leonding, Austria) and then incubated under sterile conditions on a rotary shaker (37 °C, 100/min, Unimax 1010, Heidolph, Schwabach, Germany) for three, seven, and fourteen days, respectively. For each time interval and each orthodontic appliance, five samples were prepared. For the purpose of this experiment, samples from the same time interval were combined in a new sterile vessel (100 mL total) to obtain a concentrated sample from which working concentrations were prepared. From each sample, 7 mL were taken to determine the released metal ions using an inductively coupled plasma mass spectrometer. Remainder of each sample (approximately 93 mL) was concentrated on a rotary evaporator (RV 10, IKA, Staufen, Germany). The resulting volume of each sample was concentrated 20-fold and used as stock solution to prepare the working solution (concentration range: 2.0×, 1.0×, 0.5×, 0.1×). All dilutions were prepared from a concentrated saliva stock solution using RPMI-1640 culture medium (Sigma-Aldrich-Chemie, Steinheim, Germany). Artificial saliva used as a control was prepared in the same way (concentrated 20× and diluted accordingly).

4.2.2. Cytotoxic and Prooxidant Effects of Metals in Artificial Saliva Eluted from Dental Alloys

In brief, the cell lines CAL 27, AGS, CaCo2, and HepG2 were seeded in transparent multiwell microtiter plates (96 wells) for the cytotoxic assay and in black multiwell microtiter plates for the prooxidant assay. The concentration of cells in each well was 5×10^4 cells/mL. After attachment, cells were treated with prior prepared eluates of parts of orthodontic appliances (archwires, brackets, bands, and ligatures) in four different concentrations of artificial saliva ($0.1\times$, $0.5\times$, $1.0\times$, and $2\times$) for 24 h. Treatment with artificial saliva was used as a control. Each concentration was tested in six replicates, and each experiment was repeated three times.

The cytotoxic effect of the prepared samples was determined by the neutral red method [48]. After incubation and removal of eluates, cells were washed and 50 µM neutral red was added. The excess amount of neutral red was discarded, the cells were washed with PBS buffer, and the dye accumulated in the lysosomes was extracted with an extraction solution (ethanol:glacial acetic acid:water = 50:1:49, *v/v/v*). Absorbance was measured at 540 nm, and the percentage of surviving cells was calculated using the following formula:

$$\% \text{ of cell survival} = (A_{540} \text{ nm sample}/A_{540} \text{ nm control}) \times 100$$

The prooxidant effect of the artificial saliva samples was determined by the DCFH-DA method. After treatment of the cells, a 50 µM solution of DCFH-DA was added to the cells. After 30 min of incubation, fluorescence was measured. The fluorescence intensity was proportional to the free radicals formed in the cells [49].

Sample fluorescence was normalized to cell survival. Prooxidant effect was expressed as a percentage of sample fluorescence compared to control fluorescence:

$$\% \text{ prooxidant effect} = (\text{sample fluorescence}/\text{control fluorescence}) \times 100$$

4.2.3. Effect of Dental Alloy Eluates on Hydroxyl Radical-Mediated DNA Strand Breakage

DNA damage was measured by the conversion of supercoiled φX-174 RF I double-stranded DNA into open circular and linear forms [50]. Briefly, the plasmid φX-174 RF I is in the form of supercoiled circular DNA and converts to a coiled form in the presence of the hydroxyl radical, which travels much more slowly through the gel than the supercoiled form. Hydrogen peroxide cannot induce the formation of hydroxyl radicals, but in the presence of metals, this form of this type of free radical can be formed and cause linearization of the plasmid [41,50,51]. Consequently, two bands of different intensity are formed in the gel after electrophoresis; the lower band represents the supercoiled DNA and the upper band represents the coiled (damaged) DNA. The ratio between the lower and upper bands and comparison with the negative control indicates the potential for the eluate from the dental alloy to induce DNA damage. The reaction mixtures contained 0.3 µg/mL of the plasmid, 0.3% hydrogen peroxide, and artificial saliva samples at a final concentration of $0.1\times$, $0.5\times$, $1.0\times$, and $2.0\times$. The reaction mixture was prepared in TE buffer. After incubation at 37 °C for 30 min, loading dye was added, and samples were immediately loaded into a 1% agarose gel containing 40 mM Tris, 20 mM sodium acetate, and 2 mM EDTA, and then electrophoresed in a horizontal slab-gel apparatus in Tris/acetate/EDTA gel buffer at 150 V for 60 min. After electrophoresis, the gels were stained with a 0.5 mg/mL solution of ethidium bromide for 2 h and then destained in water. The gels were then photographed under ultraviolet illumination and quantified using GelAnalyzer 19.1 software.

4.2.4. Comet Assay

The comet assay was performed under alkaline conditions as described by Azqueta and Collins [52], with some modifications. Two replicate slides were prepared per sample. Agarose gels were prepared on fully frosted slides coated with 1% and 0.6% normal melting point (NMP) agarose. Cells were seeded in Petri dishes, Φ 5 cm, at a concentration of

5×10^4 cells/mL and treated for 24 h with selected concentrations of dental alloy samples. After treatment, 10 µL of the cell suspension was mixed with 0.5% low melting point (LMP) agarose, placed on slides, and covered with a layer of 0.5% LMP agarose. The slides were immersed in freshly prepared ice-cold lysis solution (2.5 M NaCl, 100 mM Na_2EDTA, 10 mM Tris—1% Na-sarcosinate, pH 10, all components provided by Sigma Aldrich, Oakville, Canada) containing 1% Triton X-100 (Sigma Aldrich, Oakville, Canada) and 10% dimethyl sulfoxide (Kemika, Zagreb, Croatia) for 1 h. Alkaline denaturation and electrophoresis were performed at 4 °C under dim light in freshly prepared electrophoresis buffer (300 mM NaOH, 1 mM Na_2EDTA (Honeywell, Charlotte, NC, USA), pH 13.0). After 20 min of denaturation, the slides were randomly placed side by side in the horizontal gel electrophoresis tank, facing the anode. Electrophoresis at 25 V continued for an additional 20 min. After electrophoresis, the slides were gently washed three times at five-minute intervals with neutralization buffer (0.4 M Tris-HCl, pH 7.5). The slides were stained with ethidium bromide (20 µg/mL) and stored at 4 °C in humidified, sealed containers until analysis. Each slide was examined at $250\times$ magnification using a fluorescence microscope (Zeiss, Oberkochen, Germany) equipped with a 515–560 nm excitation filter and a 590 nm blocking filter. The microscope image was transferred to a computerized image analysis system (Comet Assay II, Perceptive Instruments Ltd., Bury Saint Edmunds, UK). The comet parameters analyzed were tail length and tail DNA content. A total of 100 comets per sample (50 from each of the two replicate slides) were scored. Slides were blinded prior to analysis.

4.2.5. Statistical Analysis

For the comparison of prooxidant effects and cytotoxicity, analysis of variance (ANOVA) was used together with the Student–Newman–Keuls post hoc test (using the commercial software IBM SPSS Statistics for Windows, Version 22.0 (IBM Corp., Armonk, NY, USA)). For comparison of genotoxicity, analysis of variance (ANOVA) was used together with the LSD test for the comet assay and the Mann–Whitney U test for the plasmid φX-174 RF I assay (using the Statistica data analysis software system, version 13.4.04; Tibco Software Inc., Palo Alto, CA, USA). PCA was used to determine the potential correlation between the orthodontic appliance and the analyses performed, using the Statistica software system (version 13.4.04; Tibco Software Inc., Palo Alto, CA, USA).

5. Conclusions

The effects of fixed orthodontic appliances on various cell lines are being studied intensively, and numerous clinical trials are being conducted, but each trial is different and sheds only a small light on the overall picture. Research in this area is still very thorough, useful, and intense, and, on the other hand, there is still much to discover. This type of research, using a fixed orthodontic appliance as would be used in the oral cavity and using eluates from such an appliance with known metal content and composition for toxicity research, brings us closer to more accurate in vivo conditions. In addition, this study uses cells from the tongue, stomach, liver, and colon, which are selected to mimic the passage of saliva through the gastrointestinal tract.

By deciding to conduct the study at a low pH (4.8), we focused on patients with poor oral hygiene, plaque accumulation, and altered acidity, which further compromised alloy stability. The average exposure time (for orthodontic treatments) is 18–24 months. Such long-term contact of alloys with the oral mucosa is not negligible, and it is possible that different cellular reactions may develop under real conditions. Under certain circumstances, there may be an accumulation of ions due to other intakes (food, beverages, piercing, etc.), which may lead to a significant synergistic effect and, consequently, to intense effects on human health. For this reason, the information on cytotoxicity should not be disregarded.

Author Contributions: Conceptualization, G.Č.J. and K.D.; methodology, K.D.; formal analysis, S.O. and Ž.F.; data curation, M.R.M., S.O. and G.Č.J.; writing—original draft preparation, K.D.; writing—review and editing, G.Č.J.; visualization, G.Č.J.; supervision, G.Č.J.; project administration, G.Č.J. All authors have read and agreed to the published version of the manuscript.

Funding: This research was funded by the Croatian Science Foundation within the framework of the Slovenian–Croatian bilateral project (IPS-2020-01-7418): "Determination of the occurrence, cause and harmful effects of oxidative stress caused by the use of fixed orthodontic appliances".

Institutional Review Board Statement: Not applicable.

Informed Consent Statement: Not applicable.

Data Availability Statement: The data presented in this study are available on request from the corresponding author.

Acknowledgments: The authors would like to thank Stjepan Špalj for constructive advice, help, and data processing.

Conflicts of Interest: The authors declare no conflict of interest.

References

1. Martín-Cameán, A.; Jos, Á.; Mellado-García, P.; Iglesias-Linares, A.; Solano, E.; Cameán, A.M. In Vitro and in Vivo Evidence of the Cytotoxic and Genotoxic Effects of Metal Ions Released by Orthodontic Appliances: A Review. *Environ. Toxicol. Pharmacol.* **2015**, *40*, 86–113. [CrossRef] [PubMed]

2. Wendl, B.; Wiltsche, H.; Lankmayr, E.; Winsauer, H.; Walter, A.; Muchitsch, A.; Jakse, N.; Wendl, M.; Wendl, T. Metal Release Profiles of Orthodontic Bands, Brackets, and Wires: An in Vitro Study. *J. Orofac. Orthop.* **2017**, *78*, 494–503. [CrossRef] [PubMed]

3. Brantley, W.; Berzins, D.; Iijima, M.; Tufekçi, E.; Cai, Z. Structure/Property Relationships in Orthodontic Alloys. In *Orthodontic Applications of Biomaterials*; Elsevier: Amsterdam, The Netherlands, 2017; pp. 3–38. ISBN 978-0-08-100383-1.

4. Proffit, W.R. *Contemporary Orthodontics*, 6th ed.; Elsevier: Philadelphia, IL, USA, 2018; ISBN 978-0-323-54387-3.

5. Igarashi, K.; Lee, I.K.; Schachtele, C.F. Effect of Dental Plaque Age and Bacterial Composition on the PH of Artificial Fissures in Human Volunteers. *Caries Res.* **1990**, *24*, 52–58. [CrossRef] [PubMed]

6. Ferrer, M.D.; Pérez, S.; Lopez, A.L.; Sanz, J.L.; Melo, M.; Llena, C.; Mira, A. Evaluation of Clinical, Biochemical and Microbiological Markers Related to Dental Caries. *Int. J. Environ. Res. Public Health* **2021**, *18*, 6049. [CrossRef] [PubMed]

7. Martín-Cameán, A.; Jos, A.; Cameán, A.M.; Solano, E.; Iglesias-Linares, A. Genotoxic and Cytotoxic Effects and Gene Expression Changes Induced by Fixed Orthodontic Appliances in Oral Mucosa Cells of Patients: A Systematic Review. *Toxicol. Mech. Methods* **2015**, *25*, 440–447. [CrossRef]

8. Mikulewicz, M.; Chojnacka, K. Trace Metal Release from Orthodontic Appliances by In Vivo Studies: A Systematic Literature Review. *Biol. Trace Elem. Res.* **2010**, *137*, 127–138. [CrossRef]

9. Jacoby, L.S.; Rodrigues Junior, V.D.S.; Campos, M.M.; Macedo De Menezes, L. Cytotoxic Outcomes of Orthodontic Bands with and without Silver Solder in Different Cell Lineages. *Am. J. Orthod. Dentofac. Orthop.* **2017**, *151*, 957–963. [CrossRef]

10. Ajami, S.; Dadras, S.; Faghih, Z.; Shobeiri, S.S.; Mahdian, A. In Vitro Evaluation of Immediate Cytotoxicity of Resterilised Orthodontic Bands on HGF-1 Cell Line. *Int. Orthod.* **2021**, *19*, 500–504. [CrossRef]

11. Nimeri, G.; Curry, J.; Berzins, D.; Liu, D.; Ahuja, B.; Lobner, D. Cytotoxic Evaluation of Two Orthodontic Silver Solder Materials on Human Periodontal Ligament Fibroblast Cells and the Effects of Antioxidant and Antiapoptotic Reagents. *Angle Orthod.* **2021**, *91*, 349–355. [CrossRef]

12. Wishney, M.; Mahadevan, S.; Cornwell, J.A.; Savage, T.; Proschogo, N.; Darendeliler, M.A.; Zoellner, H. Toxicity of Orthodontic Brackets Examined by Single Cell Tracking. *Toxics* **2022**, *10*, 460. [CrossRef]

13. Escobar, L.M.; Rivera, J.R.; Arbelaez, E.; Torres, L.F.; Villafañe, A.; Díaz-Báez, D.; Mora, I.; Lafaurie, G.I.; Tanaka, M. Comparison of Cell Viability and Chemical Composition of Six Latest Generation Orthodontic Wires. *Int. J. Biomater.* **2021**, *2021*, 8885290. [CrossRef]

14. Loyola-Rodríguez, J.P.; Lastra-Corso, I.; García-Cortés, J.O.; Loyola-Leyva, A.; Domínguez-Pérez, R.A.; Avila-Arizmendi, D.; Contreras-Palma, G.; González-Calixto, C. In Vitro Determination of Genotoxicity Induced by Brackets Alloys in Cultures of Human Gingival Fibroblasts. *J. Toxicol.* **2020**, *2020*, 1467456. [CrossRef]

15. Rincic Mlinaric, M.; Durgo, K.; Katic, V.; Spalj, S. Cytotoxicity and Oxidative Stress Induced by Nickel and Titanium Ions from Dental Alloys on Cells of Gastrointestinal Tract. *Toxicol. Appl. Pharmacol.* **2019**, *383*, 114784. [CrossRef]

16. Gonçalves, T.S.; Menezes, L.M.D.; Trindade, C.; Machado, M.D.S.; Thomas, P.; Fenech, M.; Henriques, J.A.P. Cytotoxicity and Genotoxicity of Orthodontic Bands with or without Silver Soldered Joints. *Mutat. Res. Genet. Toxicol. Environ. Mutagen.* **2014**, *762*, 1–8. [CrossRef]

17. Kovač, V.; Poljšak, B.; Primožič, J.; Jamnik, P. Are Metal Ions That Make up Orthodontic Alloys Cytotoxic, and Do They Induce Oxidative Stress in a Yeast Cell Model? *Int. J. Mol. Sci.* **2020**, *21*, 7993. [CrossRef]

18. Kovač, V.; Bergant, M.; Ščančar, J.; Primožič, J.; Jamnik, P.; Poljšak, B. Causation of Oxidative Stress and Defense Response of a Yeast Cell Model after Treatment with Orthodontic Alloys Consisting of Metal Ions. *Antioxidants* **2021**, *11*, 63. [CrossRef]

19. Papaioannou, P.; Sütel, M.; Hüsker, K.; Müller, W.-D.; Bartzela, T. A New Setup for Simulating the Corrosion Behavior of Orthodontic Wires. *Materials* **2021**, *14*, 3758. [CrossRef]

20. Hussain, H.D.; Ajith, S.D.; Goel, P. Nickel Release from Stainless Steel and Nickel Titanium Archwires—An in Vitro Study. *J. Oral Biol. Craniofacial Res.* **2016**, *6*, 213–218. [CrossRef]

21. Da Silveira, R.; Gonçalves, T.; De Souza Schacher, H.; De Menezes, L. Ion Release and Surface Roughness of Silver Soldered Bands with Two Different Polishing Methods: An in-Vitro Study. *J. Orthodont. Sci.* **2022**, *11*, 11. [CrossRef]

22. Petković Didović, M.; Jelovica Badovinac, I.; Fiket, Ž.; Žigon, J.; Rinčić Mlinarić, M.; Čanadi Jurešić, G. Cytotoxicity of Metal Ions Released from NiTi and Stainless Steel Orthodontic Appliances, Part 1: Surface Morphology and Ion Release Variations. *Materials* **2023**, *16*, 4156. [CrossRef]

23. Kuhta, M.; Pavlin, D.; Slaj, M.; Varga, S.; Lapter-Varga, M.; Slaj, M. Type of Archwire and Level of Acidity: Effects on the Release of Metal Ions from Orthodontic Appliances. *Angle Orthod.* **2009**, *79*, 102–110. [CrossRef] [PubMed]

24. Hwang, C.-J.; Shin, J.-S.; Cha, J.-Y. Metal Release from Simulated Fixed Orthodontic Appliances. *Am. J. Orthod. Dentofac. Orthop.* **2001**, *120*, 383–391. [CrossRef] [PubMed]

25. Hsu, M.Y.; Mina, E.; Roetto, A.; Porporato, P.E. Iron: An Essential Element of Cancer Metabolism. *Cells* **2020**, *9*, 2591. [CrossRef] [PubMed]

26. Alberto, A.F.; Smith, L.A.; Reichhardt, C.C.; Hansen, S.L.; Thornton, K.J. PSV-13 Understanding the Role of Zinc and Manganese in Proliferation and Protein Synthesis of Primary Bovine Satellite Cells. *J. Anim. Sci.* **2021**, *99*, 308. [CrossRef]

27. Francisco, L.F.V.; Baldivia, D.D.S.; Crispim, B.D.A.; Klatke, S.M.F.F.; Castilho, P.F.D.; Viana, L.F.; Santos, E.L.D.; Oliveira, K.M.P.D.; Barufatti, A. Acute Toxic and Genotoxic Effects of Aluminum and Manganese Using In Vitro Models. *Toxics* **2021**, *9*, 153. [CrossRef]

28. Chen, X.; Yu, C.; Kang, R.; Tang, D. Iron Metabolism in Ferroptosis. *Front. Cell Dev. Biol.* **2020**, *8*, 590226. [CrossRef]

29. Primožič, J.; Poljšak, B.; Jamnik, P.; Kovač, V.; Čanadi Jurešić, G.; Spalj, S. Risk Assessment of Oxidative Stress Induced by Metal Ions from Fixed Orthodontic Appliances during Treatment and Indications for Supportive Antioxidant Therapy: A Narrative Review. *Antioxidants* **2021**, *10*, 1359. [CrossRef]

30. Ardelean, L.; Reclaru, L.; Bortun, C.M.; Rusu, L.C. Assessment of Dental Alloys by Different Methods. In *Superalloys*; Aliofkhazraei, M., Ed.; IntechOpen: London, UK, 2015; ISBN 978-953-51-2212-8.

31. Downarowicz, P.; Mikulewicz, M. Trace Metal Ions Release from Fixed Orthodontic Appliances and DNAdamage in Oral Mucosa Cells by in Vivo Studies: A Literature Review. *Adv. Clin. Exp. Med.* **2017**, *26*, 1155–1162. [CrossRef]

32. Costa, M.T.; Lenza, M.A.; Gosch, C.S.; Costa, I.; Ribeiro-Dias, F. In Vitro Evaluation of Corrosion and Cytotoxicity of Orthodontic Brackets. *J. Dent. Res.* **2007**, *86*, 441–445. [CrossRef]

33. Wen, X.; Rui, W.; Jia, X.; Tang, J.; He, X. [In Vitro Cytotoxicity Study of Nickel Ion]. *Zhongguo Yi Liao Qi Xie Za Zhi* **2015**, *39*, 212–215.

34. Kanaji, A.; Orhue, V.; Caicedo, M.S.; Virdi, A.S.; Sumner, D.R.; Hallab, N.J.; Yoshiaki, T.; Sena, K. Cytotoxic Effects of Cobalt and Nickel Ions on Osteocytes in Vitro. *J. Orthop. Surg. Res.* **2014**, *9*, 91. [CrossRef]

35. Wu, Y.; Lin, J.; Wang, T.; Lin, T.; Yen, M.; Liu, Y.; Wu, P.; Chen, F.; Shih, Y.; Yeh, I. Hexavalent Chromium Intoxication Induces Intrinsic and Extrinsic Apoptosis in Human Renal Cells. *Mol. Med. Rep.* **2019**, 851–857. [CrossRef]

36. Van Breemen, R.B.; Li, Y. Caco-2 Cell Permeability Assays to Measure Drug Absorption. *Expert Opin. Drug Metab. Toxicol.* **2005**, *1*, 175–185. [CrossRef]

37. Folkard, M.; Prise, K.M.; Turner, C.J.; Michael, B.D. The Production of Single Strand and Double Strand Breaks in DNA in Aqueous Solution by Vacuum UV Photons Below 10 Ev. *Radiat. Prot. Dosim.* **2002**, *99*, 147–149. [CrossRef]

38. Prise, K.M.; Folkard, M.; Michael, B.D.; Vojnovic, B.; Brocklehurst, B.; Hopkirk, A.; Munro, I.H. Critical Energies for SSB and DSB Induction in Plasmid DNA by Low-Energy Photons: Action Spectra for Strand-Break Induction in Plasmid DNA Irradiated in Vacuum. *Int. J. Radiat. Biol.* **2000**, *76*, 881–890. [CrossRef]

39. Cherny, D.I.; Jovin, T.M. Electron and Scanning Force Microscopy Studies of Alterations in Supercoiled DNA Tertiary Structure. *J. Mol. Biol.* **2001**, *313*, 295–307. [CrossRef]

40. Vologodskii, A.V.; Cozzarelli, N.R. Conformational and Thermodynamic Properties of Supercoiled DNA. *Annu. Rev. Biophys. Biomol. Struct.* **1994**, *23*, 609–643. [CrossRef]

41. Reid, T.M.; Feig, D.I.; Loeb, L.A. Mutagenesis by Metal-Induced Oxygen Radicals. *Environ. Health Perspect.* **1994**, *102*, 57–61. [CrossRef]

42. Angelé-Martínez, C.; Goodman, C.; Brumaghim, J. Metal-Mediated DNA Damage and Cell Death: Mechanisms, Detection Methods, and Cellular Consequences. *Metallomics* **2014**, *6*, 1358–1381. [CrossRef]

43. Hafez, H.S.; Selim, E.M.N.; Kamel Eid, F.H.; Tawfik, W.A.; Al-Ashkar, E.A.; Mostafa, Y.A. Cytotoxicity, Genotoxicity, and Metal Release in Patients with Fixed Orthodontic Appliances: A Longitudinal in-Vivo Study. *Am. J. Orthod. Dentofac. Orthop.* **2011**, *140*, 298–308. [CrossRef]

44. Karlina, I.; Amtha, R.; Roeslan, B.O.; Zen, Y. The Release of Total Metal Ion and Genotoxicity of Stainless Steel Brackets: Experimental Study Using Micronucleus Assay. *Indones. Biomed. J.* **2016**, *8*, 97. [CrossRef]

45. Kishore, S.; Felicita, A.S.; Siva, S. Evaluation of Nickel Release in Blood and Periodontal Tissue with the Use of NiTi Wires, Bands and Brackets in Orthodontics–A Systematic Review. *J. Evol. Med. Dent. Sci.* **2021**, *10*, 1539–1546. [CrossRef]

46. DesMarias, T.L.; Costa, M. Mechanisms of Chromium-Induced Toxicity. *Curr. Opin. Toxicol.* **2019**, *14*, 1–7. [CrossRef] [PubMed]

47. *ISO 10993-5:2009*; Biological Evaluation of Medical Devices—Part 5: Tests for In Vitro Cytotoxicity. ISO Standard: Geneva, Switzerland, 2009.

48. Repetto, G.; Del Peso, A.; Zurita, J.L. Neutral Red Uptake Assay for the Estimation of Cell Viability/Cytotoxicity. *Nat. Protoc.* **2008**, *3*, 1125–1131. [CrossRef]

49. Wang, X.; Roper, M.G. Measurement of DCF Fluorescence as a Measure of Reactive Oxygen Species in Murine Islets of Langerhans. *Anal. Methods* **2014**, *6*, 3019–3024. [CrossRef]

50. Green, M.R.; Sambrook, J. Analysis of DNA by Agarose Gel Electrophoresis. *Cold Spring Harb. Protoc.* **2019**, *2019*, pdb-top100388. [CrossRef]

51. Vandjelovic, N.; Zhu, H.; Misra, H.P.; Zimmerman, R.P.; Jia, Z.; Li, Y. EPR Studies on Hydroxyl Radical-Scavenging Activities of Pravastatin and Fluvastatin. *Mol. Cell Biochem.* **2012**, *364*, 71–77. [CrossRef]

52. Azqueta, A.; Collins, A.R. The Essential Comet Assay: A Comprehensive Guide to Measuring DNA Damage and Repair. *Arch. Toxicol.* **2013**, *87*, 949–968. [CrossRef]

International Journal of
Molecular Sciences

Article

Anti-*Candida albicans* Effects and Mechanisms of Theasaponin E1 and Assamsaponin A

Yuhong Chen [1,2], Ying Gao [1,*], Mingan Yuan [3], Zhaisheng Zheng [3] and Junfeng Yin [1,*]

1   Key Laboratory of Tea Biology and Resources Utilization, Tea Research Institute of Chinese Academy of Agricultural Sciences, Ministry of Agriculture, 9 South Meiling Road, Hangzhou 310008, China; chenyuhong@tricaas.com
2   Graduate School of Chinese Academy of Agricultural Sciences, Beijing 100081, China
3   Jinhua Academy of Agricultural Science, Jinhua 321000, China; minganyuan@126.com (M.Y.); zzs165@163.com (Z.Z.)
*   Correspondence: yinggao@tricaas.com (Y.G.); yinjf@tricaas.com (J.Y.); Tel.: +86-571-86650031 (Y.G. & J.Y.)

Abstract: *Candida albicans* is an opportunistic human fungal pathogen, and its drug resistance is becoming a serious problem. *Camellia sinensis* seed saponins showed inhibitory effects on resistant *Candida albicans* strains, but the active components and mechanisms are unclear. In this study, the effects and mechanisms of two *Camellia sinensis* seed saponin monomers, theasaponin E1 (TE1) and assamsaponin A (ASA), on a resistant *Candida albicans* strain (ATCC 10231) were explored. The minimum inhibitory concentration and minimum fungicidal concentration of TE1 and ASA were equivalent. The time–kill curves showed that the fungicidal efficiency of ASA was higher than that of TE1. TE1 and ASA significantly increased the cell membrane permeability and disrupted the cell membrane integrity of *C. albicans* cells, probably by interacting with membrane-bound sterols. Moreover, TE1 and ASA induced the accumulation of intracellular ROS and decreased the mitochondrial membrane potential. Transcriptome and qRT-PCR analyses revealed that the differentially expressed genes were concentrated in the cell wall, plasma membrane, glycolysis, and ergosterol synthesis pathways. In conclusion, the antifungal mechanisms of TE1 and ASA included the interference with the biosynthesis of ergosterol in fungal cell membranes, damage to the mitochondria, and the regulation of energy metabolism and lipid metabolism. Tea seed saponins have the potential to be novel anti-*Candida albicans* agents.

Keywords: *Camellia sinensis* seeds; theasaponin E1; assamsaponin A; antifungal activity; transcriptome

Citation: Chen, Y.; Gao, Y.; Yuan, M.; Zheng, Z.; Yin, J. Anti-*Candida albicans* Effects and Mechanisms of Theasaponin E1 and Assamsaponin A. *Int. J. Mol. Sci.* **2023**, 24, 9350. https://doi.org/10.3390/ijms24119350

Academic Editors: Manuel Simões, Rossana Morabito and Alessia Remigante

Received: 7 April 2023
Revised: 27 April 2023
Accepted: 23 May 2023
Published: 27 May 2023

1. Introduction

Candida albicans is one of the ubiquitous commensal fungi and the most common causative pathogen found on the skin and mucous membranes such as the oral cavity, gastrointestinal tract, and vagina [1–3]. It grows as a yeast, pseudohyphae, or true hyphae depending on environmental conditions. *C. albicans* is more infectious when transformed from yeast form to hypha form [4,5]. According to the data on fungal infections in hospitals, the mortality rate of immunocompromised patients infected with *C. albicans* is estimated to be 40%, making *C. albicans* the most serious fungal invasive infectious fungal strain [6]. *C. albicans* has expanded from simple endogenous infections in hospitals to exogenous infections that may be acquired through the hands of healthcare workers, contaminated fluids and biomaterials, and inanimate environments [7]. The disinfection and sterilization of *Candida albicans* on equipment and the environment is also a potential way to prevent and control the spread of intervention. Currently, there are five common drugs used in clinical applications for the treatment of *C. albicans* infections, which are classified as azoles, polyenes, allylamines, candins, and flucytosine according to their sites of action [8].

However, more and more antibiotic-resistant strains are appearing, leading to an urgent need to explore novel antifungal drugs which are efficient, safe, and cost-effective.

Natural products are popular sources for screening antifungal drugs. Tea saponins are triterpenoids with various physiological functions such as anti-inflammatory, antiviral, anticancer, and antifungal activities, as well as gastrointestinal protection [9–16]. Previous studies have shown that tea saponins exhibited good inhibitory activity against the skin-pathogenic fungus *Microsporum audouinii* [17]. They also had an inhibitory effect on *Botrytis cinerea*, which was the main cause of the rot and mold of nectarines [15]. In particular, tea saponins had inhibitory activity against *C. albicans* [16,18]. Zhang et al. found that after treatment with tea saponins, the morphology of *C. albicans* was changed, including a shrinkage in appearance and a rupture of the cell wall and plasma membrane [16]. Li et al. explained that the induction of oxidative stress might be an important antifungal mechanism of saponins because increases in intracellular ROS and mitochondrial dysfunction were observed in *C. albicans* [18]. However, tea saponins are composed of a series of saponins with similar structures. According to differences in sapogenin, glycone, and organic acids, more than 40 monomers have been identified in *Camellia sinensis* seeds [19]. The active components in tea seed saponin mixtures and their antifungal mechanisms against *C. albicans* are still to be elucidated.

In this study, *C. albicans* ATCC 10231, which is resistant to most antifungal drugs (fluconazole, anidulafungin, itraconazole, and voriconazole), was used as the experiment strain, and the anti-*C. albicans* activities and mechanisms of two tea seed saponin monomers, theasaponin E1 (TE1) and assamsaponin A (ASA) (Figure 1), were evaluated against *C. albicans* ATCC 10231 through physiological and biochemical examinations, morphological characteristics, and transcriptome analysis. These findings will provide a theoretical basis for the development and utilization of tea seed saponins.

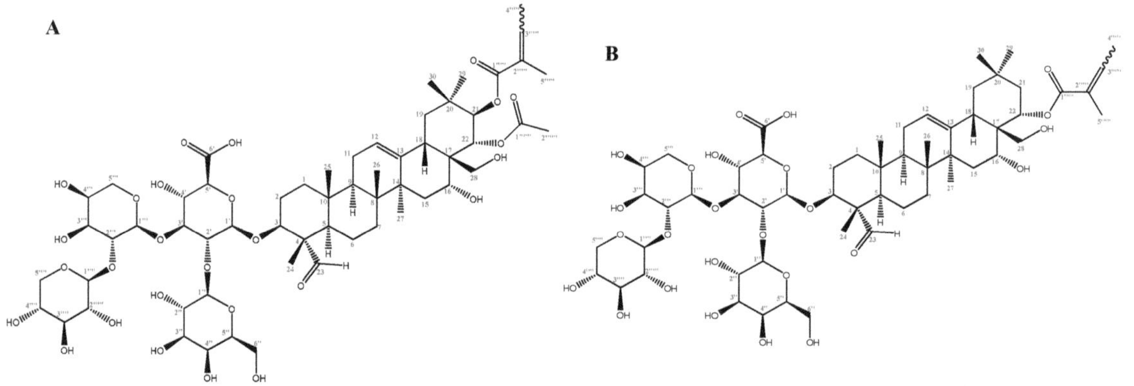

Figure 1. The structures of theasaponin E1 (**A**) and assamsaponin A (**B**).

2. Results

2.1. Antifungal Activities of TE1 and ASA

The minimum inhibitory concentrations (MICs) were determined using the CLSI broth microdilution method (Table 1). TE1 showed an obvious fungicidal activity effect on the fluconazole-resistant *C. albicans* strain (ATCC 10231), with an MIC and minimum fungicidal concentration (MFC) of 100 μM and 100 μM, respectively, representing an MFC/MIC ratio of 1 [20]. ASA showed the same effect. Time–kill curves (Figure 2A,B) showed that TE1 and ASA exhibited a good and rapid fungicidal effect within 4–10 h. At 1 MIC of TE1, the cell count of *C. albican* was reduced from 5.64 to 4.02 log CFU/mL after 2 h, and finally to 0 log CFU/mL after 10 h. At 1 MIC of ASA, the cell count of *C. albican* was reduced from 5.76 to 3.21 log CFU/mL after 2 h, and finally to 0 log CFU/mL after 4 h. This implies that ASA had a better fungicidal efficiency rate than TE1. Moreover, STYO-9/propidium iodide

(PI) staining was used to measure the cell damage (Figure 2C). SYTO-9 stains all microbes and displays green fluorescence, whereas PI merely stains microbes with compromised membranes and displays red fluorescence. Compared with the control group, the red fluorescence intensities (dead cells) of *C. albicans* treated with TE1 and ASA for 2 h were significantly increased (Figure 2C). The above results indicate that TE1 and ASA exhibited high antifungal activities against *C. albicans*.

Table 1. Antifungal activity against *C. albicans* ATCC 10231.

Compounds	MIC	MFC
AMB	0.5 µg/mL	0.5 µg/mL
CPF	0.25 µg/mL	0.5 µg/mL
FLC	>2000 µg/mL	>2000 µg/mL
Saponin mixture (purity > 96%)	125 µg/mL	250 µg/mL
TE1	100 µM	100 µM
ASA	100 µM	100 µM

AMB, amphotericin B; CPF, caspofungin; FLC, fluconazole; TE1, theasaponin E1; ASA, assamsaponin A.

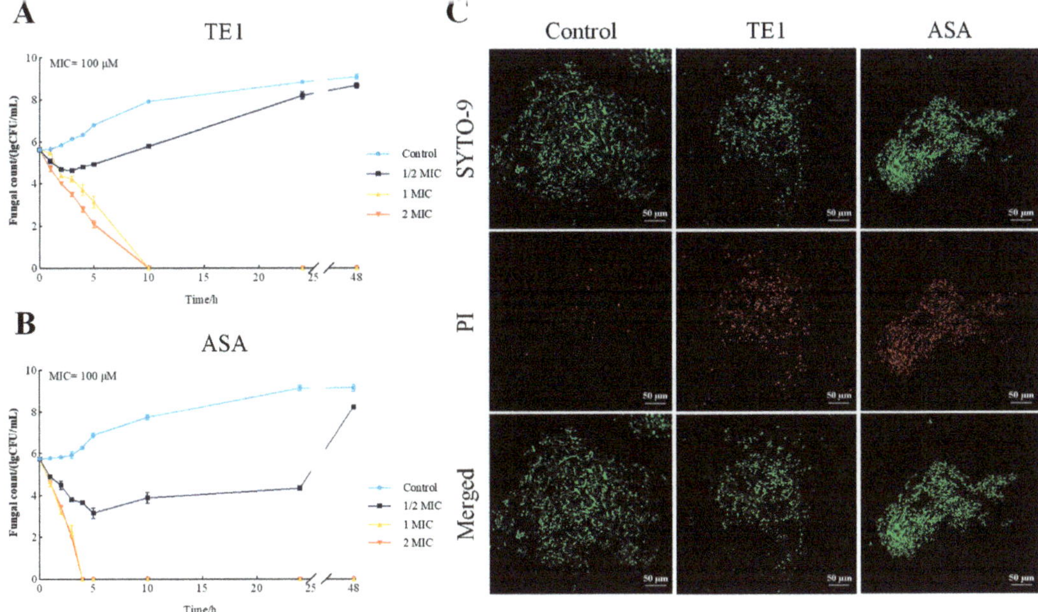

Figure 2. Antifungal activities of theasaponin E1 (TE1) and assamsaponin A (ASA). (**A**) Time-kill curves of TE1; (**B**) Time-kill curves of ASA; (**C**) Viability fluorescent staining of *C. albicans* 10231 observed by CLSM. Data are presented as (means ± SD) of three replicates.

2.2. Effects of TE1 and ASA on the Cell Wall and Membrane of C. albicans

Previous studies have demonstrated that many antibiotics display antifungal activity by impairing cell membrane integrity and permeability [21]. Scanning electron microscopy (SEM) showed that cells cultured in fresh culture medium formed a mycelium form with smooth cell walls, while cells cultured in TE1- or ASA-containing medium maintained yeast morphology and exhibited several morphological changes, such as shrinkage, indentation, and breakage (Figure 3A). As shown in Figure 3B, changes in the ultrastructure of *C. albicans* cells with a 2 h TE1 or ASA treatment were observed by transmission electron microscope (TEM). The cell walls were not clearly discernible, and many small vacuoles gathered near the cell membrane in TE1- or ASA-treated cells, indicating the loss of membrane integrity. The integrity of cell membranes was further evaluated by measuring the leakage of cellular contents such as nucleic acids and proteins. The OD260 and OD280 values of the culture medium represented the concentrations of extracellular nucleic acids and proteins, respectively. The leakage of nucleic acids (Figure 3C,D) and proteins (Figure 3E,F) was increased by TE1 and ASA in a dose- and time-dependent manner. The OD260 values of the culture medium from *C. albicans* treated with 1 MIC TE1 and ASA for 48 h were 2.12 and 2.20 times higher than the control, respectively. PI is a dye that penetrates the cell membrane of damaged cells and combines with the DNA of the nucleus to generate a red fluorescent PI complex, thereby being used to verify the integrity of the cell membrane [22]. The higher red fluorescence intensity of TE1- or ASA-treated cells provided more evidence of impaired membrane integrity and permeability (Figure 2C). The above results indicate that TE1 and ASA destroyed the structure of the fungal cell walls and membranes.

Sorbitol is used as an osmotic stabilizer of fungal growth. Cell walls cannot grow or be repaired without sorbitol [23,24]. To investigate the role of impairing cell walls in the TE1- and ASA-induced inhibition of *C. albicans*, *C. albicans* was cultured in the medium with or without sorbitol. As shown in Table 2, the MIC values of TE1 and ASA did not increase within 7 days, suggesting that impairing cell walls might not be vital for the TE1- and ASA-induced inhibition of *C. albicans*. On the contrary, the MIC of caspofungin (CPF), a medication known to display antifungal activity using sorbitol, was increased from 0.25 to 8 μg/mL.

Most of the previous literature has shown that the fungicidal activity of saponins is attributed to their interactions with sterol [25,26]. To verify whether TE1 and ASA had similar functions, different concentrations of ergosterol were added to observe their effects on the MICs of TE1 and ASA. The results (Table 3) showed after adding 250 μg/mL of ergosterol, the MICs of TE1 and ASA were increased 2-fold, while the MIC of amphotericin B (AMB, known to bind to ergosterol) was increased 64-fold. This indicates that the damage to the cell membrane by binding ergosterol might be involved in the TE1- and ASA-induced inhibition of *C. albicans*.

2.3. Effect of TE1 and ASA on Intracellular Reactive Oxygen Species (ROS) and Mitochondria

Mitochondria are essential for the initiation of intrinsic apoptosis, which is triggered by an increase in ROS. According to Figure 4A,B, ROS were generated in *C. albicans* cells in a concentration-dependent manner after TE1 and ASA treatment, respectively, suggesting that TE1 and ASA caused ROS accumulation in *C. albicans*. The JC-1 staining assay was performed to examine the effect of TE1 and ASA on the mitochondria membrane potential (MMP) (Figure 4C,D). Both compounds dose-dependently decreased MMP. A 1/4 MIC of TE1 reduced the accumulation of fluorescence by 12.33%, while a 1/4 MIC of ASA reduced it by 34.83%. There was no significant change in the fluorescence ratio between cells treated with 1/2 MIC and 1 MIC ASA ($p > 0.05$), both of which had a significant impact on the mitochondria. According to the results, TE1 and ASA induced the intracellular accumulation of ROS and a decrease in MMP, which could result in the apoptosis of *C. albicans*.

To investigate the role of ROS in the inhibitory effect of TE1 and ASA on *C. albicans*, the intracellular ROS level of *C. albicans* after treatment with N-acetyl-L-cysteine (NAC) or vitamin C (VC) was compared with that of untreated control cells (Figure 4E,F). However, the addition of antioxidants (NAC or VC) did not weaken the inhibitory effect of TE1 and ASA and could not prevent the increase in the intracellular ROS level. From these results, it was speculated that the two saponins caused oxidative damage to *C. albicans*, but it might not be the main cause of TE1- or ASA-induced cell death.

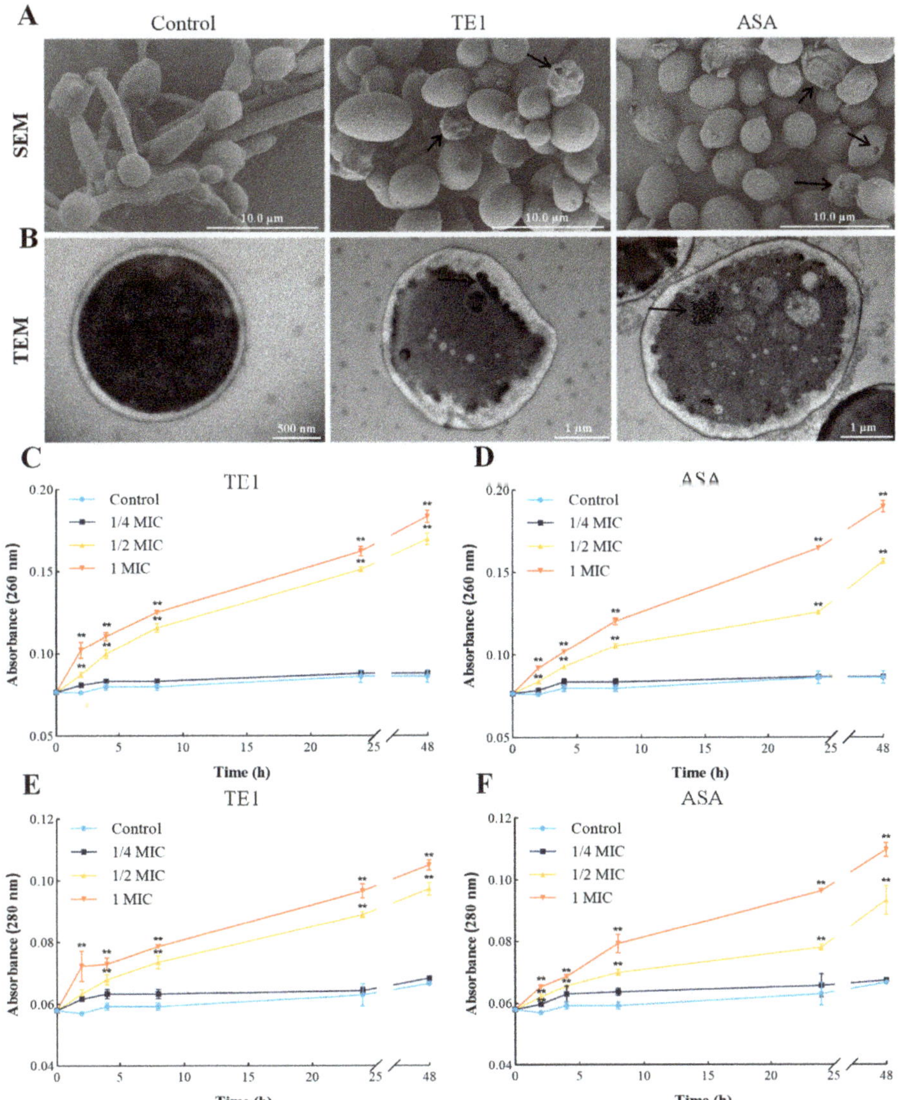

Figure 3. Theasaponin E1 (TE1) and assamsaponin A (ASA) destroyed the structure of the *C. albicans* 10231 cell wall and membrane. (**A**) Representative SEM images of *C. albicans* treated with 1 MIC of TE1 and ASA. Cells with indentations and breakages (arrows) can be seen. (**B**) Representative TEM images of *C. albicans* treated with 1 MIC of TE1 and ASA. TE1- or ASA-treated cells in which aggregations of tiny vacuoles are seen (arrows). (**C,D**) The content of nucleic acids in *C. albicans* culture medium quantified after treatment with different concentrations of TE1 and ASA. (**E,F**) The content of protein in *C. albicans* culture medium quantified after treatment with different concentrations of TE1 and ASA. Data are presented as (means ± SD) of three replicates. ** $p < 0.01$ were obtained for treated samples vs. control.

Table 2. MIC values of TE1, ASA and CPF against *C. albicans* ATCC 10231.

MIC TE1				MIC ASA				MIC CPF			
Day 2		Day 7		Day 2		Day 7		Day 2		Day 7	
(-) S	(+) S	(-) S	(+) S	(-) S	(+) S	(-) S	(+) S	(-) S	(+) S	(-) S	(+) S
100 μM	100 μM	100 μM	100 μM	100 μM	100 μM	100 μM	100 μM	0.25 μg/mL	8 μg/mL	0.5 μg/mL	8 μg/mL

TE1, theasaponin E1; ASA, assamsaponin A; S, sorbitol; (+), presence; (-) absence.

Table 3. MIC values of TE1, ASA and AMB against *C. albicans* ATCC 10231.

Compounds	Ergosterol Concentration (μg/mL)					
	0	50	100	150	200	250
TE1	100 μM	100 μM	100 μM	100 μM	100 μM	200 μM
ASA	100 μM	100 μM	100 μM	100 μM	100 μM	200 μM
AMB	0.5 μg/mL	2 μg/mL	4 μg/mL	8 μg/mL	16 μg/mL	32 μg/mL

TE1, theasaponin E1; ASA, assamsaponin A; AMB, amphotericin B.

2.4. Transcriptomic Analysis of C. albicans with TE1 and ASA Treatment

To gain insights into the possible molecular mechanism in the antifungal activity of TE1 and ASA, transcriptome analyses by RNA sequencing (RNA-seq) were performed. The Q20 of each piece of raw data was more than 97%, which indicated its accuracy and reliability and that it was worthy of further analyses (Table S1). A total of 2135 DEGs were identified between the control samples and the TE1-treated samples, including 1344 upregulated and 791 downregulated genes (Table S2). A total of 2105 DEGs were identified between the control samples and the ASA-treated samples, including 1339 upregulated and 766 downregulated genes (Table S2). In addition, only 17 DEGs were identified between the TE1-treated samples and the ASA-treated samples, which suggested that TE1 and ASA had similar fungicidal mechanisms.

Gene ontology (GO) analysis showed that the DEGs between the control samples and TE1- or ASA-treated samples mainly included two categories, which were the cellular component (CC) comprising "cell periphery", "cell wall", and "plasma membrane", and the biological process (BP), comprising "carbohydrate metabolic process", "rRNA processing", and "ribosome biogenesis" (Figure 5A,C). These results indicate that TE1 and ASA damage cell membrane function. The Kyoto Encyclopedia of Genes and Genomes (KEGG) database analysis revealed that most DEGs between the control samples and the TE1- or ASA-treated samples were related to "glycolysis/gluconeogenesis", "starch and sucrose metabolism", "glycerophospholipid metabolism", "fatty acid degradation", "steroid biosynthesis", and "DNA replication", which could be further categorized into three major pathways: carbohydrate metabolism, lipid metabolism, and replication and repair (Figure 5B,D).

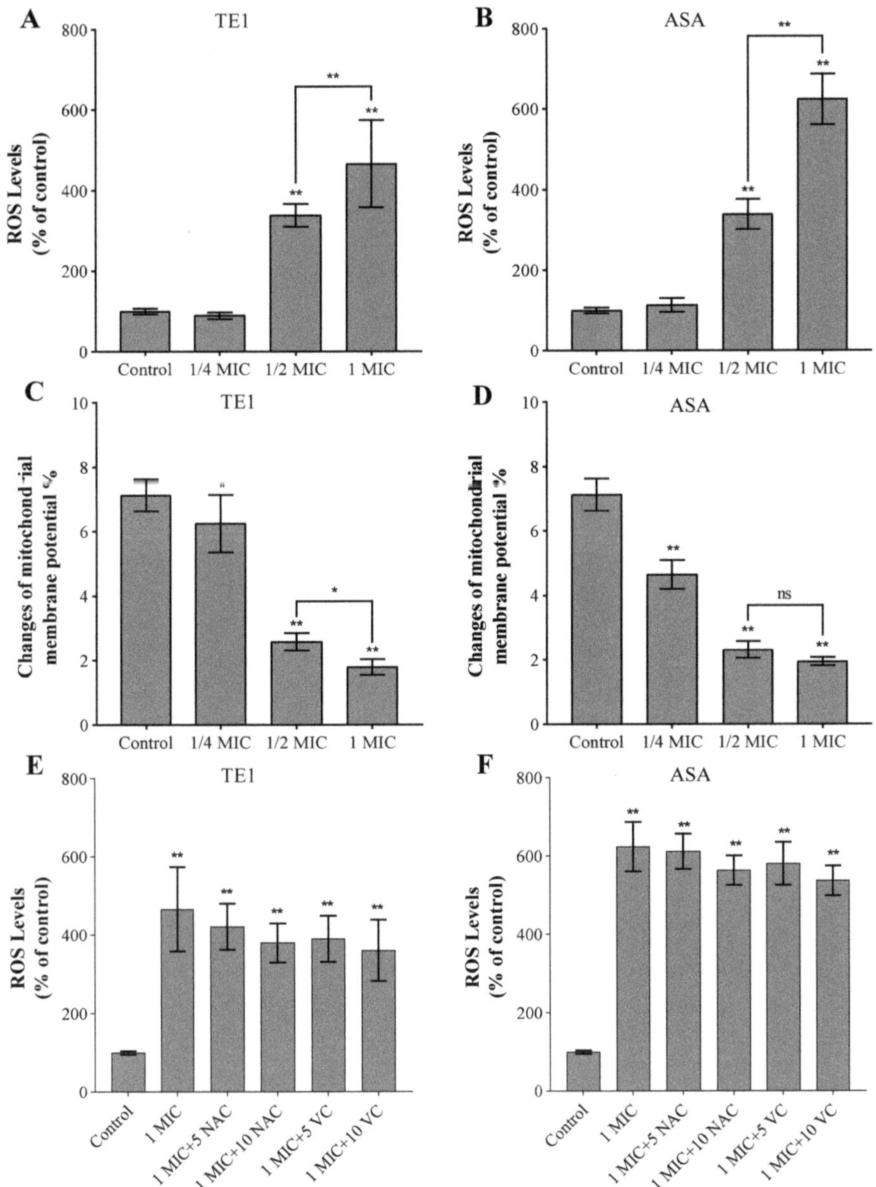

Figure 4. Effects of theasaponin E1 (TE1) and assamsaponin A (ASA) on the reactive oxygen species (ROS) production and mitochondrial membrane potentials (MMP) in *C. albicans* 10231. (**A**,**B**) The production of ROS from *C. albicans* treated with different concentrations of TE1 and ASA was determined by 2′,7′-Dichlorodihydrofluorescein diacetate (DCFH-DA); (**C**,**D**) MMP of *C. albicans* treated with various concentrations of TE1 and ASA were detected by JC-1 fluorescence. (**E**,**F**) The production of ROS from *C. albicans* treated with 1 MIC of TE1 and ASA following the addition of different concentrations (5 mM and 10 mM) of antioxidant (N-acetyl-L-cysteine, NAC or vitamin C, VC). Data are means (±SD) of three replicates. * $p < 0.05$ or ** $p < 0.01$ were obtained for treated samples vs. control, and ns indicated not significant.

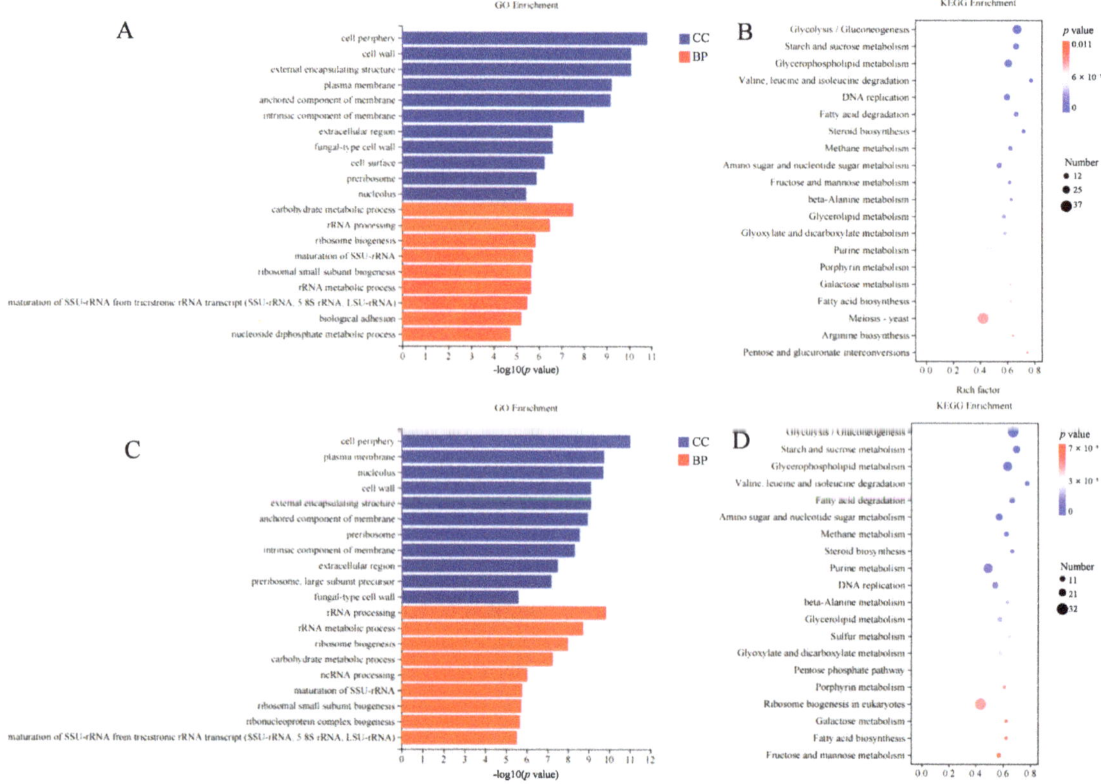

Figure 5. Transcriptome analysis of the effects of theasaponin E1 (TE1) and assamsaponin A (ASA) on *C. albicans* 10231. (**A**,**C**) DEGs of the top 20 enriched GO terms of TE1 and ASA; (**B**,**D**) KEGG enrichment analysis of DEGs of TE1 and ASA.

Quantitative real-time PCR (qRT-PCR) was used to confirm the changes in the 16 DEGs identified by RNA-seq, including genes related to glucose metabolism, ergosterol biosynthesis, fatty acid degradation, and membranes. The qRT-PCR results were consistent with the gene expression trends observed by transcriptome sequencing (Figure 6).

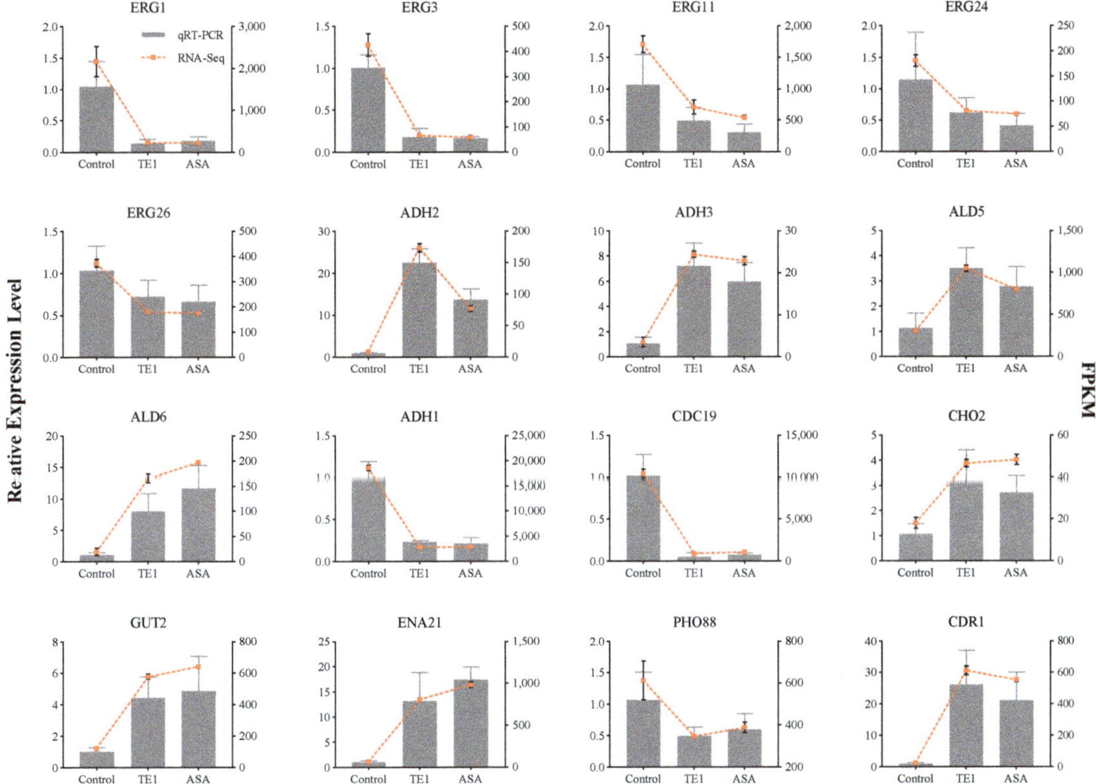

Figure 6. qRT-PCR validation of 16 DEGs. Data are means (±SD) of three replicates. FPKM: fragments per kilobase of exon model per million mapped fragments.

3. Discussion

With the increase in drug resistance in fungi, natural products are viewed as a potential alternative for controlling infectious diseases. Azole drugs are common low-toxicity antifungal drugs, but due to long-term use, some resistant strains have emerged. The biochemical basis for the development of resistance to azoles includes four aspects. First, mutations in the 14α-demethylase, which serves as the target of the action, result in a decreased affinity between azoles and the enzyme. Second, lesions in the Δ 5(6)-desaturase result in the production of 14α-methyl fecosterol instead of ergosterol. Third, the overexpression of 14α-demethylase, which attenuates the azoles-induced decrease in ergosterol enhances drug resistance [27]. The changes in these enzymes themselves may all come from mutations in the *ERG11* gene. Fourth, the overexpression of efflux pump genes leads to a decrease in the antifungal drug concentration inside the fungi and results in drug resistance. Discovering novel targets of action, inhibiting glucan synthase enzyme activities, and modifying compound configurations to enhance affinity with enzymes are possible ways to overcome and solve the issue. VT1129 and VT1161 are triazole compounds whose affinity to the Candida cytochrome P-450 enzyme (CYP51) is at least 1000 times higher than that of homologous zymosomes in the human body [28]. SCY-078 affects the surface of β-(1,3)-D-glucan synthase and achieves antifungal effects by destroying cellular walls [29,30]. T-2307 disrupts fungal mitochondrial membranes, thereby interfering with cell energy metabolism [31]. The antifungal activity of saponins is comparable to or even stronger than traditional antifungal drugs, and some saponins inhibit drug-resistant strains [32,33]. Yin et al. found that analogues of *Panax stipulcanatus* saponins have good antifungal activity against *C. albicans* and have synergistic antifungal effects against fluconazole-resistant strains [34]. Coleman et al. also found saponins against the resistant strains at low concentrations (16 and 32 µg/mL) [33]. Here, we found similar results. *C. albicans* ATCC 10231 was resistant to fluconazole, while the saponin mixture and saponin monomers (TE1 and ASA) extracted from tea seeds have antifungal effects towards it. Additionally, saponins have relatively low cytotoxicity to animal hosts [33,35]. Their high effectiveness and safety make saponins ideal alternative antifungal agents.

C. albicans, one of the most common human fungal pathogens [36], has emerged as one of the major problems for nosocomial infections, posing a fatal factor for patients with septic shock [37,38]. Triterpenoid saponins have been reported to inhibit the proliferation of various *Candida* species [39–41]. Tea saponins, which belong to triterpenoid saponins, are abundant in tea seeds and have an inhibitory effect against *C. albicans* [16,18]. In this study, two tea seed saponin monomers, TE1 and ASA were prepared from a high-purity tea seed saponin mixture. The saponin mixture (purity > 96%) along with TE1 and ASA, had inhibitory activity against *C. albicans* ATCC 10231. They all showed high antifungal efficacy within 2 h. When treated for 10 h, none of the *C. albicans* were alive. The MIC of TE1 was 100 µM (123 µg/mL) and the MIC of ASA was 100 µM (117 µg/mL), suggesting that TE1 and ASA were superior to the saponin mixture (MIC of 125 µg/mL, MFC of 250 µg/mL) (Table 1). The higher MIC of the mixture might be due to the presence of compounds without antifungal activity in the mixture, resulting in a decrease in the ratio of antifungal substances. Proteins, fats, sugars, and phenolic substances were possible residues in the purified tea saponin mixture [42]. Polysaccharides and phenols have been reported to have antifungal activity, but the biological activity of monomer components still has limitations [43,44]. The MIC of the phenolic mixture isolated from tea seeds for *C. albicans* was greater than 2 mg/mL. Interactions between substances might also affect antifungal activity. Numerous studies have reported that polysaccharides benefited the repair of damaged cell membranes [45–47]. The existence of tea seed polysaccharides might also have an antagonistic effect on saponins, resulting in a weakened antifungal effect. Further studies are needed to prove this.

The kill curve showed that the fungicidal efficiency of ASA was higher than that of TE1. According to previous reports, the biological activities of triterpenoid saponins with different structures were varied. Some modification groups, such as the type of sugar group on the C3 position, had a crucial impact on the cytotoxicity, hemolysis [48], and antifungal activity [49]. Acetyl groups in the C21 and C22 positions increased antihyperlipidemic activity [50]. A cinnamoyl group in the C22 position increased the cytotoxicity, while an acyl group in the C22 position decreased the cytotoxicity [51]. The groups in the C22 position of ASA and TE1 were an angelic acid moiety and an acetyl group, respectively. Accordingly, ASA had better antifungal activity than TE1, implying that the group type in the C22 position might also have an impact on the antifungal activity. Other studies reported that saponins damaged the integrity of membranes by binding their hydrophobic parts to cholesterol [25,26,52,53]. Therefore, the hydrophobicity of saponins may also affect their antifungal activity. Studies on the structure–activity relationship and molecular targets of saponins are needed to further explore their potential as antifungal agents.

Cell membranes have important physiological functions, including maintaining the stability of the intracellular environment, signal transduction, and material transportation [54]. The integrity of cell membranes is crucial for cell viability, and membrane damage can lead to high cytotoxicity. Numerous studies have shown that the cell membrane of fungi is a target for inhibiting fungal growth and reproduction [55]. In this study, it was found that saponins caused significant damage to cell membranes, which is consistent with previous studies [16,56]. Morphological changes in cell membrane damage were observed using SEM and TEM. The damaged cell membrane integrity was also verified by the PI staining and leakage experiments.

Ergosterol is an important component of the fungal cell membrane, which is of great significance for maintaining normal physiological functions [57]. Ergosterol biosynthesis is the most important cellular pathway targeted by antifungal compounds [58,59]. Ergosterol is synthesized on the endoplasmic reticulum by 25 enzymes in sequence. The three major classes of clinical antifungal substances include azoles, allylamines, and morpholines. Their targets are the *EGR11* gene encoding 14α-demethylase, the *ERG1* gene encoding squalene epoxidase, and the *ERG2* gene encoding sterol C8 isomerase, respectively. By interfering with the biosynthesis of ergosterol, they damage the cell membrane. Through transcriptome analysis, it was found that after saponin treatment, genes related to the ergosterol biosynthesis pathway in *C. albicans* cells, including *ERG1*, *ERG2*, *ERG3*, *ERG5*, *ERG11*, *ERG24*, *ERG26*, and *ERG251*, were downregulated, indicating that saponins inhibited ergosterol synthesis.

Previous studies demonstrated that the higher content of ergosterol was related to the higher resistance of the strain [60,61]. Jiang et al. found that the average ergosterol content in the resistant group was higher than that in the susceptible group [62]. Differences in sterol structures in cell membranes also affected the effectiveness of antifungal agents. Amphotericin B, a polyene antifungal active substance, has different recognition effects on ergosterol and cholesterol. The combination of amphotericin B and ergosterol mainly highlights the role of the cyclomethyl group at the C19 position on the sterol ring structure and also highlights the binding force of the chain double bond without affecting the methyl group at the C21 position. On the contrary, binding with cholesterol does not affect the C19 methyl group on the ring; rather, it comes into contact with the C26 and C27 methyl groups at the end of the chain [63]. Binding with sterols was one of the ways for saponins to inhibit fungi [64]. In this study, the ergosterol binding activity of the two saponins was verified. Yu et al. found that the MIC of the saponin mixture was 64 μg/mL for the *C. albicans* YEM30 strain and 78 μg/mL for the *C. albicans* CMCC98001 strain [56], indicating that the antifungal activity of saponins might be also related to the content and type of sterols on the cell membrane of the strain. The relationship between the structure of the saponins and the content and structure of the sterols in the cell membrane of the fungal strains still deserves further research.

In addition to targeting ergosterol, ROS induction plays an important role in the potential mechanisms of antifungal drugs. The overproduction of ROS may lead to cell damage and death [65]. The presence of ROS causes a decrease in mitochondrial membrane potentials, which had previously been used to determine mitochondrial function [66]. Previous studies have mostly elucidated the antifungal mechanism of saponins against *C. albicans* strains from the perspective of ROS. Li et al. found that the saponins extracted from *Camellia Oleifera* seeds affect mitochondrial dysfunction by inducing the production of reactive oxygen species, leading to the death of *C. albicans* [18]. In this study, we also investigated the role of oxidative stress pathways against *C. albicans* (Figure 4). Our results indicated that saponins caused a significant elevation in intracellular ROS levels, and a significant decrease in the mitochondrial membrane potential, which are negatively correlated. Therefore, the death of *C. albicans* may be caused by mitochondrial dysfunction, in turn caused by the accumulation of ROS through the depolarization of the mitochondrial membrane potential. This is concordant with prior studies [18]. However, supplementation with antioxidants (NAC and VC) did not decrease the level of ROS and did not restore *C. albicans'* damage potential, which differs from the previous studies. The above results indicate that the oxidative stress pathway was not the main pathway for saponins leading to *C. albicans'* death.

4. Materials and Methods

4.1. Strains and Chemicals

The strain *C. albicans* (ATCC 10231) was obtained from the Guangdong Microbial Culture Collection Center (Guangzhou, China) and grown on yeast extract–peptone–dextrose agar (YPD, Sangon Biotech (Shanghai) Co., Ltd., Shanghai, China). Before each test, inoculations of single colonies of the strain were cultured in YPD liquid media and incubated overnight at 30 °C at 200 rpm.

Amphotericin B (AMB), caspofungin (CPF), fluconazole (FLC), ergosterol (Shanghai Aladdin Biochemical Technology Co., Ltd., Shanghai, China), and sorbitol (Sigma-Aldrich Co., St. Louis, MO, USA) were used in this study.

4.2. Preparation of TE1 and ASA

Camellia sinensis seed cake was provided by the Jinhua Academy of Agricultural Science (Jinhua, China). The tea seed cake (1 kg) was crushed into powder and extracted three times with 70% MeOH (10 L) at 65 °C under reflux each for 2 h. After the removal of the solvent from the MeOH solution under reduced pressure, it was then extracted successively with hexane, ethyl acetate, and saturated n-butanol [67]. The n-butanol fraction was purified using a D101 macroporous resin absorption, and the saponin mixture was prepared by the collection and concentration of 70% ethanol elution [67]. The saponin monomers TE1 and ASA (Figure 1) were purified via a Waters Auto Purification system (2545 pump-PCM 510 Compensated pump-2767 sample manager-2489 DUV-2424 ELSD, Waters, Milford, MA, USA). For complete separation, purification, and characterization data, see Supplementary Methods.

4.3. Antifungal Susceptibility Tests

Minimum inhibitory concentrations (MICs) were determined according to the broth microdilution method of the Clinical Laboratory Standards Institute (CLSI), with slight modification [68]. Cells were diluted to a concentration of 1×10^6 CFU/mL in a YPD medium and then mixed together with different concentrations of TE1 and ASA. Then, 200 µL of solutions was added to 96-well plates (Corning Int., Corning, NY, USA) containing 3.125–400 µM of TE1 and ASA, respectively, and incubated for 24 h at 30 °C. AMB (0.125–16 µg/mL), FLC (15.625–2000 µg/mL), and CFP (0.25–8 µg/mL) were used as positive controls. DMSO and distilled water were used as negative, solvent, and growth control, respectively. Culture medium was used as a blank control. MIC was set for wells without turbidity. In addition, the MIC wells and the wells with concentrations above the MIC were plated onto YPD agar plates with 10 µL per well and incubated overnight at 30 °C. The minimum fungicidal concentration (MFC) was defined as the lowest concentration where no colony growth was observed. Each experiment was performed in triplicate.

4.4. Time–Kill Curves

The kill curve was used to investigate the fungicidal effects of TE1 and ASA. The YPD medium of fresh *C. albicans* cells (1×10^6 CFU/mL) was combined with 0, 1/2 MIC, 1 MIC, and 2 MIC of TE1 and ASA, respectively. Cells were incubated at 30 °C, 200 rpm for 48 h and sampled at different time points (0 h, 1 h, 2 h, 3 h, 4 h, 5 h, 10 h, 24 h, and 48 h). Following these time periods, we took an aliquot of 10 µL and serially diluted it with sterile PBS, and then plated it onto YPD agar plates. These were incubated for 24 h at 30 °C, after which the CFUs were counted. Curves were constructed by plotting the log10 of CFU/mL against time. Results are reported as the means of three separate experiments.

4.5. Live/Dead Fungicidal Fluorescent Imaging

Further detailed evaluation of the antifungal effects of TE1 and ASA was performed using live/dead fungicidal fluorescent imaging of confocal laser scanning microscopy (CLSM, FV1200, Olympus, Tokyo, Japan). Briefly, cells (1×10^7 CFU/mL) were treated with 1 MIC of TE1 and ASA at 30 °C, 200 rpm, for 2 h. Next, the cells were washed twice with sterile PBS and incubated with 2 µg/mL PI (Solarbio Science & Technology Co., Ltd., Beijing, China) and 5 µM SYTO-9 (Invitrogen Detection Technologies, Eugene, OR, USA) at 37 °C in darkness for 15 min. After being stained, samples were washed and resuspended in PBS. Finally, the samples were put on a glass cover slip and imaged by CLSM.

4.6. Morphology Observation of C. albicans ATCC 10231

Cells were treated with 1 MIC of TE1 and ASA at 30 °C, 200 rpm for 2 h. The ultrastructure observation of the treated *C. albicans* cells was achieved by Shiyanjia Lab on a scanning electron microscopy (SEM) and transmission electron microscope (TEM) according to the standard protocols.

4.7. Cellular Leakage Effect

When the fungal membrane was damaged, the nucleic acids and proteins leaking from the fungal culture medium could be measured. Briefly, after treatment with different concentrations (0, 1/4 MIC, 1/2 MIC, and 1 MIC) of TE1 and ASA, the *C. albicans* cell cultures were centrifuged for 10 min at 5000 rpm. Then, the suspensions were detected by a Synergy H1 microplate reader (BioTek Instruments, Inc., Winooski, VT, USA) at OD260 and OD280.

4.8. Sorbitol Protection Assay

The experimental method was conducted according to the literature reports [23,24]. The effect of TE1 and ASA on the fungal cell wall was conducted by sorbitol protection assay. The specific step was that the 0.8 M sorbitol in the absence and presence of YPD medium, and the MIC of TE1 and ASA, were determined using the broth microdilution method after 2 and 7 days. Caspofungin was used as a positive control.

4.9. Ergosterol Binding Assay

The experimental method to evaluate whether the compound binds to fungal membrane sterols through ergosterol binding experiments was conducted according to the literature reports [23,24]. In brief, with the addition of 0, 50, 100, 150, 200, and 250 μg/mL ergosterol in YPD medium, the MIC value of TE1 and ASA was determined after 48 h in accordance with the broth microdilution method described above. Amphotericin B was used as a positive control.

4.10. Reactive Oxygen Species and Membrane Potential Assays

ROS were detected by the DCFH-DA assay (Beyotime Biotech, Shanghai, China). MMP was detected by the Mitochondrial Membrane Potential Assay Kit with JC-1 (Beyotime Biotech, Shanghai, China). The *C. albicans* cells (1×10^7 CFU/mL) were cultured with various concentrations of TE1 and ASA (0, 1/4 MIC, 1/2 MIC, and 1 MIC) for 2 h at 30 °C. Then, the treated cells were washed twice with PBS and stained with DCFH-DA or JC-1 for 30 min in the dark. Subsequently, the cells were washed with PBS and resuspended in PBS post-centrifugation. Finally, the absorbance was measured using a Synergy H1 microplate reader. CCK-8 assay (Beyotime Biotech, Shanghai, China) was used to measure viable cell numbers. The percentage of control was used to express ROS. The MMP was determined as the ratio of JC-1 monomer (490/530 nm) and JC-1 aggregate (525/590 nm) fluorescence intensity. All experiments were performed in triplicate.

4.11. The Effect of the Antioxidant on the Fungicidal Activity of Saponins

The *C. albicans* cells (1×10^7 CFU/mL) were pretreated with or without antioxidants, including NAC (5 mM and 10 mM) and VC (5 mM and 10 mM) at 30 °C for 1 h, respectively. Then, they were cultured with 1 MIC of TE1 and ASA for 2 h at 30 °C, respectively. The detection method of ROS was the same as in Section 4.10.

4.12. Transcriptome Analysis

The Yeast RNA Extraction Kit (Aidlab Biotech, Beijing, China) was used for total RNA extraction. The *C. albicans* cells (1×10^7 CFU/mL) were cultured with 1 MIC of TE1 and ASA for 2 h at 30 °C, with three independent biological replicates. The RNA concentration and purity were verified. The cDNA libraries were constructed with an Illumina NovaSeq 6000 platform by Personalbio Technology (Shanghai, China) with paired-end reads. The clean reads were mapped to the *C. albicans* SC5314 genome sequence with HISAT2. The analysis of the differential expressed genes (DEGs) was performed by DESeq2 software (v1.40.1) between the TE1 treatment [69], the ASA treatment, and the control, and | log2foldchang | > 1 and *p*-value < 0.05 were used as conditions for screening DEGs. On the basis of the GO and KEGG pathways, functional annotation was performed using the GO (http://geneontology.org, accessed on 8 December 2022) and KEGG (http://www.genome.ad.jp/kegg/, accessed on 14 December 2022) databases [70].

4.13. Real-Time Quantitative Reverse Transcription PCR (qRT-PCR) Validation

Sixteen DEGs were validated by qRT-PCR with the paired primers (Table S3), with 18S rRNA as the reference gene. Total RNA was reverse transcribed to cDNA using SYBR® Premix Ex Taq™ (Monad Biotech, Shanghai, China). The PCR program was as follows: 95 °C for 10 min, followed by 40 cycles of 95 °C for 10 s, 60 °C for 10 s, and 72 °C for 15 s. The relative expression levels of the genes were normalized using the $2^{-\Delta\Delta CT}$ method.

4.14. Statistical Analysis

All data are presented as the mean ± standard deviation (three replicates). The results were analyzed with SPSS Version 22.0 using one-way ANOVA to demonstrate the significant differences ($p < 0.05$, $p < 0.01$). GraphPad Prism (v9.4.1) was used to process the data and generate the figures.

5. Conclusions

In this study, the fungicidal effects of saponin monomers isolated and purified from *Camellia sinensis* seeds, which were TE1 and ASA, were highlighted against fluconazole-resistant *Candida albicans* strains for the first time, and there were differences in antifungal activity between the monomers. TE1 and ASA bound with ergosterol on the fungal cell membrane to increase the cell membrane permeability and disrupted the cell membrane integrity of the *C. albicans* cells. Moreover, the accumulation of intracellular ROS treated with TE1 and ASA lead to mitochondrial dysfunction, affecting energy metabolism, but this may not be the main cause of *C. albicans'* death. These results suggest that tea seed saponins could be used as broad-spectrum and potential fungicides for the management of pathogenic fungi with ergosterol as the main target. In addition, the structure–activity relationship of the antifungal activity of saponins remains to be further studied.

Supplementary Materials: The supporting information can be downloaded at: https://www.mdpi.com/article/10.3390/ijms24119350/s1.

Author Contributions: Y.C. carried out the experiments, collected all the data, and worked on writing the manuscript. M.Y. and Z.Z. contributed to the sample collection. Y.G. and J.Y. designed the overall experiment and reviewed the manuscript. All authors have read and agreed to the published version of the manuscript.

Funding: This research was supported by the Innovation Project for Chinese Academy of Agricultural Sciences (CAAS-ASTIP-TRI), China Agriculture Research System of MOF and MARA (CARS-19).

Institutional Review Board Statement: Not applicable.

Informed Consent Statement: Not applicable.

Data Availability Statement: The authors declare that all data generated or analyzed during this study are included in this published article and its Supplementary Information files. Related data are available from the authors upon reasonable request.

Acknowledgments: We thank Changqiang Ke of the Chinese Academy of Sciences Shanghai Institute of Materia Medica for his guidance on the separation of saponin monomers. We thank the Shiyanjia Lab (www.Shiyanjia.com, accessed on 5 January 2023) for the SEM and TEM testing.

Conflicts of Interest: The authors declare no conflict of interest.

References

1. Erb Downward, J.R.; Falkowski, N.R.; Mason, K.L.; Muraglia, R.; Huffnagle, G.B. Modulation of post-antibiotic bacterial community reassembly and host response by *Candida albicans*. *Sci. Rep.* **2013**, *3*, 2191. [CrossRef] [PubMed]
2. da Silva Dantas, A.; Lee, K.K.; Raziunaite, I.; Schaefer, K.; Wagener, J.; Yadav, B.; Gow, N.A. Cell biology of *Candida albicans*-host interactions. *Curr. Opin. Microbiol.* **2016**, *34*, 111–118. [CrossRef] [PubMed]
3. Verma, A.; Gaffen, S.L.; Swidergall, M. Innate immunity to mucosal *Candida* Infections. *J. Fungi.* **2017**, *3*, 60. [CrossRef] [PubMed]
4. Brown, G.D.; Denning, D.W.; Gow, N.A.R.; Levitz, S.M.; Netea, M.G.; White, T.C. Hidden killers: Human fungal infections. *Sci. Transl. Med.* **2012**, *4*, 165rv13. [CrossRef]
5. Sievert, D.M.; Ricks, P.; Edwards, J.R.; Schneider, A.; Patel, J.; Srinivasan, A.; Kallen, A.; Limbago, B.; Fridkin, S.; National Healthcare Safety Network (NHSN) Team and Participating NHSN Facilities. Antimicrobial-resistant pathogens associated with healthcare-associated infections summary of data reported to the national healthcare safety network at the centers for disease control and prevention, 2009–2010. *Infect. Control Hosp. Epidemiol.* **2013**, *34*, 1–14. [CrossRef]
6. Lu, Y.; Su, C.; Liu, H.P. *Candida albicans* hyphal initiation and elongation. *Trends Microbiol.* **2014**, *22*, 707–714. [CrossRef]
7. Pfaller, M.A. Nosocomial candidiasis: Emerging species, reservoirs, and modes of transmission. *Clin. Infect. Dis.* **1996**, *22*, S89–S94. [CrossRef]
8. Zida, A.; Bamba, S.; Yacouba, A.; Ouedraogo-Traore, R.; Guiguemdé, R.T. Anti-*Candida albicans* natural products, sources of new antifungal drugs: A review. *J. Mycol. Med.* **2017**, *27*, 1–19. [CrossRef]
9. Hu, W.L.; Liu, J.X.; Ye, J.; Wu, Y.M.; Guo, Y.Q. Effect of tea saponin on rumen fermentation in vitro. *Anim. Feed Sci. Technol.* **2005**, *120*, 333–339. [CrossRef]
10. Yang, W.S.; Ko, J.; Kim, E.; Kim, J.H.; Park, J.G.; Sung, N.Y.; Kim, H.G.; Yang, S.; Rho, H.S.; Hong, Y.D.; et al. 21-O-angeloyltheasapogenol E3, a novel triterpenoid saponin from the seeds of tea plants, inhibits macrophage-mediated inflammatory responses in a NF-κB-dependent manner. *Mediat. Inflamm.* **2014**, *2014*, 658351. [CrossRef]
11. Yang, H.; Cai, R.; Kong, Z.Y.; Chen, Y.; Cheng, C.; Qi, S.H.; Gu, B. Teasaponin ameliorates murine colitis by regulating gut microbiota and suppressing the immune system response. *Front. Med.* **2020**, *7*, 584369. [CrossRef]
12. Yang, H.; Shao, X.; Yu, G.H. Effect of tea saponin on blood pressure in spontaneously hypertensive rats. *Chin. J. Clin. Healthc.* **2007**, *116*, 388–395. (In Chinese)
13. Zhao, W.H.; Li, N.; Zhang, X.R.; Wang, W.L.; Li, J.Y.; Si, Y.Y. Cancer chemopreventive theasaponin derivatives from the total tea seed saponin of *Camellia sinensis*. *J. Funct. Foods* **2015**, *12*, 192–198. [CrossRef]
14. Tsukamoto, S.; Kanegae, T.; Nagoya, T.; Shimamura, M.; Kato, T.; Watanabe, S.; Kawaguchi, M. Effects of seed saponins of *Thea sinensis* L. (Ryokucha saponin) on alcohol absorption and metabolism. *Alcohol Alcohol.* **1993**, *28*, 687–692. [CrossRef] [PubMed]
15. Yang, X.P.; Jiang, X.D.; Chen, J.J.; Zhang, S.S. Control of postharvest grey mould decay of nectarine by tea polyphenol combined with tea saponin. *Lett. Appl. Microbiol.* **2013**, *57*, 502–509. [CrossRef]
16. Zhang, X.F.; Yang, S.L.; Han, Y.Y.; Zhao, L.; Lu, G.L.; Xia, T.; Gao, L.P. Qualitative and quantitative analysis of triterpene saponins from tea seed pomace (*Camellia oleifera* Abel) and their activities against bacteria and fungi. *Molecules* **2014**, *19*, 7568–7580. [CrossRef]
17. Sagesaka, Y.M.; Uemura, T.; Suzuki, Y.; Sugiura, T.; Yoshida, M.; Yamaguchi, K.; Kyuki, K. Antimicrobial and anti-inflammatory actions of tea-leaf saponin. *Yakugaku Zasshi J. Pharm. Soc. Jpn.* **1996**, *116*, 238–243. [CrossRef]
18. Li, S.H.; Guo, K.M.; Yin, N.N.; Shan, M.Z.; Zhu, Y.; Li, Y. Teasaponin induces reactive oxygen species-mediated mitochondrial dysfunction in *Candida albicans*. *J. Chin. Pharm. Sci.* **2021**, *30*, 895–903. [CrossRef]
19. Guo, N.; Tong, T.T.; Ren, N.; Tu, Y.Y.; Li, B. Saponins from seeds of *Genus Camellia*: Phytochemistry and bioactivity. *Phytochemistry* **2018**, *149*, 42–55. [CrossRef]
20. Wiegand, I.; Hilpert, K.; Hancock, R.E.W. Agar and broth dilution methods to determine the minimal inhibitory concentration (MIC) of antimicrobial substances. *Nat. Protoc.* **2008**, *3*, 163–175. [CrossRef]
21. Qin, F.; Wang, Q.; Zhang, C.L.; Fang, C.Y.; Zhang, L.P.; Chen, H.L.; Zhang, M.; Cheng, F. Efficacy of antifungal drugs in the treatment of vulvovaginal candidiasis: A Bayesian network meta-analysis. *Infect. Drug Resist.* **2018**, *11*, 1893–1901. [CrossRef] [PubMed]
22. Niu, Y.Y.; Li, S.S.; Lin, Z.T.; Liu, M.X.; Wang, D.D.; Wang, H.; Chen, S.Z. Development of propidium iodide as a fluorescence probe for the on-line screening of non-specific DNA-intercalators in Fufang Banbianlian Injection. *J. Chromatogr. A.* **2016**, *1463*, 102–109. [CrossRef] [PubMed]
23. Turecka, K.; Chylewska, A.; Kawiak, A.; Waleron, K.F. Antifungal activity and mechanism of action of the Co(III) coordination complexes with diamine chelate ligands against reference and clinical strains of *Candida* spp. *Front. Microbiol.* **2018**, *9*, 1594. [CrossRef] [PubMed]
24. de Oliveira Filho, A.A.; de Oliveira, H.M.B.F.; de Sousa, J.P.; Meireles, D.; de Azevedo Maia, G.L.; Filho, J.M.B.; Lima, E.O. In vitro anti-*Candida* activity and mechanism of action of the flavonoid isolated from *Praxelis clematidea* against *Candida albicans* species. *J. App. Pharm. Sci.* **2016**, *6*, 66–69. [CrossRef]
25. Simons, V.; Morrissey, J.P.; Latijnhouwers, M.; Csukai, M.; Cleaver, A.; Yarrow, C.; Osbourn, A. Dual effects of plant steroidal alkaloids on *Saccharomyces cerevisiae*. *Antimicrob. Agents Chemother.* **2006**, *50*, 2732–2740. [CrossRef] [PubMed]
26. Keukens, E.A.; De, V.T.; Fabrie, C.H.; Demel, R.A.; Jongen, W.M.; De, K.B. Dual specificity of sterol-mediated glycoalkaloid induced membrane disruption. *Biochimica et Biophysica Acta (BBA) Biomembr.* **1992**, *1110*, 127–136. [CrossRef]

27. Ghannoum, M.A.; Rice, L.B. Antifungal agents: Mode of action, mechanisms of resistance, and correlation of these mechanisms with bacterial resistance. *Clin. Microbiol. Rev.* **1999**, *12*, 501–517. [CrossRef]

28. Garvey, E.P.; Hoekstra, W.J.; Schotzinger, R.J.; Sobel, J.D.; Lilly, E.A.; Fidel, P.L., Jr. Efficacy of the clinical agent VT-1161 against fluconazole-sensitive and -resistant *Candida albicans* in a murine model of vaginal candidiasis. *Antimicrob. Agents Chemother.* **2015**, *59*, 5567–5573. [CrossRef]

29. Peláez, F.; Cabello, A.; Platas, G.; Díez, M.T.; del Val, A.G.; Basilio, A.; Martán, I.; Vicente, F.; Bills, G.E.; Giacobbe, R.A.; et al. The discovery of enfumafungin, a novel antifungal compound produced by an endophytic Hormonema species biological activity and taxonomy of the producing organisms. *Syst. Appl. Microbiol.* **2000**, *23*, 333–343. [CrossRef]

30. Jiménez-Ortigosa, C.; Paderu, P.; Motyl, M.R.; Perlin, D.S. Enfumafungin derivative MK-3118 shows increased in vitro potency against clinical echinocandin-resistant *Candida* Species and *Aspergillus* species isolates. *Antimicrob. Agents Chemother.* **2014**, *58*, 1248–1251. [CrossRef]

31. Shibata, T.; Takahashi, T.; Yamada, E.; Kimura, A.; Nishikawa, H.; Hayakawa, H.; Nomura, N.; Mitsuyama, J. T-2307 causes collapse of mitochondrial membrane potential in yeast. *Antimicrob. Agents Chemother.* **2012**, *56*, 5892–5897. [CrossRef] [PubMed]

32. Zhang, J.D.; Cao, Y.B.; Xu, Z.; Sun, H.H.; An, M.M.; Yan, L.; Chen, H.S.; Gao, P.H.; Wang, Y.; Jia, X.M.; et al. In vitro and in vivo antifungal activities of the eight steroid saponins from *Tribulus terrestris* L. with potent activity against fluconazole-resistant fungal. *Biol. Pharm. Bull.* **2005**, *28*, 2211–2215. [CrossRef] [PubMed]

33. Coleman, J.J.; Okoli, I.; Tegos, G.P.; Holson, E.B.; Wagner, F.F.; Hamblin, M.R.; Mylonakis, E. Characterization of plant-derived saponin natural products against *Candida albicans*. *ACS Chem. Biol.* **2010**, *5*, 321–332. [CrossRef] [PubMed]

34. Yin, S.; Li, L.; Su, L.Q.; Li, H.F.; Zhao, Y.; Wu, Y.D.; Liu, R.N.; Zou, F.; Ni, G. Synthesis and in vitro synergistic antifungal activity of analogues of Panax stipulcanatus saponin against fluconazole-resistant Candida albicans. *Carbohydr. Res.* **2022**, *517*, 108575. [CrossRef]

35. Dawgul, M.A.; Grzywacz, D.; Liberek, B.; Kamysz, W.; Myszka, H. Activity of diosgenyl 2-amino-2-deoxy-β-D-glucopyranoside, its hydrochloride, and N,N-dialkyl derivatives against non-albicans *Candida* Isolates. *Med. Chem.* **2018**, *14*, 460–467. [CrossRef] [PubMed]

36. Kim, J.; Sudbery, P. *Candida albicans*, a major human fungal pathogen. *J. Microbiol.* **2011**, *49*, 171–177. [CrossRef]

37. Pfaller, M.A.; Diekema, D.J. Epidemiology of invasive mycoses in North America. *Crit. Rev. Microbiol.* **2010**, *36*, 1–53. [CrossRef]

38. Kullberg, B.J.; Arendrup, M.C. Invasive Candidiasis. *N. Engl. J. Med.* **2015**, *373*, 1445–1456. [CrossRef]

39. Favel, A.; Steinmetz, M.D.; Regli, P.; Vidal-Ollivier, E.; Balansard, G. In vitro antifungal activity of triterpenoid saponins. *Planta Med.* **1994**, *60*, 50–53. [CrossRef]

40. Zhang, J.D.; Xu, Z.; Cao, Y.B.; Chen, H.S.; Yan, L.; An, M.M.; Gao, P.H.; Wang, Y.; Jia, X.M.; Jiang, Y.Y. Antifungal activities and action mechanisms of compounds from *Tribulus terrestris* L. *J. Ethnopharmacol.* **2006**, *103*, 76–84. [CrossRef]

41. Li, Y.; Shan, M.Z.; Zhu, Y.; Yao, H.K.; Li, H.C.; Gu, B.; Zhu, Z.B. Kalopanaxsaponin A induces reactive oxygen species mediated mitochondrial dysfunction and cell membrane destruction in *Candida albicans*. *PLoS ONE* **2020**, *15*, e0243066. [CrossRef] [PubMed]

42. Luo, J.J.; Jiang, Z.H.; Huang, Z.A.; Chen, T.; Liao, Y.X. Resource and development of tea meal (tea bran or tea seed). *Guangzhou Chem. Ind.* **2021**, *49*, 26–27. (In Chinese)

43. Lin, Z.; Tan, X.H.; Zhang, Y.; Li, F.P.; Luo, P.; Liu, H.Z. Molecular targets and related biologic activities of fucoidan: A Review. *Mar. Drugs* **2020**, *18*, 376. [CrossRef] [PubMed]

44. Nørskov, N.P.; Bruhn, A.; Cole, A.; Nielsen, M.O. Targeted and untargeted metabolic profiling to discover bioactive compounds in seaweeds and hemp using gas and liquid chromatography-mass spectrometry. *Metabolites* **2021**, *11*, 259. [CrossRef] [PubMed]

45. Wen, Z.S.; Liu, L.J.; OuYang, X.K.; Qu, Y.L.; Chen, Y.; Ding, G.F. Protective effect of polysaccharides from *Sargassum horneri* against oxidative stress in RAW264.7 cells. *Int. J. Biol. Macromol.* **2014**, *68*, 98–106. [CrossRef] [PubMed]

46. Cui, C.; Cui, N.S.; Wang, P.; Song, S.L.; Liang, H.; Ji, A.G. Sulfated polysaccharide isolated from the sea cucumber *Stichopus japonicus* against PC12 hypoxia/reoxygenation injury by inhibition of the MAPK signaling pathway. *Cell. Mol. Neurobiol.* **2015**, *35*, 1081–1092. [CrossRef]

47. Ye, Z.P.; Wang, W.; Yuan, Q.X.; Ye, H.; Sun, Y.; Zhang, H.C.; Zeng, X.X. Box–Behnken design for extraction optimization, characterization and in vitro antioxidant activity of *Cicer arietinum* L. hull polysaccharides. *Carbohydr. Polym.* **2016**, *147*, 354–364. [CrossRef]

48. Oda, K.; Matsuda, T.; Murakami, S.; Katayama, S.; Ohgitani, T.; Yoshikawa, M. Adjuvant and haemolytic activities of 47 saponins derived from medicinal and food plants. *Biol. Chem.* **2000**, *381*, 67–74. [CrossRef]

49. Sokolov, S.S.; Volynsky, P.E.; Zangieva, O.T.; Severin, F.F.; Glagoleva, E.S.; Knorre, D.A. Cytostatic effects of structurally different ginsenosides on yeast cells with altered sterol biosynthesis and transport. *Biochimica et Biophysica Acta (BBA) Biomembr.* **2022**, *1864*, 183993. [CrossRef]

50. Yoshikawa, M.; Morikawa, K.; Yamamoto, Y.; Kato, Y.; Nagatomo, A.; Matsuda, H. Floratheasaponins A-C, acylated oleanane-type triterpene oligoglycosides with anti-hyperlipidemic activities from flowers of the tea plant (*Camellia sinensis*). *J. Nat. Prod.* **2005**, *68*, 1360–1365. [CrossRef]

51. Cui, C.J.; Zong, J.F.; Sun, Y.; Zhang, L.; Ho, C.T. Triterpenoid saponins from the *genus Camellia*: Structures, biological activities, and molecular simulation for structure-activity relationship. *Food Funct.* **2018**, *9*, 3069–3091. [CrossRef] [PubMed]

52. Bangham, A.D.; Horne, R.W. Action of saponin on biological cell membranes. *Nature* **1962**, *196*, 952–953. [CrossRef] [PubMed]

53. Seeman, P. Ultrastructure of membrane lesions in immune lysis, osmotic lysis and drug-induced lysis. *Fed. Proc.* **1974**, *33*, 2116–2124. [CrossRef] [PubMed]
54. Shi, Y.; Cai, M.J.; Zhou, L.L.; Wang, H.D. The structure and function of cell membranes studied by atomic force microscopy. *Semin. Cell Dev. Biol.* **2018**, *73*, 31–44. [CrossRef]
55. Ma, M.M.; Wen, X.F.; Xie, Y.T.; Guo, Z.; Zhao, R.B.; Yu, P.; Gong, D.M.; Deng, S.G.; Zeng, Z.L. Antifungal activity and mechanism of monocaprin against food spoilage fungi. *Food Control.* **2018**, *84*, 561–568. [CrossRef]
56. Yu, Z.L.; Wu, X.H.; He, J.H. Study on the antifungal activity and mechanism of tea saponin from *Camellia oleifera* cake. *Eur. Food. Res. Technol.* **2022**, *248*, 783–795. [CrossRef]
57. Behbehani, J.M.; Irshad, M.; Shreaz, S.; Karched, M. Anticandidal activity of capsaicin and its effect on ergosterol biosynthesis and membrane integrity of *Candida albicans*. *Int. J. Mol. Sci.* **2023**, *24*, 1046. [CrossRef]
58. Daum, G.; Lees, N.D.; Bard, M.; Dickson, R. Biochemistry, cell biology and molecular biology of lipids of Saccharomyces cerevisiae. *Yeast* **1998**, *14*, 1471–1510. [CrossRef]
59. Prasad, R.; Shah, A.H.; Rawal, M.K. Antifungals: Mechanism of action and drug resistance. *Adv. Exp. Med. Biol.* **2016**, *892*, 327–349. [CrossRef]
60. Espinoza, C.; González, M.C.R.; Mendoza, G.; Creus, A.H.; Trigos, Á.; Fernández, J.J. Exploring photosensitization as an efficient antifungal method. *Sci. Rep.* **2018**, *8*, 14489. [CrossRef]
61. Veen, M.; Lang, C. Interactions of the ergosterol biosynthetic pathway with other lipid pathways. *Biochem. Soc. Trans.* **2005**, *33*, 1178–1181. [CrossRef] [PubMed]
62. Jiang, C.; Dong, D.F.; Yu, B.Q.; Cai, G.; Wang, X.F.; Ji, Y.H.; Peng, Y.B. Mechanisms of azole resistance in 52 clinical isolates of *Candida tropicalis* in China. *J. Antimicrob. Chemother.* **2013**, *68*, 778–785. [CrossRef] [PubMed]
63. Ciesielski, F.; Griffin, D.C.; Loraine, J.; Rittig, M.; Delves-Broughton, J.; Bonev, B.B. Recognition of membrane sterols by polyene antifungals amphotericin B and natamycin, a 13C MAS NMR Study. *Front. Cell Dev. Biol.* **2016**, *4*, 57. [CrossRef]
64. Lin, F.; Wang, R. Hemolytic mechanism of dioscin proposed by molecular dynamics simulations. *J. Mol. Model.* **2009**, *16*, 107–118. [CrossRef] [PubMed]
65. Dantas Ada, S.; Day, A.; Ikeh, M.; Kos, I.; Achan, B.; Quinn, J. Oxidative stress responses in the human fungal pathogen, *Candida albicans*. *Biomolecules.* **2015**, *5*, 142–165. [CrossRef] [PubMed]
66. Suski, J.M.; Lebiedzinska, M.; Bonora, M.; Pinton, P.; Duszynski, J.; Wieckowski, M.R. Relation between mitochondrial membrane potential and ROS formation. *Methods Mol. Biol.* **2012**, *810*, 183–205. [CrossRef]
67. Wu, X.J. Separation, Identification and Quantification of Triterpene Saponins in Seeds of *Camellia sinensis*. Master's Thesis, Zhejiang University, Hangzhou, China, 2018. (In Chinese).
68. Alastruey-Izquierdo, A.; Melhem, M.S.; Bonfietti, L.X.; Rodriguez-Tudela, J.L. Susceptibility test for fungi: Clinical and laboratorial correlations in medical mycology. *Rev. Inst. Med. Trop. Sao Paulo* **2015**, *57*, 57–64. [CrossRef]
69. Love, M.I.; Huber, W.; Anders, S. Moderated estimation of fold change and dispersion for RNA-seq data with DESeq2. *Genome Biol.* **2014**, *15*, 550. [CrossRef]
70. Kanehisa, M.; Goto, S. KEGG: Kyoto encyclopedia of genes and genomes. *Nucleic Acids Res.* **2000**, *28*, 27–30. [CrossRef]

International Journal of
Molecular Sciences

Review

The Potential of Cylindromatosis (CYLD) as a Therapeutic Target in Oxidative Stress-Associated Pathologies: A Comprehensive Evaluation

Zhenzhou Huang and Yanjie Tan *

Center for Cell Structure and Function, Shandong Provincial Key Laboratory of Animal Resistance Biology, Collaborative Innovation Center of Cell Biology in Universities of Shandong, College of Life Sciences, Shandong Normal University, Jinan 250358, China; huangzhenzhou97@163.com
* Correspondence: 620036@sdnu.edu.cn

Abstract: Oxidative stress (OS) arises as a consequence of an imbalance between the formation of reactive oxygen species (ROS) and the capacity of antioxidant defense mechanisms to neutralize them. Excessive ROS production can lead to the damage of critical biomolecules, such as lipids, proteins, and DNA, ultimately contributing to the onset and progression of a multitude of diseases, including atherosclerosis, chronic obstructive pulmonary disease, Alzheimer's disease, and cancer. Cylindromatosis (CYLD), initially identified as a gene linked to familial cylindromatosis, has a well-established and increasingly well-characterized function in tumor inhibition and anti-inflammatory processes. Nevertheless, burgeoning evidence suggests that CYLD, as a conserved deubiquitination enzyme, also plays a pivotal role in various key signaling pathways and is implicated in the pathogenesis of numerous diseases driven by oxidative stress. In this review, we systematically examine the current research on the function and pathogenesis of CYLD in diseases instigated by oxidative stress. Therapeutic interventions targeting CYLD may hold significant promise for the treatment and management of oxidative stress-induced human diseases.

Keywords: oxidative stress; CYLD; disease; deubiquitination

check for
updates

Citation: Huang, Z.; Tan, Y. The Potential of Cylindromatosis (CYLD) as a Therapeutic Target in Oxidative Stress-Associated Pathologies: A Comprehensive Evaluation. *Int. J. Mol. Sci.* **2023**, *24*, 8368. https://doi.org/10.3390/ijms24098368

Academic Editors: Rossana Morabito and Alessia Remigante

Received: 29 March 2023
Revised: 25 April 2023
Accepted: 5 May 2023
Published: 6 May 2023

1. CYLD: An Overview

1.1. CYLD Structure and Function

Cylindromatosis (CYLD), belonging to the ubiquitin-specific protease (USP) family, is a deubiquitinase that selectively removes K63-linked ubiquitin chains and exhibits widespread distribution in vivo. Initially identified in familial cylindromatosis (FC), a skin appendage tumor typically manifesting on the scalp, CYLD is implicated in a genetic syndrome thought to originate from hair follicle stem cells [1]. Biggs et al. discovered the association between cylindromatosis and CYLD gene deletion on chromosome 16, subsequently determining the gene's location on the long arm of chromosome 16 (16q12-13) in 1995 [2]. Multiple familial trichoepitheliomas (MFT) and Brooke–Spiegler syndrome (BSS) also exhibit loss of heterozygosity of CYLD [3]. FC, MFT, and BSS represent overlapping phenotypes resulting from CYLD deletion [3], underscoring the significance of CYLD as a crucial tumor suppressor.

CYLD, composed of 20 exons, encodes a protein containing 956 amino acids [4]. Its N-terminal region houses three CAP-Gly domains, which interact with targets such as NEMO in the NF-kB pathway [4]. These domains, originally identified in connections between endocytic vesicles and microtubules, comprise approximately 70 hydrophobic amino acid residues [5]. Functionally, CAP-Gly domains are present in several microtubule-binding proteins, including cytoplasmic adaptor protein CLIP-170 and dynactin1, and are postulated to facilitate the attachment of proteins like microtubule-binding proteins to microtubules [6]. Additionally, a ubiquitin-specific protease (USP) catalytic domain is located at the C-terminal region. This USP domain, capable of specifically removing K63-linked ubiquitin chains, also contains a B-box domain that mediates CYLD dimerization [7]. A small zinc finger binding module, akin to the E3 ligase B box and RING finger structure, is embedded within this domain, playing a role in CYLD subcellular localization [8].

1.2. The Cap-Gly Domains of CYLD and Microtubule-Related Cellular Processes

Research has revealed that the N-terminus of CYLD harbors three Cap-Gly domains, which are evolutionarily conserved motifs consisting of approximately 70 amino acids and an abundance of glycine residues [9]. A key function of the Cap-Gly domains is that two domains near the 3' end can bind to the C-terminal EEY/F-COO(-) motifs of α-tubulin and certain microtubule-associated proteins, with the first Cap-Gly domain found to be necessary for this activity [10,11]. These domains regulate microtubule–tubulin interactions and suppress the deacetylation of tubulin by downregulating histone deacetylase-6 (HDAC6) or inhibiting its activity via trichostatin A (TSA), thereby influencing microtubule-related cell migration [11], cell cycle [12], and ciliogenesis [13]. The third CAP-Gly domain of CYLD specifically interacts with one of the two proline-rich sequences of NEMO/IKKγ in the NF-κB signaling pathway, a pathway that regulates gene expression involved in various biological processes such as development, inflammation, and tumorigenesis. CYLD-mediated NEMO deubiquitination impedes its phosphorylation of IκB, consequently inhibiting NF-κB signaling [14,15].

1.3. The USP Catalytic Structural Domain of CYLD and Deubiquitination Function

Ubiquitination has emerged as a crucial post-translational modification in diverse cellular processes, regulating protein degradation, autophagy, intracellular protein transport, DNA damage response, protein activation, and protein–protein interactions. Given that the deregulation of these processes can result in pathological conditions such as inflammatory diseases, neurodegeneration, or cancer, stringent regulation of the ubiquitin system is paramount [16]. CYLD proteins possess a USP catalytic structural domain at their C-terminus, which specifically removes Lys63 and Met-1-linked polyubiquitin chains [16] and disrupts protein interactions, leading to protein degradation by the protease system. This encompasses TNF receptor-associated factor 2 (TRAF2) and nuclear factor (NF)-κB essential modulator (NEMO), which are necessary for the canonical activation of NF-κB, and Bcl-37, which is required for noncanonical activation of NF-κB [7]. A B-box structure is embedded within the USP structural domain, and the deletion of the B-box does not significantly affect the deubiquitinase activity of CYLD. However, it disrupts CYLD intermolecular interactions, rendering the CYLD molecule incapable of localizing in the cytoplasm and binding to the ubiquitin chain complex [17].

2. CYLD and Oxidative Stress

2.1. Oxidative Stress and Associated Pathologies

Reactive oxygen species (ROS) are capable of inflicting damage upon lipids, nucleic acids, and proteins, subsequently altering their functionality [18]. Oxidative stress arises when an imbalance transpires between ROS production and the antioxidant defense mechanism [19]. A multitude of diseases, including atherosclerosis [20], chronic obstructive pulmonary disease (COPD) [21], Alzheimer's disease [22], and cancer [23], have been associated with oxidative stress. For instance, cardiovascular disease, the foremost cause of death globally, is influenced by oxidative stress [24]. Elevated ROS levels lead to diminished nitric oxide availability and vasoconstriction, thereby promoting arterial hypertension [24]. Additionally, ROS negatively impacts myocardial calcium handling, instigates arrhythmias, and exacerbates cardiac remodeling by inducing hypertrophic signaling and apoptosis [25]. ROSs have also been implicated in the formation of atherosclerotic plaques.

Oxidative stress may function as an initiator in oocyte aging and reproductive pathologies, resulting in abnormal follicular atresia, aberrant meiosis, reduced fertilization rates, delayed embryonic development, and reproductive disorders such as polycystic ovary syndrome and ovarian endometriosis cysts [26]. Traumatic brain injury (TBI), a leading cause of mortality and morbidity worldwide, induces glutamate elevation at the synapse following a severe TBI event. Excess glutamate subsequently activates corresponding NMDA and AMPA receptors, promoting excessive calcium influx into neuronal cells. This cascade generates oxidative stress, culminating in mitochondrial dysfunction, lipid peroxidation, and oxidation of proteins and DNA, ultimately resulting in neuronal cell death [27].

ROS are implicated in various oncogenic processes, including initiation, promotion, activation, and inactivation of proto-oncogenes, as well as the stability and function of tumor suppressor genes [28]. Numerous studies have demonstrated that oxidative stress influences several signaling pathways linked to cell proliferation. Key signaling proteins, such as nuclear factor erythroid 2-related factor 2, RAS/RAF, mitogen-activated protein kinases ERK1/2 and MEK, phosphatidylinositol 3-kinase, phospholipase C, and protein kinase C, are affected by oxidative stress [29,30]. Moreover, ROSs modify the expression of p53 repressor genes, a crucial factor in apoptosis [31]. Consequently, oxidative stress induces alterations in gene expression, cell proliferation, and apoptosis, playing a significant role in tumorigenesis and progression. Table 1 enumerates other diseases induced by oxidative stress.

Table 1. Oxidative Stress-Induced Diseases.

Disease	Mechanism	Reference
Alzheimer's	promotes Aβ deposition, tau hyperphosphorylation, and the subsequent loss of synapses and neurons	[22]
chronic kidney disease	antioxidant depletions and increases ROS production	[32]
periodontitis	increases ROS production	[33]
male infertility	damages sperm DNA, RNA transcripts, and telomeres	[34]
osteoporosis	diminishes bone mineral density in osteoporosis	[35]
endometriosis	causes a general inflammatory response in the abdominal cavity	[36]
vitiligo	damages melanocytes by ROS	[37]
nonalcoholic fatty liver	increases ROS production	[38]

2.2. Function of CYLD in Oxidative Stress-Related Diseases

2.2.1. Function of CYLD in Oxidative Stress-Induced Obesity-Related Nephropathy

With a rapidly increasing global prevalence of obesity, obesity has become a serious public health problem. In addition to predisposing to cardiovascular disease and diabetes [39,40], a growing number of reports suggest that obesity is also an important risk factor for kidney damage, namely obesity-related nephropathy (ORN) [40,41], which has become one of the major causes of end-stage renal disease.

Many studies have shown that oxidative stress is a characteristic of obesity [42] and is one of the main causes of kidney damage in ORN [43,44]. The imbalance between increased reactive oxygen species (ROSs) and/or decreased antioxidant activity promotes oxidative stress damage to tissues or cells [45,46]. ROS production induces glomerular and tubular damage, suggesting that ROSs play an important role in mediating renal injury, which may ultimately lead to the development of end-stage renal disease [47–49]. Therefore, reducing ROS production to ameliorate oxidative stress injury may be a new therapeutic target for ORN. IκB kinase (IKK) induces phosphorylation of CYLD. Phosphorylation serves as a mechanism to temporarily inactivate the deubiquitination activity of CYLD, thereby promoting the ubiquitination of its downstream molecules [50,51]. In one study, researchers found that oxidative stress damage was observed in the kidney tissue of ORN model mice and that IKK induced phosphorylation of CYLD, which in turn inactivated its deubiquitination activity [52]. Thus, phosphorylated CYLD instead promoted ubiquitination of Nrf2, which ultimately led to oxidative stress injury in ORN (Figure 1). These findings suggest that IKK promotes obesity-induced kidney injury via CYLD phosphorylation and that IKK inhibitors can alleviate lipid deposition and oxidative stress injury in ORN. Furthermore, IKK/CYLD/Nrf2 axis may provide a viable target for the treatment of ORN-induced kidney injury.

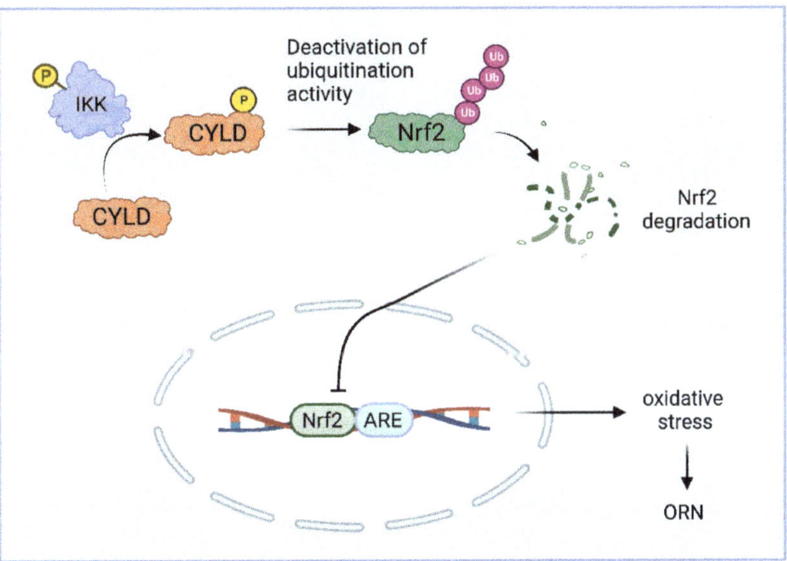

Figure 1. Schematic illustration of the mechanism through which IKK and CYLD phosphorylation mediates the activation of the Nrf2/ARE pathway, subsequently inducing oxidative stress in human kidney cells. IKK activates CYLD phosphorylation, which in turn inactivates CYLD's deubiquitination activity. This promotes the ubiquitination of Nrf2, resulting in Nrf2 protein degradation and inhibition of the Nrf2/ARE signaling pathway. Consequently, oxidative stress is exacerbated in ORN-associated kidney injury.

2.2.2. Role of CYLD in Malignant Transformation of Tumors Resulting from Oxidative Stress-Induced DNA Damage

A vast majority of human cancers present persistent DNA damage and genomic instability as pathological features [53]. The accumulation of reactive oxygen species (ROS)-induced DNA damage, leading to genomic instability, is a crucial factor in the malignant transformation of tumors [53]. Several characteristic alterations transpire during cellular transformation, including autonomous proliferation, apoptosis evasion, invasion of surrounding tissues, and tumor metastasis [54]. These properties are concomitant with the aberrant activation of nuclear factor-kappa B (NF-kB) to bolster cancer cell survival and proliferation [55].

Oxidative stress-induced DNA damage may stimulate DNAPKsome assembly and Mps1 activation, which phosphorylates c-Abl at threonine 735 (T735) and promotes its cytoplasmic translocation [56]. Persistent cytoplasmic localization of c-Abl is associated with tumor cell transformation [57]. In addition, c-Abl phosphorylates OTULIN at tyrosine [58], disrupting its binding to LUBAC. The liberated LUBAC interacts with SPATA2 and is recruited to TNF-R1sc, promoting SPATA2-CYLD interactions [54]. These interactions are essential for oxidative stress to activate IKKβ, which in turn stimulates NF-κB transcriptional activity. IKKβ also induces phosphorylation of CYLD at serine 568, subsequently activating CYLD's deubiquitination function to terminate NF-κB signaling [4].

Contrary to the prevailing notion of CYLD as a strict tumor suppressor, it initiates and terminates NF-κB activity by alternating between oncoprotein and tumor suppressor roles, respectively. Should IKKβ fail to induce the DUB activity of serine 568 phosphorylation,

CYLD will persistently exhibit oncogenic activity [54]. The ensuing dysregulation of NF-κB activity and other associated pathological changes would disrupt intracellular homeostasis, thereby favoring tumor transformation (Figure 2).

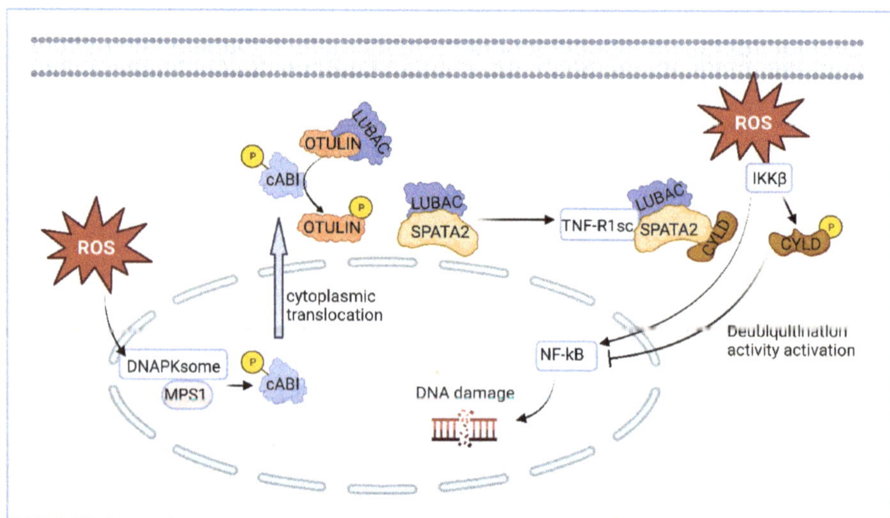

Figure 2. Schematic representation of the mechanism of CYLD function in oxidative stress-induced DNA damage. Oxidative stress-induced DNA damage may stimulate the assembly of DNAPKsome and the activation of Mps1, which phosphorylates c-Abl and promotes its cytoplasmic translocation. Also, c-Abl phosphorylates OTULIN, disrupting its binding to LUBAC. The released LUBAC interacts with SPATA2 and participates in the TNF-R1-mediated signaling pathway, promoting the interaction between CYLD and LUBAC. CYLD-LUBAC binding induces DNA damage by interacting with regulatory proteins and stimulating NF-kB activation. Additionally, oxidative stress activates IKKβ, which induces CYLD phosphorylation, in turn activating CYLD's deubiquitination function and terminating NF-κB signaling.

2.2.3. Role of CYLD in Ischemia-Reperfusion-Induced Liver Inflammation

Ischemia and reperfusion (IR)-induced liver inflammation and injury are important causes of liver dysfunction and failure after liver transplantation, resection, and hemorrhagic shock [59]. The IR-induced liver injury involves oxidative stress and endoplasmic reticulum (ER) stress-mediated inflammatory responses. Hepatic macrophages (Kupffer cells) are a key component of the liver's innate immune system and are the first line of defense in detecting invading pathogens in the liver [60]. Activated macrophages produce reactive oxygen species (ROS) and initiate activation of the tophane-like receptor 4 (TLR4) or NLRP3 inflammasome, leading to liver inflammation and injury [59,61,62]. It has been shown that the knockdown of bone marrow-specific TXNIP ameliorates IR-induced liver injury and reduces macrophage/neutrophil accumulation and pro-inflammatory mediators in IR-stressed livers [63]. Macrophage TXNIP deficiency activates the NRF2-OASL1 pathway and regulates TBK1 function in IR-induced hepatitis injury. This study found that macrophage TXNIP deficiency promoted CYLD-NADPH oxidase 4 (NOX4) interactions and enhanced nuclear factor-like 2 (NRF2) and its target gene 2′,5′ oligoadenylate synthase-like 1 (OASL1) activity, leading to IR stress-induced liver injury inhibiting Ras-GTPase-activating protein binding protein 1 (G3BP1) and TBK1-driven inflammatory response and hepatocyte death [63]. Thus, the molecular regulatory mechanisms of the macrophage TXNIP-mediated CYLD-NRF2-OASL1 pathway in the IR-stressed liver may provide potential therapeutic targets for stress-induced liver inflammation and injury (Figure 3).

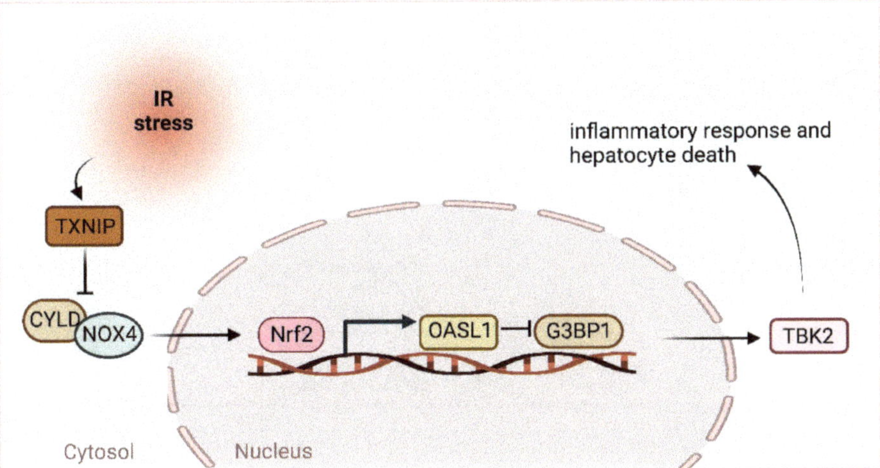

Figure 3. Schematic diagram of the mechanism of the role of CYLD in ischemia-reperfusion-induced liver inflammation. IR activates TXNIP to inhibit CYLD-NOX4 interaction, which activates NRF2 and its target gene OASL1 to suppress G3BP1 and subsequent TBK2-driven inflammatory responses and hepatocyte death.

2.2.4. CYLD Enhances Oxidative Stress in the Heart

In response to pathological stresses, such as pressure overload and other myocardial injuries, the heart initially activates an adaptive physiological hypertrophic response. However, a sustained cardiac hypertrophic response can lead to pathological cardiac hypertrophy, fibrosis, and cell death, ultimately resulting in heart failure and death [64]. Oxidative stress is a state of excessive intracellular ROS levels, which can cause DNA damage, lipid peroxidation, and protein aggregation, leading to pathological cardiomyocyte hypertrophy and death in the heart [65].

Recently, CYLD expression was found to be significantly upregulated in cardiomyocytes from hypertrophied and failing human and mouse hearts. CYLD knockout improved survival in mice and attenuated myocardial hypertrophy, fibrosis, apoptosis, oxidative stress, and dysfunction due to sustained pressure overload caused by transverse aortic constriction [52]. The most significantly altered genes revealed by sequencing and gene array analysis were those involved in free radical scavenging pathways and cardiovascular disease, including fos, jun, myc, and nuclear factor-erythroid-2-related factor 2 (Nrf2) in the heart. Nrf2 is an essential negative regulator of oxidative stress in cardiomyocytes, thereby inhibiting cardiac rational remodeling and dysfunction in different pathological settings [66–69].

This study found that CYLD knockdown enhanced the expression of mitogen-activated protein kinase (MAPK) ERK- and p38-mediated c-jun, c-fos, and c-myc that control Nrf2 expression in cardiomyocytes. The inhibition of cardiomyocyte reactive oxygen species (ROS) formation, death, and hypertrophy by CYLD deficiency was blocked by Nrf2 knockdown [52]. Thus, CYLD mediates cardiac maladaptive remodeling and dysfunction, likely by enhancing myocardial oxidative stress in response to stress overload. CYLD blocks ERK-, p38-/AP-1, and c-Myc pathways and inhibits the antioxidant capacity of Nrf2, thereby enhancing oxidative stress in the heart (Figure 4).

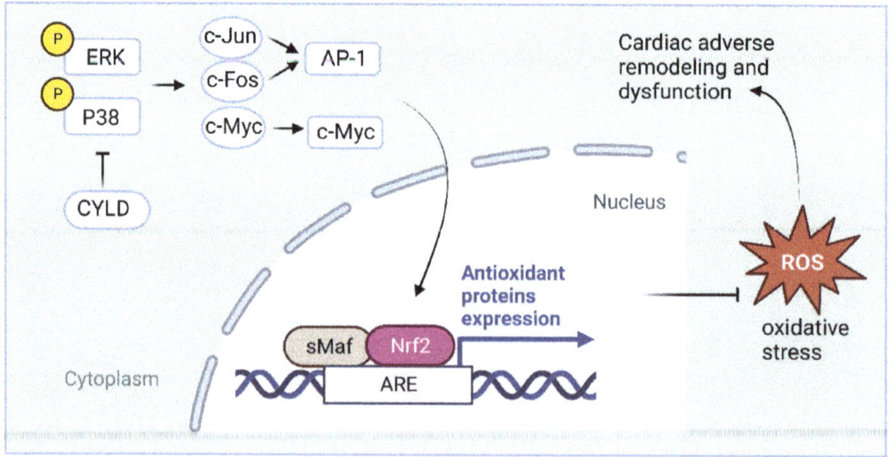

Figure 4. Schematic representation of the mechanism by which CYLD enhances oxidative stress in the heart. CYLD inhibits the antioxidant capacity of Nrf2 by blocking ERK- and p38-/AP-1, as well as c-Myc pathways, consequently enhancing oxidative stress in the heart.

2.2.5. Role of CYLD in Oxidative Stress-Induced Retinal Pigment Epithelial Cell Damage and Dysfunction

Age-related macular degeneration (AMD) is a chronic and progressive degenerative disease of the retina that ultimately results in blindness [70]. Retinal pigment epithelial (RPE) cell damage and dysfunction induced by oxidative stress are significant pathogenic factors in AMD [71]. During oxidative stress, CYLD-AS1 expression is upregulated in RPE cells. Depletion of CYLD-AS1 promotes cell proliferation and mitochondrial function, protecting RPE cells from hydrogen peroxide (H_2O_2)-induced damage. CYLD-AS1 also modulates the expression of members of the NRF2 and inflammation-related NF-κB signaling pathways, which are associated with oxidative stress. These two signaling pathways are mediated by the CYLD-AS1 interactor, miR-134-5p. CYLD-AS1 influences the oxidative stress-related inflammatory function of RPE cells by sponging miR-134-5p-mediated NRF2/NF-κB signaling pathway activity [72]. Consequently, targeting CYLD-AS1 may represent an effective strategy for the treatment of AMD-related diseases.

3. Other Biological Functions of CYLD

3.1. Function of CYLD in Ciliary Diseases

Cilia are highly specialized cellular structures that protrude from the surface of the cell membrane and are found in a wide range of organisms, from unicellular eukaryotes to vertebrates. They sense extracellular signaling molecules and regulate various cellular activities [73]. Defects in the structure and function of primary cilia cause a range of diseases, including polycystic kidney disease, microcephaly, retinal degeneration, obesity, liver dysfunction, polydactylism, neurological disorders, and malignancy, collectively known as ciliopathy [74]. CYLD plays a crucial role in matrix anchoring and assembly of primary and motor cilia in several organs. Ciliopathy phenotypes, including male sterility, impaired lung maturation, and osteoporosis, have been observed in Cyld-deficient mice. CYLD is required for ciliogenesis, and Cyld-knockout mice exhibit defects in multiple organs, including skin, kidneys, trachea, and testis [13]. Transmission electron microscopy has shown that anchoring of the matrix and proper assembly of the matrix and axon require CYLD. CYLD deubiquitinates the 70 kDa central protein (Cep70), thereby increasing the Cep70 localization of the matrix, facilitating matrix organization and anchoring to the plasma membrane. Additionally, CYLD-mediated inactivation of HDAC6 enhances microtubule protein acetylation, stabilizing axonal microtubules and promoting cilia formation [13].

Some studies have also shown that CYLD locates in the centrosome and basal body through its interaction with the centrosome protein CAP350. When the interaction between the two proteins is eliminated, matrix migration docking is damaged, resulting in cilia loss [75]. Furthermore, CENPV, a component of mitotic chromosomes associated with cytoplasmic microtubules, interacts with CYLD through the CAP-Gly structural domain and is deubiquitinated by CYLD to promote cilia formation [76]. Many studies have shown that CYLD gene deletion mice exhibit various cilia-related diseases [77–79], and the treatment of cilia diseases by CYLD warrants further investigation.

3.2. Function of CYLD in Neuronal Development

Abnormal fear memory is a hallmark of many neuropsychiatric disorders [80]. Proper neuronal activation and excitability in the basolateral amygdala (BLA) are necessary for the formation of fear memories [81]. It has been shown that Cyld knockdown impairs amygdala-dependent tone-cued fear memory [82]. Cyld is expressed in several brain regions, including the amygdala. Cyld deficiency leads to abnormal neuronal excitation, with reduced frequency of spontaneous excitatory postsynaptic currents and amplitude of microexcitatory presynaptic currents in BLA principal neurons. It has been demonstrated that CYLD deficiency disrupts neuronal activity and synaptic transmission in the BLA of mice, potentially leading to impaired fear memory. Auditory neuropathy is an important cause of hearing loss [82]. It has been shown that CYLD-KO mice have mild hearing impairment. CYLD is widely expressed and localized in cochlear tissue in vitro and in various neuronal cell models. Knockdown of CYLD reduces the length and proportion of neurite outgrowth in neuronal cells. The abnormal hearing in Cyld KO mice may be caused by a reduction in the length and number of neurite outgrowths in auditory neurons in the cochlea. This suggests that CYLD is a key protein affecting hearing. Proteomic analysis of rodent brain samples also shows that CYLD is partially enriched in purified postsynaptic densities. CYLD regulates dendritic growth and postsynaptic differentiation in mouse hippocampal neurons [82].

3.3. Function of CYLD in Vascular Disease

Arteries transport blood from the heart to other organs, and their walls are composed of three layers from the lumen to the exterior: the inner membrane, the middle membrane, and the outer mold. The endothelium consists mainly of endothelial cells, a few fibroblasts, smooth muscle cells, and sparse elastic fibers; the mesothelium consists mainly of smooth muscle cells interspersed with fibroblasts; and the outer mold is mainly collagen fibers and fibroblasts, with an elastic layer separating the mesothelium from the endothelium and outer mold [83,84]. Studies have shown that CYLD affects vascular disease in several ways.

Rac1 is a Rho family of GTPases that regulates actin cytoskeleton and adhesion rearrangement and plays a key role in cell polarization and migration [85]. CYLD deubiquitinates Rac1 and promotes Rac1 activation for endothelial cell migration and angiogenesis [86]. CYLD expression in endothelial cells (ECs) and macrophages decreases with age, exacerbating monocyte adhesion to endothelial cells and foam cell formation, triggering the development of age-related atherosclerosis. Therefore, the CYLD gene in the vasculature may be a new therapeutic target in early intervention to prevent age-related atherosclerosis formation [87].

It has been shown that CYLD mediates a pro-inflammatory phenotype of vascular smooth muscle cells (VSMC) through MAPK activation, characterized by loss of contractility, apoptosis, production of extracellular matrix and cytokines, and foam cell-like transformation, which may contribute to the development of coronary artery lesions [88]. Pulmonary arterial hypertension (PAH) is a common complication of congenital heart disease (CHD), and CYLD mediates human pulmonary artery smooth muscle cell (HPASMC) dysfunction, which regulates HPASMC phenotypic transformation, proliferation, and migration through modulation of p38 and ERK activation [89]. CYLD is a potential new therapeutic target for the prevention of PAH and pulmonary vascular remodeling in CHD-PAH.

Transdifferentiation of extravascular fibroblasts (AFs) to myofibroblasts plays a key role in atherosclerosis, postoperative restenosis, and vascular remodeling in aortic aneurysms [90]. Nicotinamide adenine dinucleotide phosphate oxidase 4 (Nox4), a member of the NADPH oxidase family, is a major source of ROS in the vascular wall. AFs have been shown to produce large amounts of NADPH oxidase-derived ROS in response to vascular injury [6]. CYLD promotes the transdifferentiation of AFs by directly binding to Nox4 through the USP structural domain, which plays a role in vascular remodeling, and CYLD can be used as a new target for vascular anti-inflammatory therapy for diseases such as abdominal aortic aneurysms [91]. In addition, CYLD-mediated deubiquitination of IκB kinase C (IKKγ), IκBα, or TNF receptor-associated factor 2 (TRAF2), leading to inhibition of NF-κB activity, is thought to resolve the vascular inflammatory response and thus inhibit vascular injury [92].

3.4. The Role of CYLD in Nephropathy

Diabetic nephropathy (DN) is a primary cause of chronic kidney disease (CKD). Podocytes, the end-differentiated epithelial cells of the glomerulus, are vital for maintaining an intact glomerular filtration barrier (GFB). Damaged podocytes are a crucial factor in the development of proteinuria and DN [93]. A contributing factor to podocyte damage is the destruction of actin and intermediate filaments resulting from mitochondrial damage [94]. RING-finger protein 166 (RNF166) is a member of the E3 ubiquitin ligase family. It has been demonstrated that RNF166 can directly interact with CYLD to regulate CYLD degradation. Overexpression of CYLD following RNF166 gene knockdown eliminated mitochondrial dysfunction and apoptosis of podocytes under high glucose stimulation due to RNF166 knockdown. Consequently, maintaining the protein level of CYLD in podocytes by inhibiting RNF166 expression or promoting CYLD expression might be a promising therapeutic strategy for DN treatment [78].

3.5. CYLD and Cancer

As a tumor suppressor gene, the expression level of CYLD plays a crucial regulatory role in tumorigenesis (Table 1). Most solid tumors exhibit infiltration of immune and inflammatory cells. Inflammation is a hallmark of cancer and plays a key role in cell transformation, invasion, metastasis, and treatment resistance [95,96]. Certain factors in CYLD regulation can also influence tumorigenesis. MicroRNAs (miRNAs) are a class of small non-coding RNAs consisting of 17–25 nucleotides. The miRNAs target the 3′-untranslated region (UTR) of mRNA or other non-coding RNAs and interact with AGO proteins to form RNA-induced silencing complexes (RISCs) that inhibit the expression or degrade target genes [97]. In non-small cell lung cancer (NSCLC), microRNA-135b (miR-135b) directly targets the 3′-untranslated region (UTR) of the deubiquitinase CYLD, thereby regulating the ubiquitination and activation of NF-κB signaling [98], promoting lung cancer cell proliferation, migration, invasion, anti-apoptosis, and angiogenesis. MicroRNA-587 (miR-587) can exacerbate NSCLC by downregulating CYLD and promoting the proliferation and migration ability of NSCLC [99]. Additionally, long non-coding RNAs (lncRNAs) can regulate malignant tumor initiation, development, and metastasis [100]. The LncRNA-LINC01260 gene can suppress NSCLC tumorigenesis through competitive endogenous RNA and ultimately inhibit NF-κB pathway activation by regulating CYLD expression, providing a potential target for NSCLC treatment [101]. Defective CYLD expression or function may profoundly impact the growth and survival of various cancer cell types, with some cancer types associated with CYLD function listed in Table 2. Therefore, CYLD may be considered a new target for cancer therapy.

Table 2. The relationship between CYLD and cancer.

Function	Cancer Relevance	Reference
phosphorylation	lymphoma, breast cancer, B-cell lymphoma	[51,102,103]
deubiquitination	prostate cancer, nasopharyngeal carcinoma, cancer of the stomach, lung cancer	[98,104–106]
mutation	basal cell salivary gland tumor, skin cancer, squamous cell carcinoma of the head and neck	[107–109]
defect	multiple myeloma, melanoma	[110,111]
transcriptional inhibition	colon cancer, liver cancer	[112,113]
regulatory microtubule/tubulin	pancreatic cancer, leukemia	[6,114]

4. CYLD Is a Potential Therapeutic Target for Disease

4.1. CYLD Is a Potential Therapeutic Target for Autism Spectrum Disorder and Parkinson's Disease

Current treatment options for autism spectrum disorder (ASD) and Parkinson's disease (PD) remain limited in their efficacy [115]. Clinical trials aimed at improving the lives of affected individuals have delivered disappointing outcomes. For instance, the Phase 3 clinical trial of arbaclofen, a GABA-B receptor agonist, failed to show significant improvements in social function in patients with ASD [21]. Likewise, the trial of the phosphodiesterase-10A inhibitor, PF-02545920, did not show promising results in improving motor symptoms in PD patients [116]. Recent studies on the CYLD have provided intriguing insights into its role in the pathophysiology of ASD and PD. While the potential of targeting CYLD as a therapeutic option is promising, several challenges must be addressed. For instance, the development of CYLD inhibitors or modulators of the pathways it is involved in may face obstacles related to off-target effects, toxicity concerns, and the complexity of the underlying molecular networks.

In the context of ASD, CYLD has been implicated in the regulation of mechanistic target of rapamycin (mTOR) signaling, synaptic α-amino-3-hydroxy-5-methyl-4-isoxazolepropionic acid (AMPA) receptor subunits, and autophagy in the hippocampus. However, the mTOR signaling pathway is complex and has a plethora of downstream targets that could be affected by the inhibition of CYLD, raising concerns about the potential off-target effects and the risk of dysregulating other physiological processes [117,118]. Similarly, the development of CYLD inhibitors for PD treatment may encounter challenges related to the PINK1/parkin pathway's multifaceted nature, which regulates mitochondrial quality control and has implications for cellular homeostasis beyond neurodegeneration. Inhibition of CYLD may influence other proteins within this pathway, potentially leading to unforeseen consequences or toxicity concerns [119].

Although no specific CYLD inhibitors have been developed to date, the search for molecules capable of modulating CYLD activity is ongoing. Some potential CYLD inhibitors, such as PR-619, have shown promising results in vitro by reducing CYLD activity and rescuing mitochondrial dysfunction in cellular models of PD [115]. Additionally, the development of small-molecule modulators targeting CYLD-related pathways, such as mTOR inhibitors like rapamycin and its analogs, has demonstrated therapeutic potential in preclinical models of ASD [120]. These findings suggest that CYLD inhibition, either directly or indirectly, could pave the way for innovative treatment options for ASD and PD.

The efficacy of CYLD-targeted therapies can be evaluated using a combination of in vitro and in vivo models, as well as clinical trials. For example, cellular models of ASD and PD can be employed to assess the impact of CYLD inhibition on neuronal connectivity, synaptic function, and mitochondrial quality control. Furthermore, animal models, such as transgenic mice, can provide valuable insights into the therapeutic potential of CYLD inhibitors in alleviating behavioral and motor deficits. Ultimately, well-designed clinical trials are necessary to establish the safety, tolerability, and efficacy of CYLD-targeted therapies in patients with ASD and PD.

4.2. Drugs Targeting Pathways Implicated in CYLD Dysfunction

The development of small molecule drugs targeting signaling pathways implicated in CYLD dysfunction has gained significant interest, as these drugs hold promise for modulating dysregulated cellular processes associated with CYLD mutations and related diseases [121,122]. CYLD, a deubiquitinating enzyme, negatively regulates pathways such as the NF-κB, Wnt/β-catenin, and JNK pathways through the removal of K63-linked ubiquitin chains from target proteins [8,123]. Small molecule inhibitors targeting these pathways function by modulating specific components of the pathways, potentially ameliorating the effects of CYLD dysfunction. For the NF-κB pathway, BMS-345541 selectively inhibits the ATP-binding site of IKK, preventing IκB phosphorylation and degradation and ultimately inhibiting NF-κB activation [124]. Bortezomib, a proteasome inhibitor, prevents IκB degradation, thereby indirectly inhibiting NF-κB activation [125,126]. In the Wnt/β-catenin pathway, LGK974 and ETC-159 function as porcupine inhibitors, blocking the secretion of Wnt proteins essential for pathway activation [127,128]. This inhibition disrupts the activation of downstream signaling components, such as Dishevelled (Dvl) and β-catenin [129]. Tankyrase inhibitors, on the other hand, stabilize axin, a negative regulator of the Wnt/β-catenin pathway, leading to decreased β-catenin levels and pathway inhibition [130]. Regarding the JNK pathway, SP600125 inhibits JNK activation by targeting the ATP-binding site of the kinase, preventing the activation of downstream targets involved in cell proliferation and survival [131].

These small molecule inhibitors have shown potential in preclinical studies and early-phase clinical trials, with some, like bortezomib, already approved for specific clinical indications [132,133]. However, their safety profiles and potential side effects, which may include gastrointestinal symptoms, fatigue, hematological toxicities, and hepatotoxicity, need to be carefully evaluated. Additionally, the context-dependent effectiveness of these drugs warrants further investigation, as factors such as disease stage, genetic background, and the presence of other molecular alterations can influence their therapeutic outcomes [134]. Continued research is essential to optimize the safety and efficacy profiles of these small molecule drugs, enabling their use as potential therapeutic options for patients with CYLD-related diseases and broadening our understanding of the intricate interplay between CYLD and its associated signaling pathways.

5. Conclusions

Oxidative stress (OS) is an imbalance between reactive oxygen species (ROS) formation and antioxidant defense mechanisms and affects the normal function of multiple tissues. Many age-related chronic diseases, such as diabetes and cardiovascular, renal, pulmonary and skeletal muscle diseases, are also directly associated with OS. Although many of the small molecules evaluated as antioxidants have shown therapeutic potential in preclinical studies, clinical trial results have been disappointing. Therefore, many studies are also trying to elucidate the potential mechanisms and role of OS in disease onset and progression and to find new therapeutic strategies to reduce OS. CYLD encodes a deubiquitinating enzyme that is a key regulator of various cellular processes, including immune response, inflammation, death and proliferation, and directly regulates several keys signaling cascades, such as NF-kB and MAPK pathways, involved in the development of multiple diseases, including cancer, poor infection control, pulmonary fibrosis, neurodevelopment and cardiovascular dysfunction. This review explores the functional and mechanistic studies of CYLD in oxidative stress-induced diseases, which provide a strong rationale for the design and testing of specific CYLD inhibitors that may have translational potential for the treatment of oxidative stress-related diseases. In conclusion, targeting CYLD and its associated pathways holds promise as a novel therapeutic strategy for oxidative stress-induced diseases. However, significant challenges remain, including the development of specific CYLD inhibitors, potential off-target effects, and toxicity concerns. Future research should focus on addressing these issues and elucidating the precise mechanisms by which CYLD modulation could ameliorate the symptoms of these complex disorders.

In addition to the loss of function in human disease tissues through gene deletion or mutation, CYLD expression can also be regulated at the RNA level through transcriptional regulation or at the protein level through post-translational modifications, if necessary. The identification of CYLD-mediated signaling pathways during disease progression will also provide a solid basis for diagnosis and facilitate the development of new tools for disease treatment. We expect that all of these approaches will help to advance antioxidant therapy and hope that this review will encourage and inform a sound approach to this worthwhile endeavor.

Author Contributions: In this review paper, Z.H. (first author) and Y.T. (corresponding author) have significantly contributed to the conceptualization, design, analysis, and interpretation of the information presented. Z.H. led the literature search, drafting, and initial editing, while Y.T. provided valuable insights, guidance, and mentorship. Both authors collaborated on refining the manuscript and have read and approved the final version. All authors have read and agreed to the published version of the manuscript.

Funding: This paper is supported by the Shandong Normal University Research Initiation Funding Project (109198).

Institutional Review Board Statement: Not applicable.

Informed Consent Statement: Not applicable.

Data Availability Statement: Not applicable.

Conflicts of Interest: The authors declare no conflict of interest.

References

1. Misago, N.; Narisawa, Y. Cytokeratin 15 expression in apocrine mixed tumors of the skin and other benign neoplasms with apocrine differentiation. *J. Dermatol.* **2006**, *33*, 2–9. [CrossRef] [PubMed]
2. Biggs, P.J.; Wooster, R.; Ford, D.; Chapman, P.; Mangion, J.; Quirk, Y.; Easton, D.F.; Burn, J.; Stratton, M.R. Familial cylindromatosis (turban tumour syndrome) gene localised to chromosome 16q12-q13: Evidence for its role as a tumour suppressor gene. *Nat. Genet.* **1995**, *11*, 441–443. [CrossRef] [PubMed]
3. Nagy, N.; Dubois, A.; Szell, M.; Rajan, N. Genetic Testing in CYLD Cutaneous Syndrome: An Update. *Appl. Clin. Genet.* **2021**, *14*, 427–444. [CrossRef] [PubMed]
4. Elliott, P.R.; Leske, D.; Wagstaff, J.; Schlicher, L.; Berridge, G.; Maslen, S.; Timmermann, F.; Ma, B.; Fischer, R.; Freund, S.M.V.; et al. Regulation of CYLD activity and specificity by phosphorylation and ubiquitin-binding CAP-Gly domains. *Cell Rep.* **2021**, *37*, 109777. [CrossRef]
5. Douanne, T.; Andre-Gregoire, G.; Thys, A.; Trillet, K.; Gavard, J.; Bidere, N. CYLD Regulates Centriolar Satellites Proteostasis by Counteracting the E3 Ligase MIB1. *Cell Rep.* **2019**, *27*, 1657–1665.e4. [CrossRef]
6. Yang, Y.; Ran, J.; Sun, L.; Sun, X.; Luo, Y.; Yan, B.; Liu, M.; Li, D.; Zhang, L.; Bao, G.; et al. CYLD Regulates Noscapine Activity in Acute Lymphoblastic Leukemia via a Microtubule-Dependent Mechanism. *Theranostics* **2015**, *5*, 656–666. [CrossRef]
7. Sun, S.C. CYLD: A tumor suppressor deubiquitinase regulating NF-kappaB activation and diverse biological processes. *Cell Death Differ.* **2010**, *17*, 25–34. [CrossRef]
8. Komander, D.; Lord, C.J.; Scheel, H.; Swift, S.; Hofmann, K.; Ashworth, A.; Barford, D. The structure of the CYLD USP domain explains its specificity for Lys63-linked polyubiquitin and reveals a B box module. *Mol. Cell* **2008**, *29*, 451–464. [CrossRef]
9. Massoumi, R. CYLD: A deubiquitination enzyme with multiple roles in cancer. *Future Oncol.* **2011**, *7*, 285–297. [CrossRef]
10. Akhmanova, A.; Steinmetz, M.O. Tracking the ends: A dynamic protein network controls the fate of microtubule tips. *Nat. Rev. Mol. Cell Biol.* **2008**, *9*, 309–322. [CrossRef]
11. Gao, J.; Huo, L.; Sun, X.; Liu, M.; Li, D.; Dong, J.T.; Zhou, J. The tumor suppressor CYLD regulates microtubule dynamics and plays a role in cell migration. *J. Biol. Chem.* **2008**, *283*, 8802–8809. [CrossRef] [PubMed]
12. Wickström, S.A.; Masoumi, K.C.; Khochbin, S.; Fässler, R.; Massoumi, R. CYLD negatively regulates cell-cycle progression by inactivating HDAC6 and increasing the levels of acetylated tubulin. *EMBO J.* **2010**, *29*, 131–144. [CrossRef]
13. Yang, Y.; Ran, J.; Liu, M.; Li, D.; Li, Y.; Shi, X.; Meng, D.; Pan, J.; Ou, G.; Aneja, R.; et al. CYLD mediates ciliogenesis in multiple organs by deubiquitinating Cep70 and inactivating HDAC6. *Cell Res.* **2014**, *24*, 1342–1353. [CrossRef] [PubMed]

14. Saito, K.; Kigawa, T.; Koshiba, S.; Sato, K.; Matsuo, Y.; Sakamoto, A.; Takagi, T.; Shirouzu, M.; Yabuki, T.; Nunokawa, E.; et al. The CAP-Gly domain of CYLD associates with the proline-rich sequence in NEMO/IKKgamma. *Structure* **2004**, *12*, 1719–1728. [CrossRef] [PubMed]

15. Hadian, K.; Griesbach, R.A.; Dornauer, S.; Wanger, T.M.; Nagel, D.; Metlitzky, M.; Beisker, W.; Schmidt-Supprian, M.; Krappmann, D. NF-κB essential modulator (NEMO) interaction with linear and lys-63 ubiquitin chains contributes to NF-κB activation. *J. Biol. Chem.* **2011**, *286*, 26107–26117. [CrossRef]

16. Mansour, M.A. Ubiquitination: Friend and foe in cancer. *Int. J. Biochem. Cell Biol.* **2018**, *101*, 80–93. [CrossRef]

17. Fiil, B.K.; Gyrd-Hansen, M. The Met1-linked ubiquitin machinery in inflammation and infection. *Cell Death Differ.* **2021**, *28*, 557–569. [CrossRef]

18. Pierzynowska, K.; Gaffke, L.; Cyske, Z.; Wegrzyn, G.; Buttari, B.; Profumo, E.; Saso, L. Oxidative Stress in Mucopolysaccharidoses: Pharmacological Implications. *Molecules* **2021**, *26*, 5616. [CrossRef]

19. Teleanu, D.M.; Niculescu, A.G.; Lungu, I.I.; Radu, C.I.; Vladacenco, O.; Roza, E.; Costachescu, B.; Grumezescu, A.M.; Teleanu, R.I. An Overview of Oxidative Stress, Neuroinflammation, and Neurodegenerative Diseases. *Int. J. Mol. Sci.* **2022**, *23*, 5938. [CrossRef]

20. Kattoor, A.J.; Pothineni, N.V.K.; Palagiri, D.; Mehta, J.L. Oxidative Stress in Atherosclerosis. *Curr. Atheroscler. Rep.* **2017**, *19*, 42. [CrossRef]

21. Barnes, P.J. Oxidative stress-based therapeutics in COPD. *Redox Biol.* **2020**, *33*, 101544. [CrossRef] [PubMed]

22. Chen, Z.; Zhong, C. Oxidative stress in Alzheimer's disease. *Neurosci. Bull.* **2014**, *30*, 271–281. [CrossRef] [PubMed]

23. Jelic, M.D.; Mandic, A.D.; Maricic, S.M.; Srdjenovic, B.U. Oxidative stress and its role in cancer. *J. Cancer Res. Ther.* **2021**, *17*, 22–28. [CrossRef] [PubMed]

24. Senoner, T.; Dichtl, W. Oxidative Stress in Cardiovascular Diseases: Still a Therapeutic Target? *Nutrients* **2019**, *11*, 2090. [CrossRef] [PubMed]

25. Bertero, E.; Maack, C. Calcium Signaling and Reactive Oxygen Species in Mitochondria. *Circ. Res.* **2018**, *122*, 1460–1478. [CrossRef]

26. Wang, L.; Tang, J.; Wang, L.; Tan, F.; Song, H.; Zhou, J.; Li, F. Oxidative stress in oocyte aging and female reproduction. *J. Cell Physiol.* **2021**, *236*, 7966–7983. [CrossRef]

27. Khatri, N.; Thakur, M.; Pareek, V.; Kumar, S.; Sharma, S.; Datusalia, A.K. Oxidative Stress: Major Threat in Traumatic Brain Injury. *CNS Neurol. Disord. Drug Targets* **2018**, *17*, 689–695. [CrossRef]

28. Sosa, V.; Moline, T.; Somoza, R.; Paciucci, R.; Kondoh, H.; ME, L.L. Oxidative stress and cancer: An overview. *Ageing Res. Rev.* **2013**, *12*, 376–390. [CrossRef]

29. Prakash, R.; Fauzia, E.; Siddiqui, A.J.; Yadav, S.K.; Kumari, N.; Singhai, A.; Khan, M.A.; Janowski, M.; Bhutia, S.K.; Raza, S.S. Oxidative Stress Enhances Autophagy-Mediated Death Of Stem Cells Through Erk1/2 Signaling Pathway—Implications For Neurotransplantations. *Stem Cell Rev. Rep.* **2021**, *17*, 2347–2358. [CrossRef]

30. Ritt, D.A.; Abreu-Blanco, M.T.; Bindu, L.; Durrant, D.E.; Zhou, M.; Specht, S.I.; Stephen, A.G.; Holderfield, M.; Morrison, D.K. Inhibition of Ras/Raf/MEK/ERK Pathway Signaling by a Stress-Induced Phospho-Regulatory Circuit. *Mol. Cell* **2016**, *64*, 875–887. [CrossRef]

31. Yang, Y.; Karsli-Uzunbas, G.; Poillet-Perez, L.; Sawant, A.; Hu, Z.S.; Zhao, Y.; Moore, D.; Hu, W.; White, E. Autophagy promotes mammalian survival by suppressing oxidative stress and p53. *Genes Dev.* **2020**, *34*, 688–700. [CrossRef] [PubMed]

32. Daenen, K.; Andries, A.; Mekahli, D.; Van Schepdael, A.; Jouret, F.; Bammens, B. Oxidative stress in chronic kidney disease. *Pediatr. Nephrol.* **2019**, *34*, 975–991. [CrossRef] [PubMed]

33. Sczepanik, F.S.C.; Grossi, M.L.; Casati, M.; Goldberg, M.; Glogauer, M.; Fine, N.; Tenenbaum, H.C. Periodontitis is an inflammatory disease of oxidative stress: We should treat it that way. *Periodontology* **2020**, *84*, 45–68. [CrossRef]

34. Bisht, S.; Faiq, M.; Tolahunase, M.; Dada, R. Oxidative stress and male infertility. *Nat. Rev. Urol.* **2017**, *14*, 470–485. [CrossRef] [PubMed]

35. Kimball, J.S.; Johnson, J.P.; Carlson, D.A. Oxidative Stress and Osteoporosis. *J. Bone Jt. Surg. Am.* **2021**, *103*, 1451–1461. [CrossRef] [PubMed]

36. Scutiero, G.; Iannone, P.; Bernardi, G.; Bonaccorsi, G.; Spadaro, S.; Volta, C.A.; Greco, P.; Nappi, L. Oxidative Stress and Endometriosis: A Systematic Review of the Literature. *Oxidative Med. Cell. Longev.* **2017**, *2017*, 7265238. [CrossRef]

37. Wang, Y.; Li, S.; Li, C. Perspectives of New Advances in the Pathogenesis of Vitiligo: From Oxidative Stress to Autoimmunity. *Med. Sci. Monit.* **2019**, *25*, 1017–1023. [CrossRef]

38. Pierantonelli, I.; Svegliati-Baroni, G. Nonalcoholic Fatty Liver Disease: Basic Pathogenetic Mechanisms in the Progression From NAFLD to NASH. *Transplantation* **2019**, *103*, e1–e13. [CrossRef] [PubMed]

39. Xia, Y.; Xie, Z.; Huang, G.; Zhou, Z. Incidence and trend of type 1 diabetes and the underlying environmental determinants. *Diabetes Metab. Res. Rev.* **2019**, *35*, e3075. [CrossRef]

40. Su, X.; Peng, D. Emerging functions of adipokines in linking the development of obesity and cardiovascular diseases. *Mol. Biol. Rep.* **2020**, *47*, 7991–8006. [CrossRef]

41. Hsu, C.Y.; McCulloch, C.E.; Iribarren, C.; Darbinian, J.; Go, A.S. Body mass index and risk for end-stage renal disease. *Ann. Intern. Med.* **2006**, *144*, 21–28. [CrossRef] [PubMed]

42. Vincent, H.K.; Taylor, A.G. Biomarkers and potential mechanisms of obesity-induced oxidant stress in humans. *Int. J. Obes.* **2006**, *30*, 400–418. [CrossRef] [PubMed]

43. Xu, T.; Sheng, Z.; Yao, L. Obesity-related glomerulopathy: Pathogenesis, pathologic, clinical characteristics and treatment. *Front. Med.* **2017**, *11*, 340–348. [CrossRef] [PubMed]
44. Shi, Y.; Wang, C.; Zhou, X.; Li, Y.; Ma, Y.; Zhang, R.; Li, R. Downregulation of PTEN promotes podocyte endocytosis of lipids aggravating obesity-related glomerulopathy. *Am. J. Physiol. Renal. Physiol.* **2020**, *318*, F589–F599. [CrossRef]
45. Tang, J.; Yan, H.; Zhuang, S. Inflammation and oxidative stress in obesity-related glomerulopathy. *Int. J. Nephrol.* **2012**, *2012*, 608397. [CrossRef]
46. Rani, V.; Deep, G.; Singh, R.K.; Palle, K.; Yadav, U.C. Oxidative stress and metabolic disorders: Pathogenesis and therapeutic strategies. *Life Sci.* **2016**, *148*, 183–193. [CrossRef]
47. Jaimes, E.A.; Hua, P.; Tian, R.X.; Raij, L. Human glomerular endothelium: Interplay among glucose, free fatty acids, angiotensin II, and oxidative stress. *Am. J. Physiol. Renal. Physiol.* **2010**, *298*, F125–F132. [CrossRef]
48. Habibi, J.; Hayden, M.R.; Sowers, J.R.; Pulakat, L.; Tilmon, R.D.; Manrique, C.; Lastra, G.; Demarco, V.G.; Whaley-Connell, A. Nebivolol attenuates redox-sensitive glomerular and tubular mediated proteinuria in obese rats. *Endocrinology* **2011**, *152*, 659–668. [CrossRef]
49. Fernandes, S.M.; Cordeiro, P.M.; Watanabe, M.; Fonseca, C.D.; Vattimo, M.F. The role of oxidative stress in streptozotocin-induced diabetic nephropathy in rats. *Arch. Endocrinol. Metab.* **2016**, *60*, 443–449. [CrossRef]
50. Reiley, W.; Zhang, M.; Wu, X.; Granger, E.; Sun, S.C. Regulation of the deubiquitinating enzyme CYLD by IkappaB kinase gamma-dependent phosphorylation. *Mol. Cell Biol.* **2005**, *25*, 3886–3895. [CrossRef]
51. Hutti, J.E.; Shen, R.R.; Abbott, D.W.; Zhou, A.Y.; Sprott, K.M.; Asara, J.M.; Hahn, W.C.; Cantley, L.C. Phosphorylation of the tumor suppressor CYLD by the breast cancer oncogene IKKepsilon promotes cell transformation. *Mol. Cell* **2009**, *34*, 461–472. [CrossRef] [PubMed]
52. Wang, H.; Lai, Y.; Mathis, B.J.; Wang, W.; Li, S.; Qu, C.; Li, B.; Shao, L.; Song, H.; Janicki, J.S.; et al. Deubiquitinating enzyme CYLD mediates pressure overload-induced cardiac maladaptive remodeling and dysfunction via downregulating Nrf2. *J. Mol. Cell. Cardiol.* **2015**, *84*, 143–153. [CrossRef]
53. Halazonetis, T.D.; Gorgoulis, V.G.; Bartek, J. An oncogene-induced DNA damage model for cancer development. *Science* **2008**, *319*, 1352–1355. [CrossRef] [PubMed]
54. Erol, A. Genotoxity-Stimulated and CYLD-Driven Malignant Transformation. *Cancer Manag. Res.* **2022**, *14*, 2339–2356. [CrossRef] [PubMed]
55. Lin, Y.; Bai, L.; Chen, W.; Xu, S. The NF-kappaB activation pathways, emerging molecular targets for cancer prevention and therapy. *Expert Opin. Ther. Targets* **2010**, *14*, 45–55. [CrossRef] [PubMed]
56. Nihira, K.; Taira, N.; Miki, Y.; Yoshida, K. TTK/Mps1 controls nuclear targeting of c-Abl by 14-3-3-coupled phosphorylation in response to oxidative stress. *Oncogene* **2008**, *27*, 7285–7295. [CrossRef]
57. Wang, J.Y. The capable ABL: What is its biological function? *Mol. Cell. Biol.* **2014**, *34*, 1188–1197. [CrossRef]
58. Westbrook, A.M.; Wei, B.; Hacke, K.; Xia, M.; Braun, J.; Schiestl, R.H. The role of tumour necrosis factor-α and tumour necrosis factor receptor signalling in inflammation-associated systemic genotoxicity. *Mutagenesis* **2012**, *27*, 77–86. [CrossRef]
59. Li, C.; Sheng, M.; Lin, Y.; Xu, D.; Tian, Y.; Zhan, Y.; Jiang, L.; Coito, A.J.; Busuttil, R.W.; Farmer, D.G.; et al. Functional crosstalk between myeloid Foxo1-β-catenin axis and Hedgehog/Gli1 signaling in oxidative stress response. *Cell Death Differ.* **2021**, *28*, 1705–1719. [CrossRef]
60. Dara, L.; Ji, C.; Kaplowitz, N. The contribution of endoplasmic reticulum stress to liver diseases. *Hepatology* **2011**, *53*, 1752–1763. [CrossRef]
61. Yue, S.; Zhu, J.; Zhang, M.; Li, C.; Zhou, X.; Zhou, M.; Ke, M.; Busuttil, R.W.; Ying, Q.L.; Kupiec-Weglinski, J.W.; et al. The myeloid heat shock transcription factor 1/β-catenin axis regulates NLR family, pyrin domain-containing 3 inflammasome activation in mouse liver ischemia/reperfusion injury. *Hepatology* **2016**, *64*, 1683–1698. [CrossRef] [PubMed]
62. Lu, L.; Yue, S.; Jiang, L.; Li, C.; Zhu, Q.; Ke, M.; Lu, H.; Wang, X.; Busuttil, R.W.; Ying, Q.L.; et al. Myeloid Notch1 deficiency activates the RhoA/ROCK pathway and aggravates hepatocellular damage in mouse ischemic livers. *Hepatology* **2018**, *67*, 1041–1055. [CrossRef] [PubMed]
63. Zhan, Y.; Xu, D.; Tian, Y.; Qu, X.; Sheng, M.; Lin, Y.; Ke, M.; Jiang, L.; Xia, Q.; Kaldas, F.M.; et al. Novel role of macrophage TXNIP-mediated CYLD-NRF2-OASL1 axis in stress-induced liver inflammation and cell death. *JHEP Rep.* **2022**, *4*, 100532. [CrossRef] [PubMed]
64. Li, J.; Ichikawa, T.; Villacorta, L.; Janicki, J.S.; Brower, G.L.; Yamamoto, M.; Cui, T. Nrf2 protects against maladaptive cardiac responses to hemodynamic stress. *Arter. Thromb. Vasc. Biol.* **2009**, *29*, 1843–1850. [CrossRef] [PubMed]
65. Li, J.; Ichikawa, T.; Janicki, J.S.; Cui, T. Targeting the Nrf2 pathway against cardiovascular disease. *Expert Opin. Ther. Targets* **2009**, *13*, 785–794. [CrossRef]
66. Li, J.; Zhang, C.; Xing, Y.; Janicki, J.S.; Yamamoto, M.; Wang, X.L.; Tang, D.Q.; Cui, T. Up-regulation of p27(kip1) contributes to Nrf2-mediated protection against angiotensin II-induced cardiac hypertrophy. *Cardiovasc. Res.* **2011**, *90*, 315–324. [CrossRef]
67. Li, S.; Wang, W.; Niu, T.; Wang, H.; Li, B.; Shao, L.; Lai, Y.; Li, H.; Janicki, J.S.; Wang, X.L.; et al. Nrf2 deficiency exaggerates doxorubicin-induced cardiotoxicity and cardiac dysfunction. *Oxidative Med. Cell. Longev.* **2014**, *2014*, 748524. [CrossRef]
68. Wang, W.; Li, S.; Wang, H.; Li, B.; Shao, L.; Lai, Y.; Horvath, G.; Wang, Q.; Yamamoto, M.; Janicki, J.S.; et al. Nrf2 enhances myocardial clearance of toxic ubiquitinated proteins. *J. Mol. Cell. Cardiol.* **2014**, *72*, 305–315. [CrossRef]

69. Tan, Y.; Ichikawa, T.; Li, J.; Si, Q.; Yang, H.; Chen, X.; Goldblatt, C.S.; Meyer, C.J.; Li, X.; Cai, L.; et al. Diabetic downregulation of Nrf2 activity via ERK contributes to oxidative stress-induced insulin resistance in cardiac cells in vitro and in vivo. *Diabetes* **2011**, *60*, 625–633. [CrossRef]

70. Hanus, J.; Anderson, C.; Wang, S. RPE necroptosis in response to oxidative stress and in AMD. *Ageing Res. Rev.* **2015**, *24*, 286–298. [CrossRef]

71. Mitter, S.K.; Song, C.; Qi, X.; Mao, H.; Rao, H.; Akin, D.; Lewin, A.; Grant, M.; Dunn, W., Jr.; Ding, J.; et al. Dysregulated autophagy in the RPE is associated with increased susceptibility to oxidative stress and AMD. *Autophagy* **2014**, *10*, 1989–2005. [CrossRef] [PubMed]

72. Luo, R.; Li, L.; Xiao, F.; Fu, J. LncRNA FLG-AS1 Mitigates Diabetic Retinopathy by Regulating Retinal Epithelial Cell Inflammation, Oxidative Stress, and Apoptosis via miR-380-3p/SOCS6 Axis. *Inflammation* **2022**, *45*, 1936–1949. [CrossRef] [PubMed]

73. Oh, E.C.; Katsanis, N. Cilia in vertebrate development and disease. *Development* **2012**, *139*, 443–448. [CrossRef] [PubMed]

74. Mitchison, H.M.; Valente, E.M. Motile and non-motile cilia in human pathology: From function to phenotypes. *J. Pathol.* **2017**, *241*, 294–309. [CrossRef] [PubMed]

75. Eguether, T.; Ermolaeva, M.A.; Zhao, Y.; Bonnet, M.C.; Jain, A.; Pasparakis, M.; Courtois, G.; Tassin, A.M. The deubiquitinating enzyme CYLD controls apical docking of basal bodies in ciliated epithelial cells. *Nat. Commun.* **2014**, *5*, 4585. [CrossRef] [PubMed]

76. Chiticariu, E.; Regamey, A.; Huber, M.; Hohl, D. CENPV Is a CYLD-Interacting Molecule Regulating Ciliary Acetylated α-Tubulin. *J. Investig. Dermatol.* **2020**, *140*, 66–74.e4. [CrossRef]

77. Jin, W.; Chang, M.; Paul, E.M.; Babu, G.; Lee, A.J.; Reiley, W.; Wright, A.; Zhang, M.; You, J.; Sun, S.C. Deubiquitinating enzyme CYLD negatively regulates RANK signaling and osteoclastogenesis in mice. *J. Clin. Investig.* **2008**, *118*, 1858–1866. [CrossRef]

78. Hongbo, M.; Yanjiao, D.; Shuo, W.; Kun, S.; Yanjie, L.; Mengmeng, L. Podocyte RNF166 deficiency alleviates diabetic nephropathy by mitigating mitochondria impairment and apoptosis via regulation of CYLD signal. *Biochem. Biophys. Res. Commun.* **2021**, *545*, 46–53. [CrossRef]

79. Ji, Y.X.; Huang, Z.; Yang, X.; Wang, X.; Zhao, L.P.; Wang, P.X.; Zhang, X.J.; Alves-Bezerra, M.; Cai, L.; Zhang, P.; et al. The deubiquitinating enzyme cylindromatosis mitigates nonalcoholic steatohepatitis. *Nat. Med.* **2018**, *24*, 213–223. [CrossRef]

80. Varinthra, P.; Ganesan, K.; Huang, S.P.; Chompoopong, S.; Eurtivong, C.; Suresh, P.; Wen, Z.H.; Liu, I.Y. The 4-(Phenylsulfanyl) butan-2-one Improves Impaired Fear Memory Retrieval and Reduces Excessive Inflammatory Response in Triple Transgenic Alzheimer's Disease Mice. *Front. Aging Neurosci.* **2021**, *13*, 615079. [CrossRef]

81. Marek, R.; Sun, Y.; Sah, P. Neural circuits for a top-down control of fear and extinction. *Psychopharmacology* **2019**, *236*, 313–320. [CrossRef] [PubMed]

82. Li, H.D.; Li, D.N.; Yang, L.; Long, C. Deficiency of the CYLD Impairs Fear Memory of Mice and Disrupts Neuronal Activity and Synaptic Transmission in the Basolateral Amygdala. *Front. Cell. Neurosci.* **2021**, *15*, 740165. [CrossRef] [PubMed]

83. Fhayli, W.; Boëté, Q.; Harki, O.; Briançon-Marjollet, A.; Jacob, M.P.; Faury, G. Rise and fall of elastic fibers from development to aging. Consequences on arterial structure-function and therapeutical perspectives. *Matrix Biol. J. Int. Soc. Matrix Biol.* **2019**, *84*, 41–56. [CrossRef] [PubMed]

84. Gialeli, C.; Shami, A.; Gonçalves, I. Extracellular matrix: Paving the way to the newest trends in atherosclerosis. *Curr. Opin. Lipidol.* **2021**, *32*, 277–285. [CrossRef]

85. Watanabe, T.; Noritake, J.; Kaibuchi, K. Regulation of microtubules in cell migration. *Trends Cell Biol.* **2005**, *15*, 76–83. [CrossRef]

86. Gao, J.; Sun, L.; Huo, L.; Liu, M.; Li, D.; Zhou, J. CYLD regulates angiogenesis by mediating vascular endothelial cell migration. *Blood* **2010**, *115*, 4130–4137. [CrossRef]

87. Imaizumi, Y.; Takami, Y.; Yamamoto, K.; Nagasawa, M.; Nozato, Y.; Nozato, S.; Takeshita, H.; Wang, C.; Yokoyama, S.; Hayashi, H.; et al. Pathophysiological significance of cylindromatosis in the vascular endothelium and macrophages for the initiation of age-related atherogenesis. *Biochem. Biophys. Res. Commun.* **2019**, *508*, 1168–1174. [CrossRef]

88. Liu, S.; Lv, J.; Han, L.; Ichikawa, T.; Wang, W.; Li, S.; Wang, X.L.; Tang, D.; Cui, T. A pro-inflammatory role of deubiquitinating enzyme cylindromatosis (CYLD) in vascular smooth muscle cells. *Biochem. Biophys. Res. Commun.* **2012**, *420*, 78–83. [CrossRef]

89. Zhou, J.J.; Li, H.; Li, L.; Li, Y.; Wang, P.H.; Meng, X.M.; He, J.G. CYLD mediates human pulmonary artery smooth muscle cell dysfunction in congenital heart disease-associated pulmonary arterial hypertension. *J. Cell. Physiol.* **2021**, *236*, 6297–6311. [CrossRef]

90. Xu, F.; Liu, Y.; Hu, W. Adventitial fibroblasts from apoE(-/-) mice exhibit the characteristics of transdifferentiation into myofibroblasts. *Cell Biol. Int.* **2013**, *37*, 160–166. [CrossRef]

91. Yu, B.; Liu, Z.; Fu, Y.; Wang, Y.; Zhang, L.; Cai, Z.; Yu, F.; Wang, X.; Zhou, J.; Kong, W. CYLD Deubiquitinates Nicotinamide Adenine Dinucleotide Phosphate Oxidase 4 Contributing to Adventitial Remodeling. *Arterioscler. Thromb. Vasc. Biol.* **2017**, *37*, 1698–1709. [CrossRef] [PubMed]

92. Takami, Y.; Nakagami, H.; Morishita, R.; Katsuya, T.; Hayashi, H.; Mori, M.; Koriyama, H.; Baba, Y.; Yasuda, O.; Rakugi, H.; et al. Potential role of CYLD (Cylindromatosis) as a deubiquitinating enzyme in vascular cells. *Am. J. Pathol.* **2008**, *172*, 818–829. [CrossRef]

93. Diez-Sampedro, A.; Lenz, O.; Fornoni, A. Podocytopathy in diabetes: A metabolic and endocrine disorder. *Am. J. Kidney Dis. Off. J. Natl. Kidney Found.* **2011**, *58*, 637–646. [CrossRef] [PubMed]

94. Wang, R.M.; Wang, Z.B.; Wang, Y.; Liu, W.Y.; Li, Y.; Tong, L.C.; Zhang, S.; Su, D.F.; Cao, Y.B.; Li, L.; et al. Swiprosin-1 Promotes Mitochondria-Dependent Apoptosis of Glomerular Podocytes via P38 MAPK Pathway in Early-Stage Diabetic Nephropathy. *Cell. Physiol. Biochem. Int. J. Exp. Cell. Physiol. Biochem. Pharmacol.* **2018**, *45*, 899–916. [CrossRef] [PubMed]
95. Grivennikov, S.I.; Karin, M. Inflammation and oncogenesis: A vicious connection. *Curr. Opin. Genet. Dev.* **2010**, *20*, 65–71. [CrossRef]
96. Grivennikov, S.I.; Greten, F.R.; Karin, M. Immunity, inflammation, and cancer. *Cell* **2010**, *140*, 883–899. [CrossRef]
97. Treiber, T.; Treiber, N.; Meister, G. Regulation of microRNA biogenesis and its crosstalk with other cellular pathways. *Nat. Rev. Mol. Cell Biol.* **2019**, *20*, 5–20. [CrossRef]
98. Zhao, J.; Wang, X.; Mi, Z.; Jiang, X.; Sun, L.; Zheng, B.; Wang, J.; Meng, M.; Zhang, L.; Wang, Z.; et al. STAT3/miR-135b/NF-κB axis confers aggressiveness and unfavorable prognosis in non-small-cell lung cancer. *Cell Death Dis.* **2021**, *12*, 493. [CrossRef]
99. Li, X.J.; Chen, L.W.; Gao, P.; Jia, Y.J. MiR-587 acts as an oncogene in non-small-cell lung carcinoma via reducing CYLD expression. *Eur. Rev. Med. Pharmacol. Sci.* **2020**, *24*, 12741–12747. [CrossRef]
100. Hennessy, E.J. Cardiovascular Disease and Long Noncoding RNAs: Tools for Unraveling the Mystery Lnc-ing RNA and Phenotype. *Circ. Cardiovasc. Genet.* **2017**, *10*, e001556. [CrossRef]
101. Chen, Y.; Lei, Y.; Lin, J.; Huang, Y.; Zhang, J.; Chen, K.; Sun, S.; Lin, X. The LINC01260 Functions as a Tumor Suppressor via the miR-562/CYLD/NF-κB Pathway in Non-Small Cell Lung Cancer. *OncoTargets Ther.* **2020**, *13*, 10707–10719. [CrossRef] [PubMed]
102. Xu, X.; Kalac, M.; Markson, M.; Chan, M.; Brody, J.D.; Bhagat, G.; Ang, R.L.; Legarda, D.; Justus, S.J.; Liu, F.; et al. Reversal of CYLD phosphorylation as a novel therapeutic approach for adult T-cell leukemia/lymphoma (ATLL). *Cell Death Dis.* **2020**, *11*, 94. [CrossRef] [PubMed]
103. Xu, X.; Wei, T.; Zhong, W.; Ang, R.; Lei, Y.; Zhang, H.; Li, Q. Down-regulation of cylindromatosis protein phosphorylation by BTK inhibitor promotes apoptosis of non-GCB-diffuse large B-cell lymphoma. *Cancer Cell Int.* **2021**, *21*, 195. [CrossRef] [PubMed]
104. Haq, S.; Sarodaya, N.; Karapurkar, J.K.; Suresh, B.; Jo, J.K.; Singh, V.; Bae, Y.S.; Kim, K.S.; Ramakrishna, S. CYLD destabilizes NoxO1 protein by promoting ubiquitination and regulates prostate cancer progression. *Cancer Lett.* **2022**, *525*, 146–157. [CrossRef]
105. Wang, L.; Lin, Y.; Zhou, X.; Chen, Y.; Li, X.; Luo, W.; Zhou, Y.; Cai, L. CYLD deficiency enhances metabolic reprogramming and tumor progression in nasopharyngeal carcinoma via PFKFB3. *Cancer Lett.* **2022**, *532*, 215586. [CrossRef] [PubMed]
106. Yan, Y.F.; Gong, F.M.; Wang, B.S.; Zheng, W. MiR-425-5p promotes tumor progression via modulation of CYLD in gastric cancer. *Eur. Rev. Med. Pharmacol. Sci.* **2017**, *21*, 2130–2136. [PubMed]
107. Rito, M.; Mitani, Y.; Bell, D.; Mariano, F.V.; Almalki, S.T.; Pytynia, K.B.; Fonseca, I.; El-Naggar, A.K. Frequent and differential mutations of the CYLD gene in basal cell salivary neoplasms: Linkage to tumor development and progression. *Mod. Pathol. Off. J. United States Can. Acad. Pathol. Inc.* **2018**, *31*, 1064–1072. [CrossRef]
108. Williams, E.A.; Montesion, M.; Alexander, B.M.; Ramkissoon, S.H.; Elvin, J.A.; Ross, J.S.; Williams, K.J.; Glomski, K.; Bledsoe, J.R.; Tse, J.Y.; et al. CYLD mutation characterizes a subset of HPV-positive head and neck squamous cell carcinomas with distinctive genomics and frequent cylindroma-like histologic features. *Mod. Pathol. Off. J. United States Can. Acad. Pathol. Inc.* **2021**, *34*, 358–370. [CrossRef]
109. Alameda, J.P.; Moreno-Maldonado, R.; Navarro, M.; Bravo, A.; Ramírez, A.; Page, A.; Jorcano, J.L.; Fernández-Aceñero, M.J.; Casanova, M.L. An inactivating CYLD mutation promotes skin tumor progression by conferring enhanced proliferative, survival and angiogenic properties to epidermal cancer cells. *Oncogene* **2010**, *29*, 6522–6532. [CrossRef]
110. van Andel, H.; Kocemba, K.A.; de Haan-Kramer, A.; Mellink, C.H.; Piwowar, M.; Broijl, A.; van Duin, M.; Sonneveld, P.; Maurice, M.M.; Kersten, M.J.; et al. Loss of CYLD expression unleashes Wnt signaling in multiple myeloma and is associated with aggressive disease. *Oncogene* **2017**, *36*, 2105–2115. [CrossRef]
111. de Jel, M.M.; Schott, M.; Lamm, S.; Neuhuber, W.; Kuphal, S.; Bosserhoff, A.K. Loss of CYLD accelerates melanoma development and progression in the Tg(Grm1) melanoma mouse model. *Oncogenesis* **2019**, *8*, 56. [CrossRef] [PubMed]
112. Hellerbrand, C.; Bumes, E.; Bataille, F.; Dietmaier, W.; Massoumi, R.; Bosserhoff, A.K. Reduced expression of CYLD in human colon and hepatocellular carcinomas. *Carcinogenesis* **2007**, *28*, 21–27. [CrossRef] [PubMed]
113. Urbanik, T.; Köhler, B.C.; Boger, R.J.; Wörns, M.A.; Heeger, S.; Otto, G.; Hövelmeyer, N.; Galle, P.R.; Schuchmann, M.; Waisman, A.; et al. Down-regulation of CYLD as a trigger for NF-κB activation and a mechanism of apoptotic resistance in hepatocellular carcinoma cells. *Int. J. Oncol.* **2011**, *38*, 121–131. [PubMed]
114. Xie, S.; Wu, Y.; Hao, H.; Li, J.; Guo, S.; Xie, W.; Li, D.; Zhou, J.; Gao, J.; Liu, M. CYLD deficiency promotes pancreatic cancer development by causing mitotic defects. *J. Cell. Physiol.* **2019**, *234*, 9723–9732. [CrossRef]
115. Pirooznia, S.K.; Wang, H.; Panicker, N.; Kumar, M.; Neifert, S.; Dar, M.A.; Lau, E.; Kang, B.G.; Redding-Ochoa, J.; Troncoso, J.C.; et al. Deubiquitinase CYLD acts as a negative regulator of dopamine neuron survival in Parkinson's disease. *Sci. Adv.* **2022**, *8*, eabh1824. [CrossRef]
116. Lenda, T.; Ossowska, K.; Berghauzen-Maciejewska, K.; Matloka, M.; Pieczykolan, J.; Wieczorek, M.; Konieczny, J. Antiparkinsonian-like effects of CPL500036, a novel selective inhibitor of phosphodiesterase 10A, in the unilateral rat model of Parkinson's disease. *Eur. J. Pharmacol.* **2021**, *910*, 174460. [CrossRef]
117. Saxton, R.A.; Sabatini, D.M. mTOR Signaling in Growth, Metabolism, and Disease. *Cell* **2017**, *168*, 960–976. [CrossRef]
118. Colombo, E.; Horta, G.; Roesler, M.K.; Ihbe, N.; Chhabra, S.; Radyushkin, K.; Di Liberto, G.; Kreutzfeldt, M.; Schumann, S.; von Engelhardt, J.; et al. The K63 deubiquitinase CYLD modulates autism-like behaviors and hippocampal plasticity by regulating autophagy and mTOR signaling. *Proc. Natl. Acad. Sci. USA* **2021**, *118*, e2110755118. [CrossRef]

119. Pickrell, A.M.; Youle, R.J. The roles of PINK1, parkin, and mitochondrial fidelity in Parkinson's disease. *Neuron* **2015**, *85*, 257–273. [CrossRef]
120. Gkogkas, C.G.; Khoutorsky, A.; Cao, R.; Jafarnejad, S.M.; Prager-Khoutorsky, M.; Giannakas, N.; Kaminari, A.; Fragkouli, A.; Nader, K.; Price, T.J.; et al. Pharmacogenetic inhibition of eIF4E-dependent Mmp9 mRNA translation reverses fragile X syndrome-like phenotypes. *Cell Rep.* **2014**, *9*, 1742–1755. [CrossRef]
121. Brummelkamp, T.R.; Nijman, S.M.; Dirac, A.M.; Bernards, R. Loss of the cylindromatosis tumour suppressor inhibits apoptosis by activating NF-kappaB. *Nature* **2003**, *424*, 797–801. [CrossRef] [PubMed]
122. Kovalenko, A.; Chable-Bessia, C.; Cantarella, G.; Israel, A.; Wallach, D.; Courtois, G. The tumour suppressor CYLD negatively regulates NF-kappaB signalling by deubiquitination. *Nature* **2003**, *424*, 801–805. [CrossRef] [PubMed]
123. Trompouki, E.; Hatzivassiliou, E.; Tsichritzis, T.; Farmer, H.; Ashworth, A.; Mosialos, G. CYLD is a deubiquitinating enzyme that negatively regulates NF-kappaB activation by TNFR family members. *Nature* **2003**, *424*, 793–796. [CrossRef] [PubMed]
124. Burke, J.R.; Pattoli, M.A.; Gregor, K.R.; Brassil, P.J.; MacMaster, J.F.; McIntyre, K.W.; Yang, X.; Iotzova, V.S.; Clarke, W.; Strnad, J.; et al. BMS-345541 is a highly selective inhibitor of I kappa B kinase that binds at an allosteric site of the enzyme and blocks NF-kappa B-dependent transcription in mice. *J. Biol. Chem.* **2003**, *278*, 1450–1456. [CrossRef] [PubMed]
125. Adams, J.; Palombella, V.J.; Sausville, E.A.; Johnson, J.; Destree, A.; Lazarus, D.D.; Maas, J.; Pien, C.S.; Prakash, S.; Elliott, P.J. Proteasome inhibitors: A novel class of potent and effective antitumor agents. *Cancer Res.* **1999**, *59*, 2615–2622.
126. Hideshima, T.; Chauhan, D.; Richardson, P.; Mitsiades, C.; Mitsiades, N.; Hayashi, T.; Munshi, N.; Dang, L.; Castro, A.; Palombella, V.; et al. NF kappa B as a therapeutic target in multiple myeloma. *J. Biol. Chem.* **2002**, *277*, 16639–16647. [CrossRef]
127. Liu, J.; Pan, S.; Hsieh, M.H.; Ng, N.; Sun, F.; Wang, T.; Kasibhatla, S.; Schuller, A.G.; Li, A.G.; Cheng, D.; et al. Targeting Wnt-driven cancer through the inhibition of Porcupine by LGK974. *Proc. Natl. Acad. Sci. USA* **2013**, *110*, 20224–20229. [CrossRef]
128. Madan, B.; Ke, Z.; Harmston, N.; Ho, S.Y.; Frois, A.O.; Alam, J.; Jeyaraj, D.A.; Pendharkar, V.; Ghosh, K.; Virshup, I.H.; et al. Wnt addiction of genetically defined cancers reversed by PORCN inhibition. *Oncogene* **2016**, *35*, 2197–2207. [CrossRef]
129. Niehrs, C. The complex world of WNT receptor signalling. *Nat. Rev. Mol. Cell Biol.* **2012**, *13*, 767–779. [CrossRef]
130. Huang, S.M.; Mishina, Y.M.; Liu, S.; Cheung, A.; Stegmeier, F.; Michaud, G.A.; Charlat, O.; Wiellette, E.; Zhang, Y.; Wiessner, S.; et al. Tankyrase inhibition stabilizes axin and antagonizes Wnt signalling. *Nature* **2009**, *461*, 614–620. [CrossRef]
131. Bennett, B.L.; Sasaki, D.T.; Murray, B.W.; O'Leary, E.C.; Sakata, S.T.; Xu, W.; Leisten, J.C.; Motiwala, A.; Pierce, S.; Satoh, Y.; et al. SP600125, an anthrapyrazolone inhibitor of Jun N-terminal kinase. *Proc. Natl. Acad. Sci. USA* **2001**, *98*, 13681–13686. [CrossRef] [PubMed]
132. Richardson, P.G.; Barlogie, B.; Berenson, J.; Singhal, S.; Jagannath, S.; Irwin, D.; Rajkumar, S.V.; Srkalovic, G.; Alsina, M.; Alexanian, R.; et al. A phase 2 study of bortezomib in relapsed, refractory myeloma. *N. Engl. J. Med.* **2003**, *348*, 2609–2617. [CrossRef] [PubMed]
133. Adams, J. The proteasome: A suitable antineoplastic target. *Nat. Rev. Cancer* **2004**, *4*, 349–360. [CrossRef] [PubMed]
134. Massagué, J. How cells read TGF-beta signals. *Nat. Rev. Mol. Cell Biol.* **2000**, *1*, 169–178. [CrossRef] [PubMed]

International Journal of
Molecular Sciences

Review

Molecular Pathology, Oxidative Stress, and Biomarkers in Obstructive Sleep Apnea

Piero Giuseppe Meliante [1], Federica Zoccali [1], Francesca Cascone [1], Vanessa Di Stefano [1], Antonio Greco [1], Marco de Vincentiis [1], Carla Petrella [2], Marco Fiore [2], Antonio Minni [1,3,*] and Christian Barbato [2,*]

1   Department of Sense Organs DOS, Sapienza University of Rome, Viale del Policlinico 155, 00161 Roma, Italy
2   Institute of Biochemistry and Cell Biology (IBBC), National Research Council (CNR), Department of Sense Organs DOS, Sapienza University of Rome, Viale del Policlinico 155, 00161 Roma, Italy
3   Division of Otolaryngology-Head and Neck Surgery, Ospedale San Camillo de Lellis, ASL Rieti-Sapienza University, Viale Kennedy, 02100 Rieti, Italy
*   Correspondence: antonio.minni@uniroma1.it (A.M.); christian.barbato@cnr.it (C.B.)

Abstract: Obstructive sleep apnea syndrome (OSAS) is characterized by intermittent hypoxia (IH) during sleep due to recurrent upper airway obstruction. The derived oxidative stress (OS) leads to complications that do not only concern the sleep-wake rhythm but also systemic dysfunctions. The aim of this narrative literature review is to investigate molecular alterations, diagnostic markers, and potential medical therapies for OSAS. We analyzed the literature and synthesized the evidence collected. IH increases oxygen free radicals (ROS) and reduces antioxidant capacities. OS and metabolic alterations lead OSAS patients to undergo endothelial dysfunction, osteoporosis, systemic inflammation, increased cardiovascular risk, pulmonary remodeling, and neurological alterations. We treated molecular alterations known to date as useful for understanding the pathogenetic mechanisms and for their potential application as diagnostic markers. The most promising pharmacological therapies are those based on N-acetylcysteine (NAC), Vitamin C, Leptin, Dronabinol, or Atomoxetine + Oxybutynin, but all require further experimentation. CPAP remains the approved therapy capable of reversing most of the known molecular alterations; future drugs may be useful in treating the remaining dysfunctions.

Keywords: obstructive sleep apnea syndrome; OSAS; OSA; oxidative stress; intermittent hypoxia; biomarkers; CPAP; cognitive

Citation: Meliante, P.G.; Zoccali, F.; Cascone, F.; Di Stefano, V.; Greco, A.; de Vincentiis, M.; Petrella, C.; Fiore, M.; Minni, A.; Barbato, C. Molecular Pathology, Oxidative Stress, and Biomarkers in Obstructive Sleep Apnea. *Int. J. Mol. Sci.* **2023**, *24*, 5478. https://doi.org/10.3390/ijms24065478

Academic Editors: Rossana Morabito and Alessia Remigante

Received: 29 January 2023
Revised: 6 March 2023
Accepted: 10 March 2023
Published: 13 March 2023

1. Introduction

Obstructive sleep apnea syndrome (OSAS) affects from 9% to 39% of the adult population, with a higher incidence in males and the elderly, and is the most common form of respiratory sleep disorder [1–4]. It is characterized by recurrent, complete, or partial upper airway obstruction due to their collapse, with consequent hypopnea or apnea, leading to hypoventilation and chronic intermittent hypoxemia (IH) and increasing blood carbon dioxide partial pressure [5]. The OSAS consequences do not only concern excessive daytime sleepiness; they also independently favor the development of cardiovascular pathologies as an independent risk factor for hypercholesterolemia and hypertension, obesity, diabetes, and neuropsychological diseases such as depression [6–12]. Therefore, OSAS patients have higher cardiovascular-related morbidity compared with non-OSAS ones [13].

The diagnosis of OSAS follows the diagnostic criteria codified by the American Academy of Sleep Medicine (AASM) after the exclusion of other pathologies that may be the cause of the apnea/hypopnea events [14]. Furthermore, it is possible to make a diagnosis of OSAS when one is faced with anamnestic data, collected by interviewing the patient or those who sleep with him, with reported episodes of falling asleep when awake, excessive daytime sleepiness, non-refreshing sleep, tiredness, or insomnia, breathing, snoring, wheezing, choking, loud snoring, and at least 5 episodes of apnea, hypopnea, or breathing-related awakenings per hour in polysomnography. Alternatively, even in the absence of anamnestic data, it is possible to diagnose OSAS when polysomnography shows 15 or more apneas, hypopneas, or awakenings related to respiratory events per hour with evidence of respiratory effort in all or part of them in polysomnography [1]. To measure the degree of OSAS, the polysomnographic apnea-hypopnea index (AHI) is used [14,15]. Depending on the severity of the disease, treatment options include surgical interventions, lifestyle modifications, continuous positive airway pressure (CPAP), oral appliances such as mandibular advancement, and hypoglossal nerve stimulation [16–18]. However, the OSAS diagnosis and treatment currently do not take into consideration what happens at the molecular and cellular level, which instead causes systemic complications related to OSAS. This review aims to recognize molecular mechanisms, markers, and potential treatments useful for counteracting the cellular damage resulting from OSAS. This could help to better understand which patients are candidates, not only for the use of CPAP or MAD (mandibular advancement device), but also for medical therapy to counteract or reverse cellular damage.

2. Results

2.1. Oxidative Stress (OS)

OS is an imbalance between reactive oxygen species (ROS) production and antioxidant capacity. The hypoxia and reoxygenation cycles cause a change in the oxidative balance, leading to the formation of ROS and a decrease in endogenous antioxidant molecules [19]. ROS react with organic molecules, impairing their functions, altering cellular metabolism, and causing cell damage. They are considered one of the main mechanisms responsible for cardiovascular complications in patients with OSAS, and their production correlates with the severity of the disease [13,20–22].

The presence of oxidative imbalance has also been observed in uvular tissues removed after uvulopalatoplasty in patients with OSAS and correlates with the severity of the disease [23].

The human organism tries to adapt to the condition of lack of oxygen with the production of specific molecules useful for cell survival in hypoxic conditions such as Hypoxia Induced Factor-1α (HIF-1α) and Vascular endothelial growth factor (VEGF) [23,24]. HIF-1 α, instead, is one of the main actors in oxygen homeostasis. Its levels, together with those of NF-kB, correlate with the severity of the disease measured by the AHI and desaturation number, and the levels of surfactant protein D (SPD) are reduced in an inverse manner [24–26]. There is no significant variation in HIF-1α levels during the day [25,26]. The alteration of circulating HIF-1α levels is chronic; it has been observed that a single night with CPAP in severe OSAS patients does not significantly modify its expression [24]. The expression of HIF-1α and NF-kB is reduced after two months of continuous nasal CPAP use, and the SPD levels increase [25]. VEGF is increased as the nocturnal oxygen saturation decreases [24,25]. OS also underlies the increase in myeloperoxidase (MPO), intracellular adhesion molecule 1 (ICAM-1), vascular cell adhesion protein (VCAM-1), L-selectin, and E-selectin [26,27]. The differences in molecular expression in OSAS patients also concern the reduction of the morning levels of the Rho-associated protein kinase (ROCK) 1 and 2 molecules and the circulating nitric oxide, as well as the expression of endothelial eNOS [28,29]. These observations were confirmed in vivo, with an observed down-regulation of eNOS and an increase in nitrotyrosine [30]. VE-cadherin cleavage is one of the mechanisms of endothelial dysfunction and in OSAS patients, the circulating plasma values of the soluble form of VE-cadherin (sVE) were increased, suggesting an augmented endothelial permeability. This mechanism appears to be associated with ROS production, and activation of HIF-1, VEGF, and tyrosine kinase pathways [31]. Another mechanism that contributes to the endothelial dysfunction in IH is the production of extracellular vesicles by red blood cells, and their pathogenetic mechanism involves decreased eNOS, increased Endothelin-1 (ET-1) production via the Erk1/2 pathway, and phosphorylation via the PI3K/AKT pathway (Figure 1 [32]. OSAS-induced endothelial dysfunction has also been related to some microRNAs, such as miR-630 in infantile OSAS and miR-30a, miR-34a-5p, and miR-193 in mouse models (Figure 1) [33–37]. CPAP is widely used in the treatment of OSAS, and its systemic benefits have also been observed at the molecular level regarding endothelial function. Myocardial damage is also linked to microRNA expression alterations such as miR-146a-5p (Figure 1) [37].

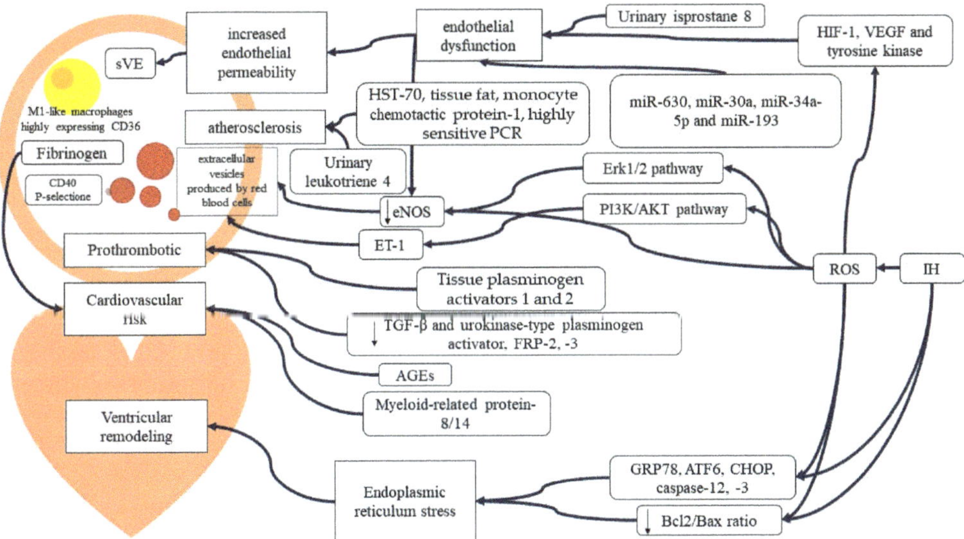

Figure 1. Mechanisms of cardiovascular risk in OSAS.

2.1.1. Mitochondrial Involvement

ROS are mainly produced at the mitochondrial level during the reoxygenation phase [27,38]. Mitochondrial damage is confirmed by the reduction of circulating mitochondrial DNA in conjunction with the worsening of OSAS [39]. Studies from mouse models showed that mitochondrial OS is also the basis of the damage to auditory hair cells, with aberrant mitochondrial morphology and overexpression of PGC-1α and Tfam mRNA [40]. The role of mitochondria in recurrent hypoxia damage is not limited to neuronal cells, and studies on genioglossus and palatine muscles have also shown damage at this level [41].

2.1.2. Inflammatory Signaling

IH leads to the formation of proinflammatory factors such as tumor necrosis factor (TNF), C-reactive protein (CRP), and interleukins (IL)-6 and -8 [6,21,42,43], and elevated values of NF-kB and TNF-α are linked to OSAS and daytime sleepiness. Cytokines related to NF-kB are also consequently increased, such as IL-8 [42,44], whereas IL-17 levels are also correlated with OSAS and its severity, as well as an inverse correlation with Vitamin D levels in enrolled patients [45].

CRP and TNF-α levels decrease after OSAS surgery, but they still remain higher than those of healthy control groups [43].

Further molecules whose expression is altered are Osteoprotegerine, Chitinase 3-like protein 1 (YKL-40), and Cardiotrophin-1 (CT-1), and their value correlates with AHI [46].

2.1.3. Peripheral Blood Cells

Various changes in peripheral blood cells have been observed. Overexpression of Toll-Like Receptor (TLR) was observed in circulating monocytes of OSAS patients [47], and the TLR 6 gene is upregulated in peripheral blood cells through DNA methylation.

In particular, the cytosine-phosphate-guanine (CPG) site number 1 is hypermethylated in patients with severe OSAS, and its methylation is reduced after at least 6 months of CPAP therapy [48]. Epigenetic studies also identified hypermethylation of interleukin 1 receptor 2 (IL1R2) and androgen receptors with increased expression of both [49].

2.1.4. Cardiovascular Implication of OSAS Inflammation

The role of inflammation has also been observed in the aortas of OSAS mouse models, evidencing an accumulation and proliferation of pro-inflammatory metabolic M1-like macrophages highly expressing CD36 and an increase in the transcription of atherogenic pathways, inflammation, and OS (Figure 1) [50]. Increased circulating fibrinogen values are associated with an elevated risk of cardiovascular events as well as OSAS, correlating high serum levels with the severity of the disease [51]. The inflammatory molecules increased in OSAS patients are numerous and also include heat shock protein 70, tissue fat, monocyte chemotactic protein-1, and highly sensitive C-reactive protein [52]. The prothrombotic state due to OSAS is also documented by the increase in tissue plasminogen activators 1 and 2 and the decrease in TGF-β and urokinase-type plasminogen activator [53]. Epigenetic dysregulation of DNA in OSAS was shown by aberrant methylation of the formyl peptide receptor (FPR) 1, 2, 3 genes, and FPR1 overexpression and the deficiency of FPR2 and FPR3 were associated with OSAS and its severity as well as with the development of diabetes mellitus and cardiovascular diseases.

Literature evidence did not show a correlation between OS and cardiovascular risk in OSAS patients [54].

Neurotrophins are proteins that regulate the nervous system; alterations of some of them, such as brain-derived neurotrophic factor (BDNF) and nerve growth factor (NGF), can also lead to cardiovascular complications. BDNF is related to cardiomyocyte contractility and its alterations with atherosclerosis and hypertension, and NGF plays a role in atrial autonomic OSAS-induced atrial fibrillation [54].

2.2. Circulating Metabolites

Previous research has shown that OSAS induces an increase in the blood levels of endocannabinoids such as anandamide and ethanolamide, which are associated with an increase in blood pressure and cardiovascular risk [55]. Likewise, an increase in the blood levels of adenosine, epinephrine, norepinephrine, and aldosterone was also observed [56,57], as was an increase in retinoids, carotenoids, and tocopherol, which increase the susceptibility to vascular pathologies [58].

Levels of saturated fatty acids and n-3 fatty acids also correlate with sleep quality, duration, and rapid eye movements. An increase in lactic acid and some fatty acids such as arabinose, arabitol, cellulose, glyceraldehyde, and threitol has been observed in these patients [59], and metabolomic studies have also shown an increase in glutamic acid, deoxy sugar, arachidonic acid, phosphatidylethanolamine, sphingomyelin, and lysophosphatidylcholine. The metabolic alterations mentioned above seem to be attributable to the hypoxia induced by OSAS [60].

In addition to the metabolites, the regulatory metabolism molecules are also impaired in OSAS patients. OSAS is independently correlated with insulin resistance and fatty liver disease, and several genes involved in cholesterol metabolism were impaired, such as malic enzyme and acetyl coenzyme A (CoA) synthetase, or acetyl-CoA carboxylase, stearoyl-CoA desaturases 1 and mitochondrial glycerol- 3-phosphate acyltransferase [61]. OSAS also induces adipose tissue inflammation and dysfunction [62].

The circulating omentin expression increased in OSAS [63]. Moreover, OSAS patients have significantly increased levels of leptin, affecting sleep, ventilation, and upper airway defenses [64,65].

2.3. Urine Molecules

Urine analysis from OSAS patients revealed high concentrations of adrenaline, noradrenaline, and homocysteine, which are associated with increased cardiovascular risk, as well as leukotriene E4. Increased levels of homovanillic acid, a metabolite of dopamine, and 3-4-dihydroxyphenylacetic acid were also measured [29,66,67]. Urinary isoprostane 8, which is associated with daytime sleepiness and OSAS, as well as being an element of endothelial damage [68], and an increase in urinary leukotriene-4 has also been observed, which is related to atherosclerosis [69,70]. The significance of these altered urine metabolites in OSAS is unknown.

2.4. Neurotransmitters

Several neurotransmitters could play a role in the pathogenesis of the disease. Histamine is responsible for the state of arousal in the central nervous system. In studies on mouse models, it also seems to have a role in the neuromuscular transmission that occurs from the hypoglossal nerve to the genioglossus, both of which are altered in OSAS patients [71].

Starting from the observation that apneas increase during REM sleep, due to motor inhibition of the cervical-cephalic muscles, several authors investigated the potential mechanism of muscular inactivation. REM sleep-related upper airway collapse is due to a change in norepinephrine and serotonin secretion by cranial nerve XII. The secretory patterns and increased distribution of α1-adrenoceptors are impaired in mouse models of IH during sleep [72].

Loss of muscle tone leading to recurrent airway collapse appears to be related to reduced activity of pharyngeal motoneurons, also due to decreased stimulation of cholinergic acetylcholine receptors. This molecule, therefore, not only acts at the level of the central nervous system with different concentrations depending on whether we are in a state of sleep or wakefulness, but its action also varies at a peripheral level and could be one of the mechanisms of muscular hypotonia pharyngeal in OSAS patients [73]. Grace et al., observed that the G-protein-coupled inwardly rectifying potassium (GIRK) channels are also involved in muscular inactivation during REM sleep. Therefore, the authors hypothesized that the potential target to counteract the loss of muscle tone is a potassium channel, and to avoid having an undesirable systemic effect, they suggested the inwardly rectifying potassium channel Kir2.4, which is expressed almost exclusively in the motor nuclei of the cranial nerves. To date, there are no clinical studies with this hypothesis.

The brain neurotrophic factor (BDNF) is a neurotrophin responsible for neuron growth, development, and plasticity. Therefore, it has a role in memory and learning mechanisms. It can cross the brain-blood barrier, and its blood level dysregulation is related to depression and cognitive decline. Despite there being no significant difference in BDNF levels between OSAS and control patients, the morning molecule expression is related to age and oxygen saturation during sleep. The proBDNF, a derived product of BDNF, relates to age and HIF-1α morning quantity [74]. BDNF and proBDNF evening concentrations are higher in patients with alteration measured with the Athens Insomnia Scale (AIS), Pittsburg Sleep Quality Index (PSQUI) positive, and lower in those with Beck Depression Inventory (BDI) positive [75]. There is also a correlation between plasmatic BDNF levels and the oxygen desaturation index, and a negative one with the AHI. It has been hypothesized that BDNF could be involved in apoptosis-related neural injury, contributing to OSAS-induced cognitive degeneration and psychiatric pathologies [76]. OSAS patients with depression have lower levels of BDNF and pro-BDNF compared with OSAS patients with no mood disturbances [76].

IH could also be responsible for peripheric neural damage. In mice models, IH-induced OS decreases BDNF and pro-BDNF expression, which increases retinal cell apoptosis [72]. BDNF expression is altered in cognitive diseases (e.g., Alzheimer's disease, Parkinson's related dementia, etc.) [77] BDNF is also involved in nociception, and some authors have hypothesized his involvement in the greater pain susceptibility of OSAS patients compared with the general population [78].

Further neurotrophins are known and are related to OSAS, such as the nerve growth factor (NGF), neurotrophins 3 and 4, and one neurotrophic factor, the glial cell-line-derived neurotrophic factor (GDNF) [79]. NGF is responsible for sympathetic neurons' maturation, differentiation, and survival. His precursor protein, the proNGF, has the opposite action [73]. Its neurological involvement in OSAS has not been observed, but it has been related to cardiac autonomic nervous system disturbances and pediatric neurogenic tonsillar inflammation and hypertrophy [54,79].

GDNF is essential for dopaminergic system development and respiratory pattern generation. It has been observed that its levels are lower in patients with OSAS [54,80].

OSAS exacerbates Parkinson's disease; sleep disturbance severity and motor dysfunction have been related to IL-6 levels. In the same study, Kaminska et al. observed a correlation between BDNF and increased sleepiness [81].

OSAS, Neurocognition and Neurofilament

Oxidative stress has repercussions in many parts of the body, including the central nervous system. Therefore, OS in OSAS is also responsible for neurocognitive dysfunction. The imbalance between oxidization and antioxidants, as previously cited, could cause neuron injury in brain regions most susceptible to hypoxia and oxidative stress, such as the hippocampus and cerebral cortex regions [82]. Different domains could be altered in OSAS patients, such as attention/vigilance, memory, global cognitive, and executive function. However, CPAP treatment seems to improve some but not all executive functions in different degrees of cognitive dysfunction. Compromised cognition could be partially reversed after CPAP. As mentioned above, several studies highlighted the association between OS and nervous system diseases such as Parkinson's disease, Alzheimer's disease, and epilepsy [54]. In this direction, identification and quantification of neuronal biomarkers of axonal damage could improve the diagnostic accuracy and the prognostic assessment in the management of neurological disease. Neurofilament light chain (NFL) is a neuronal protein, exclusively located in the neuronal cytoplasm, whose levels increase in serum and cerebrospinal fluid (CSF) proportionally to the degree of neuronal axonal damage, and it is a valuable biomarker in several neurological disorders. In OSAS patients, it could be interesting to monitor NFL variations after CPAP treatment aimed to evaluate neuronal recovery. The neuronal damage has also been confirmed by the first studies, which show an increase in the expression of neurofilament (NFL) and its correlation with the severity of the disease [83]. Further studies are needed to measure the impact of CPAP on this parameter.

2.5. Potential Therapies

2.5.1. Antioxidants

There are many potentially beneficial molecules for patients with OSAS, but most of them have not been tested on humans. Manganese superoxide dismutase is protective against cortical neuron oxidative damage by IH in mouse models [84]. Adiponectin also proved to be useful in counteracting mitochondrial damage in the genioglossus muscle of OSAS mice [46].

In addition, ROS scavenger administration with Endavarone in mouse models of IH has been tested, showing a significant reduction in cognitive impairment associated with increased brain expression of phosphorylated-cAMP response element-binding (p-CREB) [85].

The molecules tested in humans have not been studied in courts large enough to give indications for their use. Vitamin C and N-acetylcysteine (NAC) have shown interesting results in the reduction of OS in OSAS [7], and NAC reduces OS in OSAS through the reduction of peroxidized lipids and the increase in glutathione. Surprisingly, patients who received it continuously also had improvements in sleep parameters [86]. Vitamin C, on the other hand, proved to be effective in improving the endothelial function of OSAS patients in a study that took as its reference the diameter of the brachial artery, an indirect indicator of endothelial function [87]. Lastly, Leptin is both a drug capable of reducing free radicals, OS, and atherosclerosis in patients with OSAS [88].

To date, the only confirmed antioxidant therapy is CPAP itself, which has shown to be able to reverse many of the molecular alterations observed in vivo, such as eNOS, nitro-tyrosine, and NF-kB in the endothelium and circulating TNF-α [32].

2.5.2. Non-Antioxidant-Based Therapy

Estrogen-related receptor-α (ERR-α) is downregulated in OSAS patients and its ligand-binding induces the expression of fast-type muscle fibers in palatopharyngeal muscles. The interaction between estrogens and ERR-α could be a therapeutic target to reverse the muscle remodeling typical of these patients [89]. They inhibit the overexpression of HIF-1α induced by chronic IH and improve the endurance and regeneration of the genioglossus muscle in OSAS animal models [90]. Estrogens, in particular 17β-estradiol (E2) and a resveratrol dimer (RD), have a protective action against OSAS by limitation of HIF-1α action.

A pilot study in OSAS patients evidenced that Desipramine reduced airway collapse. At the same time, its anti-inflammatory properties could be beneficial in counteracting the systemic effects of OSAS, but further clinical studies are needed on a larger scale to evaluate its application [91,92].

The use of sedatives in the treatment of OSAS appears to be counterintuitive. However, it has been hypothesized that trazodone may reduce the respiratory arousal threshold and upper airway obstruction. The first phenomenon occurred in an experimental group, while the second was not significant, and the magnitude of the threshold change was not sufficient to counteract the changes due to mechanical obstruction [93].

The Phase II Pharmacotherapy of Apnea by Cannabimimetic Enhancement (PACE) has shown encouraging preliminary results for Dronabinol. A reduction in the AHI, a reduction in the feeling of sleepiness, and good satisfaction in the treated patients have been observed [94].

Sildenafil involves the inhibition of cyclin guanosine monophosphate phosphodiesterase 5, resulting in an increase in cyclic guanosine monophosphate and NO. Its experimentation in a randomized controlled trial in which it was compared with a placebo, however, showed a worsening of the disease [95].

In a randomized trial, the combination of Atomoxetine, a norepinephrine reuptake inhibitor, and antimuscarinic Oxybutynin, taken before going to sleep, was shown to be able to reduce the severity of OSAS, and further studies on larger sample sizes are needed [96].

It is important to observe that many molecules have only been tested in mouse models, such as Astragaloside IV, which showed an improvement of hypoxia-induced endothelial function [97]; Tauroursodeoxycholic acid, against hepatic damage induced by HI [98]; Pitavastatin, showing a reversal of IH-induced myocardial hypertrophy, cardiac function, perivascular fibrosis and inflammatory indices [99]; Allopurinol also showed beneficial effects in mouse models of OSAS with a reduction of lipid peroxidation and an improvement in cardiac function [100]; and for all molecules, clinical trials are needed.

2.6. Diagnostic Biomarkers

Some authors have attempted to identify OSAS biomarkers. From the studies of Fleming et al. it has emerged that altered values of glycated hemoglobin, c-reactive protein, and erythropoietin may be useful in OSAS diagnosis [101]. Variants of the 5-hydroxytryptamine receptor 2A have also been studied as markers of risk and severity of OSAS. It has been observed that some of them can be protective, while others predispose to greater severity of the disease [102].

The study of the overexpression of the genes disintegrin and metalloproteinase domain 29 (ADAM29), solute carrier family 18 (vesicular acetylcholine) member 3 (SLC18A3), and fibronectin-like domain-containing leucine-rich transmembrane protein 2 (FLRT2) showed how these latter are overexpressed in Asian subjects with OSAS and could be used to screen patients, at risk for the severe form of the disease [103].

Numerous genetic polymorphisms have been associated with the development of OSAS, but none of them have been validated in clinical trials as useful for screening or diagnosis [104–111].

MicroRNAs could be helpful for diagnosis, but we still don't have clinical validations in OSAS [112–114], such as the downregulation of miR-664a-3p [115], and dysregulation of miR-126-3p, miR-26a-5p, and miR-107, which associate with arterial hypertension in OSAS patients [116]. (Table 1)

Table 1. Potential OSAS diagnostic biomarkers. Clinical validations are needed.

Potential Biomarkers	References
glycated hemoglobin, c-reactive protein, erythropoietin	[101]
5-hydroxytryptamine receptor 2A	[102]
ADAM29, SLC18A3, FLRT2	[103]
miR-664a-3p, miR-126-3p, miR-26a-5p, miR-107	[115,116]
thioredoxin, malondialdehyde, superoxide dismutase	[117–119]
TBARS	[120,121]
FRAP	[122,123]
TAS	[124]

The most accurate OS markers in OSAS are thioredoxin, malondialdehyde, superoxide dismutase, and iron reduction [7]. Thioredoxin (TRX) concentration is a marker of disease severity as it is proportional to that of AHI and is inversely related to oxygen saturation [117,118].

Although several molecules related to OS were not significantly increased in OSAS, glutathione, 8-isoprostane, substances reactive to barbituric acid (TBARS), catalase activity, copper-zinc superoxide dismutase (SOD), and products of lipid peroxidation [119,120]. However, there is no agreement between SOD and malondialdehyde (a TBARS), because, in some studies, their levels seem to be correlated with the severity of OSAS [120,121]. The markers of the loss of antioxidant capacity in OSAS patients have been observed in several studies as the lowering of the antioxidant power of reduced iron (FRAP), the concentration of reduced iron, and the total serum antioxidant status (TAS) [122–124].

3. Discussion

Patients with OSAS often have an altered lipidic profile. The molecular imbalances above mentioned could determine alterations detectable before OSAS leads to the typical increase in total cholesterol, triglycerides, low-density lipoprotein, high-density lipoprotein, and low-density lipoprotein cholesterol [64]. The metabolic imbalance observed in the OSAS leading to an increase in glycolysis products seems to be also attributable to the action of HIF-1α which also induces glycolytic enzymes [125]. However, the list of impaired molecules in OSAS is wide, such as cardiolipin, phosphatidylcholine, phosphatidylethanolamine, bile acids, and oxylipids [126].

Since a correlation between OSAS and Vitamin D levels has been observed, its concentration measurement and possible supplementation could be useful, not to improve apneas but for the systemic damage connected to the deficiency [44].

IH causes increased production of ROS and a reduction of endogenous antioxidant molecules [19]. OS is a crucial component of dysfunctional pathologies associated with OSAS, such as obesity, hypertension, dyslipidemia, sympathetic activation, and diabetes. It has even been hypothesized that the OS produced by IH favors the development of obesity, thus favoring the development of OSAS [127]. It has been observed that the reduction of antioxidant enzymes occurs through the methylation of DNA, which is reversible with the normalization of breathing, inverting ROS production. Along with it, the chemosensory reflex of the carotid body and hypertension, which are impaired following IH, also stabilize [128].

Aldosterone levels increase with the increasing severity of OSAS. Aldosterone is a molecule related to resistant hypertension, just as OSAS is a disease related to the same disorder [57]. The interaction between the increase in all the proinflammatory molecules and the endothelium could be one of the causes of the increased cardiovascular risk in OSAS, such as TNF, interleukins, NF-kB, TLR receptors, myeloid-related protein-8/14, the accumulation of pro-inflammatory metabolic M1-lie macrophages highly expressing CD36, fibrinogen, shock protein 70, monocyte chemotactic protein 1, highly sensitive C-reactive protein, P-selectin, soluble CD40, ICAM-1, VCAM-1, L-selectin, E-selectin and MPO [6,20,42–45,48,51,52,55].

Some authors hypothesized the role of microRNAs (miR-126-3p, miR-26a-5p, and miR-107) in the diagnosis of arterial hypertension in OSAS patients [129]. In our opinion, they could be useful for better understanding the etiopathogenetic mechanisms, but they are certainly less relevant from a hypertension diagnostic point of view

In the brain, chronic IH is associated with hippocampal cortical damage. The OS produced by the cycles of ischemia and reoxygenation with ROS production is thought to mimic that of stroke. Confirmation of this phenomenon was investigated in mouse models, and an increase in ROS and OS response markers was observed. Recurrent ischemia induced for prolonged times induces an increase in the molecules produced by the action of ROS, such as oxidized proteins, peroxidized lipids, and oxidized nucleic acids, with activation of caspase 3 and neuronal cell apoptosis. Further, confirming the role of OS in neuronal degeneration following OSAS, it was observed that mice overexpressing ROS scavenger molecules were less susceptible to neuronal damage following recurrent ischemia [19]. Oxidative damage from IH also affects areas of the sleep-wake rhythm, which could aggravate the persistent feeling of sleepiness [130]. Regarding the potential neurocognitive dysfunction induced by OSAS, it should also be noted that markers of Alzheimer's disease (Aβ40, t-tau, p-tau) are increased in affected patients [131]. In the study of OSAS-related cognitive impairment, it was observed that it is associated with the expression of miR-26b and miR-207 [132].

The chronic IH that occurs in OSAS patients also leads to the formation of ROS at the cortical level, and their production increases above all during the re-oxygenation phases. ROS are considered responsible for cortical oxidative damage and the reduction of neurocognitive functions. The maximum production of ROS appears to occur at the mitochondrial level [38,83]. Mitochondrial OS is also responsible for damage to auditory sensory neuronal hair cells. The study of mouse models of hypoxia has highlighted alterations in mitochondrial morphology, and alterations in mRNA expression and leaves us with the prospect of experimenting with drugs active at this level to prevent hearing loss in OSAS patients [38]. It should be evaluated whether a mechanic counteracting airway collapse, associated with substances that reduce or reverse mitochondrial oxidative damage to cortical neurons, can alleviate neuronal damage.

Several molecules are potentially useful in counteracting OS, cardiovascular, and neuronal damage due to IH, such as NAC, Vitamin C, Leptin, Dronabinol, and a combination of Atomoxetine and Oxybutinin [7,81,82,92,94].

From a clinical point of view, however, it is also useful to know which drugs are not suggested in patients suffering from a specific pathology. Indeed, even if the randomized trial on Sildenafil showed a negative impact on OSAS, the study is useful from a clinical point of view. Patients with OSAS suffer more frequently from erectile dysfunction; therefore, it is necessary to remember in their treatment not to administer this drug as it is pejorative for their disease [93].

The only confirmed antioxidant therapy is CPAP itself, which is able to reverse many of the molecular alterations [43]. Therefore, the utility of molecular antioxidant therapies must be viewed with caution. The mechanisms underlying cell damage could be disrupted by the restoration of normal nocturnal oxygenation through CPAP. This phenomenon has been observed, for example, with the return to normal values of eNOS, nitrotyrosine, and NF-kB in the endothelium of OSAS subjects after the use of CPAP and, as noted by Ryan et al., with TNF-α [44]. Similarly, CPAP also reduces silent brain infarctions, which are usually increased in subjects with OSAS. Therefore, also in this case, the mechanical action of the positive pressure and the rebalancing of nocturnal oxygenation are sufficient to interrupt the phenomenon [55]. CPAP rebalances the expression of numerous pro-inflammatory and pro-coagulation molecules, but it is not always able to restore the levels of healthy subjects. This is the case with inhibitors of plasminogen-1 activation and TGF-β [52].

Importantly, not all damages appear to be reversible with CPAP. Aortic injury mediated by pro-inflammatory metabolic M1-like macrophages highly expressing CD36, upregulation of atherogenic pathway transcription, inflammation, and OS in mouse models, once triggered, is not reversed by a return to normal oxygenation values [51].

Despite the identification of numerous molecules whose levels are significantly increased in patients with OSAS, studies that have tried to identify diagnostic or severity biomarkers have not led to conclusive results [27]. Plasma thioredoxin (TRX) is a marker of OS, and CPAP is able to reduce its concentration after 1 month of treatment [119]. In this direction, there are several molecules that can potentially support the clinician in selecting patients for specific antioxidant therapy, such as MMP-2, -9, highly sensitive C-reactive protein, soluble receptors for advanced glycation end-products (sRAGE), and copper (Cu).

We know that cellular hypoxia induces acidosis, but an interesting observation has also been made on the role that IH has on the ability of cells to respond to it. As an effect of IH, the overproduced ROS induces the release of Ca^{2+} and the entry of Na^+ through the activation of the Na^+/Ca^{2+} exchanger, resulting in an increase in Na^+ ions that inhibit the activity of the Na^+/H^+ exchanger, leading to an accumulation of H^+ ions and acidosis [133].

4. Materials and Methods

4.1. Literature Research

We performed a narrative literature review with articles from PubMed, Embase, and the Cochrane Central Register of Controlled Trials. We considered articles concerning molecular and metabolic alterations, OS, biomarkers, and antioxidant therapy in OSAS patients. We also analyzed the bibliography of the selected manuscripts for further relevant articles.

4.2. Inclusion and Exclusion Criteria

We have considered articles in the English language without time limits. We preferred in vivo clinical analyses, but manuscripts describing animal models were not excluded if useful for understanding and completing the mechanisms analyzed. Unpublished studies were not considered for the present review.

4.3. Data Selection

As this is a narrative review, the decision regarding the inclusion of each article was addressed jointly by the authors. After a careful selection of sources, the collected evidence was discussed by the authors and summarized in this manuscript.

5. Conclusions

OSAS is not only a pathology related to sleep dysfunction but also has a significant systemic impact. OS and IH lead to impaired endothelial function, osteoporosis, metabolic alterations, systemic inflammation, cardiovascular complications, central and peripheral neuronal degeneration, and pulmonary remodeling.

To counteract systemic effects, therapies based on NAC, Vitamin C, Leptin, or a combination of Dronabinol and Atomoxetine appear to have promising results. Currently, the only approved therapy is CPAP, which is also capable of reversing many of the observed molecular alterations. Therefore, drug therapy can be useful in the treatment of all those dysfunctions that remain even after the restoration of normal nocturnal oxygenation.

OSAS patients are more frequently subject to erectile dysfunction; however, Sildenafil should not be prescribed because it worsens the underlying disease. OSAS patients are also more prone to vitamin D deficiency; therefore, this should be sought in newly diagnosed patients and corrected.

In conclusion, it is important to discover new biomarkers through innovative diagnostic tools [134,135], opening a new translational phase aimed at tuning oxidative profiles in OSAS patients.

Author Contributions: A.M. and C.B. contributed conception and design of the study; P.G.M., F.Z., F.C., V.D.S., A.G. and M.d.V. contributed to the methodology; P.G.M. and F.Z. wrote the first draft of the manuscript; supervision, C.P., M.F., C.B. and A.M.; project administration, C.B.; funding acquisition, A.M. and C.B. All authors have read and agreed to the published version of the manuscript.

Funding: 'TRANSLATIONAL BIOMEDICINE: MULTI-ORGAN PATHOLOGY AND THERAPY', n. DSB.AD007.256 (CNR), to C.B.; and Prot. N.0000625 by BANCA D'ITALIA to A.M.

Institutional Review Board Statement: Not applicable.

Informed Consent Statement: Not applicable.

Data Availability Statement: Not applicable.

Acknowledgments: Authors thank IBBC-CNR, Sapienza University of Rome, and DOS, Sapienza University, Policlinico Umberto I, Rome, Italy.

Conflicts of Interest: The authors declare no conflict of interest.

References

1. Epstein, L.J.; Kristo, D.; Strollo, P.J.J.r.; Friedman, N.; Malhotra, A.; Patil, S.P.; Ramar, K.; Rogers, R.; Schwab, R.J.; Weaver, E.M.; et al. Clinical guideline for the evaluation, management and long-term care of obstructive sleep apnea in adults. *J. Clin. Sleep Med.* **2009**, *5*, 263–276. [CrossRef] [PubMed]
2. Peppard, P.E.; Young, T.; Barnet, J.H.; Palta, M.; Hagen, E.W.; Hla, K.M. Increased prevalence of sleep-disordered breathing in adults. *Am. J. Epidemiol.* **2013**, *177*, 1006–1014. [CrossRef]
3. Senaratna, C.V.; Perret, J.L.; Lodge, C.J.; Lowe, A.J.; Campbell, B.E.; Matheson, M.C.; Hamilton, G.S.; Dharmage, S.C. Prevalence of obstructive sleep apnea in the general population: A systematic review. *Sleep Med. Rev.* **2017**, *34*, 70–81. [CrossRef]
4. Iannella, G.; Vicini, C.; Colizza, A.; Meccariello, G.; Polimeni, A.; Greco, A.; de Vincentiis, M.; de Vito, A.; Cammaroto, G.; Gobbi, R.; et al. Aging effect on sleepiness and apneas severity in patients with obstructive sleep apnea syndrome: A meta-analysis study. *Eur. Arch. Oto-Rhino-Laryngol.* **2019**, *276*, 3549–3556. [CrossRef]
5. Flemons, W.W.; Buysse, D.; Redline, S. Sleep-related breathing disorders in adults: Recommendations for syndrome definition and measurement techniques in clinical research. *Sleep* **1999**, *22*, 667–689. [CrossRef]
6. Lavie, L. Obstructive sleep apnoea syndrome—An oxidative stress disorder. *Sleep Med. Rev.* **2003**, *7*, 35–51. [CrossRef]
7. Lira, A.B.; de Sousa Rodrigues, C.F. Evaluation of oxidative stress markers in obstructive sleep apnea syndrome and additional antioxidant therapy: A review article. *Sleep Breath.* **2016**, *20*, 1155–1160. [CrossRef] [PubMed]
8. Mohit; Tomar, M.S.; Sharma, D.; Nandan, S.; Pateriya, A.; Shrivastava, A.; Chand, P. Emerging role of metabolomics for biomarker discovery in obstructive sleep apnea. *Sleep Breath.* **2022**, *2*, 1–8. [CrossRef]
9. Young, T.; Palta, M.; Dempsey, J.; Skatrud, J.; Weber, S.; Badr, S. The Occurrence of Sleep-Disordered Breathing among Middle-Aged Adults. *N. Engl. J. Med.* **1993**, *328*, 1230–1235. [CrossRef] [PubMed]
10. Nieto, F.J.; Young, T.B.; Lind, B.K.; Shahar, E.; Samet, J.M.; Redline, S.; D'Agostino, R.B.; Newman, A.B.; Lebowitz, M.D.; Pickering, T.G. Association of sleep-disordered breathing sleep apnea, and hypertension in a large community-based study. *J. Am. Med. Assoc.* **2000**, *283*, 1829–1836. [CrossRef]
11. Li, J.; Thorne, L.N.; Punjabi, N.M.; Sun, C.K.; Schwartz, A.R.; Smith, P.L.; Marino, R.L.; Rodriguez, A.; Hubbard, W.C.; O'Donnell, C.P.; et al. Intermittent hypoxia induces hyperlipidemia in lean mice. *Circ. Res.* **2005**, *97*, 698–706. [CrossRef] [PubMed]
12. Peppard, P.E.; Young, T.; Palta, M.; Skatrud, J. Prospective Study of the Association between Sleep-Disordered Breathing and Hypertension. *N. Engl. J. Med.* **2000**, *342*, 1378–1384. [CrossRef]
13. Gozal, D.; Kheirandish-Gozal, L. Cardiovascular morbidity in obstructive sleep apnea: Oxidative stress, inflammation, and much more. *Am. J. Respir. Crit. Care Med.* **2008**, *177*, 369–375. [CrossRef] [PubMed]

14. Kapur, V.K.; Auckley, D.H.; Chowdhuri, S.; Kuhlmann, D.C.; Mehra, R.; Ramar, K.; Harrod, C.G. Clinical practice guideline for diagnostic testing for adult obstructive sleep apnea: An American academy of sleep medicine clinical practice guideline. *J. Clin. Sleep Med.* **2017**, *13*, 479–504. [CrossRef] [PubMed]

15. Punjabi, N.M.; Sorkin, J.D.; Katzel, L.I.; Goldberg, A.P.; Schwartz, A.R.; Smith, P.L. Sleep-disordered breathing and insulin resistance in middle-aged and overweight men. *Am. J. Respir. Crit. Care Med.* **2002**, *165*, 677–682. [CrossRef] [PubMed]

16. Taheri, S.; Mignot, E. The genetics of sleep disorders. *Lancet Neurol.* **2002**, *1*, 242–250. [CrossRef]

17. Hukins, C.A. Obstructive sleep apnea—Management update. *Neuropsychiatr. Dis. Treat.* **2006**, *2*, 309–326. [CrossRef]

18. Lee, J.S.; Choi, H.; Lee, H.; Ahn, S.; Noh, G. Biomechanical effect of mandibular advancement device with different protrusion positions for treatment of obstructive sleep apnoea on tooth and facial bone: A finite element study. *J. Oral Rehabil.* **2018**, *45*, 948–958. [CrossRef]

19. Ferguson, A.; Ono, T.; Lowe, A.A.; Al-Majed, S.; Love, L.L.; Fleetham, J.A. A short-term controlled trial of ant adjustable oral appliance for the treatment of mild to moderate obstructive sleep apnoea. *Thorax* **1997**, *52*, 362–368. [CrossRef]

20. Xu, W.; Chi, L.; Row, B.W.; Xu, R.; Ke, Y.; Xu, B.; Luo, C.; Kheirandish, L.; Gozal, D.; Liu, R. Increased oxidative stress is associated with chronic intermittent hypoxia-mediated brain cortical neuronal cell apoptosis in a mouse model of sleep apnea. *Neuroscience* **2004**, *126*, 313–323. [CrossRef]

21. Celec, P.; Jurkovičová, I.; Buchta, R.; Bartík, I.; Gardlík, R.; Pálffy, R.; Mucska, I.; Hodosy, J. Antioxidant vitamins prevent oxidative and carbonyl stress in an animal model of obstructive sleep apnea. *Sleep Breath.* **2013**, *17*, 867–871. [CrossRef]

22. Pilkauskaite, G.; Miliauskas, S.; Sakalauskas, R. Reactive oxygen species production in peripheral blood neutrophils of obstructive sleep apnea patients. *Sci. World J.* **2013**, *2013*, 421763. [CrossRef]

23. Olszewska, E.; Rogalska, J.; Brzóska, M.M. The association of oxidative stress in the uvular mucosa with obstructive sleep apnea syndrome: A clinical study. *J. Clin. Med.* **2021**, *10*, 1132. [CrossRef]

24. Gabryelska, A.; Stawski, R.; Sochal, M.; Szmyd, B.; Białasiewicz, P. Influence of one-night CPAP therapy on the changes of HIF-1α protein in OSA patients: A pilot study. *J. Sleep Res.* **2020**, *29*, e12995. [CrossRef] [PubMed]

25. Lu, D.; Li, N.; Yao, X.; Zhou, L. Potential inflammatory markers in obstructive sleep apnea-hypopnea syndrome. *Biomol. Biomed.* **2017**, *17*, 47–53 . [CrossRef]

26. Gabryelska, A.; Szmyd, B.; Szemraj, J.; Stawski, R.; Sochal, M.; Białasiewicz, P. Patients with obstructive sleep apnea present with chronic upregulation of serum HIF-1α protein. *J. Clin. Sleep Med.* **2020**, *16*, 1761–1768. [CrossRef]

27. Schulz, R.; Hummel, C.; Heinemann, S.; Seeger, W.; Grimminger, F. Serum levels of vascular endothelial growth factor are elevated in patients with obstructive sleep apnea and severe nighttime hypoxia. *Am. J. Respir. Crit. Care Med.* **2002**, *165*, 67–70. [CrossRef] [PubMed]

28. El-Solh, A.A.; Mador, M.J.; Sikka, P.; Dhillon, R.S.; Amsterdam, D.; Grant, B.J.B. Adhesion molecules in patients with coronary artery disease and moderate-to-severe obstructive sleep apnea. *Chest* **2002**, *121*, 1541–1547. [CrossRef]

29. Aydin, Ş.; Özdemir, C.; Küçükali, C.I.; Sökücü, S.N.; Giriş, M.; Akcan, U.; Tüzün, E. Reduced peripheral blood mononuclear cell ROCK1 and ROCK2 levels in obstructive sleep apnea syndrome. *In Vivo* **2018**, *32*, 319–325. [CrossRef] [PubMed]

30. Jelic, S.; Lederer, D.J.; Adams, T.; Padeletti, M.; Colombo, P.C.; Factor, P.H.; Le Jemtel, T.H. Vascular inflammation in obesity and sleep apnea. *Circulation* **2010**, *121*, 1014–1021. [CrossRef]

31. Harki, O.; Tamisier, R.; Pépin, J.L.; Bailly, S.; Mahmani, A.; Gonthier, B.; Salomon, A.; Vilgrain, I.; Faury, G.; Briançon-Marjollet, A. VE-cadherin cleavage in sleep apnoea: New insights into intermittent hypoxia-related endothelial permeability. *Eur. Respir. J.* **2021**, *58*, 2004518. [CrossRef]

32. Peng, L.; Li, Y.; Li, X.; Du, Y.; Li, L.; Hu, C.; Zhang, J.; Qin, Y.; Wei, Y.; Zhang, H. Extracellular Vesicles Derived from Intermittent Hypoxia–Treated Red Blood Cells Impair Endothelial Function Through Regulating eNOS Phosphorylation and ET-1 Expression. *Cardiovasc. Drugs Ther.* **2020**, *35*, 901–913. [CrossRef] [PubMed]

33. Liu, K.X.; Chen, G.P.; Lin, P.L.; Huang, J.C.; Lin, X.; Qi, J.C.; Lin, Q.C. Detection and analysis of apoptosis- and autophagy-related miRNAs of mouse vascular endothelial cells in chronic intermittent hypoxia model. *Life Sci.* **2018**, *193*, 194–199. [CrossRef]

34. Khalyfa, A.; Kheirandish-Gozal, L.; Khalyfa, A.A.; Philby, M.F.; Alonso-Álvarez, M.L.; Mohammadi, M.; Bhattacharjee, R.; Terán-Santos, J.; Huang, L.; Andrade, J.; et al. Circulating plasma extracellular microvesicle MicroRNA cargo and endothelial dysfunction in children with obstructive sleep apnea. *Am. J. Respir. Crit. Care Med.* **2016**, *194*, 1116–1126. [CrossRef] [PubMed]

35. Bi, R.; Dai, Y.; Ma, Z.; Zhang, S.; Wang, L.; Lin, Q. Endothelial cell autophagy in chronic intermittent hypoxia is impaired by miRNA-30a-mediated translational control of Beclin-1. *J. Cell. Biochem.* **2019**, *120*, 4214–4224. [CrossRef]

36. Lv, X.; Wang, K.; Tang, W.; Yu, L.; Cao, H.; Chi, W.; Wang, B. miR-34a-5p was involved in chronic intermittent hypoxia-induced autophagy of human coronary artery endothelial cells via Bcl-2/beclin 1 signal transduction pathway. *J. Cell. Biochem.* **2019**, *120*, 18871–18882. [CrossRef] [PubMed]

37. Lin, G.; Huang, J.; Chen, Q.; Chen, L.; Feng, D.; Zhang, S.; Huang, X.; Huang, Y.; Lin, Q. MiR-146a-5p mediates intermittent hypoxia-induced injury in H9c2 cells by targeting XIAP. *Oxid. Med. Cell. Longev.* **2019**, *2019*, 6. [CrossRef] [PubMed]

38. Kim, Y.S.; Kwak, J.W.; Lee, K.E.; Cho, H.S.; Lim, S.J.; Kim, K.S.; Yang, H.S.; Kim, H.J. Can mitochondrial dysfunction be a predictive factor for oxidative stress in patients with obstructive sleep apnea? *Antioxid. Redox Signal.* **2014**, *21*, 1285–1288. [CrossRef]

39. Seo, Y.J.; Ju, H.M.; Lee, S.H.; Kwak, S.H.; Kang, M.J.; Yoon, J.H.; Kim, C.H.; Cho, H.J. Damage of inner ear sensory hair cells via mitochondrial loss in a murine model of sleep apnea with chronic intermittent hypoxia. *Sleep* **2017**, *40*, 9. [CrossRef]

40. Huang, H.; Jiang, X.; Dong, Y.; Zhang, X.; Ding, N.; Liu, J.; Hutchinson, S.Z.; Lu, G.; Zhang, X. Adiponectin alleviates genioglossal mitochondrial dysfunction in rats exposed to intermittent hypoxia. *PLoS ONE* **2014**, *9*, 10. [CrossRef]

41. Stl, P.S.; Johansson, B. Abnormal mitochondria organization and oxidative activity in the palate muscles of long-term snorers with obstructive sleep apnea. *Respiration* **2012**, *83*, 407–417. [CrossRef]

42. Vgontzas, A.N.; Papanicolaou, D.A.; Bixler, E.O.; Kales, A.; Tyson, K.; Chrousos, G.P. Elevation of plasma cytokines in disorders of excessive daytime sleepiness: Role of sleep disturbance and obesity. *J. Clin. Endocrinol. Metab.* **1997**, *82*, 1313–1316. [CrossRef]

43. Olszewska, E.; Pietrewicz, T.M.; Świderska, M.; Jamiołkowski, J.; Chabowski, A. A Case-Control Study on the Changes in High-Sensitivity C-Reactive Protein and Tumor Necrosis Factor-Alpha Levels with Surgical Treatment of OSAS. *Int. J. Mol. Sci.* **2022**, *23*, 14116. [CrossRef]

44. Ryan, S.; Taylor, C.T.; McNicholas, W.T. Predictors of elevated nuclear factor-κB-dependent genes in obstructive sleep apnea syndrome. *Am. J. Respir. Crit. Care Med.* **2006**, *174*, 824–830. [CrossRef] [PubMed]

45. Toujani, S.; Kaabachi, W.; Mjid, M.; Hamzaoui, K.; Cherif, J.; Beji, M. Vitamin D deficiency and interleukin-17 relationship in severe obstructive sleep apnea-hypopnea syndrome. *Ann. Thorac. Med.* **2017**, *12*, 107–113. [CrossRef]

46. Fiedorczuk, P.; Olszewska, E.; Rogalska, J.; Brzóska, M.M. Osteoprotegerin, Chitinase 3-like Protein 1, and Cardiotrophin-1 as Potential Biomarkers of Obstructive Sleep Apnea in Adults—A Case-Control Study. *Int. J. Mol. Sci.* **2023**, *24*, 2607. [CrossRef] [PubMed]

47. Akinnusi, M.; Jaoude, P.; Kufel, T.; El-Solh, A.A. Toll-like receptor activity in patients with obstructive sleep apnea. *Sleep Breath.* **2013**, *17*, 1009–1016. [CrossRef] [PubMed]

48. Huang, K.T.; Chen, Y.C.; Tseng, C.C.; Chang, H.C.; Su, M.C.; Wang, T.Y.; Lin, Y.Y.; Zheng, Y.X.; Chang, J.C.; Chin, C.H.; et al. Aberrant DNA methylation of the toll-like receptors 2 and 6 genes in patients with obstructive sleep apnea. *PLoS ONE* **2020**, *15*, 2. [CrossRef]

49. Chen, Y.C.; Chen, T.W.; Su, M.C.; Chen, C.J.; Chen, K.D.; Liou, C.W.; Tang, P.; Wang, T.Y.; Chang, J.C.; Wang, C.C.; et al. Whole genome DNA methylation analysis of obstructive sleep apnea: IL1R2, NPR2, AR, SP140 methylation and clinical phenotype. *Sleep* **2016**, *39*, 743–755. [CrossRef] [PubMed]

50. Shamsuzzaman, A.; Amin, R.S.; Calvin, A.D.; Davison, D.; Somers, V.K. Severity of obstructive sleep apnea is associated with elevated plasma fibrinogen in otherwise healthy patients. *Sleep Breath.* **2014**, *18*, 761–766. [CrossRef] [PubMed]

51. Hayashi, M.; Fujimoto, K.; Urushibata, K.; Takamizawa, A.; Kinoshita, O.; Kubo, K. Hypoxia-sensitive molecules may modulate the development of atherosclerosis in sleep apnoea syndrome. *Respirology* **2006**, *11*, 24–31. [CrossRef] [PubMed]

52. Steffanina, A.; Proietti, L.; Antonaglia, C.; Palange, P.; Angelici, E.; Canipari, R. The plasminogen system and transforming growth factor-β in subjects with obstructive sleep apnea syndrome: Effects of CPAP treatment. *Respir. Care* **2015**, *60*, 1643–1651. [CrossRef]

53. Fiedorczuk, P.; Stróżyński, A.; Olszewska, E. Is the oxidative stress in obstructive sleep apnea associated with cardiovascular complications? Systematic review. *J. Clin. Med.* **2020**, *9*, 3734. [CrossRef] [PubMed]

54. Gabryelska, A.; Turkiewicz, S.; Ditmer, M.; Sochal, M. Neurotrophins in the Neuropathophysiology, Course, and Complications of Obstructive Sleep Apnea-A Narrative Review. *Int. J. Mol. Sci.* **2023**, *24*, 1808. [CrossRef] [PubMed]

55. Engeli, S.; Blüher, M.; Jumpertz, R.; Wiesner, T.; Wirtz, H.; Bosse-Henck, A.; Stumvoll, M.; Batkai, S.; Pacher, P.; Harvey-White, J.; et al. Circulating anandamide and blood pressure in patients with obstructive sleep apnea. *J. Hypertens.* **2012**, *30*, 2345–2351. [CrossRef]

56. Findley, L.J.; Boykin, M.; Fallon, T.; Belardinelli, L. Plasma adenosine and hypoxemia in patients with sleep apnea. *J. Appl. Physiol.* **1988**, *64*, 556–561. [CrossRef]

57. Gonzaga, C.C.; Gaddam, K.K.; Ahmed, M.I.; Pimenta, E.; Thomas, S.J.; Harding, S.M.; Oparil, S.; Cofield, S.S. Calhoun DA Severity of obstructive sleep apnea is related to aldosterone status in subjects with resistant hypertension. *J. Clin. Sleep Med.* **2010**, *6*, 363–368. [CrossRef]

58. Day, R.M.; Matus, I.A.; Suzuki, Y.J.; Yeum, K.J.; Qin, J.; Park, A.M.; Jain, V.; Kuru, T.; Tang, G. Plasma levels of retinoids, carotenoids and tocopherols in patients with mild obstructive sleep apnoea. *Respirology* **2009**, *14*, 1134–1142. [CrossRef]

59. Papandreou, C. Independent associations between fatty acids and sleep quality among obese patients with obstructive sleep apnoea syndrome. *J. Sleep Res.* **2013**, *22*, 569–572. [CrossRef]

60. Xu, H.; Zheng, X.; Qian, Y.; Guan, J.; Yi, H.; Zou, J.; Wang, Y.; Meng, L.; Zhao, A.; Yin, S.; et al. Metabolomics Profiling for Obstructive Sleep Apnea and Simple Snorers. *Sci. Rep.* **2016**, *6*, srep30958. [CrossRef]

61. Lebkuchen, A.; Carvalho, V.M.; Venturini, G.; Salgueiro, J.S.; Freitas, L.S.; Dellavance, A.; Martins, F.C.; Lorenzi-Filho, G.; Cardozo, K.H.M.; Drager, L.F. Metabolomic and lipidomic profile in men with obstructive sleep apnoea: Implications for diagnosis and biomarkers of cardiovascular risk. *Sci. Rep.* **2018**, *8*, 11270. [CrossRef] [PubMed]

62. Ryan, S.; Arnaud, C.; Fitzpatrick, S.F.; Gaucher, J.; Tamisier, R.; Pépin, J.L. Adipose tissue as a key player in obstructive sleep apnoea. *Eur. Respir. Rev.* **2019**, *28*, 190006. [CrossRef] [PubMed]

63. Kurt, O.K.; Tosun, M.; Alcelik, A.; Yilmaz, B.; Talay, F. Serum omentin levels in patients with obstructive sleep apnea. *Sleep Breath.* **2014**, *18*, 391–395. [CrossRef]

64. Imayama, I.; Prasad, B. Role of leptin in obstructive sleep apnea. *Ann. Am. Thorac. Soc.* **2017**, *14*, 1607–1621. [CrossRef]

65. Schiza, S.E.; Mermigkis, C.; Bouloukaki, I. Leptin and leptin receptor gene polymorphisms and obstructive sleep apnea syndrome: Is there an association? *Sleep Breath.* **2015**, *19*, 1079–1080. [CrossRef]

66. Gharib, S.A.; Hayes, A.L.; Rosen, M.J.; Patel, S.R. A pathway-based analysis on the effects of obstructive sleep apnea in modulating visceral fat transcriptome. *Sleep* **2013**, *36*, 23–30. [CrossRef]

67. Badran, M.; Gozal, D. PAI-1: A Major Player in the Vascular Dysfunction in Obstructive Sleep Apnea? *Int. J. Mol. Sci.* **2022**, *23*, 5516. [CrossRef]

68. Jordan, W.; Berger, C.; Cohrs, S.; Rodenbeck, A.; Mayer, G.; Niedmann, P.D.; von Ahsen, N.; Rüther, E.; Kornhuber, J.; Bleich, S. CPAP-therapy effectively lowers serum homocysteine in obstructive sleep apnea syndrome. *J. Neural Transm.* **2004**, *111*, 683–689. [CrossRef]

69. Ozkan, Y.; Fırat, H.; Şimşek, B.; Torun, M.; Yardim-Akaydin, S. Circulating nitric oxide (NO), asymmetric dimethylarginine (ADMA), homocysteine, and oxidative status in obstructive sleep apnea-hypopnea syndrome (OSAHS). *Sleep Breath.* **2008**, *12*, 149–154. [CrossRef]

70. Paik, M.J.; Kim, D.K.; Nguyen, D.T.; Lee, G.; Rhee, C.S.; Yoon, I.Y.; Kim, J.W. Correlation of daytime sleepiness with urine metabolites in patients with obstructive sleep apnea. *Sleep Breath.* **2014**, *18*, 517–523. [CrossRef] [PubMed]

71. Liu, Z.L.; Wu, X.; Luo, Y.J.; Wang, L.; Qu, W.M.; Li, S.Q.; Huang, Z.L. Signaling mechanism underlying the histamine-modulated action of hypoglossal motoneurons. *J. Neurochem.* **2016**, *137*, 277–286. [CrossRef]

72. Kubin, L. Sleep-wake control of the upper airway by noradrenergic neurons, with and without intermittent hypoxia. *Prog. Brain Res.* **2014**, *209*, 255–274. [CrossRef] [PubMed]

73. Grace, K.P.; Hughes, S.W.; Shahabi, S.; Horner, R.L. K+ Channel modulation causes genioglossus inhibition in REM sleep and is a strategy for reactivation. *Respir. Physiol. Neurobiol.* **2013**, *188*, 277–288. [CrossRef] [PubMed]

74. Gabryelska, A.; Sochal, M. Evaluation of HIF-1 Involvement in the BDNF and ProBDNF Signaling Pathways among Obstructive Sleep Apnea Patients. *Int. J. Mol. Sci.* **2022**, *23*, 14876. [CrossRef]

75. Gabryelska, A.; Turkiewicz, S.; Ditmer, M.; Karuga, F.F.; Strzelecki, D.; Białasiewicz, P.; Sochal, M. BDNF and proBDNF Serum Protein Levels in Obstructive Sleep Apnea Patients and Their Involvement in Insomnia and Depression Symptoms. *J. Clin. Med.* **2022**, *11*, 7135. [CrossRef] [PubMed]

76. Wang, W.H.; He, G.P.; Xiao, X.P.; Gu, C.; Chen, H.Y. Relationship between brain-derived neurotrophic factor and cognitive function of obstructive sleep apnea/hypopnea syndrome patients. *Asian Pac. J. Trop. Med.* **2012**, *5*, 906–910. [CrossRef]

77. Andrade, A.G.; Bubu, O.M.; Varga, A.W.; Osorio, R.S. The Relationship between Obstructive Sleep Apnea and Alzheimer's Disease. *J. Alzheimer's Dis.* **2018**, *64*, S255–S270. [CrossRef]

78. Flores, K.R.; Viccaro, F.; Aquilini, M.; Scarpino, S.; Ronchetti, F.; Mancini, R.; Di Napoli, A.; Scozzi, D.; Ricci, A. Protective role of brain derived neurotrophic factor (BDNF) in obstructive sleep apnea syndrome (OSAS) patients. *PLoS ONE.* **2020** 15, e0227834. [CrossRef]

79. Miao, F.; Shuang, L.; Yin, Y.; Huo, H.; Kui, Z.; Hui, L. 7,8-Dihydroxyflavone protects retinal ganglion cells against chronic intermittent hypoxia—Induced oxidative stress damage via activation of the BDNF/TrkB signaling pathway. *Sleep Breath.* **2022**, *26*, 287–295. [CrossRef]

80. Gabryelska, A.; Sochal, M.; Turkiewicz, S.; Białasiewicz, P. Relationship between HIF-1 and circadian clock proteins in obstructive sleep apnea patients—Preliminary study. *J. Clin. Med.* **2020**, *9*, 1599. [CrossRef]

81. Kaminska, M.; O'Sullivan, M.; Mery, V.P.; Lafontaine, A.L.; Robinson, A.; Gros, P.; Martin, J.G.; Benedetti, A.; Kimoff, R.J. Inflammatory markers and BDNF in obstructive sleep apnea (OSA) in Parkinson's disease (PD). *Sleep Med.* **2022**, *90*, 258–261. [CrossRef]

82. Yang, C.; Zhou, Y.; Liu, H.; Xu, P. The Role of Inflammation in Cognitive Impairment of Obstructive Sleep Apnea Syndrome. *Brain Sci.* **2022**, *12*, 1303. [CrossRef]

83. Arslan, B.; Şemsi, R.; İriz, A.; Dinçel, A.S. The evaluation of serum brain-derived neurotrophic factor and neurofilament light chain levels in patients with obstructive sleep apnea syndrome. *Laryngoscope Investig. Otolaryngol.* **2021**, *6*, 1466–1473. [CrossRef] [PubMed]

84. Shan, X.; Chi, L.; Ke, Y.; Luo, C.; Qian, S.; Gozal, D.; Liu, R. Manganese superoxide dismutase protects mouse cortical neurons from chronic intermittent hypoxia-mediated oxidative damage. *Neurobiol. Dis.* **2007**, *28*, 206–215. [CrossRef]

85. Ling, J.; Yu, Q.; Li, Y.; Yuan, X.; Wang, X.; Liu, W.; Guo, T.; Duan, Y.; Li, L. Edaravone improves intermittent hypoxia-induced cognitive impairment and hippocampal damage in rats. *Biol. Pharm. Bull.* **2020**, *43*, 1196–1201. [CrossRef]

86. Sadasivam, K.; Patial, K.; Vijayan, V.K.; Ravi, K. Antioxidant treatment in obstructive sleep apnoea syndrome. *Indian, J. Chest Dis. Allied Sci.* **2011**, *53*, 153–162.

87. Grebe, M.; Eisele, H.J.; Weissmann, N.; Schaefer, C.; Tillmanns, H.; Seeger, W.; Schulz, R. Antioxidant vitamin C improves endothelial function in obstructive sleep apnea. *Am. J. Respir. Crit. Care Med.* **2006**, *173*, 897–901. [CrossRef]

88. MacRea, M.; Martin, T.; Zagrean, L.; Jia, Z.; Misra, H. Role of leptin as antioxidant in obstructive sleep apnea: An in vitro study using electron paramagnetic resonance method. *Sleep Breath.* **2013**, *17*, 105–110. [CrossRef] [PubMed]

89. Chen, H.H.; Lu, J.; Guan, Y.F.; Li, S.J.; Hu, T.T.; Xie, Z.S.; Wang, F.; Peng, X.H.; Liu, X.; Xu, X.; et al. Estrogen/ERR-α signaling axis is associated with fiber-type conversion of upper airway muscles in patients with obstructive sleep apnea hypopnea syndrome. *Sci. Rep.* **2016**, *6*, 27088. [CrossRef] [PubMed]

90. Zhou, J.; Liu, Y. Effects of genistein and estrogen on the genioglossus in rats exposed to chronic intermittent hypoxia may be HIF-1α dependent. *Oral Dis.* **2013**, *19*, 702–711. [CrossRef] [PubMed]

91. Jaffuel, D.; Mallet, J.P.; Dauvilliers, Y.; Bourdin, A. Is the muscle the only potential target of desipramine in obstructive sleep apnea syndrome? *Am. J. Respir. Crit. Care Med.* **2017**, *195*, 1677–1678. [CrossRef]

92. Taranto-Montemurro, L.; Sands, S.A.; Edwards, B.A.; Eckert, D.J.; White, D.P.; Wellman, A. Is the muscle the only potential target of desipramine in obstructive sleep apnea syndrome? Reply. *Am. J. Respir. Crit. Care Med.* **2017**, *195*, 1678–1679. [CrossRef] [PubMed]

93. Eckert, D.J.; Malhotra, A.; Wellman, A.; White, D.P. Trazodone increases the respiratory arousal threshold in patients with obstructive sleep apnea and a low arousal threshold. *Sleep* **2014**, *37*, 811–819. [CrossRef]

94. Carley, D.W.; Prasad, B.; Reid, K.J.; Malkani, R.; Attarian, H.; Abbott, S.M.; Vern, B.; Xie, H.; Yuan, C.; Zee, P.C. Pharmacotherapy of apnea by cannabimimetic enhancement, the PACE clinical trial: Effects of dronabinol in obstructive sleep apnea. *Sleep* **2018**, *41*, zsx184. [CrossRef] [PubMed]

95. Roizenblatt, S.; Guilleminault, C.; Poyares, D.; Cintra, F.; Kauati, A.; Tufik, S. A double-blind, placebo-controlled, crossover study of sildenafil in obstructive sleep apnea. *Arch. Intern. Med.* **2006**, *166*, 1763–1767. [CrossRef] [PubMed]

96. Taranto-Montemurro, L.; Messineo, L.; Sands, S.A.; Azarbarzin, A.; Marques, M.; Edwards, B.A.; Eckert, D.J.; White, D.P.; Wellman, A. The combination of atomoxetine and oxybutynin greatly reduces obstructive sleep apnea severity a randomized, placebo-controlled, double-blind crossover trial. *Am. J. Respir. Crit. Care Med.* **2019**, *199*, 1267–1276. [CrossRef] [PubMed]

97. Zhao, F.; Meng, Y.; Wang, Y.; Fan, S.; Liu, Y.; Zhang, X.; Ran, C.; Wang, H.; Lu, M. Protective effect of Astragaloside IV on chronic intermittent hypoxia-induced vascular endothelial dysfunction through the calpain-1/SIRT1/AMPK signaling pathway. *Front. Pharmacol.* **2022**, *13*, 920977. [CrossRef] [PubMed]

98. Hou, Y.; Yang, H.; Cui, Z.; Tai, X.; Chu, Y.; Guo, X. Tauroursodeoxycholic acid attenuates endoplasmic reticulum stress and protects the liver from chronic intermittent hypoxia induced injury. *Exp. Ther. Med.* **2017**, *14*, 2461–2468. [CrossRef]

99. Inamoto, S.; Yoshioka, T.; Yamashita, C.; Miyamura, M.; Mori, T.; Ukimura, A.; Matsumoto, C.; Matsumura, Y.; Kitaura, Y.; Hayashi, T. Pitavastatin reduces oxidative stress and attenuates intermittent hypoxia-induced left ventricular remodeling in lean mice. *Hypertens. Res.* **2010**, *33*, 579–586. [CrossRef]

100. Williams, A.L.; Chen, L.; Scharf, S.M. Effects of allopurinol on cardiac function and oxidant stress in chronic intermittent hypoxia. *Sleep Breath.* **2010**, *14*, 51–57. [CrossRef]

101. Fleming, W.E.; Holty, J.C.; Bogan, R.K.; Hwang, D.; Ferouz-Colborn, A.S.; Budhiraja, R.; Redline, S.; Mensah-Osman, E.; Osman, N.I.; Li, Q.; et al. Use of blood biomarkers to screen for obstructive sleep apnea. *Nat. Sci. Sleep* **2018**, *10*, 159–167. [CrossRef] [PubMed]

102. Wu, W.; Li, Z.; Tang, T.; Wu, J.; Liu, F.; Gu, L. 5-HTR2A and IL-6 polymorphisms and obstructive sleep apnea-hypopnea syndrome. *Biomed. Rep.* **2016**, *4*, 203–208. [CrossRef]

103. Lin, S.W.; Tsai, C.N.; Lee, Y.S.; Chu, S.F.; Chen, N.H. Gene expression profiles in peripheral blood mononuclear cells of asian obstructive sleep apnea patients. *Biomed. J.* **2014**, *37*, 60–70. [CrossRef] [PubMed]

104. Strausz, S.; Ruotsalainen, S.; Ollila, H.M.; Karjalainen, J.; Kiiskinen, T.; Reeve, M.; Kurki, M.; Mars, N.; Havulinna, A.S.; Luonsi, E.; et al. Genetic analysis of obstructive sleep apnoea discovers a strong association with cardiometabolic health. *Eur. Respir. J.* **2021**, *57*, 2003091. [CrossRef] [PubMed]

105. Li, J.; Lv, Q.; Sun, H.; Yang, Y.; Jiao, X.; Yang, S.; Yu, H.; Qin, Y. Combined Association Between ADIPOQ, PPARG, and TNF Genes Variants and Obstructive Sleep Apnea in Chinese Han Population. *Nat. Sci. Sleep* **2022**, *14*, 363–372. [CrossRef]

106. Tanizawa, K.; Chin, K. Genetic factors in sleep-disordered breathing. *Respir. Investig.* **2018**, *56*, 111–119. [CrossRef]

107. Wang, H.; Cade, B.E.; Sofer, T.; Sands, S.A.; Chen, H.; Browning, S.R.; Stilp, A.M.; Louie, T.L.; Thornton, T.A.; Johnson, W.C.; et al. Admixture mapping identifies novel loci for obstructive sleep apnea in Hispanic/Latino Americans. *Hum. Mol. Genet.* **2019**, *28*, 675–687. [CrossRef]

108. Kalra, M.; Pal, P.; Kaushal, R.; Amin, R.S.; Dolan, L.M.; Fitz, K.; Kumar, S.; Sheng, X.; Guha, S.; Mallik, J.; et al. Association of ApoE genetic variants with obstructive sleep apnea in children. *Sleep Med.* **2008**, *9*, 260–265. [CrossRef]

109. Gozal, D.; Khalyfa, A.; Capdevila, O.S.; Kheirandish-Gozal, L.; Khalyfa, A.A.; Kim, J. Cognitive function in prepubertal children with obstructive sleep apnea: A modifying role for NADPH oxidase p22 subunit gene polymorphisms? *Antioxid. Redox Signal.* **2012**, *16*, 171–177. [CrossRef]

110. Cade, B.E.; Chen, H.; Stilp, A.M.; Gleason, K.J.; Sofer, T.; Ancoli-Israel, S.; Arens, R.; Bell, G.I.; Below, J.E.; Bjonnes, A.C.; et al. Genetic associations with obstructive sleep apnea traits in Hispanic/Latino Americans. *Am. J. Respir. Crit. Care Med.* **2016**, *194*, 886–897. [CrossRef]

111. Chen, W.; Ye, J.; Han, D.; Yin, G.; Wang, B.; Zhang, Y. Association of prepro-orexin polymorphism with obstructive sleep apnea/hypopnea syndrome. *Am. J. Otolaryngol.-Head Neck Med. Surg.* **2012**, *33*, 31–36. [CrossRef]

112. Pinilla, L.; Barbé, F.; de Gonzalo-Calvo, D. MicroRNAs to guide medical decision-making in obstructive sleep apnea: A review. *Sleep Med. Rev.* **2021**, *59*, 101458. [CrossRef]

113. Li, K.; Wei, P.; Qin, Y.; Wei, Y. MicroRNA expression profiling and bioinformatics analysis of dysregulated microRNAs in obstructive sleep apnea patients. *Medicine* **2017**, *96*, e7917. [CrossRef]

114. Chen, Q.; Lin, G.; Huang, J.; Chen, G.; Huang, X.; Lin, Q. Expression profile of long non-coding RNAs in rat models of OSA-induced cardiovascular disease: New insight into pathogenesis. *Sleep Breath.* **2019**, *23*, 795–804. [CrossRef]

115. Li, K.; Chen, Z.; Qin, Y.; Wei, Y. MiR-664a-3p expression in patients with obstructive sleep apnea: A potential marker of atherosclerosis. *Medicine* **2018**, *97*, e9813. [CrossRef] [PubMed]

116. Yang, X.; Niu, X.; Xiao, Y.; Lin, K.; Chen, X. MiRNA expression profiles in healthy OSAHS and OSAHS with arterial hypertension: Potential diagnostic and early warning markers 11 Medical and Health Sciences 1102 Cardiorespiratory Medicine and Haematology. *Respir. Res.* **2018**, *19*, 194. [CrossRef]

117. Guo, Q.; Wang, Y.; Li, Q.Y.; Li, M.; Wan, H.Y. Levels of thioredoxin are related to the severity of obstructive sleep apnea: Based on oxidative stress concept. *Sleep Breath.* **2013**, *17*, 311–316. [CrossRef]

118. Takahashi, K.I.; Chin, K.; Nakamura, H.; Morita, S.; Sumi, K.; Oga, T.; Matsumoto, H.; Niimi, A.; Fukuhara, S.; Yodoi, J.; et al. Plasma thioredoxin, a novel oxidative stress marker, in patients with obstructive sleep apnea before and after nasal continuous positive airway pressure. *Antioxid. Redox Signal.* **2008**, *10*, 715–726. [CrossRef] [PubMed]

119. Alzoghaibi, M.A.; BaHammam, A.S.O. Lipid peroxides, superoxide dismutase and circulating IL-8 and GCP-2 in patients with severe obstructive sleep apnea: A pilot study. *Sleep Breath.* **2005**, *9*, 119–126. [CrossRef] [PubMed]

120. Jordan, W.; Cohrs, S.; Degner, D.; Meier, A.; Rodenbeck, A.; Mayer, G.; Pilz, J.; Rüther, E.; Kornhuber, J.; Bleich, S. Evaluation of oxidative stress measurements in obstructive sleep apnea syndrome. *J. Neural Transm.* **2006**, *113*, 239–254. [CrossRef]

121. Cofta, S.; Wysocka, E.; Piorunek, T.; Rzymkowska, M.; Batura-Gabryel, H.; Torlinski, L. Oxidative stress markers in the blood of persons with different stages of obstructive sleep apnea syndrome. *J. Physiol. Pharmacol.* **2008**, *59*, 183–190.

122. Mancuso, M.; Bonanni, E.; LoGerfo, A.; Orsucci, D.; Maestri, M.; Chico, L.; DiCoscio, E.; Fabbrini, M.; Siciliano, G.; Murri, L. Oxidative stress biomarkers in patients with untreated obstructive sleep apnea syndrome. *Sleep Med.* **2012**, *13*, 632–636. [CrossRef] [PubMed]

123. Simiakakis, M.; Kapsimalis, F.; Chaligiannis, E.; Loukides, S.; Sitaras, N.; Alchanatis, M. Lack of effect of sleep apnea on oxidative stress in obstructive sleep apnea syndrome (OSAS) patients. *PLoS ONE* **2012**, *7*, e39172. [CrossRef] [PubMed]

124. Katsoulis, K.; Kontakiotis, T.; Spanogiannis, D.; Vlachogiannis, E.; Kougioulis, M.; Gerou, S.; Daskalopoulou, E. Total antioxidant status in patients with obstructive sleep apnea without comorbidities: The role of the severity of the disease. *Sleep Breath.* **2011**, *15*, 861–866. [CrossRef]

125. Gabryelska, A.; Karuga, F.F.; Szmyd, B.; Białasiewicz, P. HIF-1α as a Mediator of Insulin Resistance, T2DM, and Its Complications: Potential Links with Obstructive Sleep Apnea. *Front. Physiol.* **2020**, *11*, 1035. [CrossRef] [PubMed]

126. Pinilla, L.; Benítez, I.D.; Santamaria-Martos, F.; Targa, A.; Moncusí-Moix, A.; Dalmases, M.; Mínguez, O.; Aguilà, M.; Jové, M.; Sol, J.; et al. Plasma profiling reveals a blood-based metabolic fingerprint of obstructive sleep apnea. *Biomed. Pharmacother.* **2022**, *145*, 112425. [CrossRef]

127. Lavie, L.; Lavie, P. Molecular mechanisms of cardiovascular disease in OSAHS: The oxidative stress link. *Eur. Respir. J.* **2009**, *33*, 1467–1484. [CrossRef]

128. Kim, J.; Bhattacharjee, R.; Snow, A.B.; Capdevila, O.S.; Kheirandish-Gozal, L.; Gozal, D. Myeloid-related protein 8/14 levels in children with obstructive sleep apnoea. *Eur. Respir. J.* **2010**, *35*, 843–850. [CrossRef]

129. Barbato, C. MicroRNA-Mediated Silencing Pathways in the Nervous System and Neurological Diseases. *Cells* **2022**, *11*, 2375. [CrossRef]

130. Liang, S.; Li, N.; Heizhati, M.; Yao, X.; Abdireim, A.; Wang, Y.; Abulikemu, Z.; Zhang, D.; Chang, G.; Kong, J.; et al. What do changes in concentrations of serum surfactant proteins A and D in OSA mean? *Sleep Breath.* **2015**, *19*, 955–962. [CrossRef]

131. Kong, W.; Zheng, Y.; Xu, W.; Gu, H.; Wu, J. Biomarkers of Alzheimer's disease in severe obstructive sleep apnea–hypopnea syndrome in the Chinese population. *Eur. Arch. Oto-Rhino-Laryngol.* **2021**, *278*, 865–872. [CrossRef]

132. Gao, H.; Han, Z.; Huang, S.; Bai, R.; Ge, X.; Chen, F.; Lei, P. Intermittent hypoxia caused cognitive dysfunction to relate to miRNAs dysregulation in hippocampus. *Behav. Brain Res.* **2017**, *335*, 80–87. [CrossRef] [PubMed]

133. Chang, H.R.; Lien, C.F.; Jeng, J.R.; Hsieh, J.C.; Chang, C.W.; Lin, J.H.; Yang, K.T. Intermittent Hypoxia Inhibits Na + -H + Exchange-Mediated Acid Extrusion Via Intracellular Na + Accumulation in Cardiomyocytes. *Cell Physiol. Biochem.* **2018**, *46*, 1252–1262. [CrossRef]

134. Fiedorczuk, P.; Polecka, A.; Walasek, M.; Olszewska, E. Potential Diagnostic and Monitoring Biomarkers of Obstructive Sleep Apnea–Umbrella Review of Meta-Analyses. *J. Clin. Med.* **2023**, *12*, 60. [CrossRef]

135. Riccardi, G.; Bellizzi, M.G.; Fatuzzo, I.; Zoccali, F.; Cavalcanti, L.; Greco, A.; Vincentiis, M.d.; Ralli, M.; Fiore, M.; Petrella, C.; et al. Salivary Biomarkers in Oral Squamous Cell Carcinoma: A Proteomic Overview. *Proteomes* **2022**, *10*, 37. [CrossRef] [PubMed]

International Journal of
Molecular Sciences

Review

Molecular Genetics of Abnormal Redox Homeostasis in Type 2 Diabetes Mellitus

Iuliia Azarova [1,2], Alexey Polonikov [3,4,]* and Elena Klyosova [2]

1 Department of Biological Chemistry, Kursk State Medical University, 3 Karl Marx Street, 305041 Kursk, Russia
2 Laboratory of Biochemical Genetics and Metabolomics, Research Institute for Genetic and Molecular Epidemiology, Kursk State Medical University, 18 Yamskaya Street, 305041 Kursk, Russia
3 Laboratory of Statistical Genetics and Bioinformatics, Research Institute for Genetic and Molecular Epidemiology, Kursk State Medical University, 18 Yamskaya Street, 305041 Kursk, Russia
4 Department of Biology, Medical Genetics and Ecology, Kursk State Medical University, 3 Karl Marx Street, 305041 Kursk, Russia
* Correspondence: polonikov@rambler.ru

Abstract: Numerous studies have shown that oxidative stress resulting from an imbalance between the production of free radicals and their neutralization by antioxidant enzymes is one of the major pathological disorders underlying the development and progression of type 2 diabetes (T2D). The present review summarizes the current state of the art advances in understanding the role of abnormal redox homeostasis in the molecular mechanisms of T2D and provides comprehensive information on the characteristics and biological functions of antioxidant and oxidative enzymes, as well as discusses genetic studies conducted so far in order to investigate the contribution of polymorphisms in genes encoding redox state-regulating enzymes to the disease pathogenesis.

Keywords: type 2 diabetes; genetic susceptibility; molecular mechanisms; oxidative stress; redox homeostasis; glutathione metabolism; antioxidant enzymes; oxidative enzymes; single nucleotide polymorphism

Citation: Azarova, I.; Polonikov, A.; Klyosova, E. Molecular Genetics of Abnormal Redox Homeostasis in Type 2 Diabetes Mellitus. *Int. J. Mol. Sci.* **2023**, 24, 4738. https://doi.org/10.3390/ijms24054738

Academic Editors: Rossana Morabito and Alessia Remigante

Received: 26 January 2023
Revised: 20 February 2023
Accepted: 24 February 2023
Published: 1 March 2023

1. Introduction

Diabetes mellitus is one of the most common metabolic diseases, affecting every tenth person on the planet [1]. According to the latest edition of the IDF Diabetes Atlas, there are more than 537 million people who suffer from diabetes in the world [2]. It has been recently predicted that 643 million people will have diabetes by 2030 (11.3% of the population) [3]. If trends continue, this number will jump to a staggering 783 million (12.2%) by 2045 [2]. A great majority of diabetics in the world suffer from type 2 diabetes mellitus, or non-insulin-dependent diabetes mellitus, a serious chronic disease that develops when the body does not produce enough insulin or is unable to use it effectively [4–6].

Epidemiological studies of T2D conducted over the past decades indicate that T2D is a heterogeneous disease determined by genetic, epigenetic, and environmental risk factors that closely interact with each other [7–9]. A huge number of genetic studies have been conducted to elucidate the molecular mechanisms of T2D, including beta-cell dysfunction, insulin resistance, imbalance in redox homeostasis, and impairment of incretin signaling; from these studies, multiple disease-associated gene polymorphisms have been identified [10–14]. Nonetheless, many aspects of the molecular mechanisms of disease pathogenesis remain poorly characterized. In particular, numerous studies have shown that oxidative stress (the imbalance caused by excess ROS or oxidants over the cell's ability to realize an effective antioxidant response) resulting from an imbalance between the production of free radicals and their neutralization by antioxidant enzymes is one of the major pathological disorders underlying the development and progression of type 2 diabetes.

However, the literature data on the molecular genetic mechanisms underlying abnormal redox homeostasis in type 2 diabetes have not so far been reviewed.

The present review summarizes current state-of-the art advances in understanding the role of abnormal redox homeostasis in the molecular mechanisms of T2D and provides comprehensive information on the characteristics and biological functions of antioxidant and oxidative enzymes, as well as discusses genetic studies conducted so far in order to investigate the contribution of polymorphisms in genes encoding redox state-regulating enzymes in disease pathogenesis. The following internet resources were used in preparing this review: Entrez gene [15], UniProt Knowledgebase [16], BRENDA Enzyme Database [17], GENATLAS [18], HUGO Gene Nomenclature Committee [19], GeneCards [20], BioGPS [21], Genotype-Tissue Expression (GTEx) Portal [22], eQTLGen Consortium [23], GWAS Catalog [24], and DisGeNET [25].

2. General Pathological Alterations in Type 2 Diabetes

It is widely agreed that chronic hyperglycemia, the primary diagnostic indicator of T2D, results from pancreatic beta-cell failure, which manifests by a gradient reduction in beta cell mass and insulin production in response to glucose [26–28]. With the loss of metabolic flexibility, beta-cells begin to oxidize fatty acids instead of glucose, which results in the production of harmful byproducts (peroxides) and decreased insulin secretion [29,30]. It is noteworthy that the damage of the endocrine part of the pancreas is caused by at least three biological phenomena: apoptosis of beta-cells, their dedifferentiation, and transformation into glucagon-producing alpha cells [31–34]. A number of studies have shown a decrease in the expression of key transcription factors (PDX1, NKX6.1, and MAFA) regulating the function of beta-cells, and discussed the subsequent disorders that are accompanied by a loss of insulin-producing ability in T2D [33,35].

The islets of Langerhans have been found to contain bihormonal cells that produce both glucagon and insulin after the conversion of beta-cells into alpha cells, as has been demonstrated by experimental studies on transgenic mice [33,36]. A shift in the phenotype of pancreatic beta-cells may serve as a mechanism of cell mass conservation [37] since alpha cells are more resistant to metabolic stress induced by excessive nutrition. Lim and co-authors observed that changes in insulin secretion in T2D are potentially reversible due to beta-cell re-differentiation: complete recovery of the first phase of secretion was observed by the eighth week of a hypocaloric diet in 87% of T2D patients with less than four years of disease duration [38]. There is definitely a "point of no return" in the progression of the disease beyond which the lost ability to produce insulin cannot be restored due to irreversible apoptotic changes in the pancreatic islets.

The development of insulin resistance, another phenomenon accompanying T2D pathogenesis, involves supra- and post-receptor mechanisms. The former are associated with hypertrophic extracellular matrix signaling, a decrease in capillary filling, and thus deterioration of local blood flow, limiting the availability of insulin and glucose in muscles and other tissues [39–41]. Insulin resistance is exacerbated by an increase in the expression of collagen and other extracellular matrix proteins, including their integrin receptors, which are in direct contact with skeletal muscle capillaries [42]. It is worth noting that the formation of insulin resistance in adipose tissue is triggered by the accumulation of extracellular matrix and fibrosis [43,44]. Given that insulin-like growth factor 1 (IGF1) is a powerful stimulator of collagen expression, hyperinsulinemia can promote collagen synthesis by adipose tissue fibroblasts, leading to insulin resistance by activating IGF1 receptors or affecting IGF1 binding protein [45]. Post-receptor mechanisms underlying insulin resistance are associated with a decrease in intracellular glucose metabolism activity, including de novo fatty acid synthesis in adipose tissue [46–48]. These effects are associated with decreased expression of the lipogenic transcription factors ChREBP-α and ChREBP-β, which are linked to GLUT4 inactivation [49–51]. The importance of controlling the metabolism of acetyl-CoA is also related to the fact that this substrate is necessary for the acetylation of proteins, including histones [52–54].

Insulin resistance and Langerhans islet dysfunction appear early in the pathogenesis of T2D, as shown in observational studies of Pima Indians, whose natural course of disease progressed over time from euglycemia to impaired glucose tolerance and T2D [55,56]. There are two points of view regarding the sequence of development of insulin resistance and beta-cell dysfunction, which is often manifested as hyperinsulinemia. The first explanation states that insulin resistance is the primary cause of hyperinsulinemia. An experimental study on mice has shown that insulin resistance appears as a result of induced disorders of the insulin signaling pathway in the liver, skeletal muscle, and adipose tissue and causes hyperinsulinemia, thereby leading to the development of T2D [57,58]. In humans, mutations in genes encoding insulin signaling proteins are also manifested by an increase in circulating insulin levels in the blood and are the causes of hereditary forms of diabetes [59,60]. A decrease in the effectiveness of insulin in relation to glycemic control is associated with the activation of the FOXO1 transcription factor in the liver [61,62] and impaired translocation of the glucose transporter GLUT4 into the membranes of skeletal myocytes [63,64]. Consequently, FOXO1 increases the expression of key enzymes of gluconeogenesis, causing the liver to produce more glucose. A decrease in GLUT4 in the membrane of skeletal myocytes reduces the uptake of glucose from the bloodstream into the cells. In the liver, insulin normally causes phosphorylation and suppression of FOXO1 by protein kinase B (Akt), which keeps FOXO1 in the cytoplasm, where this transcription factor is inactive [65–67]. However, FOXO1 expression was found to be increased in the liver of obese mice, and this transcription factor loses its sensitivity to insulin regulation [68–70]. Impact of overnutrition on the FOXO1 dysregulation is currently being extensively researched [71,72] to better understand the primary mechanism of uncontrolled gluconeogenesis in obesity [73–75]. It is thought that hyperglycemia caused by uncontrolled FOXO1 activation, combined with chronic hyperinsulinemia, can remove insulin's inhibitory effect on lipolysis in adipose tissue [69]. Activated lipolysis in abdominal adipocytes in turn increases the flow to the liver of its products—free fatty acids, which are allosteric activators of pyruvate carboxylase, a key enzyme of gluconeogenesis, and glycerol, a substrate of the same pathway [76]. Thus, a high-calorie diet induces the production of glucose by the liver under the influence of FOXO1, which is additionally stimulated by the activation of lipolysis in adipose tissue. In this context, functional beta-cell deficiency results in an inability to secrete enough insulin to compensate for the effects of FOXO1, resulting in diabetes [77,78].

An alternative hypothesis of T2D pathogenesis implies that hyperinsulinemia is the primary cause of insulin resistance. In obese people without diabetes, fasting hyperinsulinemia is found without detectable hyperglycemia, which is theoretically necessary to stimulate additional insulin secretion by beta-cells. This apparent disparity between blood glucose and insulin levels led to the hypothesis that hyperinsulinemia is the initial, primary effect of overeating and obesity [79,80] and is caused by stimulation of insulin secretion [81,82] and suppression of its degradation [83]. Insulin resistance in the liver is caused by primary hyperinsulinemia, which is associated with the suppression of signal transmission from the insulin receptor to Akt. Activation of the Akt protein appears to be sufficient to activate the mTORC1 kinase complex and the SREBP-1c transcription factor, both of which promote de novo fatty acid synthesis [84]. Furthermore, hyperinsulinemia has been found to cause activation of inflammatory signaling pathways in humans [85–87] and rats [88], which can impair insulin sensitivity in the target tissues [89]. A similar increase in glucose tolerance was observed in the knockout mice with impaired insulin secretion [90]. Thus, both beta-cell dysfunction and insulin resistance are disorders that can occur at the same time and are not mutually exclusive.

3. Environmental Risk Factors of Type 2 Diabetes

Type 2 diabetes is a multifactorial disease that develops as a result of interactions between environmental and genetic factors [91,92]. It has been estimated that the majority of the burden of type 2 diabetes is related to environmental exposures and modifiable risk factors, such as lifestyle [93,94]. High-calorie diets, high consumption of fatty foods and refined carbohydrates, hypodynamia, and smoking are modifiable risk factors for T2D, whereas gender, age, and family history of diabetes are non-modifiable risk factors.

The leading causes underlying the epidemics of T2D and obesity include urbanization, sedentary lifestyles, and unhealthy diets [26]. Undoubtedly, nutritional status is one of the most important environmental factors influencing the risk of T2D [26]. Food quality is more significant than overall food intake, as evaluated by naturally occurring polyunsaturated fatty acid content, the lack of trans-unsaturated fatty acids, and meals with a high glycemic index. An increase in dietary fat intake raises insulin levels in contrast to normal glycemia [95]. Carter P. et al. [96] observed that daily eating of fresh fruits and vegetables reduces the risk of disease by 14%. According to a study by Cooper A.J. and co-authors [97], a diet with a greater quantity of vegetables and a greater variety of both fruits and vegetables in diet is associated with a reduced risk of type 2 diabetes. It has been found that there is an inverse relationship between consumption of plant food and the risk of disease [98,99]. It is estimated that consuming even four servings of fruit each day can lower the incidence of T2D by 7% [99]. Increased consumption of fresh, plant-based foods and whole-meal breads, as well as decreased consumption of sugary drinks, high-calorie, fried foods, and white bread, are among the nutritional recommendations made by the WHO for the prevention of T2D [100]. Numerous studies have demonstrated that the Mediterranean diet lowers the incidence of T2D [101,102]. The fast-food boom is one of the factors that's caused the pandemic spike in T2D incidence in Asia (China and South Korea) over the past two decades [103]. Clinical studies, such as the Diabetes Prevention Program in the USA, the Finnish Diabetes Prevention Studies in Finland, the Da Qing IGT, and the Diabetes Study in China, have demonstrated the benefits of changing lifestyle, particularly through a healthy diet and increased physical activity.

Physical inactivity is associated with poor glucose tolerance, hyperinsulinemia, and insulin resistance, which dramatically raises the risk of T2D. The prevalence of type 2 diabetes was found to be two times higher among those who have a sedentary lifestyle than those who engage in strenuous physical activity [104–106]. Prospective studies have demonstrated a positive relationship between aerobic and power loads and the risk of T2D [107,108]. Ekelund et al. [109] calculated that the reduction in T2D risk caused by moderate and vigorous physical activity depends only on the length of training and does not depend on the amount of time spent in an inactive state. Rockette-Wagner [110] showed that long-term sedentary work can neutralize the positive effect of physical activity in terms of the T2D risk. Numerous tissues, including skeletal and cardiac muscle, the liver, and brain cells, have been found to be involved in the regulation of redox homeostasis by physical exercise [111]. A body of evidence shows that disruption in exercise-induced redox homeostasis serves as an upstream signal for transcription factor activation, followed by gene expression stimulation, increased antioxidant defense, and decreased age-related oxidative stress [112,113].

Tobacco smoking is also an environmental risk factor for T2D [114]. The number of cigarettes smoked per day was found to be dose-dependently correlated with the risk of T2D, which was found to be 45% higher in smokers than nonsmokers [115]. There is evidence that passive smoking also increases the incidence of T2D [116]. Smoking is an independent factor that worsens tissue insulin sensitivity. Tobacco use has been shown in studies to be toxic to the pancreas and to reduce insulin release by beta-cells [117].

Alcohol drinking patterns have been found to be associated with the risk of type 2 diabetes [118]. Okamura T. and colleagues discovered that heavy alcohol consumers (>280 g/week) with fatty liver disease (FLD) had a higher risk of developing type 2 diabetes than the other groups. Moderate alcohol consumers (140–280 g/week) without FLD had a significantly higher risk for type 2 diabetes, compared with minimal (<40 g/week) and light (40–140 g/week) alcohol consumers without FLD. In contrast, there was no apparent difference in the risk for incident type 2 diabetes between non-drinkers and minimal, light, or moderate alcohol consumers with FLD. Furthermore, no significant difference in the risk for T2D was found between moderate and heavy alcohol drinkers without FLD and non-drinkers or minimal, light, or moderate alcohol consumers with FLD [119].

The prevalence of type 2 diabetes, like many other multifactorial diseases, is linked to cumulative exposure to lipophilic and hydrophilic environmental pollutants such as POPs (persistent organic pollutants), exudates produced by common plastics, air pollutants, and some medicines [120,121]. Prospective studies investigating the dose-response relationship between environmental pollutants and T2D prevalence are needed to substantiate the causal role of chemical factors in disease susceptibility.

4. Genetic Factors of Type 2 Diabetes

Studies in multiethnic populations with varying T2D rates suggest that some ethnic groups may have a genetic predisposition to developing insulin resistance and diabetes when exposed to adverse conditions [122]. It is thought that between 35% and 70% of type 2 diabetes cases are genetically predisposed [123]. The genetic contribution to diabetes is demonstrated by familial aggregation [124] and a high concordance rate for T2D in monozygotic twins. In particular, the relative risk of T2D is 35–39% if one parent suffers from the disease; the risk is 60–70% if both parents have T2D; the risk of T2D in monozygotic twins is 58–65%; and the risk in heterozygotic twins is 16–30% [125,126].

Genome-wide association studies (GWAS), a hypothesis-free method created to test hundreds of thousands of genetic variants across many genomes to find those statistically associated with a specific trait or disease, have made significant progress in the last two decades in identifying the genes that cause diabetes susceptibility [127]. Numerous GWASs have identified multiple genetic loci that are linked to T2D and disease-related phenotypes [9,10,128]. According to the GWAS Catalog (https://www.ebi.ac.uk/gwas/, accessed on 25 February 2023), there have been 222 genome-wide studies on type 2 diabetes and 5348 SNP-disease associations have been detected. Furthermore, large meta-analyses of GWAS were performed [10,129] and identified 2284 SNPs associated with T2D, but only 183 and 38 loci were successfully replicated as T2D susceptibility markers once and twice, respectively [130]. Thus, a large portion of the originally discovered SNP-disease associations were poorly replicated by independent studies [131–133].

TCF7L2, HHEX, KCNJ11, CENTD2, ARAP1, FTO, HNF1B, PPARG, IGF2BP2, CDKAL1, VEGFA, SLC30A8, CDKN2A, CDKN2B, KCNQ1, UBE2E2, ANK1, ABCC8, IRS1, and *GLP2R* are the genes whose polymorphic variants showed strong associations with susceptibility to T2D and related phenotypes [134–136]. The variants most strongly associated with disease risk affect gene expression in the islets of Langerhans as well as insulin-sensitive tissues and organs such as adipose, muscle, and liver. Transcriptome studies [137,138] showed that some differentially expressed genes (DEGs) are associated with a decrease in insulin secretion (*ZMIZ1, MTNR1B, ADCY5, GIPR, C2CDC4A, CDKAL1, GCK, TCF7L2, GLIS3, THADA,* and *IGF2BP2*), while other DEGs are associated with insulin resistance (*PPARG, KLF14,* and *IRS1*) and a decrease in the incretin response (*TCF7L2, KCNQ1, GIPR, THADA, WFS1,* and MTNR1B). However, a larger portion of the identified genes are not associated with either production and/or action of insulin or glucose metabolism, and their role in disease pathogenesis remains unknown. These genes produce proteins that are involved in signaling pathways (*ACHE, ADCY5, ARAP1, ARL15, ATP8A1, BECN1, BRAF, CAMKK2, CMIP, DAAM1, DCDC2C, GATAD2A, GPSM1, GRK5, IGF2BP2, MAEA, MCC, PLEKHA1, RGS7, SHB, SUGP1, SYK,* and *TMEM132D*), antigens and receptors (*GABRA4, GIPR, GLP2R, IL17REL, IL23R, LRP12, MTNR1B, NOTCH2, NRXN3, PPARG, PTPRD, PVRL2, RASGRP1, SSR1, TGFBR3,* and *THADA*), transport proteins (*ATP8A1, EXOC6, KCNJ11, KCNQ1, SLC16A11, SLC16A13, SLC9B2, SLCO4C1, VPS26A, VPS33B, WFS1,* and *YKT6*), proteins regulating cell cycle, mitosis and apoptosis (*BCL2, CENPW, FAM58A, FAM60A, MPHOSPH9, NDUFAF6, RCCD1, RHOU,* and *ZZEF1*) and other proteins (*ANK1, FSCN3, INTS8, MACF1, SGCD, SGCG, LAMA1, LIMS2,* and *ST6GAL1*). SNP rs7903146 at the *TCF7L2* (transcription factor 7 like 2) gene is a genetic marker, the most strongly associated with T2D susceptibility, and it has been successfully replicated in twenty-two studies in various populations of the world [139–144]. Polymorphisms of *TCF7L2*, *IGF2BP2*, *CDKAL1*, *KCNQ1*, and *PPARG* genes showed a strong contribution to type 2 diabetes.

The rs7903146 polymorphism of *TCF7L2*, one of the most strongly T2D-associated loci, has been convincingly linked to diminished incretin responsiveness, elevated hepatic glucose production, and impaired insulin secretion [145,146]. It has been experimentally shown that turning off the *TCF7L2* gene in mouse beta-cells impairs glucose tolerance and reduces beta-cell mass [147]. Functional studies have confirmed the essential role of the gene in the control of glucose homeostasis through the regulation of beta-cell mass. TCF7L2 is a dual-acting transcription factor that participates in the Wnt signaling pathway, operating as an activator when CTNNB1 beta-catenin is present and as a repressor when it is not [148]. TCF7L2 also inhibits adipogenesis via NLK (Nemo-like kinase) in adipose tissue and inhibits gluconeogenesis in the liver [149].

The rs11927381 variant of the *IGF2BP2* gene is a marker that has been shown to be strongly associated with T2D susceptibility [150,151]. *IGF2BP2* encodes a protein that binds insulin-like growth factor 2 (IGF2) mRNA and was found to be associated with both beta-cell dysfunction and decreased sensitivity of peripheral tissues to its effects [152]. Associations of six polymorphisms of the *IGF2BP2* gene such as rs4402960, rs1470579, rs7640539, rs71320321, rs1374910, and rs6769511 with type 2 diabetes have been discovered and subsequently replicated by independent studies [134]. Dai and co-authors have revealed that an increase in *IGF2BP2* expression can cause a deficit of mitochondrial uncoupler protein 1 (UCP1) and hyperproduction of IGF2 because the IGF2BP2 protein facilitates the translation of insulin-like growth factor 2 (IGF2) mRNA and inhibits the synthesis of UCP1 [152]. Mice knocked out of the *IGF2BP2* gene showed that a decrease in UCP1 synthesis worsens tissue sensitivity to insulin and contributes to the development of alimentary obesity [152]. Additionally, a study [153] discovered an inverse correlation between the level of *IGF2BP2* gene expression in the blood and serum insulin concentration in T2D patients. Experiments with transgenic mice have shown that an increased level of IGF2 leads to insulin resistance, hyperglycemia, and diabetes in 30% of the animals [154]. An increase in IGF2 levels was found to cause disruption of pancreatic islets, whereas a decrease in IGF2 expression in human liver, muscles, and beta-cells reduced the risk of T2D [155].

CDKAL1 (cyclin dependent kinase 5 regulatory subunit associated protein 1 like 1) is another gene whose polymorphisms (rs4712524, rs6931514) are linked to beta-cell dysfunction [156]. This enzyme regulates cyclin-dependent kinase 5, an activator of insulin production in pancreatic beta-cells [157,158]. As part of the formation of ms2t6A37, the enzyme transfers the thiomethyl group from the S-adenosylmethionine molecule to the N6-threonylcarbamoyladenosine at position 37 of the cytosolic tRNALys (t6A37), which is required for the proper reading of lysine codons during translation of the proinsulin mRNA [159]. The accurate recognition of the AAA and AAG lysine codons during preproinsulin mRNA translation is made possible by CDKAL1-catalyzed tRNA modification [160]. The lysine residue is located at the cleavage site of proinsulin into the A-chain and C-peptide; therefore, translational errors can disrupt the formation and secretion of insulin in pancreatic beta-cells. Loss-of-function SNPs in the introns of the *CDKAL1* gene were found to be associated with defects in insulin secretion but not with obesity or insulin resistance [161,162].

The first gene described in candidate gene studies and subsequently replicated in GWAS was the *PPARG* (peroxisome proliferator activated receptor gamma) gene [163]. The gene encodes a nuclear transcription factor involved in the regulation of hundreds of genes for carbohydrate and lipid metabolism. PPARG controls the expression of lipoprotein lipase and CD36 fatty acid transporter [128]. Adipocytes with activated expression of PPARG secrete leptin and adiponectin in a balanced manner, which mediates the effects of insulin on peripheral organs. It has been shown that polymorphisms in the *PPARG* gene are involved in the development of both insulin resistance and impaired insulin secretion by pancreatic beta-cells [164]. In particular, the rs11709077 polymorphism of *PPARG* is associated with T2D, obesity, and cardiovascular diseases [165].

The *KCNQ1* gene encodes the potassium channel protein of beta-cells and is directly involved in the process of glucose-stimulated insulin secretion [166]. Polymorphism rs2237896 of the *KCNQ1* gene was found to be associated with dysfunction of beta-cells and decreased incretin response [167].

Thus, genome-wide association studies have discovered hundreds of SNPs linked to T2D. The discovered disease-associated alleles influence functional properties of pancreatic beta-cells and sensitivity of insulin-dependent tissues. However, the molecular processes by which many susceptibility genes are implicated in the pathogenesis of T2D are yet recognized and are the subject of ongoing research [168,169]. The polygenic nature of this complex disease, epistatic interactions between genes, effects of environmental factors, variable penetrance (between 10 and 40%), and high frequency of disease-related alleles with weak or moderate effects on disease phenotype make it difficult to understand the primary molecular mechanisms underlying T2D [170].

5. Glutathione Metabolism and Oxidative Enzymes as the Core of Redox Homeostasis

Cellular redox homeostasis is a crucial dynamic process that maintains the balance of reducing and oxidizing reactions within cells and governs a wide range of biological responses and events. Reactive oxygen species are required for life and play a role in practically all biological activities. Excess ROS are neutralized by antioxidant enzymes that catalyze the conversion of reactive oxygen species and their byproducts into stable harmless compounds. Redox homeostasis controls a wide range of biological responses and events by ensuring the balance between oxidation and reduction reactions within the cell [171]. The roles of reactive oxygen species in redox homeostasis and cellular signaling were reviewed in detail [172,173]. Increased levels of reactive oxygen species (ROS) damage macromolecules, impair redox signaling, and were found to be associated with beta-cell dysfunction and insulin resistance [174–176]. To reduce the harm of free radicals, cells use both enzymatic and non-enzymatic antioxidant mechanisms. Superoxide dismutase (SOD), catalase (CAT), glutathione peroxidase (GPX), and glutathione S-transferases (GST) comprise the enzymatic part of antioxidant defense, while vitamin A, ascorbic acid (vitamin C), and alpha-tocopherol (vitamin E) represent non-enzymatic antioxidants [177,178]. Supplementary Table S1 summarizes the biological functions and tissue-specific gene expression of antioxidant enzymes that fight against oxidative stress by multiple mechanisms, including the conversion of superoxide anion into hydrogen peroxide (SOD), breakdown of hydrogen peroxide (CAT, GPX), reduction of disulfide bonds into dithiol-containing ones (thioredoxin reductase), biosynthesis of glutathione (glutamate-cysteine ligase, glutathione synthase, glutathione reductase), and conjugation of xenobiotic substrates with glutathione (GST).

Glutathione-dependent enzymes constitute a crucial antioxidant system and control thiol-disulfide equilibrium. Thiol-dependent peroxidases require electrons to deactivate free radicals with the help of reduced glutathione (GSH). Glutathione is the master antioxidant, capable of protecting cellular components from reactive oxygen species, free radicals, peroxides, lipid peroxides, and heavy metals; it also participates in DNA repair, immunological response, and controls cell death [179]. Furthermore, glutathione determines the activity of several enzymes, acts as a mediator of signaling pathways, regulates the cellular life cycle and division, and promotes protein folding as well as acts as a cysteine depot and a source of vitamins C and E [174,180]. Glutathione is converted by free radical scavenging into the oxidized form known as the GSSG dimer, which is involved in glutathionylation, post-translational protein modification, and epigenetic control of gene expression [181–183].

The main metabolic pathway for glutathione biosynthesis is the gamma-glutamyl cycle, which begins with the ATP-dependent formation of gamma-glutamylcysteine from glutamate and cysteine (https://www.kegg.jp/pathway/map00480, accessed on 22 January 2022). The reaction is catalyzed by the enzyme glutamate cysteine ligase (GCL), consisting of modifying (GCLM) and catalytic (GCLC) subunits [184]. In the second step, glutathione synthetase (GSS) is used to add glycine to the cysteine residue of the γ-Glu-Cys dipeptide. GSH is then released from the cell. Since the concentration of glutathione in blood plasma is 1000 times lower than that in a cell's cytosol, the current gradient acts as a driving force for glutathione transport. Once GSH has left the cell, gamma-glutamyl transferase (GGT) replaces the glutathione-Cys-Gly moiety with an alternative amino acid (phenylalanine, leucine, or lysine). Following the hydrolysis of the dipeptide Cys-Gly by dipeptidase into cysteine and glycine, these amino acids are delivered into the cell via specialized membrane transporters [185]. Gamma-glutamyl-amino acid, a product of the GGT reaction, is also transported into the cell, where it is converted into an amino acid and oxoproline by gamma-glutamyl cyclotransferase (GGCT). With the use of a mole of ATP, oxoproline is transformed into glutamic acid by the enzyme oxoprolinase. Therefore, the cell contains all three of the amino acids required to replenish GSH. In order to assure the formation of GSH and to supply some positively charged (lysine) and non-polar (phenylalanine, leucine) amino acids into the cell, one cycle turn requires the consumption of three ATP molecules [186].

The ChaC-family of proteins [187], which catalyzes a process similar to that of the GGCT enzyme but uses glutathione itself as a substrate, is one of the new enzymes implicated in glutathione catabolism that have been discovered in recent years. This fact has led to a considerable revision of theories regarding GSH degradation and the roles of GGT and GGCT in glutathione metabolism. GGT catalyzes the hydrolysis of GSH into glutamate and cysteinylglycine (1): transpeptidation of amino acids (2):

$$GSH + H_2O \rightarrow Glu + Cys\text{-}Gly \tag{1}$$

$$GSH + AA \rightarrow \gamma\text{-}Glu\text{-}AA + Cys\text{-}Gly \tag{2}$$

Several experimental studies showed that there is no connection between GGT activity and the transport of amino acids into the cell. Glutathione is transported by membrane glutathione transporters, including the MRP protein (multidrug resistance protein) and the ABC (ATP binding cassette) family transporter [180]. As it turns out, the main function of gamma-glutamyl transferase is the reutilization of GSSG or GSH conjugates, which are released from the cell into plasma, where they undergo conversion into glutamate and Cys-Gly or dimer $(Cys\text{-}Gly)_2$ by GGT. The resulting peptides can be transported into the cell by appropriate proteins or can be hydrolyzed into cysteine and glycine, which are taken up by the cells using specific transporters [188]. Thus, glutathione enters the cell "in parts"—in the form of its individual amino acids or dipeptides, which are subsequently used for de novo synthesis of GSH.

Gamma-glutamyl cyclotransferase was first described by Meister as one of the most important enzymes of the γ-glutamyl cycle [185]. GGCT also regulates de novo glutathione synthesis through its activity against glutamylcysteine, which is also a substrate of glutathione synthase [189]. Glutamate cysteine ligase normally generates a product that is targeted toward the production of GSH, in contrast to GGCT, which has a lesser affinity for γ–glutamylcysteine [190]. Finally, under conditions of Cys deficiency, GCL can catalyze the condensation of Glu with a non-Cys amino acid, forming γ-glutamyl-AA, a potential GGCT substrate converted to the amino acid and 5-oxoproline.

The second source of GSH formation is the enzyme glutathione reductase, which catalyzes the reduction of the GSSG dimer into a functionally active monomer according to the equation:

$$GSSG + NADPH\, H^+ \rightarrow 2GSH + NADP^+$$

It is well known that viability of the cell depends on maintaining the optimal ratio between reduced to oxidized glutathione (GSH:GSSG) [191–193]. A decrease in the activity of any enzyme involved in the metabolism of glutathione may shift the balance between reduced and oxidized glutathione and encourage the buildup of free radicals in tissues. In addition, an increase in the activity of oxidative enzymes generating ROS may cause glutathione depletion.

The main source of ROS in the cell is the mitochondrial electron transport chain, along with NADPH oxidase and nitric oxide synthase [194,195]. ROS are also produced during the folding of proteins, a process of disulfide bond formation in the endoplasmic reticulum, as well as a result of the activity of cytochromes P450 in the metabolism of xenobiotics [196]. Phagocytic cells such as neutrophils and monocytes utilize free radicals to destroy internalized microorganisms, whereas in non-phagocytic cells, ROS act as signaling molecules regulating a variety of biological processes such as cell division, regeneration, cell differentiation, their apoptosis, and cytoskeleton organization [197]. The combination of increased free radical compound production and inefficient antioxidant defense enzyme function creates the foundation for the development of oxidative stress, a pathological condition thought to be responsible for the development of complications in type 2 diabetes [198–200].

The biological functions and tissue-specific gene expression of oxidant enzymes producing superoxide anion (NADPH oxidases), hydrogen peroxide (dual oxidases, monoamine oxidases, lysyl oxidase), and hypochlorous acid (myeloperoxidase) are also summarized in Supplemental Table S1.

As mentioned above, NADPH oxidase is the main oxidant enzyme responsible for the generation of superoxide anion radicals:

$$NADPH \cdot H^+ + 2O_2 = 2O_2 \cdot^- + NADP^+ + H^+$$

As a result of the dismutation reaction, O_2 turns into H_2O_2.

The NADPH oxidase consists of transmembrane catalytic core proteins such as CYBA and CYBB, four cytosolic subunits—NCF1, NCF2, and NCF4 as well as small GTP-ases RAC1/RAC2 [201,202]. Noteworthy is the fact that only CYBB, out of all the NADPH oxidase subunits, has binding domains for the cofactors NADPH, FAD, and two heme molecules. This means that CYBB is directly involved in the formation of superoxide anion.

Nitric oxide synthases (NOS1, NOS2, and NOS3) and myeloperoxidase (MPO), which have been implicated in the development of type 2 diabetes, are also important ROS-generating enzyme families [203–206]. The three isoforms of the nitric oxide synthase family members catalyze the generation of NO from L-arginine: neuronal (nNOS/NOS1), inducible (iNOS/NOS2), and endothelial (eNOS/NOS3). Impairment in NO generation is known to be associated with endothelial dysfunction, insulin resistance, and diabetes. Several studies have revealed that single nucleotide polymorphisms (SNP) such as rs2297518 of *NOS2* and rs1799983 of *NOS3* are linked to the onset of T2D and/or its microvascular complications [206–209].

In addition, decreased antioxidant defense is a common finding in type 2 diabetes, which may be partially attributed to the spontaneous glycosylation of key antioxidant enzymes such as catalase, superoxide dismutase, glutathione-S-transferases, glutathione peroxidase, glutathione reductase, and others, as well as depletion of the intracellular pool of NADPH·H+ [85,210]. It is observed that enhanced conversion of sorbitol to fructose under the action of NAD+-dependent sorbitol dehydrogenase reduces the ratio of NAD+/NADH·H+ in the cell, leading to inhibition of glycolytic enzyme glyceraldehyde-3-phosphate dehydrogenase (GAPDH), accumulation of glyceraldehyde-3-phosphate and its conversion to diacylglycerol, a known allosteric activator of protein kinase C [211]. GAPDH, along with advanced glycation end products (AGEs), serves as a potent inducer of NADPH oxidase [212].

Reactive oxygen and nitrogen species and carbonyl compounds are respectively generated under oxidative, nitrosative, and carbonyl stress conditions, which have been found to play a role in the development and progression of type 2 diabetes [85]. Dysregulation of redox homeostasis in T2D was demonstrated at both genetic and biochemical levels [204].

6. Endogenous Deficiency of Glutathione in Type 2 Diabetes

Numerous environmental factors have been identified as risk factors for type 2 diabetes, and many of these same factors are responsible for the depletion of the endogenous glutathione pool, highlighting the importance of glutathione and redox homeostasis in the development of type 2 diabetes. Glutathione is the most prevalent non-protein thiol in cells that functions as the primary reducing agent and provides antioxidant defense against oxidative cell damage [213]. The millimolar level of reduced glutathione is maintained within cells, illustrating its essential biological functions that extend beyond antioxidant defense [214]. Glutathione is required for the detoxification of xenobiotics and endogenous toxic substances, the maintenance of mitochondrial redox balance, direct antiviral defense, immune response, vitamin C and E regeneration, control of cell proliferation, apoptosis, and protein folding [213 215]. Given the wide range of important biological functions of reduced glutathione, it is predicted that a GSH deficit plays a substantial role in the onset of a number of disorders [216,217]. It is observed that individuals with type 2 diabetes exhibit impaired redox homeostasis and oxidative stress attributed to, on the one hand, a deficiency in the production of endogenous antioxidants and, on the other hand, an excess of free radicals [218]. Decreased antioxidant defense and oxidative stress are linked to key pathophysiological abnormalities underlying type 2 diabetes mellitus, such as pancreatic beta-cell dysfunction and insulin resistance [219–221]. Lutchmansingh and colleagues discovered that T2D patients have lower levels of the reduced form of glutathione and a slower rate of its synthesis in erythrocytes than healthy individuals [193]. Furthermore, several studies [193,222,223] have revealed that depletion of the endogenous glutathione pool correlated to increased blood glucose levels and contributed to disease complications. An experimental study by Stumvoll and colleagues observed that glutathione appears to have the ability to prevent beta-cell insufficiency and reduce glucose tolerance in rats fed a long-term high-glucose diet [223].

Reduced glutathione synthesis in T2D is thought to be caused by a lack of the amino acid precursors of GSH, cysteine, and glycine [224], whereas replenishing glutathione deficiency in diabetic patients leads to improvements in metabolic disorders and disease symptoms. In particular, intravenous glutathione infusions were found to improve tissue insulin resistance [225]. The induction of GSH biosynthesis by activation of the PI3K/Akt/p70S6K signaling pathway was found to significantly improve the therapeutic effect of insulin therapy in patients with T2D [226]. A study by Murakami and colleagues found a decreased activity of key enzymes responsible for glutathione biosynthesis, such as glutamate-cysteine ligase and glutathione reductase, along with decreased levels of reduced glutathione and increased levels of oxidized glutathione in the erythrocytes of type 2 diabetics [227]. In addition, some studies showed decreased rates in the membrane transport of glutathione disulfide in diabetics [228,229]. Inhibiting glutamate-cysteine ligase activity resulted in increased hydrogen peroxide accumulation and decreased insulin transcript levels in pancreatic beta-cells [230], suggesting an important role for glutathione in the regulation of the transcriptional activity of insulin.

The ability of cells in different tissues and organs to synthesize glutathione varies, and many tissues with a limited capacity to synthesize GSH require amino acid precursors from extracellular glutathione synthesized in the liver to enter the cell. The liver, skeletal muscles, and kidneys are examples of tissues and organs with significant metabolic activity and glutathione synthesis, whereas the pancreas, despite increased activity of biosynthetic processes, has a considerably lower potential to produce glutathione [231,232]. In the pancreas, expression of GGT1, GGT6, and ANPEP enzymes, which catabolize extracellular glutathione, is far more abundant than glutathione biosynthesis enzymes such as GCLC, GCLM, GSS, GSR, and GGCT (Supplementary Table S1), meaning that the pancreatic tissues need glutathione transported from the liver [233]. It is critically important that gamma-glutamyl transferase and aminopeptidase N (ANPEP), membrane-associated enzymes that catalyze the breakdown of extracellular GSH into its individual amino acids and promote its uptake into the cell, are responsible for controlling this process [234,235]. This means the pancreas is dependent on the activity of membrane-associated enzymes that catabolize GSH and provide amino acid precursors for the synthesis of glutathione in the cell. Notably, patients with T2D and prediabetes had significantly higher plasma levels of gamma-glutamyl transferase [236,237] and mRNA levels of *ANPEP* in pancreatic islets than non-diabetics [238,239]. We recently hypothesized that increased gamma-glutamyltransferase and aminopeptidase N levels reflect a cell's adaptive response to intracellular glutathione deficiency found in T2D, and that an increase in these enzyme levels is required to promote extracellular peptide degradation, providing amino acid precursors for de novo GSH biosynthesis [240]. We also suggested that glutathione deficiency may contribute to the disruption of protein folding in the endoplasmic reticulum [240], since glutathione is required for promoting the formation of disulfide bonds in the tertiary structure of proteins and correcting non-naive formed S-S-bonds [241,242]. Impaired proinsulin folding, observed in type 2 diabetes by several studies [242–246], may be attributed to reduced glutathione level in pancreatic beta-cells. The accumulation of unfolded or misfolded proinsulin molecules may trigger endoplasmic reticulum stress and the unfolded protein response—abnormalities culminating in beta-cell apoptosis in type 2 diabetes [245,247,248]. Deeper genetic and biochemical studies of the redox homeostasis system using cell lines and animal models will shed light on the mechanisms by which glutathione deficiency triggers the development and progression of type 2 diabetes mellitus.

7. Genes Encoding Antioxidant Defense Enzymes and the Risk of Type 2 Diabetes

Variation in genes encoding antioxidant and oxidative enzymes may affect their amount or function, shifting redox homeostasis towards oxidative stress and increasing the risk of type 2 diabetes mellitus. Several genetic association studies have shown that single nucleotide polymorphisms (SNP) in genes for redox state-regulating enzymes are linked with T2D susceptibility. Table 1 summarizes the results of genetic studies that identified associations between these genes and the risk of type 2 diabetes. In particular, the following polymorphisms at genes encoding antioxidant defense enzymes were found to be associated with the risk of type 2 diabetes: del/del of *GSTM1* [249–254], del/+ *GSTT1* [249–251], rs1695 [251,252,255] and rs1138272 [249] of *GSTP1*, rs12524494 of *GCLC*, rs3827715 and rs41303970 of *GCLM* [256], rs13041792 of *GSS* [240], rs2551715 of *GSR* [257], rs11546155 and rs6119534 of *GGT7* [240], rs4270 of *GGCT* [258], rs1050450 of *GPX1* [259], rs4902346 of *GPX2* [260], rs769217 of *CAT* [261], rs2234694 of *SOD1* [261], rs4880 of *SOD2* [212], and rs2536512 of *SOD3* [262]. However, much of the genetic research has focused on polymorphisms in the glutathione S-transferases M1, T1, and P1 (enzymes that used reduced glutathione for xenobiotic conjugation reactions); loss-of-function variants of these genes have been linked to an increased risk of type 2 diabetes. A limited number of studies looked into the link between T2D and genetic variations in hydrogen peroxide-metabolizing enzymes such as superoxide dismutases types 1, 2, and 3, glutathione peroxidases types 1 and 2, and catalase. The genes encoding enzymes directly involved in the metabolism of glutathione have received less attention in diabetes research. The great majority of the investigations were undertaken by our research team to explore for a link between polymorphisms at genes for glutathione metabolism and T2D susceptibility. Below, we describe the results of genetic studies that observed the associations between polymorphisms of genes for antioxidant enzymes and type 2 diabetes risk.

Glutamate cysteine ligase (GCL), also known as gamma-glutamylcysteine synthetase, is the first rate-limiting enzyme in glutathione biosynthesis from L-cysteine and L-glutamate. The enzyme is composed of two subunits: a heavy catalytic subunit (GCLC) and a light regulatory subunit (GCLM). *GCLC* gene is highly expressed in the liver whereas *GCLM* has the highest expression in pancreatic islets (Supplementary Table S1). Our recent study showed that polymorphisms such as rs12524494 in the *GCLC* gene and rs41303970 in the 5'-flanking region of the *GCLM* gene confer protection against T2D in nonsmokers [256]. The minor allele rs3827715-C *GCLM* also showed an association with a decreased risk of T2D. In addition, rs2301022 *GCLM* was associated with decreased levels of ROS, while SNPs rs7517826 and rs41303970 of the gene were associated with increased levels of total GSH in the plasma of T2D patients. These findings are supported by the eQTL analysis, which showed that the rs41303970-A allele was associated with a decreased level of

GCLM gene expression in various tissues, including the pancreas. It is known that the rs41303970-A allele *GCLM* suppresses oxidant-induced gene expression and is associated with lower plasma levels of reduced glutathione and a risk of myocardial infarction [263]. A study by Ma et al. showed that the rs41303970-A allele *GCLM* was associated with liver damage in patients with viral hepatitis B [264]. Chromatin immunoprecipitation and high-throughput sequencing data from the Roadmap Epigenomics Consortium project show that rs41303970 was located in the region of epigenetic regulation of *GCLM* gene expression in numerous tissues, including pancreatic islets. In particular, SNP rs41303970 is located at the transcription start site and was associated with an epigenetic modification in the region of the H3K4me3 promoter [265]. Moreover, rs41303970 is located in the region of the histone modifier H3K27ac [266]. In addition, rs41303970 falls into the region of the H3K9ac epigenetic modification (acetylation at the 9th lysine residue of the H3 histone protein) [267]. Thus, the observed protective effect of the rs41303970-A allele against the risk of T2D is apparently attributed to its positive effect on the *GCLM* gene expression, thereby increasing glutathione production in the cell.

Glutathione synthetase (GSS) is the enzyme responsible for the second step in glutathione biosynthesis. *GSS* is highly expressed in the liver and pancreatic islets (Supplementary Table S1). Although the *GSS* gene is polymorphic, no studies investigating the relationship between its polymorphisms and diseases have been conducted so far. The rs13041792 polymorphism of the *GSS* gene is associated with the level of protein C in blood plasma [268]. Our recent study showed that carriers of the rs13041792-G/A genotype possess an increased risk of type 2 diabetes [240]. The GTEx portal's data show that the rs13041792 polymorphism has no effect on expression of the *GSS* gene in the pancreas and other tissues. Instead, the rs13041792-A allele is associated with increased pancreatic expression of other genes such as *EDEM2*, *PROCR*, and *MYH7B*. We also observed that *GSS* gene polymorphisms correlated with redox homeostasis parameters, such as the levels of glutathione and ROS, as well as with fasting blood glucose [240]. The relationship between the variation at the *GSS* gene and susceptibility to T2D is also supported by the finding that the rs6088660-C and rs13041792-A alleles are associated with increased levels of fasting blood glucose [240]. Taken together, these findings may indicate that the *GSS* gene polymorphisms are functionally significant variants with the potential to affect glutathione metabolism. Functional effects of the SNPs might be attributed to linkage disequilibrium with variants of neighboring genes such as *EDEM2*, *MYH7B*, and *PROCR*. rs13041792 is located in the 5′-untranslated region 1.4 kb from the transcription start of the *GSS* gene and is in strong linkage disequilibrium (D′ ≥ 0.91) with SNPs in nearby or adjacent genes such as *MYH7B* (rs6120772, rs7268266, rs6120788, rs3746436, and rs3746435), *MIR499A* (rs3746444), and *TRPC4AP* (rs752075). A SNP rs3746444 of *MIR499A* is one of the polymorphisms strongly linked to the rs13041792 variant of the *GSS* gene.

Gamma-glutamyl transpeptidase type 1 (GGT1) is an enzyme localized on the surface of almost all types of epithelial cells. It plays a critical role in the regulation of levels of reactive oxygen species (ROS) by maintaining the balance between reduced glutathione and its oxidized form [269]. The *GGT1* gene is expressed at a high level in the pancreas; however, the highest expression of *GGT1* is detected in the liver (Supplementary Table S1). It is noteworthy that an increase in plasma level of gamma-glutamyltransferase is a common finding in T2D [270,271]. A few studies have been done to assess the relationships between the SNPs of *GGT1* gene and the risk of type 2 diabetes. Lee with colleagues observed a weak association between the rs4820599 SNP and the risk of type 2 diabetes [272]. Jinnouchi et al. did not find an association between SNP rs4820599 and disease risk, but it was revealed that this polymorphism was associated with the risk of diabetic retinopathy [273]. Studies of Diergaarde B. with colleagues [274] and Brand with colleagues [275] have identified an association of the rs4820599 polymorphism with the risk of pancreatic cancer and chronic pancreatitis, respectively. Two large studies have established the effect of rs4820599 polymorphism on the level of gamma-glutamyltransferase in blood [276,277].

The rs5751909-A allele is known to be associated with increased levels of *GGT1* mRNA in the pancreas and decreased levels in the blood (Supplementary Table S1).

Gamma-glutamyl transpeptidase type 5 (GGT5) is an enzyme which cleaves the gamma-glutamyl peptide bond of glutathione and glutathione S-conjugates such as leukotriene C4 [278]. It is known that GGT5 converts C4 leukotriene to D4 leukotriene, which increases smooth muscle tone and promotes plasma exudation in tissues [279]. The *GGT5* gene is expressed in almost all tissues and organs, but to the greatest extent in adipose tissue, kidneys, thyroid gland, peripheral nerves, arteries, and cardiac muscle (Supplementary Table S1). A few studies investigated the association of polymorphisms of the *GGT5* gene with human traits or diseases. Astle with colleagues found that the rs2275984-C allele was associated with an increase in granulocytes and monocytes in the white blood cells [280]. In whole blood, the rs2275984-C allele is related with lower expression of glutathione metabolism genes *GGT5*, *GGT1*, and *GSTT1*, as well as *UPB1*, *DDT*, *SUSD2*, and *SPECC1L* genes, and higher expression of the *ADORA2A* gene (Supplementary Table S2). All the above genes are located in the 22q11.23 chromosomal segment, and their co-expression seems to be under the control of common *cis*-regulatory elements-enhancers located in this region, as can be seen from the GeneHancer data available at https://www.genecards.org (accessed on 2 December 2022).

Gamma-glutamyl transpeptidase type 6 (GGT6) is a membrane-bound extracellular enzyme which cleaves gamma-glutamyl peptide bonds in glutathione and transfers it to gamma-glutamyl acceptors [184,185]. Like other gamma-glutamyltransferases, GGT6 is a key regulator of glutathione homeostasis by providing cells with amino acid substrates for GSH biosynthesis [281]. To date, no genetic association studies have been undertaken to investigate the association of *GGT6* gene polymorphisms and susceptibility to type 2 diabetes and other diseases.

Gamma-glutamyl transpeptidase 7 (GGT7) is the extracellular membrane-bound enzyme, also known as gamma-glutamyl transferase 7, that breaks gamma-glutamyl peptide bonds in the GSH molecule and transfers gamma-glutamyl compounds to acceptors [281]. GGT7, like other gamma-glutamyltransferases, is essential for maintaining glutathione homeostasis by supplying substrates for glutathione synthesis, particularly in tissues like the pancreas that produce glutathione at a low rate. However, the enzyme mitigates oxidative damage by degrading extracellular glutathione [282]. This pathway, known as the gamma-glutamyl cycle, allows cells to use the released amino acids for *de novo* GSH synthesis [184]. No genetic association studies have been done to assess the association of *GGT7* gene polymorphisms with the risk of any disease. We found in our recent study [240] that the rs6119534-T allele of the *GGT7* gene is strongly associated with a decreased risk of T2D, especially among individuals without T2D-related risk factors such as physical inactivity, smoking, stress, insufficient consumption of fresh fruits/vegetables and proteins, excessive consumption of carbohydrates, and low-fiber foods. The rs6119534-T allele is known to be associated with an increased level of the *GGT7* and *GSS* genes in skeletal muscles (Supplementary Table S2). Surprisingly, none of the SNPs are associated with the level of *GGT7* gene expression in the pancreas. Like *GSS*, the *GGT7* gene is expressed at a low level in the pancreas compared to other organs and tissues [283,284].

Table 1. Genetic association studies of enzymes involved in the regulation of redox homeostasis in type 2 diabetes mellitus.

Gene	Gene Name	Polymorphism/ SNP ID	Genotype/ Allele	Odds Ration [95% CI]	Cofactor	Population	Reference
GSTM1	glutathione S-transferase M1	Deletion	del/del	1.99 [1.30–3.05]	males	European (Russians)	[282]
			del/del	1.99 [1.46–2.71]	-	Chinese Turkish Japanese Indians Taiwanese Iranians Egyptians	[283]
			del/del	2.90 [1.76–4.78]	-	Indians	[284]
			del/del	2.042 [1.254–3.325]	-	Indians	[285]
			del/del	3.841 [2.280–6.469]	-	Turkish	[286]
			del/del	1.74 [1.13–2.69]	-	Iranians	[287]
GSTT1	glutathione S-transferase T1	Deletion	del/del	2.23 [1.22–4.09]	males	European (Russians)	[282]
			del/del	1.61 [1.19–2.17]	-	Chinese Turkish Japanese Indians Taiwanese Iranians Egyptians	[283]
			del/del	2.90 [1.76–4.78]	-	Indians	[284]
GSTP1	glutathione S-transferase P1	rs1138272	114A/V	2.85 [1.44–5.62]	males	European (Russians)	[282]
		rs1695	105I/V	1.99 [1.20–3.32]	-	Romanian	[288]
		rs1695	105I/V	2.56 [1.47–4.48]	-	Indians	[284]
		rs1695	105I/V	0.397 [0.225–0.701]	-	Indians	[285]
GCLC	glutamate cysteine ligase catalytic subunit	rs12524494	G	0.62 [0.41–0.93]	Nonsmokers	European (Russians)	[289]
GCLM	glutamate cysteine ligase modifier subunit	rs3827715	C	0.86 [0.75–0.99]	-	European (Russians)	[289]
		rs41303970	A	0.77 [0.63–0.93]	Nonsmokers	European (Russians)	[289]
GSS	glutathione synthetase	rs13041792	A	1.14 [1.01–1.29]	-	European (Russians)	[178]

Table 1. *Cont.*

Gene	Gene Name	Polymorphism/ SNP ID	Genotype/ Allele	Odds Ration [95% CI]	Cofactor	Population	Reference
GSR	glutathione reductase	rs2551715	T/T	0.33 [0.13–0.82]	BMI < 25 kg/m^2 Daily consumption of fresh fruits and vegetables	European (Russians)	[290]
GGT7	gamma-glutamyl transferase 7	rs11546155	A/A	0.42 [0.22–0.80]	-	European (Russians)	[178]
		rs6119534	T	0.85 [0.76–0.95]	-	European (Russians)	[178]
GGCT	gamma-glutamyl cyclotransferase	rs4270	T/C-C/C	0.71 [0.54–0.93]	Nonsmokers; Daily consumption of fresh fruits and vegetables	European (Russians)	[291]
GPX1	glutathione peroxidase 1	rs1050450	T/T	1.76 [1.011–3.066]	-	South Indian	[292]
GPX2	glutathione peroxidase 2	rs4902346	G/G	1.41 [1.02–1.96]	Males	European (Russians)	[293]
CAT	catalase	rs769217	T	2.94 [1.66–5.23]	-	Egyptians	[294]
SOD1	Superoxide dismutase 1	rs2234694	C	2.9 [1.84–4.6]	-	Egyptians	[294]
SOD2	superoxide dismutase 2	rs4880	C	2.434 [1.413–4.191]	-	North Indian	[145]
SOD3	superoxide dismutase 3	rs2536512	GA-AA	1.64 [1.16–2.33]	-	Chinese	[295]
RAC1	Rac family small GTPase 1	rs7784465	T/C	1.40 [1.20–1.65]	Dietary deficit of fresh fruits and vegetables; Excess of carbohydrates in food; High calorie diet; Psychological stress; Sedentary lifestyle	European (Russians)	[296]
CYBA	cytochrome b-245 alpha chain	rs4673	A/A	1.60 [1.04–2.46]	Females	European (Russians)	[297]
			T/T	1.74 [1.15–2.64]	-	Asians	[298]
			T	1.30 [1.04–1.61]	-	Non-Asians	
CYBB	cytochrome b-245 beta chain	rs5963327	T	1.7 [1.06–2.75]	Males	European (Russians)	[299]
		rs6610650	A	1.71 [CI 1.05–2.78]	Males	European (Russians)	[299]
		rs5963327	T/T	1.35 [1.05–1.73]	Females	European (Russians)	[299]
		rs6610650	A/A	1.34 [1.05–1.72]	Females	European (Russians)	[299]
NCF2	neutrophil cytosolic factor 2	rs17849502	G/T	1.42 [1.08–1.87]	BMI > 25 kg/m^2	European (Russians)	[300]
MPO	Myeloperoxidase	rs2107545	T/C	1.563 [1.166–2.096]	-	Han Chinese	[301]

Gamma-glutamyl cyclotransferase (GGCT) is an enzyme catalyzing the formation of 5-oxoproline from gamma-glutamyl dipeptides. GGCT also causes the release of cytochrome *c* from mitochondria with subsequent induction of apoptosis [285]. No genetic association studies investigated the contribution of *GGCT* gene polymorphisms to the development of diabetes. Yu. A. Bocharova found that SNP rs6462210 was associated with a decreased risk of ischemic stroke [286]. In our recent study [258], carriage of the rs4270 T/C-C/C genotypes was associated with a decreased risk of T2D, but this association occurred only in those individuals who were non-smokers and consumed a sufficient amount (approximately 400 g daily) of fresh vegetables and fruits.

Glutathione S-transferase pi 1 (GSTP1) is an enzyme conjugating a reduced glutathione with a large number of exogenous and endogenous hydrophobic electrophilic compounds [287,288] and is involved in the formation of glutathione conjugates of prostaglandins A2 and J2 [289]. The rs1695 (Ile105Val) polymorphism is the most studied variant of the *GSTP1* gene in a variety of human diseases such as type 2 diabetes [255], acute lymphoblastic leukemia [290], Hodgkin's lymphoma [291], breast cancer [292], lung cancer [293], acute pancreatitis [294], Alzheimer's disease [295], and bronchial asthma [296]. It is also known that another well-studied polymorphism rs1138272 (Ala114Val) of the *GSTP1* gene is associated with the risk of T2D [249], bronchial asthma [297], cancers [298], and Parkinson's disease [299].

Glutathione peroxidase 1 (GPX1) is a member of the glutathione peroxidase family, which catalyzes the reduction of organic hydroperoxides and hydrogen peroxide (H_2O_2) by glutathione, protecting cells from oxidative damage [300]. A decrease in GPX activity in T2D was first noted in rats with streptozotocin-induced diabetes [300,301]. The literature on associations between SNPs in glutathione peroxidase genes and T2D risk is limited and contradictory. In particular, Tanaka [302] showed that an increase in GPX expression protected beta-cells from the damaging effects of oxidative stress during hyperglycemia, while McClung [303] reported that insulin resistance correlated with increased GPX1 expression. The role of glutathione peroxidase isoforms other than GPX1 in the development of diabetes remains unknown. Huang et al. [304] showed that knockout and overexpression of *Gpx1* in mice may induce types 1 and 2 diabetes-like phenotypes. Another study [305] discovered that *GPX1* gene polymorphisms protected the kidneys from oxidative damage in type 1 diabetic patients. Liu D. et al. [306] discovered that the *GPX1* polymorphism rs1050450 was associated with an increased risk of carotid plaques in T2D patients. In ApoE/mice, a lack of functional Gpx1 accelerated diabetes-associated atherosclerosis by upregulating proinflammatory and profibrotic pathways [307].

Glutathione peroxidase 2 (GPX2) is a member of the glutathione peroxidase family, which encodes a selenium-dependent enzyme responsible for the majority of the glutathione-dependent hydrogen peroxide-reducing activity in the epithelial cells of the gastrointestinal tract. *GPX2* is characterized by the highest level of expression in the pancreas according to the data from the BioGPS project [308]. GPX2 is active in the cytosol and mitochondrial matrix where the enzyme protects all parts of the cell from the damaging effects of ROS. Substrates for GPX2 include not only hydrogen peroxide and lipid hydroperoxides, but also peroxynitrite. We observed an association between the rs4902346 polymorphism in the intron of the *GPX2* gene and an increased risk of T2D [260]. The minor allele rs4902346-G of *GPX2* is known to be associated with a decrease in the expression of the *GPX2* gene in the liver, the small intestine, subcutaneous and visceral adipose tissues, nervous tissue and skeletal muscle [309].

8. Genes for ROS-Generating Enzymes and Susceptibility to Type 2 Diabetes

Reactive oxygen and nitrogen (RNS) species, formerly believed to be by-products of cellular metabolism, are known to play important regulatory roles in a number of vital physiological processes, including cell survival, proliferation, differentiation, migration, and adhesion [310]. Superoxide anion and hydrogen peroxide produced by NADPH oxidase (NOX) play a significant role in numerous cellular signaling networks [311]. Currently, seven NOX isoenzymes are known, with tissue-specific expression patterns: NOX1, NOX2, NOX3, NOX4, NOX5, DUOX1, and DUOX2. The main subunits of NADPH oxidase are cytochrome b-245 light chain (CYBA), cytochrome b-245 heavy chain (CYBB), neutrophil cytosolic factor 1 (NCF1), neutrophil cytosolic factor 2 (NCF2), neutrophil cytosolic factor 4 (NCF4), and small GTPases (RAC1/RAC2). Cytosolic protomers such as NCF1, NCF2, and NCF4 provide activation of two transmembrane domains CYBA and CYBB, which form the catalytic core of the multienzyme complex of NADPH oxidase [311]. This complex catalyzes the one-electron reduction of molecular oxygen to superoxide, which is essential for oxidizing pathogens [312]. All ROS-producing NOX isoforms may transport electrons across membranes and produce superoxide anion and/or hydrogen peroxide, which are signaling molecules impacting on all facets of cell activity [313]. As can be seen from Table 1, several genetic studies showed that polymorphic genes encoding oxidative enzymes are associated with the risk of T2D. These polymorphisms are rs7784465 of *RAC1* [253], rs4673 of *CYBA* [314,315], rs5963327 and rs6610650 of *CYBB* [316], rs17849502 of *NCF2* [317], and rs2107545 of *MPO* [306]. Below, we describe the results of studies that observed associations between polymorphic genes for oxidative enzymes and T2D susceptibility.

Cytochrome b-245 alpha chain (CYBA) is a critical component of the membrane-bound phagocyte oxidase, which, by binding to CYBB, forms a functionally active NADPH oxidase. Several SNPs were identified in the *CYBA* gene; for instance, –242C>T (rs4673) in the fourth exon, resulting in amino acid substitution His72Tyr, 640A>G (rs1049255) in the 3-prime-untranslated region, and –930A>G (rs9932581) located in the promoter gene region at the binding site for transcription factor CEBP, affecting transcriptional activity of the *CYBA* gene [318]. There is experimental evidence that the 242C>T and 640A>G polymorphisms of the *CYBA* gene affect the binding affinity of the p22phox and gp91phox protein subunits, modulating the activity of NADPH oxidase and the generation of superoxide anions [319,320]. A study by Meijles and colleagues showed that the 242C>T polymorphism caused structural changes in p22phox that inhibited the activation of endothelial NOX2 and the oxidative response to tumor necrosis factor alpha or glucose stimulation [321]. It is known from the literature that the rs4673C allele (polymorphism C242T) is associated with endothelial dysfunction in patients with type 2 diabetes [311], the risk of coronary artery disease [322], T2D [315], peripheral neuropathy in type 1 diabetes mellitus [323], hypertension in patients with type 2 diabetes [324], lung cancer [325], bronchial asthma [326], and some other diseases. It has also been found that the rs4673-A/G genotype, compared to the wild-type G/G genotype, of the *CYBA* gene is associated with more pronounced liver damage in patients with viral hepatitis B [264]. Our recent study showed a sex-specific association of the *CYBA* rs4673 polymorphism with the risk of type 2 diabetes in women [314]. It has been suggested that the rs9932581 variant of *CYBA* modulates the transcriptional activity of the gene promoter through allele-specific binding to CEBP (the −930G allele increases the affinity for this transcription factor to interact with the gene promoter) [327]. In particular, functional studies showed that the promoter with the −930G allele variant had a 30% higher expression of the *CYBA* gene than the promoter with the −930A allele [318]. It was also found that the −930G allele was associated with increased ROS production [327,328]. SNP rs9932581 is associated with premature development of coronary atherosclerosis [329] and hypertension [318,328]. In addition, SNP rs9932581 is associated with an increased risk of renal complications in patients with type 1 diabetes [330].

Cytochrome b-245 beta chain (CYBB) is an essential component of membrane-bound phagocyte oxidase that produces the superoxide anion. CYBB is the terminal component of the respiratory chain that transfers electrons from the cytoplasmic portion of NADPH oxidase across the plasma membrane to molecular oxygen outside. In addition, CYBB functions as a voltage-gated proton channel that transmits H+ fluxes in inactive phagocytes and is involved in the regulation of cellular pH. CYBB expression is regulated by the nuclear transcription factor NF-κB [331]. The *CYBB* gene is located at chromosome X. According to the DisGeNET database, polymorphisms in the *CYBB* gene have been the subject for genetic association studies in hematological and infectious diseases. In particular, the rs6610650A allele of the *CYBB* gene is associated with a decreased risk of tuberculosis in male smokers [332]. Our study revealed sex-specific associations of rs6610650 and rs5963327 with fasting hyperglycemia and increased risk of T2D in males in the presence of risk factors such as cigarette smoking and diets with excess calories and sugar [316]. Furthermore, the above-mentioned SNPs were related with a predisposition to T2D in females, whereas rs5917471 of *CYBB* was associated with fasting hyperglycemia in males [316] but not with T2D.

Neutrophil cytosolic factor 2 (NCF2) is a neutrophil cytosolic factor (also known as p67phox), which is part of the NADPH oxidase enzyme complex and is responsible for its activation by binding to the CYBA and CYBB subunits [320]. Few studies have been done to investigate the association of *NCF2* gene polymorphisms with human diseases. It is known that rs10911363 and rs17849502 *NCF2* are associated with the risk of systemic seropositive rheumatic disease [333], rheumatoid arthritis [334], and arthritis in patients with systemic lupus erythematosus [335]. Our recent study showed that SNP rs17849502 of *NCF2* was associated with an increased risk of T2D in overweight and obese patients [317].

Neutrophil cytosolic factor 4 (NCF4) is also known as p40phox and together with other proteins (NCF4, CYBA, and CYBB) is involved in activation of NADPH oxidase [336]. SNP rs4821544 *NCF4* is associated with the risk of rheumatoid arthritis [337] and Crohn's disease [338], whereas rs5995355 is associated with the risk of colorectal cancer [339]. Xing and colleagues [340] found an increase in NCF4 expression in the neutrophils of patients with latent adult autoimmune diabetes (LADA), which contributed to oxidative damage to pancreatic beta-cells. Our recent study showed no significant associations of *NCF4* polymorphisms such as rs5995355, rs5995357, rs1883112, rs4821544, rs760519, rs729749, rs2075938, and rs2075939 with the risk of T2D [341]. Nevertheless, we revealed associations of the rs4821544-C/C and rs5995357-A/A genotypes with increased risk of coronary artery disease in females with T2D [341]. In addition, carriers of the rs4821544-C/C genotype had significantly higher levels of glycated hemoglobin and oxidized glutathione.

NADPH oxidase 1 activator (NOXA1) is a functional homologue of p67phox for the activation of NADPH oxidase in vascular smooth muscle cells (SMCs) and plays an important role in atherogenesis [291,342]. NOXA1 is capable of activating CYBB/gp91phox and NOX3 [343]. There is no data on the association of *NOXA1* gene variants with T2D risk.

NADPH oxidase 1 (NOX1) is a homologue of the catalytic subunit of the superoxide-generating phagocyte NADPH oxidase, gp91phox. The oxidase activity of the enzyme is regulated by NOXA1 and NOXO1 [344]. As can be seen from the DisGeNET database, association studies of *NOX1* gene polymorphisms were carried out on cardiovascular diseases [345].

NADPH oxidase 4 (NOX4) was originally identified as a homologue of NADPH oxidase highly expressed in the kidney [346]. In addition to the kidneys, NOX4 is expressed in smooth muscle cells and vascular endothelium, has protective properties through the modulation of endothelial nitric oxide synthase, and is also characterized by anti-inflammatory and anti-apoptotic effects [347]. It has been established that polymorphisms rs3913535, rs10765219, and rs11018670 of the *NOX4* gene are associated with the risk of severe diabetic retinopathy [348]. The rs2164521 variant of *NOX4* is associated with a decreased risk of hepatopulmonary syndrome [349].

NADPH oxidase 5 (NOX5) from the NOX family is unique in that it does not require NADPH oxidase subunits for its activation [350]. NOX5 is localized in the regions of the perinuclear and endoplasmic reticulum of cells and, after activation, is directed to the cell membrane. Vasoactive substances, growth factors, and pro-inflammatory cytokines all activate NOX5 and modulate its activity by a variety of post-translational changes [350]. NOX5 hyperactivation is linked to the onset of cardiovascular disease, kidney damage, and cancer. However, the exact pathophysiological functions of NOX5 are still unknown [350]. NOX5 was found to be activated in human diabetic nephropathy and affects the function of the filtration barrier and blood pressure through excessive formation of ROS [351]. Expression of human NOX5 in vascular smooth muscle cells and mesangial cells in mice has been shown to induce oxidative stress, glomerulosclerosis, mesangial expansion, inflammation, and fibrosis in the kidneys: processes that are known to accelerate the progression of renal failure in diabetes [352]. It is known from the literature that the *NOX5* gene is also associated with Hirschsprung's disease [353].

Rac family small GTPase 1 (RAC1) belongs to the Rho family of GTPases that regulates the redox state of the cell by controlling enzymes generating and converting reactive oxygen and nitrogen species [310]. RAC1 is an integral part of the NOX2 holoenzyme. Activation of RAC1 is required for the subsequent generation of ROS. Indeed, RAC1 is directly involved in the assembly and activation of NADPH oxidases (NOX1/2/3), which generate superoxide anion radicals, which in turn activate nuclear factor-κB (NFκB), a transcription factor regulating the expression of redox homeostasis genes such as NOX2 and superoxide dismutase 2 [310]. While NOX2 increases ROS generation, SOD2 converts highly reactive oxygen radicals into less reactive hydrogen peroxide. When redox homeostasis is shifted to the reducing state, ROS activate RAC1 and RhoA, whereas under oxidative stress, these GTPases become inactive [310]. The depletion of intracellular glutathione leads to an increase in active forms of RAC1, replenishing the GSH deficit in the cell by the antioxidant N-acetylcysteine blocks RAC1 activation [354]. Notably, acylation of RAC1 at the amino acid residue Cys178 is a critical process for RAC1-mediated remodeling of the actin cytoskeleton [355], which is important in the regulation of transmembrane glucose transport [356] and insulin secretion in pancreatic beta-cells in T2D [357]. The suppression of RAC1 protects β-cells from the damaging effects of glucose, fatty acids, and pro-inflammatory cytokines [358–360]. Interestingly, one of the main stages in the development of metabolic dysregulation of islet beta-cells in type 2 diabetes is thought to be sustained activation of RAC1 [361]. It is important to note that chronic activation of RAC1 has been found in many diseases, including cancer, neurodegenerative diseases (Parkinson's and Alzheimer's diseases), and cardiometabolic disorders, including type 2 diabetes [362,363].

In our recent study, we discovered that the RAC1 gene polymorphism rs7784465 was associated with the risk of T2D in individuals with a lower daily intake of fresh fruits and vegetables, as well as those who experienced psycho-emotional stress and physical inactivity [314]. rs10951982, another *RAC1* gene polymorphism, was found to be protective against T2D risk in subjects who did not abuse food with excessive amount of carbohydrates. It was also revealed that another SNP, rs836478 of *RAC1*, was associated with elevated levels of glycated hemoglobin and fasting blood glucose. In addition, the *RAC1* gene polymorphisms were associated with increased plasma levels of hydrogen peroxide and uric acid in T2D patients [364]. Few studies have been undertaken to assess the association between *RAC1* gene polymorphisms and other human diseases. A meta-analysis of genome-wide association studies showed that SNP rs7784465 was associated with body mass index [365].

Rac family small GTPase 2 (RAC2), like RAC1, belongs to the family of Rho GTPases that regulates cellular redox homeostasis. RAC2 is involved in the activation of NADPH oxidase (NOX2), increasing ROS production [366]. The active form of RAC2 binds to various effector proteins that regulate a variety of cellular processes, such as secretion, phagocytosis of apoptotic cells, and polarization of epithelial cells [367]. RAC2 interacts with inducible NO synthase, producing nitric oxide [368]. According to the DisGeNET database, a limited number of genetic association studies have been done to explore the association between *RAC2* gene polymorphisms and the risk of oncological, inflammatory, and hematological diseases.

In summarizing the findings of genetic studies, it should be noted that a significant proportion of established associations between redox homeostasis gene polymorphisms and type 2 diabetes have weak or moderate effects on disease risk, and thus these findings require confirmation in independent populations as well as experimental functional annotation of disease-associated alleles.

9. Conclusions

Numerous studies have demonstrated that individuals with type 2 diabetes exhibit impaired redox homeostasis and oxidative stress attributed to, on the one hand, a deficiency in the production of endogenous antioxidants, mainly glutathione, and, on the other hand, an excess of free radicals. In type 2 diabetes, increased levels of reactive oxygen species and oxidative stress were found to be associated with pathological conditions, such as beta-cell dysfunction and insulin resistance. Patients with T2D have lower levels of reduced glutathione and a slower rate of its biosynthesis than healthy individuals, and the decreased level of glutathione correlated with increased blood glucose levels and disease progression. Glutathione has also been demonstrated to inhibit beta-cell dedifferentiation and failure caused by chronic oscillating glucose consumption [34].

Polymorphisms at genes encoding antioxidant defense and oxidative enzymes have the potential to impact redox homeostasis and therefore they represent attractive markers for testing the genetic susceptibility to type 2 diabetes. Several studies have been done to evaluate whether polymorphisms in genes for glutathione S-transferases, superoxide dismutases, glutathione peroxidases, cytochrome b-245 alpha chain, and myeloperoxidase are associated with T2D susceptibility. These studies showed that variants at the oxidative stress-related genes were significantly associated with diabetes risk. Meanwhile, a limited number of studies, a significant proportion of which were carried out by our research team, looked into the link between T2D and genetic variations in glutathione-metabolizing enzymes such as glutamate cysteine ligase, glutathione synthetase, glutathione reductase, gamma-glutamyl cyclotransferase, and gamma-glutamyltransferases. Loss of function variants at genes encoding glutathione metabolizing enzymes, together with environmental factors, contribute to an intracellular glutathione deficiency in type 2 diabetes. The mechanism by which genes encoding enzymes involved in the regulation of redox homeostasis contribute to the development of type 2 diabetes mellitus is summarized in Figure 1.

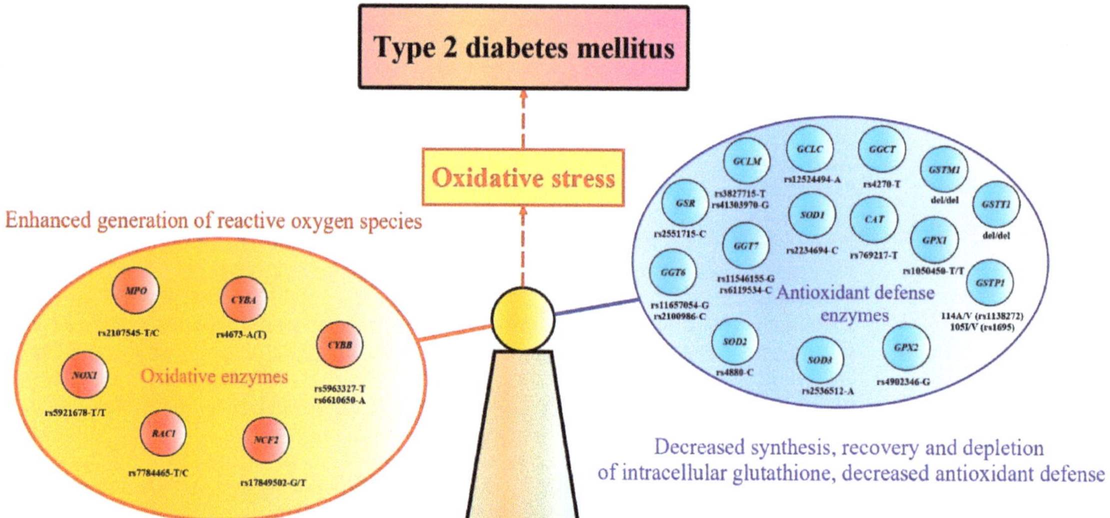

Figure 1. The relationship between genes encoding enzymes involved in the regulation of redox homeostasis and the development of type 2 diabetes. The scheme reflects the genetic causes responsible for oxidative stress in type 2 diabetes as a result of an imbalance between the generation of free radicals and their neutralization, which is attributed to the impact of polymorphisms in the genes encoding oxidative and antioxidant defense enzymes. Genotypes and alleles of the increased disease risk are shown along the genes. GSTM1, glutathione S-transferase M1; GSTT1, glutathione

S-transferase T1; GSTP1, glutathione S-transferase P1; GCLC, glutamate cysteine ligase catalytic subunit; GCLM, glutamate cysteine ligase modifier subunit; GSS, glutathione synthetase; GSR, glutathione reductase; GGT6, gamma-glutamyl transferase 6; GGT7, gamma-glutamyl transferase 7; GGCT, gamma-glutamyl cyclotransferase; GPX1, glutathione peroxidase 1; GPX2, glutathione peroxidase 2; CAT, catalase; SOD1, Superoxide dismutase; SOD2, superoxide dismutase 2; SOD3, superoxide dismutase 3; RAC1, Rac family small GTPase 1; CYBA, cytochrome b-245 alpha chain; CYBB, cytochrome b-245 beta chain; NCF2, neutrophil cytosolic factor 2; NOX1, NADPH oxidase1; MPO, Myeloperoxidase.

The most significant contribution to T2D predisposition is attributed to polymorphisms of genes such as *GCLM, GSS, GGT1, GGT7, GSTM1, GSTT1, GSTP1, GPX2, RAC1, CYBA, CYBB,* and *NCF2.* The effects of each gene polymorphism on redox homeostasis and disease risk may vary significantly across populations, and this variation is attributed to interpopulation differences in minor allele frequencies and linkage disequilibrium between SNPs, as recently demonstrated in a number of studies [240,369,370]. The link between polymorphic genes involved in redox homeostasis, such as *GCLM* (rs2301022), *GGT7* (rs6119534, rs11546155), *GSTM1* (+/del), *GSTP1* (rs1695, rs1138272), *RAC1* (rs7784465), and *CYBB* (rs591741), and type 2 diabetes differs in overweight and normal weight individuals. The impact of genetic variants of these enzymes on disease risk is triggered by environmental factors such as insufficient intake of proteins, fresh vegetables and fruits, as well as smoking and physical inactivity. Further research is needed to uncover the molecular mechanisms by which interactions between genetic and environmental factors contribute to type 2 diabetes pathogenesis via alterations in redox homeostasis. In particular, the pathophysiological relationship between glutathione deficiency and T2D development could be explained not only by the development of oxidative stress but also by other mechanisms. Since T2D-associated alleles of genes for glutathione metabolism (e.g., GSS and GGT7) correlate with expression levels of genes involved in the unfolded protein response pathway and regulation of proteostasis [240], it can be hypothesized that glutathione deficiency in beta-cells of pancreatic islets may contribute to the impaired folding of proinsulin identified in type 2 diabetes.

Understanding the importance of redox homeostasis in the pathogenesis of type 2 diabetes substantiates the need for antioxidant treatment focusing on replenishing the endogenous glutathione pool, and several clinical studies have already shown the efficacy and safety of this approach in diabetics [371–373]. However, this type of therapy for patients with type 2 diabetes has not been implemented into endocrinological practice or included in any country's national clinical recommendations.

Supplementary Materials: The following supporting information can be downloaded at: https://www.mdpi.com/article/10.3390/ijms24054738/s1.

Author Contributions: Conceptualization, A.P.; writing—original draft preparation, I.A. and A.P.; writing—review and editing, I.A., A.P. and E.K.; visualization, A.P.; supervision, A.P.; funding acquisition, I.A., A.P. and E.K. All authors have read and agreed to the published version of the manuscript.

Funding: This research was supported by the Russian Science Foundation, project number 20-15-00227.

Acknowledgments: Illustration in the text was created with yEd Graph Editor, version 3.22.

References

1. Reed, J.; Bain, S.; Kanamarlapudi, V. A Review of Current Trends with Type 2 Diabetes Epidemiology, Aetiology, Pathogenesis, Treatments and Future Perspectives. *Diabetes Metab. Syndr. Obes. Targets Ther.* **2021**, *14*, 3567–3602. [CrossRef]

2. Sun, H.; Saeedi, P.; Karuranga, S.; Pinkepank, M.; Ogurtsova, K.; Duncan, B.B.; Stein, C.; Basit, A.; Chan, J.C.; Mbanya, J.C.; et al. IDF Diabetes Atlas: Global, regional and country-level diabetes prevalence estimates for 2021 and projections for 2045. *Diabetes Res. Clin. Pract.* **2021**, *183*, 109119. [CrossRef]

3. Ogurtsova, K.; Guariguata, L.; Barengo, N.C.; Ruiz, P.L.-D.; Sacre, J.W.; Karuranga, S.; Sun, H.; Boyko, E.J.; Magliano, D.J. IDF diabetes Atlas: Global estimates of undiagnosed diabetes in adults for 2021. *Diabetes Res. Clin. Pract.* **2021**, *183*, 109118. [CrossRef]

4. *Global Report on Diabetes*; World Health Organization: Geneva, Switzerland, 2016; 86p.

5. *IDF Diabetes Atlas [Internet]*, 10th ed.; International Diabetes Federation: Brussels, Belgium, 2021.

6. Davies, M.J.; Aroda, V.R.; Collins, B.S.; Gabbay, R.A.; Green, J.; Maruthur, N.M.; Rosas, S.E.; Del Prato, S.; Mathieu, C.; Mingrone, G.; et al. Management of hyperglycaemia in type 2 diabetes, 2022. A consensus report by the American Diabetes Association (ADA) and the European Association for the Study of Diabetes (EASD). *Diabetologia* **2022**, *65*, 1925–1966. [CrossRef]

7. Kadayifci, F.Z.; Haggard, S.; Jeon, S.; Ranard, K.; Tao, D.; Pan, Y.X. Early-life Programming of Type 2 Diabetes Mellitus: Understanding the Association between Epigenetics/Genetics and Environmental Factors. *Curr. Genom.* **2019**, *20*, 453–463. [CrossRef]

8. Beulens, J.W.J.; Pinho, M.G.M.; Abreu, T.C.; Braver, N.R.D.; Lam, T.M.; Huss, A.; Vlaanderen, J.; Sonnenschein, T.; Siddiqui, N.Z.; Yuan, Z.; et al. Environmental risk factors of type 2 diabetes—An exposome approach. *Diabetologia* **2021**, *65*, 263–274. [CrossRef]

9. Laakso, M.; Silva, L.F. Genetics of Type 2 Diabetes: Past, Present, and Future. *Nutrients* **2022**, *14*, 3201. [CrossRef]

10. Fuchsberger, C.; Flannick, J.; Teslovich, T.M.; Mahajan, A.; Agarwala, V.; Gaulton, K.J.; Ma, C.; Fontanillas, P.; Moutsianas, L.; McCarthy, D.J.; et al. The genetic architecture of type 2 diabetes. *Nature* **2016**, *536*, 41–47. [CrossRef]

11. Scott, R.A.; Scott, L.J.; Mägi, R.; Marullo, L.; Gaulton, K.J.; Kaakinen, M.; Pervjakova, N.; Pers, T.H.; Johnson, A.D.; Eicher, J.D.; et al. An Expanded Genome-Wide Association Study of Type 2 Diabetes in Europeans. *Diabetes* **2017**, *66*, 2888–2902. [CrossRef]

12. Yahaya, T.O. A Review of Type 2 Diabetes Mellitus Predisposing Genes. *Curr. Diabetes Rev.* **2019**, *16*, 52–61. [CrossRef]

13. Chen, J.; Spracklen, C.N.; Marenne, G.; Varshney, A.; Corbin, L.J.; Luan, J.; Willems, S.M.; Wu, Y.; Zhang, X.; Horikoshi, M.; et al. The trans-ancestral genomic architecture of glycemic traits. *Nat. Genet.* **2021**, *53*, 840–860. [CrossRef]

14. DeForest, N.; Majithia, A.R. Genetics of Type 2 Diabetes: Implications from Large-Scale Studies. *Curr. Diabetes Rep.* **2022**, *22*, 227–235. [CrossRef]

15. Maglott, D.; Ostell, J.; Pruitt, K.; Tatusova, T. Entrez Gene: Gene-centered information at NCBI. *Nucleic Acids Res.* **2004**, *33*, D54–D58. [CrossRef]

16. UniProt Consortium. UniProt: A worldwide hub of protein knowledge. *Nucleic Acids Res.* **2019**, *47*, D506–D515. [CrossRef]

17. Placzek, S.; Schomburg, I.; Chang, A.; Jeske, L.; Ulbrich, M.; Tillack, J.; Schomburg, D. BRENDA in 2017: New perspectives and new tools in BRENDA. *Nucleic Acids Res.* **2016**, *45*, D380–D388. [CrossRef]

18. Roux-Rouquie, M.; Chauvet, M.-L.; Munnich, A.; Frezal, J. Human Genes Involved in Chromatin Remodeling in Transcription Initiation, and Associated Diseases: An Overview Using the GENATLAS Database. *Mol. Genet. Metab.* **1999**, *67*, 261–277. [CrossRef]

19. Braschi, B.; Denny, P.; Gray, K. Genenames.org: The HGNC and VGNC resources in 2019. *Nucleic Acids Res.* **2019**, *47*, D786–D792. [CrossRef]

20. Barshir, R.; Fishilevich, S.; Iny-Stein, T.; Zelig, O.; Mazor, Y.; Guan-Golan, Y.; Safran, M.; Lancet, D. GeneCaRNA: A Comprehensive Gene-centric Database of Human Non-coding RNAs in the GeneCards Suite. *J. Mol. Biol.* **2021**, *433*, 166913. [CrossRef]

21. Wu, C.; Jin, X.; Tsueng, G. BioGPS: Building your own mash-up of gene annotations and expression profiles. *Nucleic Acids Res.* **2016**, *44*, D313–D316. [CrossRef]

22. Carithers, L.J.; Carithers, L.J.; Moore, H.M. The genotype-tissue expression (GTEx) project. *Biopreservat. Biobank.* **2015**, *13*, 307–308. [CrossRef]

23. Võsa, U.; Claringbould, A.; Westra, H.J. Unraveling the polygenic architecture of complex traits using blood eQTL metaanalysis. *BioRxiv* **2018**, *2*, 447367. [CrossRef]

24. MacArthur, J.; Bowler, E.; Cerezo, M. The new NHGRI-EBI Catalog of published genome-wide association studies (GWAS Catalog). *Nucleic Acids Res.* **2017**, *45*, D896–D901. [CrossRef]

25. Piñero, J.; Saüch, J.; Sanz, F.; Furlong, L.I. The DisGeNET cytoscape app: Exploring and visualizing disease genomics data. *Comput. Struct. Biotechnol. J.* **2021**, *19*, 2960–2967. [CrossRef]

26. Zheng, Y.; Ley, S.H.; Hu, F.B. Global aetiology and epidemiology of type 2 diabetes mellitus and its complications. *Nat. Rev. Endocrinol.* **2018**, *14*, 88–98. [CrossRef]

27. Kahn, S. The importance of the β-cell in the pathogenesis of type 2 diabetes mellitus. *Am. J. Med.* **2000**, *108*, 2–8. [CrossRef]

28. Batista, T.M.; Haider, N.; Kahn, C.R. Defining the underlying defect in insulin action in type 2 diabetes. *Diabetologia* **2021**, *64*, 994–1006. [CrossRef]

29. Kitamura, Y.I.; Kitamura, T.; Kruse, J.-P.; Raum, J.C.; Stein, R.; Gu, W.; Accili, D. FoxO1 protects against pancreatic β cell failure through NeuroD and MafA induction. *Cell Metab.* **2005**, *2*, 153–163. [CrossRef]

30. Kaneto, H.; Kimura, T.; Shimoda, M.; Obata, A.; Sanada, J.; Fushimi, Y.; Matsuoka, T.-A.; Kaku, K. Molecular Mechanism of Pancreatic β-Cell Failure in Type 2 Diabetes Mellitus. *Biomedicines* **2022**, *10*, 818. [CrossRef]

31. Rahier, J.; Guiot, Y.; Goebbels, R.M.; Sempoux, C.; Henquin, J.-C. Pancreatic β-cell mass in European subjects with type 2 diabetes. *Diabetes, Obes. Metab.* **2008**, *10*, 32–42. [CrossRef]

32. Marselli, L.; Suleiman, M.; Masini, M.; Campani, D.; Bugliani, M.; Syed, F.; Martino, L.; Focosi, D.; Scatena, F.; Olimpico, F.; et al. Are we overestimating the loss of beta cells in type 2 diabetes? *Diabetologia* **2013**, *57*, 362–365. [CrossRef]

33. Brereton, M.F.; Iberl, M.; Shimomura, K.; Zhang, Q.; Adriaenssens, A.E.; Proks, P.; Spiliotis, I.I.; Dace, W.; Mattis, K.K.; Ramracheya, R.; et al. Reversible changes in pancreatic islet structure and function produced by elevated blood glucose. *Nat. Commun.* **2014**, *5*, 4639. [CrossRef]

34. Khin, P.-P.; Lee, J.-H.; Jun, H.-S. A Brief Review of the Mechanisms of β-Cell Dedifferentiation in Type 2 Diabetes. *Nutrients* **2021**, *13*, 1593. [CrossRef]

35. Spijker, H.S.; Song, H.; Ellenbroek, J.H.; Roefs, M.M.; Engelse, M.A.; Bos, E.; Koster, A.J.; Rabelink, T.J.; Hansen, B.C.; Clark, A.; et al. Loss of β-Cell Identity Occurs in Type 2 Diabetes and Is Associated With Islet Amyloid Deposits. *Diabetes* **2015**, *64*, 2928–2938. [CrossRef]

36. Talchai, C.; Xuan, S.; Lin, H.V.; Sussel, L.; Accili, D. Pancreatic β Cell Dedifferentiation as a Mechanism of Diabetic β Cell Failure. *Cell* **2012**, *150*, 1223–1234. [CrossRef]

37. Marroqui, L.; Masini, M.; Merino, B.; Grieco, F.A.; Millard, I.; Dubois, C.; Quesada, I.; Marchetti, P.; Cnop, M.; Eizirik, D.L. Pancreatic α Cells are Resistant to Metabolic Stress-induced Apoptosis in Type 2 Diabetes. *Ebiomedicine* **2015**, *2*, 378–385. [CrossRef]

38. Lim, E.L.; Hollingsworth, K.G.; Aribisala, B.S.; Chen, M.J.; Mathers, J.C.; Taylor, R. Reversal of type 2 diabetes: Normalisation of beta cell function in association with decreased pancreas and liver triacylglycerol. *Diabetologia* **2011**, *54*, 2506–2514. [CrossRef]

39. Sun, K.; Park, J.; Gupta, O.T.; Holland, W.L.; Auerbach, P.; Zhang, N.; Marangoni, R.G.; Nicoloro, S.M.; Czech, M.P.; Varga, J.; et al. Endotrophin triggers adipose tissue fibrosis and metabolic dysfunction. *Nat. Commun.* **2014**, *5*, 3485. [CrossRef]

40. Williams, A.S.; Kang, L.; Wasserman, D.H. The extracellular matrix and insulin resistance. *Trends Endocrinol. Metab.* **2015**, *26*, 357–366. [CrossRef]

41. Ahmad, K.; Lee, E.J.; Moon, J.S.; Park, S.-Y.; Choi, I. Multifaceted Interweaving Between Extracellular Matrix, Insulin Resistance, and Skeletal Muscle. *Cells* **2018**, *7*, 148. [CrossRef]

42. Kang, L.; Ayala, J.E.; Lee-Young, R.S.; Zhang, Z.; James, F.D.; Neufer, P.D.; Pozzi, A.; Zutter, M.M.; Wasserman, D.H. Diet-Induced Muscle Insulin Resistance Is Associated With Extracellular Matrix Remodeling and Interaction With Integrin α2β1 in Mice. *Diabetes* **2011**, *60*, 416–426. [CrossRef]

43. Abdennour, M.; Reggio, S.; Le Naour, G.; Liu, Y.; Poitou, C.; Aron-Wisnewsky, J.; Charlotte, F.; Bouillot, J.-L.; Torcivia, A.; Sasso, M.; et al. Association of Adipose Tissue and Liver Fibrosis With Tissue Stiffness in Morbid Obesity: Links With Diabetes and BMI Loss After Gastric Bypass. *J. Clin. Endocrinol. Metab.* **2014**, *99*, 898–907. [CrossRef] [PubMed]

44. Kusminski, C.M.; Bickel, P.E.; Scherer, P.E. Targeting adipose tissue in the treatment of obesity-associated diabetes. *Nat. Rev. Drug Discov.* **2016**, *15*, 639–660. [CrossRef] [PubMed]

45. Gealekman, O.; Gurav, K.; Chouinard, M.; Straubhaar, J.; Thompson, M.; Malkani, S.; Hartigan, C.; Corvera, S. Control of Adipose Tissue Expandability in Response to High Fat Diet by the Insulin-like Growth Factor-binding Protein-4. *J. Biol. Chem.* **2014**, *289*, 18327–18338. [CrossRef] [PubMed]

46. Czech, M.P. Insulin action and resistance in obesity and type 2 diabetes. *Nat. Med.* **2017**, *23*, 804–814. [CrossRef] [PubMed]

47. Tang, Y.; Wallace, M.; Sanchez-Gurmaches, J.; Hsiao, W.-Y.; Li, H.; Lee, P.L.; Vernia, S.; Metallo, C.M.; Guertin, D.A. Adipose tissue mTORC2 regulates ChREBP-driven de novo lipogenesis and hepatic glucose metabolism. *Nat. Commun.* **2016**, *7*, 11365. [CrossRef]

48. Eldakhakhny, B.M.; Al Sadoun, H.; Choudhry, H.; Mobashir, M. In-Silico Study of Immune System Associated Genes in Case of Type-2 Diabetes With Insulin Action and Resistance, and/or Obesity. *Front. Endocrinol.* **2021**, *12*, 641888. [CrossRef]

49. Herman, M.A.; Peroni, O.D.; Villoria, J.; Schön, M.R.; Abumrad, N.A.; Blüher, M.; Klein, S.; Kahn, B.B. A novel ChREBP isoform in adipose tissue regulates systemic glucose metabolism. *Nature* **2012**, *484*, 333–338. [CrossRef]

50. Baraille, F.; Planchais, J.; Dentin, R.; Guilmeau, S.; Postic, C. Integration of ChREBP-Mediated Glucose Sensing into Whole Body Metabolism. *Physiology* **2015**, *30*, 428–437. [CrossRef]

51. Song, Z.; Yang, H.; Zhou, L.; Yang, F. Glucose-Sensing Transcription Factor MondoA/ChREBP as Targets for Type 2 Diabetes: Opportunities and Challenges. *Int. J. Mol. Sci.* **2019**, *20*, 5132. [CrossRef]

52. Dutta, A.; Abmayr, S.M.; Workman, J.L. Diverse Activities of Histone Acylations Connect Metabolism to Chromatin Function. *Mol. Cell* **2016**, *63*, 547–552. [CrossRef]

53. McDonnell, E.; Crown, S.B.; Fox, D.B.; Kitir, B.; Ilkayeva, O.R.; Olsen, C.A.; Grimsrud, P.A.; Hirschey, M.D. Lipids Reprogram Metabolism to Become a Major Carbon Source for Histone Acetylation. *Cell Rep.* **2016**, *17*, 1463–1472. [CrossRef] [PubMed]

54. Nitsch, S.; Shahidian, L.Z.; Schneider, R. Histone acylations and chromatin dynamics: Concepts, challenges, and links to metabolism. *EMBO Rep.* **2021**, *22*, e52774. [CrossRef] [PubMed]

55. Staimez, L.R.; Deepa, M.; Ali, M.; Mohan, V.; Hanson, R.L.; Narayan, K.V. Tale of two Indians: Heterogeneity in type 2 diabetes pathophysiology. *Diabetes/Metab. Res. Rev.* **2019**, *35*, e3192. [CrossRef] [PubMed]

56. Narayan, K.M.V.; Kondal, D.; Kobes, S.; Staimez, L.R.; Mohan, D.; Gujral, U.P.; Patel, S.; Anjana, R.M.; Shivashankar, R.; Ali, M.K.; et al. Incidence of diabetes in South Asian young adults compared to Pima Indians. *BMJ Open Diabetes Res. Care* **2021**, *9*, e001988. [CrossRef]

57. Boucher, J.; Kleinridders, A.; Kahn, C.R. Insulin Receptor Signaling in Normal and Insulin-Resistant States. *Cold Spring Harb. Perspect. Biol.* **2014**, *6*, a009191. [CrossRef]

58. White, M.F.; Kahn, C.R. Insulin action at a molecular level–100 years of progress. *Mol. Metab.* **2021**, *52*, 101304. [CrossRef]

59. Parker, V.E.R.; Savage, D.B.; O'Rahilly, S.; Semple, R.K. Mechanistic insights into insulin resistance in the genetic era. *Diabet. Med.* **2011**, *28*, 1476–1486. [CrossRef]
60. Barber, T.M.; Kyrou, I.; Randeva, H.S.; Weickert, M.O. Mechanisms of Insulin Resistance at the Crossroad of Obesity with Associated Metabolic Abnormalities and Cognitive Dysfunction. *Int. J. Mol. Sci.* **2021**, *22*, 546. [CrossRef]
61. Klotz, L.-O.; Sánchez-Ramos, C.; Prieto-Arroyo, I.; Urbánek, P.; Steinbrenner, H.; Monsalve, M. Redox regulation of FoxO transcription factors. *Redox Biol.* **2015**, *6*, 51–72. [CrossRef]
62. Krafczyk, N.; Klotz, L. FOXO transcription factors in antioxidant defense. *IUBMB Life* **2021**, *74*, 53–61. [CrossRef]
63. Jeffrey, W.R.; Ryder, J.W.; Gilbert, M.; Zierath, J.R. Skeletal muscle and insulin sensitivity: Pathophysiological alterations. *Front. Biosci.* **2001**, *6*, D154–D163. [CrossRef]
64. Wasserman, D.H. Insulin, Muscle Glucose Uptake, and Hexokinase: Revisiting the Road Not Taken. *Physiology* **2022**, *37*, 115–127. [CrossRef] [PubMed]
65. Nakae, J.; Barr, V.; Accili, D. Differential regulation of gene expression by insulin and IGF-1 receptors correlates with phosphorylation of a single amino acid residue in the forkhead transcription factor FKHR. *EMBO J.* **2000**, *19*, 989–996. [CrossRef]
66. Gross, D.N.; Wan, M.; Birnbaum, M.J. The role of FOXO in the regulation of metabolism. *Curr. Diabetes Rep.* **2009**, *9*, 208–214. [CrossRef] [PubMed]
67. Kim, M.E.; Kim, D.H.; Lee, J.S. FoxO Transcription Factors: Applicability as a Novel Immune Cell Regulators and Therapeutic Targets in Oxidative Stress-Related Diseases. *Int. J. Mol. Sci.* **2022**, *23*, 11877. [CrossRef] [PubMed]
68. Qu, S.; Altomonte, J.; Perdomo, G.; He, J.; Fan, Y.; Kamagate, A.; Meseck, M.; Dong, H.H. Aberrant Forkhead Box O1 Function Is Associated with Impaired Hepatic Metabolism. *Endocrinology* **2006**, *147*, 5641–5652. [CrossRef]
69. Titchenell, P.M.; Quinn, W.J.; Lu, M.; Chu, Q.; Lu, W.; Li, C.; Chen, H.; Monks, B.R.; Chen, J.; Rabinowitz, J.D.; et al. Direct Hepatocyte Insulin Signaling Is Required for Lipogenesis but Is Dispensable for the Suppression of Glucose Production. *Cell Metab.* **2016**, *23*, 1154–1166. [CrossRef]
70. Bergman, R.N.; Iyer, M.S. Indirect Regulation of Endogenous Glucose Production by Insulin: The Single Gateway Hypothesis Revisited. *Diabetes* **2017**, *66*, 1742–1747. [CrossRef]
71. Ozcan, L.; Wong, C.C.L.; Li, G.; Xu, T.; Pajvani, U.; Park, S.K.R.; Wronska, A.; Chen, B.-X.; Marks, A.R.; Fukamizu, A.; et al. Calcium Signaling through CaMKII Regulates Hepatic Glucose Production in Fasting and Obesity. *Cell Metab.* **2012**, *15*, 739–751. [CrossRef]
72. Wu, Y.; Pan, Q.; Yan, H.; Zhang, K.; Guo, X.; Xu, Z.; Yang, W.; Qi, Y.; Guo, C.A.; Hornsby, C.; et al. Novel Mechanism of Foxo1 Phosphorylation in Glucagon Signaling in Control of Glucose Homeostasis. *Diabetes* **2018**, *67*, 2167–2182. [CrossRef]
73. Zhang, W.; Patil, S.; Chauhan, B.; Guo, S.; Powell, D.R.; Le, J.; Klotsas, A.; Matika, R.; Xiao, X.; Franks, R.; et al. FoxO1 Regulates Multiple Metabolic Pathways in the Liver: Effects on gluconeogenic, glycolytic, and lipogenic gene expression. *J. Biol. Chem.* **2006**, *281*, 10105–10117. [CrossRef] [PubMed]
74. Perry, R.J.; Camporez, J.-P.G.; Kursawe, R.; Titchenell, P.M.; Zhang, D.; Perry, C.J.; Jurczak, M.J.; Abudukadier, A.; Han, M.S.; Zhang, X.-M.; et al. Hepatic Acetyl CoA Links Adipose Tissue Inflammation to Hepatic Insulin Resistance and Type 2 Diabetes. *Cell* **2015**, *160*, 745–758. [CrossRef] [PubMed]
75. Casteras, S.; Abdul-Wahed, A.; Soty, M.; Vulin, F.; Guillou, H.; Campana, M.; Le Stunff, H.; Pirola, L.; Rajas, F.; Mithieux, G.; et al. The suppression of hepatic glucose production improves metabolism and insulin sensitivity in subcutaneous adipose tissue in mice. *Diabetologia* **2016**, *59*, 2645–2653. [CrossRef] [PubMed]
76. Schormann, N.; Hayden, K.L.; Lee, P.; Banerjee, S.; Chattopadhyay, D. An overview of structure, function, and regulation of pyruvate kinases. *Protein Sci.* **2019**, *28*, 1771–1784. [CrossRef] [PubMed]
77. Beck-Nielsen, H. The role of glycogen synthase in the development of hyperglycemia in type 2 diabetes-'To store or not to store glucose, that's the question'. *Diabetes/Metab. Res. Rev.* **2012**, *28*, 635–644. [CrossRef]
78. Nolan, C.J.; Ruderman, N.B.; Kahn, S.E.; Pedersen, O.; Prentki, M. Insulin Resistance as a Physiological Defense Against Metabolic Stress: Implications for the Management of Subsets of Type 2 Diabetes. *Diabetes* **2015**, *64*, 673–686. [CrossRef]
79. Pories, W.J.; Dohm, G.L. Diabetes: Have we got it all wrong? Hyperinsulinism as the culprit: Surgery provides the evidence? *Diabetes Care* **2012**, *35*, 2438–2442. [CrossRef]
80. Zhang, Y.; Shen, T.; Wang, S. Progression from prediabetes to type 2 diabetes mellitus induced by overnutrition. *Hormones* **2022**, *21*, 591–597. [CrossRef]
81. Erion, K.A.; Berdan, C.A.; Burritt, N.E.; Corkey, B.E.; Deeney, J.T. Chronic Exposure to Excess Nutrients Left-shifts the Concentration Dependence of Glucose-stimulated Insulin Secretion in Pancreatic β-Cells. *J. Biol. Chem.* **2015**, *290*, 16191–16201. [CrossRef]
82. Corkey, B.E.; Deeney, J.T.; Merrins, M.J. What Regulates Basal Insulin Secretion and Causes Hyperinsulinemia? *Diabetes* **2021**, *70*, 2174–2182. [CrossRef]
83. Merino, B.; Fernández-Díaz, C.M.; Parrado-Fernández, C.; González-Casimiro, C.M.; Postigo-Casado, T.; Lobatón, C.D.; Leissring, M.A.; Cózar-Castellano, I.; Perdomo, G. Hepatic insulin-degrading enzyme regulates glucose and insulin homeostasis in diet-induced obese mice. *Metabolism* **2020**, *113*, 154352. [CrossRef]
84. Caron, A.; Richard, D.; Laplante, M. The Roles of mTOR Complexes in Lipid Metabolism. *Annu. Rev. Nutr.* **2015**, *35*, 321–348. [CrossRef]

85. Yuan, T.; Yang, T.; Chen, H.; Fu, D.; Hu, Y.; Wang, J.; Yuan, Q.; Yu, H.; Xu, W.; Xie, X. New insights into oxidative stress and inflammation during diabetes mellitus-accelerated atherosclerosis. *Redox Biol.* **2019**, *20*, 247–260. [CrossRef]
86. Roden, M.; Shulman, G.I. The integrative biology of type 2 diabetes. *Nature* **2019**, *576*, 51–60. [CrossRef]
87. Janssen, J. Hyperinsulinemia and Its Pivotal Role in Aging, Obesity, Type 2 Diabetes, Cardiovascular Disease and Cancer. *Int. J. Mol. Sci.* **2021**, *22*, 7797. [CrossRef]
88. Safhi, M.M.; Anwer, T.; Khan, G.; Siddiqui, R.; Sivakumar, S.M.; Alam, M.F. The combination of canagliflozin and omega-3 fatty acid ameliorates insulin resistance and cardiac biomarkers *via* modulation of inflammatory cytokines in type 2 diabetic rats. *Korean J. Physiol. Pharmacol.* **2018**, *22*, 493–501. [CrossRef]
89. Lackey, D.E.; Olefsky, J.M. Regulation of metabolism by the innate immune system. *Nat. Rev. Endocrinol.* **2015**, *12*, 15–28. [CrossRef]
90. Flach, R.J.R.; Danai, L.V.; DiStefano, M.T.; Kelly, M.; Menendez, L.G.; Jurczyk, A.; Sharma, R.B.; Jung, D.Y.; Kim, J.H.; Kim, J.K.; et al. Protein Kinase Mitogen-activated Protein Kinase Kinase Kinase Kinase 4 (MAP4K4) Promotes Obesity-induced Hyperinsulinemia. *J. Biol. Chem.* **2016**, *291*, 16221–16230. [CrossRef]
91. Tremblay, J.; Hamet, P. Environmental and genetic contributions to diabetes. *Metabolism* **2019**, *100*, 153952. [CrossRef]
92. Pearson, E.R. Type 2 diabetes: A multifaceted disease. *Diabetologia* **2019**, *62*, 1107–1112. [CrossRef]
93. Dendup, T.; Feng, X.; Clingan, S.; Astell-Burt, T. Environmental Risk Factors for Developing Type 2 Diabetes Mellitus: A Systematic Review. *Int. J. Environ. Res. Public Health* **2018**, *15*, 78. [CrossRef] [PubMed]
94. Zhang, Y.; Pan, X.F.; Chen, J.; Xia, L.; Cao, A.; Zhang, Y.; Wang, J.; Li, H.; Yang, K.; Guo, K.; et al. Combined lifestyle factors and risk of incident type 2 diabetes and prognosis among individuals with type 2 diabetes: A systematic review and meta-analysis of prospective cohort studies. *Diabetologia.* **2020**, *63*, 21–33. [CrossRef] [PubMed]
95. Boeing, H.; Bechthold, A.; Bub, A.; Ellinger, S.; Haller, D.; Kroke, A.; Leschik-Bonnet, E.; Müller, M.J.; Oberritter, H.; Schulze, M.; et al. Critical review: Vegetables and fruit in the prevention of chronic diseases. *Eur. J. Nutr.* **2012**, *51*, 637–663. [CrossRef]
96. Carter, P.; Gray, L.J.; Troughton, J.; Khunti, K.; Davies, M.J. Fruit and vegetable intake and incidence of type 2 diabetes mellitus: Systematic review and meta-analysis. *BMJ* **2010**, *341*, c4229. [CrossRef]
97. Cooper, A.J.; Sharp, S.J.; Lentjes, M.A.; Luben, R.N.; Khaw, K.-T.; Wareham, N.J.; Forouhi, N.G. A Prospective Study of the Association Between Quantity and Variety of Fruit and Vegetable Intake and Incident Type 2 Diabetes. *Diabetes Care* **2012**, *35*, 1293–1300. [CrossRef] [PubMed]
98. Muraki, I.; Imamura, F.; Manson, J.; Hu, F.B.; Willett, W.C.; van Dam, R.; Sun, Q. Fruit consumption and risk of type 2 diabetes: Results from three prospective longitudinal cohort studies. *BMJ* **2013**, *347*, f5001, Erratum in *BMJ* **2013**, *347*, f6935. [CrossRef]
99. Yingli, F.; Fan, Y.; Zhang, X.; Hou, W.; Tang, Z. Fruit and vegetable intake and risk of type 2 diabetes mellitus: Meta-analysis of prospective cohort studies. *BMJ Open* **2014**, *4*, e005497. [CrossRef]
100. Mazzocchi, M.; Brasili, C.; Sandri, E. Trends in dietary patterns and compliance with World Health Organization recommendations: A cross-country analysis. *Public Health Nutr.* **2007**, *11*, 535–540. [CrossRef]
101. Salas-Salvadó, J.; Bulló, M.; Estruch, R.; Ros, E.; Covas, M.-I.; Ibarrola-Jurado, N.; Corella, D.; Arós, F.; Gómez-Gracia, E.; Ruiz-Gutiérrez, V.; et al. Prevention of diabetes with Mediterranean diets: A subgroup analysis of a randomized trial. *Ann. Intern. Med.* **2014**, *160*, 1–10, Erratum in *Ann. Intern. Med.* **2018**, *169*, 271–272. [CrossRef]
102. Esposito, K.; Maiorino, M.I.; Bellastella, G.; Panagiotakos, D.B.; Giugliano, D. Mediterranean diet for type 2 diabetes: Cardiometabolic benefits. *Endocrine* **2017**, *56*, 27–32. [CrossRef]
103. Saeedi, P.; Petersohn, I.; Salpea, P.; Malanda, B.; Karuranga, S.; Unwin, N.; Colagiuri, S.; Guariguata, L.; Motala, A.A.; Ogurtsova, K.; et al. Global and regional diabetes prevalence estimates for 2019 and projections for 2030 and 2045: Results from the International Diabetes Federation Diabetes Atlas, 9th edition. *Diabetes Res. Clin. Pract.* **2019**, *157*, 107843. [CrossRef] [PubMed]
104. Mohamed, H.J.B.J.; Yap, R.W.K.; Loy, S.L.; Norris, S.A.; Biesma, R.; Aagaard-Hansen, J. Prevalence and Determinants of Overweight, Obesity, and Type 2 Diabetes Mellitus in Adults in Malaysia. *Asia Pac. J. Public Health* **2014**, *27*, 123–135. [CrossRef]
105. Hussein, Z.; Taher, S.W.; Singh, H.K.G.; Swee, W.C.S. Diabetes Care in Malaysia: Problems, New Models, and Solutions. *Ann. Glob. Health* **2016**, *81*, 851–862. [CrossRef]
106. Tee, E.-S.; Yap, R.W.K. Type 2 diabetes mellitus in Malaysia: Current trends and risk factors. *Eur. J. Clin. Nutr.* **2017**, *71*, 844–849. [CrossRef]
107. Aj, S.J.; Lc, G. Effects of an educational program focused on self-care and concurrent physical training on glycemia and drug treatment of patients with diabetes mellitus. *Diabetes Updat.* **2018**, *5*, 1–7. [CrossRef]
108. Winding, K.M.; Munch, G.W.; Iepsen, U.W.; van Hall, G.; Pedersen, B.K.; Mortensen, S.P. The effect on glycaemic control of low-volume high-intensity interval training versus endurance training in individuals with type 2 diabetes. *Diabetes Obes. Metab.* **2018**, *20*, 1131–1139, Erratum in *Diabetes Obes. Metab.* **2019**, *21*, 202. [CrossRef] [PubMed]
109. Ekelund, U.; Brage, S.; Griffin, S.J.; Wareham, N.J.; the ProActive UK Research Group. Objectively Measured Moderate- and Vigorous-Intensity Physical Activity but Not Sedentary Time Predicts Insulin Resistance in High-Risk Individuals. *Diabetes Care* **2009**, *32*, 1081–1086. [CrossRef] [PubMed]
110. Rockette-Wagner, B.; Edelstein, S.; Venditti, E.M.; Reddy, D.; Bray, G.A.; Carrion-Petersen, M.L.; Dabelea, D.; Delahanty, L.M.; Florez, H.; Franks, P.W.; et al. The impact of lifestyle intervention on sedentary time in individuals at high risk of diabetes. *Diabetologia* **2015**, *58*, 1198–1202. [CrossRef] [PubMed]

111. Kruk, J.; Aboul-Enein, H.Y.; Kładna, A.; Bowser, J.E. Oxidative stress in biological systems and its relation with pathophysiological functions: The effect of physical activity on cellular redox homeostasis. *Free Radic. Res.* **2019**, *53*, 497–521. [CrossRef] [PubMed]

112. Dimauro, I.; Paronetto, M.P.; Caporossi, D. Exercise, redox homeostasis and the epigenetic landscape. *Redox Biol.* **2020**, *35*, 101477. [CrossRef]

113. Matta, L.; Fonseca, T.S.; Faria, C.C.; Lima-Junior, N.C.; De Oliveira, D.F.; Maciel, L.; Boa, L.F.; Pierucci, A.P.T.R.; Ferreira, A.C.F.; Nascimento, J.H.M.; et al. The Effect of Acute Aerobic Exercise on Redox Homeostasis and Mitochondrial Function of Rat White Adipose Tissue. *Oxidative Med. Cell. Longev.* **2021**, *2021*, 1–15. [CrossRef] [PubMed]

114. The InterAct Consortium; Spijkerman, A. M.; van der Asijsc, D.L.; Nilsson, P.M.; Ardanaz, E.; Gavrila, D.; Agudo, A.; Arriola, L.; Balkau, B.; Beulens, J.W.; et al. Smoking and Long-Term Risk of Type 2 Diabetes: The EPIC-InterAct Study in European Populations. *Diabetes Care* **2014**, *37*, 3164–3171. [CrossRef] [PubMed]

115. Willi, C.; Bodenmann, P.; Ghali, W.A.; Faris, P.D.; Cornuz, J. Active smoking and the risk of type 2 diabetes: A systematic review and meta-analysis. *JAMA* **2007**, *298*, 2654–2664. [CrossRef] [PubMed]

116. Maddatu, J.; Anderson-Baucum, E.; Evans-Molina, C. Smoking and the risk of type 2 diabetes. *Transl. Res.* **2017**, *184*, 101–107. [CrossRef] [PubMed]

117. Śliwińska-Mosson, M.; Milnerowicz, H. The impact of smoking on the development of diabetes and its complications. *Diabetes Vasc. Dis. Res.* **2017**, *14*, 265–276. [CrossRef] [PubMed]

118. Baliunas, D.O.; Taylor, B.J.; Irving, H.; Roerecke, M.; Patra, J.; Mohapatra, S.; Rehm, J. Alcohol as a risk factor for type 2 diabetes: A systematic review and meta-analysis. *Diabetes Care* **2009**, *32*, 2123–2132. [CrossRef]

119. Okamura, T.; Hashimoto, Y.; Hamaguchi, M.; Obora, A.; Kojima, T.; Fukui, M. Effect of alcohol consumption and the presence of fatty liver on the risk for incident type 2 diabetes: A population-based longitudinal study. *BMJ Open Diabetes Res. Care* **2020**, *8*, e001629. [CrossRef]

120. Zeliger, H.I. Lipophilic chemical exposure as a cause of type 2 diabetes (T2D). *Rev. Environ. Health* **2013**, *28*, 9–20. [CrossRef]

121. Kuo, C.-C.; Moon, K.; Thayer, K.A.; Navas-Acien, A. Environmental Chemicals and Type 2 Diabetes: An Updated Systematic Review of the Epidemiologic Evidence. *Curr. Diabetes Rep.* **2013**, *13*, 831–849. [CrossRef]

122. Carulli, N.; Rondinella, S.; Lombardini, S.; Canedi, I.; Loria, P.; Carulli, L. Review article: Diabetes, genetics and ethnicity. *Aliment. Pharmacol. Ther.* **2005**, *22*, 16–19. [CrossRef]

123. Stumvoll, M.; Goldstein, B.J.; van Haeften, T.W. Type 2 diabetes: Principles of pathogenesis and therapy. *Lancet* **2005**, *365*, 1333–1346. [CrossRef] [PubMed]

124. Oellgaard, J.; Gæde, P.; Rossing, P.; Rørth, R.; Køber, L.; Parving, H.-H.; Pedersen, O. Reduced risk of heart failure with intensified multifactorial intervention in individuals with type 2 diabetes and microalbuminuria: 21 years of follow-up in the randomised Steno-2 study. *Diabetologia* **2018**, *61*, 1724–1733. [CrossRef] [PubMed]

125. McCarthy, M.I.; Zeggini, E. Genetics of type 2 diabetes. *Curr. Diabetes Rep.* **2006**, *6*, 147–154. [CrossRef]

126. Kaul, N.; Ali, S. Genes, Genetics, and Environment in Type 2 Diabetes: Implication in Personalized Medicine. *DNA Cell Biol.* **2016**, *35*, 1–12. [CrossRef] [PubMed]

127. Liu, Y.; Wang, Y.; Qin, S.; Jin, X.; Jin, L.; Gu, W.; Mu, Y. Insights into Genome-Wide Association Study for Diabetes: A Bibliometric and Visual Analysis From 2001 to 2021. *Front. Endocrinol.* **2022**, *13*, 817620. [CrossRef] [PubMed]

128. Flannick, J.; Florez, J.F.J.C. Type 2 diabetes: Genetic data sharing to advance complex disease research. *Nat. Rev. Genet.* **2016**, *17*, 535–549. [CrossRef]

129. Zhao, W.; Rasheed, A.; Tikkanen, E.; Lee, J.-J.; Butterworth, A.S.; Howson, J.M.M.; Assimes, T.L.; Chowdhury, R.; Orho-Melander, M.; Damrauer, S.; et al. Identification of new susceptibility loci for type 2 diabetes and shared etiological pathways with coronary heart disease. *Nat. Genet.* **2017**, *49*, 1450–1457. [CrossRef]

130. Lyssenko, V.; Bianchi, C.; Del Prato, S. Personalized Therapy by Phenotype and Genotype. *Diabetes Care* **2016**, *39*, S127–S136. [CrossRef]

131. Zeggini, E.; Scott, L.J.; Saxena, R.; Voight, B.F.; Marchini, J.L.; Hu, T.; de Bakker, P.I.; Abecasis, G.R.; Almgren, P.; Andersen, G.; et al. Meta-analysis of genome-wide association data and large-scale replication identifies additional susceptibility loci for type 2 diabetes. *Nat. Genet.* **2008**, *40*, 638–645. [CrossRef]

132. Saxena, R.; Saleheen, D.; Been, L.F.; Garavito, M.L.; Braun, T.; Bjonnes, A.; Young, R.; Ho, W.K.; Rasheed, A.; Frossard, P.; et al. Genome-Wide Association Study Identifies a Novel Locus Contributing to Type 2 Diabetes Susceptibility in Sikhs of Punjabi Origin From India. *Diabetes* **2013**, *62*, 1746–1755. [CrossRef]

133. Cook, J.P.; Morris, A.P. Multi-ethnic genome-wide association study identifies novel locus for type 2 diabetes susceptibility. *Eur. J. Hum. Genet.* **2016**, *24*, 1175–1180. [CrossRef]

134. Langenberg, C.; Lotta, L. Genomic insights into the causes of type 2 diabetes. *Lancet* **2018**, *391*, 2463–2474. [CrossRef] [PubMed]

135. Watanabe, R.M. Physiologic Interpretation of GWAS Signals for Type 2 Diabetes. *Methods Mol Biol.* **2018**, *1706*, 323–351. [CrossRef] [PubMed]

136. Meigs, J.B. The Genetic Epidemiology of Type 2 Diabetes: Opportunities for Health Translation. *Curr. Diabetes Rep.* **2019**, *19*, 62. [CrossRef] [PubMed]

137. Segerstolpe, Å.; Palasantza, A.; Eliasson, P.; Andersson, E.-M.; Andréasson, A.-C.; Sun, X.; Picelli, S.; Sabirsh, A.; Clausen, M.; Bjursell, M.K.; et al. Single-Cell Transcriptome Profiling of Human Pancreatic Islets in Health and Type 2 Diabetes. *Cell Metab.* **2016**, *24*, 593–607. [CrossRef] [PubMed]

138. Kalai, F.Z.; Boulaaba, M.; Ferdousi, F.; Isoda, H. Effects of Isorhamnetin on Diabetes and Its Associated Complications: A Review of In Vitro and In Vivo Studies and a Post Hoc Transcriptome Analysis of Involved Molecular Pathways. *Int. J. Mol. Sci.* **2022**, *23*, 704. [CrossRef] [PubMed]

139. Juttada, U.; Kumpatla, S.; Parveen, R.; Viswanathan, V. TCF7L2 polymorphism a prominent marker among subjects with Type-2-Diabetes with a positive family history of diabetes. *Int. J. Biol. Macromol.* **2020**, *159*, 402–405. [CrossRef]

140. Mustafa, S.; Younus, D. Association of TCF7L2 rs7903146 Polymorphism with the Risk of Type 2 Diabetes Mellitus (T2DM) Among Kurdish Population in Erbil Province, Iraq. *Indian J. Clin. Biochem.* **2020**, *36*, 312–318. [CrossRef]

141. Aboelkhair, N.T.; Kasem, H.E.; Abdelmoaty, A.A.; El-Edel, R.H. TCF7L2 gene polymorphism as a risk for type 2 diabetes mellitus and diabetic microvascular complications. *Mol. Biol. Rep.* **2021**, *48*, 5283–5290. [CrossRef]

142. Elhourch, S.; Arrouchi, H.; Mekkaoui, N.; Allou, Y.; Ghrifi, F.; Allam, L.; Elhafidi, N.; Belyamani, L.; Ibrahimi, A.; Elomri, N.; et al. Significant Association of Polymorphisms in the TCF7L2 Gene with a Higher Risk of Type 2 Diabetes in a Moroccan Population. *J. Pers. Med.* **2021**, *11*, 461. [CrossRef]

143. Chaudhuri, P.; Das, M.; Lodh, I.; Goswami, R. Role of Metabolic Risk Factors, Family History, and Genetic Polymorphisms (PPARγ and TCF7L2) on Type 2 Diabetes Mellitus Risk in an Asian Indian Population. *Public Health Genom.* **2021**, *24*, 131–138. [CrossRef] [PubMed]

144. Bride, L.; Naslavsky, M.; Yamamoto, G.L.; Scliar, M.; Pimassoni, L.H.; Aguiar, P.S.; de Paula, F.; Wang, J.; Duarte, Y.; Passos Bueno, M.R.; et al. TCF7L2 rs7903146 polymorphism association with diabetes and obesity in an elderly cohort from Brazil. *Peerj* **2021**, *9*, e11349. [CrossRef] [PubMed]

145. Srinivasan, S.; Kaur, V.; Chamarthi, B.; Littleton, K.R.; Chen, L.; Manning, A.K.; Merino, J.; Thomas, M.K.; Hudson, M.; Goldfine, A.; et al. TCF7L2 Genetic Variation Augments Incretin Resistance and Influences Response to a Sulfonylurea and Metformin: The Study to Understand the Genetics of the Acute Response to Metformin and Glipizide in Humans (SUGAR-MGH). *Diabetes Care* **2018**, *41*, 554–561. [CrossRef] [PubMed]

146. Galderisi, A.; Tricò, D.; Pierpont, B.; Shabanova, V.; Samuels, S.; Man, C.D.; Galuppo, B.; Santoro, N.; Caprio, S. A Reduced Incretin Effect Mediated by the rs7903146 Variant in the *TCF7L2* Gene Is an Early Marker of β-Cell Dysfunction in Obese Youth. *Diabetes Care* **2020**, *43*, 2553–2563. [CrossRef] [PubMed]

147. Nguyen-Tu, M.-S.; Xavier, G.D.S.; Leclerc, I.; Rutter, G.A. Transcription factor-7–like 2 (TCF7L2) gene acts downstream of the Lkb1/Stk11 kinase to control mTOR signaling, β cell growth, and insulin secretion. *J. Biol. Chem.* **2018**, *293*, 14178–14189, Erratum in *J. Biol. Chem.* **2018**, *293*, 18420. [CrossRef] [PubMed]

148. Lorzadeh, S.; Kohan, L.; Ghavami, S.; Azarpira, N. Autophagy and the Wnt signaling pathway: A focus on Wnt/β-catenin signaling. *Biochim. et Biophys. Acta (BBA) Mol. Cell Res.* **2020**, *1868*, 118926. [CrossRef]

149. Nie, X.; Wei, X.; Ma, H.; Fan, L.; Chen, W.-D. The complex role of Wnt ligands in type 2 diabetes mellitus and related complications. *J. Cell. Mol. Med.* **2021**, *25*, 6479–6495. [CrossRef]

150. Huang, T.; Wang, L.; Bai, M.; Zheng, J.; Yuan, D.; He, Y.; Wang, Y.; Jin, T.; Cui, W. Influence of *IGF2BP2*, *HMG20A*, and *HNF1B* genetic polymorphisms on the susceptibility to Type 2 diabetes mellitus in Chinese Han population. *Biosci. Rep.* **2020**, *40*, BSR20193955. [CrossRef]

151. Azarova, I.E.; Klyosova, E.Y.; Lazarenko, V.A.; Konoplya, A.I.; Polonikov, A.V. rs11927381 Polymorphism and Type 2 Diabetes Mellitus: Contribution of Smoking to the Realization of Susceptibility to the Disease. *Bull. Exp. Biol. Med.* **2020**, *168*, 313–316. [CrossRef]

152. Dai, N.; Zhao, L.; Wrighting, D.; Krämer, D.; Majithia, A.; Wang, Y.; Cracan, V.; Borges-Rivera, D.; Mootha, V.K.; Nahrendorf, M.; et al. IGF2BP2/IMP2-Deficient Mice Resist Obesity through Enhanced Translation of Ucp1 mRNA and Other mRNAs Encoding Mitochondrial Proteins. *Cell Metab.* **2015**, *21*, 609–621. [CrossRef]

153. Kokkinopoulou, I.; Maratou, E.; Mitrou, P.; Boutati, E.; Sideris, D.C.; Fragoulis, E.G.; Christodoulou, M.-I. Decreased expression of microRNAs targeting type-2 diabetes susceptibility genes in peripheral blood of patients and predisposed individuals. *Endocrine* **2019**, *66*, 226–239. [CrossRef] [PubMed]

154. Livingstone, C.; Borai, A. Insulin-like growth factor-II: Its role in metabolic and endocrine disease. *Clin. Endocrinol.* **2014**, *80*, 773–781. [CrossRef]

155. Mercader, J.M.; Liao, R.G.; Bell, A.D.; Dymek, Z.; Estrada, K.; Tukiainen, T.; Huerta-Chagoya, A.; Moreno-Macías, H.; Jablonski, K.A.; Hanson, R.L.; et al. A Loss-of-Function Splice Acceptor Variant in *IGF2* Is Protective for Type 2 Diabetes. *Diabetes* **2017**, *66*, 2903–2914. [CrossRef]

156. Zeng, Y.; He, H.; Zhang, L.; Zhu, W.; Shen, H.; Yan, Y.-J.; Deng, H.-W. GWA-based pleiotropic analysis identified potential SNPs and genes related to type 2 diabetes and obesity. *J. Hum. Genet.* **2020**, *66*, 297–306. [CrossRef] [PubMed]

157. Wei, F.-Y.; Suzuki, T.; Watanabe, S.; Kimura, S.; Kaitsuka, T.; Fujimura, A.; Matsui, H.; Atta, M.G.; Michiue, H.; Fontecave, M.; et al. Deficit of tRNALys modification by Cdkal1 causes the development of type 2 diabetes in mice. *J. Clin. Investig.* **2011**, *121*, 3598–3608. [CrossRef]

158. Palmer, C.J.; Bruckner, R.J.; Paulo, J.A.; Kazak, L.; Long, J.Z.; Mina, A.I.; Deng, Z.; LeClair, K.B.; Hall, J.A.; Hong, S.; et al. Cdkal1, a type 2 diabetes susceptibility gene, regulates mitochondrial function in adipose tissue. *Mol. Metab.* **2017**, *6*, 1212–1225. [CrossRef] [PubMed]

159. May, M.J.; Kopp, E.B. NF-κB AND REL PROTEINS: Evolutionarily Conserved Mediators of Immune Responses. *Annu. Rev. Immunol.* **1998**, *16*, 225–260. [CrossRef]

160. Arragain, S.; Handelman, S.K.; Forouhar, F.; Wei, F.-Y.; Tomizawa, K.; Hunt, J.F.; Douki, T.; Fontecave, M.; Mulliez, E.; Atta, M. Identification of Eukaryotic and Prokaryotic Methylthiotransferase for Biosynthesis of 2-Methylthio-N6-threonylcarbamoyladenosine in tRNA. *J. Biol. Chem.* **2010**, *285*, 28425–28433. [CrossRef]

161. Groenewoud, M.J.; Dekker, J.M.; Fritsche, A.; Reiling, E.; Nijpels, G.; Heine, R.J.; Maassen, J.A.; Machicao, F.; Schäfer, S.A.; Häring, H.U.; et al. Variants of CDKAL1 and IGF2BP2 affect first-phase insulin secretion during hyperglycaemic clamps. *Diabetologia* **2008**, *51*, 1659–1663. [CrossRef]

162. Stančáková, A.; Pihlajamäki, J.; Kuusisto, J.; Stefan, N.; Fritsche, A.; Häring, H.; Andreozzi, F.; Succurro, E.; Sesti, G.; Boesgaard, T.W.; et al. Single-Nucleotide Polymorphism rs7754840 of *CDKAL1* Is Associated with Impaired Insulin Secretion in Nondiabetic Offspring of Type 2 Diabetic Subjects and in a Large Sample of Men with Normal Glucose Tolerance. *J. Clin. Endocrinol. Metab.* **2008**, *93*, 1924–1930. [CrossRef]

163. Li, J.; Niu, X.; Li, J.; Wang, Q. Association of PPARG Gene Polymorphisms Pro12Ala with Type 2 Diabetes Mellitus: A Meta-analysis. *Curr. Diabetes Rev.* **2019**, *15*, 277–283. [CrossRef] [PubMed]

164. Herder, C.; Rathmann, W.; Strassburger, K.; Finner, H.; Grallert, H.; Huth, C.; Meisinger, C.; Gieger, C.; Martin, S.; Giani, G.; et al. Variants of the *PPARG, IGF2BP2, CDKAL1, HHEX*, and *TCF7L2* Genes Confer Risk of Type 2 Diabetes Independently of BMI in the German KORA Studies. *Horm. Metab. Res.* **2008**, *40*, 722–726. [CrossRef] [PubMed]

165. Zia, A.; Bhatti, A.; John, P.; Kiani, A.K. Data interpretation: Deciphering the biological function of Type 2 diabetes associated risk loci. *Acta Diabetol.* **2015**, *52*, 789–800. [CrossRef] [PubMed]

166. Erfani, T.; Sarhangi, N.; Afshari, M.; Abbasi, D.; Meybodi, H.R.A.; Hasanzad, M. KCNQ1 common genetic variant and type 2 diabetes mellitus risk. *J. Diabetes Metab. Disord.* **2019**, *19*, 47–51. [CrossRef]

167. Unoki, H.; Takahashi, A.; Kawaguchi, T.; Hara, K.; Horikoshi, M.; Andersen, G.; Ng, D.P.K.; Holmkvist, J.; Borch-Johnsen, K.; Jørgensen, T.; et al. SNPs in KCNQ1 are associated with susceptibility to type 2 diabetes in East Asian and European populations. *Nat. Genet.* **2008**, *40*, 1098–1102. [CrossRef]

168. Cirillo, E.; Kutmon, M.; Hernandez, M.G.; Hooimeijer, T.; Adriaens, M.E.; Eijssen, L.M.T.; Parnell, L.D.; Coort, S.L.; Evelo, C.T. From SNPs to pathways: Biological interpretation of type 2 diabetes (T2DM) genome wide association study (GWAS) results. *PLoS ONE* **2018**, *13*, e0193515. [CrossRef]

169. Rouf, A.S.M.R.; Amin, A.; Islam, K.; Haque, F.; Ahmed, K.R.; Rahman, A.; Islam, Z.; Kim, B. Statistical Bioinformatics to Uncover the Underlying Biological Mechanisms That Linked Smoking with Type 2 Diabetes Patients Using Transcritpomic and GWAS Analysis. *Molecules* **2022**, *27*, 4390. [CrossRef]

170. Smushkin, G.; Vella, A. Genetics of type 2 diabetes. *Curr. Opin. Clin. Nutr. Metab. Care* **2010**, *13*, 471–477. [CrossRef]

171. Le Gal, K.; Schmidt, E.E.; Sayin, V.I. Cellular Redox Homeostasis. *Antioxidants* **2021**, *10*, 1377. [CrossRef]

172. Ray, P.D.; Huang, B.-W.; Tsuji, Y. Reactive oxygen species (ROS) homeostasis and redox regulation in cellular signaling. *Cell. Signal.* **2012**, *24*, 981–990. [CrossRef]

173. Lennicke, C.; Cochemé, H.M. Redox metabolism: ROS as specific molecular regulators of cell signaling and function. *Mol. Cell* **2021**, *81*, 3691–3707. [CrossRef] [PubMed]

174. Sies, H.; Belousov, V.V.; Chandel, N.S.; Davies, M.J.; Jones, D.P.; Mann, G.E.; Murphy, M.P.; Yamamoto, M.; Winterbourn, C. Defining roles of specific reactive oxygen species (ROS) in cell biology and physiology. *Nat. Rev. Mol. Cell Biol.* **2022**, *23*, 499–515. [CrossRef] [PubMed]

175. Onyango, A.N. Cellular Stresses and Stress Responses in the Pathogenesis of Insulin Resistance. *Oxidative Med. Cell. Longev.* **2018**, *2018*, 4321714. [CrossRef] [PubMed]

176. Eguchi, N.; Vaziri, N.D.; Dafoe, D.C.; Ichii, H. The Role of Oxidative Stress in Pancreatic β Cell Dysfunction in Diabetes. *Int. J. Mol. Sci.* **2021**, *22*, 1509. [CrossRef]

177. Ali, S.S.; Ahsan, H.; Zia, M.K.; Siddiqui, T.; Khan, F.H. Understanding oxidants and antioxidants: Classical team with new players. *J. Food Biochem.* **2020**, *44*, e13145. [CrossRef]

178. Lubrano, S.B.V. Enzymatic antioxidant system in vascular inflammation and coronary artery disease. *World J. Exp. Med.* **2015**, *5*, 218–224. [CrossRef]

179. Lu, J.; Holmgren, A. The thioredoxin antioxidant system. *Free Radic. Biol. Med.* **2014**, *66*, 75–87. [CrossRef]

180. Bachhawat, A.K.; Yadav, S. The glutathione cycle: Glutathione metabolism beyond the γ-glutamyl cycle. *IUBMB Life* **2018**, *70*, 585–592. [CrossRef]

181. Garcã a-Gimã©Nez, J.L.; Pallardã³, F.V. Maintenance of glutathione levels and its importance in epigenetic regulation. *Front. Pharmacol.* **2014**, *5*, 88. [CrossRef]

182. Scirè, A.; Cianfruglia, L.; Minnelli, C.; Bartolini, D.; Torquato, P.; Principato, G.; Galli, F.; Armeni, T. Glutathione compartmentalization and its role in glutathionylation and other regulatory processes of cellular pathways. *Biofactors* **2018**, *45*, 152–168. [CrossRef]

183. Parsanathan, R.; Jain, S.K. Glutathione deficiency induces epigenetic alterations of vitamin D metabolism genes in the livers of high-fat diet-fed obese mice. *Sci. Rep.* **2019**, *9*, 1–11. [CrossRef] [PubMed]

184. Lu, S.C. Glutathione synthesis. *Biochim. et Biophys. Acta (BBA) Gen. Subj.* **2013**, *1830*, 3143–3153. [CrossRef] [PubMed]

185. Meister, A. Glutathione metabolism and its selective modification. *J. Biol. Chem.* **1988**, *263*, 17205–17208. [CrossRef] [PubMed]

186. Zhou, X.; He, L.; Zuo, S.; Zhang, Y.; Wan, D.; Long, C.; Huang, P.; Wu, X.; Wu, C.; Liu, G.; et al. Serine prevented high-fat diet-induced oxidative stress by activating AMPK and epigenetically modulating the expression of glutathione synthesis-related genes. *Biochim. et Biophys. Acta (BBA) Mol. Basis Dis.* **2018**, *1864*, 488–498. [CrossRef]

187. Kumar, S.; Kaur, A.; Chattopadhyay, B.; Bachhawat, A.K. Defining the cytosolic pathway of glutathione degradation in *Arabidopsis thaliana*: Role of the ChaC/GCG family of γ-glutamyl cyclotransferases as glutathione-degrading enzymes and AtLAP1 as the Cys-Gly peptidase. *Biochem. J.* **2015**, *468*, 73–85. [CrossRef] [PubMed]

188. Kennedy, L.; Sandhu, J.K.; Harper, M.-E.; Cuperlovic-Culf, M. Role of Glutathione in Cancer: From Mechanisms to Therapies. *Biomolecules* **2020**, *10*, 1429. [CrossRef] [PubMed]

189. Janowiak, B.E.; Hayward, M.A.; Peterson, F.C.; Volkman, B.F.; Griffith, O.W. γ-Glutamylcysteine Synthetase−Glutathione Synthetase: Domain Structure and Identification of Residues Important in Substrate and Glutathione Binding. *Biochemistry* **2006**, *45*, 10461–10473. [CrossRef]

190. Oakley, A.; Yamada, T.; Liu, D.; Coggan, M.; Clark, A.G.; Board, P.G. The The identification and structural characterization of C7orf24 as gamma-glutamyl cyclotransferase. An essential enzyme in the gamma-glutamyl cycle. *J. Biol. Chem.* **2008**, *283*, 22031–22042, Erratum in *J. Biol. Chem.* **2008**, *283*, 32152. [CrossRef]

191. Shenoy, N.; Stenson, M.; Lawson, J.; Abeykoon, J.; Patnaik, M.; Wu, X.; Witzig, T. Drugs with anti-oxidant properties can interfere with cell viability measurements by assays that rely on the reducing property of viable cells. *Lab. Investig.* **2017**, *97*, 494–497. [CrossRef]

192. Panigrahy, S.K.; Bhatt, R.; Kumar, A. Reactive oxygen species: Sources, consequences and targeted therapy in type 2 diabetes. *J. Drug Target.* **2016**, *25*, 93–101. [CrossRef]

193. Lutchmansingh, F.K.; Hsu, J.W.; Bennett, F.I.; Badaloo, A.; McFarlane-Anderson, N.; Gordon-Strachan, G.M.; Wright-Pascoe, R.A.; Jahoor, F.; Boyne, M.S. Glutathione metabolism in type 2 diabetes and its relationship with microvascular complications and glycemia. *PLoS ONE* **2018**, *13*, e0198626. [CrossRef] [PubMed]

194. Zhao, R.-Z.; Jiang, S.; Zhang, L.; Yu, Z.-B. Mitochondrial electron transport chain, ROS generation and uncoupling (Review). *Int. J. Mol. Med.* **2019**, *44*, 3–15. [CrossRef] [PubMed]

195. Magnani, F.; Mattevi, A. Structure and mechanisms of ROS generation by NADPH oxidases. *Curr. Opin. Struct. Biol.* **2019**, *59*, 91–97. [CrossRef] [PubMed]

196. Sies, H.; Jones, D.P. Reactive oxygen species (ROS) as pleiotropic physiological signalling agents. *Nat. Rev. Mol. Cell Biol.* **2020**, *21*, 363–383. [CrossRef]

197. Manea, A. NADPH oxidase-derived reactive oxygen species: Involvement in vascular physiology and pathology. *Cell Tissue Res.* **2010**, *342*, 325–339. [CrossRef]

198. Pickering, R.J.; Rosado, C.J.; Sharma, A.; Buksh, S.; Tate, M.; de Haan, J.B. Recent novel approaches to limit oxidative stress and inflammation in diabetic complications. *Clin. Transl. Immunol.* **2018**, *7*, e1016. [CrossRef]

199. Bigagli, E.; Lodovici, M. Circulating Oxidative Stress Biomarkers in Clinical Studies on Type 2 Diabetes and Its Complications. *Oxidative Med. Cell. Longev.* **2019**, *2019*, 1–17. [CrossRef]

200. Teodoro, J.; Nunes, S.R.R.P.; Rolo, A.; Reis, F.; Palmeira, C.M. Therapeutic Options Targeting Oxidative Stress, Mitochondrial Dysfunction and Inflammation to Hinder the Progression of Vascular Complications of Diabetes. *Front. Physiol.* **2019**, *9*, 1857. [CrossRef]

201. Vignais, P.V. The superoxide-generating NADPH oxidase: Structural aspects and activation mechanism. *Cell. Mol. Life Sci. CMLS* **2002**, *59*, 1428–1459. [CrossRef]

202. Sahoo, S.; Meijles, D.; Pagano, P.J. NADPH oxidases: Key modulators in aging and age-related cardiovascular diseases? *Clin. Sci.* **2016**, *130*, 317–335. [CrossRef]

203. Rovira-Llopis, S.; Rocha, M.; Falcon, R.; De Pablo, C.; Alvarez, A.; Jover, A.; Hernandez-Mijares, A.; Victor, V.M. Is Myeloperoxidase a Key Component in the ROS-Induced Vascular Damage Related to Nephropathy in Type 2 Diabetes? *Antioxidants Redox Signal.* **2013**, *19*, 1452–1458. [CrossRef]

204. Banerjee, M.; Vats, P. Reactive metabolites and antioxidant gene polymorphisms in type 2 diabetes mellitus. *Indian J. Hum. Genet.* **2014**, *20*, 10–19. [CrossRef] [PubMed]

205. Król, M.; Kepinska, M. Human Nitric Oxide Synthase−Its Functions, Polymorphisms, and Inhibitors in the Context of Inflammation, Diabetes and Cardiovascular Diseases. *Int. J. Mol. Sci.* **2020**, *22*, 56. [CrossRef] [PubMed]

206. Santos, K.G.; Crispim, D.; Canani, L.H.; Ferrugem, P.T.; Gross, J.L.; Roisenberg, I. Relationship of endothelial nitric oxide synthase (*eNOS*) gene polymorphisms with diabetic retinopathy in Caucasians with type 2 diabetes. *Ophthalmic Genet.* **2011**, *33*, 23–27. [CrossRef] [PubMed]

207. Porojan, M.D.; Catana, A.; Popp, R.; Dumitrascu, D.L.; Bala, C. The role of NOS2A -954G/C and vascular endothelial growth factor +936C/T polymorphisms in type 2 diabetes mellitus and diabetic nonproliferative retinopathy risk management. *Ther. Clin. Risk Manag.* **2015**, *11*, 1743–1748. [CrossRef]

208. Priščáková, P.; Minárik, G.; Repiská, V. Candidate gene studies of diabetic retinopathy in human. *Mol. Biol. Rep.* **2016**, *43*, 1327–1345. [CrossRef]
209. Chen, F.; Li, Y.-M.; Yang, L.-Q.; Zhong, C.-G.; Zhuang, Z.-X. Association of NOS2 and NOS3 gene polymorphisms with susceptibility to type 2 diabetes mellitus and diabetic nephropathy in the Chinese Han population. *IUBMB Life* **2016**, *68*, 516–525. [CrossRef]
210. Ganjifrockwala, F.; Joseph, J.; George, G. Decreased total antioxidant levels and increased oxidative stress in South African type 2 diabetes mellitus patients. *J. Endocrinol. Metab. Diabetes S. Afr.* **2017**, *22*, 21–25. [CrossRef]
211. Heinonen, S.; Lautala, S.; Koivuniemi, A.; Bunker, A. Insights into the behavior of unsaturated diacylglycerols in mixed lipid bilayers in relation to protein kinase C activation—A molecular dynamics simulation study. *Biochim. Biophys. Acta (BBA) Biomembr.* **2022**, *1864*, 183961. [CrossRef]
212. Vats, P.; Sagar, N.; Singh, T.P.; Banerjee, M. Association of *Superoxide dismutases* (SOD1 and SOD2) and *Glutathione peroxidase 1 (GPx1)* gene polymorphisms with Type 2 diabetes mellitus. *Free Radic. Res.* **2014**, *49*, 17–24. [CrossRef]
213. Forman, H.J.; Zhang, H.; Rinna, A. Glutathione: Overview of its protective roles, measurement, and biosynthesis. *Mol. Asp. Med.* **2009**, *30*, 1–12. [CrossRef] [PubMed]
214. Pizzorno, J. Glutathione! *Integr Med.* **2014**, *13*, 8–12.
215. Diotallevi, M.; Checconi, P.; Palamara, A.T.; Celestino, I.; Coppo, L.; Holmgren, A.; Abbas, K.; Peyrot, F.; Mengozzi, M.; Ghezzi, P. Glutathione Fine-Tunes the Innate Immune Response toward Antiviral Pathways in a Macrophage Cell Line Independently of Its Antioxidant Properties. *Front. Immunol.* **2017**, *8*, 1239. [CrossRef] [PubMed]
216. Bajic, V.P.; Van Neste, C.; Obradovic, M.; Zafirovic, S.; Radak, D.; Bajic, V.B.; Essack, M.; Isenovic, E.R. Glutathione "Redox Homeostasis" and Its Relation to Cardiovascular Disease. *Oxidative Med. Cell. Longev.* **2019**, *2019*, 5028181. [CrossRef]
217. McBean, G.J.; Aslan, M.; Griffiths, H.R.; Torrão, R.C. Thiol redox homeostasis in neurodegenerative disease. *Redox Biol.* **2015**, *5*, 186–194. [CrossRef]
218. Garg, S.S.; Gupta, J. Polyol pathway and redox balance in diabetes. *Pharmacol. Res.* **2022**, *182*, 106326. [CrossRef]
219. Hurrle, S.; Hsu, W.H. The etiology of oxidative stress in insulin resistance. *Biomed J.* **2017**, *40*, 257–262. [CrossRef]
220. Gerber, P.A.; Rutter, G.A. The Role of Oxidative Stress and Hypoxia in Pancreatic Beta-Cell Dysfunction in Diabetes Mellitus. *Antioxid. Redox Signal.* **2017**, *26*, 501–518. [CrossRef]
221. Yaribeygi, H.; Farrokhi, F.R.; Butler, A.E.; Sahebkar, A. Insulin resistance: Review of the underlying molecular mechanisms. *J. Cell. Physiol.* **2018**, *234*, 8152–8161. [CrossRef]
222. Traverso, N.; Furfaro, A.L.; Nitti, M.; Marengo, B.; Domenicotti, C.; Cottalasso, D.; Marinari, U.M.; Pronzato, M.A. Impaired synthesis contributes to diabetes-induced decrease in liver glutathione. *Int. J. Mol. Med.* **2012**, *29*, 899–905. [CrossRef]
223. Zhang, J.; An, H.; Ni, K.; Chen, B.; Li, H.; Li, Y.; Sheng, G.; Zhou, C.; Xie, M.; Chen, S.; et al. Glutathione prevents chronic oscillating glucose intake-induced β-cell dedifferentiation and failure. *Cell Death Dis.* **2019**, *10*, 321. [CrossRef]
224. Sekhar, R.V. GlyNAC Supplementation Improves Glutathione Deficiency, Oxidative Stress, Mitochondrial Dysfunction, Inflammation, Aging Hallmarks, Metabolic Defects, Muscle Strength, Cognitive Decline, and Body Composition: Implications for Healthy Aging. *J. Nutr.* **2021**, *151*, 3606–3616. [CrossRef] [PubMed]
225. De Mattia, G.; Bravi, M.; Laurenti, O.; Cassone-Faldetta, M.; Armiento, A.; Ferri, C.; Balsano, F. Influence of reduced glutathione infusion on glucose metabolism in patients with non—Insulin-dependent diabetes mellitus. *Metabolism* **1998**, *47*, 993–997. [CrossRef] [PubMed]
226. Kim, J.H.; Moon, M.K.; Kim, S.W.; Shin, H.D.; Hwang, Y.H.; Ahn, C.; Lee, H.K. Glutathion S-Transferase M1 Gene Polymorphism is Associated with Type 2 Diabetic Nephropathy. *Korean Diabetes J.* **2005**, *29*, 315–321.
227. Murakami, K.; Takahito, K.; Ohtsuka, Y.; Fujiwara, Y.; Shimada, M.; Kawakami, Y. Impairment of glutathione metabolism in erythrocytes from patients with diabetes mellitus. *Metabolism* **1989**, *38*, 753–758. [CrossRef]
228. Yoshida, K.; Hirokawa, J.; Tagami, S.; Kawakami, Y.; Urata, Y.; Kondo, T. Weakened cellular scavenging activity against oxidative stress in diabetes mellitus: Regulation of glutathione synthesis and efflux. *Diabetologia* **1995**, *38*, 201–210. [CrossRef] [PubMed]
229. Samiec, P.S.; Drews-Botsch, C.; Flagg, E.W.; Kurtz, J.C.; Sternberg, P.; Reed, R.L.; Jones, D.P. Glutathione in Human Plasma: Decline in Association with Aging, Age-Related Macular Degeneration, and Diabetes. *Free Radic. Biol. Med.* **1998**, *24*, 699–704. [CrossRef] [PubMed]
230. Fu, J.; Cui, Q.; Yang, B.; Hou, Y.; Wang, H.; Xu, Y.; Wang, D.; Zhang, Q.; Pi, J. The impairment of glucose-stimulated insulin secretion in pancreatic β-cells caused by prolonged glucotoxicity and lipotoxicity is associated with elevated adaptive antioxidant response. *Food Chem. Toxicol.* **2017**, *100*, 161–167. [CrossRef] [PubMed]
231. Neuschwander-Tetri, B.A.; Presti, M.E.; Wells, L.D. Glutathione Synthesis in the Exocrine Pancreas. *Pancreas* **1997**, *14*, 342–349. [CrossRef]
232. Wallig, M.A. Xenobiotic Metabolism, Oxidant Stress and Chronic Pancreatitis. *Digestion* **1998**, *59*, 13–24. [CrossRef]
233. Lu, S.C. Dysregulation of glutathione synthesis in liver disease. *Liver Res.* **2020**, *4*, 64–73. [CrossRef]
234. Githens, S. Glutathione metabolism in the pancreas compared with that in the liver, kidney, and small intestine. *Int. J. Pancreatol.* **1991**, *8*, 97–109. [CrossRef] [PubMed]
235. Inoue, M. Glutathionists in the battlefield of gamma-glutamyl cycle. *Arch. Biochem. Biophys.* **2016**, *595*, 61–63. [CrossRef] [PubMed]

236. Noordam, R.; Smit, R.A.; Postmus, I.; Trompet, S.; Van Heemst, D. Assessment of causality between serum gamma-glutamyltransferase and type 2 diabetes mellitus using publicly available data: A Mendelian randomization study. *Leuk. Res.* **2016**, *45*, 1953–1960. [CrossRef] [PubMed]

237. Nano, J.; Muka, T.; Ligthart, S.; Hofman, A.; Murad, S.D.; LA Janssen, H.; Franco, O.H.; Dehghan, A. Gamma-glutamyltransferase levels, prediabetes and type 2 diabetes: A Mendelian randomization study. *Leuk. Res.* **2017**, *46*, 1400–1409. [CrossRef] [PubMed]

238. Locke, J.M.; Hysenaj, G.; Wood, A.R.; Weedon, M.N.; Harries, L.W. Targeted Allelic Expression Profiling in Human Islets Identifies *cis*-Regulatory Effects for Multiple Variants Identified by Type 2 Diabetes Genome-Wide Association Studies. *Diabetes* **2014**, *64*, 1484–1491. [CrossRef]

239. Ding, L.; Fan, L.; Xu, X.; Fu, J.; Xue, Y. Identification of core genes and pathways in type 2 diabetes mellitus by bioinformatics analysis. *Mol. Med. Rep.* **2019**, *20*, 2597–2608. [CrossRef]

240. Azarova, I.; Klyosova, E.; Polonikov, A. The Link between Type 2 Diabetes Mellitus and the Polymorphisms of Glutathione-Metabolizing Genes Suggests a New Hypothesis Explaining Disease Initiation and Progression. *Life* **2021**, *11*, 886. [CrossRef]

241. Arolas, J.L.; Aviles, F.X.; Chang, J.-Y.; Ventura, S. Folding of small disulfide-rich proteins: Clarifying the puzzle. *Trends Biochem. Sci.* **2006**, *31*, 292–301. [CrossRef]

242. Tsunoda, S.; Avezov, E.; Zyryanova, A.; Konno, T.; Mendes-Silva, L.; Melo, E.P.; Harding, H.P.; Ron, D. Intact protein folding in the glutathione-depleted endoplasmic reticulum implicates alternative protein thiol reductants. *Elife* **2014**, *3*, e03421. [CrossRef]

243. Costes, S.; Langen, R.; Gurlo, T.; Matveyenko, A.V.; Butler, P.C. β-Cell Failure in Type 2 Diabetes: A Case of Asking Too Much of Too Few? *Diabetes* **2013**, *62*, 327–335. [CrossRef] [PubMed]

244. Sun, J.; Cui, J.; He, Q.; Chen, Z.; Arvan, P.; Liu, M. Proinsulin misfolding and endoplasmic reticulum stress during the development and progression of diabetes ☆. *Mol. Asp. Med.* **2015**, *42*, 105–110. [CrossRef] [PubMed]

245. Liu, M.; Weiss, M.A.; Arunagiri, A.; Yong, J.; Rege, N.; Sun, J.; Haataja, L.; Kaufman, R.J.; Arvan, P. Biosynthesis, structure, and folding of the insulin precursor protein. *Diabetes, Obes. Metab.* **2018**, *20*, 28–50. [CrossRef] [PubMed]

246. Arunagiri, A.; Haataja, L.; Pottekat, A.; Pamenan, F.; Kim, S.; Zeltser, L.M.; Paton, A.W.; Paton, J.C.; Tsai, B.; Itkin-Ansari, P.; et al. Proinsulin misfolding is an early event in the progression to type 2 diabetes. *Elife* **2019**, *8*, e44532. [CrossRef] [PubMed]

247. Scheuner, D.; Kaufman, R.J. The Unfolded Protein Response: A Pathway That Links Insulin Demand with β-Cell Failure and Diabetes. *Endocr. Rev.* **2008**, *29*, 317–333, Erratum in *Endocr. Rev.* **2008**, *29*, 631. [CrossRef]

248. Moreno-Gonzalez, I.; Edwards, G., III; Salvadores, N.; Shahnawaz, M.; Diaz-Espinoza, R.; Soto, C. Molecular interaction between type 2 diabetes and Alzheimer's disease through cross-seeding of protein misfolding. *Mol. Psychiatry* **2017**, *22*, 1327–1334. [CrossRef] [PubMed]

249. Azarova, I.; Bushueva, O.; Konoplya, A.; Polonikov, A. Glutathione S-transferase genes and the risk of type 2 diabetes mellitus: Role of sexual dimorphism, gene-gene and gene-smoking interactions in disease susceptibility. *J. Diabetes* **2018**, *10*, 398–407. [CrossRef] [PubMed]

250. Tang, S.-T.; Wang, C.-J.; Tang, H.-Q.; Zhang, Q.; Wang, Y. Evaluation of glutathione S-transferase genetic variants affecting type 2 diabetes susceptibility: A meta-analysis. *Gene* **2013**, *530*, 301–308. [CrossRef]

251. Mastana, S.S.; Kaur, A.; Hale, R.; Lindley, M.R. Influence of glutathione S-transferase polymorphisms (GSTT1, GSTM1, GSTP1) on type-2 diabetes mellitus (T2D) risk in an endogamous population from north India. *Mol. Biol. Rep.* **2013**, *40*, 7103–7110. [CrossRef] [PubMed]

252. Banerjee, M.; Bid, H.; Konwar, R.; Saxena, M.; Chaudhari, P.; Agrawal, C. Association of glutathione S-transferase (GSTM1, T1 and P1) gene polymorphisms with type 2 diabetes mellitus in north Indian population. *J. Postgrad. Med.* **2010**, *56*, 176–181. [CrossRef]

253. Gönül, N.; Kadioglu, E.; Kocabaş, N.A.; Özkaya, M.; Karakaya, A.E.; Karahalil, B. The role of GSTM1, GSTT1, GSTP1, and OGG1 polymorphisms in type 2 diabetes mellitus risk: A case–control study in a Turkish population. *Gene* **2012**, *505*, 121–127. [CrossRef] [PubMed]

254. Moasser, E.; Kazemi-Nezhad, S.R.; Saadat, M.; Azarpira, N. Study of the association between glutathione S-transferase (GSTM1, GSTT1, GSTP1) polymorphisms with type II diabetes mellitus in southern of Iran. *Mol. Biol. Rep.* **2012**, *39*, 10187–10192. [CrossRef]

255. Stoian, A.; Bănescu, C.; Bălaşa, R.I.; Moţăţăianu, A.; Stoian, M.; Moldovan, V.G.; Voidăzan, S.; Dobreanu, M. Influence of *GSTM1*, *GSTT1*, and *GSTP1* Polymorphisms on Type 2 Diabetes Mellitus and Diabetic Sensorimotor Peripheral Neuropathy Risk. *Dis. Markers* **2015**, *2015*, 1–10. [CrossRef] [PubMed]

256. Azarova, I.; Klyosova, E.; Lazarenko, V.; Konoplya, A.; Polonikov, A. Genetic variants in glutamate cysteine ligase confer protection against type 2 diabetes. *Mol. Biol. Rep.* **2020**, *47*, 1–13. [CrossRef] [PubMed]

257. Azarova, I.E.; Klyosova, E.Y.; Polonikov, A.V. Polymorphic variants of glutathione reductase–new genetic markers of predisposition to type 2 diabetes mellitus. *Ter. arkhiv* **2021**, *93*, 1164–1170. [CrossRef] [PubMed]

258. Azarova, I.E.; Klyosova, E.Y.; Churilin, M.I.; Samgina, T.A.; Konoplya, A.I.; Polonikov, A.V. Genetic and biochemical investigation of the gamma-glutamylcyclotransferase role in predisposition to type 2 diabetes mellitus. *Ecol. Genet.* **2020**, *18*, 215–228. [CrossRef]

259. Ramprasath, T.; Murugan, P.S.; Kalaiarasan, E.; Gomathi, P.; Rathinavel, A.; Selvam, G.S. Genetic association of Glutathione peroxidase-1 (GPx-1) and NAD(P)H:Quinone Oxidoreductase 1(NQO1) variants and their association of CAD in patients with type-2 diabetes. *Mol. Cell. Biochem.* **2011**, *361*, 143–150. [CrossRef]

260. Azarova, I.E.; Klyosova, E.; Samgina, T.A.; Bushueva, O.; Yu Azarova, V.A.; Konoplya, A.I.; Polonikov, A.V. Polymorphic variant in gpx2 gene (rs4902346) and predisposition to type 2 diabetes mellitus. *Med. Genet.* **2020**, *19*, 17–27. (In Russian) [CrossRef]

261. Ghattas, M.H.; Abo-Elmatty, D.M. Association of Polymorphic Markers of the Catalase and Superoxide Dismutase Genes with Type 2 Diabetes Mellitus. *DNA Cell Biol.* **2012**, *31*, 1598–1603. [CrossRef]

262. Yang, Y.; Xie, X.; Jin, A. Genetic polymorphisms in extracellular superoxide dismutase Leu53Leu, Arg213Gly, and Ala40Thr and susceptibility to type 2 diabetes mellitus. *Genet. Mol. Res.* **2016**, *15*, gmr15048418. [CrossRef]

263. Nakamura, S.-I.; Kugiyama, K.; Sugiyama, S.; Miyamoto, S.; Koide, S.-I.; Fukushima, H.; Honda, O.; Yoshimura, M.; Ogawa, H. Polymorphism in the 5′-Flanking Region of Human Glutamate-Cysteine Ligase Modifier Subunit Gene Is Associated With Myocardial Infarction. *Circulation* **2002**, *105*, 2968–2973. [CrossRef] [PubMed]

264. Ma, N.; Liu, W.; Zhang, X.; Gao, X.; Yu, F.; Guo, W.; Meng, Y.; Gao, P.; Zhou, J.; Yuan, M.; et al. Oxidative Stress-Related Gene Polymorphisms Are Associated With Hepatitis B Virus-Induced Liver Disease in the Northern Chinese Han Population. *Front. Genet.* **2020**, *10*, 1290. [CrossRef] [PubMed]

265. Santos-Rosa, H.; Schneider, R.; Bernstein, B.E.; Karabetsou, N.; Morillon, A.; Weise, C.; Schreiber, S.L.; Mellor, J.; Kouzarides, T. Methylation of Histone H3 K4 Mediates Association of the Isw1p ATPase with Chromatin. *Mol. Cell* **2003**, *12*, 1325–1332, Erratum in *Mol. Cell.* **2018**, *70*, 983. [CrossRef] [PubMed]

266. Creyghton, M.P.; Cheng, A.W.; Welstead, G.G.; Kooistra, T.; Carey, B.W.; Steine, E.J.; Hanna, J.; Lodato, M.A.; Frampton, G.M.; Sharp, P.A.; et al. Histone H3K27ac separates active from poised enhancers and predicts developmental state. *Proc. Natl. Acad. Sci. USA* **2010**, *107*, 21931–21936. [CrossRef] [PubMed]

267. Tjeertes, J.V.; Miller, K.M.; Jackson, S.P. Screen for DNA-damage-responsive histone modifications identifies H3K9Ac and H3K56Ac in human cells. *EMBO J.* **2009**, *28*, 1878–1889. [CrossRef] [PubMed]

268. Tang, W.; Basu, S.; Kong, X.; Pankow, J.S.; Aleksic, N.; Tan, A.; Cushman, M.; Boerwinkle, E.; Folsom, A.R. Genome-wide association study identifies novel loci for plasma levels of protein C: The ARIC study. *Blood* **2010**, *116*, 5032–5036. [CrossRef]

269. Long, Y.; Jia, D.; Wei, L.; Yang, Y.; Tian, H.; Chen, T. Liver-Specific Overexpression of Gamma-Glutamyltransferase Ameliorates Insulin Sensitivity of Male C57BL/6 Mice. *J. Diabetes Res.* **2017**, *2017*, 2654520. [CrossRef]

270. Sabanayagam, C.; Shankar, A.; Li, J.; Pollard, C.; Ducatman, A. Serum gamma-glutamyl transferase level and diabetes mellitus among US adults. *Eur. J. Epidemiology* **2009**, *24*, 369–373. [CrossRef]

271. Zhao, W.; Tong, J.; Liu, J.; Liu, J.; Li, J.; Cao, Y. The Dose-Response Relationship between Gamma-Glutamyl Transferase and Risk of Diabetes Mellitus Using Publicly Available Data: A Longitudinal Study in Japan. *Int. J. Endocrinol.* **2020**, *2020*, 5356498. [CrossRef]

272. Lee, Y.S.; Cho, Y.; Burgess, S.; Smith, G.D.; Relton, C.L.; Shin, S.-Y.; Shin, M.-J. Serum gamma-glutamyl transferase and risk of type 2 diabetes in the general Korean population: A Mendelian randomization study. *Hum. Mol. Genet.* **2016**, *25*, 3877–3886. [CrossRef]

273. Jinnouchi, H.; Morita, K.; Tanaka, T.; Kajiwara, A.; Kawata, Y.; Oniki, K.; Saruwatari, J.; Nakagawa, K.; Otake, K.; Ogata, Y.; et al. Interactive effects of a common γ-glutamyltransferase 1 variant and low high-density lipoprotein-cholesterol on diabetic macro- and micro-angiopathy. *Cardiovasc. Diabetol.* **2015**, *14*, 1–11. [CrossRef] [PubMed]

274. Diergaarde, B.; Brand, R.; Lamb, J.; Cheong, S.Y.; Stello, K.; Barmada, M.M.; Feingold, E.; Whitcomb, D.C. Pooling-Based Genome-Wide Association Study Implicates Gamma-Glutamyltransferase 1 (GGT1) Gene in Pancreatic Carcinogenesis. *Pancreatology* **2010**, *10*, 194–200. [CrossRef] [PubMed]

275. Brand, H.; Diergaarde, B.; O'Connell, M.R.; Whitcomb, D.C.; Brand, R.E. Variation in the γ-Glutamyltransferase 1 Gene and Risk of Chronic Pancreatitis. *Pancreas* **2013**, *42*, 836–840. [CrossRef] [PubMed]

276. Yuan, X.; Waterworth, D.; Perry, J.R.; Lim, N.; Song, K.; Chambers, J.C.; Zhang, W.; Vollenweider, P.; Stirnadel, H.; Johnson, T.; et al. Population-Based Genome-wide Association Studies Reveal Six Loci Influencing Plasma Levels of Liver Enzymes. *Am. J. Hum. Genet.* **2008**, *83*, 520–528. [CrossRef]

277. Cho, Y.-G.; Park, K.H.; Kim, C.-W.; Hur, Y.-I. The Relationship between Serum Gamma-glutamyltransferase Level and Overweight in Korean Urban Children. *Korean J. Fam. Med.* **2011**, *32*, 182–188. [CrossRef]

278. Wickham, S.; West, M.B.; Cook, P.F.; Hanigan, M.H. Gamma-glutamyl compounds: Substrate specificity of gamma-glutamyl transpeptidase enzymes. *Anal. Biochem.* **2011**, *414*, 208–214. [CrossRef]

279. Shin, D.M.S.; Yon, D.K. Apparently healthy adults with high serum gamma-glutamyl transferase levels are at increased risk of asthma development in the near future: A Korean nationwide cohort study. *Eur. Rev. Med. Pharmacol. Sci.* **2022**, *26*, 3958–3966. [CrossRef]

280. Astle, W.J.; Elding, H.; Jiang, T.; Allen, D.; Ruklisa, D.; Mann, A.L.; Mead, D.; Bouman, H.; Riveros-Mckay, F.; Kostadima, M.A.; et al. The Allelic Landscape of Human Blood Cell Trait Variation and Links to Common Complex Disease. *Cell* **2016**, *167*, 1415–1429. [CrossRef]

281. Heisterkamp, N.; Groffen, J.; Warburton, D.; Sneddon, T.P. The human gamma-glutamyltransferase gene family. *Hum. Genet.* **2008**, *123*, 321–332. [CrossRef]

282. Whitfield, J.B. Gamma Glutamyl Transferase. *Crit. Rev. Clin. Lab. Sci.* **2001**, *38*, 263–355. [CrossRef]

283. GTEx Consortium. The GTEx Consortium atlas of genetic regulatory effects across human tissues. *Science* **2020**, *369*, 1318–1330. [CrossRef] [PubMed]

284. Uhlén, M.; Fagerberg, L.; Hallström, B.M.; Lindskog, C.; Oksvold, P.; Mardinoglu, A.; Sivertsson, Å.; Kampf, C.; Sjöstedt, E.; Asplund, A.; et al. Proteomics. Tissue-Based Map of the Human Proteome. *Science* **2015**, *347*, 1260419. [CrossRef]

285. Masuda, Y.; Maeda, S.; Watanabe, A.; Sano, Y.; Aiuchi, T.; Nakajo, S.; Itabe, H.; Nakaya, K. A novel 21-kDa cytochrome c-releasing factor is generated upon treatment of human leukemia U937 cells with geranylgeraniol. *Biochem. Biophys. Res. Commun.* **2006**, *346*, 454–460. [CrossRef]

286. Bocharova, J.A.; Azarova, J.E.; Klyosova EYu Drozdova, E.L.; Solodilova, M.A.; Polonikov, A.V. Gene of gamma-glutamylcyclotransferase, a key enzyme of glutathione catabolism, and predisposition to ischemic stroke: Association analysis and functional annotation of gene polymorphisms. *Med. Genet.* **2020**, *19*, 32–39. (In Russian) [CrossRef]

287. Singh, R.R.; Mohammad, J.; Orr, M.; Reindl, K.M. Glutathione S-Transferase pi-1 Knockdown Reduces Pancreatic Ductal Adenocarcinoma Growth by Activating Oxidative Stress Response Pathways. *Cancers* **2020**, *12*, 1501. [CrossRef] [PubMed]

288. Fujikawa, Y.; Mori, M.; Tsukada, M.; Miyahara, S.; Sato-Fukushima, H.; Watanabe, E.; Murakami-Tonami, Y.; Inoue, H. Pi-Class Glutathione S-transferase (GSTP1)-Selective Fluorescent Probes for Multicolour Imaging with Various Cancer-Associated Enzymes. *Chembiochem.* **2022**, *23*, e202200443. [CrossRef]

289. Bogaards, J.J.P.; Venekamp, J.C.; van Bladeren, P.J. Stereoselective Conjugation of Prostaglandin A$_2$ and Prostaglandin J$_2$ with Glutathione, Catalyzed by the Human Glutathione S-Transferases A1-1, A2-2, M1a-1a, and P1-1. *Chem. Res. Toxicol.* **1997**, *10*, 310–317. [CrossRef]

290. Al-Eitan, L.N.; Rababa'H, D.M.; Alkhatib, R.Q.; Khasawneh, R.H.; Aljarrah, O.A. GSTM1 and GSTP1 Genetic Polymorphisms and Their Associations With Acute Lymphoblastic Leukemia Susceptibility in a Jordanian Population. *J. Pediatr. Hematol.* **2016**, *38*, e223–e229. [CrossRef]

291. Hohaus, S.; Di Russo, A.; Di Febo, A.; Massini, G.; D'Alo', F.; Guidi, F.; Mansueto, G.; Voso, M.T.; Leone, G. Glutathione S-transferase P1 Genotype and Prognosis in Hodgkin's Lymphoma. *Clin. Cancer Res.* **2005**, *11*, 2175–2179. [CrossRef]

292. Lu, S.; Wang, Z.; Cui, D.; Liu, H.; Hao, X. Glutathione S-transferase P1 Ile105Val polymorphism and breast cancer risk: A meta-analysis involving 34,658 subjects. *Breast Cancer Res. Treat.* **2010**, *125*, 253–259. [CrossRef]

293. Li, X.-M.; Yu, X.-W.; Yuan, Y.; Pu, M.-Z.; Zhang, H.-X.; Wang, K.-J.; Han, X.-D. Glutathione S-transferase P1, gene-gene interaction, and lung cancer susceptibility in the Chinese population: An updated meta-analysis and review. *J. Cancer Res. Ther.* **2015**, *11*, 565. [CrossRef] [PubMed]

294. Martins, F.D.O.; Gomes, B.C.; Rodrigues, A.S.; Rueff, J. Genetic Susceptibility in Acute Pancreatitis: Genotyping of GSTM1, GSTT1, GSTP1, CASP7, CASP8, CASP9, CASP10, LTA, TNFRSF1B, and TP53 Gene Variants. *Pancreas* **2017**, *46*, 71–76. [CrossRef] [PubMed]

295. Wang, M.; Li, Y.; Lin, L.; Song, G.; Deng, T. GSTM1 Null Genotype and GSTP1 Ile105Val Polymorphism Are Associated with Alzheimer's Disease: A Meta-Analysis. *Mol. Neurobiol.* **2015**, *53*, 1355–1364. [CrossRef] [PubMed]

296. Minelli, C.; Granell, R.; Newson, R.; Rose-Zerilli, M.; Torrent, M.; Ring, S.M.; Holloway, J.; Shaheen, S.; Henderson, J. Glutathione-S-transferase genes and asthma phenotypes: A Human Genome Epidemiology (HuGE) systematic review and meta-analysis including unpublished data. *Leuk. Res.* **2009**, *39*, 539–562. [CrossRef]

297. MacIntyre, E.A.; Brauer, M.; Melén, E.; Bauer, C.P.; Bauer, M.; Berdel, D.; Bergström, A.; Brunekreef, B.; Chan-Yeung, M.; Klümper, C.; et al. *GSTP1* and *TNF* Gene Variants and Associations between Air Pollution and Incident Childhood Asthma: The Traffic, Asthma and Genetics (TAG) Study. *Environ. Health Perspect.* **2014**, *122*, 418–424. [CrossRef]

298. Huang, S.-X.; Wu, F.-X.; Luo, M.; Ma, L.; Gao, K.-F.; Li, J.; Wu, W.-J.; Huang, S.; Yang, Q.; Liu, K.; et al. The Glutathione S-Transferase P1 341C>T Polymorphism and Cancer Risk: A Meta-Analysis of 28 Case-Control Studies. *PLoS ONE* **2013**, *8*, e56722. [CrossRef]

299. Kelada, S.N.; Stapleton, P.L.; Farin, F.M.; Bammler, T.K.; Eaton, D.L.; Smith-Weller, T.; Franklin, G.M.; Swanson, P.D.; Longstreth, W.; Checkoway, H. Glutathione S-transferase M1, T1, and P1 Polymorphisms and Parkinson's Disease. *Neurosci. Lett.* **2002**, *337*, 5–8. [CrossRef]

300. Awasthi, Y.C.; Sharma, R.; Cheng, J.; Yang, Y.; Sharma, A.; Singhal, S.S.; Awasthi, S. Role of 4-hydroxynonenal in stress-mediated apoptosis signaling. *Mol. Asp. Med.* **2003**, *24*, 219–230. [CrossRef]

301. Mastrocola, R.; Restivo, F.; Vercellinatto, I.; Danni, O.; Brignardello, E.; Aragno, M.; Boccuzzi, G. Oxidative and nitrosative stress in brain mitochondria of diabetic rats. *J. Endocrinol.* **2005**, *187*, 37–44. [CrossRef]

302. Tanaka, Y.; Tran, P.O.T.; Harmon, J.; Robertson, R.P. A role for glutathione peroxidase in protecting pancreatic β cells against oxidative stress in a model of glucose toxicity. *Proc. Natl. Acad. Sci. USA* **2002**, *99*, 12363–12368. [CrossRef]

303. McClung, J.P.; Roneker, C.A.; Mu, W.; Lisk, D.J.; Langlais, P.; Liu, F.; Lei, X.G. Development of insulin resistance and obesity in mice overexpressing cellular glutathione peroxidase. *Proc. Natl. Acad. Sci. USA* **2004**, *101*, 8852–8857. [CrossRef] [PubMed]

304. Huang, J.-Q.; Zhou, J.-C.; Wu, Y.-Y.; Ren, F.-Z.; Lei, X.G. Role of glutathione peroxidase 1 in glucose and lipid metabolism-related diseases. *Free Radic. Biol. Med.* **2018**, *127*, 108–115. [CrossRef] [PubMed]

305. Mohammedi, K.; Patente, T.A.; Bellili-Muñoz, N.; Driss, F.; Le Nagard, H.; Fumeron, F.; Roussel, R.; Hadjadj, S.; Corrêa-Giannella, M.L.; Marre, M.; et al. Glutathione peroxidase-1 gene (GPX1) variants, oxidative stress and risk of kidney complications in people with type 1 diabetes. *Metabolism* **2015**, *65*, 12–19. [CrossRef] [PubMed]

306. Liu, D.; Liu, L.; Hu, Z.; Song, Z.; Wang, Y.; Chen, Z. Evaluation of the oxidative stress–related genes *ALOX5, ALOX5AP, GPX1, GPX3* and *MPO* for contribution to the risk of type 2 diabetes mellitus in the Han Chinese population. *Diabetes Vasc. Dis. Res.* **2018**, *15*, 336–339. [CrossRef]

307. Lewis, P.; Stefanovic, N.; Pete, J.; Calkin, A.; Giunti, S.; Thallas-Bonke, V.; Jandeleit-Dahm, K.A.; Allen, T.J.; Kola, I.; Cooper, M.E.; et al. Lack of the Antioxidant Enzyme Glutathione Peroxidase-1 Accelerates Atherosclerosis in Diabetic Apolipoprotein E–Deficient Mice. *Circulation* **2007**, *115*, 2178–2187. [CrossRef]

308. Wu, C.; Orozco, C.; Boyer, J.; Leglise, M.; Goodale, J.; Batalov, S.; Hodge, C.L.; Haase, J.; Janes, J.; Huss, J.W., 3rd; et al. BioGPS: An extensible and customizable portal for querying and organizing gene annotation resources. *Genome Biol.* **2009**, *10*, R130. [CrossRef]

309. DeFronzo, R.A. Banting Lecture. From the Triumvirate to the Ominous Octet: A New Paradigm for the Treatment of Type 2 Diabetes Mellitus. *Diabetes* **2009**, *58*, 773–795. [CrossRef]

310. Hobbs, G.A.; Zhou, B.; Cox, A.D.; Campbell, S.L. Rho GTPases, oxidation, and cell redox control. *Small GTPases* **2014**, *5*, e28579. [CrossRef]

311. Osmenda, G.; Matusik, P.T.; Sliwa, T.; Czesnikiewicz-Guzik, M.; Skupien, J.; Malecki, M.T.; Siedlinski, M. Nicotinamide adenine dinucleotide phosphate (NADPH) oxidase p22phox subunit polymorphisms, systemic oxidative stress, endothelial dysfunction, and atherosclerosis in type 2 diabetes mellitus. *Pol. Arch. Intern. Med.* **2021**, *131*, 447–454. [CrossRef]

312. Taylor, J.P.; Tse, H.M. The role of NADPH oxidases in infectious and inflammatory diseases. *Redox Biol.* **2021**, *48*, 102159. [CrossRef]

313. Chen, F.; Wang, Y.; Barman, S.; Fulton, D. Enzymatic regulation and functional relevance of NOX5. *Curr. Pharm. Des.* **2015**, *21*, 5999–6008. [CrossRef]

314. Azarova, I.E.; Klyosova EYu Samgina, T.A.; Sakali SYu Kolomoets, I.I.; Azarova, V.A.; Konoplya, A.I.; Polonikov, A.V. Role of cyba gene polymorphisms in pathogenesis of type 2 diabetes mellitus. *Med. Genet.* **2019**, *18*, 37–48. (In Russian) [CrossRef]

315. Sun, Q.; Yin, Y.; Zhu, Z.; Yan, Z. Association of the C242T polymorphism in the NAD(P)H oxidase P22 phox gene with type 2 diabetes mellitus risk: A meta-analysis. *Curr. Med. Res. Opin.* **2013**, *30*, 415–422. [CrossRef] [PubMed]

316. Azarova, I.E.; Klyosova, E.Y.; Kolomoets, I.I.; Azarova, V.A.; Ivakin, V.E.; Konoplya, A.I.; Polonikov, A.V. Polymorphisms of the Gene Encoding Cytochrome b-245 Beta Chain of NADPH Oxidase: Relationship with Redox Homeostasis Markers and Risk of Type 2 Diabetes Mellitus. *Russ. J. Genet.* **2020**, *56*, 856–862. [CrossRef]

317. Azarova, I.E.; Klyosova, E.Y.; Kolomoets, I.I.; Polonikov, A.V. Polymorphic Variants of the Neutrophil Cytosolic Factor 2 Gene: Associations with Susceptibility to Type 2 Diabetes Mellitus and Cardiovascular Autonomic Neuropathy. *Russ. J. Genet.* **2022**, *58*, 593–602. [CrossRef]

318. Moreno, M.U.; José, G.S.; Orbe, J.; Páramo, J.; Beloqui, O.; Díez, J.; Zalba, G. Preliminary characterisation of the promoter of the human p22^{phox} gene: Identification of a new polymorphism associated with hypertension. *FEBS Lett.* **2003**, *542*, 27–31, Erratum in *FEBS Lett.* **2010**, *584*, 4709. [CrossRef] [PubMed]

319. Schirmer, M.; Hoffmann, M.; Kaya, E.; Tzvetkov, M.; Brockmoller, J. Genetic polymorphisms of NAD(P)H oxidase: Variation in subunit expression and enzyme activity. *Pharm. J.* **2007**, *8*, 297–304. [CrossRef]

320. Belambri, S.A.; Rolas, L.; Raad, H.; Hurtado-Nedelec, M.; Dang, P.M.-C.; El-Benna, J. NADPH oxidase activation in neutrophils: Role of the phosphorylation of its subunits. *Eur. J. Clin. Investig.* **2018**, *48*, e12951. [CrossRef]

321. Meijles, D.N.; Fan, L.M.; Ghazaly, M.M.; Howlin, B.; Krönke, M.; Brooks, G.; Li, J.-M. p22 phox C242T Single-Nucleotide Polymorphism Inhibits Inflammatory Oxidative Damage to Endothelial Cells and Vessels. *Circulation* **2016**, *133*, 2391–2403. [CrossRef]

322. Fang, S.; Wang, L.; Jia, C. Association of p22phox gene C242T polymorphism with coronary artery disease: A meta-analysis. *Thromb. Res.* **2010**, *125*, e197–e201. [CrossRef]

323. Snahnicanova, Z.; Mendelova, A.; Grendar, M.; Holubekova, V.; Kostkova, M.; Pozorciakova, K.; Jancinová, M.; Kasubova, I.; Vojtkova, J.; Durdik, P.; et al. Association of Polymorphisms in CYBA, SOD1, and CAT Genes with Type 1 Diabetes and Diabetic Peripheral Neuropathy in Children and Adolescents. *Genet. Test. Mol. Biomark.* **2018**, *22*, 413–419. [CrossRef] [PubMed]

324. Petrovič, D. Association of the −262C/T polymorphism in the catalase gene promoter and the C242T polymorphism of the NADPH oxidase P22phox gene with essential arterial hypertension in patients with diabetes mellitus type 2. *Clin. Exp. Hypertens.* **2013**, *36*, 36–39. [CrossRef]

325. Yamamoto, Y.; Kiyohara, C.; Suetsugu-Ogata, S.; Hamada, N.; Nakanishi, Y. Biological interaction of cigarette smoking on the association between genetic polymorphisms involved in inflammation and the risk of lung cancer: A case-control study in Japan. *Oncol. Lett.* **2017**, *13*, 3873–3881. [CrossRef] [PubMed]

326. Holla, L.I.; Kaňková, K.; Znojil, V. Haplotype Analysis of the NADPH Oxidase p22phox Gene in Patients with Bronchial Asthma. *Int. Arch. Allergy Immunol.* **2008**, *148*, 73–80. [CrossRef] [PubMed]

327. José, G.S.; Moreno, M.U.; Oliván, S.; Beloqui, O.; Fortuño, A.; Díez, J.; Zalba, G. Functional Effect of the p22 ^{phox} −930 $^{A/G}$ Polymorphism on p22 ^{phox} Expression and NADPH Oxidase Activity in Hypertension. *Hypertension* **2004**, *44*, 163–169. [CrossRef]

328. Moreno, M.U.; José, G.S.; Fortuño, A.; Beloqui, O.; Díez, J.; Zalba, G. The C242T CYBA polymorphism of NADPH oxidase is associated with essential hypertension. *J. Hypertens.* **2006**, *24*, 1299–1306. [CrossRef]

329. Niemiec, P.; Nowak, T.; Iwanicki, T.; Krauze, J.; Górczyńska-Kosiorz, S.; Grzeszczak, W.; Ochalska-Tyka, A.; Żak, I. The −930A>G polymorphism of the CYBA gene is associated with premature coronary artery disease. A case–control study and gene–risk factors interactions. *Mol. Biol. Rep.* **2014**, *41*, 3287–3294. [CrossRef]

330. Patente, T.A.; Mohammedi, K.; Bellili-Muñoz, N.; Driss, F.; Sanchez, M.; Fumeron, F.; Roussel, R.; Hadjadj, S.; Corrêa-Giannella, M.L.; Marre, M.; et al. Allelic variations in the CYBA gene of NADPH oxidase and risk of kidney complications in patients with type 1 diabetes. *Free Radic. Biol. Med.* **2015**, *86*, 16–24. [CrossRef]

331. Frazão, J.B.; Thain, A.; Zhu, Z.; Luengo, M.; Condino-Neto, A.; Newburger, P.E. Regulation of *CYBB* Gene Expression in Human Phagocytes by a Distant Upstream NF-κB Binding Site. *J. Cell. Biochem.* **2015**, *116*, 2008–2017. [CrossRef]

332. Liu, Q.; Wang, J.; Sandford, A.J.; Wu, J.; Wang, Y.; Wu, S.; Ji, G.; Chen, G.; Feng, Y.; Tao, C.; et al. Association of CYBB polymorphisms with tuberculosis susceptibility in the Chinese Han population. *Infect. Genet. Evol.* **2015**, *33*, 169–175. [CrossRef]

333. Acosta-Herrera, M.; Kerick, M.; González-Serna, D.; Wijmenga, C.; Franke, A.; Gregersen, P.K.; Padyukov, L.; Worthington, J.; Vyse, T.J.; Alarcón-Riquelme, M.E.; et al. Genome-wide meta-analysis reveals shared new *loci* in systemic seropositive rheumatic diseases. *Ann. Rheum. Dis.* **2018**, *78*, 311–319. [CrossRef] [PubMed]

334. Gutierrez-Achury, J.; Zorro, M.M.; Ricaño-Ponce, I.; Zhernakova, D.V.; Diogo, D.; Raychaudhuri, S.; Franke, L.; Trynka, G.; Wijmenga, C.; Zhernakova, A. Functional implications of disease-specific variants in loci jointly associated with coeliac disease and rheumatoid arthritis. *Hum. Mol. Genet.* **2015**, *25*, 180–190. [CrossRef] [PubMed]

335. Yu, B.; Chen, Y.; Wu, Q.; Li, P.; Shao, Y.; Zhang, J.; Zhong, Q.; Peng, X.; Yang, H.; Hu, X.; et al. The association between single-nucleotide polymorphisms of NCF2 and systemic lupus erythematosus in Chinese mainland population. *Clin. Rheumatol.* **2010**, *30*, 521–527. [CrossRef] [PubMed]

336. Sumimoto, H. Structure, regulation and evolution of Nox-family NADPH oxidases that produce reactive oxygen species. *FEBS J.* **2008**, *275*, 3249–3277, Erratum in *FEBS J.* **2008**, *275*, 3984. [CrossRef] [PubMed]

337. Zhang, T.-P.; Li, R.; Huang, Q.; Pan, H.-F.; Ye, D.-Q.; Li, X.-M. Association of NCF2, NCF4, and CYBA Gene Polymorphisms with Rheumatoid Arthritis in a Chinese Population. *J. Immunol. Res.* **2020**, *2020*, 8528976. [CrossRef]

338. Roberts, R.L.; Hollis-Moffatt, J.; Gearry, R.; Kennedy, M.; Barclay, M.; Merriman, T. Confirmation of association of IRGM and NCF4 with ileal Crohn's disease in a population-based cohort. *Genes Immun.* **2008**, *9*, 561–565. [CrossRef]

339. Ryan, B.M.; Zanetti, K.A.; Robles, A.I.; Schetter, A.J.; Goodman, J.; Hayes, R.D.; Huang, W.-Y.; Gunter, M.J.; Yeager, M.; Burdette, L.; et al. Germline variation in *NCF4*, an innate immunity gene, is associated with an increased risk of colorectal cancer. *Int. J. Cancer* **2013**, *134*, 1399–1407. [CrossRef]

340. Xing, Y.; Lin, Q.; Tong, Y.; Zhou, W.; Huang, J.; Wang, Y.; Huang, G.; Li, Y.; Xiang, Z.; Zhou, Z.; et al. Abnormal Neutrophil Transcriptional Signature May Predict Newly Diagnosed Latent Autoimmune Diabetes in Adults of South China. *Front. Endocrinol.* **2020**, *11*, 581902. [CrossRef]

341. Azarova, I.E. NCF4 gene polymorphism, level of glutathione and glycated hemoglobin in type 2 diabetics with coronary artery disease. *Med. Genet.* **2021**, *20*, 37–47. (In Russian) [CrossRef]

342. Vendrov, A.E.; Sumida, A.; Canugovi, C.; Lozhkin, A.; Hayami, T.; Madamanchi, N.R.; Runge, M.S. NOXA1-dependent NADPH oxidase regulates redox signaling and phenotype of vascular smooth muscle cell during atherogenesis. *Redox Biol.* **2019**, *21*, 101063. [CrossRef]

343. Ueno, N.; Takeya, R.; Miyano, K.; Kikuchi, H.; Sumimoto, H.; Ueno, N.; Takeya, R.; Miyano, K.; Kikuchi, H.; Sumimoto, H. The NADPH oxidase Nox3 constitutively produces superoxide in a p22phox-dependent manner: Its regulation by oxidase organizers and activators. *J. Biol. Chem.* **2005**, *280*, 23328–23339. [CrossRef] [PubMed]

344. Valente, A.J.; El Jamali, A.; Epperson, T.K.; Gamez, M.J.; Pearson, R.W.; Clark, R.A. NOX1 NADPH oxidase regulation by the NOXA1 SH3 domain. *Free Radic. Biol. Med.* **2007**, *43*, 384–396. [CrossRef] [PubMed]

345. Gimenez, M.; Schickling, B.M.; Lopes, L.R.; Miller, F.J. Nox1 in cardiovascular diseases: Regulation and pathophysiology. *Clin. Sci.* **2015**, *130*, 151–165. [CrossRef] [PubMed]

346. Kawahara, T.; Ritsick, D.; Cheng, G.; Lambeth, J.D. Point Mutations in the Proline-rich Region of p22 Are Dominant Inhibitors of Nox1- and Nox2-dependent Reactive Oxygen Generation. *J. Biol. Chem.* **2005**, *280*, 31859–31869. [CrossRef]

347. Schroeder, K.; Zhang, M.; Benkhoff, S.; Mieth, A.; Pliquett, R.; Kosowski, J.; Kruse, C.; Luedike, P.; Michaelis, U.R.; Weissmann, N.; et al. Nox4 Is a Protective Reactive Oxygen Species Generating Vascular NADPH Oxidase. *Circ. Res.* **2012**, *110*, 1217–1225. [CrossRef]

348. Meng, W.; Shah, K.P.; Pollack, S.; Toppila, I.; Hebert, H.L.; McCarthy, M.I.; Groop, L.; Ahlqvist, E.; Lyssenko, V.; Agardh, E.; et al. A genome-wide association study suggests new evidence for an association of the NADPH Oxidase 4 (*NOX4*) gene with severe diabetic retinopathy in type 2 diabetes. *Acta Ophthalmol.* **2018**, *96*, e811–e819. [CrossRef]

349. Roberts, K.E.; Kawut, S.M.; Krowka, M.J.; Brown, R.S.; Trotter, J.F.; Shah, V.; Peter, I.; Tighiouart, H.; Mitra, N.; Handorf, E.; et al. Genetic Risk Factors for Hepatopulmonary Syndrome in Patients With Advanced Liver Disease. *Gastroenterology* **2010**, *139*, 130–139. [CrossRef]

350. Touyz, R.M.; Anagnostopoulou, A.; Rios, F.; Montezano, A.C.; Camargo, L.L. NOX5: Molecular biology and pathophysiology. *Exp. Physiol.* **2019**, *104*, 605–616. [CrossRef]

351. Holterman, C.E.; Thibodeau, J.-F.; Towaij, C.; Gutsol, A.; Montezano, A.C.; Parks, R.J.; Cooper, M.E.; Touyz, R.M.; Kennedy, C.R. Nephropathy and Elevated BP in Mice with Podocyte-Specific NADPH Oxidase 5 Expression. *J. Am. Soc. Nephrol.* **2014**, *25*, 784–797. [CrossRef]

352. Jha, J.C.; Banal, C.; Okabe, J.; Gray, S.P.; Hettige, T.; Chow, B.S.; Thallas-Bonke, V.; De Vos, L.; Holterman, C.E.; Coughlan, M.T.; et al. NADPH Oxidase Nox5 Accelerates Renal Injury in Diabetic Nephropathy. *Diabetes* **2017**, *66*, 2691–2703. [CrossRef]

353. Shin, J.-G.; Seo, J.Y.; Seo, J.-M.; Kim, D.-Y.; Oh, J.-T.; Park, K.-W.; Kim, H.-Y.; Kim, J.-H.; Shin, H.D. Association analysis of NOX5 polymorphisms with Hirschsprung disease. *J. Pediatr. Surg.* **2019**, *54*, 1815–1819. [CrossRef] [PubMed]

354. Nagase, M.; Ayuzawa, N.; Kawarazaki, W.; Ishizawa, K.; Ueda, K.; Yoshida, S.; Fujita, T. Oxidative Stress Causes Mineralocorticoid Receptor Activation in Rat Cardiomyocytes. *Hypertension* **2012**, *59*, 500–506. [CrossRef] [PubMed]

355. Johnson, J.; Erickson, J.W.; Cerione, R.A. New Insights into How the Rho Guanine Nucleotide Dissociation Inhibitor Regulates the Interaction of Cdc42 with Membranes. *J. Biol. Chem.* **2009**, *284*, 23860–23871. [CrossRef] [PubMed]

356. Sylow, L.; Jensen, T.E.; Kleinert, M.; Højlund, K.; Kiens, B.; Wojtaszewski, J.; Prats, C.; Schjerling, P.; Richter, E.A. Rac1 Signaling Is Required for Insulin-Stimulated Glucose Uptake and Is Dysregulated in Insulin-Resistant Murine and Human Skeletal Muscle. *Diabetes* **2013**, *62*, 1865–1875. [CrossRef]

357. Kalwat, M.; Thurmond, D.C. Signaling mechanisms of glucose-induced F-actin remodeling in pancreatic islet β cells. *Exp. Mol. Med.* **2013**, *45*, e37. [CrossRef]

358. Subasinghe, W.; Syed, I.; Kowluru, A. Phagocyte-like NADPH oxidase promotes cytokine-induced mitochondrial dysfunction in pancreatic β-cells: Evidence for regulation by Rac1. *Am. J. Physiol. Integr. Comp. Physiol.* **2011**, *300*, R12–R20. [CrossRef]

359. Newsholme, P.; Morgan, D.; Rebelato, E.; Oliveira-Emilio, H.C.; Procopio, J.; Curi, R.; Carpinelli, A. Insights into the critical role of NADPH oxidase(s) in the normal and dysregulated pancreatic beta cell. *Diabetologia* **2009**, *52*, 2489–2498. [CrossRef]

360. Syed, I.; Kyathanahalli, C.N.; Jayaram, B.; Govind, S.; Rhodes, C.J.; Kowluru, R.A.; Kowluru, A. Increased phagocyte-like NADPH oxidase and ROS generation in type 2 diabetic ZDF rat and human islets: Role of Rac1-JNK1/2 signaling pathway in mitochondrial dysregulation in the diabetic islet. *Diabetes* **2011**, *60*, 2843–2852. [CrossRef]

361. Kowluru, A.; Kowluru, R.A. RACking up ceramide-induced islet β-cell dysfunction. *Biochem. Pharmacol.* **2018**, *154*, 161–169. [CrossRef]

362. Marei, H.; Malliri, A. Rac1 in human diseases: The therapeutic potential of targeting Rac1 signaling regulatory mechanisms. *Small GTPases* **2016**, *8*, 139–163. [CrossRef]

363. Aguilar, B.J.; Zhu, Y.; Lu, Q. Rho GTPases as therapeutic targets in Alzheimer's disease. *Alzheimer's Res. Ther.* **2017**, *9*, 1–10. [CrossRef] [PubMed]

364. Azarova, I.; Klyosova, E.; Polonikov, A. Association between RAC1 gene variation, redox homeostasis and type 2 diabetes mellitus. *Eur. J. Clin. Investig.* **2022**, *52*, e13792. [CrossRef] [PubMed]

365. Pulit, S.L.; Stoneman, C.; Morris, A.P.; Wood, A.R.; Glastonbury, C.A.; Tyrrell, J.; Yengo, L.; Ferreira, T.; Marouli, E.; Ji, Y.; et al. Meta-analysis of genome-wide association studies for body fat distribution in 694 649 individuals of European ancestry. *Hum. Mol. Genet.* **2019**, *28*, 166–174. [CrossRef]

366. Pick, E. Role of the Rho GTPase Rac in the activation of the phagocyte NADPH oxidase: Outsourcing a key task. *Small GTPases* **2014**, *5*, e27952. [CrossRef] [PubMed]

367. Beckers, C.M.L.; van Hinsbergh, V.W.M.; Amerongen, G.P.V.N. Driving Rho GTPase activity in endothelial cells regulates barrier integrity. *Thromb. Haemost.* **2010**, *103*, 40–55. [CrossRef] [PubMed]

368. Jyoti, A.; Singh, A.K.; Dubey, M.; Kumar, S.; Saluja, R.; Keshari, R.S.; Verma, A.; Chandra, T.; Kumar, A.; Bajpai, V.K.; et al. Interaction of Inducible Nitric Oxide Synthase with Rac2 Regulates Reactive Oxygen and Nitrogen Species Generation in the Human Neutrophil Phagosomes: Implication in Microbial Killing. *Antioxidants Redox Signal.* **2014**, *20*, 417–431. [CrossRef] [PubMed]

369. Kinney, N.; Kang, L.; Bains, H.; Lawson, E.; Husain, M.; Husain, K.; Sandhu, I.; Shin, Y.; Carter, J.K.; Anandakrishnan, R.; et al. Ethnically biased microsatellites contribute to differential gene expression and glutathione metabolism in Africans and Europeans. *PLoS ONE* **2021**, *16*, e0249148. [CrossRef] [PubMed]

370. Polonikov, A.; Bocharova, I.; Azarova, I.; Klyosova, E.; Bykanova, M.; Bushueva, O.; Polonikova, A.; Churnosov, M.; Solodilova, M. The Impact of Genetic Polymorphisms in Glutamate-Cysteine Ligase, a Key Enzyme of Glutathione Biosynthesis, on Ischemic Stroke Risk and Brain Infarct Size. *Life* **2022**, *12*, 602. [CrossRef] [PubMed]

371. Lasram, M.M.; Dhouib, I.B.; Annabi, A.; El Fazaa, S.; Gharbi, N. A review on the possible molecular mechanism of action of N-acetylcysteine against insulin resistance and type-2 diabetes development. *Clin. Biochem.* **2015**, *48*, 1200–1208. [CrossRef]

372. El-Hafidi, M.; Franco, M.; Ramírez, A.R.; Sosa, J.S.; Flores, J.A.P.; Acosta, O.L.; Salgado, M.C.; Cardoso-Saldaña, G. Glycine Increases Insulin Sensitivity and Glutathione Biosynthesis and Protects against Oxidative Stress in a Model of Sucrose-Induced Insulin Resistance. *Oxidative Med. Cell. Longev.* **2018**, *2018*, 2101562. [CrossRef]

373. Kalamkar, S.; Acharya, J.; Madathil, A.K.; Gajjar, V.; Divate, U.; Karandikar-Iyer, S.; Goel, P.; Ghaskadbi, S. Randomized Clinical Trial of How Long-Term Glutathione Supplementation Offers Protection from Oxidative Damage and Improves HbA1c in Elderly Type 2 Diabetic Patients. *Antioxidants* **2022**, *11*, 1026. [CrossRef] [PubMed]

Review

The Yin and Yang Effect of the Apelinergic System in Oxidative Stress

Benedetta Fibbi [1,2], Giada Marroncini [1], Laura Naldi [1] and Alessandro Peri [1,2,*]

1 "Pituitary Diseases and Sodium Alterations" Unit, AOU Careggi, 50139 Florence, Italy
2 Endocrinology, Department of Experimental and Clinical Biomedical Sciences "Mario Serio", University of Florence, 50139 Florence, Italy
* Correspondence: alessandro.peri@unifi.it; Tel.: +39-05-5794-9275

Abstract: Apelin is an endogenous ligand for the G protein-coupled receptor APJ and has multiple biological activities in human tissues and organs, including the heart, blood vessels, adipose tissue, central nervous system, lungs, kidneys, and liver. This article reviews the crucial role of apelin in regulating oxidative stress-related processes by promoting prooxidant or antioxidant mechanisms. Following the binding of APJ to different active apelin isoforms and the interaction with several G proteins according to cell types, the apelin/APJ system is able to modulate different intracellular signaling pathways and biological functions, such as vascular tone, platelet aggregation and leukocytes adhesion, myocardial activity, ischemia/reperfusion injury, insulin resistance, inflammation, and cell proliferation and invasion. As a consequence of these multifaceted properties, the role of the apelinergic axis in the pathogenesis of degenerative and proliferative conditions (e.g., Alzheimer's and Parkinson's diseases, osteoporosis, and cancer) is currently investigated. In this view, the dual effect of the apelin/APJ system in the regulation of oxidative stress needs to be more extensively clarified, in order to identify new potential strategies and tools able to selectively modulate this axis according to the tissue-specific profile.

Keywords: apelin; APJ; apelinergic system; oxidative stress

Citation: Fibbi, B.; Marroncini, G.; Naldi, L.; Peri, A. The Yin and Yang Effect of the Apelinergic System in Oxidative Stress. *Int. J. Mol. Sci.* **2023**, *24*, 4745. https://doi.org/10.3390/ijms24054745

Academic Editors: Rossana Morabito and Alessia Remigante

Received: 26 January 2023
Revised: 20 February 2023
Accepted: 23 February 2023
Published: 1 March 2023

1. Physiology of the Apelin/APJ System

Apelin is a biologically active neuropeptide that was first isolated in 1998 from bovine stomach extracts [1,2] and identified as the endogenous ligand for the orphan receptor APJ, which was characterized in 1993 as a seven transmembrane G-protein coupled receptor (GPCR) with high affinity (homology of 40–50% in the hydrophobic transmembrane region) with the angiotensin II receptor type 1a [3–5].

In humans, apelin is encoded by the APLN gene, which is located on the long arm of X chromosome (Xq25-q26.1) and encodes the 77-aminoacid precursor peptide pre-pro-apelin [2], whose enzymatic hydrolysis originates several active peptide fragments able to activate APJ by their common C-terminal sequence [6]. Apelin isoforms have 12, 13, 15, 16, 17, 19, 28, 31, 36, or 55 aminoacids and display a distinct receptor binding affinity [6], with apelin-13 representing the most effective activator of APJ [7], followed by apelin-17 and apelin-36 [8].

As a GPCR, APJ interacts with G proteins (mainly Gi/o and Gq/11), leading to the modulation of several different signaling pathways after ligand binding. Specifically, via Gi/o, the apelin/APJ system activates the phospho-inositide 3-kinase (PI3K)/AKT (also named protein kinase B, PKB) and protein kinase C (PKC)/extracellular signal-regulated kinase 1/2 (ERK 1/2) pathways, thereby being involved in the regulation of apoptosis, cell proliferation, neuroinflammation, and oxidative stress [9,10]. Moreover, Gi/o is implicated in the downregulation of protein kinase A (PKA) by inhibiting cAMP production [10]. Upregulation of phospholipase C beta (PLCβ) by Gq/11 triggers the generation of diacylglycerol (DAG) and inositol 1,4,5-triphosphate (IP3), which lead to the initiation of the

PKC cascade and the intracellular release of Ca^{2+}, respectively [10]. Both AKT activation and increase of intracellular Ca^{2+} induces nitric oxide synthase (NOS), thus promoting vasodilation. Binding of apelin to APJ can also result in the autophosphorylation of the receptor through G protein-coupled receptor kinase (GRK). This event initiates a β-arrestin-mediated response involving the desensitization and clathrin-dependent internalization of APJ, which can activate G protein-independent signaling pathways [10,11]. Finally, APJ has also been shown to activate G13 in human umbilical vein endothelial cells, leading to histone deacetylases (HDAC) type 4 and 5 inactivation, activation of myocyte enhancer factor-2 (MEF2) and expression of MEF2 target gene Kruppel-like factor 2 (KLF2) [12] (Figure 1).

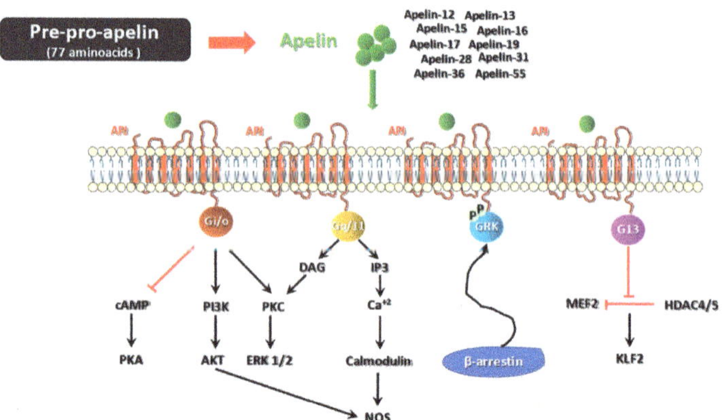

Figure 1. Intracellular pathways modulated by the apelin/APJ system. The 77-aminoacid precursor peptide, pre-pro-apelin, is cleaved in active fragments (apelin-12, apelin-13, apelin-15, apelin-16, apelin-17, apelin-19, apelin-28, apelin-31, apelin-36, apelin-55), which bind the apelin receptor APJ. By interacting with G proteins, apelin/APJ is able to modulate different signaling pathways: inhibition of cAMP generation and protein kinase A (PKA) and activation of phospho-inositide 3-kinase (PI3K)/AKT through Gi/o; activation of protein kinase C (PKC)-dependent extracellular signal-regulated kinase 1/2 (ERK1/2) through Gi/o or Gq/11; initiation of the intracellular release of Ca^{2+} by Gq/11 and inositol 1,4,5-triphosphate (IP3) synthesis; autophosphorylation of APJ through G protein-coupled receptor kinase (GRK) and initiation of the β-arrestin-mediated internalization of the receptor; activation of G13 and inactivation of histone deacetylases (HDAC) type 4 and 5, determining the activation of myocyte enhancer factor-2 (MEF2) and expression of MEF2 target gene Kruppel-like factor 2 (KLF2). Both AKT activation and increase of intracellular Ca^{2+} induces nitric oxide synthase (NOS).

APJ and apelin are both highly conserved among species and widely expressed in rodents and human tissues, including lung, heart, spinal cord, brain, placenta, endocrine (thyroid, parathyroid, adrenal, pituitary), gastrointestinal and urinary apparatuses, bone marrow, skeletal and smooth muscles, and adipose tissue, among others [13]. The capability of APJ to interact with several G proteins according to cell types (i.e., heterologous signaling) and stimulate different intracellular pathways explains the variety of biological effects potentially mediated by the apelin/APJ system: vasodilation and lowering of blood pressure [14], increase of cardiac contractility and heart rate [15,16], control of pituitary hormone release, drinking behavior and body fluid homeostasis [17], neuroendocrine stress response [18], food intake and appetite regulation [17,19], glucose metabolism and insulin sensitivity [20], promotion of cell proliferation, migration and angiogenesis, and regulation of gastrointestinal and immune functions [21]. The type and magnitude of downstream events may be cell type- and context-dependent, since apelin isoforms induce different APJ trafficking [22,23]. It is worth noting that although apelin-13 and apelin-36 are able to promote the internalization of APJ, only apelin-13-internalized receptors can be rapidly recovered to the cell surface [24]. Conversely, apelin receptors internalized after binding to apelin-36 represent a target for lysosomal degradation [25].

Elabela is a micropeptide recognized as the second endogenous ligand for APJ [26]. Its involvement in physiological and pathological conditions will not be discussed in this review.

2. Apelin/APJ System and Oxidative Stress

Oxidative stress is a condition characterized by an excessive accumulation of reactive oxygen species (ROS) in cells and tissues, which overwhelms the normal dynamic homeostasis and the ability of a biological system to detoxify them. The imbalance between ROS and antioxidants exerts harmful effects on several cellular structures (proteins, lipids, and nucleic acids) [27] and processes (protein phosphorylation, transcriptional factors activation, apoptosis, differentiation, and immunity) [28], thus leading to cell and tissue damage. In this view, oxidative stress is considered as one of the underlying mechanisms of the onset and/or progression of several diseases (i.e., cancer, diabetes, metabolic disorders, atherosclerosis, and cardiovascular diseases) [29].

Mitochondria are the major intracellular site of energy metabolism regulation and therefore they are heavily involved in ROS production [30]. Both enzymatic and non-enzymatic (oxygen reaction with organic compounds, cell exposure to ionizing radiations, mitochondrial respiration) reactions participate in ROS generation from both endogenous (inflammation, ischemia, immune cell activation, infections, cancer, aging) and exogenous (chemical drugs and solvents, smoke, radiations, alcohol) sources [31,32].

As a consequence of apelin/APJ system characterization and evidence of its involvement in the regulation of many intracellular pathways and cell functions, it was not long before there was a demonstration of a close link between this axis and oxidative stress. In fact, not only the myocardial APLN gene expression and protein secretion have been shown to be upregulated by hypoxia via activation of hypoxia-inducible factor (HIF) [33], but the crucial role of apelin in regulating oxidative stress-related processes was also revealed in many tissues and pathological conditions.

Although the activation of apelin/APJ-associated signaling pathways is secondary to both ROS-dependent and ROS-independent stimuli, this review focused on the role of the apelinergic system in oxidative stress-mediated pathologic conditions.

2.1. Oxidative Stress-Linked Hypertension, Atherosclerosis, and Pre-Eclampsia

In the vascular system, apelin and APJ are expressed by endothelial and vascular smooth muscle cells (VSMCs), where they are implicated in a complex regulation of blood vessels' function [34]. Under physiologic conditions, apelin binding to APJ results in vasodilation and transient hypotension by modulating both NO synthesis (via PI3K/Akt and IP3/Ca^{2+} pathways) and the renin–angiotensin–aldosterone system (RAAS). Indeed, its counterregulatory role against angiotensin II-dependent vasopressor stimulation [35–38] is at least in part secondary to the upregulation of angiotensin converting enzyme 2 (ACE2), which is a negative modulator of RAAS [39,40].

Oxidative stress and vascular NO bioavailability imbalance represent the major etiopathogenetic factors of vascular injury and hypertension [41], with angiotensin II and RAAS acting as crucial triggers of ROS production (e.g., by angiotensin II-induced activity of mitochondrial NADPH oxidase 4, NOX4, which is the upstream signaling molecule of ERK) and endothelial NOS (eNOS) inhibition [42–44]. As expected, based on the endothelium-dependent vasodilative properties of apelin, different isoforms of this peptide were demonstrated to mitigate hypertension in in vivo models, with apelin-12 exhibiting the greater effect on blood pressure lowering after intraperitoneal injection of apelin-12, apelin-13, and apelin-36 in anesthetized rats. The absence of a significant antihypertensive effect in APJ-deficient mice suggests that apelin binding to an intact endothelial APJ is required for its vasodilative action [45].

Oxidative stress is the major trigger in the initiation and progression of atherosclerosis. Through the upregulation of selected genes [46], ROS promote mitogenicity and inhibit apoptosis of VSMCs, which contributes to the recruitment of circulant inflammatory cells and the production of extracellular matrix and cytokines, thus participating both in early- and late-stage atherogenesis [47]. Apelin/APJ has been demonstrated to be involved in the development of hypercholesterolemia-associated atherosclerosis similarly to angiotensin II/AT1, which promotes endothelial dysfunction and myosin light chain phosphorylation in VSMCs [48–51]. By exerting an opposite role on RAAS function to that previously described, apelin-13 is able to activate the ERK-Jagged-1/Notch3-cyclin D1 pathway [52], NOX4 expression and NOX4-derived ROS generation, and oxidative stress-linked proliferation in VSMCs [53]. Accordingly, APJ deficiency can prevent oxidative stress-induced atherosclerosis and protect blood vessels from atherosclerotic plaques [53,54]. In parallel, apelin/APJ-dependent activation of ERK increases the endothelial expression of intercellular adhesion molecule-1 (ICAM-1) and vascular cell adhesion molecule-1 (VCAM-1), and the release of monocyte chemoattractant protein-1 (MCP-1) through the NF-κB/JNK signaling pathway, thus leading to monocyte recruitment and adhesion to endothelial cells [55,56]. Hence, apelin/APJ and oxidative stress seem to be involved in early atherogenesis via the activation of the NOX4-ROS-NF-κB/ERK signaling pathway in VSMCs and in the endothelium.

Abnormal proliferation and migration of VSMCs results in a large number of cells able to penetrate the endothelial layer, deposit in the arterial intima, and secrete bone morphogenetic proteins, which can promote the spontaneous calcification of plaques in late-staged atherosclerosis [57–59]. Abnormal apoptosis of mouse aortic vascular smooth muscle cells (MOVAS) secondary to intracellular oxidative stress has been closely related to vascular calcifications through ERK and PI3K/AKT pathways, which both affect MOVAS osteogenic differentiation and calcium deposition [5,60,61]. Zhang et al. have recently reported that apelin-13 significantly reduced high glucose-induced proliferation, invasion, and osteoblastic differentiation of MOVAS—therefore suppressing vascular calcification processes—by inhibiting ROS-mediated DNA damage and regulating ERK and PI3K/AKT pathways [62]. However, apelin's ability to abrogate the development of atherosclerosis by increasing NO bioavailability and antagonizing angiotensin II cellular signaling was also described [63].

The exact pathophysiological mechanism of pre-eclampsia (PE) is not clearly defined, but abnormal placentation with angiogenic factors levels disproportion and placental insufficiency, increased inflammation, and oxidative stress are known to exert critical roles [64,65]. Adipokines including resistin, adiponectin, and apelin are released even from the placenta during pregnancy [66], and a significant decrease in circulating apelin levels has been demonstrated in PE women compared to normal pregnancies [67–70]. Circulating apelin decreases in the middle of pregnancy and rises again in the third trimester in healthy pregnancy [71]. Hence, maternal concentrations of apelin lower than expected may play a key role in the etiology of PE. Circulating apelin concentrations showed a significant negative correlation with mean arterial blood pressure, proteinuria, serum soluble fms-like tyrosine kinase-1 (sFlt-1, a soluble form of VEGF/PLGF receptors which acts as an effective scavenger of VEGF and PLGF and sensitizes maternal endothelium to proinflammatory cytokines, thus inducing endothelial dysfunction and multiorgan damage), soluble endoglin (sEng, that acts as a limiting factor for eNOS activity), and IFN-γ levels in PE compared to control women [72]. Furthermore, a positive correlation of apelin levels with serum placental growth factor (PLGF), VEGF and IL-10 levels, and superoxide dismutase (SOD) and catalase activities was also recognized [72]. However, apelin administration significantly improved sFlt-1 and sEng values in the treated group. These results, which are in line with previous reports stating that inflammation is one of the mechanisms of PE by inducing placental ischemia and endothelial dysfunction [73,74], also strengthen the effect of apelin in the pathogenesis of PE.

2.2. Oxidative Stress and Diabetic Microvascular Complications

The role of the apelinergic system on the endothelial function accounts for its close association with diabetic microvascular complications, which have, in oxidative stress, one of the underlying pathogenetic mechanisms [75]. In the kidney of diabetic mice, apelin was able to restore antioxidant enzymes' activity and reduce oxidative stress, thus preventing chronic injury [76] and progression of diabetic nephropathy [77]. Moreover, the evidence of apelin-induced inhibition of ROS generation in an in vitro model of cortical neurons supports the hypothesis of its positive effect in preventing the occurrence of diabetic neuropathy [78]. Nevertheless, the mRNA levels of APJ, apelin, and VEGF are all upregulated in the vascular tissue membrane in proliferative diabetic retinopathy [79], and apelin/APJ was demonstrated to be involved in retinal neoangiogenesis by promoting the expression of VEGF [80,81]. Hence, apelin is supposed to exert a pathogenetic effect in the onset of diabetic retinopathy, and the inhibition of the apelinergic system has been proposed as an effective tool to prevent it.

2.3. Oxidative Stress and Cardiac Function

The role of apelin/APJ in myocardial homeostasis and pathology is uncertain and data from literature are conflicting.

On the one hand, it was linked to a protecting effect against ventricular hypertrophy in murine models, where apelin was reported to reduce oxidative stress induced by hydrogen peroxide or 5-hydroxytryptamine [82], and endoplasmic reticulum stress [83]. Similarly, in a model of ischemia-induced heart failure, apelin was proved to reduce ROS production and to ameliorate cardiac dysfunction and RAAS hyperactivation-associated fibrosis, via inhibiting the PI3K/Akt signaling pathway [84].

Peripheral and coronary vasodilatation and improved cardiac output were observed even in patients affected by chronic heart failure after apelin injection [85].

Contrastingly, an increased expression of cardiac myosin and β-MHC (β-myosin heavy chain) mRNA was observed in normotensive rats 15 days after chronic infusion of apelin-13 into the paraventricular nucleus, thus indicating a role of the peptide in the induction of cardiac hypertrophy [86].

2.4. Oxidative Stress and Ischemia/Reperfusion Injury

Ischemia/reperfusion (I/R) injury (IRI) consists of the paradoxical exacerbation of cellular dysfunction and death after restoration of blood flow to previously ischemic tissues. Oxidative stress and inflammation secondary to hypoxia-induced production of ROS are the main determinants of cellular and tissue damage [87], which are sustained by activation of matrix metalloproteinase enzymes and degradation of the extracellular matrix and tight junction proteins around endothelial vascular cells [88].

During I/R in in vivo models, apelin is able to protect myocardiocytes against oxidative stress and inhibit mitochondrial oxidative damage and lipid peroxidation by activating eNOS and reperfusion injury salvage kinase (RISK) [89,90]. Hemodynamically, it results in reduced left ventricular preload and afterload, improved cardiac contractility [91], and reduced infarct size [92]. The effect of apelin-13 on post-myocardial infarction repair is partially mediated by an increase of myocardial progenitor cells in the infarcted hearts [93].

Epidemiological data show that diabetes is the most important risk factor for cardiovascular diseases and IRI, with a 2–6-fold increased mortality compared to non-diabetic conditions [94]. Results from animal models showed that heart failure was more severe in diabetic IRI rats compared to non-diabetic IRI rats, and that apelin overexpression significantly decreased injury size and heart weight index and improved cardiac function [95]. Upregulation of PPARα (a well-known modulator of lipid metabolism, antioxidant defense, mitochondrial and endothelial functions, atherosclerosis, and inflammation) and inhibition of apoptosis (enhanced Bcl-2 levels and decreased Bax and cleaved caspase-3 levels) and oxidative stress via the PI3K and p38MAPK pathways has been characterized as the major determinant of apelin's cardio-protective effects [96,97].

In lungs, IRI often occurs after pulmonary oedema or acute respiratory distress syndrome [98]. Apelin-13 administration to lung IRI rats resulted in a mild damage of alveolar structures, a reduced number of erythrocytes and inflammatory cells, and lower inflammatory cytokines (IL-1β, IL-6 and TNF-α) expression levels [99]. These morphological and molecular changes observed in tissues were associated with an increase of PaO_2 and a decrease of $PaCO_2$ compared to non-apelin-treated IRI rats, thus suggesting that apelin/APJ could minimize IRI by improving lung oxygenation and peroxidation. Finally, apelin-induced expression of uncoupling protein 2 (UCP2), an anionic carrier located on mitochondria which increases SOD activity and improves cell survival in a reduced ROS environment [100], could imply a direct effect of apelin/APJ in ameliorating mitochondrial damage [99].

In the brain, ischemia-induced injury is not only considered as an outcome of inadequate oxygen supply, but it has also been related to an excessive amount of ROS, which lead to cellular and protein dysfunction [101–103] and tissue disruption [104]. The degradation of the extracellular matrix secondary to IRI-associated oxidative stress leads to blood–brain barrier (BBB) destruction and vasogenic edema [105], which is a severe consequence of ischemic brain stroke, resulting in a 5% mortality rate [106–108]. The subsequent reperfusion contributes to cerebral oedema by initiating the activation of several destructing signaling pathways, including inflammatory responses, alteration of cellular receptors, ion imbalance, oxidative stress, changes in water channel expression, activation of proteinase enzymes, as well as changing tight junction proteins expression [109–112]. Apelin-13's ability to significantly decrease brain IRI is mediated by different mechanisms. Gholamzadeh et al. showed that, in mice, oxidative stress markers increased due to ischemia, and that the injection of apelin-13 only 5 min before the onset of reperfusion could significantly reduce vasogenic cerebral oedema and protect BBB integrity [113]. Apelin-13 administration also decreased the expression of endothelin-1 receptor type B [113], whose up-regulation in astrocytes and endothelial cells is associated with metalloproteinase activation [114]. By the activation of ERK1/2 intracellular pathway, the apelinergic system inhibited the production of ROS and increased SOD activity [115]. In parallel, apelin-13 was able to inhibit the ROS-mediated inflammatory response of ischemic stroke by activating the phosphorylation level of AMP-activated protein kinase (AMPK) and the expression of nuclear factor erythroid 2-related factor 2 (Nrf2) [116]. AMPK signaling was also reported to participate in the antiapoptotic role of apelin-13 in ischemic stroke [117]. Conversely, apelin-36-mediated decrease of Bax and caspase-3 levels associated with IRI was related to the PI3K/Akt pathway [118], inhibition of ER stress/unfolded protein response (UPR) activation induced by brain I/R injury [119], and SK1/JNK/caspase-3 apoptotic pathway [120], whereas apelin-12 neuroprotection after ischemia was associated with restrainment of the c-Jun N-terminal kinase (JNK) and p38MAPK signaling pathways of apoptosis-related MAPKs family [121].

Autophagy is a homeostatic process involved in the lysosomal-dependent degradation and elimination of damaged and/or misfolded proteins and organelles. It is negatively modulated by the AMPK/mammalian target of the rapamycin (mTOR) axis [122], and apelin-13 was suggested to attenuate traumatic brain-associated IRI by suppressing autophagy [123].

Finally, apelin/APJ reduced renal IRI by promoting the activity of the mitochondrial enzymes SOD, catalase, and glutathione peroxidase, and decreasing the formation of hydroxyl radicals and malondialdehyde [124].

2.5. Oxidative Stress, Obesity, and Insulin Resistance

Data from in vivo obesity models suggest that apelin may function as an adipokine [125–127]. Serum levels of this neuropeptide positively correlate with insulin resistance and obesity [125–127], and inflammation (particularly by TNF-α production) and oxidative stress have been proposed as the link between apelin/APJ and insulin resistance [128]. In skeletal muscle, apelin enhances the expression of mitochondrial biogenesis markers and enzymes (e.g., citrate synthase, β-hydroxyacyl-CoA dehydrogenase, cytochrome c oxidase) and the content of proteins involved in the assembly of mitochondrial respiratory chain complexes [129,130]. In adipocytes, the apelin/APJ axis prevents the generation of ROS by stimulating the expression of antioxidant enzymes (through MAPK/ERK and AMPK pathways) and inhibiting the expression of pro-oxidant enzymes [131].

The direct effect of insulin on the adipocytic production of apelin is supported by the statistical association among different markers of adiposity, related risk factors, and apelin expression from rat subcutaneous and retroperitoneal adipose tissue [132]. On the other hand, the correlation between apelin mRNA levels and markers of hepatic oxidative stress highlighted a possible role of the apelinergic system in obesity-induced liver oxidative steatosis and dysfunction [132]. Accordingly, exogenous apelin injection restored glucose tolerance and increased glucose utilization in peripheral tissues in high fat diet mice with hyperinsulinemia, hyperglycemia, and obesity [133].

2.6. Oxidative Stress and Aging

Apelin has been reported to be downregulated with age in different tissues, and its absence accelerates the onset and progression of aging. Again, oxidative stress is considered to be the link between apelin and the aging process [134]. Specifically, increasing evidence has shown that the apelinergic system participates in autophagy [135,136] and alleviates oxidative stress [82,131,137], which contributes to the development of aging.

2.7. Oxidative Stress in the Central Nervous System

Apelin and APJ mRNAs are widely expressed in neuronal cell bodies and fibers throughout the entire central nervous system (CNS), such as in the thalamus, subthalamic nucleus, pituitary gland, hippocampus, basal forebrain, frontal and piriform cortex, striatum, corpus callosum, substantia nigra, olfactory tract, amygdala, central gray matter, spinal cord, and cerebellum [34,138,139].

This broad localization fits with the huge impact of the apelinergic system in neuroprotection, which goes through several mechanisms: suppression of oxidative stress, inhibition of apoptosis and excitotoxicity, and modulation of inflammatory responses and autophagy. Interestingly, these different processes are frequently interconnected and regulated by the same intracellular pathways [45]. Indeed, apelin's beneficial properties on ethanol-induced memory impairment and neuronal injury of rats are sustained by inhibitory effects on hippocampal oxidative stress, apoptosis, and neuroinflammation [140]. Specifically, the administration of apelin-13 was observed to increase antioxidant enzymes' activity and glutathione concentration, reduce lipid peroxidation and the number of active caspase-3 positive cells, and attenuate TNF-α production and glial fibrillary acidic protein (GFAP) as a neuroinflammation mediators [140].

The regulatory role of apelin/APJ in neuroinflammation is exerted by suppressing the activity of microglia, astrocytes, and other inflammatory cells [141,142]. Microglia are the innate immune cells of the CNS, able to both eliminate pathogens and cell debris and contribute to neuronal regeneration after tissue damage through the acquisition of different activated phenotypes: M1 cells produce pro-inflammatory cytokines and ROS, causing cytotoxic effects, whereas M2 cells synthetize anti-inflammatory cytokines and stimulate tissue repair [143]. In an in vivo model of ischemic stroke, apelin-13 reduced the expression of pro-inflammatory cytokines and chemokines (IL-1β, TNF-α, macrophage inflammatory protein 1α or MIP-1α, monocyte chemoattractant protein 1 or MCP-1) produced by M1 microglia and increased the expression of the M2-derived anti-inflammatory cytokine IL-10 [144]. The shift of microglial M1 polarization toward the M2 phenotype may be sustained by the blockage of STAT3 signal [145]. The activation of the brain-derived neurotrophic factor (BDNF)-Tyrosine Kinase receptor B (TrkB) signaling pathway, and the inhibition of the NF-κB pathway and endoplasmic reticulum (ER) stress-associated AMPK/TXNIP/NLRP3 inflammasome are other targets of apelin-mediated suppression of neuroinflammation, resulting in improvement of cognitive dysfunction, depressive-like behavior, and early brain injury subarachnoid hemorrhage [142,146,147]. The overactivation of ER stress induced by ROS, with the aim to remove the damaged elements, induces calcium release and reinforces oxidative stress, which promote further microglia activation and leukocyte infiltration into the brain, which subsequently trap in a vicious circle to exacerbate brain injury after stroke [148,149]. In this view, apelin-13 activates AMPK and degradation of TXNIP, which suppresses the overactivation of ER stress and reduces the level of NLRP3 [150].

Excitotoxicity is a complex process of neuronal sufferance and death triggered by the excessive levels of neurotransmitters, which result in a pathologic stimulation of specific receptors. Glutamate neurotoxicity (GNT) is a condition characterized by time-dependent damage of several cell components driven by a massive cell influx of calcium ions and activation of enzymes, including phospholipases, endonucleases, and proteases such as calpain [151,152]. Among the neuroprotective effects of the apelinergic system, the inhibition of excitotoxicity by the activation of pro-survival pathways (i.e., PI3K/Akt and PKC/ERK1/2) and the regulation of N-Methyl-D-aspartic acid (NMDA) receptor activity [153–156] have also been described.

Patients who have undergone thoracic and abdominal aortic surgery are frequently faced with nerve injury induced by spinal cord ischemia, which is driven by ROS-induced neuronal apoptosis, neuroinflammation, and autophagy [157]. The intraperitoneal injection of apelin-13 exerted spinal cord protection and recovery of motor function in rats by suppressing autophagy, oxidative stress, and mitochondrial dysfunction [158].

Recent evidence suggests that GnRH neurons are targets of apelin-associated neuroprotection. APJ signaling pathway activation via either apelin-13 or transient overexpression is able to increase GnRH neurons proliferation after H_2O_2 exposure and hypoxia, and to stimulate the conversion of G0/G1 to S phase through AKT and ERK-1/2 kinase pathways activation [159]. Therefore, the expression and activation of the apelin/APJ system in GnRH neurons might support a protective mechanism against oxidative stress-induced cell death. Furthermore, the observation of a promoting effect of the apelinergic system on GnRH release in embryonic stem cell-derived GnRH neurons supports the hypothesis of its pro-differentiating role during developmental stages [159].

The antioxidative stress effects of apelin/APJ prompted the research to evaluate its potential correlation with neurodegenerative diseases.

Alzheimer's disease (AD) is the most prevalent form of dementia in the elderly, characterized by intracellular neurofibrillary tangles (NFTs) and extracellular amyloid beta (Aβ) protein deposits that contribute to senile plaques and progressive neurodegeneration [160]. The neuronal loss that appears in the cerebral cortex and in the hippocampus as a consequence of mitochondrial dysfunction and ROS production is an early event in AD and anticipates senile plaques appearance [161,162]. By activating glycogen synthase kinase-3 (GSK-3) and c-Jun N-terminal kinase (JNK)/p38MAPK, oxidative stress induces Tau phosphorylation and beta-site amyloid precursor protein cleaving enzyme 1 (BACE1) expression, and therefore promotes the production of NFTs and Aβ [163–165]. Moreover, dysregulation of intracellular calcium exerts a crucial role in the regulation of familial Alzheimer's proteins (PSEN1 and PSEN2) and Aβ, which results in altered calcium signaling, loss of synapses, and memory impairment [166]. In this scenario, serum apelin-13 has been shown to be lower in AD patients compared to control subjects [167] and its exogenous administration attenuates Aβ-induced memory deficit in Aβ-treated animals [168]. Subsequent molecular studies revealed that the apelinergic system participates in the pathophysiology of AD via regulating Tau and Aβ [146]. Again, the intracellular mechanisms involved in this complex regulation are multiple: (i) activation of PI3K/AKT phosphorylates and inactivates GSK3β, thus suppressing Tau hyperphosphorylation and Aβ accumulation [169]; (ii) inhibition of Aβ-induced autophagy through mTOR signaling pathway [168]; (iii) inhibition of the synthesis of inflammatory mediators, especially TNF-α and IL-1β [170]; (iv) improvement of cell survival and inhibition of neuronal apoptosis through reduction of cytochrome c, increase of caspase-3, and suppression of intracellular calcium release [170,171]; (v) modulation of excitotoxicity [153]. The effects of the apelinergic system on multiple mechanisms involved in AD pathogenesis make apelin a potential therapeutic agent in AD.

In Parkinson's Disease (PD), the progressive loss of dopaminergic neurons in the substantia nigra is secondary to the accumulation of misfolded α-synuclein (α-Syn) in cytoplasmic inclusions named Lewy bodies [172,173]. Dysfunction of parkin, a key part of a multiprotein E3 ubiquitin ligase complex which destroys malformed proteins in neurons, is associated with the pathogenesis of PD [174], which is sustained by oxidative stress, microglia activation, and excessive neuroinflammation [175]. The dysregulation of PI3K/Akt and MAPKs cascades is implicated in the imbalance between cellular anti-apoptotic and pro-apoptotic pathways [175]. As in AD, apelin/APJ axis activation was linked to inhibition of apoptosis and dopaminergic neuronal loss, activation of antioxidants and autophagy, prevention of excessive neuroinflammation, suppression of endoplasmic reticulum stress, and glutamate-induced excitotoxicity. In in vitro models of SH-SY5Y cells, apelin-13 pre-treatment preserved the mitochondrial membrane potential, inhibited the release of cytochrome c and cleaved-caspase 3, and reduced ROS production, thus improving cell viability via PI3K-induced Akt activation [8,22]. AMPK/mTOR-dependent activation of autophagy [22] and ERK1/2-mediated attenuation of ER stress [176] contribute to apelin 13 protection against dopaminergic neurodegeneration. Downregulation of ROS and prevention of SH-SY5Y apoptosis was also described for apelin-36 [120]. In agreement with in vitro observations, different in vivo studies confirmed the neuroprotective role of apelin isoforms. Apelin-36 was able to prevent dopamine depletion in the striatum, at least partially via improving antioxidant cellular mechanisms (including SOD and glutathione) and downregulating inducible NOS and nitrated α-Syn expression [120], whereas apelin-13 markedly improved cognitive impairments in 6-OHDA-treated animals [177].

2.8. Oxidative Stress and Osteoporosis

High levels of oxidative stress and mitochondrial dysfunction are key regulators of bone marrow mesenchymal stem cells (BMSCs) survival and bone formation [178]. ROS overproduction associated with aging and estrogen deficiency determines the establishment of a "pro-osteoporotic" microenvironment, which alters the commitment of BMSCs and shifts their differentiation from the osteogenic to the adipogenic line [179]. Furthermore, intracellular ROS accumulation promotes BMSCs apoptosis [180,181], induces loss of function and apoptosis in osteoblasts [182], and increases osteoclastic bone resorption [183], thus contributing to the development of osteoporosis. Upon mitochondrial damage, mitophagy (a unique form of autophagy) selectively removes damaged mitochondria and prevents their accumulation and oxidative stress aggravation [184]. Hence, its activation in BMSCs contributes to promoting osteogenic function at the expense of adipogenic commitment [185–190]. As expected, based on the essential role of adipokines in bone homeostasis, the apelin/APJ system is a potential therapeutic tool in the treatment of osteoporosis. Endogenous apelin is highly expressed during osteogenesis in human BMSCs [191], whereas both apelin and APJ are downregulated in distal femurs of ovariectomy-induced osteoporotic rats [192]. Accordingly, serum apelin-13 in osteoporotic patients was significantly lower than in osteopenia and normal subjects [193]. Molecular and cellular studies demonstrated that apelin is able to stimulate proliferation and to suppress apoptosis of the osteoblastic cell line MC3T3-E1 [194] and to prevent mitochondrial ROS accumulation [195] and mitophagy [192] in BMSCs via the AMPK pathway [196].

2.9. Drug-Induced Oxidative Stress

Cisplatin, a broad-spectrum chemotherapeutic drug which affects DNA replication and inhibits cell division, is burdened by cardio- and ototoxicity [197,198]. Oxidative stress-dependent apoptosis of cardiomyocytes, which are limitedly able to regenerate, results in irreversible cisplatin-induced cardiomyopathy [199,200]. Oxidation resistance is recognized as a key cellular event in the protective effects of apelin-13 in cisplatin-exposed cardiomyocytes, where it efficaciously blocks the mitochondrial apoptosis pathway by inhibiting

ROS-mediated DNA damage and p53 phosphorylation and regulating MAPKs and AKT pathways [201]. In the cochlea, excessive ROS production and mitochondrial dysfunction induced by cisplatin are key contributors of cochlear hair cells (HCs) [202–204]. Down-regulation of apelin expression has been related to cisplatin-induced damage to HCs, and exogenous apelin's otoprotective effect against cisplatin-induced injury is closely associated with its ability to inhibit ROS production and mitochondrial dysfunction, which are known to potentiate cisplatin-induced apoptosis, via deregulation of JNK signaling [205]. Most recently, apelin-13 administration was demonstrated to reduce nephrotoxicity induced by cisplatin by triggering oxidative stress and inflammation [206].

Bupivacaine is a commonly used local anesthetic which may cause cardiotoxicity via inhibition of PI3K/AKT signaling [207], respiratory chain complexes I, III, and IV [208], and carnitine palmitoyl transferase [209]. As a result, cardiac energy metabolism is altered, and cardiac arrest may occur. In a rat model, apelin-13 treatment reduced bupivacaine-induced oxidative stress, attenuated mitochondrial morphological change and DNA damage, and enhanced mitochondrial energy metabolism through modulation of AMPK cascade, ultimately reversing bupivacaine-induced cardiotoxicity [210].

2.10. Oxidative Stress and Cancer

Cancer cells show a great ability to adapt their functions to perturbation of cellular homeostasis, particularly the imbalanced redox status secondary to local hypoxia and high metabolism. The theory of ROS rheostat predicts a fine regulation of ROS production and scavenging pathways to potentiate the antioxidant capacity of neoplastic cells and allow oxidative stress levels compatible with intracellular activities, even if higher than in normal cells [211]. Accordingly, an increased expression of ROS scavengers and low oxidative stress levels were described as crucial for the survival of pre-neoplastic foci in breast and liver cancer stem cells [212,213]. Indeed, oxidative stress is involved in the regulation of several cell functions, which are deregulated in cancer (i.e., cell growth, excitability, cytoskeleton remodeling and migration, autophagy, exocytosis and endocytosis, hormone signaling, necrosis, and apoptosis) [214,215], in the promotion of genomic instability and/or transcriptional errors [216], and in the activation of pro-survival and pro-metastatic pathways [215]. Consequently, the three steps of carcinogenesis (initiation, promotion, progression), local invasiveness and metastatization, and the resistance to treatment are strongly conditioned by the imbalance between ROS and antioxidant production [217]. As strong inducers of ROS generation, chemotherapy and radiotherapy are often unable to definitivly cure cancer: antineoplastic drugs and radiations may eliminate the bulk of cancer cells, but the upregulation of antioxidants in the presence of high ROS levels and ROS-dependent accumulation of DNA mutations are mechanisms that spare cancer stem cells and lead to therapeutic failure [211]. In this very complex scenario, antioxidant inhibitors (e.g., glutathione, HSP90, thioredoxin, enzyme poly-ADP-ribose polymerase or PARP) are considered a promising therapeutic tool in cancer treatment in association with radiotherapy or chemotherapy [211].

In the last 15 years, the role of the apelinergic system in tumorigenesis and cancer progression emerged from several studies and it has been proposed as a novel therapeutic target for different malignant tumors [218]. The apelin/APJ axis is upregulated in glioblastoma, esophageal squamous cell carcinoma, cholangiocarcinoma, and lymphoma, and it has been associated with carcinogenesis [218–221]. Furthermore, serum apelin levels were correlated with shorter survival, higher incidence of cancer recurrence and resistance to anticancer drugs in some human solid tumors, such as gastric cancer, lung adenocarcinoma, and breast cancer [222].

Hypoxia caused by the hypermorphosis of tumor cells was shown to promote apelin expression [223] via increased ROS-dependent hypoxia inducible factors (HIFs) activation [224], even in cancer stem cells [225]. The promoting effect of apelin/APJ in oxidative stress-associated cancer proliferation was reported in gastric adenocarcinoma cells [160] and melanoma [219], where apelin stimulated cancer cells survival and accelerated tumor growth in addition to allowing intratumoral lymphatic capillary and lymphnode metastatization. In several cancers, apelin may also protect cancer cells from apoptosis [226,227] and may play a role in mediating differentiation of mesenchymal stem cells into cancer stem cells, whose self-renewal is facilitated by activating signaling pathways such as wnt/β-catenin and Jagged/Notch [222]. In breast cancer, increased apelin levels were found to be an independent predictor of HER-2/neu expression and breast cancer phenotype, which accounts for 30% of breast carcinomas and is associated with a more aggressive tumor behavior [228].

The role of apelin/APJ signaling in angiogenesis is also well recognized in different cancers [222]. Growing evidence has suggested that apelin induces the maturation of tumor blood capillaries [229] and stimulates the proliferation of smooth muscle cells by modifying cyclin D1 expression and favoring the progression of cell cycle [230].

3. Conclusions

The apelin/APJ system may exert opposite effects on oxidative stress-mediated processes in different tissues and pathologic conditions (Table 1) by promoting prooxidant or antioxidant mechanisms (Figure 2). These contradictory functions, which can be explained by the existence of multiple isoforms of apelin, the activation of different APJ-coupled G proteins and signaling pathways, and context-dependent APJ trafficking, make the apelinergic axis a double-edged sword in regulating oxidative stress-associated diseases. In this view, a full comprehension of the complex role of apelin/APJ in ROS-related physiologic and pathologic processes is crucial, as well as to identify innovative therapeutic tools based on APJ inhibition or activation.

Table 1. Biological actions of the apelinergic system oxidative stress-related diseases.

Oxidative Stress-Related Effects the Apelinergic System in Different Organs and Tissues	
Cardiovascular system	
Vasodilation and blood pressure lowering [35–40,45]	- Induction of NO synthesis - RAAS modulation by counterregulating angiotensin II-dependent vasopressor stimulation by upregulating ACE2
Promotion of early atherogenesis [48–56]	- Promotion of endothelial dysfunction and myosin light chain phosphorylation in VSMCs - Activation of oxidative stress-linked proliferation in VSMCs - Induction of endothelial expression of ICAM-1 and VCAM-1 and release of MCP-1
Suppression of vascular calcification processes [62,63]	- Reduction of high glucose-induced proliferation, invasion, and osteoblastic differentiation of MOVAS - Increase of NO bioavailability
Prevention of diabetic microvascular complications [75–81]	- Inhibition of oxidative stress in kidney and neurons - Inhibition of retinal neoangiogenesis
Improvement of cardiac function [82–85]	- Inhibition of RAAS hyperactivation-associated fibrosis - Coronary vasodilation
Protection of myocardiocytes against IRI and reduction of infarct size in diabetic and non-diabetic patients [89–93,95–97]	- Inhibition of oxidative stress - Increase of myocardial progenitor cells in the infarcted hearts - Upregulation of PPARα - Inhibition of apoptosis
Protection of myocardiocytes against cisplatin-induced injury [201]	- Inhibition of mitochondrial apoptosis
Protection of myocardiocytes against bupivacaine-induced injury [210]	- Prevention of DNA damage - Enhanced mitochondrial energy metabolism

Table 1. *Cont.*

Oxidative Stress-Related Effects the Apelinergic System in Different Organs and Tissues	
Lung	
Restrainment of IRI-associated damage after pulmonary oedema or acute respiratory distress syndrome [99,100]	- Reduction of inflammatory infiltrates and proinflammatory cytokines secretion - Upregulation of UCP2 and SOD activity - Improvement of cell survival
Placenta	
Low apelin levels correlate with the etiopathogenesis of pre-eclampsia by inducing placental ischemia and endothelial dysfunction [67–72]	- Reduced synthesis of angiogenetic factors - Induction of a pro-inflammatory microenvironment
Central nervous system	
Protection of neurons and BBB against IRI [113–123]	- Reduction of vasogenic cerebral oedema - Protection of BBB integrity - Inhibition of inflammatory response - Inhibition of apoptosis - Inhibition of autophagy (traumatic brain-associated IRI)
Neuroprotection [45,140–177]	- Inhibition of apoptosis - Inhibition of excitotoxicity - Inhibition of neuroinflammation - Inhibition of autophagy
Kidney	
Protection of renal cells against IRI [124]	- Induction of the activity of the mitochondrial enzymes SOD, catalase, and glutathione peroxidase
Protection of renal cells against cisplatin-induced toxicity [206]	- Inhibition of inflammatory response
Adipose tissue, skeletal muscle, and liver	
Amelioration of insulin sensitivity [125–133]	- Increased expression of mitochondrial biogenesis markers and enzymes - Increased glucose utilization - Reduced liver steatosis and dysfunction
Bone	
Maintenance of bone health [192,194–196]	- Stimulation of osteoblast proliferation - Suppression of osteoblast apoptosis - Prevention of mitophagy in BMSCs
Inner ear	
Protection of cochlear cells against cisplatin-induced injury [205]	- Inhibition of cochlear cells apoptosis
Cancer cells	
Increased serum apelin levels correlate with shorter survival, higher incidence of cancer recurrence, and resistance to anticancer drugs [160,218–230]	- Increased cell proliferation - Increased cell survival and inhibition of cell apoptosis - Increased intratumoral lymphatic capillaries and lymph nodes metastasization - Differentiation of mesenchymal stem cells into cancer stem cells - Maintenance of cancer stem cells self-renewal - Increased angiogenesis

NO: nitric oxide; RAAS: renin–angiotensin–aldosterone system; ACE2: angiotensin converting enzyme 2; VSMCs: vascular smooth muscle cells; ICAM-1: intercellular adhesion molecule-1; VCAM-1: vascular cell adhesion molecule-1; MCP-1: monocyte chemoattractant protein-1; MOVAS: mouse aortic vascular smooth muscle cells; IRI: ischemia/reperfusion injury; UCP2: uncoupling protein 2; SOD: superoxide dismutase; BBB: blood–brain barrier; BMSCs: bone marrow stromal cells.

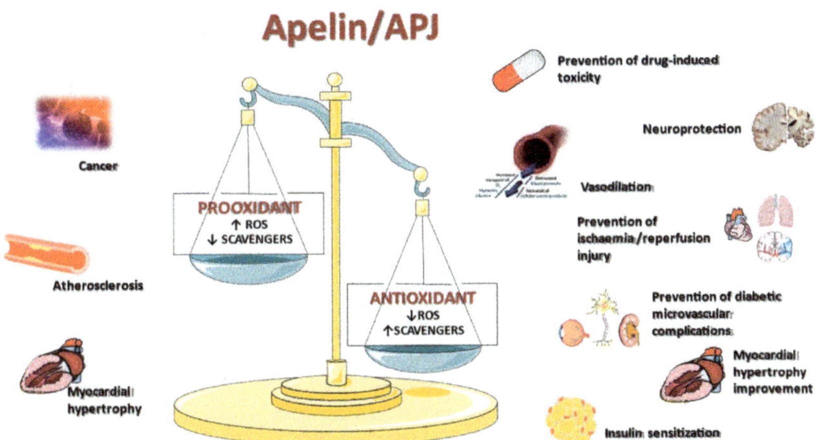

Figure 2. Prooxidant and antioxidant functions of the apelin/APJ system.

Author Contributions: Conceptualization, B.F., G.M., L.N. and A.P.; writing—original draft preparation, B.F.; writing—review and editing, G.M., L.N. and A.P.; supervision, A.P. All authors have read and agreed to the published version of the manuscript.

Funding: This research received no external funding.

Conflicts of Interest: The authors declare no conflict of interest.

References

1. Tatemoto, K.; Hosoya, M.; Habata, Y.; Fujii, R.; Kakegawa, T.; Zou, M.X.; Kawamata, Y.; Fukusumi, S.; Hinuma, S.; Kitada, C.; et al. Isolation and characterization of a novel endogenous peptide ligand for the human APJ receptor. *Biochem. Biophys. Res.Commun.* **1998**, *251*, 471–476. [CrossRef]
2. Lee, D.K.; Cheng, R.; Nguyen, T.; Fan, T.; Kariyawasam, A.P.; Liu, Y.; Osmond, D.H.; George, S.R.; O'Dowd, B.F. Characterization of apelin, the ligand for the APJ receptor. *J. Neurochem.* **2000**, *74*, 34–41. [CrossRef]
3. O'Dowd, B.F.; Heiber, M.; Chan, A.; Heng, H.H.; Tsui, L.C.; Kennedy, J.L.; Shi, X.; Petronis, A.; George, S.R.; Nguyen, T. A human gene that shows identity with the gene encoding the angiotensin receptor is located on chromosome 11. *Gene* **1993**, *136*, 355–360. [CrossRef]
4. Durham, A.L.; Speer, M.Y.; Scatena, M.; Giachelli, C.M.; Shanahan, C.M. Role of smooth muscle cells in vascular calcification: Implications in atherosclerosis and arterial stiffness. *Cardiovasc. Res.* **2018**, *114*, 590–600. [CrossRef] [PubMed]
5. Luo, X.L.; Liu, J.Q.; Zhou, H.; Chen, L.X. Apelin/APJ system: A critical regulator of vascular smooth muscle cell. *J. Cell Physiol.* **2018**, *233*, 5180–5188. [CrossRef] [PubMed]
6. Antushevich, H.; Wojcik, M. Apelin in disease. *Clin. Chim. Acta* **2018**, *483*, 241–248. [CrossRef] [PubMed]
7. Mughal, A.; O'Rourke, S.T. Vascular effects of apelin: Mechanisms and therapeutic potential. *Pharmacol. Ther.* **2018**, *190*, 139–147. [CrossRef]
8. Pouresmaeili-Babaki, E.; Esmaeili-Mahani, S.; Abbasnejad, M.; Ravan, H. Protective effect of neuropeptide apelin-13 on 6-hydroxydopamine-induced neurotoxicity in SH-SY5Y dopaminergic cells: Involvement of its antioxidant and antiapoptotic properties. *Rejuv. Res.* **2018**, *21*, 162–167. [CrossRef]
9. Kurowska, P.; Barbe, A.; Rozycka, M.; Chmielinska, J.; Dupont, J.; Rak, A. Apelin in reproductive physiology and pathology of different species: A critical review. *Int. J. Endocrinol.* **2018**, *2018*, 9170480. [CrossRef]
10. Tian, Y.; Chen, R.; Jiang, Y.; Bai, B.; Yang, T.; Liu, H. The protective effects and mechanisms of Apelin/APJ system on ischemic stroke: A promising therapeutic target. *Front. Neurol.* **2020**, *11*, 75. [CrossRef]
11. Wu, Y.; Wang, X.; Zhou, X.; Cheng, B.; Li, G.; Bai, B. Temporal expression of apelin/apelin receptor in ischemic stroke and its therapeutic potential. *Front. Mol. Neurosc.* **2017**, *10*, 1. [CrossRef] [PubMed]
12. Kang, Y.; Kim, J.; Anderson, J.P.; Wu, J.; Gleim, S.R.; Kundu, R.K.; McLean, D.L.; Kim, J.D.; Park, H.; Jin, S.W.; et al. Apelin-APJ signaling is a critical regulator of endothelial MEF2 activation in cardiovascular development. *Circ. Res.* **2013**, *113*, 22–31. [CrossRef]
13. Medhurst, A.D.; Jennings, C.A.; Robbins, M.J.; Davis, R.P.; Ellis, C.; Winborn, K.Y.; Lawrie, K.W.M.; Hervieu, G.; Riley, G.; Bolaky, J.E.; et al. Pharmacological and immunohistochemical characterization of the APJ receptor and its endogenous ligand apelin. *J. Neurochem.* **2003**, *84*, 1162–1172. [CrossRef] [PubMed]
14. Bennett, M.R.; Sinha, S.; Owens, G.K. Vascular smooth muscle cells in atherosclerosis. *Circ. Res.* **2016**, *118*, 692–702. [CrossRef]

15. Cheng, X.; Cheng, X.S.; Pang, C.C. Venous dilator effect of apelin, an endogenous peptide ligand for the orphan APJ receptor, in conscious rats. *Eur. J. Pharmacol.* **2003**, *470*, 171–175. [CrossRef]
16. Reaux, A.; De Mota, N.; Skultetyova, I.; Lenkei, Z.; El Messari, S.; Gallatz, K.; Corvol, P.; Palkovits, M.; Llorens-Cortès, C. Physiological role of a novel neuropeptide, apelin, and its receptor in the rat brain. *J. Neurochem.* **2001**, *77*, 1085–1096. [CrossRef]
17. Taheri, S.; Murphy, K.; Cohen, M.; Sujkovic, E.; Kennedy, A.; Dhillo, W.; Dakin, C.; Sajedi, A.; Ghatei, M.; Bloom, S. The effects of centrally administered apelin-13 on food intake, water intake and pituitary hormone release in rats. *Biochem. Biophys. Res. Commun.* **2002**, *291*, 1208–1212. [CrossRef]
18. O'Carroll, A.; Lolait, S.J.; Harris, L.E.; Pope, G.R. The apelin receptor APJ: Journey from an orphan to a multifaceted regulator of homeostasis. *J. Endocrinol.* **2013**, *219*, R13–R35. [CrossRef] [PubMed]
19. O'Shea, M.; Hansen, M.J.; Tatemoto, K.; Morris, M.J. Inhibitory effect of apelin-12 on nocturnal food intake in the rat. *Nutr. Neurosci.* **2003**, *6*, 163–167. [CrossRef]
20. Kazemi, F.; Zahedias, S. Effects of exercise training on adipose tissue apelin expression in streptozotocin-nicotinamide induced diabetic rats. *Gene* **2018**, *662*, 97–102. [CrossRef] [PubMed]
21. Wang, G.; Anini, Y.; Wei, W.; Qi, X.; O'Carroll, A.; Mochizuki, T.; Wang, H.; Hellmich, M.R.; Englander, E.W.; Greeley, G.H., Jr. Apelin, a new enteric peptide: Localization in the gastrointestinal tract, ontogeny, and stimulation of gastric cell proliferation and of cholecystokinin secretion. *Endocrinology* **2004**, *145*, 1342–1348. [CrossRef]
22. Chen, P.; Wang, Y.; Chen, L.; Song, N.; Xie, J. Apelin-13 protects dopaminergic neurons against rotenone-induced neurotoxicity through the AMPK/mTOR/ULK1 mediated autophagy activation. *Int. J. Mol. Sci.* **2020**, *21*, 8376. [CrossRef] [PubMed]
23. Chapman, N.A.; Dupre, D.J.; Rainey, J.K. The apelin receptor: Physiology, pathology, cell signaling, and ligand modulation of a peptide-activated class A GPCR. *Biochem. Cell Biol.* **2014**, *92*, 431–440. [CrossRef] [PubMed]
24. Zhou, N.; Zhang, X.; Fan, X.; Argyris, E.; Fang, J.; Acheampong, E.; DuBois, G.C.; Pomerantz, R.J. The N-terminal domain of APJ, a CNS-based coreceptor for HIV-1, is essential for its receptor function and coreceptor activity. *Virology* **2003**, *317*, 84–94. [CrossRef] [PubMed]
25. Lee, D.K.; Ferguson, S.S.; George, S.R.; O'Dowd, B.F. The fate of the internalized apelin receptor is determined by different isoforms of apelin mediating differential interaction with beta-arrestin. *Biochem. Biophys. Res. Commun.* **2010**, *395*, 185–189. [CrossRef]
26. Sharma, M.; Prabhavalkar, K.S.; Bhatt, L.K. Elabela peptide: An emerging target in therapeutics. *Curr. Drug Targets* **2022**, *23*, 1304–1318.
27. Wu, J.Q.; Kosten, T.R.; Zhang, X.Y. Free radicals, antioxidant defense system, and schizophrenia. *Prog.Neuropsychopharmacol. Biol. Psychiatry* **2013**, *46*, 200–206. [CrossRef] [PubMed]
28. Rajendran, P.; Nandakumar, N.; Rengarajan, T.; Palaniswami, R.; Gnanadhas, E.N.; Lakshminarasaiah, U. Antioxidants and human diseases. *Clin. Chim. Acta.* **2014**, *436*, 332–347. [CrossRef]
29. Taniyama, Y.; Griendling, K.K. Reactive oxygen species in the vasculature. *Hypertension* **2003**, *42*, 1075–1081. [CrossRef]
30. Orrenius, S. Reactive oxygen species in mitochondria-mediated cell death. *Drug Metab. Rev.* **2007**, *39*, 443–455. [CrossRef]
31. Pizzino, G.; Irrera, N.; Cucinotta, M.; Pallio, G.; Mannino, F.; Arcoraci, V.; Squadrito, F.; Altavilla, D.; Bitto, A. Oxidative Stress: Harms and Benefits for Human Health. *Oxid. Med. Cell Longev.* **2017**, *2017*, 8416763. [CrossRef] [PubMed]
32. Kinjo, T.; Higashi, H.; Uno, K.; Kuramoto, N. Apelin/Apelin Receptor System: Molecular Characteristics, Physiological Roles, and Prospects as a Target for Disease Prevention and Pharmacotherapy. *Curr. Mol. Pharmacol.* **2021**, *14*, 210–219. [CrossRef]
33. Ronkainen, V.; Ronkainen, J.J.; Hänninen, S.L.; Leskinen, H.; Ruas, J.L.; Pereira, T.; Poellinger, L.; Vuolteenaho, O.; Tavi, P. Hypoxia inducible factor regulates the cardiac expression and secretion of apelin. *FASEB J.* **2007**, *21*, 1821–1830. [CrossRef]
34. Kleinz, M.J.; Davenport, A.P. Emerging roles of apelin in biology and medicine. *Pharmacol. Ther.* **2005**, *107*, 198–211. [CrossRef]
35. Ishida, J.; Hashimoto, T.; Hashimoto, Y.; Nishiwaki, S.; Iguchi, T.; Harada, S.; Sugaya, T.; Matsuzaki, H.; Yamamoto, R.; Shiota, N.; et al. Regulatory roles for APJ, a seven-transmembrane receptor related to angiotensin-type 1 receptor in blood pressure in vivo. *J. Biol. Chem.* **2004**, *279*, 26274–26279. [CrossRef]
36. Zhong, J.C.; Huang, D.Y.; Liu, G.F.; Jin, H.Y.; Yang, Y.M.; Li, Y.F.; Song, X.H.; Du, K. Effects of all-trans retinoic acid on orphan receptor APJ signaling in spontaneously hypertensive rats. *Cardiovasc. Res.* **2005**, *65*, 743–750. [CrossRef] [PubMed]
37. Zhong, J.C.; Yu, X.Y.; Huang, Y.; Yung, L.M.; Lau, C.W.; Lin, S.G. Apelin modulates aortic vascular tone via endothelial nitric oxidase phosphorylation pathway in diabetic mice. *Cardiovasc. Res.* **2007**, *74*, 388–395. [CrossRef]
38. Gurzu, B.; Petrescu, B.C.; Costuleanu, M.; Petrescu, G. Interactions between apelin and angiotensin II on rat portal vein. *J. Renin. Angiotensin. Aldosterone. Syst.* **2006**, *7*, 212–216. [CrossRef] [PubMed]
39. Sato, T.; Suzuki, T.; Watanabe, H.; Kadowaki, A.; Fukamizu, A.; Liu, P.P.; Kimura, A.; Ito, H.; Penninger, J.M.; Imai, Y.; et al. Apelin is a positive regulator of ACE2 in failing hearts. *J. Clin. Investig.* **2013**, *123*, 5203–5211. [CrossRef]
40. Sabry, M.M.; Mahmoud, M.M.; Shoukry, H.S.; Rashed, L.; Kamar, S.S.; Ahmed, M.M. Interactive effects of apelin, renin-angiotensin system and nitric oxide in treatment of obesity-induced type 2 diabetes mellitus in male albino rats. *Arch. Physiol. Biochem.* **2019**, *125*, 244–254. [CrossRef] [PubMed]
41. Sinha, N.; Dabla, P. Oxidative stress and antioxidants in hypertension-a current review. *Curr. Hypertens. Rev.* **2015**, *11*, 132–142. [CrossRef]

42. Yoshida, K.; Kobayashi, N.; Ohno, T.; Fukushima, H.; Matsuoka, H. Cardioprotective effect of angiotensin II type 1 receptor antagonist associated with bradykinin-endothelial nitric oxide synthase and oxidative stress in dahl salt-sensitive hypertensive rats. *J. Hypertens.* **2007**, *25*, 1633–1642. [CrossRef]
43. Oidor-Chan, V.H.; Hong, E.; Pérez-Severiano, F.; Montes, S.; Torres-Narváez, J.C.; Del Valle-Mondragón, L.; Pastelín-Hernández, G.; Sánchez-Mendoza, A. Fenofibrate plus metformin produces cardioprotection in a type 2 diabetes and acute myocardial infarction model. *PPAR Res.* **2016**, *2016*, 8237264. [CrossRef] [PubMed]
44. Sowers, J.R. Insulin resistance and hypertension. *Am. J. Physiol. Heart Circ. Physiol.* **2004**, *286*, H1597–H1602. [CrossRef] [PubMed]
45. Zhou, Q.; Cao, J.; Chen, L. Apelin/APJ system: A novel therapeutic target for oxidative stress-related inflammatory diseases. *Int. J. Mol. Med.* **2016**, *37*, 1159–1169. [CrossRef] [PubMed]
46. Wassmann, S.; Nickenig, G. Pathophysiological regulation of the AT1-receptor and implications for vascular disease. *J. Hypertens. Suppl.* **2006**, *24*, S15–S21. [CrossRef]
47. Grootaert, M.O.J.; Bennett, M.R. Vascular smooth muscle cells in atherosclerosis: Time for a re-assessment. *Cardiovasc. Res.* **2021**, *117*, 2326–2339. [CrossRef]
48. Sato, K.; Kihara, M.; Hashimoto, T.; Matsushita, K.; Koide, Y.; Tamura, K.; Hirawa, N.; Toya, Y.; Fukamizu, A.; Umemura, S. Alterations in renal endothelial nitric oxide synthase expression by salt diet in angiotensin type-1a receptor gene knockout mice. *J. Am. Soc. Nephrol.* **2004**, *15*, 1756–1763. [CrossRef]
49. Pueyo, M.E.; Arnal, J.F.; Rami, J.; Michel, J.B. Angiotensin II stimulates the production of NO and peroxynitrite in endothelial cells. *Am. J. Physiol.* **1998**, *274*, C214–C220. [CrossRef]
50. Kihara, M.; Sato, K.; Hashimoto, T.; Imai, N.; Toya, Y.; Umemura, S. Expression of endothelial nitric oxide synthase is suppressed in the renal vasculature of angiotensinogen-gene knockout mice. *Cell Tissue Res.* **2006**, *323*, 313–320. [CrossRef]
51. Ramchandran, R.; Takezako, T.; Saad, Y.; Stull, L.; Fink, B.; Yamada, H.; Dikalov, S.; Harrison, D.G.; Moravec, C.; Karnik, S.S. Angiotensinergic stimulation of vascular endothelium in mice causes hypotension, bradycardia, and attenuated angiotensin response. *Proc. Natl. Acad. Sci. USA* **2006**, *103*, 19087–19092. [CrossRef] [PubMed]
52. Li, L.; Li, L.; Xie, F.; Zhang, Z.; Guo, Y.; Tang, G.; Lv, D.; Lu, Q.; Chen, L.; Li, J. Jagged-1/Notch3 signaling transduction pathway is involved in apelin-13-induced vascular smooth muscle cells proliferation. *Acta Biochim. Biophys. Sin.* **2013**, *45*, 875–881. [CrossRef] [PubMed]
53. Hashimoto, T.; Kihara, M.; Imai, N.; Yoshida, S.; Shimoyamada, H.; Yasuzaki, H.; Ishida, J.; Toya, Y.; Kiuchi, Y.; Hirawa, N.; et al. Requirement of Apelin-Apelin Receptor System for Oxidative Stress-Linked Atherosclerosis. *AJP* **2007**, *171*, 1705–1712. [CrossRef]
54. Lv, D.; Li, H.; Chen, L. Apelin and APJ, a novel critical factor and therapeutic target for atherosclerosis. *Acta Biochim. Biophys. Sin.* **2013**, *45*, 527–533. [CrossRef] [PubMed]
55. Mao, X.H.; Tao, S.; Zhang, X.H.; Li, F.; Qin, X.P.; Liao, D.F.; Li, L.F.; Chen, L.X. Apelin-13 promotes monocyte adhesion to human umbilical vein endothelial cell mediated by phosphatidylinositol 3-kinase signaling pathway. *Prog. Biochem. Biophys.* **2011**, *38*, 1162–1170. [CrossRef]
56. Lu, Y.; Zhu, X.; Liang, G.; Cui, R.; Liu, Y.; Wu, S.; Liang, Q.; Liu, G.; Jiang, Y.; Liao, X.; et al. Apelin-APJ induces ICAM-1, VCAM-1 and MCP-1 expression via NF-κB/JNK signal pathway in human umbilical vein endothelial cells. *Amino Acids* **2012**, *43*, 2125–2136. [CrossRef]
57. Ruiz, E.; Gordillo-Moscoso, A.; Padilla, E.; Redondo, S.; Rodriguez, E.; Reguillo, F.; Briones, A.M.; van Breemen, C.; Okon, E.; Tejerina, T. Human vascular smooth muscle cells from diabetic patients are resistant to induced apoptosis due to high Bcl-2 expression. *Diabetes* **2006**, *55*, 1243–1251. [CrossRef]
58. Ruiz, E.; Redondo, S.; Gordillo-Moscoso, A.; Tejerina, T. Pioglitazone induces apoptosis in human vascular smooth muscle cells from diabetic patients involving the transforming growth factor-beta/activin receptor-like kinase-4/5/7/Smad2signaling pathway. *J. Pharmacol. Exp. Ther.* **2007**, *321*, 431–438. [CrossRef]
59. Liberman, M. Oxidant generation predominates around calcifying foci and enhances progression of aortic valve calcification. *Arterioscler. Thromb. Vasc. Biol.* **2008**, *28*, 463–470. [CrossRef]
60. Nakagawa, Y.; Ikeda, K.; Akakabe, Y.; Koide, M.; Uraoka, M. Paracrine osteogenic signals via bone morphogenetic protein-2 accelerate the atheroscle-rotic intimal calcification in vivo. *Arterioscler. Thromb. Vasc. Biol.* **2010**, *30*, 1908–1915. [CrossRef]
61. Newby, A.C.; Zaltsman, A.B. Fibrous cap formation or destruction–the critical importance of vascular smooth muscle cell proliferation, migration and matrix formation. *Cardiovasc. Res.* **1999**, *41*, 345–360. [CrossRef] [PubMed]
62. Zhang, P.; Wang, A.; Yang, H.; Ai, L.; Zhang, H.; Wang, Y.; Bi, Y.; Fan, H.; Gao, J.; Zhang, H.; et al. Apelin-13 attenuates high glucose-induced calcification of MOVAS cells by regulating MAPKs and PI3K/AKT pathways and ROS-mediated signals. *Biomed. Pharmacother.* **2020**, *128*, 110271. [CrossRef] [PubMed]
63. Chun, H.J.; Ali, Z.A.; Kojima, Y.; Kundu, R.K.; Sheikh, A.Y.; Agrawal, R.; Zheng, L.; Leeper, N.J.; Pearl, N.E.; Patterson, A.J.; et al. Apelin signaling antagonizes Ang II effects in mouse models of atherosclerosis. *J. Clin. Investig.* **2008**, *118*, 3343–3354. [CrossRef]
64. Staff, A.C. The two-stage placental model of preeclampsia: An update. *J. Reprod. Immunol.* **2019**, *134–135*, 1–10. [CrossRef]
65. Daskalakis, G.; Papapanagiotou, A. Serum markers for the prediction of preeclampsia. *J. Neurol. Neurophysiol.* **2015**, *6*, 264.
66. Cobellis, L.; De Falco, M.; Mastrogiacomo, A.; Giraldi, D.; Dattilo, D.; Scaffa, C.; Colacurci, N.; De Luca, A. Modulation of apelin and APJ receptor in normal and preeclampsia-complicated placentas. *Histol. Histopathol.* **2007**, *22*, 1–8. [PubMed]

67. Deniz, R.; Baykus, Y.; Ustebay, S.; Ugur, K.; Yavuzkir, S.; Aydin, S. Evaluation of elabela, apelin and nitric oxide findings in maternal blood of normal pregnant women, pregnant women with pre-eclampsia, severe pre-eclampsia and umbilical arteries and venules of newborns. *J. Obstet. Gynaecol. Res.* **2019**, *39*, 907–912. [CrossRef]
68. Gürlek, B.; Yılmaz, A.; Durakoğlugil, M.E.; Karakaş, S.; Kazaz, I.M.; Önal, Ö.; Şatıroğlu, Ö. Evaluation of serum apelin-13 and apelin-36 concentrations in preeclamptic pregnancies. *J. Obstet. Gynaecol. Res.* **2020**, *46*, 58–65. [CrossRef]
69. Inuzuka, H.; Nishizawa, H.; Inagaki, A.; Suzuki, M.; Ota, S.; Miyamura, H.; Miyazaki, J.; Sekiya, T.; Kurahashi, H.; Udagawa, Y. Decreased expression of apelin in placentas from severe pre-eclampsia patients. *Hypertens. Pregnancy* **2013**, *32*, 410–421. [CrossRef]
70. Bortoff, K.D.; Qiu, C.; Runyon, S.; Williams, M.A.; Maitra, R. Decreased maternal plasma apelin concentrations in preeclampsia. *Hypertens. Pregnancy* **2012**, *31*, 398–404. [CrossRef]
71. Van Mieghem, T.; Doherty, A.; Baczyk, D.; Drewlo, S.; Baud, D.; Carvalho, J. Apelin in normal pregnancy and pregnancies complicated by placental insufficiency. *Reprod. Sci.* **2016**, *23*, 1037–1043. [CrossRef]
72. Hamza, R.Z.; Diab, A.A.A.; Zahra, M.H.; Asalah, A.K.; Moursi, S.M.M.; Al-Baqami, N.M.; Al-Salmi, F.A.; Attia, M.S. Correlation between Apelin and Some Angiogenic Factors in the Pathogenesis of Preeclampsia: Apelin-13 as Novel Drug for Treating Preeclampsia and Its Physiological Effects on Placenta. *Int. J. Endocrinol.* **2021**, *15*, 5017362. [CrossRef] [PubMed]
73. Murthi, P.; Pinar, A.A.; Dimitriadis, E.; Samuel, C.S. Inflammasomes—A molecular link for altered immunoregulation and inflammation mediated vascular dysfunction in Preeclampsia. *Int. J. Mol. Sci.* **2020**, *21*, 1406. [CrossRef]
74. Armistead, B.; Kadam, L.; Drewlo, S.; Kohan-Ghadr, H. The role of NFκB in healthy and preeclamptic placenta: Trophoblasts in the spotlight. *Int. J. Mol. Sci.* **2020**, *21*, 1775. [CrossRef] [PubMed]
75. Ryu, S.; Ornoy, A.; Samuni, A.; Zangen, S.; Kohen, R. Oxidative stress in Cohen diabetic rat model by high-sucrose, low-copper diet: Inducing pancreatic damage and diabetes. *Metabolism* **2008**, *57*, 1253–1261. [CrossRef] [PubMed]
76. Nishida, M.; Okumura, Y.; Oka, T.; Toiyama, K.; Ozawa, S.; Itoi, T.; Hamaoka, K. The role of apelin on the alleviative effect of Angiotensin receptor blocker in unilateral ureteral obstruction-induced renal fibrosis. *Nephron. Extra* **2012**, *2*, 39–47. [CrossRef] [PubMed]
77. Day, R.T.; Cavaglieri, R.C.; Feliers, D. Apelin retards the progression of diabetic nephropathy. *Am. J. Physiol. Renal. Physiol.* **2013**, *304*, F788–F800. [CrossRef] [PubMed]
78. Zeng, X.J.; Yu, S.P.; Zhang, L.; Wei, L. Neuroprotective effect of the endogenous neural peptide apelin in cultured mouse cortical neurons. *Exp. Cell Res.* **2010**, *316*, 1773–1783. [CrossRef]
79. Tao, Y.; Lu, Q.; Jiang, Y.R.; Qian, J.; Wang, J.Y.; Gao, L.; Jonas, J.B. Apelin in plasma and vitreous and in fibrovascular retinal membranes of patients with proliferative diabetic retinopathy. *Investig. Ophthalmol. Vis. Sci.* **2010**, *51*, 4237–4242. [CrossRef]
80. Lu, Q.; Feng, J.; Jiang, Y.R. The role of apelin in the retina of diabetic rats. *PLoS ONE* **2013**, *8*, e69703. [CrossRef]
81. Saint-Geniez, M.; Masri, B.; Malecaze, F.; Knibiehler, B.; Audigier, Y. Expression of the murine msr/apj receptor and its ligand apelin is upregulated during formation of the retinal vessels. *Mech. Dev.* **2002**, *110*, 183–186. [CrossRef] [PubMed]
82. Foussal, C.; Lairez, O.; Calise, D.; Pathak, A.; Guilbeau-Frugier, C.; Valet, P.; Parini, A.; Kunduzova, O. Activation of catalase by apelin prevents oxidative stress-linked cardiac hypertrophy. *FEBS Lett.* **2010**, *584*, 2363–2370. [CrossRef] [PubMed]
83. Ceylan-Isik, A.F.; Kandadi, M.R.; Xu, X.; Hua, Y.; Chicco, A.J.; Ren, J.; Nair, S. Apelin administration ameliorates high fat diet-induced cardiac hypertrophy and contractile dysfunction. *J. Mol. Cell Cardiol.* **2013**, *63*, 4–13. [CrossRef] [PubMed]
84. Zhong, S.; Guo, H.; Wang, H.; Xing, D.; Lu, T.; Yang, J.; Wang, C. Apelin-13 alleviated cardiac fibrosis via inhibiting the PI3K/Akt pathway to attenuate oxidative stress in rats with myocardial infarction-induced heart failure. *Biosci. Rep.* **2020**, *40*, BSR20200040. [CrossRef]
85. Japp, A.; Cruden, N.; Barnes, G.; Van Gemeren, N.; Mathews, J.; Adamson, J.; Johnston, N.; Denvir, M.; Megson, I.; Flapan, A. Acute cardiovascular effects of apelin in humans: Potential role in patients with chronic heart failure. *Circulation* **2010**, *121*, 1818–1827. [CrossRef]
86. Zhang, F.; Sun, H.J.; Xiong, X.Q.; Chen, Q.; Li, Y.; Kang, Y.; Wang, J.; Gao, X.; Zhu, G. Apelin-13 and APJ in paraventricular nucleus contribute to hypertension via sympathetic activation and vasopressin release in spontaneously hypertensive rats. *Acta Physiol.* **2014**, *212*, 17–27. [CrossRef]
87. Cowled, P.; Fitridge, R. Pathophysiology of Reperfusion Injury. In *Mechanisms of Vascular Disease: A Reference Book for Vascular Specialists*; University of Adelaide Press: Adelaide, Australia, 2011; ISBN 978-0-9871718-2-5.
88. Lakhan, S.E.; Kirchgessner, A.; Tepper, D.; Aidan, L. Matrix metalloproteinases and blood-brain barrier disruption in acute ischemic stroke. *Front. Neurol.* **2013**, *4*, 32. [CrossRef]
89. Zeng, X.J.; Zhang, L.K.; Wang, H.X.; Lu, L.Q.; Ma, L.Q.; Tang, C.S. Apelin protects heart against ischemia/reperfusion injury in rat. *Peptides* **2009**, *30*, 1144–1152. [CrossRef]
90. Simpkin, J.C.; Yellon, D.M.; Davidson, S.M.; Lim, S.Y.; Wynne, A.M.; Smith, C.C. Apelin-13 and apelin-36 exhibit direct cardioprotective activity against ischemia-reperfusion injury. *Basic Res. Cardiol.* **2007**, *102*, 518–528. [CrossRef]
91. Ashley, E.A.; Powers, J.; Chen, M.; Kundu, R.; Finsterbach, T.; Caffarelli, A.; Deng, A.; Eichhorn, J.; Mahajan, R.; Agrawal, R. The endogenous peptide apelin potently improves cardiac contractility and reduces cardiac loading in vivo. *Cardiovasc. Res.* **2005**, *65*, 73–82. [CrossRef]
92. Xu, W.; Yu, H.; Ma, R.; Ma, L.; Liu, Q.; Shan, H.; Wu, C.; Zhang, R.; Zhou, Y.; Shan, H. Apelin protects against myocardial ischemic injury by inhibiting dynamin-related protein 1. *Oncotarget* **2017**, *8*, 100034. [CrossRef]

93. Li, L.; Zeng, H.; Chen, J.X. Apelin-13 increases myocardial progenitor cells and improves repair post-myocardial infarction. *Am. J. Phys. Heart Circ. Phys.* **2012**, *303*, H605–H618.
94. Kupai, K.; Szabó, R.; Veszelka, M.; Awar, A.A.; Török, S.; Csonka, A.; Baráth, Z.; Pósa, A.; Varga, C. Consequences of exercising on ischemia-reperfusion injury in type 2 diabetic Goto-Kakizaki rat hearts: Role of the HO/NOS system. *Diabetol. Metab. Syndr.* **2015**, *7*, 85. [CrossRef] [PubMed]
95. An, S.; Wang, X.; Shi, H.; Zhang, X.; Meng, H.; Li, W.; Chen, D.; Ge, J. Apelin protects against ischemia-reperfusion injury in diabetic myocardium via inhibiting apoptosis and oxidative stress through PI3K and p38-MAPK signaling pathways. *Aging* **2020**, *12*, 25120–25137. [CrossRef] [PubMed]
96. Robinson, E.; Grieve, D.J. Significance of peroxisome proliferator-activated receptors in the cardiovascular system in health and disease. *Pharmacol. Ther.* **2009**, *122*, 246–263. [CrossRef] [PubMed]
97. Kim, T.; Yang, Q. Peroxisome-proliferator-activated receptors regulate redox signaling in the cardiovascular system. *World J. Cardiol.* **2013**, *5*, 164–174. [CrossRef] [PubMed]
98. Weyker, P.D.; Webb, C.A.; Kiamanesh, D.; Flynn, B.C. Lung ischemia reperfusion injury: A bench-to-bedside review. *Semin. Cardiothorac. Vasc. Anesth.* **2013**, *17*, 28–43. [CrossRef] [PubMed]
99. Xia, F.; Chen, H.; Jin, Z.; Fu, Z. Apelin-13 protects the lungs from ischemia-reperfusion injury by attenuating inflammatory and oxidative stress. *Hum. Exp. Toxicol.* **2021**, *40*, 685–694. [CrossRef]
100. Singh, P.K.; Gari, M.; Choudhury, S.; Shukla, A.; Gangwar, N.; Garg, S.K. Oleic acid prevents isoprenaline-induced cardiac injury: Effects on cellular oxidative stress, inflammation and histopathological alterations. *Cardiovasc. Toxicol.* **2020**, *20*, 28–48. [CrossRef]
101. Beckman, J.S.; Beckman, T.W.; Chen, J.; Marshall, P.A.; Freeman, B.A. Apparent hydroxyl radical production by peroxynitrite: Implications for endothelial injury from nitric oxide and superoxide. *Proc. Natl. Acad. Sci. USA* **1990**, *87*, 1620–1624. [CrossRef]
102. Gu, Z.; Kaul, M.; Yan, B.; Kridel, S.J.; Cui, J.; Strongin, A.; Smith, J.W.; Liddington, R.C.; Lipton, S.A. S-nitrosylation of matrix metalloproteinases: Signaling pathway to neuronal cell death. *Science* **2002**, *297*, 1186–1190. [CrossRef]
103. Chen, H.; Yoshioka, H.; Kim, G.S.; Jung, J.E.; Okami, N.; Sakata, H.; Maier, C.M.; Narasimhan, P.; Goeders, C.E.; Chan, P.H. Oxidative stress in ischemic brain damage: Mechanisms of cell death and potential molecular targets for neuroprotection. *Antioxid. Redox. Signal.* **2011**, *14*, 1505–1517. [CrossRef] [PubMed]
104. Chan, P.H. Reactive oxygen radicals in signaling and damage in the ischemic brain. *J. Cereb. Blood Flow Metab.* **2002**, *21*, 2–14. [CrossRef] [PubMed]
105. Kim, J.; Ko, A.R.; Hyun, H.W.; Kang, T.C. ETB receptor-mediated MMP-9 activation induces vasogenic edema via ZO-1 protein degradation following status epilepticus. *Neuroscience* **2015**, *304*, 355–367. [CrossRef] [PubMed]
106. Lochhead, J.J.; McCaffrey, G.; Quigley, C.E.; Finch, J.; DeMarco, K.M.; Nametz, N.; Davis, T.P. Oxidative stress increases blood–brain barrier permeability and induces alterations in occludin during hypoxia–reoxygenation. *J. Cereb. Blood Flow Metab.* **2010**, *30*, 1625–1636. [CrossRef]
107. Thor´en, M.; Azevedo, E.; Dawson, J.; Egido, J.A.; Falcou, A.; Ford, G.A.; Holmin, S.; Mikulik, R.; Ollikainen, J.; Wahlgren, N. Predictors for cerebral edema in acute ischemic stroke treated with intravenous thrombolysis. *Stroke* **2017**, *48*, 2464–2471. [CrossRef]
108. Amani, H.; Habibey, R.; Shokri, F.; Hajmiresmail, S.J.; Akhavan, O.; Mashaghi, A.; Pazoki-Toroudi, H. Selenium nanoparticles for targeted stroke therapy through modulation of inflammatory and metabolic signaling. *Sci. Rep.* **2019**, *9*, 6044. [CrossRef]
109. Yang, Y.; Thompson, J.F.; Taheri, S.; Salayandia, V.M.; McAvoy, T.A.; Hill, J.W.; Yang, Y.; Estrada, E.Y.; Rosenberg, G.A. Early inhibition of MMP activity in ischemic rat brain promotes expression of tight junction proteins and angiogenesis during recovery. *J. Cereb. Blood Flow Metab.* **2013**, *33*, 1104–1114. [CrossRef]
110. Michinaga, S.; Nagase, M.; Matsuyama, E.; Yamanaka, D.; Seno, N.; Fuka, M.; Yamamoto, Y.; Koyama, Y. Amelioration of cold injury-induced cortical brain edema formation by selective endothelin ETB receptor antagonists in mice. *PLoS ONE* **2014**, *9*, e102009. [CrossRef] [PubMed]
111. Chu, H.; Yang, X.; Huang, C.; Gao, Z.; Tang, Y.; Dong, Q. Apelin-13 protects against ischemic blood-brain barrier damage through the effects of aquaporin-4. *Cerebrovasc. Dis.* **2017**, *44*, 10–25. [CrossRef]
112. Yuan, M.; Ge, M.; Yin, J.; Dai, Z.; Xie, L.; Li, Y.; Liu, X.; Peng, L.; Zhang, G.; Si, J. Isoflurane post-conditioning down-regulates expression of aquaporin 4 in rats with cerebral ischemia/reperfusion injury and is possibly related to bone morphogenetic protein 4/Smad1/5/8 signaling pathway. *Biomed. Pharm.* **2018**, *97*, 429–438. [CrossRef]
113. Gholamzadeh, R.; Ramezani, F.; Tehrani, P.M.; Aboutaleb, N. Apelin-13 attenuates injury following ischemic stroke by targeting matrix metalloproteinases (MMP), endothelin- B receptor, occludin/claudin-5 and oxidative stress. *J. Chem. Neuroanat.* **2021**, *118*, 102015. [CrossRef] [PubMed]
114. Kim, J.E.; Ryu, H.; Kang, T. Status epilepticus induces vasogenic edema via tumor necrosis factor-α/endothelin-1-mediated two different pathways. *PLoS ONE* **2013**, *8*, e74458. [CrossRef] [PubMed]
115. Wu, G.; Li, L.; Liao, D.; Wang, Z. Protective effect of Apelin-13 on focal cerebral ischemia-reperfusion injury in rats. *Nan Fang Yi Ke Da Xue Xue Bao* **2015**, *35*, 1335–1339. [PubMed]
116. Duan, J.; Cui, J.; Yang, Z.; Guo, C.; Cao, J.; Xi, M.; Weng, Y.; Yin, Y.; Wang, Y.; Wei, G.; et al. Neuroprotective effect of Apelin 13 on ischemic stroke by activating AMPK/GSK-3beta/Nrf2 signaling. *J. Neuroinflammation* **2019**, *16*, 24. [CrossRef] [PubMed]
117. Yang, Y.; Zhang, X.J.; Li, L.T.; Cui, H.Y.; Zhang, C.; Zhu, C.H.; Miao, J.Y. Apelin-13 protects against apoptosis by activating AMP-activated protein kinase pathway in ischemia stroke. *Peptides* **2016**, *75*, 96–100. [CrossRef]

118. Gu, Q.; Zhai, L.; Feng, X.; Chen, J.; Miao, Z.; Ren, L.; Qian, X.; Yu, J.; Li, Y.; Xu, X.; et al. Apelin-36, a potent peptide, protects against ischemic brain injury by activating the PI3K/Akt pathway. *Neurochem. Int.* **2013**, *63*, 535–540. [CrossRef]
119. Qiu, J.; Wang, X.; Wu, F.; Wan, L.; Cheng, B.; Wu, Y.; Bai, B. Low dose of Apelin-36 attenuates ER stress-associated apoptosis in rats with Ischemic stroke. *Front. Neurol.* **2017**, *8*, 556. [CrossRef]
120. Zhu, J.; Gao, W.; Shan, X.; Wang, C.; Wang, H.; Shao, Z.; Dou, S.; Jiang, Y.; Wang, C.; Cheng, B. Apelin-36 mediates neuroprotective effects by regulating oxidative stress, autophagy and apoptosis in MPTP-induced Parkinson's disease model mice. *Brain Res.* **2020**, *1726*, 146493. [CrossRef]
121. Liu, D.R.; Hu, W.; Chen, G.Z. Apelin-12 exerts neuroprotective effect against ischemia-reperfusion injury by inhibiting JNK and P38MAPK signaling pathway in mouse. *Eur. Rev. Med. Pharmacol. Sci.* **2018**, *22*, 3888–3895.
122. Curry, D.W.; Stutz, B.; Andrews, Z.B.; Elsworth, J.D. Targeting AMPK signaling as a neuroprotective strategy in Parkinson's disease. *J. Park. Dis.* **2018**, *8*, 161–181. [CrossRef]
123. Bao, H.J.; Zhang, L.; Han, W.C.; Dai, D.K. Apelin-13 attenuates traumatic brain injury-induced damage by suppressing autophagy. *Neurochem. Res.* **2015**, *40*, 89–97. [CrossRef]
124. Bircan, B.; Cakir, M.; Kirbag, S.; Gul, H.F. Effect of apelin hormone on renal ischemia/reperfusion induced oxidative damage in rats. *Ren. Fail.* **2016**, *38*, 1122–1128. [CrossRef]
125. Boucher, J.; Masri, B.; Daviaud, D.; Gesta, S.; Guigné, C.; Mazzucotelli, A.; Castan-Laurell, I.; Tack, I.; Knibiehler, B.; Carpéné, C.; et al. Apelin, a newly identified adipokine up-regulated by insulin and obesity. *Endocrinology* **2005**, *146*, 1764–1771. [CrossRef]
126. GorhedeWinzell, M.; Magnusson, C.; Ahren, B. The APJ receptor is expressed in pancreatic islets and its ligand, apelin, inhibits insulin secretion in mice. *Regul. Pept.* **2005**, *131*, 12–17. [CrossRef]
127. Wei, L.; Hou, X.; Tatemoto, K. Regulation of apelin mRNA expression by insulin and glucocorticoids in mouse 3T3-L1 adipocytes. *Regul. Pept.* **2005**, *132*, 27–32. [CrossRef]
128. Daviaud, D.; Boucher, J.; Gesta, S.; Dray, C.; Guigne, C.; Quilliot, D.; Ayav, A.; Ziegler, O.; Carpene, C.; Saulnier-Blache, J.; et al. TNF alpha up-regulates apelin expression in human and mouse adipose tissue. *FASEB J.* **2006**, *20*, 1528–1530. [CrossRef] [PubMed]
129. Frier, B.C.; Williams, D.B.; Wright, D.C. The effects of apelin treatment on skeletal muscle mitochondrial content. *Am. J. Physiol. Regul. Integr. Comp. Physiol.* **2009**, *297*, R1761–R1768. [CrossRef] [PubMed]
130. Attané, C.; Foussal, C.; Le Gonidec, S.; Benani, A.; Daviaud, D.; Wanecq, E.; Guzmán-Ruiz, R.; Dray, C.; Bezaire, V.; Rancoule, C. Apelin treatment increases complete fatty acid oxidation, mitochondrial oxidative capacity, and biogenesis in muscle of insulin-resistant mice. *Diabetes* **2012**, *61*, 310–320. [CrossRef] [PubMed]
131. Than, A.; Zhang, X.; Leow, M.K.; Poh, C.L.; Chong, S.K.; Chen, P. Apelin attenuates oxidative stress in human adipocytes. *J. Biol. Chem.* **2014**, *289*, 3763–3774. [CrossRef]
132. Milagro, F.I.; Campion, J.; Martinez, J.A. Weight gain induced by high-fat feeding involves increased liver oxidative stress. *Obesity* **2006**, *14*, 1118–1123. [CrossRef] [PubMed]
133. Dray, C.; Knauf, C.; Daviaud, D.; Waget, A.; Boucher, J.; Buléon, M.; Cani, P.D.; Attané, C.; Guigné, C.; Carpéné, C. Apelin stimulates glucose utilization in normal and obese insulin-resistant mice. *Cell Metab.* **2008**, *8*, 437–445. [CrossRef] [PubMed]
134. Rai, R.; Ghosh, A.K.; Eren, M.; Mackie, A.R.; Levine, D.C.; Kim, S.; Cedernaes, J.; Ramirez, V.; Procissi, D.; Smith, L.H.; et al. Downregulation of the apelinergic axis accelerates aging, whereas its systemic restoration improves the mammalian healthspan. *Cell Rep.* **2017**, *21*, 1471–1480. [CrossRef] [PubMed]
135. Yao, F.; Lv, Y.; Zhang, M.; Xie, W.; Tan, Y.; Gong, D.; Cheng, H.; Liu, D.; Li, L.; Liu, X.; et al. Apelin-13 impedes foam cell formation by activating Class III PI3K/Beclin-1-mediated autophagic pathway. *Biochem. Biophys. Res. Comm.* **2015**, *466*, 637–643. [CrossRef]
136. Xie, F.; Liu, W.; Feng, F.; Li, X.; He, L.; Lv, D.; Qin, X.; Li, L.; Li, L.; Chen, L. Apelin-13 promotes cardiomyocyte hypertrophy via PI3K-Akt-ERK1/2-p70S6K and PI3K-induced autophagy. *Acta Biochim. Biophys. Sin.* **2015**, *47*, 969–980. [CrossRef] [PubMed]
137. Busch, R.; Strohbach, A.; Pennewitz, M.; Lorenz, F.; Bahls, M.; Busch, M.C.; Felix, S.B. Regulation of the endothelial apelin/APJ system by hemodynamic fluid flow. *Cell. Signal.* **2015**, *27*, 1286–1296. [CrossRef]
138. Cheng, B.; Chen, J.; Bai, B.; Xin, Q. Neuroprotection of apelin and its signaling pathway. *Peptides* **2012**, *37*, 171–173. [CrossRef]
139. O'Donnell, L.A.; Agrawal, A.; Sabnekar, P.; Dichter, M.A.; Lynch, D.R.; Kolson, D.L. Apelin, an endogenous neuronal peptide, protects hippocampal neurons against excitotoxic injury. *J. Neurochem.* **2007**, *102*, 1905–1917. [CrossRef]
140. Mohseni, F.; Garmabi, B.; Khaksari, M. Apelin-13 attenuates spatial memory impairment by anti-oxidative, anti-apoptosis, and anti-inflammatory mechanism against ethanol neurotoxicity in the neonatal rat hippocampus. *Neuropeptides* **2021**, *87*, 102130. [CrossRef]
141. Xin, Q.; Cheng, B.; Pan, Y.; Liu, H.; Yang, C.; Chen, J.; Bai, B. Neuroprotective effects of apelin-13 on experimental ischemic stroke through suppression of inflammation. *Peptides* **2015**, *63*, 55–62. [CrossRef]
142. Xu, W.; Li, T.; Gao, L.; Zheng, J.; Yan, J.; Zhang, J.; Shao, A. Apelin-13/APJ system attenuates early brain injury via suppression of endoplasmic reticulum stress-associated TXNIP/NLRP3 inflammasome activation and oxidative stress in a AMPK-dependent manner after subarachnoid hemorrhage in rats. *J. Neuroinflamm.* **2019**, *16*, 247. [CrossRef] [PubMed]
143. Cherry, J.D.; Olschowka, J.A.; O'Banion, M.K. Neuroinflammation and M2 microglia: The good, the bad, and the inflamed. *J. Neuroinflamm.* **2014**, *11*, 98. [CrossRef]
144. Chen, D.; Lee, J.; Gu, X.; Wei, L.; Yu, S.P. Intranasal delivery of Apelin-13 is neuroprotective and promotes angiogenesis after ischemic stroke in mice. *ASN Neuro* **2015**, *7*, 1759091415605114. [CrossRef] [PubMed]

145. Zhou, S.; Guo, X.; Chen, S.; Xu, Z.; Duan, W.; Zeng, B. Apelin-13 regulates LPS-induced N9 microglia polarization involving STAT3 signaling pathway. *Neuropeptides* **2019**, *76*, 101938. [CrossRef]
146. Luo, H.; Xiang, Y.; Qu, X.; Liu, H.; Liu, C.; Li, G.; Han, L.; Qin, X. Apelin-13 suppresses neuroinflammation against cognitive deficit in a streptozotocin-induced rat model of Alzheimer's disease through activation of BDNF-TrkB signaling pathway. *Front. Pharmacol.* **2019**, *10*, 395. [CrossRef]
147. Zhang, Z.X.; Li, E.; Yan, J.P.; Fu, W.; Shen, P.; Tian, S.W.; You, Y. Apelin attenuates depressive-like behaviour and neuroinflammation in rats co-treated with chronic stress and lipopolysaccharide. *Neuropeptides* **2019**, *77*, 101959. [CrossRef] [PubMed]
148. Keep, R.F.; Hua, Y.; Xi, G. Intracerebral haemorrhage: Mechanisms of injury and therapeutic targets. *Lancet Neurol.* **2012**, *11*, 720–731. [CrossRef] [PubMed]
149. Xie, Y.; Guo, H.; Wang, L.; Xu, L.; Zhang, X.; Yu, L.; Liu, Q.; Li, Y.; Zhao, N.; Zhao, N.; et al. Human albumin attenuates excessive innate immunity via inhibition of microglial Mincle/Syk signaling in subarachnoid hemorrhage. *Brain Behav. Immun.* **2017**, *60*, 346–360. [CrossRef]
150. Li, Y.; Li, J.; Li, S.; Li, Y.; Wang, X.; Liu, B.; Fu, Q.; Ma, S. Curcumin attenuates glutamate neurotoxicity in the hippocampus by suppression of ER stress-associated TXNIP/NLRP3 inflammasome activation in a manner dependent on AMPK. *Toxicol. Appl. Pharmacol.* **2015**, *286*, 53–63. [CrossRef]
151. Jaiswal, M.K.; Zech, W.D.; Goos, M.; Leutbecher, C.; Ferri, A.; Zippelius, A.; Carrì, M.T.; Nau, R.; Keller, B.U. Impairment of mitochondrial calcium handling in a mtSOD1 cell culture model of motoneuron disease. *BMC Neurosci.* **2009**, *10*, 64. [CrossRef]
152. Manev, H.; Favaron, M.; Guidotti, A.; Costa, E. Delayed increase of Ca2+ influx elicited by glutamate: Role in neuronal death. *Mol. Pharmacol.* **1989**, *36*, 106–112. [PubMed]
153. Cook, D.R.; Gleichman, A.J.; Cross, S.A.; Doshi, S.; Ho, W.; Jordan-Sciutto, K.L.; Lynch, D.R.; Kolson, D.L. NMDA receptor modulation by the neuropeptide apelin: Implications for excitotoxic injury. *J. Neurochem.* **2011**, *118*, 1113–1123. [CrossRef]
154. Khaksari, M.; Aboutaleb, N.; Nasirinezhad, F.; Vakili, A.; Madjd, Z. Apelin-13 protects the brain against ischemic reperfusion injury and cerebral edema in a transient model of focal cerebral ischemia. *J. Mol. Neurosci.* **2012**, *48*, 201–208. [CrossRef]
155. Ishimaru, Y.; Sumino, A.; Kajioka, D.; Shibagaki, F.; Yamamuro, A.; Yoshioka, Y.; Maeda, S. Apelin protects against NMDA-induced retinal neuronal death via an APJ receptor by activating Akt and ERK1/2, and suppressing TNF-alpha expression in mice. *J. Pharmacol. Sci.* **2017**, *133*, 34–41. [CrossRef] [PubMed]
156. Shibagaki, F.; Ishimaru, Y.; Sumino, A.; Yamamuro, A.; Yoshioka, Y.; Maeda, S. Systemic administration of an apelin receptor agonist prevents NMDA-induced loss of retinal neuronal cells in mice. *Neurochem. Res.* **2020**, *45*, 752–759. [CrossRef]
157. Song, W.; Sun, J.; Su, B.; Yang, R.; Dong, H.; Xiong, L. Ischemic post-conditioning protects the spinal cord from ischemia–reperfusion injury via modulation of redox signaling. *J. Thoracic. Cardiovasc. Surg.* **2013**, *146*, 688–695. [CrossRef]
158. Xu, Z.; Li, Z. Experimental Study on the Role of Apelin-13 in Alleviating Spinal Cord Ischemia Reperfusion Injury Through Suppressing Autophagy. *Drug Des. Devel. Ther.* **2020**, *14*, 1571–1581. [CrossRef]
159. Şişli, H.B.; Hayal, T.B.; Şenkal, S.; Kıratlı, B.; Sağraç, D.; Seçkin, S.; Özpolat, M.; Şahin, F.; Yılmaz, B.; Doğan, A. Apelin Receptor Signaling Protects GT1-7 GnRH Neurons Against Oxidative Stress In Vitro. *Cell. Mol. Neurobiol.* **2022**, *42*, 753–775. [CrossRef]
160. Selkoe, D.J. The molecular pathology of Alzheimer's disease. *Neuron* **1991**, *6*, 487–498. [CrossRef]
161. Gilgun-Sherki, Y.; Melamed, E.; Offen, D. Oxidative stress induced-neurodegenerative diseases: The need for antioxidants that penetrate the blood brain barrier. *Neuropharmacol* **2001**, *40*, 959–975. [CrossRef]
162. Nunomura, A.; Perry, G.; Aliev, G.; Hirai, K.; Takeda, A.; Balraj, E.K.; Jones, P.K.; Ghanbari, H.; Wataya, T.; Shimohama, S. Oxidative damage is the earliest event in Alzheimer disease. *J. Neuropathol. Exp. Neurol.* **2001**, *60*, 759–767. [CrossRef] [PubMed]
163. Lovell, M.A.; Xiong, S.; Xie, C.; Davies, P.; Markesbery, W.R. Induction of hyperphosphorylated tau in primary rat cortical neuron cultures mediated by oxidative stress and glycogen synthase kinase-3. *J. Alzheimer's Dis.* **2004**, *6*, 659–671. [CrossRef] [PubMed]
164. Sahara, N.; Murayama, M.; Lee, B.; Park, J.M.; Lagalwar, S.; Binder, L.I.; Takashima, A. Active c-Jun N-terminal kinase induces caspase cleavage of tau and additional phosphorylation by GSK-3β is required for tau aggregation. *Eur. J. Neurosci.* **2008**, *27*, 2897–2906. [CrossRef] [PubMed]
165. Tamagno, E.; Parola, M.; Bardini, P.; Piccini, A.; Borghi, R.; Guglielmotto, M.; Santoro, G.; Davit, A.; Danni, O.; Smith, M. β-Site APP cleaving enzyme up-regulation induced by 4-hydroxynonenal is mediated by stress-activated protein kinases pathways. *J. Neurochem.* **2005**, *92*, 628–636. [CrossRef] [PubMed]
166. Supnet, C.; Bezprozvanny, I. The dysregulation of intracellular calcium in Alzheimer disease. *Cell Calcium* **2010**, *47*, 183–189. [CrossRef] [PubMed]
167. Eren, N.; Deni, Z.; Yildiz, Z.; Mu, F.; Go, N.; Gu, L.; Karabiyik, T. P200-Levels of apelin-13 and total oxidant/antioxidant status in sera of Alzheimer patients. *Turk. J. Biochem.* **2012**, *4*, 201–205.
168. Aminyavari, S.; Zahmatkesh, M.; Farahmandfar, M.; Khodagholi, F.; Dargahi, L.; Zarrindast, M.R. Protective role of Apelin-13 on amyloid beta25-35-induced memory deficit; Involvement of autophagy and apoptosis process. *Prog. Neuro-Psychopharmacol. Biol. Psychiatry* **2018**, *89*, 322–334. [CrossRef] [PubMed]
169. Yang, C.; Li, X.; Gao, W.; Wang, Q.; Zhang, L.; Li, Y.; Li, L.; Zhang, L. Cornel Iridoid Glycoside Inhibits Tau Hyperphosphorylation via Regulating Cross-Talk Between GSK-3beta and PP2A Signaling. *Front. Pharmacol.* **2018**, *9*, 682. [CrossRef] [PubMed]
170. Masoumi, J.; Abbasloui, M.; Parvan, R.; Mohammadnejad, D.; Pavon-Djavid, G.; Barzegari, A.; Abdolalizadeh, J. Apelin, a promising target for Alzheimer disease prevention and treatment. *Neuropeptides* **2018**, *70*, 76–86. [CrossRef]

171. Samandari-Bahraseman, M.R.; Elyasi, L. Apelin-13 protects human neuroblastoma SH-SY5Y cells against amyloid-beta induced neurotoxicity: Involvement of antioxidant and antiapoptotic properties. *J. Basic Clin. Physiol. Pharmacol.* **2022**, *33*, 599–605. [CrossRef]

172. Pollanen, M.S.; Bergeron, C.; Weyer, L. Deposition of detergent resistant neurofilaments into Lewy body fibrils. *Brain Res.* **1992**, *603*, 121–124. [CrossRef]

173. Riess, O.; Jakes, R.; Kruger, R. Genetic dissection of familial Parkinson's disease. *Mol. Med. Today* **1998**, *4*, 438–444. [CrossRef] [PubMed]

174. Yang, F.; Jiang, Q.; Zhao, J.; Ren, Y.; Sutton, M.D.; Feng, J. Parkin stabilizes microtubules through strong binding mediated by three independent domains. *J. Biol. Chem.* **2005**, *280*, 17154–17162. [CrossRef] [PubMed]

175. Jha, S.K.; Jha, N.K.; Kar, R.; Ambasta, R.K.; Kumar, P. p38 MAPK and PI3K/AKT signalling cascades in Parkinson's disease. *Int. J. Mol. Cell. Med.* **2015**, *4*, 67–86. [PubMed]

176. Jiang, Y.; Liu, H.; Ji, B.; Wang, Z.; Wang, C.; Yang, C.; Pan, Y.; Chen, J.; Cheng, B.; Bai, B.O. Apelin13 attenuates ER stress-associated apoptosis induced by MPP+ in SHSY5Y cells. *Int. J. Mol. Med.* **2018**, *42*, 1732–1740. [PubMed]

177. Haghparast, E.; Esmaeili-Mahani, S.; Abbasnejad, M.; Sheibani, V. Apelin-13 ameliorates cognitive impairments in 6-hydroxydopamineinduced substantia nigra lesion in rats. *Neuropeptides* **2018**, *68*, 28–35. [CrossRef]

178. Cai, W.J.; Chen, Y.; Shi, L.X.; Cheng, H.R.; Banda, I.; Ji, Y.H.; Wang, Y.T.; Li, X.M.; Mao, Y.X.; Zhang, D.F.; et al. AKT-GSK3β signaling pathway regulates mitochondrial dysfunction-associated OPA1 cleavage contributing to osteoblast apoptosis: Preventative effects of hydroxytyrosol. *Oxidative Med. Cell. Longev.* **2019**, *2019*, 4101738. [CrossRef] [PubMed]

179. Manolagas, S.C. From estrogen-centric to aging and oxidative stress: A revised perspective of the pathogenesis of osteoporosis. *Endocr. Rev.* **2010**, *31*, 266–300. [CrossRef]

180. Banfi, G.; Iorio, E.L.; Corsi, M.M. Oxidative stress, free radicals and bone remodeling. *Clin. Chem. Lab. Med.* **2008**, *46*, 1550–1555. [CrossRef]

181. Domazetovic, V.; Marcucci, G.; Falsetti, I.; Bilia, A.R.; Vincenzini, M.T.; Brandi, M.L.; Iantomasi, T. Blueberry juice antioxidants protect osteogenic activity against oxidative stress and improve long-term activation of the mineralization process in human osteoblast-like SaOS-2 cells: Involvement of SIRT1. *Antioxidants* **2020**, *9*, 125. [CrossRef]

182. Zhen, Y.F.; Wang, G.D.; Zhu, L.Q.; Tan, S.P.; Zhang, F.Y.; Zhou, X.Z.; Wang, X.D. P53 dependent mitochondrial permeability transition pore opening is required for dexamethasone-induced death of osteoblasts. *J. Cell. Physiol.* **2014**, *229*, 1475–1483. [CrossRef] [PubMed]

183. Menale, C.; Robinson, L.J.; Palagano, E.; Rigoni, R.; Erreni, M.; Almarza, A.J.; Strina, D.; Mantero, S.; Lizier, M.; Forlino, A.; et al. Absence of dipeptidyl peptidase 3 increases oxidative stress and causes bone loss. *J. Bone Miner. Res.* **2019**, *34*, 2133–2148. [CrossRef] [PubMed]

184. Wang, L.; Qi, H.; Tang, Y.; Shen, H.M. Post-translational modifications of key machinery in the control of mitophagy. *Trends Biochem. Sci.* **2020**, *45*, 58–75. [CrossRef]

185. Zhang, F.; Peng, W.; Zhang, Y.; Dong, W.; Wu, J.; Wang, T.; Xie, Z. P53 and Parkin coregulate mitophagy in bone marrow mesenchymal stem cells to promote the repair of early steroid-induced osteonecrosis of the femoral head. *Cell Death Dis.* **2020**, *11*, 42. [CrossRef]

186. Fan, P.; Yu, X.Y.; Xie, X.H.; Chen, C.H.; Zhang, P.; Yang, C.; Peng, X.; Wang, Y.T. Mitophagy is a protective response against oxidative damage in bone marrow mesenchymal stem cells. *Life Sci.* **2019**, *229*, 36–45. [CrossRef] [PubMed]

187. Li, Q.; Gao, Z.; Chen, Y.; Guan, M.X. The role of mitochondria in osteogenic, adipogenic and chondrogenic differentiation of mesenchymal stem cells. *Protein Cell* **2017**, *8*, 439–445. [CrossRef]

188. Shen, Y.; Wu, L.; Qin, D.; Xia, Y.; Zhou, Z.; Zhang, X.; Wu, X. Carbon black suppresses the osteogenesis of mesenchymal stem cells: The role of mitochondria. *Part. Fibre Toxicol.* **2018**, *15*, 16. [CrossRef]

189. Jing, X.; Du, T.; Yang, X.; Zhang, W.; Wang, G.; Liu, X.; Li, T.; Jiang, Z. Desferoxamine protects against glucocorticoid-induced osteonecrosis of the femoral head via activating HIF-1α expression. *J. Cell. Physiol.* **2020**, *235*, 9864–9875. [CrossRef]

190. Feng, X.; Yin, W.; Wang, J.; Feng, L.; Kang, Y.J. Mitophagy promotes the stemness of bone marrow-derived mesenchymal stem cells. *Exp. Biol. Med.* **2020**, *246*, 97–105. [CrossRef]

191. Hang, K.; Ye, C.; Xu, J.; Chen, E.; Wang, C.; Zhang, W.; Ni, L.; Kuang, Z.; Ying, L.; Xue, D.; et al. Apelin enhances the osteogenic differentiation of human bone marrow mesenchymal stem cells partly through Wnt/β-catenin signaling pathway. *Stem Cell Res. Ther.* **2019**, *10*, 189. [CrossRef]

192. Chen, L.; Shi, X.; Xie, J.; Weng, S.; Xie, Z.; Tang, J.; Yan, D.; Wang, B.; Fang, K.; Hong, C.; et al. Apelin-13 induces mitophagy in bone marrow mesenchymal stem cells to suppress intracellular oxidative stress and ameliorate osteoporosis by activation of AMPK signaling pathway. *Free Radical. Biol. Med.* **2021**, *163*, 356–368. [CrossRef] [PubMed]

193. Liu, S.; Wang, W.; Yin, L.; Zhu, Y. Influence of Apelin-13 on osteoporosis in Type-2 diabetes mellitus: A clinical study. *Pak. J. Med. Sci.* **2018**, *34*, 159–163. [CrossRef] [PubMed]

194. Tang, S.Y.; Xie, H.; Yuan, L.Q.; Luo, X.H.; Huang, J.; Cui, R.R.; Zhou, H.D.; Wu, X.P.; Liao, E.Y. Apelin stimulates proliferation and suppresses apoptosis of mouse osteoblastic cell line MC3T3-E1 via JNK and PI3-K/Akt signaling pathways. *Peptides* **2007**, *28*, 708–718. [CrossRef]

195. Zeng, X.; Yu, S.P.; Taylor, T.; Ogle, M.; Wei, L. Protective effect of apelin on cultured rat bone marrow mesenchymal stem cells against apoptosis. *Stem Cell Res.* **2012**, *8*, 357–367. [CrossRef]

196. Yan, J.; Wang, A.; Cao, J.; Chen, L. Apelin/APJ system: An emerging therapeutic target for respiratory diseases. *Cell. Mol. Life Sci.* **2020**, *77*, 2919–2930. [CrossRef] [PubMed]

197. Patane, S. Cardiotoxicity: Cisplatin and long-term cancer survivors. *Int. J. Cardiol.* **2014**, *175*, 201–202. [CrossRef]

198. Paken, C.D.; Govender, M.; Pillay, V.; Sewram, A. Review of Cisplatin-Associated Ototoxicity. *Semin. Hear.* **2019**, *40*, 108–121. [CrossRef]

199. El-Awady, S.E.; Moustafa, Y.M.; Abo-Elmatty, D.M.; Radwan, A. Cisplatin-induced cardiotoxicity: Mechanisms and cardioprotective strategies. *Eur. J. Pharmacol.* **2011**, *650*, 335–341. [CrossRef]

200. Ferroni, P.; Della-Morte, D.; Palmirotta, R.; McClendon, M.; Testa, G.; Abete, P. Platinum-based compounds and risk for cardiovascular toxicity in the elderly: Role of the antioxidants in chemoprevention. *Rejuvenation Res.* **2011**, *14*, 293–308. [CrossRef]

201. Zhang, P.; Yi, L.; Meng, G.; Zhang, H.; Sun, H.; Cui, L. Apelin-13 attenuates cisplatin-induced cardiotoxicity through inhibition of ROSmediated DNA damage and regulation of MAPKs and AKT pathways. *Free Radical. Res.* **2017**, *51*, 449–459. [CrossRef]

202. Kamogashira, T.; Fujimoto, C.; Yamasoba, T. Reactive oxygen species, apoptosis, and mitochondrial dysfunction in hearing loss. *Biomed. Res. Int.* **2015**, *2015*, 617207. [CrossRef]

203. Callejo, A.; Sedo-Cabezon, L.; Juan, I.D.; Llorens, J. Cisplatin-Induced Ototoxicity: Effects, Mechanisms and Protection Strategies. *Toxics* **2015**, *3*, 268–293. [CrossRef] [PubMed]

204. Majumder, P.; Duchen, M.R.; Gale, J.E. Cellular glutathione content in the organ of Corti and its role during ototoxicity. *Front. Cell. Neurosci.* **2015**, *9*, 143. [CrossRef]

205. Yin, H.; Zhang, H.; Kong, Y.; Wang, C.; Guo, Y.; Gao, Y.; Yuan, L.; Yang, X.; Chen, J. Apelin protects auditory cells from cisplatin-induced toxicity in vitro by inhibiting ROS and apoptosis. *Neurosci. Lett.* **2020**, *728*, 134948. [CrossRef]

206. Topcu, A.; Saral, S.; Mercantepe, T.; Akyildiz, K.; Tumkaya, L.; Yilmaz, A. The effects of apelin-13 against cisplatin-induced nephrotoxicity in rats. *Drug Chem. Toxicol.* **2023**, *46*, 47. [CrossRef]

207. Fettiplace, M.R.; Kowal, K.; Ripper, R.; Young, A.; Lis, K.; Rubinstein, I.; Bonini, M.; Minshall, R.; Weinberg, G. Insulin signaling in bupivacaine-induced cardiac toxicity: Sensitization during recovery and potentiation by lipid emulsion. *Anesthesiology* **2016**, *124*, 428–442. [CrossRef] [PubMed]

208. Cela, O.; Piccoli, C.; Scrima, R.; Quarato, G.; Marolla, A.; Cinnella, G.; Dambrosio, M.; Capitanio, N. Bupivacaine uncouples the mitochondrial oxidative phosphorylation, inhibits respiratory chain complexes I and III and enhances ROS production: Results of a study on cell cultures. *Mitochondrion* **2010**, *10*, 487–496. [CrossRef] [PubMed]

209. Weinberg, G.L.; Palmer, J.W.; Vade Boncouer, T.R.; Zuechner, M.B.; Edelman, G.; Hoppel, C.L. Bupivacaine inhibits acylcarnitine exchange in cardiac mitochondria. *Anesthesiology* **2000**, *92*, 523–528. [CrossRef]

210. Ye, Y.; Cai, Y.; Xia, E.; Shi, K.; Jin, Z.; Chen, H.; Xia, F.; Xia, Y.; Papadimos, T.J.; Xu, X.; et al. Apelin-13 Reverses Bupivacaine-Induced Cardiotoxicity via the Adenosine Monophosphate–Activated Protein Kinase Pathway. *Anesth. Analg.* **2021**, *133*, 1048–1059. [CrossRef]

211. Gorrini, C.; Harris, I.S.; Mak, T.W. Modulation of oxidative stress as an anticancer strategy. *Nat. Rev. Drug Discov.* **2013**, *12*, 931–947. [CrossRef]

212. Diehn, M.; Cho, R.W.; Lobo, N.A.; Kalisky, T.; Dorie, M.J.; Kulp, A.N.; Qian, D.; Lam, J.S.; Ailles, L.E.; Wong, M.; et al. Association of reactive oxygen species levels and radioresistance in cancer stem cells. *Nature* **2009**, *458*, 780–783. [CrossRef] [PubMed]

213. Kim, H.M.; Haraguchi, N.; Ishii, H.; Ohkuma, M.; Okano, M.; Mimori, K.; Eguchi, H.; Yamamoto, H.; Nagano, H.; Sekimoto, M.; et al. Increased CD13 expression reduces reactive oxygen species, promoting survival of liver cancer stem cells via an epithelial-mesenchymal transition-like phenomenon. *Ann. Surg. Oncol.* **2012**, *19*, 539–548. [CrossRef] [PubMed]

214. Hanahan, D.; Weinberg, R.A. Hallmarks of cancer: The next generation. *Cell* **2011**, *144*, 646–674. [CrossRef] [PubMed]

215. Zelenka, J.; Koncosova, M.; Ruml, T. Targeting of stress response pathways in the prevention and treatment of cancer. *Biotechnol. Adv.* **2018**, *36*, 583–602. [CrossRef] [PubMed]

216. Marnett, L.J. Oxyradicals and DNA damage. *Carcinogenesis* **2000**, *21*, 361–370. [CrossRef]

217. Panieri, E.; Santoro, M.M. ROS homeostasis and metabolism: A dangerousliason in cancer cells. *Cell Death Dis.* **2016**, *7*, e2253. [CrossRef]

218. Yang, Y.; Lv, S.Y.; Ye, W.; Zhang, L. Apelin/APJ system and cancer. *Clin. Chim. Acta* **2016**, *457*, 112–116. [CrossRef]

219. Berta, J.; Hoda, M.A.; Laszlo, V.; Rozsas, A.; Garay, T.; Torok, S.; Grusch, M.; Berger, W.; Paku, S.; Renyi-Vamos, F.; et al. Apelin promotes lymphangiogenesis and lymph node metastasis. *Oncotarget* **2014**, *5*, 4426–4437. [CrossRef]

220. Hall, C.; Ehrlich, L.; Venter, J.; O'Brien, A.; White, T.; Zhou, T.; Dang, T.; Meng, F.; Invernizzi, P.; Bernuzzi, F.; et al. Inhibition of the apelin/apelin receptor axis decreases cholangiocarcinoma growth. *Cancer Lett.* **2017**, *386*, 179–188. [CrossRef]

221. Diakowska, D.; Markocka-Maczka, K.; Nienartowicz, M.; Rosińczuk, J.; Krzystek-Korpacka, M. Assessment of apelin, apelin receptor, resistin, and adiponectin levels in the primary tumor and serum of patients with esophageal squamous cell carcinoma. *Adv. Clin. Exp. Med.* **2019**, *28*, 671–678. [CrossRef]

222. Masoumi, J.; Jafarzadeh, A.; Khorramdelazad, H.; Abbasloui, M.; Abdolalizadeh, J.; Jamali, N. Role of Apelin/APJ axis in cancer development and progression. *Adv. Med. Sci.* **2020**, *65*, 202–213. [CrossRef] [PubMed]

223. Sorli, S.C.; Le Gonidec, S.; Knibiehler, B.; Audigier, Y. Apelin is a potent activator of tumourneoangiogenesis. *Oncogene* **2007**, *26*, 7692–7699. [CrossRef] [PubMed]

224. Han, S.; Wang, G.; Qi, X.; Lee, H.M.; Englander, E.W.; Greeley, G.H., Jr. A possible role for hypoxia-induced apelin expression in enteric cell proliferation. *Am. J. Physiol. Regul. Integr. Comp. Physiol.* **2008**, *294*, R1832–R1839. [CrossRef]

225. Liu, J.; Wang, Z. Increased oxidative stress as a selective anti cancer therapy. *Oxidative Med. Cell. Longev.* **2015**, *2015*, 294303. [CrossRef] [PubMed]

226. Picault, F.X.; Chaves-Almagro, C.; Projetti, F.; Prats, H.; Masri, B.; Audigier, Y. Tumour co-expression of apelin and its receptor is the basis of an autocrine loop involved in the growth of colon adenocarcinomas. *Eur. J. Cancer* **2014**, *50*, 663–674. [CrossRef]

227. Harford-Wright, E.; Andre-Gregoire, G.; Jacobs, K.A.; Treps, L.; Le Gonidec, S.; Leclair, H.M.; Gonzalez-Diest, S.; Roux, Q.; Guillonneau, F.; Loussouarn, D.; et al. Pharmacological targeting of apelin impairs glioblastoma growth. *Brain* **2017**, *140*, 2939–2954. [CrossRef]

228. Grupinska, J.; Budzyn, M.; Brezinski, J.J.; Gryszczynska, B.; Kasprzak, M.P.; Kycler, W.; Leporowska, E.; Iskra, M. Association between clinicopathological features of breast cancer with adipocytokine levels and oxidative stress markers before and after chemotherapy. *Biomed. Rep.* **2021**, *14*, 30. [CrossRef]

229. Muto, J.; Shirabe, K.; Yoshizumi, T.; Ikegami, T.; Aishima, S.; Ishigami, K.; Yonemitsu, Y.; Ikeda, T.; Soejima, Y.; Maehara, Y. The apelin-APJ system induces tumor arteriogenesis in hepato- cellular carcinoma. *Anticancer Res.* **2014**, *34*, 5313–5320.

230. Yoshiya, S.; Shirabe, K.; Imai, D.; Toshima, T.; Yamashita, Y.; Ikegami, T.; Okano, S.; Yoshizumi, T.; Kawanaka, H.; Maehara, Y. Blockade of the apelin-APJ system promotes mouse liver regeneration by activating Kupffer cells after partial hepatectomy. *J. Gastroenterol.* **2015**, *50*, 573–582. [CrossRef]

International Journal of
Molecular Sciences

MDPI

Review

Manipulation of Oxidative Stress Responses by Non-Thermal Plasma to Treat Herpes Simplex Virus Type 1 Infection and Disease

Julia Sutter [1], Peter J. Bruggeman [2], Brian Wigdahl [1], Fred C. Krebs [1] and Vandana Miller [1,*]

1   Center for Molecular Virology and Gene Therapy, Institute for Molecular Medicine and Infectious Disease, Department of Microbiology and Immunology, Drexel University College of Medicine, Philadelphia, PA 19129, USA
2   Department of Mechanical Engineering, University of Minnesota, Minneapolis, MN 55455, USA
*   Correspondence: vam54@drexel.edu

Abstract: Herpes simplex virus type 1 (HSV-1) is a contagious pathogen with a large global footprint, due to its ability to cause lifelong infection in patients. Current antiviral therapies are effective in limiting viral replication in the epithelial cells to alleviate clinical symptoms, but ineffective in eliminating latent viral reservoirs in neurons. Much of HSV-1 pathogenesis is dependent on its ability to manipulate oxidative stress responses to craft a cellular environment that favors HSV-1 replication. However, to maintain redox homeostasis and to promote antiviral immune responses, the infected cell can upregulate reactive oxygen and nitrogen species (RONS) while having a tight control on antioxidant concentrations to prevent cellular damage. Non-thermal plasma (NTP), which we propose as a potential therapy alternative directed against HSV-1 infection, is a means to deliver RONS that affect redox homeostasis in the infected cell. This review emphasizes how NTP can be an effective therapy for HSV-1 infections through the direct antiviral activity of RONS and via immunomodulatory changes in the infected cells that will stimulate anti-HSV-1 adaptive immune responses. Overall, NTP application can control HSV-1 replication and address the challenges of latency by decreasing the size of the viral reservoir in the nervous system.

Keywords: oxidative stress; immunomodulation; reactive oxygen and nitrogen species; redox homeostasis; antioxidant; innate immunity; adaptive immunity; antiviral therapy; plasma

check for
updates

Citation: Sutter, J.; Bruggeman, P.J.; Wigdahl, B.; Krebs, F.C.; Miller, V. Manipulation of Oxidative Stress Responses by Non-Thermal Plasma to Treat Herpes Simplex Virus Type 1 Infection and Disease. *Int. J. Mol. Sci.* **2023**, *24*, 4673. https://doi.org/10.3390/ijms24054673

Academic Editors: Rossana Morabito and Alessia Remigante

Received: 21 December 2022
Revised: 16 February 2023
Accepted: 24 February 2023
Published: 28 February 2023

1. Introduction to HSV-1 Infection

As a prevalent pathogen and global health concern, herpes simplex virus type 1 (HSV-1) is among one of the most studied human herpesviruses. HSV-1 has long served as a useful research tool in providing information that has enhanced the understanding of viruses. It also has a large global footprint in which more than 70% of the world's population under the age of 50 harbors an HSV-1 infection [1]. Most individuals who are infected with HSV-1 are asymptomatic because the viral genome remains transcriptionally dormant within the peripheral nervous system during latent infection, allowing lifelong persistence of the virus. Replication of this latent virus can be reactivated by external and internal stress stimuli, resulting in productive infection in neurons and re-infection of epithelial cells in the mucosal epithelium innervated by the infected neuron. Reactivation results in clinical signs of infection in the form of cold sores around the mouth, eyes, and genitalia. Although these lesions can be discomforting to the patient, most cases of HSV-1 infection are mild [2,3]. In rare cases, HSV-1 can cause serious disease by spreading to the brain to cause encephalitis [4] and to the eye to cause keratitis [5]. There is also increasing evidence that chronic HSV-1 infection can contribute to the development of neurodegenerative diseases such as Alzheimer's Disease later in life [6–8] (Figure 1).

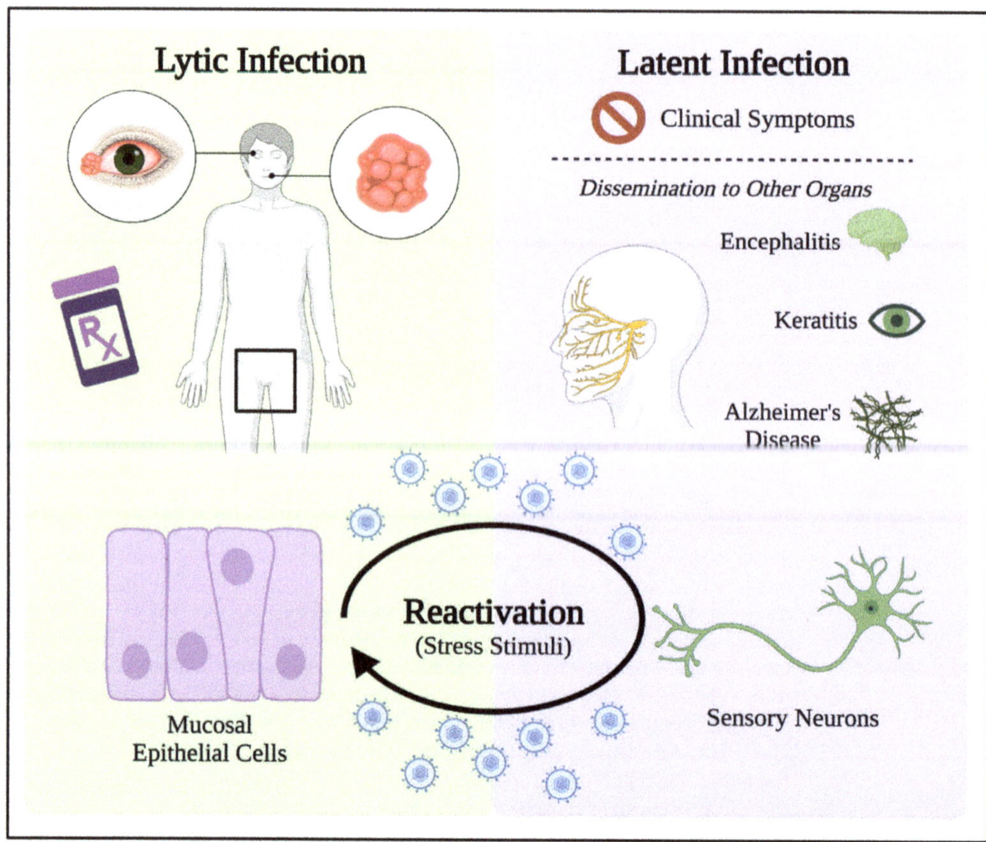

Figure 1. The HSV-1 lifecycle is characterized by the cycle between lytic infection and latent infection over the course of a patient's life. Infection with HSV-1 is known for causing cold sores that can appear around the mouth, eyes, and genitalia. These clinical manifestations are characteristic of lytic infection, when HSV-1 is actively replicating and producing new viruses in mucosal epithelial cells in these areas. It is at this point in infection where antiviral therapies are administered to alleviate symptoms and shorten the course of virus infection. As clinical symptoms disappear, HSV-1 virions travel to nearby nerves and establish viral latency in peripheral sensory neurons. Latent infection is characterized by negligible viral gene expression, which prevents detection by the host immune system. During the life of the infected individual, HSV-1 lytic infection can become reactivated by stress stimuli. Reactivation can lead to the recurrence of lesions caused by productive, lytic infection, as well as an increased risk of more serious disease.

For patients who experience severe clinical symptoms of HSV-1 infection, treatment options, while available, are limited and sometimes ineffective. Antiviral therapies are typically administered upon the appearance of clinical symptoms after the initial acquisition of the virus or when the virus is reactivated from latency. The first line of defense against these symptoms is nucleoside analogs, the most common being acyclovir. As a prodrug, acyclovir becomes activated through phosphorylation by virus-encoded thymidine kinases (TK) to the monophosphate form and further phosphorylation by other thymidylate kinases, converting it to acyclovir trisphosphate [9–11]. This form of acyclovir then competes with guanine nucleotides and binds the growing DNA strand during viral DNA replication in the nucleus, thus inhibiting the synthesis of the viral DNA. This leads to the selective inhibition of the viral DNA polymerase [11]. Penciclovir is another nucleoside analog prodrug prescribed to patients with a similar mechanism of action as acyclovir, but with an even higher affinity for the viral DNA [10,11].

Due to their mechanism of action, nucleoside analogs are effective during acute infection but do not affect the latent virus residing in neurons, allowing HSV-1 to persist in its host. This persistence can lead to the development of viral mutants upon reactivation, which give rise to drug-resistant strains [9,11,12]. Additionally, by failing to eliminate the virus, these antiviral therapies do not prevent future recurrent acute infection and transmission of HSV-1 that occurs through direct contact with the resulting lesions [1,3]. These limitations highlight the need for new therapeutic approaches that effectively address viral latency and reactivation by targeting other available viral and cellular components involved in HSV-1 pathogenesis. These include structural components of the virion structure and cellular processes manipulated by HSV-1 to facilitate productive infection of the cell.

Reduction-oxidation (redox) homeostasis, which involves the maintenance of intracellular reactive oxygen and nitrogen species (RONS), is one of these processes that is significantly disrupted during HSV-1 infection in both epithelial cells (host cells for lytic infection) and neurons (latently infected cells). Disruption of redox homeostasis by HSV-1 in both cell types allows the increase in RONS generation, leading to greater RONS concentrations in spite of secondary reactions and enzymatic pathways. This increase in RONS, as a result, contributes to their activity in modulating cellular signaling pathways, the downregulation of cellular antiviral responses, and the promotion of an environment favorable for HSV-1 infection [13]. This review discusses how the development and progression of HSV-1 to chronic infection is closely tied to the alteration of the redox state in the host cell and subsequent reductions in the cell's antiviral defense against HSV-1. Since maintaining redox homeostasis is critical in cells, we discuss how alterations in cellular redox by the virus may be harnessed as a therapeutic target for HSV-1 infections.

2. HSV-1 Disrupts Cellular Redox Homeostasis during Infection

RONS are potent intracellular signaling molecules with abilities to stimulate various cell signaling pathways, including those that regulate cell proliferation and apoptosis [14–19]. RONS also modulate immune responses against invading pathogens through the manipulation of cell signaling, or by influencing the production of proinflammatory mediators to promote an antiviral environment in and around the cell. Cellular RONS often participate in antimicrobial killing of invading pathogens during phagocytosis [20]. These functions of RONS are designed to maintain the health and integrity of the cell. HSV-1 is an obligate intracellular pathogen that relies heavily on host cellular machinery for its replication. To create an intracellular environment that favors productive replication (or sustained latent infection), HSV-1 affects multiple cellular processes, including redox homeostasis.

HSV-1 Can Hijack Cellular RONS Generation to Cause Oxidative Stress

To maintain redox homeostasis, RONS are constantly produced and destroyed in a healthy cell. The functions of RONS are attributed to their abilities to react with macromolecular structures. RONS are reactive species derived from oxygen and/or nitrogen-containing compounds, some containing an unpaired electron in the outer shell [21,22]. Therefore, these species can modify macromolecules leading to the modulation of signaling pathways. These actions are characteristic of primary RONS such as nitric oxide, hydrogen peroxide, and superoxide [15]. In the cell, primary RONS are reactive species that are typically generated during normal metabolic processes that occur in the mitochondria [23–30], endoplasmic reticulum (ER) [31,32], and peroxisome [33,34]. Alternatively, these RONS can be generated via specialized enzymes present in the cell such as nicotinamide adenine dinucleotide phosphate (NADPH) oxidases (NOX) [35–37] or nitric oxide synthases (NOS) [38–40]. These primary RONS may alter signaling cascades by reversible modifications on target macromolecular structures. Overall, primary RONS exhibit a weak damaging potential. Due to their metastable nature, primary RONS can participate in reactions with one another to form secondary RONS (e.g., peroxynitrite and hypochlorous acid). Unlike primary RONS, secondary RONS are not tightly regulated by the cell and can inflict damage when they accumulate [41,42]. This accumulation and subsequent damage of the cell by RONS is referred to as oxidative stress.

Oxidative stress is often induced by viruses to overwhelm the infected cell and create an environment that favors its replication. For example, respiratory syncytial virus (RSV) upregulates RONS in airways, while simultaneously diminishing the host's antioxidant system [21,43]. Similarly, upregulation of serum markers of lipid peroxidation in human immunodeficiency virus type 1 (HIV-1)-infected patients suggests increased oxidative stress and a link between RONS dysregulation and viral pathogenesis [44,45]. Oxidative stress in infected cells can damage cellular macromolecules and impair immune responses against the virus. However, the viral genome is also at risk for modification and mutation, leading to the formation of viral variants that may have enhanced virulence [46].

3. HSV-1 Lytic Infection Manipulates Oxidative Stress in the Cell

Like RSV and HIV-1, HSV-1 induces intracellular oxidative stress conditions during infection [47,48]. HSV-1 infection results in two distinct outcomes: lytic infection and latent infection. Lytic infection is characterized by active viral gene expression and results in the assembly of progeny virions that are released into the extracellular environment to infect nearby uninfected cells. During a lytic infection, the high level of replication results in the destruction of the host cell. This type of HSV-1 infection occurs in mucosal epithelial cells, usually at the site of infection or reactivation. During a period of acute infection characterized by lytic replication, patients present with cold sores and are responsive to antiviral drugs administered to alleviate these clinical symptoms [2,3]. Lytic infection depends on multiple viral and cellular components to establish and sustain infection in epithelial cells, impacting the entire cell and utilizing cellular processes to create an environment suitable for productive HSV-1 replication. One way this occurs is through the manipulation of oxidative stress responses by the virus, generating RONS that can interfere with cellular signaling and responses.

3.1. HSV-1 Alters Cellular Redox Homeostasis Early in Infection

HSV-1 is an enveloped virus, composed of an encapsulated genome surrounded by a host cell-derived lipid bilayer decorated with glycoproteins [49,50]. In the initial steps of a lytic infection, these envelope glycoproteins facilitate attachment and subsequent entry into target epithelial cells through interactions with cellular heparan sulfate (HS) [5,51–53], a cell surface proteoglycan involved in signaling, host defense, and metabolic processes [54]. Upon glycoprotein-receptor binding, cytoskeletal rearrangements are triggered to allow membrane fusion between the viral envelope and host cell membrane, facilitating the release of the protein capsid and the enclosed viral genome into the cytoplasm [5,51–53,55].

In establishing a redox environment that favors HSV-1 replication, two envelope proteins have secondary roles in recalibrating oxidative stress responses in the cell. Glycoprotein B (gB), which mediates attachment to HS, also induces ER stress by interacting with the ER lumen domain protein kinase R-like ER kinase (PERK), a signaling transducer involved in the ER stress response to misfolded proteins during viral infections. To prevent ER-mediated oxidative stress and subsequent apoptosis, gB interacts with PERK to maintain homeostasis in the ER [56]. HSV-1 glycoprotein J (gJ), which does this during binding and entry, is also implicated in viral-induced oxidative stress by promoting RONS accumulation through the adenosine triphosphate (ATP) synthase molecule [57,58].

Following fusion of the viral and cellular membranes, compartmentalization of the encapsulated viral genome into host endosomes allows trafficking of the viral contents through the cytoplasm toward the nucleus in a pH-dependent process. The lower pH of the endosome facilitates the release of the capsid into the cytosol [59]. HSV-1 tegument proteins, which are packaged in the virion, are simultaneously released into the cytoplasm where they can regulate cellular processes in favor of HSV-1 replication and subsequent pathogenesis [55,60–62]. Like the envelope proteins, HSV-1 tegument proteins influence the redox state of the cell to prevent clearance of the virus. Virus protein 16 (VP16), a tegument protein released into the cytoplasm following viral entry, manipulates HSV-1 by downregulating nuclear factor-κB (NF-κB) immune signaling, which is known to be responsive to RONS [60,63]. HSV-1 can also manipulate oxidative stress responses in the cell through infected cell protein 0 (ICP0), which is expressed as an immediate-early (IE) protein but is also packaged in the virus. By acting as an ubiquitin ligase, ICP0 can selectively target host cell signaling proteins to promote their degradation and loss of function [64]. The actions of these tegument proteins are critical in downregulating the antiviral response toward HSV-1, creating an intracellular environment suitable for productive replication. HSV-1 continues to hijack cellular machinery, such as the endocytic pathway, to allow the trafficking of the viral genome into the nucleus through the nuclear pore complex (NPC). Additionally, tegument proteins facilitate the shedding of the viral protein capsid and release of the genome into the nucleus [55,60,62,65–68].

Upon entry into the nucleus, the HSV-1 genome is still at risk of detection and subsequent clearance by the cell. Therefore, HSV-1 continues to employ tactics to dodge these attempts at cellular immune detection by manipulating redox homeostasis. By upregulating the production of RONS in the cell, HSV-1 can modify key cell signaling proteins to dampen and negate innate detection mechanisms. For example, HSV-1 interferes with the cyclic GMP-AMP synthase/stimulator of interferon genes (cGas/STING), a nuclear DNA sensing pathway that induces potent type 1 interferon (IFN) responses. By directly modifying the STING protein via RONS production, HSV-1 can dampen the IFN response [69]. HSV-1 also depletes the cellular antioxidant glutathione peroxidase 4 (GPX4) during infection. Since GPX4 is an activator of cGAS/STING, its depletion by HSV-1 can prevent the activation of this detection pathway. Without detection of HSV-1 by the cell, viral-induced oxidative stress accumulates, which can be in the form of lipid peroxidation by-products, and further overwhelms cellular machinery that maintains redox balance [70]. This control that HSV-1 holds over antioxidant concentrations can further be imposed through the Nrf2-antioxidant response element (Nrf2-ARE) pathway, which mediates expression of antioxidant genes. Through its ability to modify the Nrf2 transcription factors, HSV-1 can control the redox state in the cell to favor its replication [71].

3.2. HSV-1 Maniplates the Cell Environment to Facilitate Viral Gene Expression and Assembly

As a DNA virus, HSV-1 diverts the cellular machinery in the nucleus to replicate its own viral genome and produce new progeny viruses. Once the cell replication machinery is taken over by HSV-1, the virus begins to express its genes in three distinct but overlapping phases: immediate early (IE), early (E), and late (L). All phases of viral gene expression involve viral and cellular factors [9,72]. First, IE genes encode α proteins, which play important roles in enhancing HSV-1 infection by promoting virus replication and dampening antiviral responses in the host cell [50,73–75]. The IE phase includes the expression of the E3 ubiquitin ligase ICP0 (also packaged in the virus as a tegument protein), which induces oxidative stress and targets host cell proteins involved in the antiviral defense for degradation [64]. ICP27 is also encoded by IE genes and inhibits cell protein translation alongside tegument protein virion host shutoff (vhs) [63]. IE gene products also initiate a transcriptional cascade, resulting in the expression of downstream E genes that mediate HSV-1 DNA replication [73,76–79] and L genes that encode protein components used to assemble new viral particles [65,72,73]. L genes also encode ICP34.5 that further modulates the oxidative stress responses in cells by directly binding to Beclin 1, an autophagy-stimulating protein. Through this interaction, ICP34.5 can directly prevent autophagy of the cell, which is often activated by virus-induced oxidative stress as a means to eliminate HSV-1 from the cell [55,80].

During the later events in viral replication, HSV-1 must evade cell detection mechanisms and interfere with signaling to negate anti-HSV-1 immune responses [81]. Interference with antiviral responses is achieved by the manipulation of redox homeostasis in the cell by HSV-1. Specifically, HSV-1 can induce oxidative stress conditions to modify key signaling proteins, resulting in their degradation or loss of function [13,82]. In addition to the ubiquitin activity of ICP0, carbonylation is another type of post-translational modification that can divert cellular proteins from their roles in signal transduction. Carbonylation is described as the introduction of ketones and aldehydes to amino acid side chains. These modifications are recognized as indicators of oxidative stress during viral infections. Proteins with these modifications are inactivated and targeted for degradation in the proteasome. As a result, HSV-1 has adopted carbonylation to target key host cell proteins, such as apoptotic proteins, that would otherwise interrupt productive replication [46,64,83,84].

Cellular mechanisms are also manipulated by HSV-1 to promote virus assembly. Cellular proteins are preferentially degraded over viral proteins by HSV-1 induced expression of virus-induced chaperone-enriched (VICE) domains in the nucleus. In addition, heat shock protein 27 (Hsp27), a molecular chaperone that aids in protein folding, was found to be associated with VICE domains in the nucleus during HSV-1 infection. It is likely that the increased presence of Hsp27 is correlated with promoting viral protein folding in preparation for assembly [64,85].

Following HSV-1 genome replication, the virus begins the transcription of L genes that encode structural viral components required to assemble progeny virions. This assembly process is complex, requiring several viral proteins to form the capsid structure and initiate insertion of the newly replicated genome into the nascent virus particle [86–89]. Once the genome is packaged into the viral capsid, viral tegument proteins assist the progeny virion in budding from the inner nuclear membrane into the perinuclear space to gain a temporary lipid envelope [87,89]. Virions then de-envelope as they bud into the cytoplasm and are most likely taken up by the trans-Golgi network (TGN) where they are assumed to obtain their tegument protein layers. The TGN is proposed to serve as the site of secondary envelopment, in which virion envelopes derive from vesicles that bud from the TGN to the plasma membrane. The assembled viruses then use the host cell microtubule system for transport to the plasma membrane where the fully assembled virions egress the host cell via exocytosis [90–93] (Figure 2).

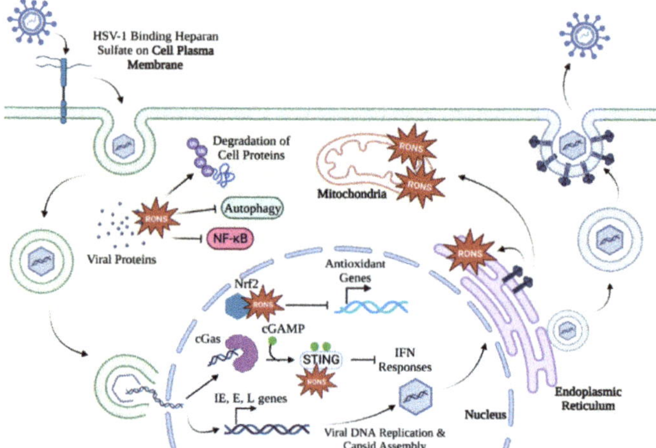

Figure 2. During infection of an epithelial cell, HSV-1 depends on the cell for endosome trafficking and its replication machinery. Due to this dependence, HSV-1 infection is invasive and must continuously avoid immune clearance from the cell. To overcome this, HSV-1 manipulates the cellular redox environment by upregulating oxidative stress responses to overwhelm the cell's capacity for balancing redox, crafting an environment that promotes HSV-1 replication. This allows the virus to avoid sensing by immune receptors and the subsequent innate responses directed against HSV-1.

4. Latent Infection Induces Oxidative Stress for HSV-1 Persistence in Patients

The alternative outcome of HSV-1 replication is a latent infection. Latency is a state in which virus gene expression and replication are silenced to escape cellular immune detection and clearance, and to allow for long-term persistence of the viral genome in the infected cell. During acute HSV-1 infection, progeny viruses produced by lytic infection of epithelial cells leave the mucosal epithelium, enter the trigeminal nerve, and latently infect the cell bodies of sensory neurons. Since genes associated with replication and viral assembly are expressed to a much lesser extent, patients exhibit no clinical symptoms of HSV-1 infection [50]. The ability of HSV-1 to remain transcriptionally dormant and evade immune system surveillance contributes to viral persistence, allowing lifelong infection with HSV-1 and its continued spread in the general population. While the exact mechanisms by which HSV-1 establishes and maintains latency are not fully understood, evidence suggests HSV-1 can manipulate redox homeostasis within the neurons to promote their survival and continue to escape immune detection [47,82].

Latent infection of neurons begins with mechanisms that parallel those involved in lytic infection but quickly diverges as the virus enters the neuron. Like lytic infection of epithelial cells, initial attachment of the HSV-1 virion to neurons involves interactions between the envelope glycoproteins and cell surface proteoglycans, including HS and nectin-1 [66,94]. Following attachment to these molecules, entry into the neuron occurs via a pH-independent manner that results in the fusion of the virus envelope with the neuronal plasma membrane, a process mediated by the host cell meshwork of filamentous actin [94]. Rather than trafficking via endocytic pathways, the released encapsulated genome uses the neuronal cytoskeleton to initiate retrograde axonal transport from the entry point on the axon. This transport mechanism involves microtubule networks running along the axon to the cell body where latency is established [61,94].

When HSV-1 reaches the cell body of the neuron, its genome rapidly condenses into heterochromatin via epigenetic mechanisms, making genes associated with lytic infection inaccessible to the host transcription machinery. Instead, the condensation of the genome facilitates the expression of a single transcript, the latency-associated transcript (LAT), expressed under a neuron-specific promoter [55,65,95–100]. In addition to silencing replication-associated genes [101], LAT expression is responsible for upregulating cellular genes that promote neuron survival and HSV-1 persistence [97,102].

4.1. HSV-1 Latency Results in Long-Term Oxidative Damage in Neurons

The nervous system, where HSV-1 resides during latency, is highly susceptible to oxidative stress. In particular, the trigeminal ganglia at the base of the brain are characterized by high oxygen uptake and generally low levels of antioxidants [8,103]. Therefore, during HSV-1 latency, cellular genes induced by the LAT may promote the onset of oxidative stress. Because the brain is rich in polyunsaturated fatty acids, oxidative stress may cause lipid peroxidation reactions, which are known to interfere with cell signaling pathways [15,16]. In fact, HSV-1 infection of neurons is correlated with the release of hydroxynonenal (HNE) and malondialdehyde (MDA), which are by-products of RONS-mediated lipid peroxidation. During HSV-1 establishment of latency, these products can contribute to the downregulation of immune pathways such as NF-κB to negate host innate immune responses and upregulate pathways such as c-Jun N-terminal kinase/mitogen-activated protein kinase (JNK/MAPK) to promote survival of the infected cell [47,104]. Manipulation of these cellular processes contributes to reduced immune visibility and neuron survival, which facilitates the persistence of HSV-1 in the latent stage. To further promote persistent infection, LAT offers protection to the condensed and mostly silent genome of HSV-1 within a circularized episome in the neuron until its eventual reactivation [66,96,105].

Persistence of HSV-1 in the neuron and the associated chronic oxidative stress can result in damage to infected neural tissues. Although latency allows the evasion of HSV-1 from immune clearance, the host still tries to protect itself. Dissemination of HSV-1 to neural tissues can lead to oxidative stress from microglia cells, a common cell type in the nervous system involved in protective immune responses. They are cited as being major cytokine producers and influencing excessive inflammatory responses to HSV-1 infection. Over time, these stressful conditions, in some cases, can lead to serious disease such as encephalitis in the brain, resulting in neuronal death and irreversible brain tissue damage [4,47,106–108]. Although the exact mechanisms for disease development are unknown, Alzheimer's Disease and other neurodegenerative conditions have been associated with HSV-1 infection [6–8]. These potential disease outcomes highlight the importance of redox homeostasis and how critical it is in determining the fate of HSV-1 infection and efficiency of the cell to overcome it.

4.2. Oxidative Stress in the Reactivation from Latency

Reactivation from latency is a hallmark of HSV-1 infection and occurs periodically throughout the lifetime of an infected patient. This process involves the escape of HSV-1 virions from neurons and re-infection of the innervated epithelial cells within the mucosal epithelium via anterograde transport. Upon re-infection of epithelial cells, the ensuing productive infection of epithelial cells results in the re-appearance of clinical symptoms [2,65,105].

The process of reactivation has long been associated with an individual's response to external and internal stress stimuli that trigger the escape of HSV-1 from neurons. Regardless of the stimulus, HSV-1 takes advantage of the stress stimuli to escape immune detection in order to re-infect epithelial cells within the mucosal epithelium, resulting in the re-emergence of clinical symptoms [105]. While stimuli involved in cellular stress are known to trigger reactivation, the exact mechanisms by which this process occurs are poorly understood. Overall, reactivation is critical for the lifecycle and persistence of HSV-1. Due to the involvement of oxidative stress conditions in maintaining latency and triggering reactivation, the latently infected neurons are subjected to long-term oxidative damage [8,109].

Oxidative Stress as Stimulus in Models for HSV-1 Reactivation

Both in vitro and in vivo models of HSV-1 infection have been used to demonstrate the effects of external stimuli in HSV-1 reactivation from latency. Many of these external stimuli are tied to the induction of oxidative stress responses that push the potential of reactivation of latent HSV-1. Early studies of HSV-1 reactivation focused on the delivery of electrical stimulation of the trigeminal nerve [110]. A commonly used reactivation stimulus is ultraviolet (UV) light, which causes oxidative stress responses in cells through the generation of RONS [111]. High intensities of UV-B light were found to be sufficient to reactivate HSV-1 replication in a coculture system involving HaCaT cells (a neurofibrillary keratinocyte cell line susceptible t1o HSV-1 in vitro), and PC12 neuronal cells (a cell line susceptible to HSV-1 in vitro) [112]. UV light has also been a tool for reactivation in latently infected mice [113]. Relative to mechanism of action, damage to cellular macromolecules by UV light exposure can impact enzyme functions, create lipid peroxidation by-products, and damage nucleic acids [114–116]. However, oxidative stress was also shown to aid in reactivation through the administration of sodium arsenite, an inducer of heat shock, and gramicidin D, a toxin that interacts with membrane lipids, both of which are inducers of oxidative stress. Using ICP0-null mutant strains of HSV-1, these agents were capable of reactivating replication in absence of the ICP0 tegument protein, which is implicated in promoting lytic gene expression [117]. Other triggers have been attributed to emotional stress or hormonal imbalances [105,110,118]. This includes adrenergic hormones involved in the body's stress response such as epinephrine and norepinephrine, which have been shown to reactivate HSV-1 in vivo using a latent rabbit infection model [110]. These studies demonstrate that exposure to harmful stress stimuli can cause the re-emergence of infection and clinical disease via the reactivation of acute HSV-1 infection.

The reactivation of HSV-1 replication can also be affected by the onset of oxidative stress caused by intracellular stimuli. Reductions in the temperature or superinfection with cytomegalovirus (another member of the *Herpesviridae* family known to induce oxidative stress) can activate HSV-1 infection from latently infected cells [119–122]. Overall.

the redox state of cells involved in HSV-1 replication can contribute to reactivation of HSV-1 replication.

5. Cellular Mechanisms That Control RONS Concentrations

RONS have dual roles in cells. RONS are naturally produced by the cell and aid in the maintenance of homeostasis. When tightly controlled, RONS are beneficial to the cell, aiding in cell survival, signaling, and immune responses. However, when this control of RONS concentrations is lost, which occurs during HSV-1 infection, RONS can damage cellular macromolecules and dysregulate cellular processes.

To regulate the concentrations of RONS and protect the cell from oxidative damage, cells utilize an antioxidant system composed of enzymatic and non-enzymatic defenses. These defenses are typically used to prevent the accumulation of primary RONS by converting them into less reactive molecules [21,42,44]. For instance, superoxide dismutases (SOD) remove superoxide radicals that are produced during metabolism, generating hydrogen peroxide (another primary RONS and an important signaling molecule) and molecular oxygen, both of which can readily be used by the cell [14,42,123]. Then, to prevent the subsequent accumulation of hydrogen peroxide, catalases convert hydrogen peroxide to water and oxygen [14,41,57]. In addition, glutathione (GSH), in conjunction with other non-enzymatic antioxidants, also controls RONS concentrations in the cell as one of its numerous functions. As a co-substrate for glutathione peroxidase (GPx), GSH limits peroxides through the formation of glutathione disulfide (GSSG), which is then reduced back to GSH, making the balance between GSH and GSSG a useful indicator of the antioxidant capacity in a cell [44,69,124]. Other non-enzymatic scavengers for RONS include vitamins, carotenoids, flavonoids, and melatonin [46] (Figure 3).

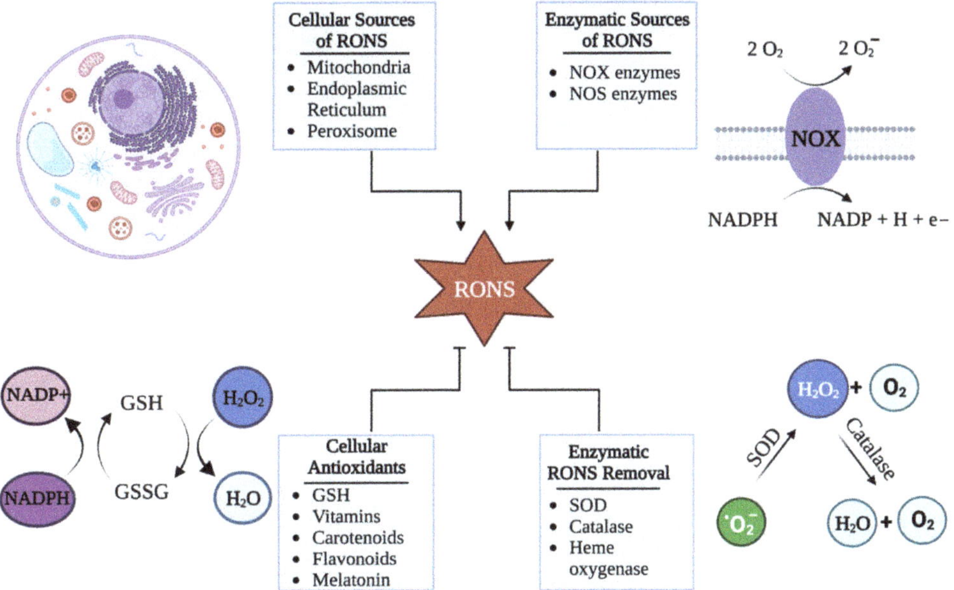

Figure 3. In a cell, RONS can be produced by various organelles during normal metabolic processes including cellular respiration in the mitochondria, protein folding in the ER, and in the peroxisome. Alternatively, enzymes such as NOX and NOS can generate RONS by transporting free electrons across membranes to reduce molecular oxygen. To avoid the accumulation of RONS, cells can oxidize RONS through antioxidants or enzymatically remove them by converting RONS into non-reactive products.

5.1. Cells Control the Upregulation of RONS in Response to HSV-1

The tightly regulated generation and destruction of RONS in the cell contributes to the antiviral defense mechanisms employed by the cell against HSV-1. In response to HSV-1 infection, cells try to maintain a stable redox environment to prevent the virus from hijacking the cell.

RONS can serve as potent signaling molecules by modifying key cellular proteins involved in signaling pathways and making them integral components of the cellular innate immune system. NOX enzymes, involved in RONS production, influence many toll-like receptor (TLR)-stimulated pathways that can activate or inhibit cellular responses to a pathogen [42]. TLRs are pattern recognition receptors (PRRs) located on the plasma membrane, within endosomes, and in the cytoplasm of cells. TLRs recognize conserved motifs within pathogen structures and upregulate innate immune responses that promote pathogen clearance from a cell. Specifically, TLR2 and TLR4 reside on the plasma membrane of cells, while TLR9 is located in endosomal compartments [42,55,69]. TLR pathogen recognition upregulates innate immune responses that promote pathogen clearance from a cell. Upon stimulation of these TLRs by binding viral components, NOX expression is reported to be upregulated, inducing the production of RONS [42]. Additionally, TLRs promote mitochondrial reactive oxygen species (mROS) generation to aid in immune responses [125]. Lastly, retinoic acid inducible gene 1 (RIG-1)-like receptors, another class of PRRs, have been implicated in their ability to mediate NOX expression by directly sensing foreign DNA and RNA molecules in the cytoplasm [42].

PRRs play active roles in sensing and activation of the host cell immune response, including the generation of RONS. HSV-1 envelope glycoproteins are recognized by PRRs such as TLR2 and $\alpha v\beta 3$ integrin on the plasma membrane. Specifically, gH/gL glycoprotein heterodimer was sufficient for this activation and subsequent immune response [126,127]. Intracellularly, endosomal TLR3 and TLR9, located on the ER, have also been shown to recognize HSV-1 structural components and synergize with TLR2 to elicit an inflammatory response and produce RONS [104,128]. In the cell, DNA/RNA sensors such as cGas, RIG-1, and γ-interferon inducible protein 16 (IFNI16) sense HSV-1 during the replication cycle to trigger RONS production and promote innate immune responses [104,127–129].

RONS produced during viral infection can lead to oxidative damage to cellular macromolecules, which can prompt immune responses toward the invading virus. Mitochondrial DNA (mtDNA) is a nucleic acid susceptible to RONS-mediated damage from virus-induced oxidative stress due to the lack of DNA repair mechanisms present in the mitochondria. DNA packing proteins that bind the mtDNA to maintain its structure and control the accessibility of genes can also be subjected to oxidative damage by RONS. Overall, damage to mtDNA serves as an antiviral sensor during infection and promotes the stimulation of innate immune responses [130] (Figure 4).

5.2. Cellular RONS Mediate Innate Immune Responses against HSV-1

Following recognition of HSV-1 by innate sensors, the cell is prompted to create an antiviral environment to prevent efficient viral replication and assembly. Activation of the NF-κB pathway, known to be influenced by RONS, results in the production of proinflammatory mediators, primarily type 1 IFN, which promote inflammation and T cell responses. These molecules are important mediators in antiviral defenses [126,127,130,131]. While not a direct inducer of NF-κB, micromolar concentrations of hydrogen peroxide are shown to be sufficient for NF-κB activation and function [132]. In the absence of stimulation, NF-κB resides in the cytoplasm bound to the IκBα inhibitor to prevent its translocation into the nucleus. The dynein motor complex protein, LC8, is a protein that can bind and modulate the activity of the IκBα inhibitor bound to NF-κB. Following activation of NF-κB by TNF-α, subsequent RONS production was found to oxidize LC8, releasing it from IκBα, which is then subjected to phosphorylation and ubiquitination [133]. As a result, NF-κB is released, allowing the transcription factor to translocate into the nucleus and promote expression of its target genes. NF-κB responsive genes are known to induce proinflammatory responses including the production of cytokines [15,132,134]. Thioredoxin, a group of redox proteins with a known role in signaling, was also found to modulate NF-κB activation through increased DNA binding activity in bone marrow dendritic cells [135]. Additionally, NF-κB knockout experiments demonstrated that HSV-1 replication and production of viral-induced RONS increased in the absence of NF-κB, highlighting the efficiency of the pathway as an immediate immune response toward HSV-1 while also implicating its susceptibility to redox signaling [81]. Additionally, NF-κB and Nrf2-Keap1 pathways have been shown to counteract each other by competing for the same co-activator during transcription. This is a mechanism used by the cell to regulate these pathways and to control redox levels and inflammatory responses to maintain cellular homeostasis [136,137].

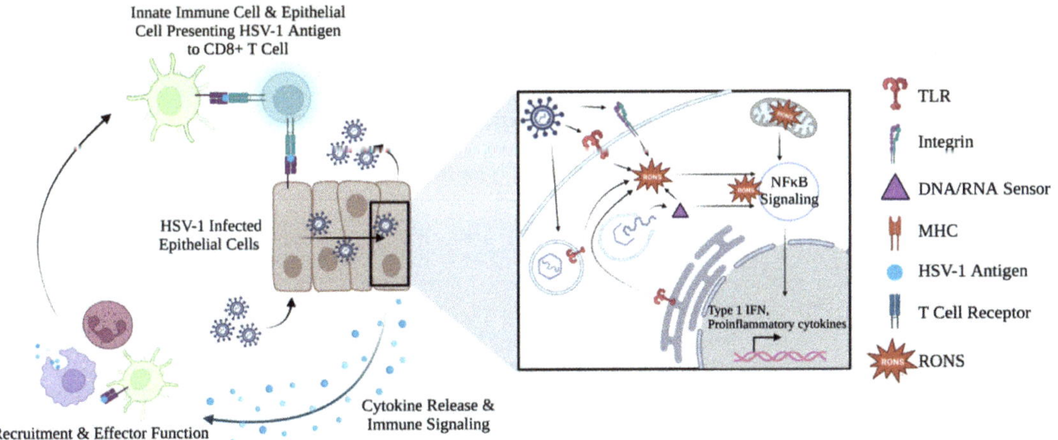

Figure 4. Overview of how cellular immune responses are stimulated during HSV-1 lytic infection. Inside the infected cell, PRRs involving TLRs, DNA/RNA sensors, and integrins sense components of HSV-1. Upon activation of sensors by HSV-1 or HSV-1-induced oxidative stress, immune signaling pathways such as NF-κB can be stimulated to promote the expression of proinflammatory cytokines. When released by the infected cell, these cytokines aid in innate immune cell (e.g., macrophage, dendritic cell, neutrophil) recruitment to promote phagocytosis and uptake of HSV-1 antigens for presentation to the immune system and stimulation of an adaptive anti-HSV-1 immune response (e.g., CD8+ T cells). Alternatively, epithelial cells can also present CD8+ T cells with HSV-1 antigens to prompt elimination of the infected cell.

As HSV-1 infection progresses, accumulation of RONS can result in the activation of inflammasomes [127] and autophagy [80] within the infected cell. Intracellular accumulation of RONS in response to HSV-1 infection activates the NOD-, LRR- and pyrin domain-containing protein 3 (NLRP3) inflammasome, resulting in the secretion of proinflammatory cytokines interleukin (IL)-1 β and IL-18, along with caspase-8 [42,125]. Once PRRs within the infected cell recognize HSV-1 components, immune signaling cascades are activated to promote the expression of proinflammatory mediators to attract innate immune cells. These recruited cell types mediate the destruction of pathogens during phagocytosis by phagocytes (e.g., neutrophils and macrophages). These cell types are capable of engulfing viruses or viral components into intracellular phagosomes. Within these phagosomes, oxidative stress conditions, via upregulation of RONS generation, are created to mediate destruction of the phagocytosed pathogen. To generate sufficient quantities of RONS for this function, mitochondria are recruited to the phagosome through signaling cascades to produce mROS and supply electrons in the form of NADPH [138]. As a result, NOX2 can generate superoxide radicals in large quantities, which then dismutate into hydrogen peroxide. Hydrogen peroxide is then consumed by phagosomal myeloperoxidase (MPO) to produce hypochlorous acid, a secondary RONS, through secondary reactions. Other reactive species, such as hydroxyl radicals and singlet oxygen, are also involved in this process [139,140]. This process is commonly referred to as a respiratory burst, known to mediate destruction to the pathogen structure and activate additional signaling cascades that can promote the pathogen's clearance [22,42,140].

5.3. Evidence of RONS Involvement in Adaptive Immune Response against HSV-1

Immune-associated RONS also have roles in the induction of adaptive immune responses. As a novel sensor for HSV-1, TLR3 is linked to the induction of antigen presentation by dendritic cells, which are involved in the induction of a robust CD8+ T cell response against HSV-1. TLR3-deficient mice demonstrate impaired HSV-1-specific CD8+ T cell responses and loss of control over HSV-1 infection in epithelial cells [127]. TLR3 is a sensor that can become activated by oxidative stress induced during viral infections [141]. Therefore, its role in promoting CD8+ T cell responses could be influenced by RONS. Nitric oxide was also observed to modulate the adaptive immune response by regulating the proliferation of lymphocytes in response to HSV-1, while also recruiting immune cells for antigen presentation [39]. These findings indicate that RONS can facilitate adaptive antiviral responses and are critical mediators of the cell-based immune response against HSV-1.

6. Examination of RONS as Antiviral Agents

The potent antiviral capabilities of RONS have garnered interest in the development of RONS cocktails as antiviral therapies. RONS have demonstrated roles in the cellular antiviral response, and some species are currently utilized as the basis of disinfectant agents for the decontamination of surfaces.

By itself, ozone inactivates a multitude of viruses. For example, ozone application to cell-free poliovirus-1, which is a single-stranded RNA virus, was shown to modify the protein sequences within the viral capsid, resulting in the impairment of viral adsorption. Furthermore, ozone damaged the viral RNA genome, leading to its inactivation [142]. Similarly, a 3-h application of ozone to cell-free HSV-1 resulted in 90% inhibition of viral infection. Not only did ozone treatment reduce viral infectivity, it was also shown to induce cytokine expression in the infected cell that promoted an innate immune response [143]. Ozone can enter the liquid phase and interact with other RONS such as nitrogen dioxide to further enhance its antiviral effect. Specifically, the concurrent presence of ozone and nitrogen dioxide in liquid media coincided with the enhanced generation of secondary RONS, such as dinitrogen pentoxide [144]. As a secondary species, dinitrogen pentoxide can rapidly accumulate and contribute to the oxidative damage of viruses. Like ozone, dinitrogen pentoxide acts as an antiviral agent in plants by decreasing viral lesions and inducing antiviral immunity in plants [145].

As a common disinfectant, hydrogen peroxide is known to be a powerful antimicrobial agent against viruses and microorganisms. Hydrogen peroxide has antiviral effects against both cell-free and intracellular viruses. Specifically, hydrogen peroxide is virucidal against many viruses spread through contaminated surfaces in healthcare, veterinary, and public facilities. These include feline calicivirus (FCV), transmissible gastroenteritis coronavirus (TGEV), avian influenza virus (AIV), and swine influenza virus (SwIV) [146]. Additionally, hydrogen peroxide was effective for decontamination of human severe acute respiratory syndrome-coronavirus-2 (SARS-CoV-2), with enhanced viral inactivation under acidic conditions through the co-presence of acidic compounds [147]. In a study involving the surface decontamination of influenza virus, a 2-min exposure to hydrogen peroxide vapor resulted in 99% viral inactivation [148]. The virucidal action of hydrogen peroxide was also reported to increase with the co-application of other RONS like sodium nitrite against FCV. This was mostly mediated through the formation of peroxynitrite, a secondary RONS. The activity of peroxynitrite was confirmed with a reduction in the pH of the exposed medium and reduced virucidal activity following the administration of ascorbic acid, a scavenger for peroxynitrite [149]. Hydrogen peroxide can also result in the formation of hypochlorous acid, a reactive chlorinated species, in a reaction catalyzed by MPO. Hypochlorous acid is commonly produced in leukocytes to cause oxidative stress conditions during phagocytosis [150]. Hypochlorous acid is also used as a disinfectant and is shown to inactivate viruses [151]. Some studies have also investigated hydrogen peroxide as a possible vaccine supplement, given its potent signaling capabilities. Interestingly, in a hydrogen peroxide-inactivated west nile virus (WNV) vaccine, specific CD8+ T cell and antibody-mediated responses were observed, resulting in immune memory upon re-infection with WNV [152].

Singlet oxygen is a short-lived RONS known to directly oxidize lipids containing carbon double bonds. Singlet oxygen is most effective against enveloped viruses (viruses enclosed by lipid envelopes). Singlet oxygen inactivates viruses by disrupting the viral envelope, which compromises its capacity to mediate entry into a target cell. In a singlet oxygen inactivated pseudorabies virus (PRV) vaccine, oxidized lipids within the virion structure also elicited a strong antibody-mediated response, while reducing PRV infectivity [153]. Singlet oxygen has also been reported to modify key amino acids in protein structures within the FCV capsid. This includes amino acids that contain double bonds in their side chains and are susceptible to disulfide bond formation [149] As a therapy, singlet oxygen generation is typically seen during photodynamic therapy (PDT) to inactivate viruses like SARS-CoV-2. However, it must be delivered to target cells immediately after generation for maximum therapeutic effect [154].

Endogenously, nitric oxide is a primary RONS involved in cellular antiviral defense. Following viral sensing by cellular PRRs, immune signaling pathways (e.g., NF-κB) are activated to promote the expression of proinflammatory cytokines that enhance the recruitment of innate immune cells to the site of viral infection. Additionally, immune signaling pathways involved in antiviral defense can further upregulate the expression of iNOS to promote nitric oxide generation [155]. Nitric oxide also has a critical role in modulating immune responses during early HSV-1 infection as inhibition of nitric oxide resulted in higher levels of HSV-1 replication and increased disease pathology in brain tissues [156]. Given its role in the antiviral defense, some studies have explored the use of exogenous nitric oxide as an antiviral therapy. Recently, this activity was demonstrated through the application of nitric oxide plasma activated water (NO-PAW), which allowed the delivery of gaseous nitric oxide into water, to samples of SARS-CoV-2 infected cells. Treatment of infected cells with NO-PAW resulted in the suppression of spike protein expression, the entry protein for SARS-CoV-2, along with other key viral proteins. In addition, NO-PAW was also able to upregulate the antiviral gene response in cells [157].

7. NTP as a Method for Controlled RONS Delivery

Although RONS are produced by and used by the cell in its antiviral defense, these species are often not produced in quantities sufficient to effectively control viral infections. Therefore, strategies that boost RONS and their effects have been explored as stand-alone antiviral therapies against viral infections. Many of the potentially therapeutic RONS cannot be produced artificially due to their stringent generation requirements and short half-lives. Furthermore, clinical approaches that use chemistry-based methods to generate RONS are often limited by the short half-lives of chemically active products and the identification of effective means to deliver them. These limitations can be addressed in therapeutic approaches involving the application of non-thermal plasma (NTP).

The field of plasma medicine is defined by explorations of NTP as a tool for the controllable delivery of RONS to biological targets. Also referred to as cold atmospheric plasma (CAP), gas plasma, or low temperature plasma (LTP), NTP has emerged as a relatively safe therapeutic tool with applications in wound healing, cancer, and infectious disease [158]. As the fourth state of matter, NTP is defined as a partially ionized gas composed of chemical, electrical, radiative, and thermal components (Figure 5). During the generation of NTP, the ionized gas produced at ambient temperature and pressure, with the application of electric fields, can excite and ionize electrons to produce RONS [158–161]. The quantities of RONS produced are regulated by changing applied electric fields, voltage, frequency, duty cycle, time, and distance of application, depending on the device. With their controllable delivery via NTP, these highly reactive species have the ability to modify macromolecular structures, induce intracellular signaling cascades, cause immunogenic cell death (ICD), and induce immune responses in a biological target [162]. While NTP is composed of multiple components, the ability to generate multiple species of RONS (some of which cannot be produced naturally by the cell or synthesized by chemical-based approaches) is critical to its ability to cause biological effects, including the inactivation of many types of viruses [163].

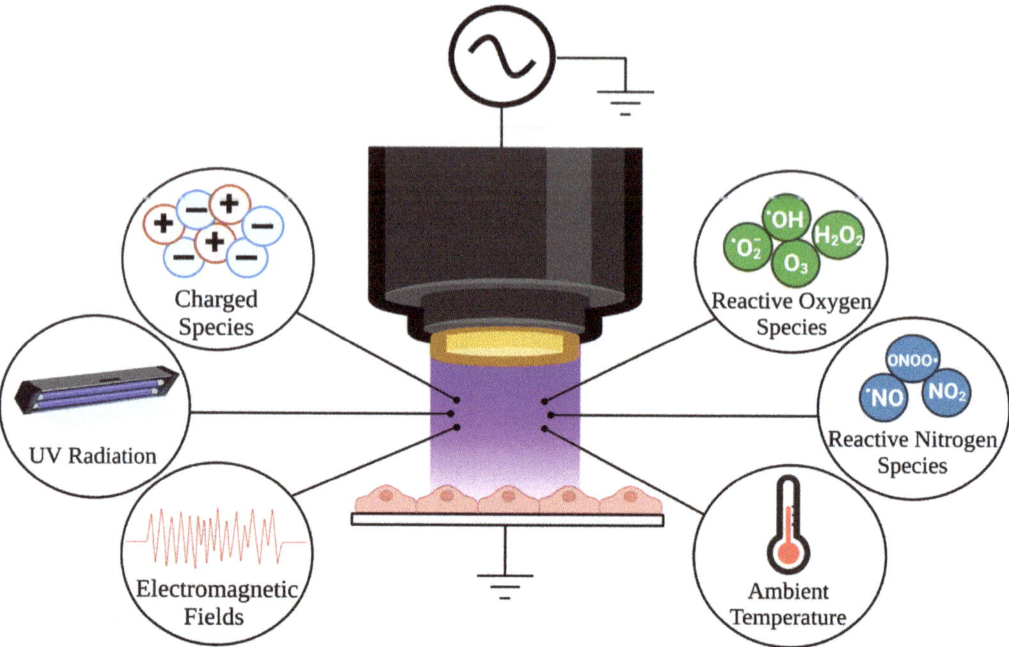

Figure 5. NTP (DBD electrode shown) is a partially ionized gas that generates charged species (electrons, and positive- and negative-charged ions), radiation, electromagnetic fields, and reactive species generated at ambient temperatures. Through collisions between electrons from NTP, and atoms and molecules in the gas and liquid phase, RONS are generated. Following their generation, RONS such as superoxide, nitric oxide, singlet oxygen, ozone, and hydrogen peroxide are delivered to biological targets, which are said to mediate much of NTP's antiviral effect. The exact composition of RONS delivered can depend on the plasma device and application modality.

7.1. NTP Adds an Additional Layer of Oxidative Stress to Control HSV-1 Infection

Redox homeostasis is an important determinant in the survival of cells. During viral infections, this homeostasis is challenged by oxidative stress induced by the cell to protect itself and by the virus to promote its replication and pathogenesis. Therefore, the redox state of the cell during a viral infection determines the likelihood of the cell being able to control the viral infection. Given the effective hijacking mechanisms of viruses, infections usually overwhelm the cell and allow the virus to utilize its machinery for its replication, which downregulates cellular DNA replication. With NTP and its controllable delivery of antiviral RONS, this redox state can be shifted back in the cell's favor to promote interference with virus replication.

While RONS are important mediators of antiviral activity, other components of NTP also contribute to its antiviral activity. By themselves, electric fields and UV radiation can kill microorganisms and can act as stand-alone therapies for treating infections [164–167]. UV light also induces oxidative stress [111,168,169] and can promote the generation and delivery of RONS to an infected cell. Additionally, cells exposed to electric fields and UV light induce the generation of RONS, activating key cellular signaling pathways in response to macromolecular damage and apoptosis [160,170,171]. Pulsed electric fields, which cause membrane permeabilization and are the basis of electroporation, also induce both extracellular and intracellular RONS in cells [172]. Although NTP can sometimes cause slightly elevated temperatures in biological targets, the negligible temperature increases caused by NTP application have little to no impact on virus inactivation [173]. It is important to note that no single component is produced in enough quantities to have a stand-alone effect. It is hypothesized that the different components of NTP work synergistically to produce the observed antiviral response. Therefore these subtherapeutic amounts of each component, working together, make NTP a unique tool for inactivating viruses [160].

7.2. Specialized Devices Allow Controllable Delivery of RONS

NTP is generated and delivered to a biological sample through specialized devices. In research and clinical settings, two main types of devices are used. Dielectric barrier discharges (DBD) devices allow the direct application of NTP to samples. These devices consist of at least one electrode encased in a dielectric material to which a high voltage is applied. The counter electrode could be the biological substrate as shown in Figure 5 for the example of a so-called floating DBD. NTP-generated RONS are directly deposited onto samples across small gap distances between the sample and the DBD electrode [158,174,175]. In addition to RONS, the direct interaction between the plasma and the sample delivers ions, UV photons, and high electric fields to the biological sample. In contrast, plasma jets are designed with a tube-like configuration that houses an electrode. Unlike DBD devices, plasma jets generate NTP at a larger distance from the target. The components of NTP are directed onto the target by the flow of typical inert gases such as helium or argon [158,176]. Plasma jets can operate in two modes: with and without direct contact of the plasma to the biological sample. When there is direct contact with the biological substrate, the interactions with a plasma jet are similar to interactions with DBD plasma as shown in Figure 5. When there is no direct interaction of plasma with the substrate, the substrate is not subjected to high electric fields and ionic species and biological interactions are mainly due to RONS. Although both types of NTP devices generate RONS, the composition of RONS delivered to the sample varies considerably between devices. Furthermore, the composition and quantity of NTP-generated RONS can be altered in a device-specific manner through adjustments in voltage, frequency, exposure time, electrode distance to the target, and type of working gas. The parameters that mediate the delivery of RONS are specific to the type and design of the device and application modality [160,163].

The composition of the RONS generated is also highly dependent on the power supply utilized to generate the NTP, as well as the sample being exposed. In the gas phase, the working gas and the parameters that result in NTP generation can influence the composition of RONS produced. As these species are delivered to a target, the interactions with the liquid-phase (typically cell culture medium or interstitial fluids in in vitro or in vivo applications) can result in the formation of secondary species [177]. Among the various RONS that can be produced, ozone (O_3), singlet oxygen ($^1O_2$), hydroxy radicals (OH), hydrogen peroxide (H_2O_2), nitric oxide (NO), nitrogen dioxide (NO_2^-), peroxynitrite ($ONOO^-$), and dinitrogen pentoxide (N_2O_5) have been implicated in pathogen inactivation [149,163]. The capability to generate and deliver these RONS controllably makes NTP a potential antiviral agent that can be used effectively to treat infections by viruses such as HSV-1.

8. NTP as a Virucidal and Antiviral Agent

The antiviral activity of NTP-associated RONS has been clearly documented and summarized by Mohamed et al. [163]. Due to NTP's ability to inactivate viruses and reduce their infectivity, NTP can be a promising therapeutic preventative measure for infections by human and non-human viral pathogens. NTP also has the ability to inactivate the plant virus, Potato Virus Y, which is a common drinking water contaminant. Studies involving NTP and this virus demonstrated NTP-mediated viral inactivation and proved NTP to be an environmental-friendly and safe decontamination tool for use in irrigation systems [178]. NTP was also shown to be an effective decontamination tool for foodborne viruses such as norovirus (NoV), a single-stranded nonenveloped virus found in fecal-contaminated produce and drinking fluids. Not only was its inactivation found to be consistent with increasing exposure times to NTP, but RONS were speculated to be the dominant effectors in its antiviral mechanism [179].

The animal virus FCV, a surrogate for human NoV, has also been the focus of research to test the antiviral efficacy of NTP. Using an argon plasma jet, FCV inactivation was more apparent with shorter exposure distances and increasing power. A correlation between NTP exposure and the oxidation of FCV viral capsid mediated by the presence of NTP-generated RONS was suggested. Specifically, modifications of the capsid and viral inactivation were proposed to be influenced by the presence of RNS, ozone, singlet oxygen, and the formation of peroxynitrite [173]. The antiviral effect of NTP against FCV has been shown using a variety of plasma devices and operational conditions, including remote plasma treatments with the effluent of a 2-dimensional (2D)-DBD plasma, which delivered mainly long-lived RONS to the virus. When NTP was applied directly, FCV inactivation could be induced by the actions of ozone in the gas-phase. In contrast, when NTP exposure was indirect, through application of liquid enriched in NTP RONS, NOx species were the dominant effectors. Additionally, pH changes secondary to peroxynitrite formation played a role in FCV inactivation [180]. Other studies that focused on FCV inactivation also highlighted ozone and peroxynitrite as the key antiviral agents in NTP function by producing oxidative damage to key amino acid residues within the viral capsid of FCV [144,181].

Few studies have investigated NTP as a potential therapeutic alternative for viruses that infect humans and cause chronic disease. Hepatitis B virus (HBV), a viral pathogen associated with a chronic liver disease, is spread through contact with bodily fluids. NTP, generated by a DBD device, was applied to blood samples infected with HBV. Key HBV antigens were susceptible to RONS-mediated damage following NTP exposure, leading to inactivation of the virus that increased with longer durations of exposure to NTP [182]. The antiviral effect of NTP was also studied using cells infected with HIV-1. NTP exposure of HIV-1-infected cells reduced HIV-1 infectivity, and impaired virus-cell fusion and viral assembly. Additionally, it was speculated that NTP exposure induced cellular factors that promoted an antiviral environment in macrophages, preventing further viral entry [183]. Similarly, viral entry mechanisms were abolished by NTP application to SARS-CoV-2. Specifically, the viral spike protein was found to be altered in its conformation and impaired in its ability to bind to cellular receptors. NTP exposure decreased the infectivity of SARS-CoV-2, and even resulted in the disruption of cell membranes which led to oxidative damage of viral RNA inside the cell [184].

The antiviral effect of NTP on HSV-1 was investigated in a model for herpes keratitis, in which explanted HSV-1-infected human cornea cells were indirectly exposed to NTP. In these experiments, cell culture medium was exposed to NTP and then applied to the HSV-1-infected cells, resulting in an 80% reduction in viral infectivity and increased antiviral activity with longer durations of NTP exposure. Importantly, minimal host cell cytotoxicity was observed [185]. An increase in 8-oxodeoxyguanosine (8-OHdg), a marker for oxidative DNA damage correlated with increased virus inactivation without damage to cells [186].

9. NTP as a Therapeutic Alternative for HSV-1 Infection

Oral, ocular, and genital lesions are considered hallmarks for HSV-1 infection and appear when the virus is actively replicating in mucosal epithelial cells [2,3]. In an envisioned NTP-based therapeutic approach, NTP will be applied directly to lesions produced by active HSV-1 infection. The application of NTP to HSV-1 lesions will deliver NTP effectors to the local area of the epithelium that includes cell-free virus, cells undergoing productive infection, and uninfected cells that may be targets for HSV-1 infection. Direct application of NTP to these lesions will likely reduce virus infectivity through oxidative damage to virus components, including envelope proteins and the DNA genome. NTP may also indirectly have an antiviral effect by altering a host cell's capacity to support productive infection and preventing infection of uninfected cells by altering cell surface molecules that participate in viral binding and entry.

9.1. Direct Effects of NTP-Generated RONS on HSV-1 Infection

Studies to date have focused on the short-term antiviral effects of NTP using cell-free HSV-1 and HSV-1 infected cells, with changes in viral infectivity as the measure of the antiviral effect of NTP. The in vitro antiviral effect of NTP on HSV-1 infection was demonstrated through the application of NTP-treated cell culture medium to HSV-1 infected cells (indirect NTP) [185]. There are multiple mechanisms through which NTP could directly affect HSV-1.

NTP is proposed to act as an antiviral agent through the controllable delivery of RONS, which are known to interact and modify macromolecular structures. Proteins, lipids, and nucleic acids in enveloped viruses such as HSV-1 are likely susceptible to RONS-mediated damage. For example, cysteine, a sulfur-containing amino acid, is prone to disulfide bond modification by RONS [187]. Additionally, MPO, an enzyme produced by cells during phagocytosis, can convert hydrogen peroxides into ions that have been implicated in the oxidation of tyrosine residues [20,188]. Given the abundance of proteins within the HSV-1 virion structure (e.g., envelope, capsid, and tegument proteins), RONS generated by NTP likely impair HSV-1 infectivity through protein modification. RONS can also affect carbon double bonds in lipid-based structures, resulting in lipid peroxidation. The viral envelope, which is important for protecting HSV-1 from the extracellular environment and for mediating fusion with target cells, can be disrupted by oxidative stress conditions through lipid modification [153]. Lastly, nucleic acids are prime targets for RONS-mediated damage. Specifically, RONS-induced modifications are typically observed within the sugar backbone of RNA and DNA molecules. Base pairing can also be disrupted by RONS, leading to the introduction of mutations in the viral genome that can impair HSV-1 infectivity [20]. Although not proven, some studies have alluded to RONS-mediated damage of viral macromolecules as the mechanism underlying the antiviral activity of NTP. This conclusion was supported by measurements of 8-OHdg, which is a marker for oxidative DNA damage. Following NTP exposure of HSV-1-infected ocular cells, increases in 8-OHdg correlated with the inactivation of HSV-1 [186].

Given the complexity of the HSV-1 structure, many viral components can be targets for damage caused by NTP RONS. As previously described, this damage will take the form of modifications to viral macromolecular components, leading to impairment of their functions. Due to the location of the envelope and envelope proteins on the exterior of the virus particle, reductions in virus infectivity attributable to NTP exposure are most likely to result from oxidative damage to the viral glycoproteins that mediate viral entry and modifications of lipids within the viral envelope that protect the encapsulated genome. Due to its chemical stability and hydrophilicity, hydrogen peroxide is an example of a RONS that can penetrate membranes [189]. As a result, hydrogen peroxide may have the ability to pass through the HSV-1 lipid envelope and gain access to the interior of the HSV-1 virion structure, causing oxidative damage to viral proteins and DNA contained within. These effects may also contribute to reductions in infectivity.

9.2. Effects of NTP-Induced Oxidative Stress on HSV-1 Replication

Redox control in a cell is important for cellular homeostasis and preventing damaging oxidative stress conditions. In the context of HSV-1, there is a clear correlation between infection and the induction of oxidative stress [85]. Part of this stress response is created by the cell. RONS are often produced to trigger intracellular signaling cascades that result in the promotion of an antiviral environment that is unfavorable for HSV-1 infection [104,127,128]. Meanwhile, HSV-1 can induce its own oxidative stress response that overcomes the cell's immune defense and enhances its own pathogenesis. This often involves the degradation of host proteins, recruitment of cellular machinery for its replication, and interference with signaling pathways through the modification of macromolecules [44,47,57,69]. Overall, HSV-1-induced oxidative stress aims to impair the cell's ability to promote its immune clearance, allowing it to replicate or persist in a dormant state within the infected cell.

Induction of oxidative stress by HSV-1 can be observed through the recruitment of VICE domains, the localization of heat shock proteins, and the activation of the unfolded protein response (UPR) in the ER. To promote the assembly of proteins for viral replication, HSV-1 induces ER stress to target cellular proteins for degradation to promote the proper assembly of virions [64,85]. This results in the accumulation of oxidized proteins and causes further dampening of the immune response towards HSV-1. In a model for diet-induced obesity, proteins associated with ER stress and the activation of the UPR were downregulated in NTP-exposed adipocytes, both in vitro and in vivo, suggesting that NTP may inhibit ER stress [190]. This ability of NTP to inhibit ER stress may allow the restoration of redox homeostasis in cells subjected to HSV-1 infection. NTP is also implicated in the increased expression of antioxidants that detoxify accumulating RONS concentrations, contributing to efforts in normalizing cellular redox levels. NTP exposure of HaCaT cells upregulated Nrf2 signaling, a transcription factor in the Nrf2-Keap1 pathway involved in the transcription of antioxidant genes as early as 20 s post-exposure to NTP and up to 24 h later [191,192]. Additionally, hydrogen peroxide, a key component of NTP, partially decreases Keap1, a negative regulator for Nrf2 expression [191]. Furthermore, NTP increases GSH levels, another antioxidant important in hydrogen peroxide neutralization [193]. This suggests that application of NTP has the potential to promote regulation of RONS concentrations in the cell, compromising the ability of HSV-1 to manipulate oxidative stress in the host cell.

Although NTP has a demonstrated antiviral effect on HSV-1, the mechanism by which RONS generated by the application of NTP to infected cells contributes to the antiviral effect of NTP is unclear. Based on reports cited above, it may be speculated that NTP exposure of HSV-1-infected cells could potentially stabilize the redox balance in a cell, diminishing oxidative stress mechanisms that favor replication.

10. NTP as an Immunomodulatory Agent for Treatment of HSV-1 Infection

In addition to its antiviral activity, NTP has potential as an immunotherapy for HSV-1 infection. RONS, as one of the main effectors of NTP, have demonstrated immunomodulatory effects on cancer cells, inducing immunogenic cell death (ICD), characterized by cell death that elicits an immune response [194,195]. The induction of ICD is normally accompanied by the display of stress-associated molecular patterns (SAMPs) on the cell surface [194–199]. These include calreticulin (CRT) and molecular chaperones such as Hsp90. In particular, Hsp90 was cleaved following exposure to NTP, and kinases that aid in its normal function were degraded [197]. Additionally, there is release of other molecules including ATP, high mobility group box protein 1 (HMGB1), and proinflammatory cytokines in response to NTP exposure. The emission of these SAMPs enhanced the functions of innate immune cells, including antigen presentation, facilitated phagocytosis, and promoted an inflammatory response within tumor environments [194–196,198]. This resulted in T helper cell activation as evidence of stimulated adaptive immune responses [200].

Literature regarding immunomodulatory changes, or SAMPs, caused by NTP in virus-infected cells is sparse. Our work in this area demonstrated the immunomodulatory effect of NTP in the context of HIV-1 infection using J-Lat cells. These cells, which contain an integrated latent HIV-1 genome that is not capable of supporting multiple rounds of replication, serve as a model for latently-infected T lymphocytes. NTP exposure of J-Lat cells resulted in emission of molecules, characteristic of SAMPs that are typically associated with ICD in tumor cells. Furthermore, there was upregulated phagocytosis by antigen presenting cells, suggesting an adjuvant effect of NTP. There was also evidence of neoepitope generation that may increase the breadth of adaptive immune responses [201]. Similar mechanisms may be activated in HSV-1-infected cells that trigger signaling pathways involved in the cellular immune defense [162].

Enhancement of Anti-HSV-1 Host Immune Responses by NTP

Innate and adaptive immune responses play integral roles in the host defense against HSV-1 infection. As previously mentioned, TLRs recognize viral components within the HSV-1 virion on the plasma membrane and within the cytoplasm [126–128,202]. Upon sensing HSV-1 components, TLRs induce immune signaling pathways that promote the transcription of antiviral mediators by the infected cell. These mediators include type 1 IFNs and other proinflammatory cytokines that are typically induced during viral infections. Type 1 IFNα, in particular, induce the expression of interferon stimulated genes (ISG), the products of which attract patrolling innate immune cells to the site of acute HSV-1 infection [13,55,63,126,128]. This trafficking of immune cells (e.g., neutrophils, macrophages, and dendritic cells) elicits their effector innate functions, which include phagocytosis and subsequent antigen presentation to adaptive immune cells (e.g., T cells, B cells) [203–205]. As a result, both T cells and B cells become activated, stimulating specific adaptive responses toward HSV-1. Specifically, CD8+ T cells, which we propose will be the main mediators of the NTP-stimulated immune response, can secrete antiviral cytokines toward invading pathogens and directly kill infected cells to inhibit viral replication [205]. Additionally, activation of B cells promotes the production and secretion of antibodies against HSV-1 antigens to prevent spread to neighboring susceptible cells [206]. To evade these immune responses, HSV-1 may escape the mucosal epithelium to establish latency in neurons. By silencing its genome expression, HSV-1 avoids sensing by the host immune system [4]. With the application of NTP to HSV-1 lesions, these responses can be stimulated or enhanced, shortening the course of acute infection and forcing HSV-1 into viral latency. Due to this shortened replication cycle in the epithelium from HSV-1's evasion tactics and the stimulated adaptive immune response, fewer virions will be produced that can travel into the nervous system, decreasing the number of latently infected cells. Over time, this can decrease the frequency and number of reactivation events leading to the recurrence of acute replication or viral reactivation.

HSV-1 has developed many ways to overcome the cellular sensing and immune clearance tactics to persist in its host. As previously mentioned, this includes the manipulation of the cell's redox state which can influence immune signaling and responses toward HSV-1. Application of NTP may be effective in overcoming viral evasion strategies through the generation of NTP-associated RONS that will enhance and stimulate more robust antiviral responses in the cell. Signaling pathways, such as NF-κB, which induce type 1 IFN responses, are upregulated by NTP exposure [104,126–128]. In addition, the application of NTP allows the emission of SAMPs on the cell surface that recruit innate immune cells [201]. These changes can lead to the aforementioned cellular innate and adaptive immune responses directed against HSV-1. The presence of NTP-generated RONS overcomes HSV-1's mechanisms of suppressing host cell responses. By controlling the cellular redox environment, NTP could prove to be an effective immunotherapy against HSV-1 infection.

11. Overlapping Roles for Oxidative Stress in Treating HSV-1 Infection with NTP

The use of NTP as a treatment for HSV-1 infection will require an integrated understanding of the roles played by oxidative stress during infection and treatment. Oxidative stress has multiple functions and effects in immune responses mounted by cells against infection, in HSV-1 replication in host cells, and in cellular and viral responses to NTP application (Figure 6). Some of the effects of RONS and modulated oxidative stress are unique to immunomodulation, infection, or NTP application. However, some roles are common to two or all three aspects of a putative treatment. Overlapping roles for RONS and oxidative stress may be advantageous or detrimental to treatment effectiveness. For example, the boost in cellular RONS by NTP application to an infected cell may augment RONS-mediated cellular mechanisms already promoting effective innate and adaptive immune responses to infection. On the other hand, oxidative damage to macromolecules and organelles during productive HSV-1 infection may be further increased by NTP application. The application of an NTP-based treatment of HSV-1 infection (or other infections or diseases) will need to be conducted at a dose that comprehensively considers the beneficial and detrimental effects of RONS and oxidative stress.

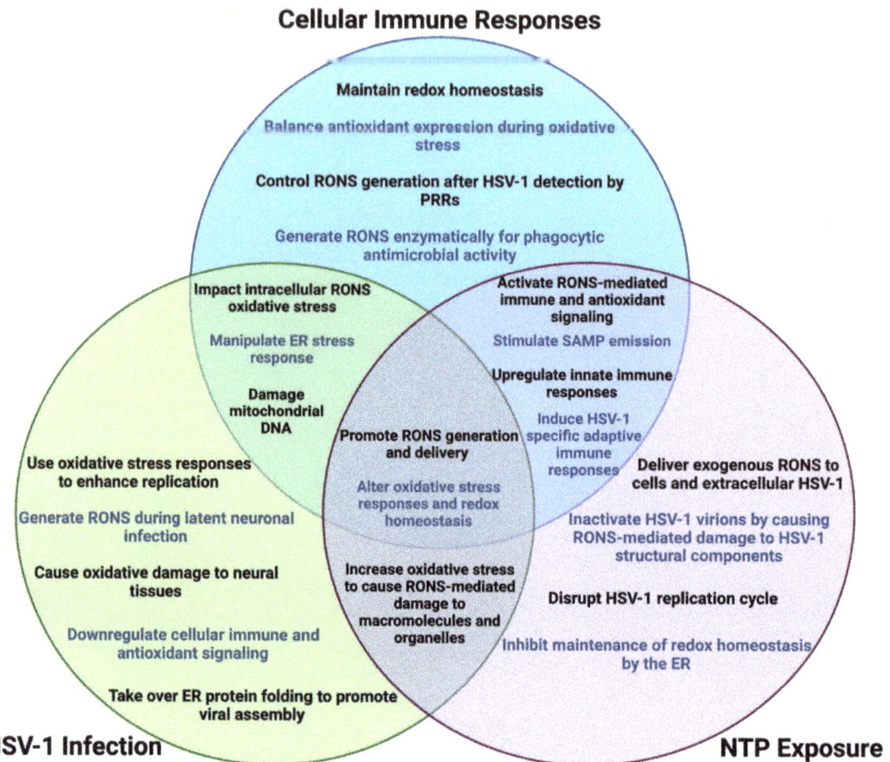

Figure 6. Summary of the roles and effects of oxidative stress in cellular immune responses, HSV-1 infection of host cells, and NTP exposure. The Venn diagram illustrates the unique and overlapping roles of RONS and oxidative stress in each aspect of a putative NTP-based treatment for HSV-1 infection.

Once considerations of NTP dose are satisfied, the potential net effects of immunomodulation induced by NTP application to an HSV-1-associated lesion are the promotion of more effective innate antiviral responses in infected cells, induction of innate protective responses in nearby uninfected cells, and recruitment of immune cells that will participate in a more effective adaptive response to infection (Figure 7). By controlling the cellular redox environment and modulating both innate and adaptive responses against HSV-1, NTP could prove to be an effective immunotherapy against HSV-1 infection.

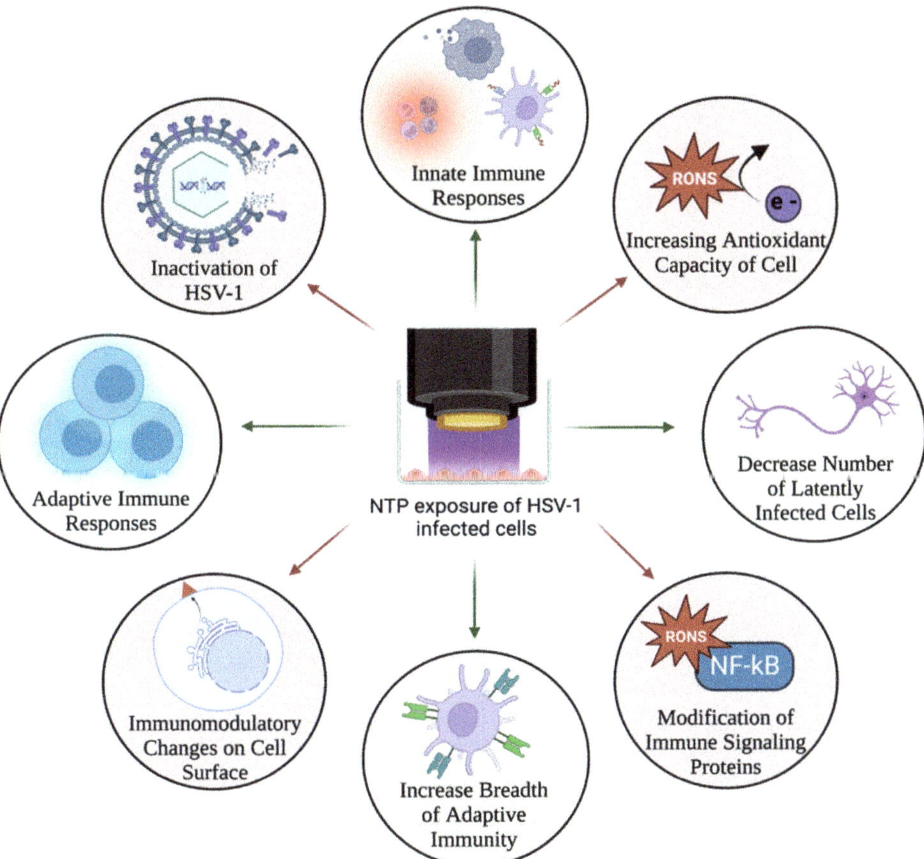

Figure 7. NTP is a potential antiviral and immunomodulatory agent for HSV-1 infection. We propose that the exposure of HSV-1-infected cells to NTP will directly (red) result in the inactivation of HSV-1 virion components and promote the display of SAMPs on the cell surface. Additionally, NTP-generated RONS can directly modify signaling proteins to activate immune signaling and balance redox levels through antioxidant transcription. Based on these effects, we believe NTP will indirectly (green) promote and expand the breadth of stimulation of both innate and adaptive immune responses against HSV-1. In return, this will decrease the amount of virus that can travel into the nervous system to establish viral latency in peripheral ganglionic neurons, ultimately reducing the frequency of reactivation over time.

12. Conclusions

HSV-1 continues to be a global health concern, with high infection rates worldwide and its ability to cause lifelong infection. Although the current antiviral therapies are effective in moderately reducing the severity of symptoms associated with acute infection, they are ineffective with respect to curing a patient after primary HSV-1 infectious due to their inability to address viral latency. Therefore, HSV-1 persists in its host, risking dissemination of the virus to other organs and causing other acute and chronic diseases.

For HSV-1, the manipulation of redox homeostasis is an evasion strategy to overcome immune clearance from the cell and to craft a cellular environment favorable for viral replication. On the other hand, maintenance of redox homeostasis by the cell is crucial in controlling the immune response directed toward HSV-1 and preventing damaging oxidative damage by RONS. Therefore, control over the redox balance within an infected cell is one of the key factors involved in determining the outcome of HSV-1 infection. Given the antiviral properties of RONS in the cell and in commercial disinfectant agents, RONS have the potential to be harnessed as antiviral therapies.

NTP technology is an inexpensive, innovative method of producing RONS responsible for inducing desired biological effects. It has the potential to act as a multi-faceted therapy effective against HSV-1 disease. Decreases in HSV-1 infectivity attributed to the direct antiviral effect of NTP on cell-free virus will decrease the overall viral burden in the lesion, thereby decreasing the clinical symptoms of infection and accelerate resolution of the lesion. An added benefit of reduced titers in the lesion will be a smaller pool of latently infected neurons that serve as reservoirs of long-term infection. The immunomodulatory effects of NTP exposure will augment local innate antiviral effects and boost adaptive immune responses involving HSV-1-specific CD8+ T lymphocytes. The combined antiviral and immunomodulatory activities of NTP are hypothesized to provide short-term relief for acute infection as well as long-term immunological control over reactivation from latently infected neurons, as indicated by reductions in or full control of recurrent lesions in HSV-1-infected individuals.

Of course, the potential of an NTP-based therapy must be fully explored along a developmental path that leads to clinical use. Future studies of the use of NTP as a therapy for HSV-1 infection will include examinations of how NTP can be used to treat lesions, a common manifestation of human HSV-1 infection. Using a preclinical murine model of HSV-1 infection, the antiviral and immunomodulatory effects of NTP on HSV-1 infection will be examined in the context of innate and adaptive, HSV-1-specific immune responses. These models will provide the insights necessary to develop this novel therapeutic as a non-invasive, pain-free, and effective clinical approach against HSV-1 infection.

Author Contributions: Conceptualization, J.S., B.W., F.C.K. and V.M.; writing—original draft preparation, J.S.; writing—review and editing, J.S., P.J.B., B.W., F.C.K. and V.M.; visualization, J.S. and V.M.; supervision, V.M. All authors have read and agreed to the published version of the manuscript.

Funding: The preparation of this review was supported by a grant awarded to V.M. and F.C.K. by the Drexel Coulter Translational Research Partnership Program.

Institutional Review Board Statement: Not applicable.

Informed Consent Statement: Not applicable.

Data Availability Statement: Not applicable.

Conflicts of Interest: The authors declare no conflict of interest.

References

1. Herpes Simplex Virus. Available online: https://www.who.int/news-room/fact-sheets/detail/herpes-simplex-virus (accessed on 16 December 2022).
2. Egan, K.P.; Allen, A.G.; Wigdahl, B.; Jennings, S.R. Modeling the pathology, immune responses, and kinetics of HSV-1 replication in the lip scarification model. *Virology* **2018**, *514*, 124–133. [CrossRef] [PubMed]
3. Whitley, R.; Baines, J. Clinical management of herpes simplex virus infections: Past, present, and future. *F1000Research* **2018**, *7*, 1726. [CrossRef] [PubMed]
4. Bradshaw, M.J.; Venkatesan, A. Herpes Simplex Virus-1 Encephalitis in Adults: Pathophysiology, Diagnosis, and Management. *NeuroTherapeutics* **2016**, *13*, 493–508. [CrossRef]
5. Shah, A.; Farooq, A.V.; Tiwari, V.; Kim, M.-J.; Shukla, D. HSV-1 infection of human corneal epithelial cells: Receptor-mediated entry and trends of re-infection. *Mol. Vis.* **2010**, *16*, 2476–2486. [PubMed]
6. Cairns, D.M.; Rouleau, N.; Parker, R.N.; Walsh, K.G.; Gehrke, L.; Kaplan, D.L. A 3D human brain–like tissue model of herpes-induced Alzheimer's disease. *Sci. Adv.* **2020**, *6*, eaay8828. [CrossRef]
7. Itzhaki, R.F. Herpes simplex virus type 1 and Alzheimer's disease: Increasing evidence for a major role of the virus. *Front. Aging Neurosci.* **2014**, *6*, 202. [CrossRef]
8. Santana, S.; Sastre, I.; Recuero, M.; Bullido, M.J.; Aldudo, J. Oxidative Stress Enhances Neurodegeneration Markers Induced by Herpes Simplex Virus Type 1 Infection in Human Neuroblastoma Cells. *PLoS ONE* **2013**, *8*, e75842. [CrossRef]

9. Ibáñez, F.J.; Farías, M.A.; Gonzalez-Troncoso, M.P.; Corrales, N.; Duarte, L.F.; Retamal-Díaz, A.; Gonzalez, P.A. Experimental Dissection of the Lytic Replication Cycles of Herpes Simplex Viruses in vitro. *Front. Microbiol.* **2018**, *9*, 2406. [CrossRef]

10. Cernik, C.; Gallina, K.; Brodell, R.T. The Treatment of Herpes Simplex InfectionsAn Evidence-Based Review. *Arch. Intern. Med.* **2008**, *168*, 1137–1144. [CrossRef]

11. Bacon, T.H.; Levin, M.J.; Leary, J.J.; Sarisky, R.T.; Sutton, D. Herpes Simplex Virus Resistance to Acyclovir and Penciclovir after Two Decades of Antiviral Therapy. *Clin. Microbiol. Rev.* **2003**, *16*, 114–128. [CrossRef]

12. Jiang, Y.-C.; Feng, H.; Lin, Y.-C.; Guo, X.-R. New strategies against drug resistance to herpes simplex virus. *Int. J. Oral Sci.* **2016**, *8*, 1–6. [CrossRef]

13. Su, C.; Zhan, G.; Zheng, C. Evasion of host antiviral innate immunity by HSV-1, an update. *Virol. J.* **2016**, *13*, 1–9. [CrossRef]

14. Laurent, A.; Nicco, C.; Chéreau, C.; Goulvestre, C.; Alexandre, J.; Alves, A.; Lévy, E.; Goldwasser, F.; Panis, Y.; Soubrane, O.; et al. Controlling Tumor Growth by Modulating Endogenous Production of Reactive Oxygen Species. *Cancer Res.* **2005**, *65*, 948–956. [CrossRef] [PubMed]

15. Zhang, J.; Wang, X.; Vikash, V.; Ye, Q.; Wu, D.; Liu, Y.; Dong, W. ROS and ROS-Mediated Cellular Signaling. *Oxidative Med. Cell. Longev.* **2016**, *2016*, 4350965. [CrossRef]

16. Son, Y.; Cheong, Y.K.; Kim, N.H.; Chung, H.T.; Kang, D.G.; Pae, H.O. Mitogen-Activated Protein Kinases and Reactive Oxygen Species: How Can ROS Activate MAPK Pathways? *J. Signal Transduct.* **2011**, *2011*, 792639. [CrossRef]

17. Goldkorn, T.; Balaban, N.; Matsukuma, K.; Chea, V.; Gould, R.; Last, J.; Chan, C.; Chavez, C. EGF-Receptor Phosphorylation and Signaling Are Targeted by H2O2 Redox Stress. *Am. J. Respir. Cell Mol. Biol.* **1998**, *19*, 786–798. [CrossRef]

18. Matsuzawa, A.; Saegusa, K.; Noguchi, T.; Sadamitsu, C.; Nishitoh, H.; Nagai, S.; Koyasu, S.; Matsumoto, K.; Takeda, K.; Ichijo, H. ROS-dependent activation of the TRAF6-ASK1-p38 pathway is selectively required for TLR4-mediated innate immunity. *Nat. Immunol.* **2005**, *6*, 587–592. [CrossRef]

19. Forman, H.J.; Ursini, F.; Maiorino, M. An overview of mechanisms of redox signaling. *J. Mol. Cell. Cardiol.* **2014**, *73*, 2–9. [CrossRef]

20. Fang, F.C. Antimicrobial reactive oxygen and nitrogen species: Concepts and controversies. *Nat. Rev. Microbiol.* **2004**, *2*, 820–832. [CrossRef]

21. Komaravelli, N.; Casola, A. Respiratory Viral Infections and Subversion of Cellular Antioxidant Defenses. *J. Pharm. Pharm.* **2014**, *5*, 1000141.

22. Jha, N.; Ryu, J.J.; Choi, E.H.; Kaushik, N.K. Generation and Role of Reactive Oxygen and Nitrogen Species Induced by Plasma, Lasers, Chemical Agents, and Other Systems in Dentistry. *Oxidative Med. Cell. Longev.* **2017**, *2017*, 1–13. [CrossRef] [PubMed]

23. Zhao, R.Z.; Jiang, S.; Zhang, L.; Yu, Z.B. Mitochondrial electron transport chain, ROS generation and uncoupling (Review). *Int. J. Mol. Med.* **2019**, *44*, 3–15. [CrossRef] [PubMed]

24. Kussmaul, L.; Hirst, J. The mechanism of superoxide production by NADH:ubiquinone oxidoreductase (complex I) from bovine heart mitochondria. *Proc. Natl. Acad. Sci. USA* **2006**, *103*, 7607–7612. [CrossRef] [PubMed]

25. Kushnareva, Y.; Murphy, A.N.; Andreyev, A.Y. Complex I-mediated reactive oxygen species generation: Modulation by cytochrome c and NAD(P)+ oxidation–reduction state. *Biochem. J.* **2002**, *368*, 545–553. [CrossRef]

26. Quinlan, C.L.; Goncalves, R.L.; Hey-Mogensen, M.; Yadava, N.; Bunik, V.I.; Brand, M.D. The 2-Oxoacid Dehydrogenase Complexes in Mitochondria Can Produce Superoxide/Hydrogen Peroxide at Much Higher Rates Than Complex I. *J. Biol. Chem.* **2014**, *289*, 8312–8325. [CrossRef]

27. Quinlan, C.L.; Orr, A.L.; Perevoshchikova, I.V.; Treberg, J.R.; Ackrell, B.A.; Brand, M.D. Mitochondrial Complex II Can Generate Reactive Oxygen Species at High Rates in Both the Forward and Reverse Reactions. *J. Biol. Chem.* **2012**, *287*, 27255–27264. [CrossRef]

28. Quinlan, C.L.; Perevoshchikova, I.V.; Hey-Mogensen, M.; Orr, A.L.; Brand, M.D. Sites of reactive oxygen species generation by mitochondria oxidizing different substrates. *Redox Biol.* **2013**, *1*, 304–312. [CrossRef]

29. Mailloux, R.J. An Update on Mitochondrial Reactive Oxygen Species Production. *Antioxidants* **2020**, *9*, 472. [CrossRef]

30. Zhang, Y.; Bharathi, S.S.; Beck, M.E.; Goetzman, E.S. The fatty acid oxidation enzyme long-chain acyl-CoA dehydrogenase can be a source of mitochondrial hydrogen peroxide. *Redox Biol.* **2019**, *26*, 101253. [CrossRef]

31. Cao, S.S.; Kaufman, R.J. Endoplasmic Reticulum Stress and Oxidative Stress in Cell Fate Decision and Human Disease. *Antioxid. Redox Signal.* **2014**, *21*, 396–413. [CrossRef]

32. Ozgur, R.; Uzilday, B.; Iwata, Y.; Koizumi, N.; Turkan, I. Interplay between the unfolded protein response and reactive oxygen species: A dynamic duo. *J. Exp. Bot.* **2018**, *69*, 3333–3345. [CrossRef]

33. Dansen, T.B.; Wirtz, K.W.A. The Peroxisome in Oxidative Stress. *IUBMB Life* **2001**, *51*, 223–230.

34. Fransen, M.; Nordgren, M.; Wang, B.; Apanasets, O. Role of peroxisomes in ROS/RNS-metabolism: Implications for human disease. *Biochim. Biophys. Acta (BBA) -Mol. Basis Dis.* **2012**, *1822*, 1363–1373. [CrossRef]

35. Bedard, K.; Krause, K.-H. The NOX Family of ROS-Generating NADPH Oxidases: Physiology and Pathophysiology. *Physiol. Rev.* **2007**, *87*, 245–313. [CrossRef]

36. Rastogi, R.; Geng, X.; Li, F.; Ding, Y. NOX Activation by Subunit Interaction and Underlying Mechanisms in Disease. *Front. Cell. Neurosci.* **2017**, *10*, 301. [CrossRef]

37. Sumimoto, H. Structure, regulation and evolution of Nox-family NADPH oxidases that produce reactive oxygen species. *FEBS J.* **2008**, *275*, 3249–3277. [CrossRef]

38. Förstermann, U.; Sessa, W.C. Nitric oxide synthases: Regulation and function. *Eur. Heart J.* **2012**, *33*, 829–837. [CrossRef]

39. Croen, K.D. Evidence for antiviral effect of nitric oxide. Inhibition of herpes simplex virus type 1 replication. *J. Clin. Investig.* **1993**, *91*, 2446–2452. [CrossRef]

40. Feng, C.; Li, J.; Zheng, H. Deciphering mechanism of conformationally controlled electron transfer in nitric oxide synthases. *Front. Biosci.* **2018**, *23*, 1803–1821. [CrossRef]

41. Weidinger, A.; Kozlov, A.V. Biological Activities of Reactive Oxygen and Nitrogen Species: Oxidative Stress versus Signal Transduction. *Biomolecules* **2015**, *5*, 472–484. [CrossRef]

42. Dumas, A.; Knaus, U.G. Raising the 'Good' Oxidants for Immune Protection. *Front. Immunol.* **2021**, *12*, 698042. [CrossRef] [PubMed]

43. Garofalo, R.P.; Kolli, D.; Casola, A. Respiratory Syncytial Virus Infection: Mechanisms of Redox Control and Novel Therapeutic Opportunities. *Antioxid. Redox Signal.* **2013**, *18*, 186–217. [CrossRef] [PubMed]

44. Reshi, M.L.; Su, Y.C.; Hong, J.-R. RNA Viruses: ROS-Mediated Cell Death. *Int. J. Cell Biol.* **2014**, *2014*, 467452. [CrossRef] [PubMed]

45. Mebrat, Y.; Amogne, W.; Mekasha, A.; Gleason, R.L., Jr.; Seifu, D. Lipid Peroxidation and Altered Antioxidant Profiles with Pediatric HIV Infection and Antiretroviral Therapy in Addis Ababa, Ethiopia. *J. Trop. Pediatr.* **2017**, *63*, 196–202. [CrossRef]

46. Rehman, Z.U.; Meng, C.; Sun, Y.; Safdar, A.; Pasha, R.H.; Munir, M.; Ding, C. Oxidative Stress in Poultry: Lessons from the Viral Infections. *Oxidative Med. Cell. Longev.* **2018**, *2018*, 5123147. [CrossRef]

47. Kavouras, J.H.; Prandovszky, E.; Valyi-Nagy, K.; Kovacs, S.K.; Tiwari, V.; Kovács, M.; Shukla, D.; Valyi-Nagy, T. Herpes simplex virus type 1 infection induces oxidative stress and the release of bioactive lipid peroxidation by-products in mouse P19N neural cell cultures. *J. NeuroVirology* **2007**, *13*, 416–425. [CrossRef]

48. Sebastiano, M.; Chastel, O.; de Thoisy, B.; Eens, M.; Costantini, D. Oxidative stress favours herpes virus infection in vertebrates: A meta-analysis. *Curr. Zool.* **2016**, *62*, 325–332. [CrossRef]

49. Laine, R.F.; Albecka, A.; van de Linde, S.; Rees, E.J.; Crump, C.M.; Kaminski, C.F. Structural analysis of herpes simplex virus by optical super-resolution imaging. *Nat. Commun.* **2015**, *6*, 5980. [CrossRef]

50. Kukhanova, M.K.; Korovina, A.N.; Kochetkov, S.N. Human herpes simplex virus: Life cycle and development of inhibitors. *Biochem. (Moscow)* **2014**, *79*, 1635–1652. [CrossRef]

51. Hilterbrand, A.T.; Daly, R.E.; Heldwein, E.E. Contributions of the Four Essential Entry Glycoproteins to HSV-1 Tropism and the Selection of Entry Routes. *Mbio* **2021**, *12*, e00143-21. [CrossRef]

52. Madavaraju, K.; Koganti, R.; Volety, I.; Yadavalli, T.; Shukla, D. Herpes Simplex Virus Cell Entry Mechanisms: An Update. *Front. Cell. Infect. Microbiol.* **2021**, *10*, 617578. [CrossRef]

53. Choudhary, S.; Marquez, M.; Alencastro, F.; Spors, F.; Zhao, Y.; Tiwari, V. Herpes Simplex Virus Type-1 (HSV-1) Entry into Human Mesenchymal Stem Cells Is Heavily Dependent on Heparan Sulfate. *J. Biomed. Biotechnol.* **2011**, *2011*, 1–11. [CrossRef]

54. Sarrazin, S.; Lamanna, W.C.; Esko, J.D. Heparan sulfate proteoglycans. *Cold Spring Harb. Perspect. Biol.* **2011**, *3*, a004952. [CrossRef]

55. Danastas, K.; Miranda-Saksena, M.; Cunningham, A.L. Herpes Simplex Virus Type 1 Interactions with the Interferon System. *Int. J. Mol. Sci.* **2020**, *21*, 5150. [CrossRef]

56. Mulvey, M.; Arias, C.; Mohr, I. Maintenance of Endoplasmic Reticulum (ER) Homeostasis in Herpes Simplex Virus Type 1-Infected Cells through the Association of a Viral Glycoprotein with PERK, a Cellular ER Stress Sensor. *J. Virol.* **2007**, *81*, 3377–3390. [CrossRef] [PubMed]

57. Molteni, C.G.; Principi, N.; Esposito, S. Reactive oxygen and nitrogen species during viral infections. *Free. Radic. Res.* **2014**, *48*, 1163–1169. [CrossRef]

58. Aubert, M.; Chen, Z.; Lang, R.; Chung, H.D.; Fowler, C.; Derek, D.S.; Keith, R.J. The Antiapoptotic Herpes Simplex Virus Glycoprotein J Localizes to Multiple Cellular Organelles and Induces Reactive Oxygen Species Formation. *J. Virol.* **2008**, *82*, 617–629. [CrossRef]

59. Tebaldi, G.; Pritchard, S.M.; Nicola, A.V. Herpes Simplex Virus Entry by a Nonconventional Endocytic Pathway. *J. Virol.* **2020**, *94*, e01910-20. [CrossRef]

60. Fan, D.; Wang, M.; Cheng, A.; Jia, R.; Yang, Q.; Wu, Y.; Zhu, D.; Zhao, X.; Chen, S.; Liu, M.; et al. The Role of VP16 in the Life Cycle of Alphaherpesviruses. *Front. Microbiol.* **2020**, *11*, 1910. [CrossRef]

61. Bearer, E.L.; Breakefield, X.O.; Schuback, D.; Reese, T.S.; LaVail, J.H. Retrograde axonal transport of herpes simplex virus: Evidence for a single mechanism and a role for tegument. *Proc. Natl. Acad. Sci. USA* **2000**, *97*, 8146–8150. [CrossRef]

62. Xu, X.; Che, Y.; Li, Q. HSV-1 tegument protein and the development of its genome editing technology. *Virol. J.* **2016**, *13*, 1–7. [CrossRef] [PubMed]

63. Kurt-Jones, E.A.; Orzalli, M.H.; Knipe, D.M. Innate Immune Mechanisms and Herpes Simplex Virus Infection and Disease. In *Cell Biology of Herpes Viruses*; Springer: Berlin/Heidelberg, Germany, 2017; Volume 223, pp. 49–75. [CrossRef]

64. Weller, S.K. Herpes Simplex Virus Reorganizes the Cellular DNA Repair and Protein Quality Control Machinery. *PLoS Pathog.* **2010**, *6*, e1001105. [CrossRef] [PubMed]

65. Harkness, J.M.; Kader, M.; DeLuca, N.A. Transcription of the Herpes Simplex Virus 1 Genome during Productive and Quiescent Infection of Neuronal and Nonneuronal Cells. *J. Virol.* **2014**, *88*, 6847–6861. [CrossRef] [PubMed]

66. Zhao, J.; Qin, C.; Liu, Y.; Rao, Y.; Feng, P. Herpes Simplex Virus and Pattern Recognition Receptors: An Arms Race. *Front. Immunol.* **2021**, *11*, 613799. [CrossRef] [PubMed]

67. Copeland, A.M.; Newcomb, W.W.; Brown, J.C. Herpes Simplex Virus Replication: Roles of Viral Proteins and Nucleoporins in Capsid-Nucleus Attachment. *J. Virol.* **2009**, *83*, 1660–1668. [CrossRef]

68. Shahin, V.; Hafezi, W.; Oberleithner, H.; Ludwig, Y.; Windoffer, B.; Schillers, H.; Kühn, J.E. The genome of HSV-1 translocates through the nuclear pore as a condensed rod-like structure. *J. Cell Sci.* **2006**, *119*, 23–30. [CrossRef]

69. Tao, L.; Lemoff, A.; Wang, G.; Zarek, C.; Lowe, A.; Yan, N.; Reese, T.A. Reactive oxygen species oxidize STING and suppress interferon production. *Elife* **2020**, *9*, e57837. [CrossRef]

70. Jia, M.; Qin, D.; Zhao, C.; Chai, L.; Yu, Z.; Wang, W.; Tong, L.; Lv, L.; Wang, Y.; Rehwinkel, J.; et al. Redox homeostasis maintained by GPX4 facilitates STING activation. *Nat. Immunol.* **2020**, *21*, 727–735. [CrossRef]

71. Zhang, L.; Wang, J.; Wang, Z.; Li, Y.; Wang, H.; Liu, H. Upregulation of nuclear factor E2-related factor 2 (Nrf2) represses the replication of herpes simplex virus type 1. *Virol. J.* **2022**, *19*, 1–8. [CrossRef]

72. Honess, R.W.; Roizman, B. Regulation of Herpesvirus Macromolecular Synthesis I. Cascade Regulation of the Synthesis of Three Groups of Viral Proteins. *J. Virol.* **1974**, *14*, 8–19. [CrossRef]

73. Packard, J.E.; Dembowski, J.A. HSV-1 DNA Replication—Coordinated Regulation by Viral and Cellular Factors. *Viruses* **2021**, *13*, 2015. [CrossRef]

74. Rodríguez, M.C.; Dybas, J.M.; Hughes, J.; Weitzman, M.D.; Boutell, C. The HSV-1 ubiquitin ligase ICP0: Modifying the cellular proteome to promote infection. *Virus Res.* **2020**, *285*, 198015. [CrossRef]

75. Tang, S.; Patel, A.; Krause, P.R. Hidden regulation of herpes simplex virus 1 pre-mRNA splicing and polyadenylation by virally encoded immediate early gene ICP27. *PLoS Pathog.* **2019**, *15*, e1007884. [CrossRef]

76. Fierer, D.S.; Challberg, M.D. Purification and characterization of UL9, the herpes simplex virus type 1 origin-binding protein. *J. Virol.* **1992**, *66*, 3986–3995. [CrossRef]

77. Muylaert, I.; Tang, K.-W.; Elias, P. Replication and Recombination of Herpes Simplex Virus DNA. *J. Biol. Chem.* **2011**, *286*, 15619–15624. [CrossRef]

78. Weller, S.K.; Coen, D.M. Herpes Simplex Viruses: Mechanisms of DNA Replication. *Cold Spring Harb. Perspect. Biol.* **2012**, *4*, a013011. [CrossRef]

79. Schumacher, A.J.; Mohni, K.N.; Kan, Y.; Hendrickson, E.A.; Stark, J.M.; Weller, S.K. The HSV-1 Exonuclease, UL12, Stimulates Recombination by a Single Strand Annealing Mechanism. *PLoS Pathog.* **2012**, *8*, e1002862. [CrossRef]

80. Cavignac, Y.; Esclatine, A. Herpesviruses and autophagy: Catch me if you can! *Viruses* **2010**, *2*, 314–333. [CrossRef]

81. Marino-Merlo, F.; Papaianni, E.; Frezza, C.; Pedatella, S.; De Nisco, M.; Macchi, B.; Grelli, S.; Mastino, A. NF-κB-Dependent Production of ROS and Restriction of HSV-1 Infection in U937 Monocytic Cells. *Viruses* **2019**, *11*, 428. [CrossRef]

82. Valyi-Nagy, T.; Dermody, T.S. Role of oxidative damage in the pathogenesis of viral infections of the nervous system. *Histol. Histopathol.* **2005**, *20*, 957–967. [CrossRef]

83. Castro-Acosta, R.M.; Rodríguez-Limas, W.A.; Valderrama, B.; Ramírez, O.T.; Palomares, L.A. Effect of metal catalyzed oxidation in recombinant viral protein assemblies. *Microb. Cell Factories* **2014**, *13*, 25. [CrossRef] [PubMed]

84. England, K.; Odriscoll, C.; Cotter, T.G. Carbonylation of glycolytic proteins is a key response to drug-induced oxidative stress and apoptosis. *Cell Death Differ.* **2003**, *11*, 252–260. [CrossRef] [PubMed]

85. Mathew, S.S.; Bryant, P.W.; Burch, A.D. Accumulation of oxidized proteins in Herpesvirus infected cells. *Free. Radic. Biol. Med.* **2010**, *49*, 383–391. [CrossRef] [PubMed]

86. Heming, J.D.; Conway, J.F.; Homa, F.L. Herpesvirus Capsid Assembly and DNA Packaging. In *Cell Biology of Herpes Viruses*; Springer: Berlin/Heidelberg, Germany, 2017; Volume 223, pp. 119–142. [CrossRef]

87. Roos, W.H.; Radtke, K.; Kniesmeijer, E.; Geertsema, H.; Sodeik, B.; Wuite, G.J.L. Scaffold expulsion and genome packaging trigger stabilization of herpes simplex virus capsids. *Proc. Natl. Acad. Sci. USA* **2009**, *106*, 9673–9678. [CrossRef] [PubMed]

88. Albright, B.S.; Nellissery, J.; Szczepaniak, R.; Weller, S.K. Disulfide Bond Formation in the Herpes Simplex Virus 1 UL6 Protein Is Required for Portal Ring Formation and Genome Encapsidation. *J. Virol.* **2011**, *85*, 8616–8624. [CrossRef]

89. Yang, K.; Baines, J. Selection of HSV capsids for envelopment involves interaction between capsid surface components pUL31, pUL17, and pUL25. *Proc. Natl. Acad. Sci. USA* **2011**, *108*, 14276. [CrossRef]

90. Leuzinger, H.; Ziegler, U.; Schraner, E.M.; Fraefel, C.; Glauser, D.L.; Heid, I.; Ackermann, M.; Mueller, M.; Wild, P. Herpes Simplex Virus 1 Envelopment Follows Two Diverse Pathways. *J. Virol.* **2005**, *79*, 13047–13059. [CrossRef]

91. Sandbaumhüter, M.; Döhner, K.; Schipke, J.; Binz, A.; Pohlmann, A.; Sodeik, B.; Bauerfeind, R. Cytosolic herpes simplex virus capsids not only require binding inner tegument protein pUL36 but also pUL37 for active transport prior to secondary envelopment. *Cell. Microbiol.* **2012**, *15*, 248–269. [CrossRef]

92. Ahmad, I.; Wilson, D.W. HSV-1 Cytoplasmic Envelopment and Egress. *Int. J. Mol. Sci.* **2020**, *21*, 5969. [CrossRef]

93. Pasdeloup, D.; McElwee, M.; Beilstein, F.; Labetoulle, M.; Rixon, F.J. Herpesvirus Tegument Protein pUL37 Interacts with Dystonin/BPAG1 To Promote Capsid Transport on Microtubules during Egress. *J. Virol.* **2013**, *87*, 2857–2867. [CrossRef]

94. Miranda-Saksena, M.; Denes, C.E.; Diefenbach, R.J.; Cunningham, A.L. Infection and Transport of Herpes Simplex Virus Type 1 in Neurons: Role of the Cytoskeleton. *Viruses* **2018**, *10*, 92. [CrossRef]

95. Wang, Q.-Y.; Zhou, C.; Johnson, K.E.; Colgrove, R.C.; Coen, D.M.; Knipe, D.M. Herpesviral latency-associated transcript gene promotes assembly of heterochromatin on viral lytic-gene promoters in latent infection. *Proc. Natl. Acad. Sci. USA* **2005**, *102*, 16055–16059. [CrossRef]

96. Deshmane, S.L.; Fraser, N.W. During latency, herpes simplex virus type 1 DNA is associated with nucleosomes in a chromatin structure. *J. Virol.* **1989**, *63*, 943–947. [CrossRef]

97. Millhouse, S.; Wigdahl, B. Molecular circuitry regulating herpes simplex virus type 1 latency in neurons. *J. Neurovirology* **2000**, *6*, 6–24. [CrossRef]

98. Kenny, J.J.; Millhouse, S.; Wotring, M.; Wigdahl, B. Upstream Stimulatory Factor Family Binds to the Herpes Simplex Virus Type 1 Latency-Associated Transcript Promoter. *Virology* **1997**, *230*, 381–391. [CrossRef]

99. Millhouse, S.; Kenny, J.J.; Quinn, P.G.; Lee, V.; Wigdahl, B. ATF/CREB elements in the herpes simplex virus type 1 latency-associated transcript promoter interact with members of the ATF/CREB and AP-1 transcription factor families. *J. Biomed. Sci.* **1998**, *5*, 451–464. [CrossRef]

100. Kenny, J.J.; Krebs, F.C.; Hartle, H.T.; Gartner, A.E.; Chatton, B.; Leiden, J.M.; Hoeffler, J.P.; Weber, P.C.; Wigdahl, B. Identification of a Second ATF/CREB-like Element in the Herpes Simplex Virus Type 1 (HSV-1) Latency-Associated Transcript (LAT) Promoter. *Virology* **1994**, *200*, 220–235. [CrossRef]

101. Nicoll, M.P.; Hann, W.; Shivkumar, M.; Harman, L.E.R.; Connor, V.; Coleman, H.M.; Proença, J.; Efstathiou, S. The HSV-1 Latency-Associated Transcript Functions to Repress Latent Phase Lytic Gene Expression and Suppress Virus Reactivation from Latently Infected Neurons. *PLoS Pathog.* **2016**, *12*, e1005539. [CrossRef]

102. Knipe, D.M.; Raja, P.; Lee, J.S. Clues to mechanisms of herpesviral latent infection and potential cures. *Proc. Natl. Acad. Sci. USA* **2015**, *112*, 11993–11994. [CrossRef]

103. Butterfield, D.A.; Reed, T.; Newman, S.F.; Sultana, R. Roles of amyloid β-peptide-associated oxidative stress and brain protein modifications in the pathogenesis of Alzheimer's disease and mild cognitive impairment. *Free. Radic. Biol. Med.* **2007**, *43*, 658–677. [CrossRef]

104. Gonzalez-Dosal, R.; Horan, K.A.; Rahbek, S.H.; Ichijo, H.; Chen, Z.; Mieyal, J.J.; Hartmann, R.; Paludan, S.R. HSV Infection Induces Production of ROS, which Potentiate Signaling from Pattern Recognition Receptors: Role for S-glutathionylation of TRAF3 and 6. *PLoS Pathog.* **2011**, *7*, e1002250. [CrossRef] [PubMed]

105. Grinde, B. Herpesviruses: Latency and reactivation—viral strategies and host response. *J. Oral Microbiol.* **2013**, *5*, 22766. [CrossRef] [PubMed]

106. Hu, S.; Sheng, W.S.; Schachtele, S.J.; Lokensgard, J.R. Reactive oxygen species drive herpes simplex virus (HSV)-1-induced proinflammatory cytokine production by murine microglia. *J. Neuroinflammation* **2011**, *8*, 123. [CrossRef]

107. Schachtele, S.J.; Hu, S.; Little, M.R.; Lokensgard, J.R. Herpes simplex virus induces neural oxidative damage via microglial cell Toll-like receptor-2. *J. Neuroinflammation* **2010**, *7*, 35. [CrossRef]

108. Marques, C.P.; Cheeran, M.C.-J.; Palmquist, J.M.; Hu, S.; Lokensgard, J.R. Microglia are the major cellular source of inducible nitric oxide synthase during experimental herpes encephalitis. *J. NeuroVirology* **2008**, *14*, 229–238. [CrossRef]

109. Doll, J.R.; Hoebe, K.; Thompson, R.L.; Sawtell, N.M. Resolution of herpes simplex virus reactivation in vivo results in neuronal destruction. *PLoS Pathog.* **2020**, *16*, e1008296. [CrossRef]

110. Webre, J.M.; Hill, J.M.; Nolan, N.M.; Clement, C.; McFerrin, H.E.; Bhattacharjee, P.S.; Hsia, V.; Neumann, D.M.; Foster, T.P.; Lukiw, W.J.; et al. Rabbit and Mouse Models of HSV-1 Latency, Reactivation, and Recurrent Eye Diseases. *J. Biomed. Biotechnol.* **2012**, *2012*, 1–18. [CrossRef]

111. Heck, D.E.; Vetrano, A.M.; Mariano, T.M.; Laskin, J.D. UVB Light Stimulates Production of Reactive Oxygen Species: Unexpected Role for Catalase. *J. Biol. Chem.* **2003**, *278*, 22432–22436. [CrossRef]

112. Hogk, I.; Kaufmann, M.; Finkelmeier, D.; Rupp, S.; Burger-Kentischer, A. An In Vitro HSV-1 Reactivation Model Containing Quiescently Infected PC12 Cells. *BioResearch Open Access* **2013**, *2*, 250–257. [CrossRef]

113. Shimeld, C.; Easty, D.L.; Hill, T.J. Reactivation of Herpes Simplex Virus Type 1 in the Mouse Trigeminal Ganglion: An In Vivo Study of Virus Antigen and Cytokines. *J. Virol.* **1999**, *73*, 1767–1773. [CrossRef]

114. Brem, R.; Macpherson, P.; Guven, M.; Karran, P. Oxidative stress induced by UVA photoactivation of the tryptophan UVB photoproduct 6-formylindolo[3,2-b]carbazole (FICZ) inhibits nucleotide excision repair in human cells. *Sci. Rep.* **2017**, *7*, 1–9. [CrossRef]

115. Rastogi, R.P.; Richa, U.; Kumar, A.; Tyagi, M.B.; Sinha, R.P. Molecular Mechanisms of Ultraviolet Radiation-Induced DNA Damage and Repair. *J. Nucleic Acids* **2010**, *2010*, 1–32. [CrossRef]

116. Ruzza, P.; Honisch, C.; Hussain, R.; Siligardi, G. Free Radicals and ROS Induce Protein Denaturation by UV Photostability Assay. *Int. J. Mol. Sci.* **2021**, *22*, 6512. [CrossRef]

117. Preston, C.M.; Nicholl, M.J. Induction of Cellular Stress Overcomes the Requirement of Herpes Simplex Virus Type 1 for Immediate-Early Protein ICP0 and Reactivates Expression from Quiescent Viral Genomes. *J. Virol.* **2008**, *82*, 11775–11783. [CrossRef]

118. Stoeger, T.; Adler, H. "Novel" Triggers of Herpesvirus Reactivation and Their Potential Health Relevance. *Front. Microbiol.* **2019**, *9*, 3207. [CrossRef]

119. Wigdahl, B.; Scheck, A.C.; Ziegler, R.J.; De Clercq, E.; Rapp, F. Analysis of the herpes simplex virus genome during in vitro latency in human diploid fibroblasts and rat sensory neurons. *J. Virol.* **1984**, *49*, 205–213. [CrossRef]

120. Wigdahl, B.L.; Isom, H.C.; Rapp, F. Repression and activation of the genome of herpes simplex viruses in human cells. *Proc. Natl. Acad. Sci. USA* **1981**, *78*, 6522–6526. [CrossRef]

121. Wigdahl, B.L.; Scheck, A.C.; De Clercq, E.; Rapp, F. High Efficiency Latency and Activation of Herpes Simplex Virus in Human Cells. *Science* **1982**, *217*, 1145–1146. [CrossRef]

122. Wigdahl, B.L.; Ziegler, R.J.; Sneve, M.; Rapp, F. Herpes simplex virus latency and reactivation in isolated rat sensory neurons. *Virology* **1983**, *127*, 159–167. [CrossRef]

123. Indo, H.P.; Yen, H.-C.; Nakanishi, I.; Matsumoto, K.-I.; Tamura, M.; Nagano, Y.; Matsui, H.; Gusev, O.; Cornette, R.; Okuda, T.; et al. A mitochondrial superoxide theory for oxidative stress diseases and aging. *J. Clin. Biochem. Nutr.* **2015**, *56*, 1–7. [CrossRef]

124. Nencioni, L.; Sgarbanti, R.; Amatore, D.; Checconi, P.; Celestino, I.; Limongi, D.; Anticoli, S.; Palamara, A.T.; Garaci, E. Intracellular Redox Signaling as Therapeutic Target for Novel Antiviral Strategy. *Curr. Pharm. Des.* **2011**, *17*, 3898–3904. [CrossRef] [PubMed]

125. Bulua, A.C.; Simon, A.; Maddipati, R.; Pelletier, M.; Park, H.; Kim, K.-Y.; Sack, M.N.; Kastner, D.L.; Siegel, R.M. Mitochondrial reactive oxygen species promote production of proinflammatory cytokines and are elevated in TNFR1-associated periodic syndrome (TRAPS). *J. Exp. Med.* **2011**, *208*, 519–533. [CrossRef] [PubMed]

126. Gianni, T.; Leoni, V.; Campadelli-Fiume, G. Type I interferon and NF-κB activation elicited by herpes simplex virus gH/gL via αvβ3 integrin in epithelial and neuronal cell lines. *J. Virol.* **2013**, *87*, 13911–13916. [CrossRef]

127. Paludan, S.R.; Bowie, A.G.; Horan, K.A.; Fitzgerald, K.A. Recognition of herpesviruses by the innate immune system. *Nat. Rev. Immunol.* **2011**, *11*, 143–154. [CrossRef] [PubMed]

128. Rasmussen, S.B.; Jensen, S.B.; Nielsen, C.; Quartin, E.; Kato, H.; Chen, Z.; Silverman, R.H.; Akira, S.; Paludan, S.R. Herpes simplex virus infection is sensed by both Toll-like receptors and retinoic acid-inducible gene- like receptors, which synergize to induce type I interferon production. *J. Gen. Virol.* **2009**, *90*, 74–78. [CrossRef]

129. Sun, L.; Wu, J.; Du, F.; Chen, X.; Chen, Z.J. Cyclic GMP-AMP Synthase Is a Cytosolic DNA Sensor That Activates the Type I Interferon Pathway. *Science* **2013**, *339*, 786–791. [CrossRef]

130. West, A.P.; Khoury-Hanold, W.; Staron, M.; Tal, M.C.; Pineda, C.M.; Lang, S.M.; Bestwick, M.; Duguay, B.A.; Raimundo, N.; MacDuff, D.A.; et al. Mitochondrial DNA stress primes the antiviral innate immune response. *Nature* **2015**, *520*, 553–557. [CrossRef]

131. Liu, H.; Zhang, H.; Iles, K.E.; Rinna, A.; Merrill, G.; Yodoi, J.; Torres, M.; Forman, H.J. The ADP-stimulated NADPH oxidase activates the ASK-1/MKK4/JNK pathway in alveolar macrophages. *Free. Radic. Res.* **2006**, *40*, 865–874. [CrossRef]

132. Lingappan, K. NF-κB in Oxidative Stress. *Curr. Opin. Toxicol.* **2018**, *7*, 81–86. [CrossRef]

133. Jung, Y.; Kim, H.; Min, S.H.; Rhee, S.G.; Jeong, W. Dynein light chain LC8 negatively regulates NF-kappaB through the redox-dependent interaction with IkappaBalpha. *J. Biol. Chem.* **2008**, *283*, 23863–23871. [CrossRef]

134. Chandel, N.S.; Trzyna, W.; McClintock, D.; Schumacker, P. Role of Oxidants in NF-κB Activation and TNF-α Gene Transcription Induced by Hypoxia and Endotoxin. *J. Immunol.* **2000**, *165*, 1013. [CrossRef]

135. Muri, J.; Thut, H.; Feng, Q.; Kopf, M. Thioredoxin-1 distinctly promotes NF-κB target DNA binding and NLRP3 inflammasome activation independently of Txnip. *ELife* **2020**, *9*, e53627. [CrossRef]

136. Bellezza, I.; Giambanco, I.; Minelli, A.; Donato, R. Nrf2-Keap1 signaling in oxidative and reductive stress. *Biochim. Biophys. Acta (BBA) -Mol. Cell Res.* **2018**, *1865*, 721–733. [CrossRef]

137. Cuadrado, A.; Martín-Moldes, Z.; Ye, J.; Lastres-Becker, I. Transcription Factors NRF2 and NF-κB Are Coordinated Effectors of the Rho Family, GTP-binding Protein RAC1 during Inflammation. *J. Biol. Chem.* **2014**, *289*, 15244–15258. [CrossRef]

138. Geng, J.; Sun, X.; Wang, P.; Zhang, S.; Wang, X.; Wu, H.; Hong, L.; Xie, C.; Li, X.; Zhao, H.; et al. Kinases Mst1 and Mst2 positively regulate phagocytic induction of reactive oxygen species and bactericidal activity. *Nat. Immunol.* **2015**, *16*, 1142–1152. [CrossRef]

139. Babior, B.M.; Takeuchi, C.; Ruedi, J.; Gutierrez, A.; Wentworth, P., Jr. Investigating antibody-catalyzed ozone generation by human neutrophils. *Proc. Natl. Acad. Sci. USA* **2003**, *100*, 3031–3034. [CrossRef]

140. Winterbourn, C.C.; Hampton, M.B.; Livesey, J.H.; Kettle, A.J. Modeling the Reactions of Superoxide and Myeloperoxidase in the Neutrophil Phagosome. *J. Biol. Chem.* **2006**, *281*, 39860–39869. [CrossRef]

141. Wang, M.; Lu, M.; Zhang, C.; Wu, X.; Chen, J.; Lv, W.; Sun, T.; Qiu, H.; Huang, S. Oxidative stress modulates the expression of toll-like receptor 3 during respiratory syncytial virus infection in human lung epithelial A549 cells. *Mol. Med. Rep.* **2018**, *18*, 1867–1877. [CrossRef]

142. Roy, D.; Wong, P.K.; Engelbrecht, R.S.; Chian, E.S. Mechanism of enteroviral inactivation by ozone. *Appl. Environ. Microbiol.* **1981**, *41*, 718–723. [CrossRef]

143. Petry, G.; Rossato, L.G.; Nespolo, J.; Kreutz, L.C.; Bertol, C.D. In Vitro Inactivation of Herpes Virus by Ozone. *Ozone Sci. Eng.* **2014**, *36*, 249–252. [CrossRef]

144. Moldgy, A.; Nayak, G.; Aboubakr, H.A.; Goyal, S.M.; Bruggeman, P.J. Inactivation of virus and bacteria using cold atmospheric pressure air plasmas and the role of reactive nitrogen species. *J. Phys. D Appl. Phys.* **2020**, *53*, 434004. [CrossRef]

145. Tsukidate, D.; Takashima, K.; Sasaki, S.; Miyashita, S.; Kaneko, T.; Takahashi, H.; Ando, S. Activation of plant immunity by exposure to dinitrogen pentoxide gas generated from air using plasma technology. *PLoS ONE* **2022**, *17*, e0269863. [CrossRef] [PubMed]

146. Goyal, S.M.; Chander, Y.; Yezli, S.; Otter, J.A. Evaluating the virucidal efficacy of hydrogen peroxide vapour. *J. Hosp. Infect.* **2014**, *86*, 255–259. [CrossRef] [PubMed]

147. Mileto, D.; Mancon, A.; Staurenghi, F.; Rizzo, A.; Econdi, S.; Gismondo, M.; Guidotti, M. Inactivation of SARS-CoV-2 in the Liquid Phase: Are Aqueous Hydrogen Peroxide and Sodium Percarbonate Efficient Decontamination Agents? *ACS Chem. Health Saf.* **2021**, *28*, 260–267. [CrossRef]

148. Rudnick, S.N.; McDevitt, J.J.; First, M.W.; Spengler, J.D. Inactivating influenza viruses on surfaces using hydrogen peroxide or triethylene glycol at low vapor concentrations. *Am. J. Infect. Control.* **2009**, *37*, 813–819. [CrossRef]

149. Aboubakr, H.A.; Gangal, U.; Youssef, M.M.; Goyal, S.M.; Bruggeman, P.J. Inactivation of virus in solution by cold atmospheric pressure plasma: Identification of chemical inactivation pathways. *J. Phys. D: Appl. Phys.* **2016**, *49*, 204001. [CrossRef]
150. Andrés, C.M.C.; de la Lastra, J.M.P.; Juan, C.A.; Plou, F.J.; Pérez-Lebeña, E. Hypochlorous Acid Chemistry in Mammalian Cells—Influence on Infection and Role in Various Pathologies. *Int. J. Mol. Sci.* **2022**, *23*, 10735. [CrossRef]
151. Block, M.S.; Rowan, B.G. Hypochlorous Acid: A Review. *J. Oral Maxillofac. Surg.* **2020**, *78*, 1461–1466. [CrossRef]
152. Pinto, A.K.; Richner, J.M.; Poore, E.A.; Patil, P.P.; Amanna, I.J.; Slifka, M.K.; Diamond, M.S. A Hydrogen Peroxide-Inactivated Virus Vaccine Elicits Humoral and Cellular Immunity and Protects against Lethal West Nile Virus Infection in Aged Mice. *J. Virol.* **2013**, *87*, 1926–1936. [CrossRef]
153. Zeng, L.; Wang, M.-D.; Ming, S.-L.; Li, G.-L.; Yu, P.-W.; Qi, Y.-L.; Jiang, D.-W.; Yang, G.-Y.; Wang, J.; Chu, B.-B. An effective inactivant based on singlet oxygen-mediated lipid oxidation implicates a new paradigm for broad-spectrum antivirals. *Redox Biol.* **2020**, *36*, 101601. [CrossRef]
154. Kipshidze, N.; Yeo, N.; Kipshidze, N. Photodynamic therapy for COVID-19. *Nat. Photonics* **2020**, *14*, 651–652. [CrossRef]
155. Garren, M.R.; Ashcraft, M.; Qian, Y.; Douglass, M.; Brisbois, E.J.; Handa, H. Nitric oxide and viral infection: Recent developments in antiviral therapies and platforms. *Appl. Mater. Today* **2020**, *22*, 100887. [CrossRef]
156. Gamba, G.; Cavalieri, H.; Courreges, M.C.; Massouh, E.J.; Benencia, F. Early inhibition of nitric oxide production increases HSV-1 intranasal infection. *J. Med. Virol.* **2004**, *73*, 313–322. [CrossRef]
157. Kaushik, N.K.; Bhartiya, P.; Kaushik, N.; Shin, Y.; Nguyen, L.N.; Park, J.S.; Kim, D.; Choi, E.H. Nitric-oxide enriched plasma-activated water inactivates 229E coronavirus and alters antiviral response genes in human lung host cells. *Bioact. Mater.* **2023**, *19*, 569–580. [CrossRef]
158. Von Woedtke, T.; Emmert, S.; Metelmann, H.-R.; Rupf, S.; Weltmann, K.-D. Perspectives on cold atmospheric plasma (CAP) applications in medicine. *Phys. Plasmas* **2020**, *27*, 070601. [CrossRef]
159. Von Woedtke, T.; Schmidt, A.; Bekeschus, S.; Wende, K.; Weltmann, K.-D. Plasma Medicine: A Field of Applied Redox Biology. *In Vivo* **2019**, *33*, 1011–1026. [CrossRef]
160. Lin, A.; Chernets, N.; Han, J.; Alicea, Y.; Dobrynin, D.; Fridman, G.; Freeman, T.A.; Fridman, A.; Miller, V. Non-Equilibrium Dielectric Barrier Discharge Treatment of Mesenchymal Stem Cells: Charges and Reactive Oxygen Species Play the Major Role in Cell Death. *Plasma Process. Polym.* **2015**, *12*, 1117–1127. [CrossRef]
161. Kang, K.A.; Piao, M.J.; Eom, S.; Yoon, S.-Y.; Ryu, S.; Kim, S.B.; Yi, J.M.; Hyun, J.W. Non-thermal dielectric-barrier discharge plasma induces reactive oxygen species by epigenetically modifying the expression of NADPH oxidase family genes in keratinocytes. *Redox Biol.* **2020**, *37*, 101698. [CrossRef]
162. Graves, D.B. The emerging role of reactive oxygen and nitrogen species in redox biology and some implications for plasma applications to medicine and biology. *J. Phys. D Appl. Phys.* **2012**, *45*, 263001. [CrossRef]
163. Mohamed, H.; Nayak, G.; Rendine, N.; Wigdahl, B.; Krebs, F.C.; Bruggeman, P.J.; Miller, V. Non-Thermal Plasma as a Novel Strategy for Treating or Preventing Viral Infection and Associated Disease. *Front. Phys.* **2021**, *9*, 683118. [CrossRef]
164. Grahl, T.; Märkl, H. Killing of microorganisms by pulsed electric fields. *Appl. Microbiol. Biotechnol.* **1996**, *45*, 148–157. [CrossRef] [PubMed]
165. Biasin, M.; Bianco, A.; Pareschi, G.; Cavalleri, A.; Cavatorta, C.; Fenizia, C.; Galli, P.; Lessio, L.; Lualdi, M.; Tombetti, E.; et al. UV-C irradiation is highly effective in inactivating SARS-CoV-2 replication. *Sci. Rep.* **2021**, *11*, 6260. [CrossRef] [PubMed]
166. Tseng, C.-C.; Li, C.-S. Inactivation of Viruses on Surfaces by Ultraviolet Germicidal Irradiation. *J. Occup. Environ. Hyg.* **2007**, *4*, 400–405. [CrossRef] [PubMed]
167. Ruiz-Fernández, A.R.; Rosemblatt, M.; Perez-Acle, T. Nanosecond pulsed electric field (nsPEF) and vaccines: A novel technique for the inactivation of SARS-CoV-2 and other viruses? *Ann. Med.* **2022**, *54*, 1749–1756. [CrossRef] [PubMed]
168. Kulms, D.; Zeise, E.; Pöppelmann, B.; Schwarz, T. DNA damage, death receptor activation and reactive oxygen species contribute to ultraviolet radiation-induced apoptosis in an essential and independent way. *Oncogene* **2002**, *21*, 5844–5851. [CrossRef]
169. Gomes, A.A.; Silva-Júnior, A.C.T.; Oliveira, E.B.; Asad, L.M.B.O.; Reis, N.C.S.C.; Felzenszwalb, I.; Kovary, K.; Asad, N.R. Reactive oxygen species mediate lethality induced by far-UV in Escherichia coli cells. *Redox Rep.* **2005**, *10*, 91–95. [CrossRef]
170. Jiang, C.; Oshin, E.A.; Guo, S.; Scott, M.; Li, X.; Mangiamele, C.; Heller, R. Synergistic Effects of an Atmospheric-Pressure Plasma Jet and Pulsed Electric Field on Cells and Skin. *IEEE Trans. Plasma Sci.* **2021**, *49*, 3317–3324. [CrossRef]
171. Sheikh, M.S.; Antinore, M.J.; Huang, Y.; Fornace, A.J. Ultraviolet-irradiation-induced apoptosis is mediated via ligand independent activation of tumor necrosis factor receptor 1. *Oncogene* **1998**, *17*, 2555–2563. [CrossRef]
172. Pakhomova, O.N.; Khorokhorina, V.A.; Bowman, A.M.; Rodaitė-Riševičienė, R.; Saulis, G.; Xiao, S.; Pakhomov, A.G. Oxidative effects of nanosecond pulsed electric field exposure in cells and cell-free media. *Arch. Biochem. Biophys.* **2012**, *527*, 55–64. [CrossRef]
173. Aboubakr, H.A.; Williams, P.; Gangal, U.; Youssef, M.M.; El-Sohaimy, S.A.A.; Bruggeman, P.J.; Goyal, S.M. Virucidal Effect of Cold Atmospheric Gaseous Plasma on Feline Calicivirus, a Surrogate for Human Norovirus. *Appl. Environ. Microbiol.* **2015**, *81*, 3612–3622. [CrossRef]
174. Kalghatgi, S.; Kelly, C.M.; Cerchar, E.; Torabi, B.; Alekseev, O.; Fridman, A.; Friedman, G.; Azizkhan-Clifford, J. Effects of Non-Thermal Plasma on Mammalian Cells. *PLoS ONE* **2011**, *6*, e16270. [CrossRef]
175. Brandenburg, R. Dielectric barrier discharges: Progress on plasma sources and on the understanding of regimes and single filaments. *Plasma Sources Sci. Technol.* **2017**, *26*, 053001. [CrossRef]

176. Winter, J.; Brandenburg, R.; Weltmann, K.-D. Atmospheric pressure plasma jets: An overview of devices and new directions. *Plasma Sources Sci. Technol.* **2015**, *24*, 064001. [CrossRef]

177. Lietz, A.M.; Kushner, M.J. Air plasma treatment of liquid covered tissue: Long timescale chemistry. *J. Phys. D Appl. Phys.* **2016**, *49*, 425204. [CrossRef]

178. Filipić, A.; Primc, G.; Zaplotnik, R.; Mehle, N.; Gutierrez-Aguirre, I.; Ravnikar, M.; Mozetič, M.; Žel, J.; Dobnik, D. Cold Atmospheric Plasma as a Novel Method for Inactivation of Potato Virus Y in Water Samples. *Food Environ. Virol.* **2019**, *11*, 220–228. [CrossRef]

179. Ahlfeld, B.; Li, Y.; Boulaaba, A.; Binder, A.; Schotte, U.; Zimmermann, J.L.; Morfill, G.; Klein, G. Inactivation of a Foodborne Norovirus Outbreak Strain with Nonthermal Atmospheric Pressure Plasma. *Mbio* **2015**, *6*, e02300-14. [CrossRef]

180. Nayak, G.; Aboubakr, H.A.; Goyal, S.M.; Bruggeman, P.J. Reactive species responsible for the inactivation of feline calicivirus by a two-dimensional array of integrated coaxial microhollow dielectric barrier discharges in air. *Plasma Process. Polym.* **2017**, *15*, 1700119. [CrossRef]

181. Yamashiro, R.; Misawa, T.; Sakudo, A. Key role of singlet oxygen and peroxynitrite in viral RNA damage during virucidal effect of plasma torch on feline calicivirus. *Sci. Rep.* **2018**, *8*, 1–13. [CrossRef]

182. Shi, X.-M.; Zhang, G.-J.; Wu, X.-L.; Peng, Z.-Y.; Zhang, Z.-H.; Shao, X.-J.; Chang, Z.-S. Effect of Low-Temperature Plasma on Deactivation of Hepatitis B Virus. *IEEE Trans. Plasma Sci.* **2012**, *40*, 2711–2716. [CrossRef]

183. Volotskova, O.; Dubrovsky, L.; Keidar, M.; Bukrinsky, M. Cold Atmospheric Plasma Inhibits HIV-1 Replication in Macrophages by Targeting Both the Virus and the Cells. *PLoS ONE* **2016**, *11*, e0165322. [CrossRef]

184. Jin, T.; Xu, Y.; Dai, C.; Zhou, X.; Xu, Q.; Wu, Z. Cold atmospheric plasma: A non-negligible strategy for viral RNA inactivation to prevent SARS-CoV-2 environmental transmission. *AIP Adv.* **2021**, *11*, 085019. [CrossRef] [PubMed]

185. Alekseev, O.; Donovan, K.; Limonnik, V.; Azizkhan-Clifford, J. Nonthermal Dielectric Barrier Discharge (DBD) Plasma Suppresses Herpes Simplex Virus Type 1 (HSV-1) Replication in Corneal Epithelium. *Transl. Vis. Sci. Technol.* **2014**, *3*, 2. [CrossRef] [PubMed]

186. Brun, P.; Vono, M.; Venier, P.; Tarricone, E.; Deligianni, V.; Martines, E.; Zuin, M.; Spagnolo, S.; Cavazzana, R.; Cardin, R.; et al. Disinfection of Ocular Cells and Tissues by Atmospheric-Pressure Cold Plasma. *PLoS ONE* **2012**, *7*, e33245. [CrossRef] [PubMed]

187. Fourquet, S.; Guerois, R.; Biard, D.; Toledano, M.B. Activation of NRF2 by Nitrosative Agents and H2O2 Involves KEAP1 Disulfide Formation. *J. Biol. Chem.* **2010**, *285*, 8463–8471. [CrossRef]

188. Khan, A.A.; Alsahli, M.A.; Rahmani, A.H. Myeloperoxidase as an Active Disease Biomarker: Recent Biochemical and Pathological Perspectives. *Med. Sci.* **2018**, *6*, 33. [CrossRef]

189. Laporte, A.; Lortz, S.; Schaal, C.; Lenzen, S.; Elsner, M. Hydrogen peroxide permeability of cellular membranes in insulin-producing cells. *Biochim. Biophys. Acta (BBA) -Biomembr.* **2019**, *1862*, 183096. [CrossRef]

190. Kang, S.U.; Kim, H.J.; Kim, D.H.; Han, C.H.; Lee, Y.S.; Kim, C.-H. Nonthermal plasma treated solution inhibits adipocyte differentiation and lipogenesis in 3T3-L1 preadipocytes via ER stress signal suppression. *Sci. Rep.* **2018**, *8*, 2277. [CrossRef]

191. Schmidt, A.; Dietrich, S.; Steuer, A.; Weltmann, K.-D.; von Woedtke, T.; Masur, K.; Wende, K. Non-thermal Plasma Activates Human Keratinocytes by Stimulation of Antioxidant and Phase II Pathways. *J. Biol. Chem.* **2015**, *290*, 6731–6750. [CrossRef]

192. Li, Y.; Choi, E.H.; Han, I. Regulation of Redox Homeostasis by Nonthermal Biocompatible Plasma Discharge in Stem Cell Differentiation. *Oxidative Med. Cell. Longev.* **2019**, *2019*, 1–15. [CrossRef]

193. Bekeschus, S.; von Woedtke, T.; Kramer, A.; Weltmann, K.-D.; Masur, K. Cold Physical Plasma Treatment Alters Redox Balance in Human Immune Cells. *Plasma Med.* **2013**, *3*, 267–278. [CrossRef]

194. Lin, A.G.; Xiang, B.; Merlino, D.J.; Baybutt, T.R.; Sahu, J.; Fridman, A.; Snook, A.E.; Miller, V. Non-thermal plasma induces immunogenic cell death in vivo in murine CT26 colorectal tumors. *OncoImmunology* **2018**, *7*, e1484978. [CrossRef]

195. Freund, E.; Liedtke, K.R.; van der Linde, J.; Metelmann, H.-R.; Heidecke, C.-D.; Partecke, L.-I.; Bekeschus, S. Physical plasma-treated saline promotes an immunogenic phenotype in CT26 colon cancer cells in vitro and in vivo. *Sci. Rep.* **2019**, *9*, 1–18. [CrossRef]

196. Bekeschus, S.; Clemen, R.; Nießner, F.; Sagwal, S.K.; Freund, E.; Schmidt, A. Medical Gas Plasma Jet Technology Targets Murine Melanoma in an Immunogenic Fashion. *Adv. Sci.* **2020**, *7*, 1903438. [CrossRef]

197. Bekeschus, S.; Lippert, M.; Diepold, K.; Chiosis, G.; Seufferlein, T.; Azoitei, N. Physical plasma-triggered ROS induces tumor cell death upon cleavage of HSP90 chaperone. *Sci. Rep.* **2019**, *9*, 1–10. [CrossRef]

198. Bekeschus, S.; Rödder, K.; Fregin, B.; Otto, O.; Lippert, M.; Weltmann, K.-D.; Wende, K.; Schmidt, A.; Gandhirajan, R.K. Toxicity and Immunogenicity in Murine Melanoma following Exposure to Physical Plasma-Derived Oxidants. *Oxid. Med. Cell. Longev.* **2017**, *2017*, 4396467. [CrossRef]

199. Kroemer, G.; Galassi, C.; Zitvogel, L.; Galluzzi, L. Immunogenic cell stress and death. *Nat. Immunol.* **2022**, *23*, 487–500. [CrossRef]

200. Bekeschus, S.; Rödder, K.; Schmidt, A.; Stope, M.B.; von Woedtke, T.; Miller, V.; Fridman, A.; Weltmann, K.-D.; Masur, K.; Metelmann, H.-R.; et al. Cold physical plasma selects for specific T helper cell subsets with distinct cells surface markers in a caspase-dependent and NF-κB-independent manner. *Plasma Process. Polym.* **2016**, *13*, 1144–1150. [CrossRef]

201. Mohamed, H.; Clemen, R.; Freund, E.; Lackmann, J.-W.; Wende, K.; Connors, J.; Haddad, E.K.; Dampier, W.; Wigdahl, B.; Miller, V.; et al. Non-thermal plasma modulates cellular markers associated with immunogenicity in a model of latent HIV-1 infection. *PLoS ONE* **2021**, *16*, e0247125. [CrossRef]

202. Reske, A.; Pollara, G.; Krummenacher, C.; Katz, D.R.; Chain, B.M. Glycoprotein-Dependent and TLR2-Independent Innate Immune Recognition of Herpes Simplex Virus-1 by Dendritic Cells. *J. Immunol.* **2008**, *180*, 7525–7536. [CrossRef]

203. Bedoui, S.; Greyer, M. The role of dendritic cells in immunity against primary herpes simplex virus infections. *Front. Microbiol.* **2014**, *5*, 533. [CrossRef]
204. Tumpey, T.M.; Chen, S.H.; Oakes, J.E.; Lausch, R.N. Neutrophil-mediated suppression of virus replication after herpes simplex virus type 1 infection of the murine cornea. *J. Virol.* **1996**, *70*, 898–904. [CrossRef] [PubMed]
205. Zhang, J.; Liu, H.; Bin Wei, B. Immune response of T cells during herpes simplex virus type 1 (HSV-1) infection. *J. Zhejiang Univ. B* **2017**, *18*, 277–288. [CrossRef] [PubMed]
206. Upasani, V.; Rodenhuis-Zybert, I.; Cantaert, T. Antibody-independent functions of B cells during viral infections. *PLoS Pathog.* **2021**, *17*, e1009708. [CrossRef] [PubMed]

International Journal of
Molecular Sciences

Article

Molecular Mechanisms of Oxidative Stress Relief by CAPE in ARPE−19 Cells

Changjie Ren [1,†], Peiran Zhou [1,†], Mingliang Zhang [1], Zihao Yu [1], Xiaomin Zhang [1], Joyce Tombran-Tink [2], Colin J. Barnstable [2,*] and Xiaorong Li [1,*]

[1]   Tianjin Key Laboratory of Retinal Functions and Diseases, Tianjin Branch of National Clinical Research Center for Ocular Disease, Eye Institute and School of Optometry, Tianjin Medical University Eye Hospital, Tianjin 300384, China
[2]   Department of Neural and Behavioral Sciences, Penn State College of Medicine, Hershey, PA 0850, USA
*   Correspondence: cbarnstable@psu.edu (C.J.B.); lixiaorong@tmu.edu.cn (X.L.)
†   These authors contributed equally to this work.

Abstract: Caffeic acid phenylethyl ester (CAPE) is an antioxidative agent originally derived from propolis. Oxidative stress is a significant pathogenic factor in most retinal diseases. Our previous study revealed that CAPE suppresses mitochondrial ROS production in ARPE−19 cells by regulating UCP2. The present study explores the ability of CAPE to provide longer-term protection to RPE cells and the underlying signal pathways involved. ARPE−19 cells were given CAPE pretreatment followed by t-BHP stimulation. We used in situ live cell staining with CellROX and MitoSOX to measure ROS accumulation; Annexin V-FITC/PI assay to evaluate cell apoptosis; ZO−1 immunostaining to observe tight junction integrity in the cells; RNA-seq to analyze changes in gene expression; q-PCR to validate the RNA-seq data; and Western Blot to examine MAPK signal pathway activation. CAPE significantly reduced both cellular and mitochondria ROS overproduction, restored the loss of ZO−1 expression, and inhibited apoptosis induced by t-BHP stimulation. We also demonstrated that CAPE reverses the overexpression of immediate early genes (IEGs) and activation of the p38-MAPK/CREB signal pathway. Either genetic or chemical deletion of UCP2 largely abolished the protective effects of CAPE. CAPE restrained ROS generation and preserved the tight junction structure of ARPE−19 cells against oxidative stress-induced apoptosis. These effects were mediated via UCP2 regulation of p38/MAPK-CREB-IEGs pathway.

Keywords: CAPE; oxidative stress; ARPE−19 cells; UCP2

Citation: Ren, C.; Zhou, P.; Zhang, M.; Yu, Z.; Zhang, X.; Tombran-Tink, J.; Barnstable, C.J.; Li, X. Molecular Mechanisms of Oxidative Stress Relief by CAPE in ARPE−19 Cells. *Int. J. Mol. Sci.* **2023**, *24*, 3565. https://doi.org/10.3390/ijms24043565

Academic Editors: Rossana Morabito and Alessia Remigante

Received: 16 December 2022
Revised: 20 January 2023
Accepted: 31 January 2023
Published: 10 February 2023

1. Introduction

Oxidative stress is involved in the pathogenesis of diverse retinal diseases, such as age-related macular degeneration (AMD), diabetic retinopathy (DR), glaucoma, retinal vascular occlusion, and inherited retinal diseases [1]. Reactive oxygen species (ROS) derived from molecular oxygen are produced by redox reactions or electronic excitation, which include free radicals, hydrogen peroxide, and oxygen ions from the byproducts of oxygen metabolism [2]. The richness of mitochondria in the retina generates sufficient ATP via oxidative metabolism to maintain its phototransduction and neurotransmission functions [3]. An imbalance between the oxidation and antioxidation systems in the retina can enhance the accumulation of ROS, and excessive levels of ROS lead to oxidative impairment of multiple types of retinal cells including retinal pigment epithelium (RPE) cells [4].

The RPE is a polarized monolayer of cells located between the neural retina and Bruch's membrane [5]. In addition to absorbing light, transferring heat, metabolizing Vitamin A, and phagocytosing photoreceptor outer segments, the RPE also has a major function in transporting nutrients from the choriocapillaris to photoreceptors, which is indispensable for maintaining visual function [6]. Under normal physiological conditions the retina demands a higher oxygen supply. Intensive oxygen metabolism, continual exposure to light, high concentrations of polyunsaturated fatty acids, and the presence of photosensitizers increase ROS production in the retina [7]. The pivotal anatomy location and physiological function render RPE cells particularly vulnerable to oxidative injury.

Mitochondria play a crucial role in energy production through oxidative phosphorylation. Extensive literature supports the idea that mitochondria are the main source of ROS generation [8–10]. The proton gradient across the inner mitochondrial membrane is a key driving force for mitochondrial ROS production, and this gradient can be modulated by members of the mitochondrial uncoupling protein (UCP) family. Among the UCPs, UCP2 is expressed in most tissues and appears to be under the tightest regulation, making it a prime candidate as a therapeutic target to alleviate numerous diseases involving ROS across many tissues, including the eye. Multiple lines of evidence suggest that UCP2 suppresses oxidative stress to protect retinal cells [11–13]. In RPE cells, various factors are able to reduce oxidative damage by enhancing the expression or activity of UCP2. More direct evidence for the role of UCP2 comes from experiments in which the expression of UCP2 was enhanced by the neuroprotective pigment epithelium-derived factor (PEDF) and allowed the RPE cells to better resist oxidative stress [14–16]. Increased expression of UCP2 improved the ability of aged RPE cells to resist oxidative damage, and UCP2 was predicted to be a new protective target for RPE cells in neurodegenerative retinal diseases [17].

Caffeic Acid Phenylethyl Ester (CAPE) was originally defined as a constituent of the propolis of honeybee hives with a broad spectrum of biological properties including anti-pathogenic, anti-carcinogenic, anti-inflammatory, and immunomodulation and wound-healing activities [18]. A growing corpus of evidence suggests CAPE has a beneficial effect in some experimental retinal disease models. CAPE protects retinal neurons and attenuates inflammatory responses in a rat model of optic nerve crush [19]. CAPE treatment also decreases the oxidative stress levels in the retinas of the streptozotocin-induced diabetic rat model [20]. In our previous work, we found that CAPE is a potent regulator of acute oxidative stress. It acts primarily through UCP2 to protect retinal ganglion cells in an ischemia/reperfusion mouse model and resists LPS-induced oxidative damage in the adult retinal pigment epithelium (ARPE−19) cell line [21]. However, the ability of CAPE to provide longer-term protective effects and its underlying mechanism of action, in addition to lowering mitochondrial membrane potential, has not been elucidated yet.

In the present study, we examined the protective role of CAPE in t-BHP-induced oxidative injury in ARPE−19 cells and found that CAPE effectively combats t-BHP stimulation-induced cellular ROS generation for at least 12 h. We further explored the mechanisms of CAPE's activity by analyzing and verifying RNA-seq data and found that CAPE could reverse abnormal expression of IEGs to physiological levels. Moreover, we demonstrated this regulation by CAPE was partly via the UCP2-mediated down-streaming signaling pathway. Based on these data, we propose that CAPE is a promising drug candidate to defend RPE cells against oxidative stress-induced cell death and barrier dysfunction in retinal diseases.

2. Results

2.1. Natural Compounds with Antioxidant Properties Inhibited t-BHP Induced ROS Production in ARPE−19 Cells

To determine whether the actions of CAPE in reducing oxidative stress were unique, we compared the effects of CAPE with a series of structurally related compounds. Chlorogenic Acid (CGA) [22], Trans-Cinnamic Acid (TCA) [23], Isoquercetin (IQT) [24], Curcumin (CUM) [25], and CAPE are natural compounds containing polyphenol structures, all of which have reported antioxidant properties. As shown in Figure 1A, we pretreated ARPE−19 cells with each of the compounds (20 µM) for 1 h, followed by stimulation with 200 µM t-BHP for 2 h, and then determined mitochondrial ROS levels using the MitoSOX assay. In Figure 1B, quantification of the fluorescence intensity of ROS (normalized to the control group) indicated that t-BHP dramatically increased mitochondrial ROS generation

(9.82 $\pm$ 0.66-fold, $p < 0.0001$). Among the compounds tested, only CUM (2.51 $\pm$ 0.62-fold, $p < 0.0001$), and CAPE (0.47 $\pm$ 0.27-fold, $p < 0.0001$) inhibited mitochondrial ROS generation. while CGA, TCA, and IQT did not show significant inhibition.

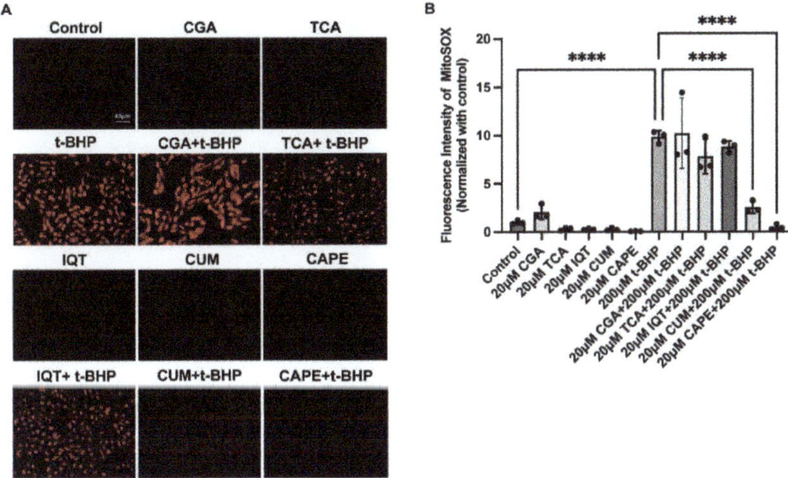

Figure 1. Treatment of ARPE−19 cells with CGA, TCA, IQT, CUM, or CAPE leads to inhibition of t-BHP-induced mitochondrial and cellular ROS generation. (**A**) Immunofluorescence microscopic images of live ARPE−19 cells stained with MitoSOX (red). Scale bar = 40 µm; (**B**) Quantification of MitoSOX immunofluorescence intensity; (**C**). Confocal images of live ARPE−19 cells stained with MitoSOX (red), CellROX (violet), and Hoechst (blue). Scale bar = 10 µm; Quantification of MitoSOX (**D**) or CellROX (**E**) immunofluorescence intensity. **** $p < 0.0001$. The data are presented as the mean $\pm$ SD, normalized to control, $n = 3$.

We next tested whether CAPE blocked the production of both mitochondrial and cellular ROS. Treatment of ARPE−19 cells with 200 µM t-BHP at different time points (2 h, 6 h, 12 h) induced an increase in both mitochondrial (determined using MitoSox) and cellular (determined using CellROX) ROS production in a time-dependent manner (Figure 1C). By 12 h cells were dying, and nuclear staining was apparent. As shown in Figure 1D–E, both mitochondrial ROS (2.58 $\pm$ 0.23, $p < 0.0001$) and cellular ROS (9.72 $\pm$ 1.20-fold, $p < 0.0001$) were significantly increased after 12 h t-BHP stimulation compared to the control. CAPE significantly decreased both t-BHP-induced mitochondrial and cellular ROS production (1.23 $\pm$ 0.13, $p < 0.0001$ and 0.48 $\pm$ 0.01-fold, $p < 0.0001$, respectively) to near control levels.

2.2. CAPE Prevents t-BHP-Induced Apoptosis in ARPE−19 Cells

To test the effects of CAPE on apoptosis we pretreated ARPE-19cells with 20 µM CAPE for 1 h, followed by 200 µM t-BHP stimulation for 12 h (Figure 2). Apoptosis was detected using an Annexin V-FITC/PI apoptosis detection kit and flow cytometry (Figure 2A,B). This assay measured both early apoptotic cells (lower right square) and late apoptotic cells (upper right square). The total percentage of apoptosis cells increased from 1.95 $\pm$ 0.22% (control) to 6.15 $\pm$ 1.03% after t-BHP stimulation, and this was significantly decreased by CAPE pretreatment (3.06 $\pm$ 0.65%). Correspondingly, the percentage of live cells (lower left square) was greater in the CAPE + t-BHP group (96.13 $\pm$ 0.61%) compared with the t-BHP group (91.60 $\pm$ 0.26%). In addition, apoptosis was monitored visually via Annexin V-FITC/PI staining (Figure 2C,D) to identify different cell populations. Positive staining of both early and late apoptotic cells was significantly increased after t-BHP stimulation, but CAPE pretreatment markedly reduced both groups of t-BHP induced apoptosis.

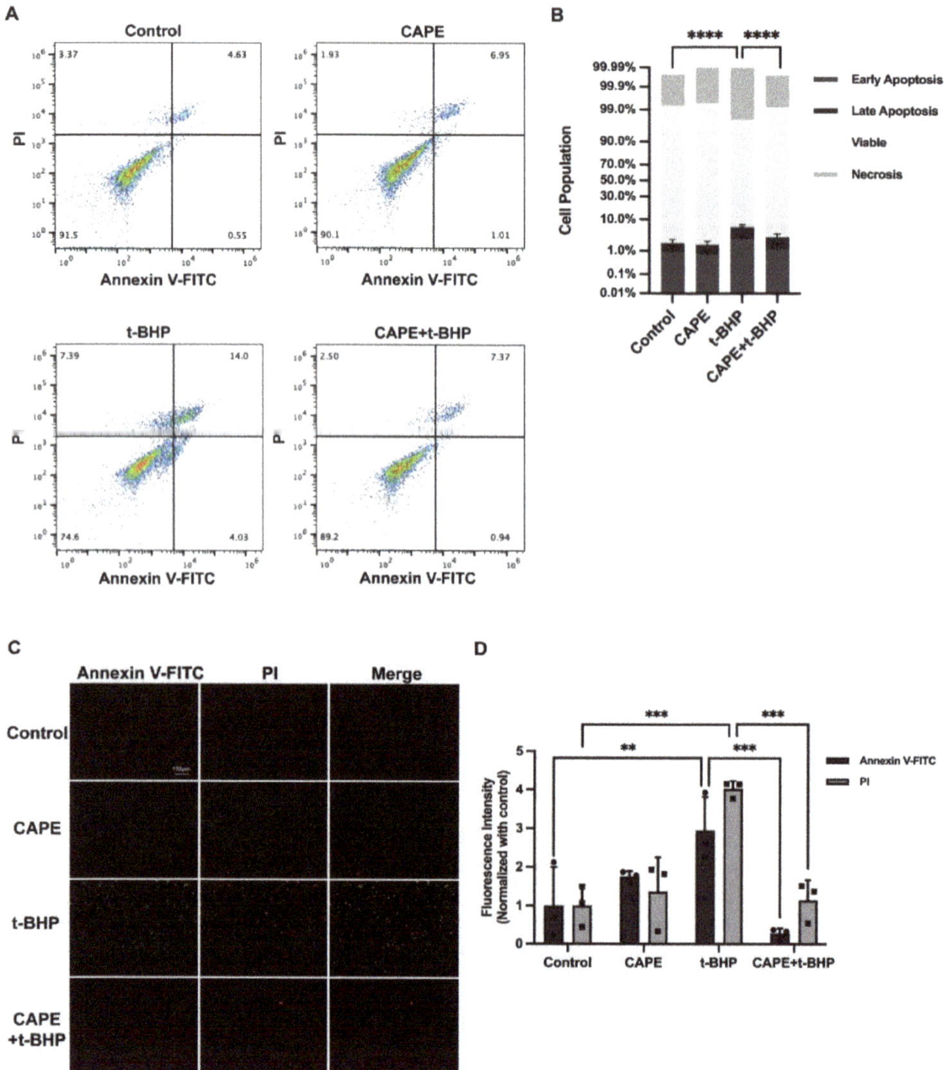

Figure 2. CAPE suppressed apoptosis in ARPE−19 cells exposed to t-BHP stimulation. (**A**) Flow cytometry after staining with both Annexin V-FITC and PI to assess; (**B**) Quantification of cell population in each treatment group. The proportion of non-apoptotic cells (lower left square: Annexin V-FITC$^-$/PI$^-$), early apoptotic cells (lower right square: Annexin V-FITC$^+$/PI$^-$), late apoptotic/necrotic cells (upper right square: Annexin V-FITC$^+$/PI$^+$) and dead cells (upper left square: Annexin V-FITC$^-$/PI$^+$); (**C**) Immunofluorescence microscopic images of ARPE−19 cells double-labeled with Annexin V-FITC (green) and PI (red): live cells (Annexin V-FITC$^-$/PI$^-$), early apoptotic cells (Annexin V-FITC$^+$/PI$^-$), late apoptotic cells (Annexin V-FITC$^+$/PI$^+$), dead cells (Annexin V-FITC$^-$/PI$^+$). Scale bar = 100μm; (**D**) Image quantification of immunofluorescence intensity for Annexin V-FITC/PI. ** $p < 0.01$, *** $p < 0.001$, **** $p < 0.0001$. Values represent the mean $\pm$ SD, normalized with control, $n = 3$.

2.3. CAPE Inhibits t-BHP Induced Disruption of Tight Junction Protein ZO−1 in ARPE−19 Cells

To determine whether CAPE treatment could maintain the barrier functions of RPE cells, we examined the tight junction protein ZO−1. ARPE−19 cells were stimulated by 200 µM t-BHP (2 h, 6 h, or 12 h) with or without CAPE pretreatment, and ZO−1 as an indicator of cell hyperpermeability was evaluated by immunofluorescence (Figure 3A,B). In contrast with the control, t-BHP significantly disturbed the integrity of ZO−1 in a time-dependent manner: the loss of ZO−1 could be observed at 2 h (0.80 ± 0.05-fold, $p < 0.01$), and worsened at 6 h (0.24 ± 0.06-fold, $p < 0.0001$), while at 12 h (0.21 ± 0.02-fold, $p < 0.0001$) almost no ZO−1 positive staining detected. Pretreatment by CAPE, however, preserved ZO−1 expression and morphology (0.85 ± 0.05-fold, $p < 0.01$) against t-BHP injury.

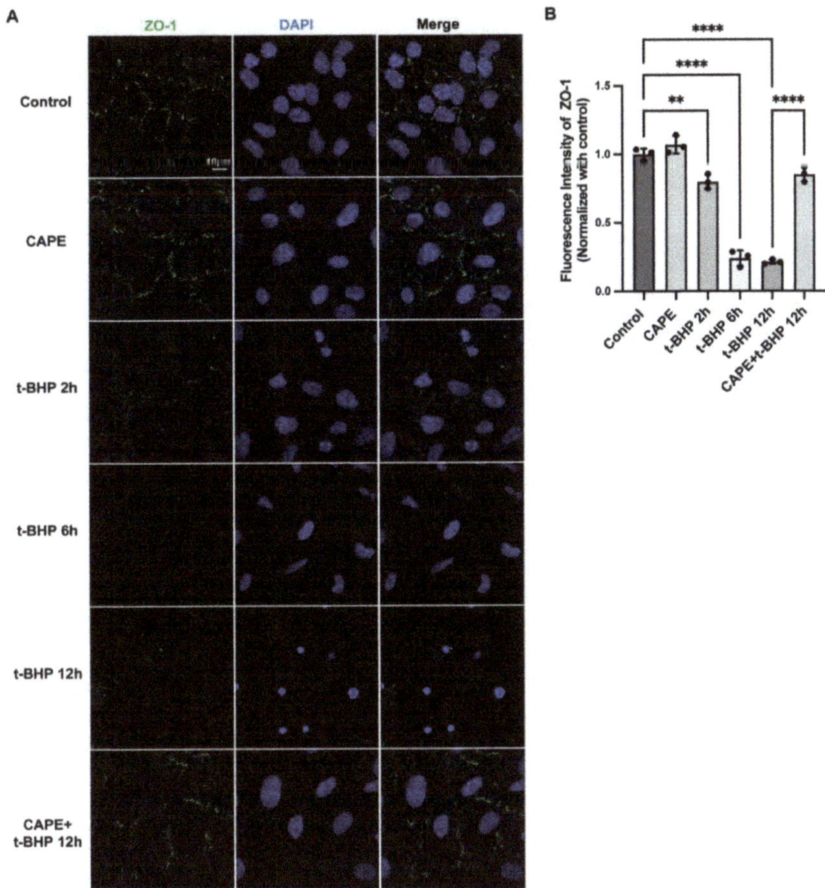

Figure 3. CAPE preserves expression of the ZO−1 tight junction protein in ARPE−19 cells exposed to t-BHP. (**A**) Confocal images of live ARPE−19 cells stained with ZO−1 (green) and DAPI (blue). Scale bar = 10 µm; (**B**) Quantification of immunofluorescence intensity of ZO−1 expression in ARPE−19 cells. ** $p < 0.01$, **** $p < 0.0001$. The data are presented as the mean $\pm$ SD, normalized to control, $n = 3$.

2.4. CAPE Modulates Oxidative Injury induced Gene Expression Profile Changes in ARPE−19 Cells

After establishing that CAPE treatment attenuated t-BHP-induced oxidative injury in ARPE−19 cells, we next studied whether this treatment also showed more widespread effects. To examine this, we carried out RNA-seq on the CAPE pretreated oxidative stress in vitro model of ARPE−19 cells and 17,157 genes were detected in total. We found after t-BHP stimulation that the expression of 6316 genes was altered. With criteria Q value < 0.05, |log2 Fold Change| > 1, 2245 genes were upregulated and 2188 genes downregulated when compared to the control (Figure 4A,C). When the t-BHP group was compared to the CAPE + t-BHP group, there were 2097 genes upregulated and 2077 genes downregulated (Figure 4B,D). Among the differentially expressed genes (DEGs), 2459 showed overlapping expression changes in the t-BHP and CAPE + t-BHP groups. Expressions of 1974 genes were uniquely regulated by t-BHP and 1708 by CAPE + t-BHP compared to controls (Figure 4E). An unbiased cluster analysis of the overlapping 2459 DEGs suggested that multiple genes regulated by t-BHP were reversed in the direction of their expression by CAPE treatment (Figure 4F). The top 20 most extremely changed genes, and the effect of CAPE, are listed (Figure 4G). Although many groups of genes were affected by t-BHP and restored to near normal by CAPE, we focused on a group of immediate early genes (IEGs) as these could potentially have pleiotropic effects on later gene expression changes. Expression levels of ARC, EGR1, FOS, JUNB, JUND, MYC, and ZNF268 were identified in each of the treatment groups compared to controls (Figure 4H). ARC, EGR1, FOS, JUNB, and JUND were significantly upregulated under t-BHP stimulation and showed a negative relationship with CAPE treatment. Interestingly, other IEGs like MYC and ZNF268, t-BHP but not significantly influenced by CAPE pretreatment.

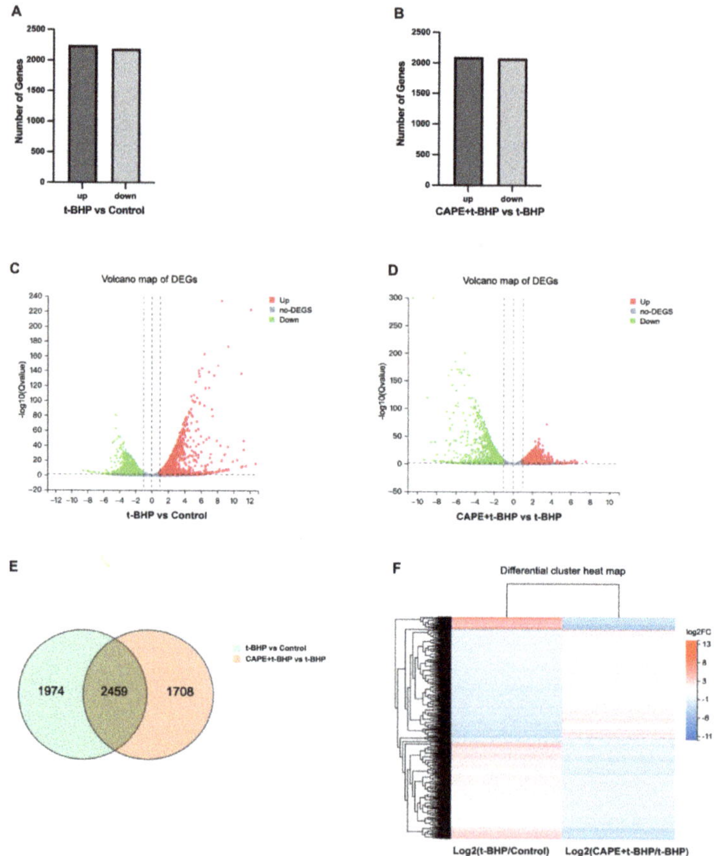

Figure 4. *Cont.*

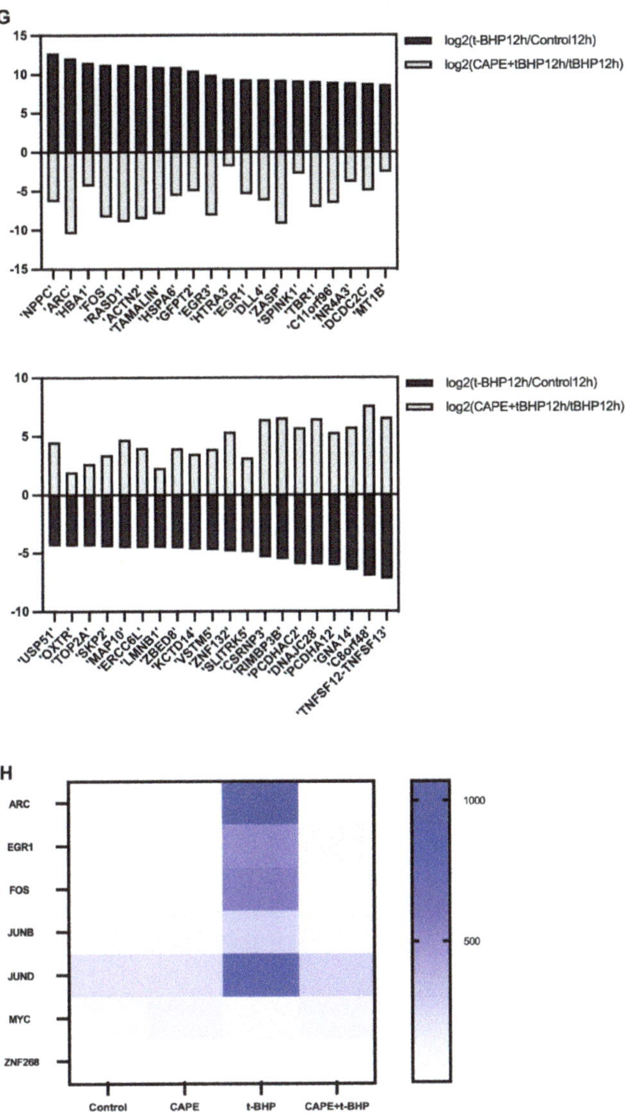

Figure 4. RNA-seq analysis of altered gene expression in CAPE pretreated ARPE−19 cells under t-BHP stimulation. (**A,B**) Overall changes in regulated DEGs after CAPE or t-BHP treatment (Q value < 0.05, |log2 Fold Change| > 1); (**C,D**) Volcano plots of DEGs after CAPE or t-BHP treatment: X-axis: fold change of the difference after conversion to log2, Y-axis: significance after conversion to log10, Red: DEG upregulated, Blue: DEG down-regulated, Gray: non-DEG (Q value < 0.05, |log2 Fold Change| > 1); (**E**) Venn of DEGs among groups. Each circle represents a group of gene sets. Areas superimposed by different circles represent the intersection of gene sets between the treatment groups. The non-overlapping sections of the circles indicate genes that are uniquely regulated in the indicated groups, (Q value < 0.05, |log2 Fold Change| > 1); (**F**) Heatmap of DEGs clusters between t-BHP/Control group and CAPE + t-BHP/t-BHP group (Q value < 0.05, |log2 Fold Change| > 1); (**G**) Top 20 most changed genes in both directions within 2459 overlapped DEGs among groups; (**H**) Heat map of IEGs expression in each group presented by the average transcripts per million (TPM) value.

2.5. GO Enrichment Analyses of DEGs

GO analysis was performed to cluster the DEGs (Figure S1), which were classified into three enriched categories: cellular component (Figure 5) molecular function, and biological process. In the cellular components' enrichment, the top five highest categories in the identified upregulated DEGs in the t-BHP vs. control group were nucleus, centrosome, nucleoplasm, centriole, and mitochondrial matrix (Figure 5A). Correspondingly, DEGs, where CAPE pretreatment blocked upregulation by t-BHP, were also involved in the cellular components categories listed for the t-BHP vs. Control group (Figure 5C). Likewise, the downregulated DEGs in the t-BHP vs. control group enrichment (top five highest categories: nucleus, cytoplasm, nucleoplasm, cytosol, and extracellular exosome) correlated with the upregulated genes in the CAPE + t-BHP vs. t-BHP group (Figure 5B,D). Detailed results of the analysis are reported in Figures S2 and S3. Overall, these results show a widespread ability of CAPE to reverse the effects of the oxidative stressor t-BHP on gene expression.

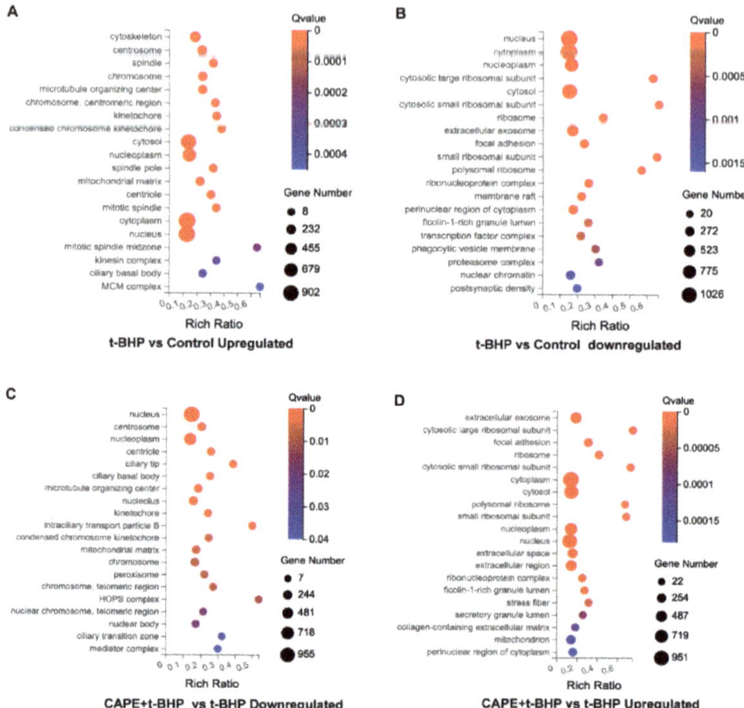

Figure 5. The top 20 GO enrichment of DEGs in cellular components. Upregulated (**A**) or Downregulated (**B**) t-BHP vs. control enrichment; Downregulated (**C**) or Upregulated (**D**) CAPE + t-BHP vs. t-BHP. *X*-axis: enrichment ratio (number of genes annotated to an entry in the selected gene set to the total number of genes annotated to the entry in the species, calculated as Rich Ratio = Term Candidate Gene Num/Term Gene Num). Size of circles: the number of DEGs annotated to a GO Term. Circle color: enriched significance. The redder the color, the smaller the significance value. Q-value $\leq$ 0.05 are considered as significant enrichment.

2.6. qPCR Validation of Changes in the Immediate Early Genes (IEGs) in the RNA-seq Data

Since IEGs can have substantial downstream effects, we investigated these further. We first confirmed the selected IEG results of the RNA-seq experiments using qPCR. The stimulation of ARPE−19 cells with t-BHP (200 µM) increases expression of almost all IEGs, including ARC (8141 ± 826.7-fold, $p < 0.0001$), EGR1 (539.8 ± 64.61-fold, $p < 0.0001$), FOS (1231 ± 241.5-fold, $p < 0.0001$), JUNB (13.24 ± 1.810-fold, $p < 0.0001$), and JUND

(10.43 ± 1.391-fold, $p < 0.0001$). The expressions of these were all reversed by CAPE to near control levels (Figure 6A–E). It is worth noting that the levels of expression of several of these IEGs were reversed, by several to even thousands of folds. However, not all IEGs were regulated by CAPE. For example, the mRNA levels of MYC (3.364 ± 0.4077-fold, $p < 0.001$) and ZNF268 (2.240 ± 0.311-fold, $p < 0.001$) were upregulated after oxidative stress in the RPE cells, but CAPE did not reverse their expression levels (Figure 6F,G).

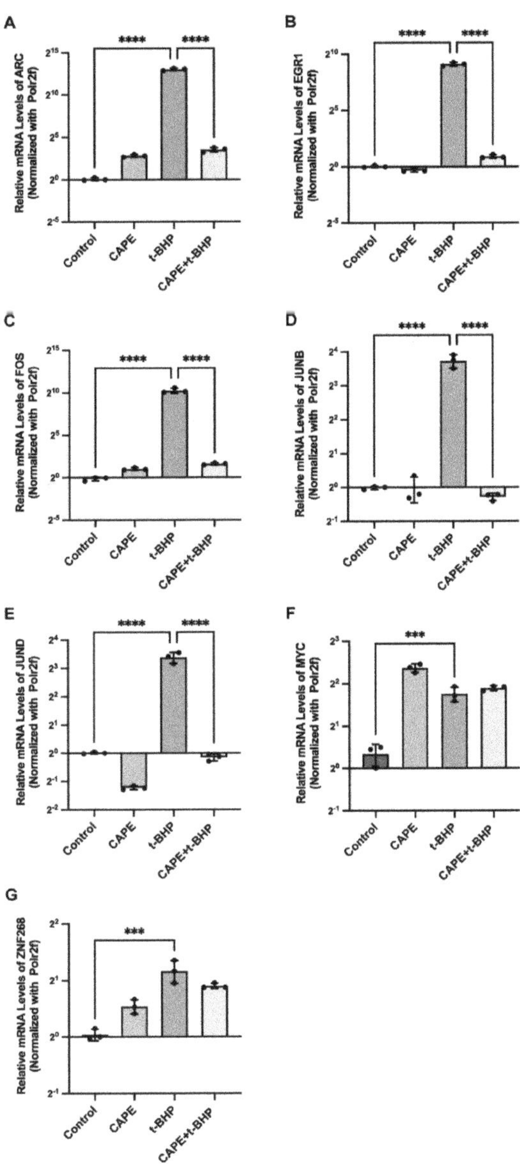

Figure 6. Pretreatment of ARPE−19 cells with CAPE before exposure to oxidative stress preserves normal expression of several IEGs. Relative mRNA expression levels of ARC (**A**), EGR1 (**B**), FOS (**C**), JUNB (**D**), JUND (**E**), MYC (**F**), ZNF268 (**G**) with each treatment. Experiments were repeated 3 times. IEGs expression was calculated by the $2^{-\triangle\triangle Ct}$ method and normalized to the internal reference Polr2f. *** $p < 0.001$, **** $p < 0.0001$.

2.7. Effects of CAPE on t-BHP-Induced Expression of IEGs

We also analyzed the time course of IEGs' expression after treatment with CAPE. Upregulation in expression of ARC, EGR1, FOS, or JUNB was detected at 4 h after exposure to t-BHP, peaking at 12 h, with expression differences among the IEGs (Figure 7). The effect of CAPE, however, was not immediate in reversing the t-BHP effect on these genes and only did so by 8 h and 12 h. Expression of JUND showed significant increases between 2 h−24 h in the oxidative stress environment with a peak at 4 h. Unlike the other IEGs, CAPE inhibited t-BHP-induced JUND overexpression as early as the 2 h time point.

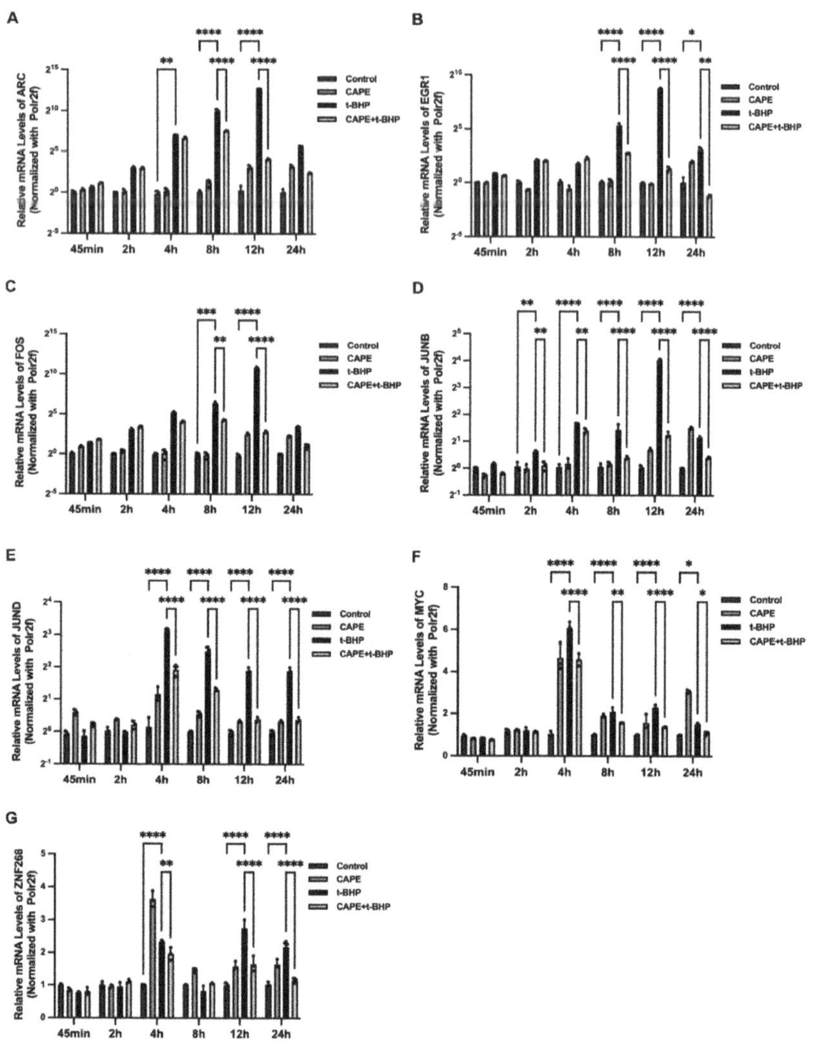

Figure 7. Changes in expression of IEGs correlate with CAPE pretreatment under varying t-BHP stimulation time. Relative mRNA expression levels of ARC (**A**), EGR1 (**B**), FOS (**C**), JUNB (**D**), JUND (**E**), MYC (**F**), ZNF268 (**G**) over 24 h. Experiments were repeated 3 times. IEGs expression was calculated by the $2^{-\triangle\triangle Ct}$ method and normalized to Polr2f. * $p < 0.05$, ** $p < 0.01$, *** $p < 0.001$, **** $p < 0.0001$.

2.8. UCP2 Mediates CAPE Regulation of IEGs

In previous studies, we showed that CAPE reduces acute ROS production by activating on mitochondrial uncoupling protein UCP2. To determine whether UCP2 modulated CAPE effects on IEGs expression we used both UCP2 siRNA and Genipin, a pharmacological inhibitor of UCP2, to inhibit UCP2 expression or activation. After UCP2 siRNA transfection, expression of UCP2 mRNA was significantly inhibited, and CAPE effects on balancing ARC, EGR1, FOS, JUNB, and JUND expression during oxidative stress were aborted (Figure 8A–E). A similar trend was observed when UCP2 was pharmacologically inhibited by Genipin (Figure 9A–E). Both methods suggest that the actions of CAPE on IEG expression are mediated by UCP2 (Figures 8F and 9F).

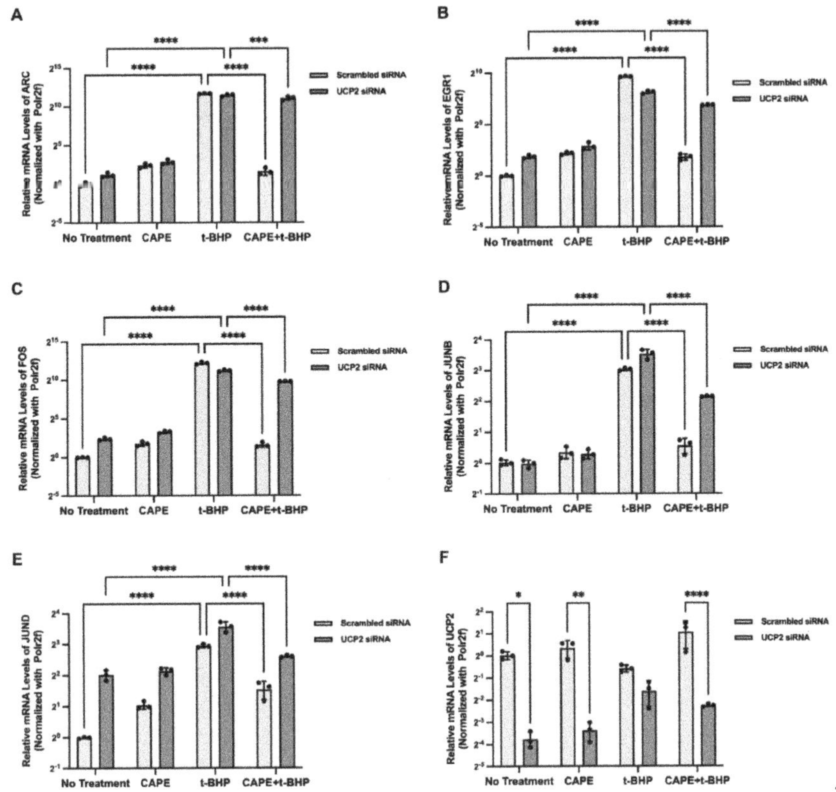

Figure 8. IEGs expression in ARPE−19 cells after UCP2 siRNA transfection. Relative mRNA expression levels of ARC (**A**), EGR1(**B**), FOS (**C**), JUNB (**D**), and JUND (**E**) in each group after 12 h t-BHP stimulation. The relative UCP2 mRNA expression levels among groups (**F**). Experiments were repeated 3 times. Relative mRNA expression was calculated using the $2^{-\triangle\triangle Ct}$ method and normalized to Polr2f. * $p < 0.05$, ** $p < 0.01$, *** $p < 0.001$, **** $p < 0.0001$.

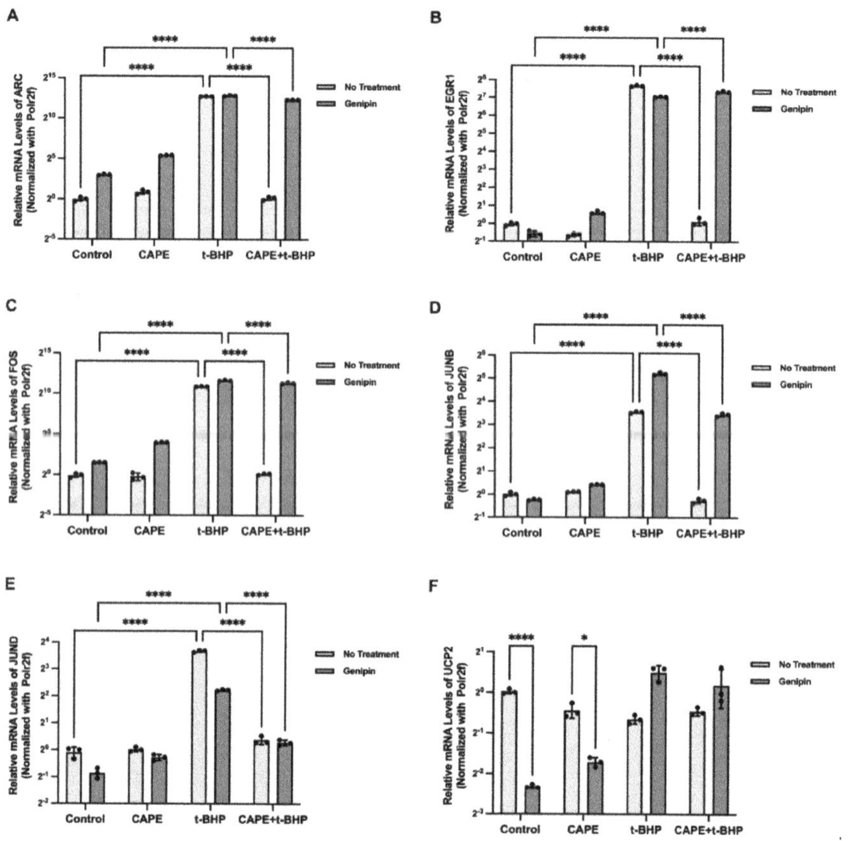

Figure 9. IEGs expression in ARPE−19 cells with Genipin treatment. The relative mRNA expression levels of ARC (**A**), EGR1(**B**), FOS (**C**), JUNB (**D**), and JUND (**E**) in each group after 12h t-BHP stimulation. The relative UCP2 mRNA expression levels among groups (**F**). Experiments were repeated 3 times. Relative mRNA expression was calculated by the $2^{-\triangle\triangle Ct}$ method and normalized to Polr2f. * $p < 0.05$, **** $p < 0.0001$.

2.9. UCP2 Deficiency Interrupts the Inhibitory Actions of CAPE on p38 MAPK/CREB Activation during Oxidative Stress

To investigate mechanisms transmitting UCP2 signals on IEG expression we examined control by the p38MAPK/CREB pathway which is intimately associated with IEGs. The phosphorylation levels of p38 and CREB were significantly upregulated in the oxidative stress environment, and activation of this pathway was inhibited by CAPE (Figure 10A,B). Conversely, knockdown or pharmacological inhibition of UCP2 expression blocked the inhibitory effects of CAPE on p38/MAPK-CREB activation (Figure 10C–F).

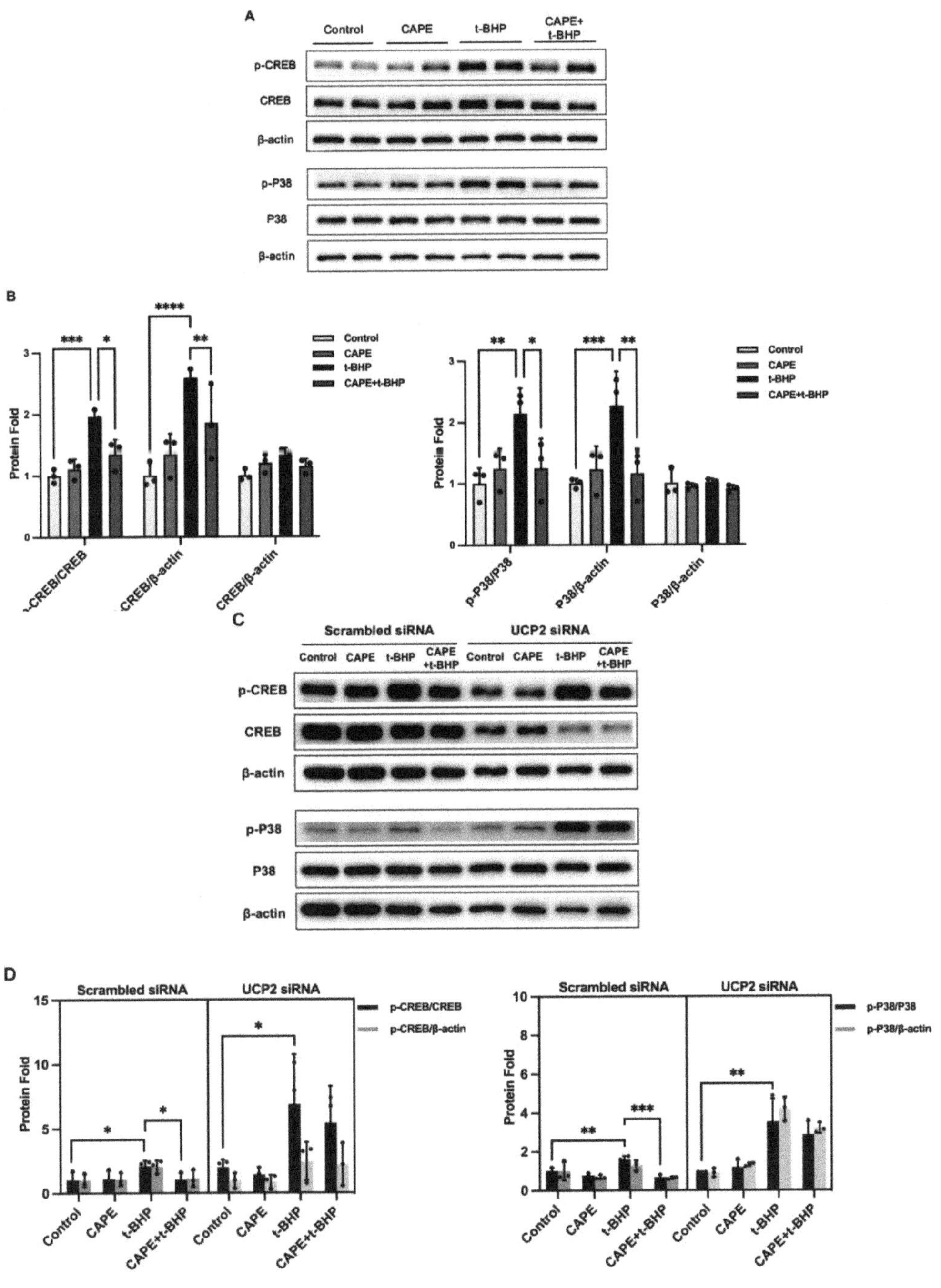

Figure 10. *Cont.*

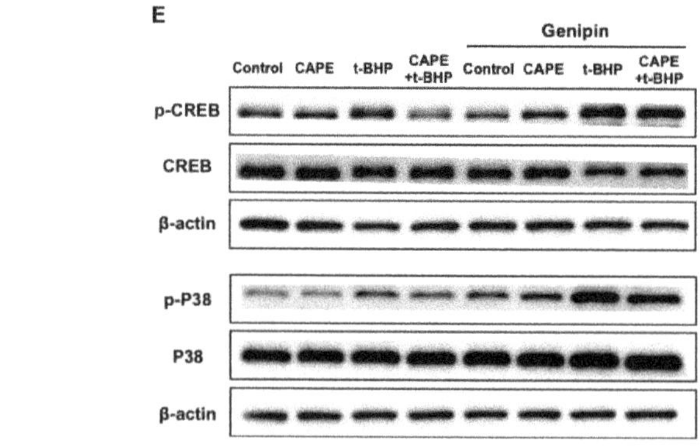

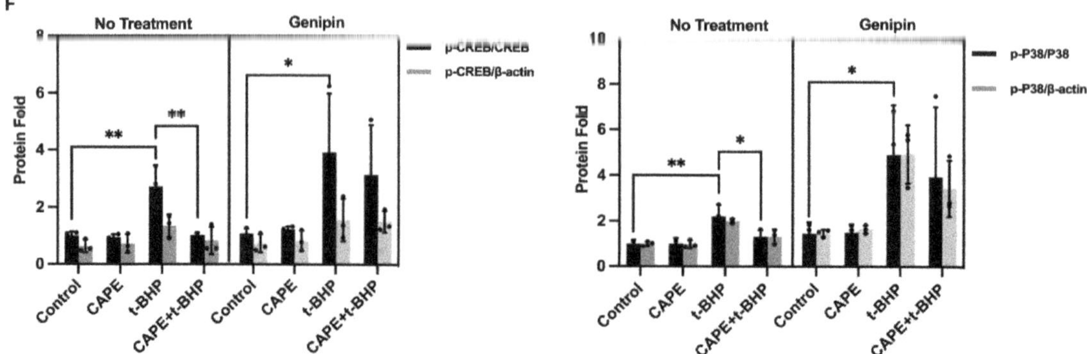

Figure 10. p38 MAPK/CREB signaling pathway is controlled by CAPE and during oxidative stress. p38 MAPK and CREB activation after ARPE−19 cells were treated with CAPE after exposure to t-BHP(**A**), UCP2 siRNA (**C**) Genipin (**E**). Densitometric quantification of relative protein (**B,D,F**). Experiments were repeated 3 times. β-actin was used as a loading reference. * $p < 0.05$, ** $p < 0.01$, *** $p < 0.001$, **** $p < 0.0001$.

3. Discussion

Oxidative stress is closely linked to several neurodegenerative ophthalmic diseases, including AMD, glaucoma, and diabetic retinopathy [26]. Increased oxidative stress and decreased antioxidant levels may contribute synergistically to disease development. As a tissue with high metabolic demand and ROS production, retinal cells, especially the RPE cells, are susceptible to oxidative insults [27]. The melanin pigment granules in RPE cells are the primary sites of light energy absorption, and the RPE cells are in close proximity to the highly oxygen-containing choroidal blood supply, which makes them particularly vulnerable to oxidative stress. Development of therapeutics that reduce oxidative insults to halt or delay retinal disease progression are therefore intensely pursued.

Compounds with antioxidant properties can mitigate the effects of oxidative insults [28]. CAPE, a major ingredient of propolis, and is responsible for most of its antioxidant benefits [29]. Polyphenols with hydroxyl groups in the catechol ring of CAPE provide strong antioxidant properties and affect a range of biological pathways [30,31], which has been demonstrated in many cell types against oxidative stimulus [32–34]. Among a series of structural mimetics, we found that CAPE had the greatest impact on the generation of mitochondrial ROS. Though unlikely, we cannot exclude the possibility that the weak activity seen with some of the compounds tested was unique to the cell type tested. CAPE was able to inhibit production of both cellular and mitochondrial ROS generation during oxidative stress induced by t-BHP (Figure 1). This study, as a confirmation and extension of our earlier results [21], suggests that CAPE has actions in addition to the general anti-oxidant properties of its polyphenol structure.

ROS production is thought to trigger a series of biochemical events that ultimately contribute to apoptosis [35]. Such cell death in the RPE may be critical in the development of ARMD. In neurons, CAPE diminishes apoptosis dysregulation to preserve the number of neurons by suppressing inflammation, mitochondrial cytochrome c release, and ROS production [31]. In an optic nerve crush rat model, attenuated inflammatory responses in Müller cells can protect retinal ganglion cells from death [19]. CAPE treatment decreases oxidative stress in retinas of the STZ-induced diabetic rat model [20], protects 661 W cells from oxidative injury-induced cell death by promoting expression of the antioxidative marker HO−1, and enhances electroretinography responses in dim-reared albino rats [36]. Similarly, in our study CAPE prevented oxidative stress-induced apoptosis in ARPE−19 cells as measured by Annexin V-FITC/PI labeling (Figure 2).

Loss of the junction protein ZO−1, and subsequent breakdown of the blood–retinal barrier, are found in a number of retinal diseases, particularly diabetic retinopathy [37]. CAPE reversed the oxidative injury-induced ZO−1 loss in a time-dependent manner and restored ARPE−19 cell morphology and tight junction structure, which suggests that CAPE is likely to preserve the barrier function of the RPE, and probably in the retinal vascular network where junction disassembly and leakage are features of retinal diseases (Figure 3). Thus, the above result points to a promising role for CAPE in retinal neuroprotection by maintaining RPE layer integrity.

To elucidate downstream molecular mechanisms of CAPE, we compared global ARPE−19 gene expression in the oxidative stress versus CAPE treatment and control groups and identified a substantial number of genes whose expression was regulated by oxidative stress and reversed by CAPE back to near control levels (Figure 4). An important cluster of genes responding in this way includes a group of IEGs (ARC, EGR1, FOS, JUNB, JUND) as validated in qPCR (Figure 5). IEGs are a group of genes expressed early in response to distinct extracellular stimuli such as growth factors, cytokines, tumor promoters, hormones, and stress. These include the AP−1 gene family (FOS and JUN), EGR, and MYC. Most of these encode transcription factors that regulate genes involved in normal cell growth and differentiation, intracellular information transmission, and energy metabolism [38]. ARC, EGR1, and FOS showed a large quantitative response (up to hundreds or thousands-fold elevation) when exposed to oxidative insults, while JUNB and JUND transcription induction showed smaller increases (up to ~10-fold elevation) over control levels then showed a declining pattern with increased exposure time. These results suggest that IEGs' regulation is rapid and large in response to oxidative stress but refractory with increased exposure time, while the effect of CAPE is to reverse these changes. Other well-characterized IEGs like ZNF268 and MYC were increased by oxidative stress, but not regulated by CAPE.

IEGs are known to be critical for the structure, function, and plasticity of both excitatory and inhibitory synapses across various cell types. Distinct IEGs are influenced in a pathway- and stimulus-specific manner on synapse development and plasticity to regulate the function of neurons [39]. For example, hyperglycemia induces abnormal IEG expression and is involved in prothrombotic and proinflammatory pathways in the retina vasculature [40], inflammatory marker expression in microglia, and astrocytes in STZ diabetic mouse retinas [41]. Involvement of IEGs in retinal degenerative conditions may lend to development of new and unique therapies. Transcripts of IEGs are rapidly induced in response to acute stress or proliferation-inducing signals. As transcription factors, rapid IEGs' expression responds to cellular stimulation and participates in cellular signaling pathways. The protein products of IEGs can form dimers and regulate apoptosis in both nonneuronal and neuronal cells. In this study, we report a selective and abnormally sustained induction of IEGs in cultured ARPE−19 cells under oxidative insults. In addition, we describe the lack of induction of IEGs after CAPE pretreatment that is relatively resistant to apoptosis. As has been suggested by others, a prolonged time course of IEGs' expression in this instance would allow for the formation of heterodimers with a series of successively transcribed gene products, potentially controlling a complex set of regulatory events for multiple target genes [42].

IEG activation is highly complex and involves various promoter elements and transcription factors. Serum response-(SREs) and cAMP-response elements (CREs) are typically found in the promoters of IEGs. The cAMP-responsive element-binding protein (CREB) has been the most studied in regulating neuronal behaviors. The activation of IEGs promoters occurs via phosphorylation of mitogen-activated protein kinase (MAPK), protein kinase A (PKA), RhoA-actin, phosphor-inositide 3 kinase (PI3K), nuclear factor-kappa B (NF-kB) or other stress-induced kinases [43]. Generally, inducible IEG expression is triggered by MAPK cascades: the RAS-MAPK pathway, which promotes ERK1/2 activation by growth factors and mitogens, or the p38-MAPK pathway, which is activated by stress. Enrichment analysis of DEGs showed multiple cellular components, biological processes, and molecular functions were under the regulation of oxidative stress and CAPE in opposite directions. For instance, the abnormal MAPK tyrosine phosphorylase activity after oxidative injury was reversed by CAPE (Figure S2).

We recently reported that a key action of CAPE on ROS production was via activation of mitochondrial UCP2 [21], the main source of intrinsic ROS generation. It is possible to achieve the optimal balance between energy production and ROS production by carefully titrating UCP2 activity using a wide variety of regulatory compounds [44]. We now show that oxidative insults promote phosphorylation of p38-MAPK and CREB, while CAPE suppressed their activation and that inhibition of UCP2 expression or activity aborts CAPE's effect on this signal pathway. Our finding agrees with previous findings that oxidative stress-induced RPE cell injury is associated with the p38-MAPK signaling pathway [45]. Proof of principle experiments in preclinical models have shown the neuroprotective effects of p38-MAPK inhibitors, and UCP2 is a negative regulator of oxidative insult via p38-MAPK signaling networks [46,47]. These results denoted the mechanism of RPE vulnerability to oxidative injury and CAPE increasing its survival.

At present we do not know whether the activity of UCP2 controlling ROS levels is sufficient to regulate p38-MAPK, CREB, and transcription of IEGs and other genes, or whether UCP2 activity is also linked to other mitochondrial pathways. Understanding these mechanisms to target IEGs expression and function remains an untapped area in retinal neurodegenerative diseases [48]. Our results further support the hypothesis that UCP2 is a key molecule regulating oxidative stress-induced injury in ARPE−19 cells, and further studies may lead to new treatments for retinal neurodegenerative diseases.

4. Materials and Methods

4.1. Antibodies and Reagents

Chemicals. CGA, TCA, IQT, and CUM were purchased from Thermo Fisher Scientific (Waltham, MA, USA) unless otherwise noted. The tert-Butyl hydroperoxide solution (t-BHP) was obtained from Sigma-Aldrich (St. Louis, MO, USA). Caffeic acid phenethyl ester (CAPE) was purchased from Santa Cruz Biotechnology (Dallas, TX, USA). UCP2 inhibitor Genipin was obtained from MedChem Express (Monmouth Junction, NJ, USA). The primary antibodies used were as follows: CREB, Phospho-CREB, p38, Phospho-p38, β-actin (Cell Signaling Technology, Danvers, MA, USA); ZO−1 (Proteintech, Rosemont, IL, USA). Information on other reagents and antibodies is shown in Supplementary Table S1.

4.2. Human Retinal Pigment Epithelial (ARPE−19) Cells Culture and Treatment

ARPE−19 cells were purchased from the American Type Culture Collection (Manassas, VA, USA), and cultured in T25 flasks using 10%FBS, 1% penicillin-streptomycin(p/S) solution in DMEM/F12 basic culture medium (Gibco, Grand Island, NY, USA) at 37 °C and 5% CO_2.

In all experiments, ARPE−19 cells used were 4–10 passages. Cells were seeded in 6-well (2×10^5 cells/well), 12-well (1×10^5 cells/well), or 96-well (5×10^3 cells/well) plate. After 12 h, the confluence reached 80–90%, and cells were washed with DPBS (Gibco, Grand Island, NY, USA). Cells were starved in Fetal Bovine Serum (FBS)-free DMEM/F12 (without phenol red) culture medium (Gibco, Grand Island, NY, USA) for 6 h. CAPE or t-BHP was also diluted in FBS-free DMEM/F12 (without phenol red) culture medium. ARPE−19 cells were pretreated with 200 μM Genepin for 2 h. After 2 h 20 μM CAPE treatment, ARPE−19 cells were stimulated with 200 μM t-BHP accordingly. Other reagents (CGA, TCA, IQT, and CUM) were diluted and applied in the same way as CAPE.

4.3. RNA-Sequencing

ARPE19 cells were treated with 20 μM CAPE for 2 h and subsequently stimulated with 200 μM t-BHP for 12 h. RNA was extracted and purified with EZ-press RNA Purification Kit (EZBioscience, Roseville, MN, USA) following the manufacturer's instructions. After analyzing RNA sample quality, the cDNA libraries were synthesized and sequenced by MGI Tech (Shenzhen, China) with the BGISEQ platform. In brief, the quality of the RNA samples was assessed by an Agilent Bioanalyzer (Agilent, Santa Clara, CA, USA). Single-strand circular DNA libraries were prepared with MGI easy PCR free library preparation kit (MGI Tech, Shenzhen, China) to generate single-strand long fragment DNA and form a DNA nano ball (DNB) structure with rolling circle amplification (RCA) technology. Each DNB was loaded on flowcell (FCL) and sequenced using MGI-seq 2000 (MGI Tech, Shenzhen, China) with combinatorial prober anchor synthesis (cPAS) technique to obtain Fastq data.

All samples were tested using the DNBSEQ platform. This project used the filtering software SOAPnuke (v1.5.2) developed by BGI independently for filtering with the "Parameters -l 15 -q 0.2 -*n* 0.05" option. After quality control, the filtered clean reads were aligned to the reference sequence (Homo_sapiens, NCBI, Reference Genome Version: GCF_.39_GRCh38.p13). Hierarchical Indexing for Spliced Alignment of Transcripts (HISAT2, v2.0.4) software was used for mapping RNA-seq reads. Perl script was used for sequencing saturation analysis. For expression quantification, we used Bowtie2 (v2.2.5) to map the clean reads to the reference gene sequence (transcriptome), and then used RSEM (v1.2.8) to calculate the gene expression level of each sample. Each group has two replicate samples. The PossionDis method (Parameters: Fold Change >= 2 and FDR <= 0.001) was applied to screen differentially expressed genes between two samples, and the DEseq2 method (Parameters: Q value or Adjusted *p*-value <= 0.05) was applied to describe the genes detection of different samples with a different expression. According to the results of differential gene detection, the R package heatmap was used to perform hierarchical clustering analysis on the union set differential genes. The average transcripts per million (TPM) value of each group was counted for further analysis. The significant levels of terms and pathways were corrected by FDR-adjusted *p*-value (q-value), with a rigorous threshold (q $\leq$ 0.05) by Bonferroni as previously described [49]. All analyses were performed on the Dr. Tom analysis system constructed by MGI.

GO enrichment analysis includes three categories: molecular function, cellular component, and biological process. Functional classification is conducted based on DEGs test results. There are sub-categories for each level under each major category. GO annotation of DEGs is classified and mapped. According to the GO annotation results and the official classification, functionally classify the DEGs. At the same time, the phyper function in R software is used for enrichment analysis, calculating the *p*-value, and the Q-value was obtained by correction of the *p*-value. Generally, functions with Q-value <= 0.05 are considered significant enrichment.

4.4. Quantitative Real-Time PCR (qPCR)

Cells were pretreated with or without 20 μM CAPE for 2 h and stimulated with 200 μM t-BHP for 45 min, 2 h, 4 h, 8 h, 12 h, and 24 h in sequence. After RNA extraction and purification, final RNA concentrations were determined using a NanoDrop 2000 Spectrophotometer (Thermo Fisher Scientific, Waltham, MA, USA). The cDNA was synthesized with Reverse Transcription Master Mix (EZBioscience, Roseville, MN, USA) according to the manufacturer's protocol. Primers were purchased from Sangon Biotech (Shanghai, China). The sequence information is listed in Supplementary Table S2. For quantitative real-time PCR we used EZ-press One Step qRT-PCR Kit (EZBioscience, Roseville, MN, USA). Samples in triplicate were run in 384-well plates on Light Cycler 480 II System (Roche, Basel, Switzerland). The relative expression level for each gene was calculated by the $2^{-\Delta\Delta Ct}$ method and normalized to Polr2f. Genes were considered upregulated or downregulated if $p < 0.05$.

4.5. Western Blot

After treatment, culture medium was aspirated and cells washed with 1 × PBS (Solarbio, Beijing, China). Cells were lysed by adding 100 µL RIPA lysis buffer containing 1% proteinase inhibitor (Target Mol, Shanghai, China) and 1% phosphatase inhibitor (Beyotime, Shanghai, China) per well of 6-well plate. Cells were immediately scraped off the plate and transferred to a microcentrifuge tube. Tubes were placed on ice and samples sonicated for 10–15 s to complete cell lysis. Samples were centrifuged at 12,000× g for 20 min, and the supernatants were collected. Protein was quantified by a BCA kit (Solarbio, Beijing, China) according to the manufacturer's instructions. An amount of 20 ul 5×SDS-PAGE loading buffer (Solarbio, Beijing, China) was added to 80 ul sample supernatant per tube and boiled at 100 °C for 5 min, then 20 ug protein of each sample was loaded, electrophoresed, and transferred onto PVDF (Roche, Basel, Switzerland) membranes, and blocked with QickBlock™ Western Block Buffer (Beyotime, Shanghai, China) at room temperature (RT) for 15 min. The primary antibodies p38 (1:1000), p-p38 (1:1000), CREB (1:1000), p-CREB (1:1000), or β-actin (1:2000) were incubated overnight at 4 °C. The membranes were washed in 1 × TBST buffer. After which, the membranes were incubated with horseradish peroxidase (HRP)-linked second antibody (1:2000) for 2 h at RT. Washing membranes after incubation and adding enhanced chemiluminescence (ECL) reagents to detect immunoblots. The reactions were visualized on Tanon 4800 (Tanon, Shanghai, China) and semi-quantified by densitometry measurement obtained using ImageJ (Version 2.1.0, National Institutes of Health, Bethesda, MD, USA) with normalization to internal control β-actin.

4.6. Transient Transfection

ARPE19 cells were transfected with UCP2 siRNA or Scrambled siRNA (Santa Cruz Biotechnology, Dallas, TX, USA) at 70–80% confluence by using the jetPRIME (Polyplus Transfection, Illkirch, France) in accordance with the manufacturer's protocol. Briefly, 200 µL jetPRIME buffer containing 50 nM siRNA and 10 µL jetPRIME reagent was added to one well in a 6-well plate and transfected for 36 h. The silencing efficiency was evidenced by qPCR. The detailed siRNA information is provided in Supplementary Table S3.

4.7. Immunolabeling

Cells were seeded in a 35 mm glass bottom dish (Cellvis, Mountain View, CA, USA) at 80–90% confluence. After treatment, cells were washed with 37 °C pre-warmed HBSS. Cells were fixed with 4% PFA for 15 min at 37 °C, followed by rinsing 3 times with 1 × PBS for 5 min each. Cells were permeabilized with 0.2% TritonX−100 in 1 × PBS for 10 min at 37 °C, followed by rinsing 3 times with 1 × PBS for 3 min each. Cells were incubated with 5% goat serum in 1×PBS blocking buffer at room temperature for 1 h. Primary antibody ZO−1(1:2000) was diluted in 1%BSA in 1 × PBS antibody dilution buffer and incubated at 4 °C overnight. After washing 3 times with 1 × PBS for 5 min each, samples were incubated with fluorochrome-labeled goat anti-rabbit secondary antibody (1:500) (Alexa Fluor 488, Abcam, Boston, MA, USA) at RT for 2 h. After washing 3 times with 1 × PBS for 5 min each, samples were mounted with Fluoroshield™ with DAPI (Sigma-Aldrich, St. Louis, MO, USA). Images were taken and processed using a confocal microscope (Zeiss, Baden-Württemberg, Germany).

4.8. Reactive Oxygen Species (ROS) Determination

To measure ROS level in cells, we detected mitochondrial ROS by applying MitoSOX™ Red mitochondrial superoxide indicator (Invitrogen, Carlsbad, CA, USA) at ~510/580 nm, and detected cellular oxidative stress with CellROX™ Deep Red Reagent (Invitrogen, Carlsbad, CA, USA) at ~644/665 nm in live cells according to the manuals. Briefly, after treatment we loaded 5μM MitoSOX™ working solution to cover cells and incubated for 10 min at 37 °C, protected from light. Similarly, for CellROX™ we loaded a 5μM working solution and incubated cells for 30 min at 37 °C, protected from light. Counterstained nuclei with Hochest 33,342 (Beyotime, Shanghai, China) for 10 min at 37 °C, protected from light. We mounted cells in warm buffer for live cell imaging with confocal microscope (Zeiss, Baden-Württemberg, Germany).

4.9. Analysis of Apoptosis

Apoptotic ARPE−19 cells were identified by staining with Annexin V-fluorescein isothiocyanate (FITC) and Propidium iodide (PI) using Annexin V-FITC/PI Apoptosis Detection Kit (Beyotime, Shanghai, China). After stimulation, cells were treated with 100 μL of binding buffer containing 1 μL of Annexin V-FITC and 10 μg/mL PI for 15 min in the dark, after which 400 μL of binding buffer was added and subjected to flow cytometry; Or, after staining directly observe the fluorescence under microscope (Zeiss, Baden-Württemberg, Germany) according to the manufacturer's manual. Flow cytometry analysis was performed using the FACSCanto II cytometer (BD FACSCanto II, San Diego, CA, USA). Analysis of data was performed using the FACSDiva software (BD, San Diego, CA, USA).

4.10. Statistical Analysis

Statistical comparisons were all conducted in Prism 9 software (Version 9.3.1, Graph-Pad Software, La Jolla, CA, USA). All data in the study were expressed as mean ± standard deviation (SD). The Shapiro–Wilk test was performed to verify the normal distribution of variables. Data were evaluated for statistical differences amongst groups using student's 2 tailed-t-test and a two-way analysis of variance (ANOVA) followed by Bonferroni's post hoc comparison. The p values < 0.05 were considered statistically significant.

5. Conclusions

In conclusion, our results reveal a potent protective role of CAPE against oxidative stress-induced apoptosis in ARPE−19 cells and show that this effect was mediated by UCP2 via regulating CREB/p38-MAPK/IEGs signal. Since systemic application of CAPE has been shown to prevent ischemia-reperfusion damage to retinal ganglion cells in mice, this compound is likely to be a potent protective agent for a variety of retinal degenerative diseases and may have strong clinical potential as a therapeutic agent.

Supplementary Materials: The following supporting information can be downloaded at: https://www.mdpi.com/article/10.3390/ijms24043565/s1.

Author Contributions: Conceptualization, C.J.B., M.Z. and C.R.; methodology, C.J.B., M.Z., J.T.-T. and C.R.; validation, C.R.; formal analysis, C.R., P.Z., and Z.Y.; writing—original draft preparation, C.R.; writing—review and editing, C.J.B., J.T.-T.; supervision, M.Z., X.Z., C.J.B. and X.L.; funding acquisition, M.Z., X.Z. and X.L. All authors have read and agreed to the published version of the manuscript.

Funding: This work was supported by the National Natural Science Foundation of China (81900894, 82171085, 81870675), the Tianjin Postgraduate Research Innovation Project (2021YJSB271), Independent and open project of Tianjin Key Laboratory of retinal function and disease (2019tjswmm004).

Institutional Review Board Statement: Not applicable.

Informed Consent Statement: Not applicable.

Data Availability Statement: The data presented in this study are available on request from the corresponding author.

Conflicts of Interest: The authors declare no conflict of interest.

References

1. Nishimura, Y.; Hara, H.; Kondo, M.; Hong, S.; Matsugi, T. Oxidative Stress in Retinal Diseases. *Oxid. Med. Cell Longev.* **2017**, *2017*, 4076518. [CrossRef] [PubMed]
2. Sies, H.; Jones, D.P. Reactive oxygen species (ROS) as pleiotropic physiological signalling agents. *Nat. Rev. Mol. Cell Biol.* **2020**, *21*, 363–383. [CrossRef]
3. Gu, L.; Kwong, J.M.; Caprioli, J.; Piri, N. DNA and RNA oxidative damage in the retina is associated with ganglion cell mitochondria. *Sci. Rep.* **2022**, *12*, 8705. [CrossRef] [PubMed]
4. Abokyi, S.; Shan, S.-W.; Lam, C.H.-I.; Catral, K.P.; Pan, F.; Chan, H.H.-L.; To, C.-H.; Tse, D.Y.-Y. Targeting Lysosomes to Reverse Hydroquinone-Induced Autophagy Defects and Oxidative Damage in Human Retinal Pigment Epithelial Cells. *Int. J. Mol. Sci.* **2021**, *22*, 9042. [CrossRef]
5. Lakkaraju, A.; Umapathy, A.; Tan, L.X.; Daniele, L.; Philp, N.J.; Boesze-Battaglia, K.; Williams, D.S. The cell biology of the retinal pigment epithelium. *Prog. Retin. Eye Res.* **2020**, *78*, 100846. [CrossRef]
6. Datta, S.; Cano, M.; Ebrahimi, K.; Wang, L.; Handa, J.T. The impact of oxidative stress and inflammation on RPE degeneration in non-neovascular AMD. *Prog. Retin. Eye Res.* **2017**, *60*, 201–218. [CrossRef] [PubMed]
7. He, Y.; Ge, J.; Burke, J.M.; Myers, R.L.; Dong, Z.Z.; Tombran-Tink, J. Mitochondria impairment correlates with increased sensitivity of aging RPE cells to oxidative stress. *J. Ocul. Biol. Dis. Inform.* **2010**, *3*, 92–108. [CrossRef]
8. Foo, J.; Bellot, G.; Pervaiz, S.; Alonso, S. Mitochondria-mediated oxidative stress during viral infection. *Trends Microbiol.* **2022**, *30*, 679–692. [CrossRef]
9. Barnstable, C.J.; Zhang, M.; Tombran-Tink, J. Uncoupling Proteins as Therapeutic Targets for Neurodegenerative Diseases. *Int. J. Mol. Sci.* **2022**, *23*, 5672. [CrossRef]
10. Zhang, B.; Pan, C.; Feng, C.; Yan, C.; Yu, Y.; Chen, Z.; Guo, C.; Wang, X. Role of mitochondrial reactive oxygen species in homeostasis regulation. *Redox Rep.* **2022**, *27*, 45–52. [CrossRef]
11. Hass, D.T.; Barnstable, C.J. Uncoupling proteins in the mitochondrial defense against oxidative stress. *Prog. Retin. Eye Res.* **2021**, *83*, 100941. [CrossRef] [PubMed]
12. Rangarajan, S.; Locy, M.L.; Chanda, D.; Kurundkar, A.; Kurundkar, D.; Larson-Casey, J.L.; Londono, P.; Bagchi, R.A.; Deskin, B.; Elajaili, H.; et al. Mitochondrial uncoupling protein-2 reprograms metabolism to induce oxidative stress and myofibroblast senescence in age-associated lung fibrosis. *Aging Cell* **2022**, *21*, e13674. [CrossRef]
13. Pan, J.A.; Zhang, H.; Lin, H.; Gao, L.; Zhang, H.L.; Zhang, J.F.; Wang, C.Q.; Gu, J. Irisin ameliorates doxorubicin-induced cardiac perivascular fibrosis through inhibiting endothelial-to-mesenchymal transition by regulating ROS accumulation and autophagy disorder in endothelial cells. *Redox Biol.* **2021**, *46*, 102120. [CrossRef]
14. Hanus, J.; Kolkin, A.; Chimienti, J.; Botsay, S.; Wang, S. 4-Acetoxyphenol Prevents RPE Oxidative Stress-Induced Necrosis by Functioning as an NRF2 Stabilizer. *Invest. Ophthalmol. Vis. Sci.* **2015**, *56*, 5048–5059. [CrossRef] [PubMed]
15. Watson, M.A.; Wong, H.-S.; Brand, M.D. Use of S1QELs and S3QELs to link mitochondrial sites of superoxide and hydrogen peroxide generation to physiological and pathological outcomes. *Biochem. Soc. Trans.* **2019**, *47*, 1461–1469. [CrossRef]
16. He, Y.; Leung, K.W.; Ren, Y.; Pei, J.; Ge, J.; Tombran-Tink, J. PEDF improves mitochondrial function in RPE cells during oxidative stress. *Invest. Ophthalmol. Vis. Sci.* **2014**, *55*, 6742–6755. [CrossRef]
17. He, Y.; Wang, X.; Liu, X.; Ji, Z.; Ren, Y. Decreased uncoupling protein 2 expression in aging retinal pigment epithelial cells. *Int. J. Ophthalmol.* **2019**, *12*, 375–380. [CrossRef]
18. Szliszka, E.; Czuba, Z.P.; Domino, M.; Mazur, B.; Zydowicz, G.; Krol, W. Ethanolic extract of propolis (EEP) enhances the apoptosis- inducing potential of TRAIL in cancer cells. *Molecules* **2009**, *14*, 738–754. [CrossRef]
19. Jia, Y.; Jiang, S.; Chen, C.; Lu, G.; Xie, Y.; Sun, X.; Huang, L. Caffeic acid phenethyl ester attenuates nuclear factor-κB-mediated inflammatory responses in Müller cells and protects against retinal ganglion cell death. *Mol. Med. Rep.* **2019**, *19*, 4863–4871. [CrossRef] [PubMed]
20. Şahin, A.; Kaya, S.; Baylan, M. The effects of caffeic acid phenethyl ester on retina in a diabetic rat model. *Cutan. Ocul. Toxicol.* **2021**, *40*, 268–273. [CrossRef] [PubMed]
21. Zhang, M.; Wang, L.; Wen, D.; Ren, C.; Chen, S.; Zhang, Z.; Hu, L.; Yu, Z.; Tombran-Tink, J.; Zhang, X.; et al. Neuroprotection of retinal cells by Caffeic Acid Phenylethyl Ester (CAPE) is mediated by mitochondrial uncoupling protein UCP2. *Neurochem. Int.* **2021**, *151*, 105214. [CrossRef]
22. Liang, N.; Kitts, D.D. Role of Chlorogenic Acids in Controlling Oxidative and Inflammatory Stress Conditions. *Nutrients* **2015**, *8*, 16. [CrossRef]
23. Lee, B.-H.; Choi, H.-S.; Hong, J. Roles of anti- and pro-oxidant potential of cinnamic acid and phenylpropanoid derivatives in modulating growth of cultured cells. *Food Sci. Biotechnol.* **2022**, *31*, 463–473. [CrossRef]

24. Chagas, M.D.S.S.; Behrens, M.D.; Moragas-Tellis, C.J.; Penedo, G.X.M.; Silva, A.R.; Gonçalves-De-Albuquerque, C.F. Flavonols and Flavones as Potential Anti-Inflammatory, Antioxidant, and Antibacterial Compounds. *Oxidative Med. Cell. Longev.* **2022**, *2022*, 9966750. [CrossRef]

25. Sathyabhama, M.; Dharshini, L.C.P.; Karthikeyan, A.; Kalaiselvi, S.; Min, T. The Credible Role of Curcumin in Oxidative Stress-Mediated Mitochondrial Dysfunction in Mammals. *Biomolecules* **2022**, *12*, 1405. [CrossRef]

26. Shi, X.; Li, P.; Liu, H.; Prokosch, V. Oxidative Stress, Vascular Endothelium, and the Pathology of Neurodegeneration in Retina. *Antioxidants* **2022**, *11*, 543. [CrossRef]

27. Subramaniam, M.D.; Iyer, M.; Nair, A.P.; Venkatesan, D.; Mathavan, S.; Eruppakotte, N.; Kizhakkillach, S.; Chandran, M.K.; Roy, A.; Gopalakrishnan, A.V.; et al. Oxidative stress and mitochondrial transfer: A new dimension towards ocular diseases. *Genes Dis.* **2020**, *9*, 610–637. [CrossRef] [PubMed]

28. Chandrasekaran, P.R.; Madanagopalan, V.G. Role of Curcumin in Retinal Diseases—A review. *Graefes Arch. Clin. Exp. Ophthalmol.* **2022**, *260*, 1457–1473. [CrossRef] [PubMed]

29. Wan, T.; Wang, Z.; Luo, Y.; Zhang, Y.; He, W.; Mei, Y.; Xue, J.; Li, M.; Pan, H.; Li, W.; et al. FA-97, a New Synthetic Caffeic Acid Phenethyl Ester Derivative, Protects against Oxidative Stress-Mediated Neuronal Cell Apoptosis and Scopolamine-Induced Cognitive Impairment by Activating Nrf2/HO-1 Signaling. *Oxid. Med. Cell. Longev.* **2019**, *2019*, 8239642. [CrossRef] [PubMed]

30. Olgierd, B.; Kamila, Ż.; Anna, B.; Emilia, M. The Pluripotent Activities of Caffeic Acid Phenethyl Ester. *Molecules* **2021**, *26*, 1335. [CrossRef]

31. Balaha, M.; De Filippis, B.; Cataldi, A.; di Giacomo, V. CAPE and Neuroprotection: A Review. *Biomolecules* **2021**, *11*, 176. [CrossRef] [PubMed]

32. Li, D.; Wang, X.; Huang, Q.; Li, S.; Zhou, Y.; Li, Z. Cardioprotection of CAPE-oNO2 against myocardial ischemia/reperfusion induced ROS generation via regulating the SIRT1/eNOS/NF-κB pathway in vivo and in vitro. *Redox Biol.* **2017**, *15*, 62–73. [CrossRef] [PubMed]

33. Soares, V.E.M.; Carmo, T.I.T.D.; dos Anjos, F.; Wruck, J.; Maciel, S.F.V.D.O.; Bagatini, M.D.; Silva, D.T.D.R.E. Role of inflammation and oxidative stress in tissue damage associated with cystic fibrosis: CAPE as a future therapeutic strategy. *Mol. Cell. Biochem.* **2021**, *477*, 39–51. [CrossRef] [PubMed]

34. Marin, E.H.; Paek, H.; Li, M.; Ban, Y.; Karaga, M.K.; Shashidharamurthy, R.; Wang, X. Caffeic acid phenethyl ester exerts apoptotic and oxidative stress on human multiple myeloma cells. *Invest. New Drugs* **2018**, *37*, 837–848. [CrossRef]

35. Kang, Q.; Yang, C. Oxidative stress and diabetic retinopathy: Molecular mechanisms, pathogenetic role and therapeutic implications. *Redox Biol.* **2020**, *37*, 101799. [CrossRef]

36. Chen, H.; Tran, J.-T.A.; Anderson, R.E.; Mandal, N.A. Caffeic acid phenethyl ester protects 661W cells from H2O2-mediated cell death and enhances electroretinography response in dim-reared albino rats. *Mol. Vis.* **2012**, *18*, 1325–1338.

37. Zhang, C.; Xie, H.; Yang, Q.; Yang, Y.; Li, W.; Tian, H.; Lu, L.; Wang, F.; Xu, J.; Gao, F.; et al. Erythropoietin protects outer blood-retinal barrier in experimental diabetic retinopathy by up-regulating ZO-1 and occludin. *Clin. Exp. Ophthalmol.* **2019**, *47*, 1182–1197. [CrossRef] [PubMed]

38. Yuan, L.; Fung, T.S.; He, J.; Chen, R.A.; Liu, D.X. Modulation of viral replication, apoptosis and antiviral response by induction and mutual regulation of EGR and AP-1 family genes during coronavirus infection. *Emerg. Microbes Infect.* **2022**, *11*, 1717–1729. [CrossRef]

39. Kim, S.; Kim, H.; Um, J.W. Synapse development organized by neuronal activity-regulated immediate-early genes. *Exp. Mol. Med.* **2018**, *50*, 1–7. [CrossRef]

40. Karthikkeyan, G.; Nareshkumar, R.N.; Aberami, S.; Sulochana, K.N.; Vedantham, S.; Coral, K. Hyperglycemia induced early growth response-1 regulates vascular dysfunction in human retinal endothelial cells. *Microvasc. Res.* **2018**, *117*, 37–43. [CrossRef] [PubMed]

41. Sun, L.; Wang, R.; Hu, G.; Liu, H.; Lv, K.; Duan, Y.; Shen, N.; Wu, J.; Hu, J.; Liu, Y.; et al. Single cell RNA sequencing (scRNA-Seq) deciphering pathological alterations in streptozotocin-induced diabetic retinas. *Exp. Eye Res.* **2021**, *210*, 108718. [CrossRef] [PubMed]

42. Rickhag, M.; Teilum, M.; Wieloch, T. Rapid and long-term induction of effector immediate early genes (BDNF, Neuritin and Arc) in peri-infarct cortex and dentate gyrus after ischemic injury in rat brain. *Brain Res.* **2007**, *1151*, 203–210. [CrossRef]

43. Guo, H.; Golczer, G.; Wittner, B.S.; Langenbucher, A.; Zachariah, M.; Dubash, T.D.; Hong, X.; Comaills, V.; Burr, R.; Ebright, R.Y.; et al. NR4A1 regulates expression of immediate early genes, suppressing replication stress in cancer. *Mol. Cell* **2021**, *81*, 4041–4058.e15. [CrossRef] [PubMed]

44. Didier, S.; Sauvé, F.; Domise, M.; Buée, L.; Marinangeli, C.; Vingtdeux, V. AMP-activated Protein Kinase Controls Immediate Early Genes Expression Following Synaptic Activation Through the PKA/CREB Pathway. *Int. J. Mol. Sci.* **2018**, *19*, 3716. [CrossRef]

45. Chan, C.-M.; Huang, D.-Y.; Sekar, P.; Hsu, S.-H.; Lin, W.-W. Reactive oxygen species-dependent mitochondrial dynamics and autophagy confer protective effects in retinal pigment epithelial cells against sodium iodate-induced cell death. *J. Biomed. Sci.* **2019**, *26*, 40. [CrossRef]

46. Gupta, A.K.; Roy, S.; Das, P.K. Antileishmanial effect of the natural immunomodulator genipin through suppression of host negative regulatory protein UCP2. *J. Antimicrob. Chemother.* **2020**, *76*, 135–145. [CrossRef] [PubMed]

47. Raghavan, S.; Kundumani-Sridharan, V.; Kumar, S.; White, C.W.; Das, K.A.-O. Thioredoxin Prevents Loss of UCP2 in Hyperoxia via MKK4-p38 MAPK-PGC1α Signaling and Limits Oxygen Toxicity. *Am. J. Respir. Cell Mol. Biol.* **2022**, *66*, 323–336. [CrossRef]

48. Healy, S.; Khan, P.; Davie, J.R. Immediate early response genes and cell transformation. *Pharmacol. Ther.* **2013**, *137*, 64–77. [CrossRef]
49. Gao, P.; You, M.; Li, L.; Zhang, Q.; Fang, X.; Wei, X.; Zhou, Q.; Zhang, H.; Wang, M.; Lu, Z.; et al. Salt-Induced Hepatic Inflammatory Memory Contributes to Cardiovascular Damage Through Epigenetic Modulation of SIRT3. *Circulation* **2022**, *145*, 375–391. [CrossRef] [PubMed]

International Journal of
Molecular Sciences

MDPI

Article

Polymorphisms in Genes Encoding VDR, CALCR and Antioxidant Enzymes as Predictors of Bone Tissue Condition in Young, Healthy Men

Ewa Jówko [1,*], Barbara Długołęcka [1], Igor Cieśliński [2] and Jadwiga Kotowska [1]

[1] Department of Physiology and Biochemistry, Faculty of Physical Education and Health in Biała Podlaska, Józef Piłsudski University of Physical Education in Warsaw, 00-968 Warsaw, Poland
[2] Department of Sports and Training Sciences, Faculty of Physical Education and Health in Biała Podlaska, Józef Piłsudski University of Physical Education in Warsaw, 00-968 Warsaw, Poland
* Correspondence: ewa.jowko@awf.edu.pl; Tel.: +48-608-074-393

Abstract: The aim of the study was to assess significant predictors of bone mineral content (BMC) and bone mineral density (BMD) in a group of young, healthy men at the time of reaching peak bone mass. Regression analyses showed that age, BMI and practicing combat sports and team sports at a competitive level (trained vs. untrained group; TR vs. CON, respectively) were positive predictors of BMD/BMC values at various skeletal sites. In addition, genetic polymorphisms were among the predictors. In the whole population studied, at almost all measured skeletal sites, the *SOD2* AG genotype proved to be a negative predictor of BMC, while the *VDR FokI* GG genotype was a negative predictor of BMD. In contrast, the *CALCR* AG genotype was a positive predictor of arm BMD. ANOVA analyses showed that, regarding *SOD2* polymorphism, the TR group was responsible for the significant intergenotypic differences in BMC that were observed in the whole study population (i.e., lower BMC values of leg, trunk and whole body were observed in AG TR compared to AA TR). On the other hand, higher BMC at L1–L4 was observed in the *SOD2* GG genotype of the TR group compared to in the same genotype of the CON group. For the *FokI* polymorphism, BMD at L1–L4 was higher in AG TR than in AG CON. In turn, the *CALCR* AA genotype in the TR group had higher arm BMD compared to the same genotype in the CON group. In conclusion, *SOD2*, *VDR FokI* and *CALCR* polymorphisms seem to affect the association of BMC/BMD values with training status. In general, at least within the *VDR FokI* and *CALCR* polymorphisms, less favorable genotypes in terms of BMD (i.e., *FokI* AG and *CALCR* AA) appear to be associated with a greater BMD response to sports training. This suggests that, in healthy men during the period of bone mass formation, sports training (combat and team sports) may attenuate the negative impact of genetic factors on bone tissue condition, possibly reducing the risk of osteoporosis in later age.

Keywords: athletes; bone mineral density; bone mineral content; osteoporosis; SOD2; GPx

check for
updates

Citation: Jówko, E.; Długołęcka, B.; Cieśliński, I.; Kotowska, J. Polymorphisms in Genes Encoding VDR, CALCR and Antioxidant Enzymes as Predictors of Bone Tissue Condition in Young, Healthy Men. *Int. J. Mol. Sci.* **2023**, *24*, 3373. https://doi.org/10.3390/ijms24043373

Academic Editors: Rossana Morabito and Alessia Remigante

Received: 29 December 2022
Revised: 30 January 2023
Accepted: 2 February 2023
Published: 8 February 2023

1. Introduction

Osteoporosis is a common metabolic bone disease with reduced BMD and microarchitectural deterioration of bone tissue, which leads to an increase in bone fragility and fracture risk. This disease is not limited to postmenopausal women. There is increasing attention being paid to osteoporosis in older men. Despite a lower prevalence of osteoporosis in men than in women, men have higher morbidity and mortality rates after fracture [1].

As a multifactorial disease, osteoporosis is influenced by various environmental and genetic factors. One of the general strategies for making the skeleton more resistant to fracture is to maximize peak bone mass by the age of 30, since bone formation is predominant until the age of 25 [2]. Among environmental factors, physical activity habits play a pivotal role in bone mass and structure characteristics [3]. Even in a population of physically active young adult men, with a BMD at the upper end of the normal range, higher levels of physical activity were associated with higher bone mass [4].

Physical activity promotes changes in the bone metabolism through a direct effect (via mechanical force) or an indirect effect (promoted by hormonal factors) [5,6]. Mechanical loadings, including compression, strain and fluid shear, are the stimuli playing essential roles in osteoblast differentiation and mineralization, as well as in maintaining proper high bone mass and density [7]. It can also be assumed that particularly dynamic, weight-bearing sports with short, high and multidimensional loads have strong effects on bone formation, independent of training quantity [8,9].

Apart from the type of physical activity, a proper diet with an adequate intake of calcium and vitamin D is a prerequisite for achieving the highest possible peak bone mass. On the other hand, absorption and metabolism of these dietary compounds, and, in turn, their influence on BMD, varies depending on genetic factors. The vitamin D receptor (*VDR*) gene polymorphisms *ApaI, BsmI, TaqI* and *FokI* are most frequently studied as potential factors predicting osteoporosis risk in middle-aged and older populations [10–12]. Additionally, the receptor of calcitonin (*CALCR*) gene polymorphism (C1377T *AluI*), related to calcium and phosphate homeostasis, was found to affect BMD [13], with ambiguous results reported in studies on men from European countries [14,15]. Moreover, associations between collagen type I (*COLIA1*) gene polymorphism (+1245G/T) and BMD were reported in postmenopausal women [16].

Although physical activity may modify the relationship between genetics and bone mineralization, few studies have evaluated the association between *VDR* polymorphisms and BMD in athletes, with limited numbers of participants and conflicting results [17,18]. Furthermore, no data are available regarding the importance of *CALCR* or *COLIA1* gene polymorphisms as potential predictors of BMD in young athletes in the phase of peak bone mass development.

It has also been suggested that dynamic formation and resorption of bone tissue throughout a person's life may depend on nonenzymatic and enzymatic antioxidant potential (superoxide dismutase, various peroxidases, catalase) [19,20]. Animal studies have shown a positive correlation between BMD and superoxide dismutase (SOD) activity in the blood [21]. In turn, lower blood glutathione peroxidase (GPx) activity was observed in a group of older women and men with osteoporosis compared to in healthy people in the same age range [22].

It is known that enzymatic antioxidant potential can be determined by genetic factors, such as polymorphism in genes encoding enzymatic proteins [23,24]. Some data in the literature indicate that polymorphism of genes encoding antioxidant enzymes (among others, *SOD2 Ala-9Val* and *GPx Pro198Leu*) may affect BMD in the middle-aged and elderly population [20,25,26]. However, there are no reports of potential links between the polymorphisms mentioned above and BMD or BMC in the younger population in the available literature.

Finally, exercise training is suggested to modulate the expression of genes encoding antioxidant enzymes in a positive manner, thus, resulting in an increase in enzymatic antioxidant potential [27]. Therefore, it is interesting to evaluate whether common polymorphisms in genes encoding antioxidant enzymes can predict BMD values in healthy young men. Furthermore, the next question is whether these polymorphisms can modulate the response of bone tissue formation (BMD, BMC values) to high-impact exercise training in professional athletes.

Therefore, the aim of the study was to assess which factors are the most potent predictors of BMC and BMD values in a group of young, healthy men during the period of reaching peak bone mass. The factors taken into account were (1) trained vs. untrained status; (2) anthropometric variables: age, BMI; (3) the gene polymorphisms mentioned above: *VDR* (*ApaI*, rs 7975232; *BsmI*, rs1544410; *FokI* rs 2228570), *COLIA1* (rs 1800012), *CALCR* (rs 1801197) and antioxidant enzymes (*SOD1*, rs2234694; *SOD2*, rs4880; and *GPx* rs1050450); (4) blood levels of bone metabolism markers: calcium, phosphates, vitamin

D (25-OH D), osteocalcin, alkaline phosphatase; (5) blood prooxidant-antioxidant homeostasis: total antioxidant capacity, uric acid, lipid hydroperoxides, superoxide dismutase, glutathione peroxides.

2. Results

The anthropometric characteristics of participants in the CON and TR groups are shown in Table 1. No intergroup differences were found in these parameters ($p > 0.05$).

Table 1. Anthropometric data, regional and total body bone mineral content (BMC) (g) and bone mineral density (BMD) (g/cm^2) in the control group (CON, n = 87) and in the trained group (TR, n = 94).

		CON (n = 87)	TR (n = 94)	ANOVA *p*-Value
Anthropometric data	age (years)	21.18 ± 1.75	20.96 ± 2.26	$p = 0.48$
	height (cm)	182.28 ± 5.28	180.78 ± 8.16	$p = 0.16$
	body mass (kg)	81.10 ± 9.73	80.38 ± 11.56	$p = 0.66$
	BMI (kg/m^2)	24.39 ± 2.56	24.54 ± 2.53	$p = 0.71$
L1–L4	BMC	83.87 ± 11.04	92.20 ± 14.28	$p = 0.00004$
	BMD	1.14 ± 0.10	1.20 ± 0.12	$p = 0.0008$
arm	BMC	217.81 ± 29.50	219.73 ± 41.63	$p = 0.73$
	BMD	0.90 ± 0.06	0.92 ± 0.08	$p = 0.030$
leg	BMC	641.04 ± 89.14	648.96 ± 118.16	$p = 0.62$
	BMD	1.49 ± 0.15	1.54 ± 0.16	$p = 0.06$
trunk	BMC	2609.18 ± 337.41	2677.09 ± 463.92	$p = 0.28$
	BMD	1.21 ± 0.10	1.25 ± 0.11	$p = 0.025$
whole body	BMC	3079.59 ± 371.80	3123.20 ± 479.68	$p = 0.51$
	BMD	1.29 ± 0.10	1.31 ± 0.11	$p = 0.075$
	Z-score (min; max)	1.01 ± 0.84 (−0.7; 2.8)	1.31 ± 0.96 (−0.6; 3.6)	$p = 0.027$
	T-score (min; max)	0.93 ± 0.82 (−0.7; 2.8)	1.17 ± 0.99 (−0.7; 3.5)	$p = 0.076$

Data are mean $\pm$ SD; L1–L4—lumbar spine.

Individuals from the CON and TR groups differed significantly in BMC value of the lumbar spine ($p = 0.00004$), whereas the groups did not differ in BMC of other points of the skeleton (Table 1). In turn, significantly higher BMD values (lumbar spine, $p = 0.0008$; arm, $p = 0.03$; trunk, $p = 0.025$) were seen in the TR group as compared to the CON group (Table 1). This significant difference was also apparent for the Z-score ($p = 0.027$), whereas only an increasing trend in the T-score ($p = 0.076$) was observed in the TR group as compared to CON group (Table 1). None of our participants had an abnormal T-score, i.e., below −1.

Table 2 presents the level of biochemical parameters in the blood (with reference range for population studied). In the TR group, significantly lower plasma TAC levels ($p = 0.002$), as well as plasma UA concentration ($p = 0.0001$), were shown in comparison to in the CON group (Table 2). No significant differences were observed between the groups in the level of other parameters in the blood. In both groups (TR and CON), the mean concentration of 25-OH D in serum was within the reference values (30–50 ng/mL). We also measured the correlations between biochemical parameters (data are not shown). In the whole study population, the only significant correlation was seen between TAC and UA ($r = 0.54$, $p < 0.00001$). Furthermore, when all participants were divided into two subgroups, taking into account individual vitamin D status (serum 20-OH D), a negative correlation between serum 25-OH D and OC was observed in the subgroup with 25-OH D < 30 ng/mL ($r = -0.32$; $p < 0.01$) but not in the subgroup with 25-OH D $\geq$ 30 ng/mL.

Table 2. Blood levels of biochemical parameters (mean $\pm$ SD) in the control group (CON, n = 87) and in the trained group (TR, n = 94).

Variables	Reference Range	CON (n = 87)	TR (n = 94)	ANOVA p-Value
TAC [mmol/L]	1.30–1.77	1.72 $\pm$ 0.18	1.64 $\pm$ 0.15	$p = 0.002$
GPx [U/g Hb]	27.5–73.6	60.1 $\pm$ 26.1	63.8 $\pm$ 26.0	$p = 0.33$
SOD [U/g Hb]	1102–1601	1507.4 $\pm$ 509.1	1518.7 $\pm$ 468.2	$p = 0.88$
LOOHs [μmol/L]	-	3.06 $\pm$ 1.65	2.72 $\pm$ 1.79	$p = 0.17$
ALP [U/L]	40–129	82.4 $\pm$ 23.5	76.9 $\pm$ 22.0	$p = 0.21$
Ca [mg/dL]	8.60–10.30	8.05 $\pm$ 0.89	8.31 $\pm$ 0.53	$p = 0.11$
UA [mg/dL]	3.40–7.00	5.55 $\pm$ 1.10	4.94 $\pm$ 0.96	$p = 0.0001$
P [mg/dL]	2.60–4.50	3.67 $\pm$ 0.55	3.66 $\pm$ 0.82	$p = 0.92$
T [ng/mL]	2.2–10.5	6.10 $\pm$ 1.45	5.86 $\pm$ 1.38	$p = 0.25$
OC [ng/mL]	<5.0	26.31 $\pm$ 10.01	24.26 $\pm$ 10.00	$p = 0.18$
25-OH D [ng/mL]	30–50	40.29 $\pm$ 28.11	36.18 $\pm$ 10.69	$p = 0.20$

TAC—total antioxidant capacity, SOD—superoxide dismutase, GPx—glutathione peroxidase, LOOH—lipid hydroperoxides, ALP—alkaline phosphatase, Ca—calcium, UA—uric acid, P—phosphates, T—testosterone, OC—osteocalcin, 25-OH D—25-OH vitamin D.

Genotype frequencies for subjects in the CON and TR groups are shown in Table 3. In both the CON and TR group, the genotype frequencies within *GPx* Pro198Leu polymorphism differed significantly ($p = 0.0008$ and $p = 0.005$, respectively) from those expected under the Hardy–Weinberg equilibrium, with the TT genotype being absent in our study population (and no differences between the CON and TR groups). In the case of *CALCR* polymorphism, it was revealed that, in the TR group (but not the CON group), the genotype distribution did not follow the Hardy–Weinberg distribution ($p = 0.00006$). Moreover, significant differences in genotype frequencies within *CALCR* polymorphism were found between groups, with GG being more frequent in the TR than the CON group ($p = 0.003$). For other SNPs in the genes analyzed in our study, the χ^2 test confirmed that the observed frequencies did not deviate from the Hardy–Weinberg equilibrium, with no significant intergroup differences in genotype frequencies (Table 3).

The predictors of BMC and BMD, as evaluated by regression models, are presented in Tables 4 and 5, respectively. Age was a positive predictor of BMC at two points of the skeleton, L1–L4 ($p < 0.01$) and arm ($p < 0.05$), whereas BMI positively affected BMC in almost all the analyzed areas ($p < 0.001$), except for in L1–L4 (Table 4). Although the status "TR vs. CON" (i.e., trained vs. untrained) was not a predictor of BMC values (Table 4), all kinds of sports disciplines studied (i.e., MMA, soccer, handball and wrestling) positively affected the BMC value at L1–L4 ($p < 0.05$ for MMA and soccer; $p < 0.001$ for handball and wrestling). In turn, handball was the positive predictor of BMC in all measured body parts ($p < 0.001$). Moreover, two polymorphisms of genes related to enzymatic antioxidant defense were factors influencing BMC (Table 4). Within *GPx* Pro198Leu polymorphism, genotype CT was a negative predictor of BMC at L1–L4 ($p < 0.05$); however, ANOVA analyses did not reveal any differences in BMC values between CC and CT genotypes either in the whole study population or in the TR and CON groups. In turn, amongst the three genotypes of *SOD2* Ala-9 Val polymorphism, AG was the independent factor that had a negative impact on BMC values in all body parts except the arm ($p < 0.05$), with GG being a negative predictor in non-significant manner (Table 4).

Table 3. The distribution of genotypes for gene polymorphisms in the control group (CON) and in the trained group (TR) (n (%)).

Polymorphism	CON (n = 87)	TR (n = 94)	TR vs. CON
SOD1 A-39 C			
AA	78 (89.7)	87 (92.5)	$\chi^2 = 0.18$
AC	9 (10.3)	7 (7.5)	$p = 0.67$
CC	0	0	
HWE	$p = 0.58$	$p = 0.30$	
SOD2 Ala-9 Val			
AA	26 (29.9)	23 (24.5)	$\chi^2 = 1.24$
AG	40 (46.0)	42 (44.7)	$p = 0.54$
GG	21 (24.1)	29 (30.8)	
HWE	$p = 0.57$	$p = 0.40$	
GPx Pro198Leu			
CC	39 (44.8)	49 (52.1)	$\chi^2 = 0.69$
CT	48 (55.2)	45 (47.9)	$p = 0.40$
TT	0	0	
HWE	$p = 0.0008$	$p = 0.005$	
VDR ApaI			
AA	26 (29.9)	23 (24.5)	$\chi^2 = 1.36$
AC	37 (42.5)	48 (51.0)	$p = 0.51$
CC	24 (27.6)	23 (24.5)	
HWE	$p = 0.22$	$p = 0.94$	
VDR BsmI			
AA	16 (18.4)	10 (10.7)	$\chi^2 = 2.21$
AG	35 (40.2)	41 (43.6)	$p = 0.33$
GG	36 (41.4)	43 (45.7)	
HWE	$p = 0.21$	$p = 0.88$	
VDR FokI			
AA	15 (17.2)	20 (21.3)	$\chi^2 = 1.45$
AG	44 (50.6)	51 (54.2)	$p = 0.48$
GG	28 (32.2)	23 (24.5)	
HWE	$p = 0.87$	$p = 0.49$	
CALCR			
AA	44 (50.6)	58 (61.7)	$\chi^2 = 11.57$
AG	38 (43.6)	21 (22.3)	$p = 0.003$
GG	5 (5.8)	15 (16.0)	
HWE	$p = 0.51$	$p = 0.00006$	
COLIA1			
AA	2 (2.3)	8 (8.5)	$\chi^2 = 3.68$
AC	22 (25.3)	19 (20.2)	$p = 0.16$
CC	63 (72.4)	67 (71.3)	
HWE	$p = 0.74$	$p = 0.52$	

HWE—Hardy–Weinberg equilibrium. χ^2 chi-square (between groups TR vs. CON), *p*-values.

In the case of BMC at L1–L4 in the whole studied group of subjects (both control and trained), a significantly lower value ($p = 0.03$) was found in the SOD2 AG genotype as compared to in genotypes AA and GG (Figure 1A; with main effect of *SOD2*, $p = 0.02$). Moreover, when this parameter was analyzed in the CON and TR groups, BMC at L1–L4 was significantly higher in GG TR than in GG CON ($p = 0.02$; Figure 1B; with main effect of *SOD2*, $p = 0.01$ and TR vs. CON, $p = 0.000000001$).

Table 4. Regression models of BMC.

Predictors / Response Variables	BMC L1–L4 (Std.Err.)	BMC Arm (Std.Err.)	BMC Leg (Std.Err.)	BMC Trunk (Std.Err.)	BMC Total (Std.Err.)
Age	1.561 ** (0.469)	2.860 * (1.309)			
BMI		5.864 *** (1.000)	12.547 *** (2.969)	51.801 *** (11.369)	60.093 *** (12.106)
GPx CT	−3.727 * (1.831)				
SOD2 AG	−5.299 * (2.208)		−36.960 * (16.191)	−153.853 * (63.413)	−144.693 * (67.997)
SOD2 GG	−0.746 (2.456)		−8.323 (18.066)	−57.214 (70.974)	−23.763 (75.905)
MMA	6.897 * (3.270)				
Soccer	5.300 * (2.404)				
Handball	19.239 *** (3.914)	53.485 *** (10.099)	142.667 *** (28.281)	647.764 *** (113.529)	555.423 *** (117.715)
Wrestling	14.187 *** (2.883)				
R^2	0.310	0.456	0.386	0.381	0.354
Adjusted R^2	0.271	0.422	0.351	0.346	0.322
RSE	11.522	29.223	84.699	332.110	356.180
F Statistic	7.927 ***	13.169 ***	11.01 6***	10.806 ***	10.909 ***

Std.Err.: standard error; RSE: residual standard error; MMA: mixed martial arts. * $p < 0.05$; ** $p < 0.01$; *** $p < 0.001$.

Table 5. Regression models of selected variables as predictors of BMD.

Predictors / Response Variables	BMD L1–L4 (Std.Err.)	BMD Arm (Std.Err.)	BMD Leg (Std.Err.)	BMD Trunk (Std.Err.)	BMD Total (Std.Err.)
TR vs. CON	0.095 *** (0.025)	0.045 ** (0.016)			
Age		0.006 * (0.002)	0.012 * (0.006)	0.008 * (0.004)	0.009 * (0.004)
BMI	0.011 *** (0.003)	0.011 *** (0.002)	0.014 ** (0.005)	0.011 *** (0.003)	0.012 *** (0.003)
FokI GG	−0.060 ** (0.022)		−0.065 * (0.031)	−0.052 * (0.021)	−0.051 * (0.021)
FokI AG	−0.020 (0.020)		−0.037 (0.028)	−0.026 (0.019)	−0.026 (0.019)
T				0.012 ** (0.005)	0.011 ** (0.005)
CALCR AG		0.020 * (0.009)			
CALCR GG		0.009 (0.013)			
OC			−0.004 *** (0.001)	−0.002 ** (0.001)	−0.002 ** (0.001)
MMA		0.034* (0.015)			
Soccer					
Handball		0.091 *** (0.018)	0.194 *** (0.045)	0.146 *** (0.030)	0.121 *** (0.030)
Wrestling		0.054 *** (0.014)		0.046* (0.024)	
R^2	0.326	0.470	0.330	0.351	0.318
Adjusted R^2	0.283	0.440	0.288	0.309	0.275
RSE	0.097	0.053	0.133	0.089	0.089
F Statistic	7.632 ***	15.575 ***	7.740 ***	8.479 ***	7.333 ***

Std.Err.: standard error; RSE: residual standard error; MMA: mixed martial arts. * $p < 0.05$; ** $p < 0.01$; *** $p < 0.001$.

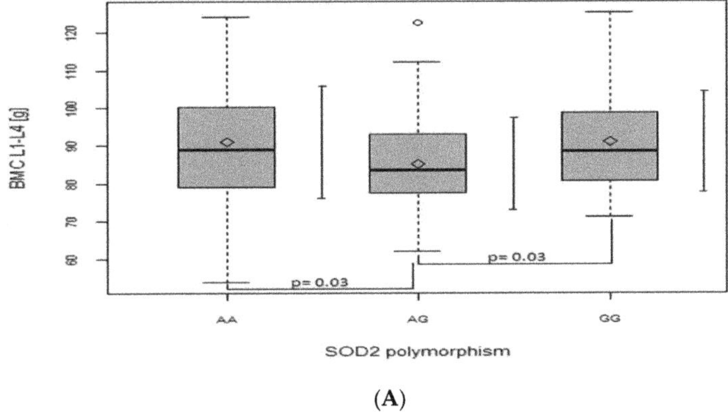

(A)

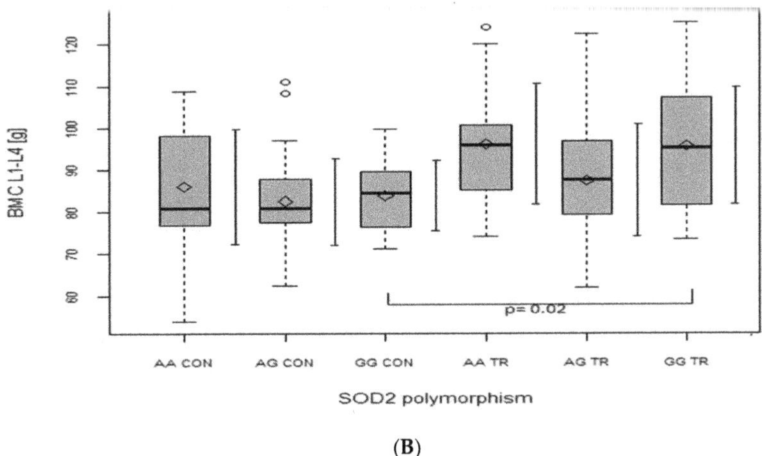

(B)

Figure 1. The effect of SOD2 polymorphism on bone mineral content (BMC) (g) of the lumbar spine (L1–L4). (**A**) In the whole group; main effects: SOD2 (p = 0.02; F = 4.23). (**B**) In control (CON) and trained (TR) group; main effects: SOD2 (p = 0.01; F = 4.67). TR vs. CON (p = 0.000000001; F = 18.1). Key: ◊—mean; ○—outliers value; boxplots—median Q1–Q4 and mean ± SD.

Regarding the BMC of the leg (Figure 2A), trunk (Figure 3A) and whole body (Figure 4A), when analyzed in all the participants together (from both the CON and TR groups), the AG genotype had lower values in comparison with the AA genotype (leg: p = 0.006, main effect of *SOD2*, p = 0.008; trunk: p = 0.002, main effect of *SOD2*, p = 0.002; whole body: p = 0.003, main effect of *SOD2*, p = 0.004). After dividing the whole studied group into the TR and CON groups, the BMC of the leg (Figure 2B), trunk (Figure 3B) and whole body (Figure 4B) was lower in AG TR than AA TR (leg: p = 0.03, main effect of *SOD2*, p = 0.008; trunk: p = 0.025, main effect of *SOD2*, p = 0.002; whole body: p = 0.057, main effect of *SOD2*, p = 0.004).

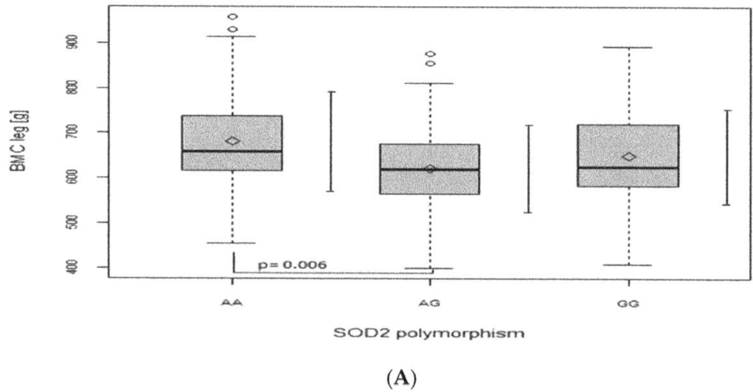

(A)

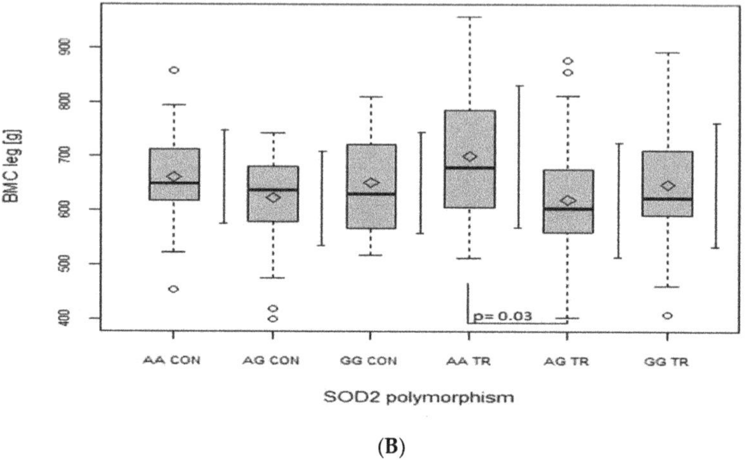

(B)

Figure 2. The effect of SOD2 polymorphism on bone mineral content (BMC) (g) of the leg. (**A**) In the whole group of studied participants; main effects: SOD2 ($p = 0.008$; F = 4.99). (**B**) In control (CON) and trained (TR) group; main effects: SOD2 ($p = 0.008$; F = 4.95). Key: ◇—mean; ○—outliers value; boxplots—median Q1–Q4 and mean $\pm$ SD.

As shown in Table 5, the factor that affected BMD to the greatest extent was practicing sports (TR vs. CON status), which concerned BMD at L1–L4 ($p < 0.001$) and the arm ($p < 0.01$). A positive impact of MMA on the arm's BMD value was found ($p < 0.05$). Wrestling had a positive impact on the BMD of the arm and trunk ($p < 0.001$ and $p < 0.05$; respectively), whereas handball positively affected the BMD of the arm, leg, trunk and total body ($p < 0.001$).

Similar to BMC, the subsequent predictors of BMD were age and BMI (Table 5). Age positively affected BMD in all the body parts studied ($p < 0.05$) besides L1–L4, whereas BMI was a positive predictor of all BMD values ($p < 0.01$ for leg; $p < 0.001$ for L1–L4, arm, trunk and total BMD). Among the biochemical parameters, serum testosterone was a positive factor of both the trunk and total BMD ($p < 0.01$). In contrast, BMD was negatively affected by serum OC ($p < 0.001$ for the leg; $p < 0.01$ for the trunk and total body).

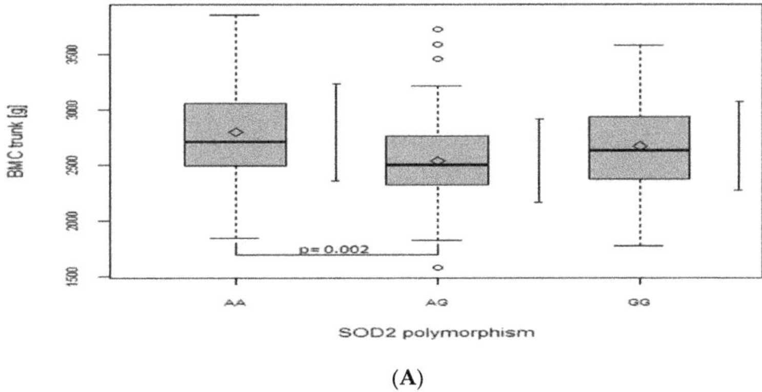

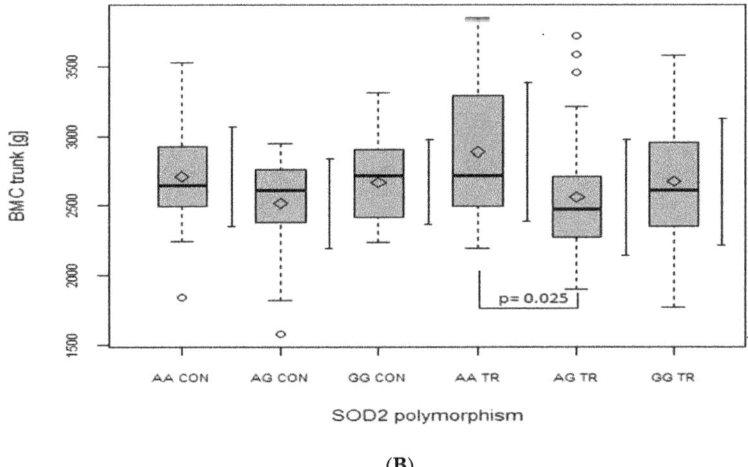

Figure 3. The effect of SOD2 polymorphism on bone mineral content (BMC) (g) of the trunk. (**A**) In the whole group; main effects: SOD2 ($p = 0.002$; F = 6.24). (**B**) In control (CON) and trained (TR) group; main effects: SOD2 ($p = 0.002$; F = 6.22). Key: ◊—mean; ○—outliers value; boxplots—median Q1–Q4 and mean ± SD.

Apart from testosterone and OC, none of the biochemical parameters was a predictor of BMC/BMD values (Tables 4 and 5). Of all the polymorphisms examined, one of the polymorphisms of gene coding the receptor of vitamin D (*FokI VDR*) and the polymorphism of gene coding the receptor of calcitonin (*CALCR*) turned out to be predictors of BMD values (Table 5). Of the three genotypes within *VDR FokI* polymorphism, GG was found to be a negative predictor of BMD at L1–L4 ($p < 0.01$), the leg, trunk and total body ($p < 0.05$, Table 5), with AG being a negative predictor in a non-significant manner. On the other hand, as indicated from ANOVA analyses, BMD at L1–L4 was significantly higher in AG TR than in AG CON ($p = 0.01$; with main effect of TR vs. CON, $p = 0.001$ and main effect of *FokI*, $p = 0.047$; Figure 5). In addition to this, the only main effect of TR vs. CON (without any *FokI* effect) was seen in the BMD of the trunk ($p < 0.05$), while no significant effects were found in the BMD of the leg or total body.

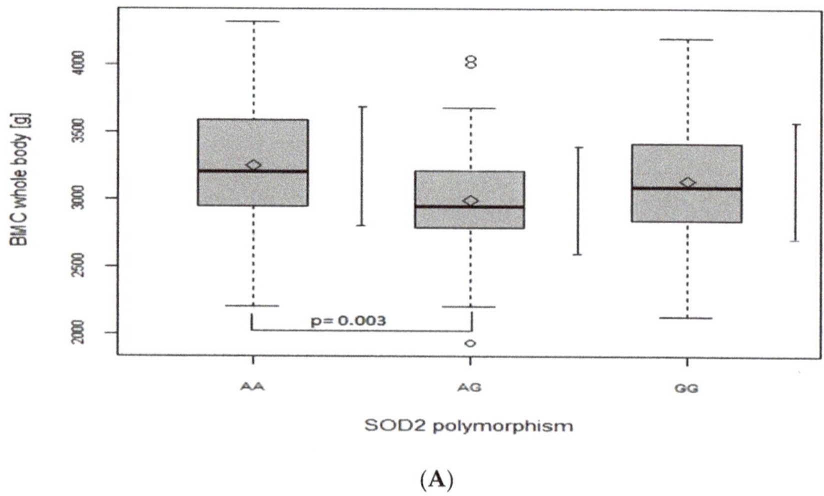

(A)

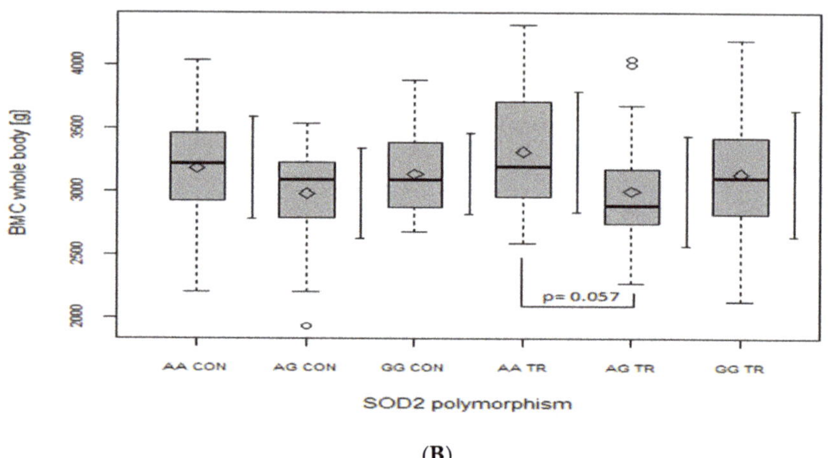

(B)

Figure 4. The effect of SOD2 polymorphism on bone mineral content (BMC) (g) of the whole body. (**A**): in the whole group; main effects: SOD2 ($p = 0.004$; F = 5.71). (**B**): in control (CON) and trained (TR) group; main effects: SOD2 ($p = 0.004$; F = 5.65). Key: ◊—mean; ○—outliers value; boxplots—median Q1–Q4 and mean ± SD.

As for *CALCR* polymorphism, the AG genotype was a positive predictor of BMD value, but only in the case of the arm ($p < 0.05$, Table 5). The same was true for the GG genotype, but in a non-significant manner (Table 5). In turn, as shown in Figure 6, the arm's BMD was found to be higher in AA TR as compared to in AA CON ($p = 0.02$; with main effects of TR vs. CON, $p = 0.039$ and the interaction of *CALCR* x TR vs. CON, $p = 0.022$).

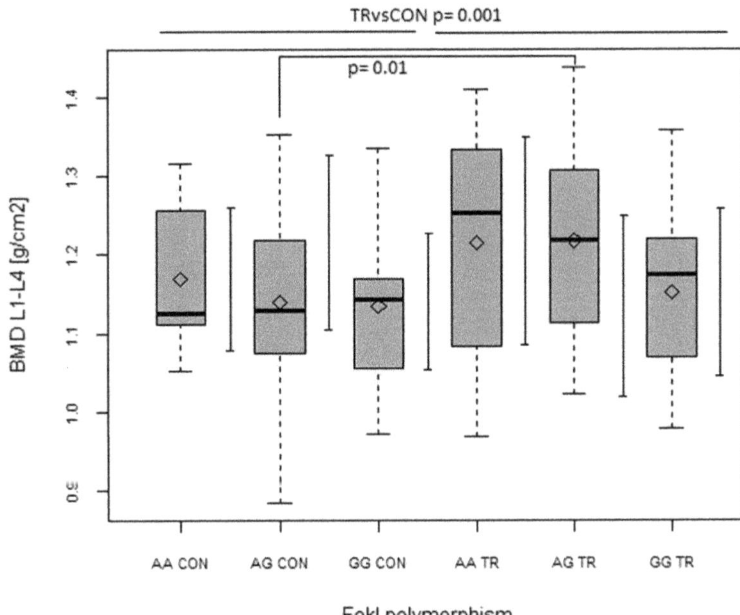

Figure 5. The effect of FokI polymorphism on bone mineral density (BMD) (g/cm^2) of the lumbar spine (L1–L4) in control (CON) and trained (TR) group. Main effects: FokI (p = 0.047; F = 2.95). TR vs. CON (p = 0.001; F = 10.46). Key: $\Diamond$—mean; boxplots—median Q1–Q4 and mean $\pm$ SD.

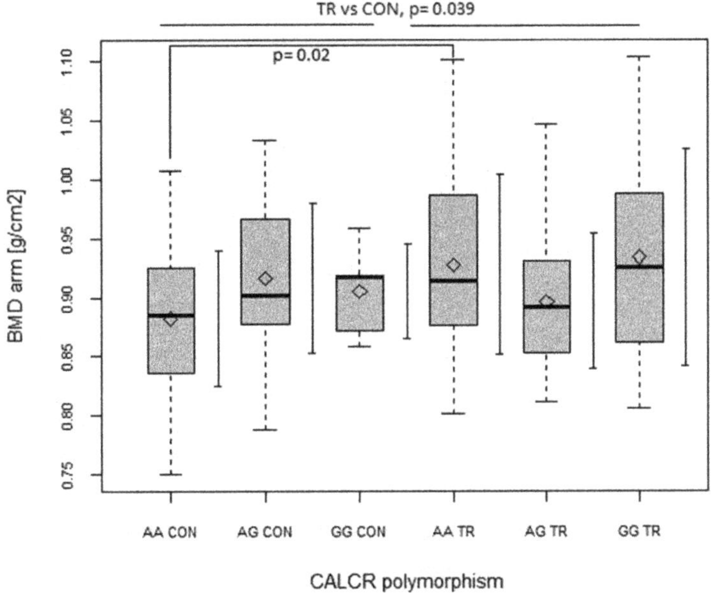

Figure 6. The effect of CALCR polymorphism on bone mineral density (BMD) (g/cm^2) of the arm in control (CON) and trained (TR) group. Main effects: CALCR NS TR vs. CON (p = 0.039; F = 4.31). CALCR x TR vs. CON (p = 0.022; F = 3.88). Key: $\Diamond$—mean; boxplots—median Q1–Q4 and mean $\pm$ SD.

3. Discussion

The most relevant finding of our study is that the common polymorphisms within the genes encoding antioxidant enzyme proteins (*SOD2* and *GPx*) are the predictors of BMC in young, healthy men. Furthermore, *SOD2*, *FokI VDR* and *CALCR* polymorphisms seem to modulate the associations of BMC/BMD values with training status (trained vs. untrained).

As confirmed by our regression analyses, BMI and age seem to be positive predictors of BMC and BMD (Tables 4 and 5), which is in line with the finding from other studies on the young, physically active, male population [4,28]. Indeed, it has been reported that, within a healthy weight range, there is a direct positive relationship between BMI and bone density [29]. In turn, positive relationships between age and bone parameters indicate that the skeleton is not fully developed. Thus, our participants were in the phase of peak bone mass development.

The main purpose of our study was to evaluate genetic factors as predictors of bone mass in young men engaged in competitive sports. We asked athletes from combat and team sports to take part in the current study, since these sports have been previously reported to have the highest positive stimulatory effects on bone formation [8]. Our results showed that the most evident positive influence of sports training on bone parameters was found at the lumbar spine (Tables 1, 4 and 5), possibly due to high amounts of trabecular bone that is more metabolically active than cortical bone [29]. On the other hand, when the relationship between sports and bone parameters in other areas of the skeleton was considered, only handball proved to be a positive predictor of both BMC and BMD at almost all the sites measured, including total body (Tables 4 and 5). Furthermore, handball, wrestling and MMA (but not soccer) were positive predictors of arm BMD, which is compatible with the nature/specificity of these sports. It confirms previous suggestions that the level of the bone adaptation through the exercise seems to be dependent on the overload, and it seems to be specific for the spots that are submitted to major stress [29].

Meanwhile, in the current study, a lack of significant influence of soccer on any BMD value was quite unexpected since the loading pattern of this sport is considered to result in high-impact forces on the bones of the lower limbs and to improve osteogenesis [28]. Instead, in our study, the only positive impact of soccer regarded BMC at the lumbar spine. The least marked effect of soccer, compared to the other disciplines studied, on bone parameters may be related to the lowest mean value of body mass in soccer as compared to other sports and those in the control group. As we aimed to evaluate all predictors of crude bone mass parameters, the analyses were not adjusted to anthropometric indices, which may be a limitation of the study. Another reason for these differences between the sports studied regarding the effect on BMD might have resulted from the different competitive levels of our athletes representing different sports, with the lowest competitive level being represented by the soccer players. However, to confirm the above-mentioned issues, further research is needed.

A study on the elderly population indicated that oxidative stress may be an independent risk factor for osteoporosis, supporting the hypothesis that links oxidative stress (resulting from a decrease in enzymatic antioxidant potential) with the etiology and physiopathology of this disease [22]. On the other hand, regular physical activity/sports training is considered as an important factor that reinforces antioxidant potential (especially enzymatic) and alleviates resting levels of oxidative stress parameters [27]. In our study, no intergroup differences were found in blood levels of enzymatic antioxidant defense and oxidative stress parameters. On the other hand, the decrease in nonenzymatic antioxidant potential (plasma TAC) found in our TR group compared to CON group was probably due to a decrease in plasma UA (Table 2) as a result of an adaptation to training, which is in line with previous findings on both endurance and speed-power athletes [30].

None of the parameters of prooxidant-antioxidant status was associated with bone mass parameters. On the other hand, our study is the first to find the association of bone parameters (BMC) with polymorphisms within genes encoding antioxidant enzymes in a young, healthy population.

Among the three polymorphisms studied, the relationship between *SOD2* polymorphism and BMC values was the most evident in our population, which is not surprising since SOD2 is considered as the key mitochondrial enzyme in the oxidative stress metabolism pathway [20].

So far, the only study on a European population (postmenopausal women) showed no association of SNP rs4880 of the *SOD2* gene to lumbar and femoral BMD [26]. In turn, in another cross-sectional study on Asian Indians, the *SOD2* G allele was found to be significantly more frequent in osteoporotic individuals [20].

Consistent with Botre's findings [20], our results also suggest that the G allele of *SOD2* gene may be disadvantageous for bone health, as GG (although not significantly) and primarily AG genotypes were generated by our regression model as a negative predictor of BMC. Moreover, ANOVA analyses revealed that, in our whole population, heterozygote AG had lower BMC values than both homozygote AA (Figures 1–4, panel A) and GG (Figure 1A). However, when BMC values (at leg, trunk and total) were analyzed separately in two groups (TR and CON), only the TR group had significant discrepancies between genotypes (Figures 2–4, panel B). It may suggest that the AG genotype blunts or slows down the rate of BMC increase induced by sports training. However, it does not seem to apply to the lumbar spine, since BMC at L1–L4 was higher in the GG genotype in the TR group as compared to the same genotype of the CON group (Figure 1B), probably due to the better response of the lumbar spine to exercise as compared to other regions of the skeleton [29].

It is difficult to explain why variant G [20] or genotype AG (our study, especially in trained persons) is associated with worse bone tissue condition in comparison to the A allele/AA genotype. It has been previously reported that the G allele, in comparison with the A allele, is related to reinforcement of the enzymatic antioxidant defense as a result of more effective transport of SOD2 protein to mitochondria and an increase in SOD2 activity [24]. In line with the above, are our previous studies on healthy men who underwent chronic endurance training [31] or wrestlers [32], in which the *SOD2* GG genotype was associated with lower oxidative stress and muscle damage parameters as compared to two other genotypes.

The explanation for a negative impact of *SOD2* GG or AG genotypes (but not the AA genotype) on bone tissue condition is possibly an imbalance between SOD and GPx activity. In the elderly Mexican population, a decrease in GPx activity, along with an increase in the SOD/GPx ratio, was seen in persons with osteoporosis [22]. As explained, greater SOD activity with respect to GPx results in an increase in H_2O_2 levels, which favors the differentiation of osteoblastic cells to osteoclasts and inhibits the differentiation of osteoblastic cells to osteoblasts, thus, propitiating an accentuated diminution of BMD [22]. In our study on a young population, no association was seen between BMD/BMC and antioxidant enzymes activities. However, variations in the coding region of antioxidant enzyme genes may not necessarily affect the protein's activity but may affect its subcellular localization [33]. Therefore, it cannot be excluded that diminished BMC in our athletes carrying AG genotypes may result from some imbalance between SOD and GPx adaptation to exercise. Finally, not only the AG genotype within *SOD2* gene, but also the CT genotype within *GPx* gene, was a negative predictor of BMC in our participants (Table 4). In the study of Mlakar et al. [25] on the elderly population, the TT genotype of the *GPx* gene was related to decreased BMD of the lumbar spine, femoral neck and total hip. Furthermore, the T allele was found to be associated with reduced GPx activity in a dose-dependent manner [34]. It must be mentioned that we did not find the TT genotype in our study population, which is in line with the results of the study on other populations [35]. As we described previously [36], the lack of TT (Leu/Leu) genotype in our study population (with deviation from Hardy–Weinberg equilibrium) may point to the prevalence of the C (Pro) allele in the healthy, physically fit population.

Among different factors, androgens take part in building the skeleton of young men and help to prevent bone loss in elderly men [37] by having antiapoptotic effects on osteocytes and osteoblasts and apoptotic effects on osteoclasts [38]. The results of our study on a young, male population confirmed the above-mentioned issue, since serum testosterone concentration was found to be a positive predictor of both trunk and total BMD.

The process of formation and resorption happens alternatively in bone development during growth and aging [2]. Serum OC is considered a specific marker of osteoblast function, as its levels have been shown to correlate with bone formation rates [39]. Furthermore, OC has been shown to be sensitive to alterations in bone metabolism due to physical exercise [40]. However, in our study, no differences in serum OC (which was at levels observed in a previous study in a young, male population [40]) were found between the CON and TR groups. Instead, our results revealed that serum OC was a negative predictor of BMD at leg, trunk and the whole body (Table 5), which was surprising in our young, healthy men. On the other hand, serum OC may be considered as the marker of overall bone turnover [39]. It has been suggested that an increase in the levels of bone turnover markers is followed by a decrease in BMD after menopause but also by an increase in BMD during puberty [41]. Therefore, in our young people in the phase of increasing their peak bone mass, an increase in serum OC with lower BMD may indicate more active bone turnover during the formation of peak BMD.

Bone turnover is mentioned to be influenced by vitamin D and calcium status. A study on a large, representative sample of healthy adolescents [42] showed that boys with high vitamin D status had significantly lower serum osteocalcin than those with low and moderate vitamin D status.

In our study, calcium intake was at a moderate level (mean values: 822.1 ± 360.3 and 792.2 ± 436.3 mg/day for CON and TR groups, respectively; data of dietary analysis are not shown in this paper). Additionally, the mean values of 25-OH D concentration in serum were within the reference ranges. Moreover, no relationship was found between areal BMD and serum Ca or serum 25-OH D concentration. However, taking into account individual vitamin D status (serum 25-OH D values) among the whole study population, vitamin D deficiency/insufficiency (below 30 ng/mL [38]) concerned 21% and 33% of the participants from the TR and CON groups, respectively. In addition, only in the subgroup with 25-OH D < 30 ng/mL, we observed a reverse correlation between serum OC and 25-OH D, which confirms previous findings [42].

Serum 25-OH D was not a predictor of BMD values in the present study. However, it must be pointed out that the polymorphisms of the *VDR* gene are well-known, important factors affecting vitamin D bioavailability and/or calcium absorption and, as a result, bone mass [43,44]. Among *VDR* polymorphisms measured in our study, only *VDR FokI* start codon polymorphism was associated with BMD values.

Our regression analysis proved the *FokI* genotypes to be independently related to BMD values, with the GG genotype (and the AG genotype, but not significantly) being a negative predictor of BMD at almost all sites except for the arm (Table 5).

However, the results of previous studies are conflicting since the genotype–phenotype relationship is not uniform [45,46]. Our finding that the G allele seems to be unfavorable to bone tissue condition is in opposition to previous results on Swedish adolescent boys [46], where the FF (GG) genotype was related to higher BMD compared to ff (AA). Additionally, the A–G transition is generally mentioned as resulting in the synthesis of a smaller VDR protein with increased biological activity [47]. These discrepancies should be clarified in further studies, but one explanation could be a possible interaction of the effects of the *FokI VDR* polymorphism and physical activity/exercise training on bone metabolism [47]. In our study, the AG genotype of the TR group had higher spine BMD than the AG genotype of the CON group (Figure 5). In addition, a clear intergroup difference (but not significant) was also seen in the case of the AA genotype (Figure 5). Thus, our results indicate that genotypes considered as being less favorable for BMD [45,46] appear to be more susceptible to changes induced by sports training.

Our findings are in line with a previous study by Laaksonen et al. [45] on Finnish adolescent boys, in which the AG genotype was associated with higher forearm BMD compared to the GG genotype, with these inter-genotype differences being significant only after adjusting for physical activity levels. In addition, in another investigation on elderly people involved in resistance training [48], the AG genotype was related to the highest responses in BMD in the femoral neck. Finally, a Brazilian study on adolescent soccer players also found higher total BMC and BMD in the AG genotype compared to the GG genotype [49].

In addition to *VDR FokI* polymorphism, the AluI polymorphism in *CALCR* gene was also associated with BMD values in our participants. This polymorphism (T to C substitution, which causes substitution of a leucine amino acid in 463 site with proline) has been reported to alter the secondary structure and, as a result, to affect the biological activity of the receptor of calcitonin, playing an important role in bone and calcium homeostasis [13]. As reported, a lack of proline residue could decrease the calcitonin hormone bond to the calcitonin receptor [50]. Accordingly, we found that the GG (CC) genotype (i.e., homozygous proline) of the *CALCR* gene was more frequent in the TR than the CON group (Table 3), which may suggest that this genotype is beneficial for athletes in combat and team sports. Additionally, both genotypes with the G allele (GG and AG) were positive predictors of arm BMD, although only AG reached a significant level ($p < 0.05$), probably because of the low frequency of the GG genotype in our participants.

The results of studies evaluating the association between *CALCR* polymorphism and BMD are inconsistent, with different genotypes indicated as unfavorable, i.e., GG in middle-aged and older men [14] and AA in postmenopausal women [13], while no association was found in a population of young, healthy men [15]. Rather, our results are in line with the findings of the study on postmenopausal women of Caucasian origin [51], in which significantly higher BMD at the femoral neck was seen in heterozygous subjects (AG) compared with in the homozygous leucine (AA) and proline (GG) genotypes. Additionally, a heterozygote was related to decreased fracture risk in that population [51]. However, it must be noted that, similar to our study, there were few subjects with the GG genotype. On the other hand, it has been suggested that heterozygotes are the most beneficial for bone health, as expression of both alleles of the receptor confers an advantage over homozygotes [50,51].

Although, in our study, the GG genotype of *CALCR* was overrepresented in the TR group, and the AG genotype was found to be a positive predictor of arm BMD, ANOVA indicated higher arm BMD in the AA genotype of the TR group than in the AA genotype of the CON group (Figure 6). Thus, even though the G allele seemed to be a positive predictor of arm BMD in our subjects, our results described above indicated that practicing combat (wrestling, MMA) and team sports (handball) improves arm BMD, and this improvement was more evident in the genotype with the lowest bone density (i.e., AA), as indicated in another study [13].

4. Material and Methods

4.1. Subjects

The study was carried out on 181 young male volunteers during the period of reaching peak bone mass (20–23 years) in the spring of 2018. Taking into account their physical activity level, they were divided into two groups: control (n = 87) and trained (n = 94). The control group consisted of untrained students from the Faculty of Physical Education and Health in Biała Podlaska who regularly participated in practical classes included in the study curriculum for 3 years and declared that they did not perform regular physical activity outside of the physical education classes at university (5 h per week). Exclusion criteria were as follows: a recent leg injury/stress fracture, a history of musculoskeletal diseases, current smokers, chronic steroid treatment, taking any medications and dietary supplements less than 3 months before the study.

The trained group included men practicing different kinds of sports in local sports clubs, such as soccer (n = 44), wrestling (n = 22), handball (n = 13) and mixed martial arts (MMA, n = 15). Soccer players (representing 4th league soccer clubs) and handball players (representing 1st league handball clubs), as well as MMA fighters (at a collegiate level), were students from the Faculty of Physical Education and Health in Biała Podlaska, whereas the wrestlers (at a national level) were current or previous students from the Sports School in Radom. All these athletes declared long training experience (7.5 ± 2.5 years) and weekly training loads of 10–12 h.

All the participants provided their written, informed consent to take part in the study. The study was in compliance with the Helsinki Declaration. The protocol of the study was approved by the Local Ethics Committee at the University of Physical Education in Warsaw (no. SKE 01-24/2015).

4.2. BMC and BMD Measurements

The height (cm) of the athletes that participated in the study was measured with an accuracy of 1 mm by using an anthropometric set in a barefoot position with feet placed on the ground on one level, heels joined, the knees stretched and upright. The weight (kg) was measured with an electronic scale with clothes as thin as possible on the athletes with an accuracy of 100 g.

Bone mineral content BMC (g) and bone mineral density BMD (g/cm^2) were obtained from a whole-body scan with the use of dual-energy X-ray absorptiometry (DEXA) on a HORIZON Ci device (USA). The following parts of the body were taken into account: the lumbar spine (L_1–L_4), the arm and the leg (averaged values of the left and the right limb), as well as trunk and total body. For the total-body scan, participants were asked to lie in the supine position, centered within the scan field. The hands were placed on the sides of the legs in the prone position, while the legs were straight and strapped together.

4.3. Blood Sampling and Biochemical Analyses

Fasting blood samples were taken from the ulnar vein in the morning (at 7:00 a.m.). All the subjects had not eaten any food for 10–12 h and had not exercised for 24 h prior to blood drawing. The samples were collected using test tubes with EDTA (for the whole blood, erythrocyte and plasma analyses, as well as DNA isolation) and without anticoagulants (for separation of serum). In order to form a blood clot, the samples were exposed at room temperature and then centrifuged (for 10 min at $3000\times g$ at a temperature of 4 °C) to separate serum. Additionally, the portion of the blood with EDTA was centrifuged to separate erythrocytes and plasma. Subsequently, the erythrocytes were washed three times with a cold, isotonic saline solution. Erythrocytes, plasma, serum and whole blood were frozen and stored at −80 °C until analysis.

The measured biochemical parameters were activity of superoxide dismutase (SOD) in erythrocytes, activity of glutathione peroxidase (GPx) in whole blood and total antioxidant capacity (TAC) of plasma, as well as plasma concentration of uric acid (UA), lipid hydroperoxides (LOOHs) and phosphates (P). Serum was analyzed for the activity of alkaline phosphatase (ALP), as well the concentration of calcium (Ca), osteocalcin (OC), testosterone (T) and 25-OH vitamin D (25-OH D).

SOD and GPx activities were determined with commercially available kits (RANSOD cat. no. SD 125 and RANSEL cat. no. RS 505, respectively; Randox, Crumlin, UK). The antioxidant enzyme activities were measured at 37 °C and expressed in U/g Hb. Hemoglobin was assessed using a standard cyanmethemoglobin method with a diagnostic kit (cat. no. HG 1539; Randox, Crumlin, UK). The enzymatic activities were measured at 37 °C and expressed in U/g Hb.

The total antioxidant capacity of plasma (TAC) to scavenge ABTS radicals was measured using a chromogenic method with a commercially available kit (cat. no. NX 2332; Randox, Crumlin, UK). The antioxidant capacity of samples was expressed as millimoles per liter of Trolox equivalents (6-hydroxy-2,5,7,8-tetramethylchroman-2-carboxylic acid).

LOOHs levels were determined as described previously [52]; the assay was based on the reaction of a chromogenic reagent, N-methyl-2-phenylindole, with malondialdehyde and 4-hydroxyalkenals at 45 °C. As a result, a stable chromophore was formed with maximum absorbance at 586 nm.

Serum levels of UA (cat. no. K6581–100), Ca (cat. no. W6504–100), P (cat. no. F6516–100) and ALP activity (cat. no. F6406–075) were determined with commercially available kits (Alpha Diagnostics, Poland) using multicalibrator (cat. no. K6504–03), normal and pathological control serum (cat. nos. S6590–05 and S6591-05, respectively; Alpha Diagnostics, Poland). These biochemical parameters (as well as SOD, GPx, TAC and LOOHs) were measured spectrophotometrically at 37 °C using an automatic biochemical analyzer A15 (Bio-Systems S.A., Montcada I Reixac, Spain) by a certified laboratory diagnostician (K.J.) to minimize as much as possible the influence of inter-assay variation.

The concentration of osteocalcin, testosterone and 25-OH vitamin D was measured with the ELISA method using diagnostic sets (Cormay, Poland) at a local, certified, clinical, diagnostic laboratory. All commercially available kits listed above were IVD (in vitro diagnostic).

4.4. Genotyping

Genomic DNA for genotyping was isolated from peripheral venous blood using a QIAamp DNA Blood Mini Kit (Qiagen GmbH, Hilden, Germany). Concentration of DNA was determined with Picodrop microliter spectrophotometer (PicoDrop, UK). The following gene polymorphisms were genotyped with commercially available TaqMan kit (Applied Biosystems, Foster City, CA, USA): vitamin D receptor (*VDR*), i.e., *ApaI* (rs 7975232, C_28977635_10, cat. no. 4351379), *BsmI* (rs1544410, C_8716062_20, cat. no. 4351379) and *FokI* (rs 2228570, C_12060045_20, cat. no. 4351379); type 1 collagen *COLIA1* (rs 1800012, C_7477170_30, cat. no. 4351379); calcitonin receptor *CALCR* (rs 1801197, C_2541576_1, cat. no. 4351379); and antioxidant enzymes, i.e., *SOD1* A-39 C (rs2234694, C_34770911_10, cat. no. 4351379), *SOD2* Ala-9Val (rs4880, C_8709053_10, cat. no. 4351379) and *GPx* Pro198Leu (rs1050450, C_175686987_10, cat. no. 4351379). Genotyping for all gene polymorphisms was carried out by a 10 μL PCR reaction on DNA (50 ng) using a TaqMan PCR Master Mix (5 μL, Applied Biosystems) and fluorescent 5'-exonuclease TaqMan SNP assays (0.5 μL, Applied Biosystems) with FAM and VIC fluorophore-labeled probes. Real-time PCR was performed on Rotor Gene (Qiagen GmbH, Hilden, Germany) according to the following protocol recommended by the manufacturer of TaqMan Assays (Applied Biosystems): an initial 5 min at 95 °C followed by 40–45 cycles of 15 s at 95 °C, 30 s at 60 °C, 30 s at 72 °C and, finally, 8 min at 72 °C. For quality control, positive and negative controls and blinded duplicate samples were run. Sample images of TaqMan analysis results for each of the polymorphisms tested are included in the Supplementary Materials (Figure S1).

4.5. Statistical Analysis

Anthropometric data, biochemical parameters and the parameters of bone tissue condition (BMC and BMD at different points of body skeleton) in both groups (trained and control) were analyzed with one-way ANOVA using Statistica version 13.3 software package (StatSoft, Krakow, Poland). Moreover, relationships between biochemical parameters were analyzed on the basis of Pearson's coefficients of linear correlation. The normal distribution of all variables was confirmed with the Shapiro–Wilk test and visual inspection (quantile distribution plots). All values were reported as mean ± standard deviation (SD). The level of statistical significance was set at $p < 0.05$.

For each SNP, deviation of the genotype frequencies from those expected under the Hardy–Weinberg equilibrium was assessed in both groups with chi-square (χ^2) test [53]. Genotype frequencies in CON and TR groups were compared using a likelihood ratio (χ^2 test).

Multiple regression was the basic data analysis tool [54]. The model was built in an iterative "brute force" way using r1071 R package [55] by testing all combinations of predictors (in this case $2^{18} - 1$) in order to avoid the influence of the order of substitution of predictors on the model quality. The model with the lowest Akaike information criterion (AIC) was selected.

5. Conclusions

Age, BMI, training status, serum testosterone and osteocalcin, as well as genetic factors, are the predictors of bone mass parameters in young, healthy men. Practicing combat sports (wrestling, MMA) and team sports (handball) is an independent positive predictor of BMD/BMC values at various skeletal sites. The main finding of this study is that, at almost all measured skeletal sites, the *SOD2* AG and *FokI VDR* GG genotypes seem to be negative predictors of BMC and BMD, respectively, while the *CALCR* AG genotype appears as a positive predictor of arm BMD values. Furthermore, these three genetic polymorphisms seem to modulate the response of bone mass parameters to sports training. Overall, at least within the *VDR FokI* and *CALCR* polymorphisms, less favorable genotypes in terms of BMD (i.e., *FokI* AG and *CALCR* AA) appear to be associated with a greater BMD response to sports training. This suggests that, in healthy men during the period of bone mass formation, sports training (combat and team sports) may attenuate the negative impact of genetic factors on bone tissue condition, possibly reducing the risk of osteoporosis in later age.

Supplementary Materials: The following supporting information can be downloaded at: https://www.mdpi.com/article/10.3390/ijms24043373/s1.

Author Contributions: E.J. and B.D. conceived and designed the research. B.D. and J.K. conducted the experiments. E.J., B.D. and I.C. analyzed the data. E.J. and I.C. interpreted the results of the experiments. E.J. wrote the manuscript. All authors have read and agreed to the published version of the manuscript.

Funding: This study was conducted within research project no. DS 248 of the Faculty of Physical Education and Health in Biała Podlaska, the Józef Piłsudski University of Physical Education in Warsaw, and was supported by the Polish Ministry of Science and Higher Education. The experimental data were obtained at the Regional Centre of Research and Development of the Faculty of Physical Education and Health in Biała Podlaska, Poland (in the Therapy and Musculoskeletal System Diagnostics Laboratory, as well as the following laboratories: Physiological, Biochemical and Molecular Diagnosis).

Institutional Review Board Statement: The protocol of the study was approved by the Research Ethics Committee, University of Physical Education, Warsaw, Poland (SKE 01-24/2015) prior to the enrolment of the participants.

Informed Consent Statement: Informed consent was obtained from all subjects involved in the study.

Data Availability Statement: The data presented in this study are available on request from the corresponding author E.J.

Acknowledgments: We wish to thank Karol Kowieski for his contribution in conducting the experiments.

Conflicts of Interest: The authors declare no conflict of interest.

Abbreviations

25-OH D	25-OH vitamin D
ALP	alkaline phosphatase
BMC	bone mineral content
BMD	bone mineral density
Ca	calcium
CALCR	receptor of calcitonin
COLIA1	collagen type I
GPx	glutathione peroxidase
LOOHs	lipid hydroperoxides
MMA	mixed martial arts
OC	osteocalcin
P	phosphates
SOD1	supeoxide dismutase 1
SOD2	superoxide dismutase 2
T	testosterone
TAC	total antioxidant capacity
UA	uric acid
VDR	vitamin D receptor

References

1. Khosla, S.; Amin, S.; Orwoll, E. Osteoporosis in men. *Endocr. Rev.* **2008**, *29*, 441–464. [PubMed]
2. Arazi, H.; Eghbali, E. Effects of Different Types of Physical Training on Bone Mineral Density in Men and Women. *J. Osteopor. Phys. Act.* **2017**, *5*, 3. [CrossRef]
3. Gentil, P.; de Lima Lins, T.C.; Lima, R.M.; de Abreu, B.S.; Grattapaglia, D.; Bottaro, M.; de Oliveira, R.J.; Pereira, R.W. Vitamin-d-receptor genotypes and bone-mineral density in postmenopausal women: Interaction with physical activity. *J. Aging Phys. Act.* **2009**, *17*, 31–45. [CrossRef] [PubMed]
4. Ruffing, J.A.; Cosman, F.; Zion, M.; Tendy, S.; Garrett, P.; Lindsay, R.; Nieves, J.W. Determinants of bone mass and bone size in a large cohort of physically active young adult men. *Nutr. Metab.* **2006**, *3*, 14. [CrossRef] [PubMed]
5. Walsh, J.S. Normal bone physiology, remodelling and its hormonal regulation. *Surgery* **2015**, *33*, 1–6. [CrossRef]
6. Ocarino, N.M.; Serakides, R. Effect of the physical activity on normal bone and on the osteoporosis prevention and treatment. *Rev. Bras. Med. Esporte* **2006**, *12*, 164–168. [CrossRef]
7. Tong, X.; Chen, X.; Zhang, S.; Huang, M.; Shen, X.; Xu, J.; Zou, J. The Effect of Exercise on the Prevention of Osteoporosis and Bone Angiogenesis. *Biomed. Res. Int.* **2019**, *2019*, 8171897. [CrossRef]
8. Hinrich, T.; Chae, E.H.; Lehmann, R.; Allolio, B.; Platen, P. Bone Mineral Density in Athletes of Different Disciplines: A Cross Sectional Study. *Open Sport. Sci. J.* **2010**, *3*, 129–133. [CrossRef]
9. Długołęcka, B.; Jówko, E. Effects of Weight-Bearing and Weight-Supporting Sports on Bone Mass in Males. *Pol. J. Sport Tour.* **2022**, *29*, 9–14. [CrossRef]
10. Ralston, S.H.; de Crombrugghe, B. Genetic regulation of bone mass and susceptibility to osteoporosis. *Genes Dev.* **2006**, *20*, 2492–2506. [CrossRef]
11. Zhang, L.; Yin, X.; Wang, J.; Xu, D.; Wang, Y.; Yang, J.; Tao, Y.; Zhang, S.; Feng, X.; Yan, C. Associations between VDR gene polymorphisms and osteoporosis risk and bone mineral density in postmenopausal women: A systematic review and meta-analysis. *Sci. Rep.* **2018**, *8*, 981. [CrossRef] [PubMed]
12. Kow, M.; Akam, E.; Singh, P.; Singh, M.; Cox, N.; Bhatti, J.S.; Tuck, S.P.; Francis, R.M.; Datta, H.; Mastana, S. Vitamin D receptor (VDR) gene polymorphism and osteoporosis risk in White British men. *Ann. Hum. Biol.* **2019**, *46*, 430–433. [CrossRef] [PubMed]
13. Masi, L.; Brandi, M.L. Calcitonin and calcitonin receptors. *Clin. Cases Miner. Bone Metab.* **2007**, *4*, 117–122. [PubMed]
14. Braga, V.; Sangalli, A.; Malerba, G.; Mottes, M.; Mirandola, S.; Gatti, D.; Rossini, M.; Zamboni, M.; Adami, S. Relationship among VDR (BsmI and FokI), COLIA1, and CTR polymorphisms with bone (BsmI and FokI), COLIA1, and CTR polymorphisms with bone mass, bone turnover markers, and sex hormones in men. *Calcif. Tissue Int.* **2002**, *70*, 457–462. [CrossRef] [PubMed]
15. Charopoulos, I.; Trovas, G.; Stathopoulou, M.; Kyriazopoulos, P.; Galanos, A.; Dedoussis, G.; Antonogiannakis, E.; Lyritis, G.P. Lack of association between vitamin D and calcitonin receptor gene polymorphisms and forearm bone values of young Greek males. *J. Musculoskelet. Neuronal Interact.* **2008**, *8*, 196–203. [PubMed]
16. Wu, J.; Yu, M.; Zhou, Y. Association of collagen type I alpha 1 +1245G/T polymorphism and osteoporosis risk in post-menopausal women: A meta-analysis. *Int. J. Rheum. Dis.* **2017**, *20*, 903–910. [CrossRef]
17. Nakamura, O.; Ishii, T.; Ando, Y.; Amagai, H.; Oto, M.; Imafuji, T.; Tokuyama, K. Potential role of vitamin D receptor gene polymorphism in determining bone phenotype in young male athletes. *J. Appl. Physiol.* **2002**, *93*, 1973–1979. [CrossRef]
18. Puthucheary, Z.; Skipworth, J.R.; Rawal, J.; Loosemore, M.; Van Someren, K.; Montgomery, H.E. Genetic influences in sport and physical performance. *Sports Med.* **2011**, *41*, 845–859. [CrossRef]

19. Maggio, D.; Barabani, M.; Pierandrei, M.; Polidori, M.C.; Catani, M.; Mecocci, P.; Senin, U.; Pacifici, R.; Cherubini, A. Marked decrease in plasma antioxidants in aged osteoporotic women: Results of a cross-sectional study. *J. Clin. Endocrinol. Metabol.* **2003**, *88*, 1523–1537. [CrossRef]

20. Botre, C.; Shahu, A.; Adkar, N.; Shouche, Y.; Ghaskadbi, S.; Ashma, R. Superoxide Dismutase 2 Polymorphisms and Osteoporosis in Asian Indians: A Genetic Association Analysis. *Cell Mol. Biol. Lett.* **2015**, *20*, 685–697. [CrossRef]

21. Zhang, Y.B.; Zhong, Z.M.; Hou, G.; Jiang, H.; Chen, J.T. Involvement of oxidative stress in age-related bone loss. *J. Surg. Res.* **2011**, *169*, 37–42. [CrossRef] [PubMed]

22. Sánchez-Rodríguez, M.A.; Ruiz-Ramos, M.; Correa-Muñoz, E.; Mendoza-Núñez, V.M. Oxidative stress as a risk factor for osteoporosis in elderly Mexicans as characterized by antioxidant enzymes. *BMC Musculoskelet. Disord.* **2007**, *8*, 124. [CrossRef] [PubMed]

23. Bastaki, M.; Huen, K.; Manzanillo, P.; Chande, N.; Chen, C.; Balmes, J.R.; Tager, I.B.; Holland, N. Genotype-activity relationship for Mn-superoxide dismutase, glutathione peroxidase 1 and catalase in humans. *Pharmacogenet. Genom.* **2006**, *16*, 279–286. [CrossRef] [PubMed]

24. Bresciani, G.; Cruz, I.B.; de Paz, J.A.; Cuevas, M.J.; Gonzalez-Gallego, J. The MnSOD Ala16Val SNP: Relevance to human diseases and interaction with environmental factors. *Free Rad. Res.* **2013**, *47*, 781–792. [CrossRef] [PubMed]

25. Mlakar, S.J.; Osredkar, J.; Prezelj, J.; Marc, J. The antioxidant enzyme GPX1 gene polymorphisms are associated with low BMD and increased bone turnover markers. *Dis. Markers* **2010**, *29*, 71–80. [CrossRef]

26. Mlakar, S.J.; Osredkar, J.; Prezelj, J.; Marc, J. Antioxidant enzymes GSR, SOD1, SOD2, and CAT gene variants and bone mineral density values in postmenopausal women: A genetic association analysis. *Menopause* **2012**, *19*, 368–376. [CrossRef] [PubMed]

27. Gomez-Cabrera, M.C.; Domenech, E.; Viña, J. Moderate exercise is an antioxidant: Upregulation of antioxidant genes by training. *Free Radic. Biol. Med.* **2008**, *44*, 126–131. [CrossRef] [PubMed]

28. Hagman, M. Bone mineral density in lifelong trained male football players compared with young and elderly untrained men. *J. Sport Health Sci.* **2018**, *7*, 159–168. [CrossRef]

29. Cadore, E.L.; Brentano, M.A.; Kruel, L.F.M. Effects of the physical activity on the bone mineral density and bone remodelation. *Rev. Bras. Med. Esporte* **2005**, *11*, 373–379. [CrossRef]

30. Fragala, M.S.; Bi, C.; Chaump, M.; Kaufman, H.W.; Kroll, M.H. Associations of aerobic and strength exercise with clinical laboratory test values. *PLoS ONE* **2017**, *12*, e0180840. [CrossRef]

31. Jówko, E.; Gromisz, W.; Sadowski, J.; Cieśliński, I.; Kotowska, J. SOD2 gene polymorphism may modulate biochemical responses to a 12-week swimming training. *Free Rad. Biol. Med.* **2017**, *113*, 571–579. [CrossRef] [PubMed]

32. Jówko, E.; Gierczuk, D.; Cieśliński, I.; Kotowska, J. SOD2 gene polymorphism and response of oxidative stress parameters in young wrestlers to a three-month training. *Free Rad. Res.* **2017**, *51*, 506–516.

33. Crawford, A.; Fassett, R.G.; Geraghty, D.P.; Kunde, D.A.; Ball, M.J.; Robertson, I.K.; Coombes, J.S. Relationships between single nucleotide polymorphisms of antioxidant enzymes and disease. *Gene* **2012**, *501*, 89–103. [PubMed]

34. Ravn-Haren, G.; Olsen, A.; Tjønneland, A.; Dragsted, L.O.; Nexø, B.A.; Wallin, H.; Overvad, K.; Raaschou-Nielsen, O.; Vogel, U. Associations between GPX1 Pro198Leu polymorphism, erythrocyte GPX activity, alcohol consumption and breast cancer risk in a prospective cohort study. *Carcinogenesis* **2006**, *27*, 820–825. [CrossRef] [PubMed]

35. Wickremasinghe, D.; Peiris, H.; Chandrasena, L.G.; Senaratne, V.; Perera, R. Case control feasibility study assessing the association between severity of coronary artery disease with Glutathione Peroxidase-1 (GPX-1) and GPX-1 polymorphism (Pro198Leu). *BMC Cardiovasc. Disord.* **2016**, *16*, 111. [CrossRef] [PubMed]

36. Kotowska, J.; Jówko, E. Effect of Gene Polymorphisms in Antioxidant Enzymes on Oxidative-Antioxidative Status in Young Men. *Pol. J. Sport Tour.* **2020**, *27*, 7–13. [CrossRef]

37. Mohammadi, Z.; Fayyazbakhsh, F.; Ebrahimi, M.; Amoli, M.M.; Khashayar, P.; Dini, M.; Zadeh, R.N.; Keshtkar, A.; Barikani, H.R. Association between vitamin D receptor gene polymorphisms (Fok1 and Bsm1) and osteoporosis: A systematic review. *J. Diabetes Metab. Disord.* **2014**, *13*, 98. [CrossRef]

38. Goolsby, M.A.; Boniqui, N. Bone Health in Athletes. *Sports Health* **2017**, *9*, 108–117.

39. Jagtap, V.R.; Ganu, J.V. Serum osteocalcin: A specific marker for bone formation in postmenopausal osteoporosis. *Int. J. Pharm. Bio. Sci.* **2011**, *1*, 510–517.

40. Alghadir, A.H.; Gabr, S.A.; Al-Eisa, E. Physical activity and lifestyle effects on bone mineral density among young adults: Sociodemographic and biochemical analysis. *J. Phys. Ther. Sci.* **2015**, *27*, 2261–2270.

41. Szulc, P.; Delmas, P.D. Biochemical Markers of Bone Turnover in Osteoporosis. In *Osteoporosis*, 3rd ed.; Chapter 63; Marcus, R., Feldman, D., Nelson, D., Rosen, C., Eds.; Academic Press, Elsevier Inc.: Cambridge, MA, USA, 2008; pp. 1519–1545.

42. Cashman, K.D.; Hill, T.R.; Cotter, A.A.; Boreham, C.A.; Dubitzky, W.; Murray, L.; Strain, J.; Flynn, A.; Robson, P.J.; Wallace, J.M.; et al. Low vitamin D status adversely affects bone health parameters in adolescents. *Am. J. Clin. Nutr.* **2008**, *87*, 1039–1044. [CrossRef] [PubMed]

43. Ames, S.K.; Ellis, K.J.; Gunn, S.K.; Copeland, K.C.; Abrams, S.A. Vitamin D receptor gene FokI polymorphism predicts calcium absorption and bone mineral density in children. *J. Bone Miner. Res.* **1999**, *14*, 740–746.

44. Tanabe, R.; Kawamura, Y.; Tsugawa, N.; Haraikawa, M.; Sogabe, N.; Okano, T.; Hosoi, T.; Goseki-Sone, M. Effects of Fok-I polymorphism in vitamin D receptor gene on serum 25-hydroxyvitamin D, bone-specific alkaline phosphatase and calcaneal quantitative ultrasound parameters in young adults. *Asia Pac. J. Clin. Nutr.* **2015**, *24*, 329–335. [PubMed]

45. Laaksonen, M.M.; Kärkkäinen, M.U.; Outila, T.A.; Rita, H.J.; Lamberg-Allardt, C.J. Vitamin D receptor gene start codon polymorphism (FokI) is associated with forearm bone mineral density and calcaneal ultrasound in Finnish adolescent boys but not in girls. *J. Bone Miner. Metab.* **2004**, *22*, 479–485.

46. Strandberg, S.; Nordström, P.; Lorentzon, R.; Lorentzon, M. Vitamin D receptor start codon polymorphism (FokI) is related to bone mineral density in healthy adolescent boys. *J. Bone Miner. Metab.* **2003**, *21*, 109–113. [CrossRef]

47. Tajima, O.; Ashizaw, N.; Ishii, T.; Amagai, H.; Mashimo, T.; Liu, L.J.; Saitoh, S.; Tokuyama, K.; Suzuki, M. Interaction of the effects between vitamin D receptor polymorphism and exercise training on bone metabolism. *J. Appl. Physiol.* **2000**, *88*, 1271–1276. [CrossRef]

48. Rabon-Stith, K.M.; Hagberg, J.M.; Phares, D.A.; Kostek, M.C.; Delmonico, M.J.; Roth, S.M.; Ferrell, R.E.; Conway, J.M.; Ryan, A.S.; Hurley, B.F. Vitamin D receptor FokI genotype influences bone mineral density response to strength training, but not aerobic training. *Exp. Physiol.* **2005**, *90*, 653–661. [CrossRef]

49. Diogenes, M.E.; Bezerra, F.F.; Cabello, G.M.; Cabello, P.H.; Mendonça, L.M.; Oliveira, A.V., Jr.; Donangelo, C.M. Vitamin D receptor gene FokI polymorphisms influence bone mass in adolescent football (soccer) players. *Eur. J. Appl. Physiol.* **2010**, *108*, 31–38. [CrossRef]

50. Dehghan, M.; Pourahmad-Jaktaji, R.; Farzaneh, Z. Calcitonin Receptor AluI (rs1801197) and Taq Calcitonin Genes Polymorphism in 45-and Over 45-year-old Women and their Association with Bone Density. *Acta Inform. Med.* **2016**, *24*, 239–243. [CrossRef]

51. Taboulet, J.; Frenkian, M.; Frendo, J.L.; Feingold, N.; Jullienne, A.; de Vernejoul, M.C. Calcitonin receptor polymorphism is associated with a decreased fracture risk in post-menopausal women. *Hum. Mol. Genet.* **1998**, *7*, 2129–2133. [CrossRef] [PubMed]

52. Gérard-Monnier, D.; Erdelmeier, I.; Régnard, K.; Moze-Henry, N.; Yadan, J.C.; Chaudière, J. Reactions of N-methyl-2-phenylindole with malondialdehyde and 4-hydroxyalkenals. Mechanistic aspects of the colorimetric assay of lipid peroxidation. *Chem. Res. Toxicol.* **1998**, *11*, 1176–1183. [CrossRef] [PubMed]

53. Graffelman, J. Exploring Diallelic Genetic Markers: The Hardy Weinberg Package. *J. Stat. Softw.* **2015**, *64*, 1–23.

54. Biecek, P. *Analiza Danych z Programem R. Modele Liniowe z Efektami Stałymi, Losowymi i Mieszanymi*; Wydawnictwo Naukowe PWN: Warsaw, Poland, 2013. (In Polish)

55. Meyer, D.; Dimitriadou, E.; Hornik, K.; Weingessel, A.; Leisch, F. *Misc Functions of the Department of Statistics, Probability Theory Group (Formerly: E1071); R Package Version 1.7-3*. 2019, e1071; TU Wien: Vienna, Austria, 2023; Available online: https://CRAN.R-project.org/package=e1071 (accessed on 5 December 2019).

International Journal of
Molecular Sciences

Review

Significance of Singlet Oxygen Molecule in Pathologies

Kazutoshi Murotomi [1,†], Aya Umeno [2,†], Mototada Shichiri [3,*], Masaki Tanito [2] and Yasukazu Yoshida [4]

1   Biomedical Research Institute, National Institute of Advanced Industrial Science and Technology (AIST), Tsukuba 305-8566, Japan
2   Department of Ophthalmology, Shimane University Faculty of Medicine, Izumo 693-8501, Japan
3   Biomedical Research Institute, National Institute of Advanced Industrial Science and Technology (AIST), Ikeda 563-8577, Japan
4   LG Japan Lab Inc., Yokohama 220-0011, Japan
*   Correspondence: mototada-shichiri@aist.go.jp; Tel.: +81-72-751-8234
†   These authors contributed equally to this work.

Abstract: Reactive oxygen species, including singlet oxygen, play an important role in the onset and progression of disease, as well as in aging. Singlet oxygen can be formed non-enzymatically by chemical, photochemical, and electron transfer reactions, or as a byproduct of endogenous enzymatic reactions in phagocytosis during inflammation. The imbalance of antioxidant enzymes and antioxidant networks with the generation of singlet oxygen increases oxidative stress, resulting in the undesirable oxidation and modification of biomolecules, such as proteins, DNA, and lipids. This review describes the molecular mechanisms of singlet oxygen production in vivo and methods for the evaluation of damage induced by singlet oxygen. The involvement of singlet oxygen in the pathogenesis of skin and eye diseases is also discussed from the biomolecular perspective. We also present our findings on lipid oxidation products derived from singlet oxygen-mediated oxidation in glaucoma, early diabetes patients, and a mouse model of bronchial asthma. Even in these diseases, oxidation products due to singlet oxygen have not been measured clinically. This review discusses their potential as biomarkers for diagnosis. Recent developments in singlet oxygen scavengers such as carotenoids, which can be utilized to prevent the onset and progression of disease, are also described.

Keywords: reactive oxygen species; singlet oxygen; biomarkers; lipid peroxidation

check for
updates

Citation: Murotomi, K.; Umeno, A.; Shichiri, M.; Tanito, M.; Yoshida, Y. Significance of Singlet Oxygen Molecule in Pathologies. *Int. J. Mol. Sci.* **2023**, *24*, 2739. https://doi.org/10.3390/ijms24032739

Academic Editors: Rossana Morabito and Alessia Remigante

Received: 11 December 2022
Revised: 22 January 2023
Accepted: 26 January 2023
Published: 1 February 2023

1. Introduction

Oxygen can be recognized as a "double-edged sword" because aerobic organisms cannot live without it, even though the oxidative metabolism produces toxic by-products. Most deleterious phenomena arise because of free radicals. Hence, for survival, the elimination of highly reactive oxygen species (ROS) is important for living organisms. The human body is also constantly threatened by oxidative injury and radicals. Over the course of evolution, humans have developed defense systems to counteract oxidative damage, including the transformation and/or elimination of ROS, as a part of normal physiological homeostasis. When this homeostatic balance is disturbed by stress caused by environmental or radiation exposure, it is termed oxidative stress, that is, a state in which the homeostatic balance between oxidation reactions and antioxidant defenses is lost, leading to the oxidation of vital biomolecules and, finally, the onset of disease.

Thousands of studies have been reported over the past decade on the role of ROS in cell and tissue injury. ROS are generally accepted to be involved in various pathologies, including inflammation, asthma, muscular dystrophy, dementia, anaphylaxis, rheumatoid arthritis, reperfusion injury following ischemic stroke or heart attacks, cardiac toxicity of anti-cancer drugs, carcinogenicity of various chemicals, and smoking [1–5].

In this review, we focus on singlet oxygen ($^1O_2$) in ROS. However, we exclude photodynamic therapy (PDT), in which $^1O_2$ is stimulated by the light irradiation of a photosensitizing agent to induce a therapeutic effect on cancer [6], because many interesting review articles on PDT have already been published in recent years. The molecular aspects of (1) the mechanism of $^1O_2$ production in vivo, (2) methods for detecting $^1O_2$ and its oxidation-modification products, (3) implications in the pathogenesis of skin and eye diseases, and (4) compounds that scavenge $^1O_2$ are presented. Furthermore, we present our findings on the measurement of $^1O_2$-mediated oxidation products in diabetes, bronchial asthma, and ocular diseases. Although $^1O_2$ has been known to potentially be involved in the pathogenesis of certain diseases, lipid peroxidation products and other products produced by $^1O_2$ have not been actively measured clinically. In this review, we examine the usefulness of lipid peroxidation products produced by $^1O_2$ in diagnosing and understanding the pathophysiology of diseases.

2. Chemical Properties and Possible Production of Singlet Oxygen In Vivo

ROS are redox-active intermediates that are formed by the chemical, photochemical, or biochemical reduction in oxygen that, at least partially, triggers a chain of oxidative reactions. ROS are grouped into non-radical and radical species, the former including $^1O_2$ and hydrogen peroxide (H_2O_2), and the latter including superoxide anions ($O_2^{\bullet-}$), hydroxyl radicals ($HO^\bullet$), hydroperoxyl radicals ($HO_2^\bullet$), alkoxyl radicals ($RO^\bullet$), and peroxyl radicals ($ROO^\bullet$) [7]. $^1O_2$ is generated by energy transfer via chemical or photochemical pathways from an activated species to the oxygen molecule (triplet oxygen, $^3O_2$) (Figure 1). Other species are formed by a series of reactions initiated by $^1O_2$, or by certain one-electron reduction processes [7]. Therefore, ROS have been actively studied over the past three decades to determine whether they are beneficial or harmful.

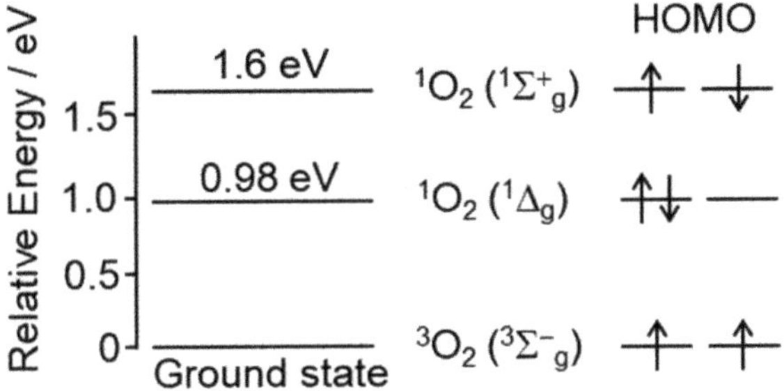

Figure 1. Relationship between energy levels of ground state triplet oxygen molecule $^3O_2$ and singlet oxygen molecule $^1O_2$. The ground state oxygen molecule, triplet oxygen $^3O_2$ ($^1\Sigma^-_g$), has two unpaired and spin-parallel electrons in π^* antibonding orbitals. $^3O_2$ ($^3\Sigma^-_g$) in the ground state receives energy and is excited to the singlet state, forming singlet oxygen ($^1O_2$). $^1O_2$ is the spin-flipped electron state of $^3O_2$ ($^3\Sigma^-_g$). There are two states of $^1O_2$: $^1O_2$ ($^1\Sigma^+_g$) and $^1O_2$ ($^1\Delta_g$). $^1O_2$ ($^1\Sigma^+_g$) and $^1O_2$ ($^1\Delta_g$) have energies that are 1.6 eV and 0.98 eV higher than the ground state $^3O_2$ ($^3\Sigma^-_g$), respectively. The lifetime of $^1O_2$ ($^1\Sigma^+_g$) is a few picoseconds, and it is rapidly converted to $^1O_2$ ($^1\Delta_g$). Since the lifetime of $^1O_2$ ($^1\Delta_g$) is several microseconds, which is significantly longer than that of $^1O_2$ ($^1\Sigma^+_g$), the $^1O_2$ generated in vivo can be $^1O_2$ ($^1\Delta_g$). The arrows indicate the direction of the electron spin of the highest occupied molecular orbital (HOMO).

High-energy radiations, such as X- and γ-rays, generate radical ROS directly by the ionization of H_2O or oxygen, whereas various chemical and photochemical reactions produce $^1O_2$ (Figure 2A), which in turn creates other ROS [8]. In the energy transfer mechanism, a photoactivated chromophore causes intersystem crossing to the triplet state, and triplet–triplet annihilation transfers part of the electronic energy to oxygen to generate $^1O_2$ [9,10]. Since the energy of most photoactivated molecules is higher than the excitation energy of $^1O_2$ (0.98 eV) (Figure 1), $^1O_2$ is readily produced by endogenous chromophores activated by ultraviolet (UV) irradiation [11]. Hatz et al. showed that the lifetime of $^1O_2$ generated by pulsed laser irradiation of a photosensitizer (5,10,15,20-tetrakis(N-methyl-4-pyridyl)-21H, 23H-phine (TMPyP)) incorporated into the nucleus of HeLa cells is about

3 µs. They also found that the generated $^1O_2$ diffuses from the formation point to a sphere radius of about 100 nm [12,13]. On the other hand, Liang et al. created a system that generates organelle-specific $^1O_2$ by irradiating 660 nm light using a photosensitizer that is locally expressed in the membrane, cytosol, endoplasmic reticulum, mitochondria, and nucleus [14]. Using this system, they observed differences in the irradiation energy required to induce cell death depending on the organelle in which the photosensitizer is expressed, as well as differences in the type of cell death (early apoptosis, necrosis, or late apoptosis) [14]. This result indicates that $^1O_2$ generated in intracellular organelles may not spread to other organelles. These reports indicate that $^1O_2$ generated by intracellular photosensitizers may induce cell death by oxidizing the components of the organelle containing the photosensitizer, rather than diffusing widely within the cell.

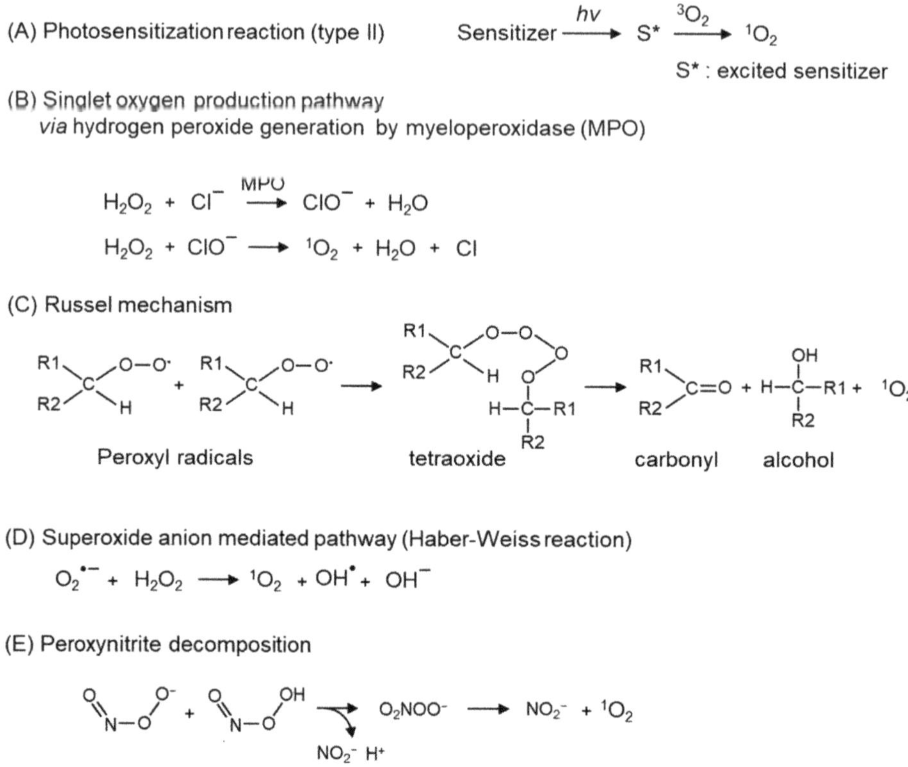

(A) Photosensitization reaction (type II)

$$\text{Sensitizer} \xrightarrow{hv} S^* \xrightarrow{^3O_2} {}^1O_2$$

S^* : excited sensitizer

(B) Singlet oxygen production pathway
via hydrogen peroxide generation by myeloperoxidase (MPO)

$$H_2O_2 + Cl^- \xrightarrow{MPO} ClO^- + H_2O$$

$$H_2O_2 + ClO^- \longrightarrow {}^1O_2 + H_2O + Cl$$

(C) Russel mechanism

(D) Superoxide anion mediated pathway (Haber-Weiss reaction)

$$O_2{}^{\bullet-} + H_2O_2 \longrightarrow {}^1O_2 + OH^\bullet + OH^-$$

(E) Peroxynitrite decomposition

Figure 2. Major $^1O_2$ production mechanisms. $^1O_2$ production pathways; (**A**) photochemical reaction, (**B**) reactions mediated by hydrogen peroxide produced by myeloperoxidase (MPO), (**C**) decomposition of peroxyl radicals (Russel mechanism), (**D**) reaction mediated by superoxide anion ($O_2{}^{\bullet-}$), (**E**) pathway via peroxynitrite decomposition.

The secretion of myeloperoxidase (MPO) from phagocytes generates hypochlorite (ClO^-), which reacts with H_2O_2 to form $^1O_2$ (Figure 2B) [15]. In leukocytes, $^1O_2$ generated via peroxidases and nicotinamide adenine dinucleotide phosphate (NADPH) oxidase critically contributes to an antimicrobial role [16]. Another endogenous pathway of $^1O_2$ is the degradation of lipid peroxides or ROO$^\bullet$ via the Russel mechanism (Figure 2C) [17,18]. ROO$^\bullet$ and lipid peroxides, which are the substrates of the Russel mechanism, are derived from the peroxidation of polyunsaturated fatty acids (PUFA) by auto-oxidation and enzymatic-oxidation, such as cytochrome c, lactoperoxidase [19], and lipoxygenases [20]. In addition, the light-independent generation pathways of $^1O_2$ include reactions involving

$O_2{}^{\bullet-}$ (Figure 2D) [21], peroxynitrite (Figure 2E) [22,23], and ozone [24–26]. Atmospheric particulates also catalyze the $^1O_2$ generation [27,28].

3. Damage to Biomolecules by $^1O_2$-Mediated Oxidation

Notably, because $^1O_2$ has electrophilic properties, it attacks the π-bonds of the compound to form hydroperoxide by the ene reaction or endoperoxide by the 1,4-addition reaction (Figure 3A). $^1O_2$ oxidatively modifies biomolecules, such as amino acids (Figure 3B) [29,30], nucleic acids (Figure 3C) [31,32], and lipids (Figure 3D) [33,34], either by a direct reaction or by the induction of ROS.

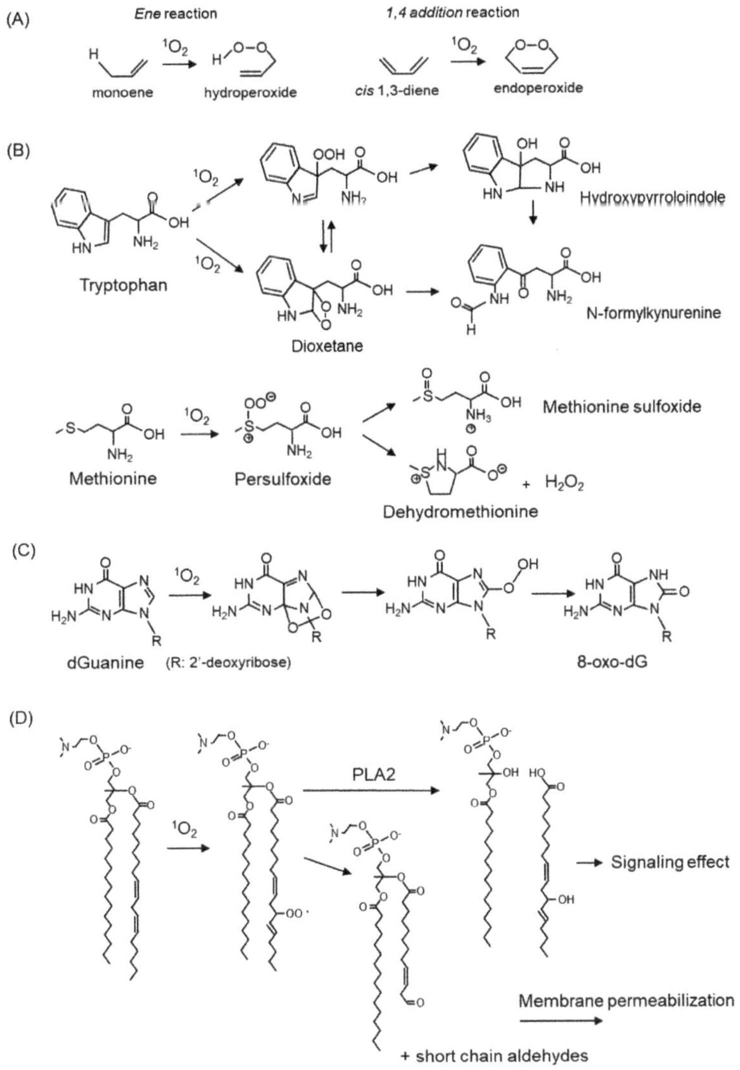

Figure 3. $^1O_2$-mediated oxidation of amino acid, nucleic acid, and lipids. (**A**) Oxygenation reaction by $^1O_2$. (**B**) Reaction products of tryptophan and methionine with $^1O_2$. (**C**) Formation of 8-oxo-2′-deoxyguanosine (8-oxo-dG) by $^1O_2$ oxidation to deoxyguanosine (dG). (**D**) Oxidative modification of fatty acids in membrane phospholipids by $^1O_2$.

Amino acids, which are components of proteins, are the targets of $^1O_2$. Among the 20 natural amino acids, tryptophan, histidine, tyrosine, and the sulfur-containing amino acids cysteine and methionine, are susceptible to reaction with $^1O_2$ because of their structural properties. The reaction products of tryptophan and methionine with $^1O_2$ are shown in Figure 3B. Tryptophan reacts with $^1O_2$ to form a hydroperoxide at the C3 position as an intermediate, which undergoes ring closure to form hydroxypyrroloindole, and finally, N-formylkynurenine [35]. Tryptophan reacts with $^1O_2$ to form C2-C3 cross-linked dioxetanes as intermediates, followed by the conversion to N-formylkynurenine [35]. Methionine reacts with $^1O_2$ to form methionine sulfoxide and the cyclic product dehydromethionine via a persulfoxide intermediate [30]. Oxidative modifications to amino acids cause alterations in protein properties, resulting in cellular damage accompanied by reduced enzyme activity and disruption of the cell structure.

$^1O_2$ reacts specifically with the guanine portion of DNA bases converting to 7,8-dihydro-8-oxo-2′-deoxy-guanin (8-oxo-dG) (Figure 3C) [36]. 8-oxo-dG is a potential mutagenic substance because it pairs with adenine to cause a G → T transversion, and it is widely recognized as one of the indicators for evaluating oxidative DNA damage [37]. The details are discussed below in Section 5.1.4. on skin cancer.

Phospholipids in the membrane react with $^1O_2$, resulting in the oxidative modification of fatty acids (Figure 3D). Oxidized fatty acids are cleaved and serve as signaling molecules involved in carcinogenesis and skin aging. The cleavage of short-chain aldehydes from fatty acids also results in membrane instability and increased permeability. The oxidation products derived from linoleic acid and cholesterol by $^1O_2$ are described below.

$^1O_2$ also oxidatively modifies steroids [38], vitamins [39,40], carbohydrates, terpenes [41], and flavonoids [42]. In addition, $^1O_2$ contributes to the stimulation of stress-activated kinases and the regulation of gene expression [43].

4. Detection Methods of $^1O_2$ In Vivo and In Vitro

The oxidative modification of biomolecules by $^1O_2$ has been related in several diseases, including skin and eye diseases [44]. The development of methods to precisely detect and quantify $^1O_2$ contributes not only to an understanding of its roles in physiological and pathological conditions, but also in cancer therapy, by understanding the mechanism of PDT. Detection methods for $^1O_2$ have been developed to verify $^1O_2$ production in living systems [45]. The detection of $^1O_2$ is mainly performed using the following methods: spectroscopic measurement of near-infrared (NIR) luminescence [46,47], electron spin resonance (ESR) with sterically hindered amine [48,49], fluorescence measurement by reaction with fluorescence probes [50,51], and the measurement of oxidation products produced by the reaction of $^1O_2$ with biomolecules [52,53]. These methods were utilized for in vitro and in vivo experiments.

4.1. Direct Detection of $^1O_2$

The measurement of $^1O_2$ NIR luminescence is a reliable method for the direct spectroscopic detection of $^1O_2$, as $^1O_2$ emission (1270 nm) can be excited by an argon laser [54]. This luminescence emission was first used for the time-resolved detection of $^1O_2$ in solution by Krasnovsky in 1976. This measurement method has been used as a standard technique for $^1O_2$ formation yields, lifetimes, and deactivation constants in various solutions. The in vivo detection of $^1O_2$ luminescence has been attempted previously, and it has been reported that $^1O_2$ in a suspension of leukemia cells [55,56] and red cell ghost [57] can be detected using a sophisticated near-infrared photomultiplier device. However, this luminescence signal is weak because the inactivation of $^1O_2$ is dominated by nonradiative pathways; typical phosphorescence yields are of the order of 10^{-5} to 10^{-7} [58]. In addition, these methods require the usage of deuterium oxide (D_2O) to extend the lifetime of $^1O_2$ and remove the 1270 nm emission absorption due to H_2O [59]. This difficulty in detecting $^1O_2$ in vivo can be attributed to the short lifetime of $^1O_2$ in cells and tissues and the lack of sufficient sensitive detectors at NIR wavelengths.

ESR spectroscopy, which has been proposed for paramagnetic materials and organic compounds with unpaired electrons, has been used for the detection of free radicals [60]. To detect $^1O_2$, a method was developed for measuring nitroxide radicals produced by the reaction of $^1O_2$ with sterically hindered secondary amine probes such as 2,2,6,6-tetramethylpiperidine (TEMP) (Figure 4A) [61–63] and hydroxy-TEMP (Figure 4B) [64]. This method has been applied to the in vitro $^1O_2$ scavenging activity of various substances in solution, and there is considerable evidence demonstrating the $^1O_2$ scavenging effects of various substances using this method [48,65,66]. However, ESR spectroscopy is not considered suitable for the detection of intracellular $^1O_2$, which may be because the time resolution in ESR measurements is not suitable for the short lifetime of $^1O_2$ in cells existing in $^1O_2$ quenching molecules.

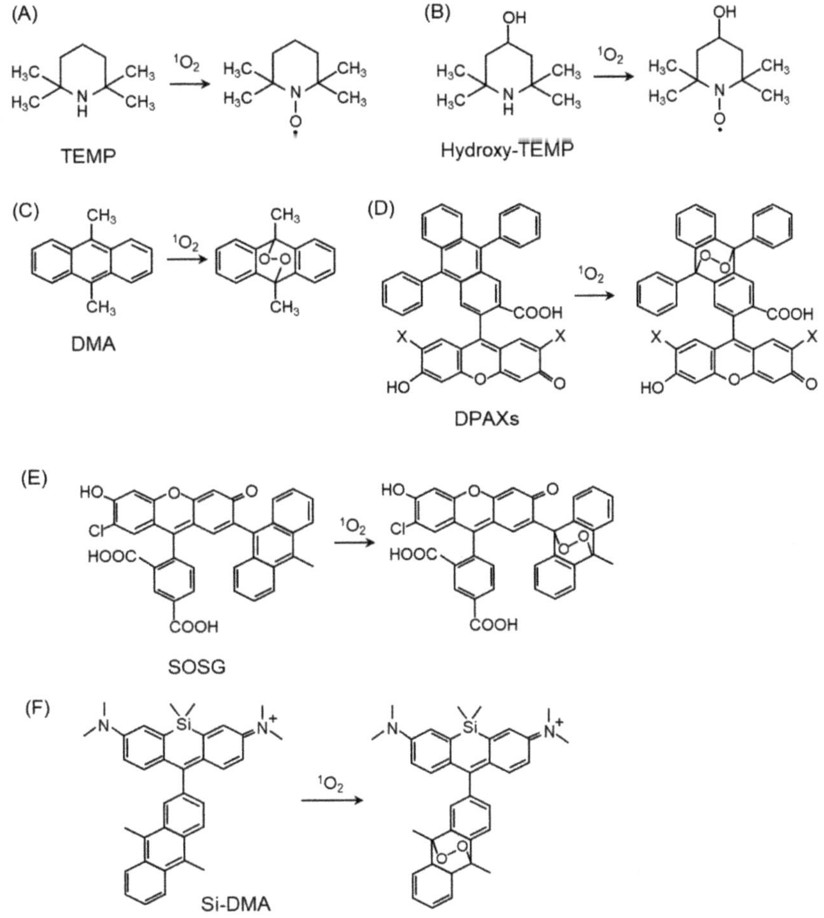

Figure 4. Structures and oxidation modification of probes for $^1O_2$ detection. (**A**) 2,2,6,6-tetramethylpiperidine (TEMP), (**B**) Hydroxy-TEMP, (**C**) 9,10-dimethylanthracene (DMA), (**D**) 9-[2-(3-Carboxy-9,10-diphenyl)anthryl]-6-hydroxy-3H-xanthen-3-ones (DPAXs), (**E**) Singlet Oxygen Sensor Green (SOSG), (**F**) silicone-containing rhodamine-9,10-dimethylanthracene (Si-DMA).

4.2. $^1O_2$ Detection by Fluorescent Probes

Fluorescent probes are excellent sensors for the detection of ROS because of their high sensitivity, simplicity of data acquisition, and high spatial resolution in microscopic imaging [67–69]. Fluorescent probes, including 9,10-dimethylanthracene (DMA) (Figure 4C) [68,70,71] and 9-[2-(3-Carboxy-9,10-diphenyl)anthryl]-6-hydroxy-3H-xanthen-3-ones (DPAXs) (Figure 4D) [72], have been developed for the detection of $^1O_2$ in neutral or basic aqueous solutions. These probes react specifically and rapidly with $^1O_2$ to form stable endoperoxides with a high rate constant. Singlet Oxygen Sensor Green (SOSG) (Figure 4E) is a fluorescent probe for in vitro detection because of its high $^1O_2$ selectivity [73–75]. To apply fluorescence probes in biological samples, the probes must penetrate the cell membrane and be localized in the cell. Recently, a far-red fluorescent probe consisting of DMA and silicon-containing rhodamine (Si-rhodamine) moieties, namely Si-DMA (Figure 4F), was developed for detecting $^1O_2$ in the mitochondria. The advantages of Si-DMA are its cell-permeable ability and increased sensitivity to specifically detect mitochondrial $^1O_2$ by the Si-rhodamine moiety [76]. It is important to precisely measure intracellular $^1O_2$ generation to clarify the characteristics of fluorescence probes in terms of their signal-to-noise ratio and dynamic changes. We have demonstrated that intracellular $^1O_2$ can be quantitatively measured using Si-DMA in living cells, as time-lapse imaging using Si-DMA provides an apparent signal-to-noise ratio in the treatments of a $^1O_2$ generator and quencher [77]. Other fluorescent probes have been used to detect intracellular $^1O_2$ using nanoparticles, including SOSG-based nanoprobes [50], super-pH-resolved nanosensors encoding SOSG [78], and biocompatible polymeric nanosensors encapsulating SOSG within their hydrophobic core [79]. The use of these fluorescence probes will improve our understanding of the $^1O_2$ generation mechanism and the biological function of $^1O_2$ in the physiological and pathological conditions of cultured cells.

4.3. Detection of $^1O_2$-Mediated Peroxidation Products

$^1O_2$ generated in vivo quickly reacts with biomolecules, including lipids, proteins, and nuclei, and the subsequent comparatively stable oxidized products remain in the $^1O_2$ generation site. This chemical property has been used for the development of in vivo ROS detection methods, and these oxidation products are widely used as oxidative stress biomarkers because they can be detected stably and correlate with oxidative stress status in vivo [80,81].

The oxidation product of linoleic acid, hydroxyoctadecadienoic acid (HODE), contains six isomers. $^1O_2$ reacts with molecules containing double bonds to form hydroperoxides as primary products, which are subsequently reduced to hydroxides (Figure 5). $^1O_2$ produces 9-, 10-, 12-, and 13-(Z,E)-hydroperoxyoctadecadienoic acid (HPODE) from linoleates; however, only 10- and 12-(Z,E)-HPODE are specific products by $^1O_2$ because 9- and 13-(Z,E)-HPODE are also formed by both free radical- and enzyme-mediated oxidation. The chemical structures of 10- and 12-(Z,E)-HPODE differ significantly from those of 9- and 13-(Z,E)-HPODE: the latter contains a conjugated diene, while the former does not. These structural features cause differences in physiological functions; for example, 10- and 12-(Z,E)-HODE cause adaptive responses to UV-derived oxidative damage in cultured cells, whereas 9- and 13-(Z,E)-HODE do not [82,83].

Cholesterol is an important lipid that constitutes biological membranes and undergoes oxidative modification to produce 7-, 24-, and 27-hydroxycholesterol. Cholesterol, as well as unsaturated fatty acids, are substrates for the oxidative reaction of $^1O_2$. When $^1O_2$ reacts with cholesterol, it forms 5α-hydroperoxide (cholesterol 5α-OOH), 6α-hydroperoxide (cholesterol 6α-OOH), and 6β-hydroperoxide (cholesterol 6β-OOH) (Figure 6). Cholesterol 5α-OOH is produced in larger amounts than 6α/β-OOH. Therefore, cholesterol 5α-OOH could be used as a biomarker for oxidative reactions involving $^1O_2$ in vivo. Furthermore, the Hock cleavage of cholesterol 5α-OOH converts to cholesterol 5,6-secosterol. Cholesterol 5,6-secosterol, also called ateronal, is involved in the development of cardiovascular diseases [84] and neurodegeneration [85].

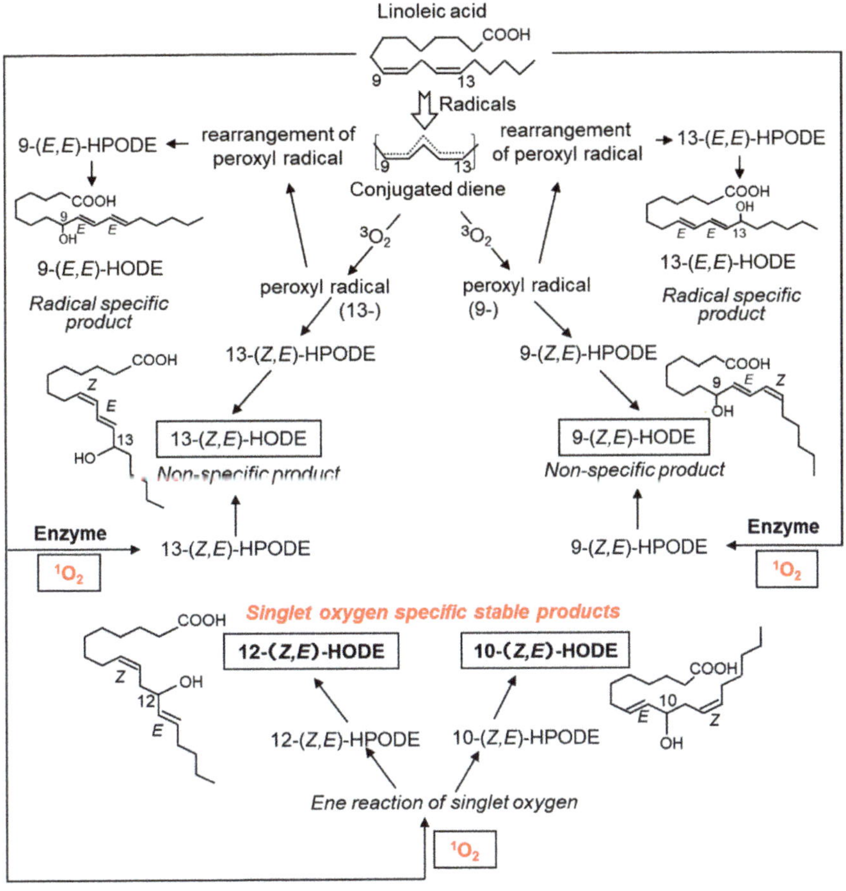

Figure 5. Structures and formation mechanism of hydroxyoctadecadienoic acid (HODE), an oxidation product derived from linoleic acid. $^1O_2$ oxidizes linoleic acid to produce 13-(Z,E)-HODE, 9-(Z,E)-HODE, 12-(Z,E)-HODE, and 10-(Z,E)-HODE. 13-(Z,E)-HODE and 9-(Z,E)-HODE are also produced by ROS other than $^1O_2$ and by enzymatic oxidation reactions via lipid oxidases. 9-(E,E)-HODE and 13-(E,E)-HODE are produced in a radical-specific manner.

We have developed a method to comprehensively measure lipid peroxidation and proposed the measurement of HODEs and hydroxycholesterol [86–91]. In this method, lipid components are extracted from biological samples after reduction and saponification. These pretreatments result in the measurement of both free and esterified forms of hydroperoxides and hydroxides as free hydroxides. HODEs and hydroxycholesterol are analyzed by liquid chromatography–tandem mass spectrometry (LC-MS/MS) and gas chromatography–mass spectrometry (GC-MS), respectively. Biological samples to be measured with this method include plasma, urine, erythrocytes, as well as cultured cells and tissues. This method allows for the measurement of not only radical-mediated oxidation products, but also $^1O_2$-specific oxidation products using the HODEs isomer as an indicator [89–91], in other words, indirect in vivo $^1O_2$ detection is possible. We have determined the relationship between $^1O_2$-mediated oxidation and the early diagnosis of diabetes in humans and mice based on the production of 10- and 12-(Z,E)-HODEs [83,92,93], which will be discussed in more detail later. $^1O_2$ measurements using 10- and 12-(Z,E)-HODEs as indicators can evaluate $^1O_2$ generation not only in living cells, but also in humans and animals in vivo, and they can precisely analyze the involvement of $^1O_2$ generation and diseases.

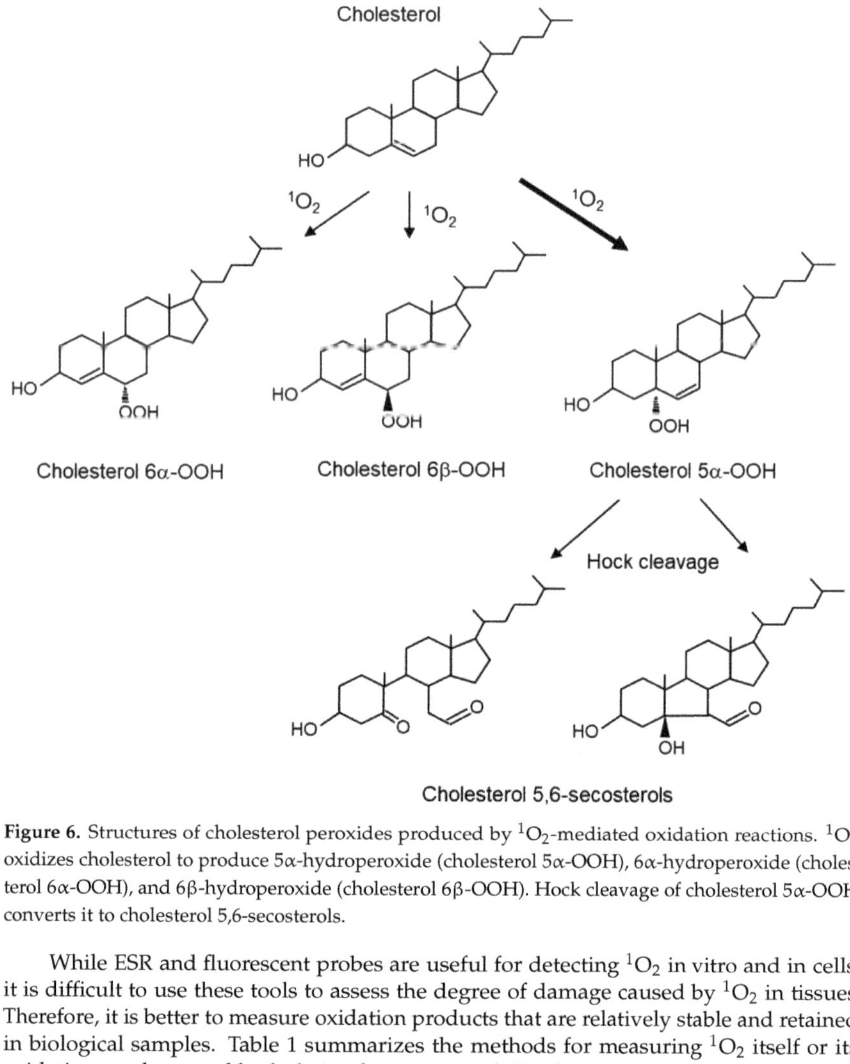

Figure 6. Structures of cholesterol peroxides produced by $^1O_2$-mediated oxidation reactions. $^1O_2$ oxidizes cholesterol to produce 5α-hydroperoxide (cholesterol 5α-OOH), 6α-hydroperoxide (cholesterol 6α-OOH), and 6β-hydroperoxide (cholesterol 6β-OOH). Hock cleavage of cholesterol 5α-OOH converts it to cholesterol 5,6-secosterols.

While ESR and fluorescent probes are useful for detecting $^1O_2$ in vitro and in cells, it is difficult to use these tools to assess the degree of damage caused by $^1O_2$ in tissues. Therefore, it is better to measure oxidation products that are relatively stable and retained in biological samples. Table 1 summarizes the methods for measuring $^1O_2$ itself or its oxidation products and both their advantages and disadvantages. Table 2 summarizes reports of the measurements of linoleic acid- or cholesterol-derived oxidation products produced by $^1O_2$-mediated oxidation reactions in human and animal samples. $^1O_2$-derived lipid oxidation products have been measured in the blood and tissue samples obtained from patients with borderline diabetes, glaucoma, alcoholism, and atherosclerosis. In animal experiments, cholesterol 5α-hydroperoxide in the skin tissue was measured following UVA irradiation in mice. These results are important to understand the contribution of $^1O_2$ to the pathogenesis of various disorders.

Table 1. Characterization of $^1O_2$ detection methods in vivo and in vitro.

Detection Method of $^1O_2$	Principle	Pros/Cons	Ref.
Near-infrared luminescence	$^1O_2$ emission at 1270 nm	• A standard technique for the $^1O_2$ yields, lifetimes, and deactivation constants in vitro. • Weak signal (phosphorescence yields are of the order 10^{-5} to 10^{-7}) • Requirement of deuterium oxide (D_2O) for longer lifetime ➢ Applied for in vitro detection	[54,57–59]
ESR spectroscopy	Measurement of nitroxide radicals produced by the reaction of $^1O_2$ with sterically hindered secondary amine probes	• Applied for in vitro $^1O_2$ scavenging activity in solution • Time resolution is longer for the short lifetime of $^1O_2$ in cells ➢ Applied almost for in vitro detection	[48,60,61,65,66]
Fluorescent probes 9,10-dimethylanthracene (DMA)		• Detection of $^1O_2$ in neutral or basic aqueous solutions • React rapidly with $^1O_2$ in a high rate constant • Impermeable cell membrane ➢ Applied for in vitro detection	[67–69] [68,70–72]
9-[2-(3-Carboxy-9,10-diphenyl)anthryl]-6-hydroxy-3H-xanthen-3-ones (DPAXs)	React with $^1O_2$ and form stable endoperoxides with high-fluorescence quantum yield (high sensitivity, simplicity of data acquisition, and high spatial resolution in microscopic imaging)		
Singlet Oxygen Sensor Green (SOSG)		• Commercially available highly selective $^1O_2$ indicator • Impermeable cell membrane ➢ Applied for in vitro detection	[73–75]
SOSG-based nanosensors NanoSOSG SPR-SOSG PAM-SOSG		• Using nanoparticles for detection of intracellular $^1O_2$ • Permeable cell membrane • Visualization of $^1O_2$ signal at the subcellular level ➢ Applied for in vitro and intracellular detection	[50,78,79]
Si-DMA		• Cell-permeable and localization of mitochondria • Specifically detect mitochondrial $^1O_2$ • Relatively quantitative ➢ Applied for intracellular detection	[76,77]
$^1O_2$-mediated peroxidation products 10- and 12-(Z,E)-HODEs	Products mediated by the reaction of $^1O_2$ with linoleic acid	• Relatively stable • Abundant in living organisms (cell membrane) • Can be used to evaluate in vivo oxidative damage ➢ Applied for in vivo and in vitro detection	[83,89–93]
Cholesterol 5α-OOH Cholesterol 6α-OOH Cholesterol 6β-OOH	Products mediated by the reaction of $^1O_2$ with cholesterol	• Relatively stable • Can be used to evaluate in vivo oxidative damage ➢ Applied for in vivo and in vitro detection	[52]

Table 2. Report on the results of the determination of linoleic acid- or cholesterol-derived peroxides produced by $^1O_2$-mediated oxidation reactions using human and animal samples.

$^1O_2$-Mediated Oxidation Product	Sample Species	Disease/Model	Ref.
10- and/or 12-(*Z,E*)-HODE	Human	Borderline diabetes/oral glucose tolerance test	[92,93]
	Human	High serum levels in patients with primary open-angle glaucoma	[94]
	Pig	Increased by kidney tissue damage	[95]
	Mouse	High in the plasma of model mice with prediabetes	[83,96]
	Mouse	Increased in the lungs of model mice with asthma and positively correlated with MPO activity and nerve growth factor (NGF)	[97]
Cholesterol 5α-hydroperoxide	Human	Alcoholic patients	[98]
	Rat	Pheophorubide a and visible light irradiation	[99]
	Mouse	UVA irradiation of hairless mice	[100]
	Mouse	Effect of beta-carotene on UVA irradiation of hairless mice	[101,102]
Cholesterol 5,6-secosterol	Human	Cholesterol 5,6-secosterol detected in atherosclerotic plaques	[103]
	Human	Cortex in patients with Alzheimer's disease	[104]
	Rat	Higher in the plasma of amyotrophic lateral sclerosis (ALS) rats than before disease onset	[85]

5. Diseases Involving $^1O_2$ and Their Pathophysiology

5.1. Skin Diseases

The skin is the organ with the largest area exposed to sunlight. The skin is regularly exposed to sunlight and is at a high risk of oxidative stress. UV radiation from the sun generates ROS in keratinocytes, the major cells of the epidermis and the outermost layer of the skin [105]. UV exposure also increases the activities of cyclooxygenase and lipoxygenase, which produce eicosanoids from arachidonic acid. UV exposure of the skin causes erythema and acute inflammation, and long-term UV exposure causes photo-carcinogenesis [106] and skin aging [107]. Lipid oxidation on the skin surface is important in relation to sunburn, hyperpigmentation, wrinkle formation, freckles, atopic dermatitis, acne, and skin cancer.

UVA (320–400 nm) penetrates the dermis, whereas UVB (280–320 nm) only attacks the epidermis. Both UVA and UVB produce ROS and free radicals in the presence of photosensitizers (Figure 7). In type I reactions, an excited photosensitizer reacts with an organic compound to produce $O_2^{\bullet-}$ and lipid peroxyl radical (LOO$^\bullet$). In a type II reaction, the excited photosensitizer reacts with a triplet oxygen molecule, producing $^1O_2$ primarily by energy transfer and, to a minor extent, $O_2^{\bullet-}$ by electron transfer.

5.1.1. Photosensitizers

Porphyrins such as protoporphyrin IX (Figure 8A) [108] have been examined as endogenous photosensitizers, and their derivatives [8] and 5-aminolevulinic acid [109], a component of porphyrins, have been investigated for application as PDT in therapeutic modalities against diseases such as cancer. Interestingly, coproporphyrin (Figure 8B) produced by *Propionibacterium acnes*, an acne-causing bacterium, also acts as a photosensitizer and generates $^1O_2$, which is associated with skin inflammation [110,111]. Hemin (Figure 8C) [112] and chlorophyll (Figure 8D) [113], which are metal complexes of porphyrin and similar compounds, also exhibit photosensitizing effects. Hemin is an iron-containing protoporphyrin IX, and chlorophyll is a chemical responsible for absorbing light energy in the light reaction of photosynthesis. Pheophytin (Figure 8E), from which magnesium ions are removed during chlorophyll degradation, and pheophorbide (Figure 8F), from which the phytyl group is eliminated from pheophytin, also act as photosensitizers [114,115].

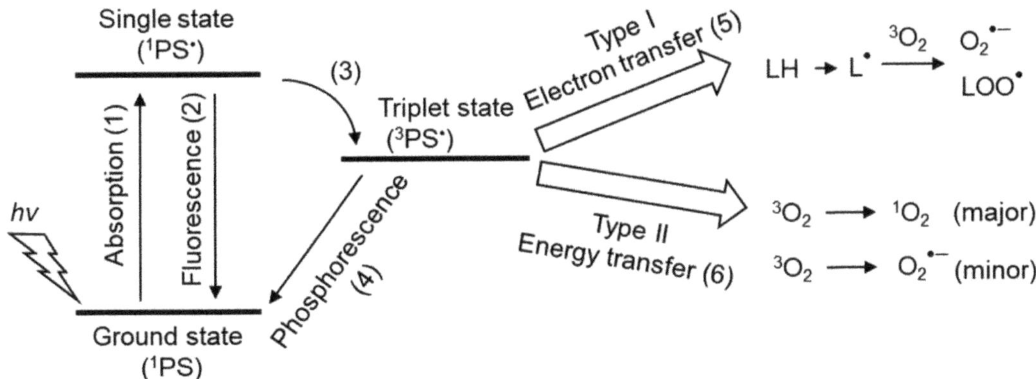

Figure 7. Photosensitizers and the mechanisms of ^{1}O$_2$ production. (1) Photosensitizers (PS) absorb photons (*hv*) from light and transform them to the excited singlet state (^{1}PS$^\bullet$). (2) ^{1}PS$^\bullet$ returns to the ground singlet state (^{1}PS) by losing energy via fluorescence emission. (3) ^{1}PS$^\bullet$ was converted to the long-lived triplet state (^{3}PS$^\bullet$). (4) ^{3}PS$^\bullet$ can return to the ground singlet state (^{1}PS) via light emission (phosphorescence). (5) ^{3}PS$^\bullet$ forms organic radicals (L$^\bullet$) from organic compounds (LH) through the transfer of electrons in the type I photochemical reaction. L$^\bullet$ affects oxygen (^{3}O$_2$) to produce ROS (superoxide anion O$_2^{\bullet-}$, lipid peroxyl radical (LOO$^\bullet$). O$_2^{\bullet-}$ leads to the production of H$_2$O$_2$ and hydroxyl radicals (OH$^\bullet$), resulting in the induction of radical chain reactions. (6) In type II photochemical reactions, the energy of ^{3}PS$^\bullet$ is transferred to ^{3}O$_2$ to mainly produce ^{1}O$_2$, but O$_2^{\bullet-}$ is produced as a minor product via electron transfer from ^{3}PS$^\bullet$.

In addition to porphyrins, tryptophan and tryptophan metabolites (kynurenine and 3-hydroxykynurenine) (Figure 8G) [116] and riboflavin (Figure 8H) [117,118] are endogenous photosensitizers. Riboflavin, a water-soluble vitamin B2, is distributed in the skin, eyes, brain, and blood vessels and increases the production of ^{1}O$_2$. Cholesta-5,7,9(11)-trien-3beta-ol (9-DDHC) (Figure 8I), a metabolite of 7-dehydrocholesterol (7-DHC), is a photosensitizer involved in rare cholesterol metabolism disorders [119]. Psoralen (Figure 8J), found in citrus fruits, is a dietary photosensitizer that causes photosensitivity [120,121].

5.1.2. Lipids in the Skin and Lipid Peroxidation Products Mediated by ^{1}O$_2$

The lipid composition of the epidermis is a mixture of free fatty acids, ceramide, and cholesterol at approximately 30–35% each [122]. The fatty acids in the epidermis include linoleic acid (21%), oleic acid (15%), palmitic acid (14%), stearic acid (11%), and arachidonic acid (6%) [123]. In contrast, the lipid composition of sebum is triacylglycerol (45–50%), wax ester (25%), squalene (Sq) (12%), and free fatty acids (10%) [122]. The fatty acid composition of sebum is palmitic acid (22%), palmitoleic acid (21%), oleic acid (15%), myristic acid (12%), and myristoleic acid (5%) [124]. The lipid composition of the sebum differs significantly from that of the epidermis.

Squalene (Sq) is a triterpene compound characteristically present in sebum, and six Sq monohydroperoxides (2-, 3-, 6-, 7-, 10-, and 11-OOH-Sq) are produced in similar amounts by the peroxidation of ^{1}O$_2$ (Figure 9) [125]. In contrast, 2- and 3-OOH-Sq were recently reported to be produced by free radical oxidation [125]. Considering these results, 6-, 7-, 10-, and 11-OOH-Sq may be the specific products of ^{1}O$_2$. SqOOHs are also associated with wrinkle formation [126] and inflammatory acne [127]. As SqOOHs are unstable, further photooxidation was revealed to produce 2-OOH-3-(1,2-dioxane)-Sq (Figure 9) as a secondary oxidation product [128]. SqOOH concentrations in sebum collected from skin exposed to tobacco heating products or electronic cigarette aerosols are lower than that in sebum from skin exposed to tobacco smoke, indicating consumer hygiene and cosmetic benefits [129].

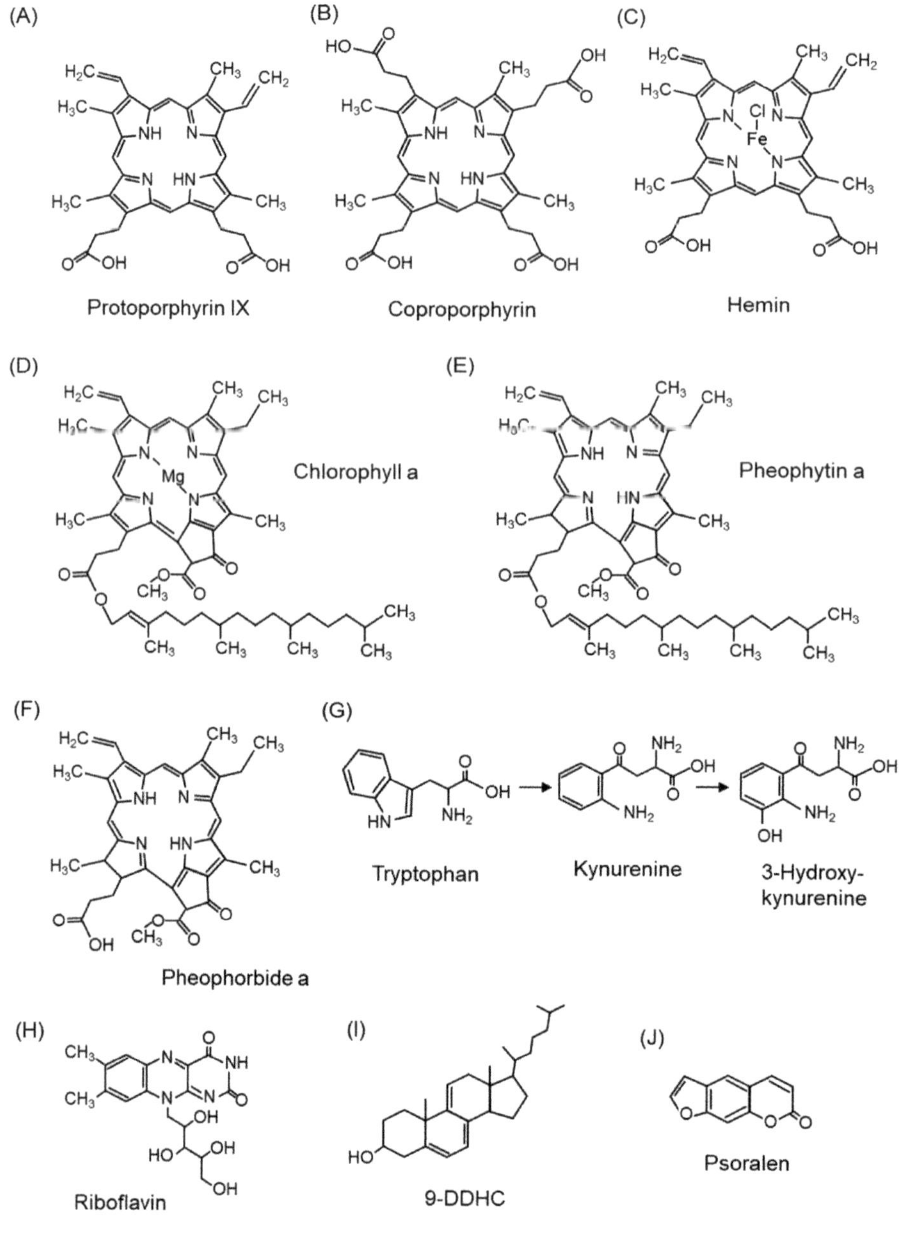

Figure 8. The structures of photosensitizers. (**A**) Protoporphyrin IX, (**B**) coproporphyrin, (**C**) hemin, (**D**) chlorophyll a, (**E**) pheophytin a, (**F**) pheophorbide a, (**G**) tryptophan, (**H**) riboflavin, (**I**) cholesta-5,7,9(11)-trien-3beta-ol (9-DDHC), (**J**) psoralen.

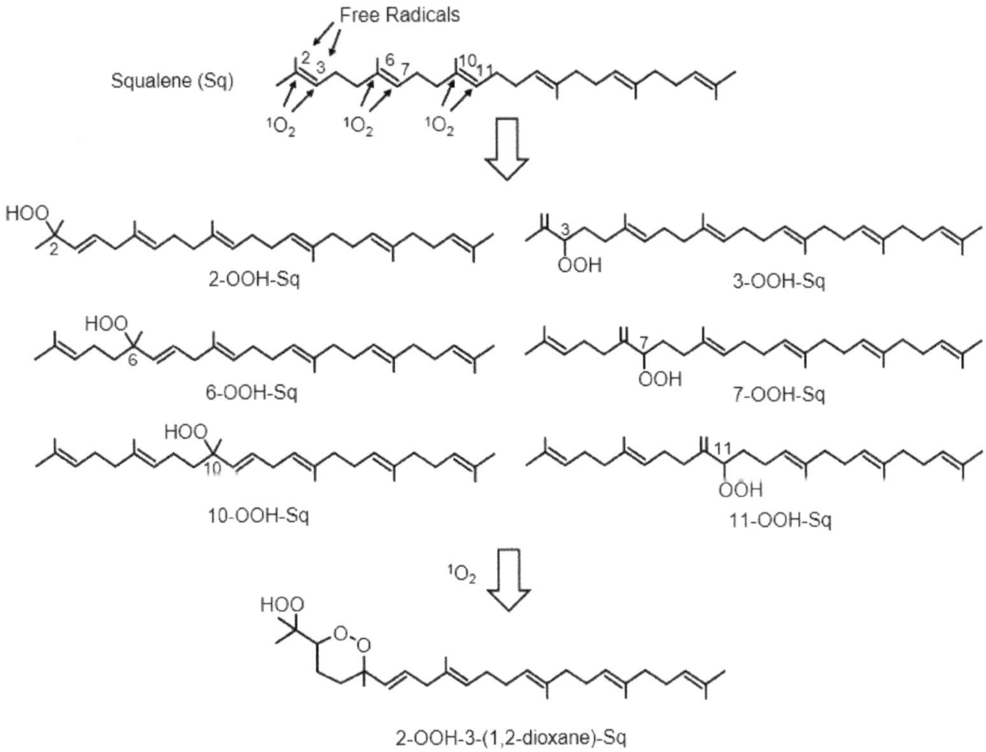

Free Radicals

Squalene (Sq)

$^1O_2$ $^1O_2$ $^1O_2$

HOO — 2-OOH-Sq

OOH — 3-OOH-Sq

HOO — 6-OOH-Sq

OOH — 7-OOH-Sq

HOO — 10-OOH-Sq

OOH — 11-OOH-Sq

$^1O_2$

HOO — O-O — 2-OOH-3-(1,2-dioxane)-Sq

Figure 9. The structures of squalene and its peroxidation products. Squalene (Sq) is converted by $^1O_2$ to six monohydroperoxides (2-, 3-, 6-, 7-, 10-, and 11-OOH-Sq). The monohydroperoxides of Sq are further photo-oxidized to 2-OOH-3-(1,2-dioxane)-Sq.

Table 3 shows the results of SqOOH measurements in human and animal sebum samples. The fact that SqOOHs, which are specifically produced by $^1O_2$, can be detected using sebum, which can be collected without severe invasion, makes it possible to assess the effects of pollutants, sunlight, and UV. In the future, techniques for the measurement of the SqOOH content in sebum could be further developed in the field of skin diseases, including cosmetology.

As mentioned above, $^1O_2$ peroxidation products from cholesterol are caused by ene reactions to 5,6 double bonds to form cholesterol 5α-OOH, 6α-OOH, and 6β-OOH (Figure 6). However, cholesterol 5α-OOH is more abundant than cholesterol 6α- and 6β-OOH [52].

PUFAs in the skin are also target molecules for peroxidation by $^1O_2$. UVA irradiation of hairless mice has also been reported to increase 10-HPODE and 12-HPODE levels (Figure 5), which are $^1O_2$-specific lipid hydroperoxides of linoleic acid [130].

Table 3. Report on the results of the determination of squalene-derived peroxides produced by oxidation reactions mediated by $^1O_2$ using human and animal samples.

$^1O_2$-Mediated Oxidation Product	Sample Species	Disease/Model	Ref.
Squalene monohydroperoxides	Human	Increased by exposure of human skin to tobacco smoke	[131]
	Human	Sunlight exposure increases SqOOH in sebum	[132]
	Human	UV irradiation to human sebum	[133–135]
	Human	SqOOH reduction in sebum by application of cosmetics	[136]
	Human	Human skin surface lipids	[137]
	Human	Increased SqOOH in the scalp due to high dandruff	[138]
	Human	Detection using human fingerprints	[139]
	Human	Increase in SqOOH in the sebum due to ambient dust and ozone	[140]
	Human	Less SqOOH in sebum exposed to tobacco heating products and electronic cigarettes compared to cigarette smoke exposure	[129]
	Pig	Cigarette smoke exposure to pig skin	[141]
	Rabbit	UVA irradiation of rabbit ears	[127]
	Guinea pig	Carotene reduces the increase in SqOOH caused by UV irradiation of the skin	[142]
2-OOH-3-(1,2-dioxane)-squalene	Human	Identification of cyclic peroxides of SqOOH	[128]

5.1.3. Skin Aging

The main clinical sign of skin aging is the dysfunction of the dermis, with ultrastructural disruption of elastic fibers [143]. This is due to an imbalance in collagen remodeling relative to proteolysis that occurs in the extracellular matrix. UVA-irradiated cultured human fibroblasts and human epidermis have been shown to induce the expression of collagenase, a protease that degrades elastic fibers in the dermis [144]. Moreover, the addition of cholesterol oxidants derived from the $^1O_2$ oxidation reaction to mouse fibroblasts causes them to accumulate in lipid rafts, as well as upregulates matrix metalloprotease-9 (MMP-9) activity [145]. The UVA irradiation of cells under protoporphyrin administration results in the upregulation of MMP-9 activity with the formation of cholesterol oxide [145]. This suggests that cholesterol oxidation products produced by $^1O_2$ may alter the structure of lipid rafts, resulting in the activation of signaling pathways that induce MMP-9 expression, leading to skin aging.

5.1.4. Skin Cancer

DNA nucleobases are targets of oxidation reactions [44]. $^1O_2$, as well as highly reactive $HO^\bullet$, are oxidants that can damage cellular DNA [146–148]. Since UVA is among the carcinogens to which organisms are exposed, it is important to understand the mechanisms that cause DNA damage [149]. UVA in sunlight is a potential source of oxidative DNA damage in the skin [150,151]. As discussed in the previous chapter, exposing skin to UVA radiation produces $^1O_2$. The reactivity of $^1O_2$ to nucleobases tends to occur in the order guanine > cytosine > adenine > uracil > thymine [152].

The introduction of an oxo group into the C8 of guanine and the addition of a hydrogen atom to the nitrogen of N7 yields 8-oxo-dG (Figure 3C). The presence of 8-oxo-dG in DNA causes problems in the S phase of the cell cycle. In the S phase, cells must replicate an exact copy of the genome. Unlike most other types of DNA damage, 8-oxo-dG does not stop the replication process but rather creates a point mutation. Normally, guanine pairs with cytosine (C), whereas 8-oxo-dG mimics thymine (T), thus forming a pair with adenine (A) [153]. Furthermore, the A:8-oxo-dG mispair does not cause helix distortion in the DNA backbone, thus avoiding correction by the error detection mechanism of replication polymerase [154]. This results in mutations that are essentially C:G→A:T mutations. The presence of C:G→A:T mutations in many cancers emphasizes the importance of 8-oxo-dG in cancer pathogenesis [155]. The amount of 8-oxo-dG produced is approximately 10^3/cell per day in normal cells, but increases to 10^5/cell in cancer cells [156]. 8-oxo-dG is also markedly increased in human skin cells after UVA irradiation [157]. There are reports that 8-oxo-dG is involved in photocarcinogenesis [158,159]. It has also been reported that 8-oxo-dG is increased in epidermal cells after chronic broadband UVB irradiation in Ogg1 knockout mice, which are unable to eliminate oxidized bases and are susceptible to skin cancer [160]. However, cyclobutane pyrimidine dimers observed in UV-irradiated DNA are also known to correlate with skin cancer development [161]. It is thought that DNA mutation due to $^1O_2$ generation is not the only cause of UV-induced carcinogenesis, but that various mutagenic molecules are involved in the pathogenesis in a complex manner.

Furthermore, 8-oxo-dG has been used as a biomarker to evaluate the degree of oxidative stress, and its application has been attempted as a risk assessment for many diseases [162].

5.1.5. Porphyria

Porphyria is a rare genetic disorder caused by a deficiency in any of the eight enzymes involved in the heme metabolism, resulting in the accumulation of the photosensitizing substance porphyrin or its precursors [163]. Porphyrin accumulation in the skin also causes photosensitivity, resulting in burn-like skin lesions [164]. The nine porphyria are divided into "cutaneous porphyria" and "acute porphyria". Among cutaneous porphyria, photosensitivity symptoms begin to appear shortly after birth for congenital erythropoietic porphyria (CEP) and hepato-erythropoietic porphyria (HEP), around age 5–6 for erythropoietic protoporphyria (EPP), and after age 50 for late-onset porphyria (PCT). Exposure to sunlight causes pain, itching, redness, and swelling, and, in severe cases, blister-like burns. Porphyrins accumulated in skin tissue become excited by the absorption of light and induce the production of $^1O_2$, resulting in tissue and vascular damage due to complement activation. In addition, the release of histamine, kinins, and chemotactic factors is thought to cause skin damage [165]. Currently, there is no cure for porphyria, and photoprotection is the only way to treat this disease. Beta-carotene, a scavenger of $^1O_2$, is widely used to treat photosensitivity caused by EPP [166,167], but it has also been reported to be less useful for photosensitivity associated with PCT [167,168].

The $^1O_2$ produced via porphyrins causes protein oxidation and aggregation [169,170]. In a porphyria model mouse, in which porphyrin accumulation was induced by 3,5-diethoxycarbonyl-1,4-dihydrocollidine, protein aggregation in liver tissue was observed upon safelight exposure compared to non-exposure [171]. Moreover, overexposure to sunlight in EPP and PCT is associated with liver dysfunction, cirrhosis, gallstones, and acute liver failure, and the relationship between hepatic damage in porphyria and $^1O_2$ production by irradiation is becoming clearer.

5.1.6. Smith–Lemli–Opitz Syndrome and Statin-Induced Skin Disorders

Smith–Lemli–Opitz syndrome (SLOS) is caused by mutations in the *DHCR7* gene, which encodes 7-dehydrocholesterol (7-DHC) reductase, involved in the final step of cholesterol biosynthesis [172,173]. SLOS shows a cholesterol deficiency and increased 7-DHC in the plasma and tissues [173]. A variety of symptoms are recognized in SLOS, including growth retardation, microcephaly, intellectual disability, characteristic facial features, and external malformations; however, hypersensitivity to ultraviolet light is recognized as a symptom of SLOS [174]. Although 7-DHC itself does not absorb UVA and is not a direct source of photosensitivity in SLOS, 9-DDHC (Figure 8I), a 7-DHC metabolite found in the plasma of SLOS patients, is highly absorbable to UVA [119]. Furthermore, UVA exposure to 9-DDHC generates $^1O_2$, providing a mechanism for the pathogenesis of photosensitivity in SLOS [119].

In addition, some statins inhibit 7-DHC reductase [175]; thus, the accumulation of 9-DDHC may also be involved in statin-induced skin photosensitivity [176] and cataracts [177].

5.1.7. Skin Disorders Caused by Exogenous Photosensitizers

Pheophorbide (Figure 8F), a degradation product of chlorophyll (Figure 8D), is a photosensitizing substance [178]. Pheophorbide is sometimes found in food, and the exposure of humans [179] and animals [180] to light after ingesting this compound can cause dermatitis due to photosensitivity. Cholesterol 5α-OOH (Figure 6) was detected in rat skin administered with pheophorbide and exposed to visible light [99].

The thiopurine prodrugs azathioprine and 6-mercaptopurine are widely prescribed for the treatment of leukemia and autoimmune diseases [181]. Although these drugs have shown therapeutic efficacy, long-term administration has been reported to increase the risk of basal cell carcinoma and squamous cell carcinoma by 10- and 65- to 250-fold, respectively [182]. These adverse effects are induced by sunlight exposure. Thiopurine derivatives absorb UVA, which is believed to be responsible for producing $^1O_2$.

Ibuprofen and ketoprofen, both marketed as nonsteroidal anti-inflammatory drugs with antipyretic and analgesic properties, also produce $^1O_2$ due to their photosensitizing effects and have been associated with adverse skin symptoms, including photosensitivity [183]. Some commonly used pharmaceutical compounds whose side effects include skin disorders associated with photosensitivity may also produce $^1O_2$.

5.1.8. Diagnosis and Evaluation of Skin Diseases by Measuring $^1O_2$-Mediated Products

In skin diseases, the measurement of $^1O_2$-derived lipid peroxidation products in sebum, especially SqOOHs, may be useful in the diagnosis of skin aging and photosensitivity. On the other hand, while the diagnosis of porphyria and SLOS can be made by measuring porphyrin and 7-DHC, respectively, the measurement of $^1O_2$-derived lipid peroxidation products in sebum may be useful in elucidating the pathogenesis and understanding a patient's disease status. In skin cancer, 8-oxo-dG may be useful in assessing the risk of DNA mutations induced by UV, but it should be evaluated in combination with the analysis of compounds produced by UV irradiation, such as cyclobutane pyrimidine dimers.

5.2. Ophthalmological Diseases

Since sunlight directly impacts the skin and eyes, oxidative stress is involved not only in the skin, but also in the eyes. In this section, we focus on eye structures, light permeation, and dysfunctions, followed by diseases related to $^1O_2$-mediated oxidative stress.

5.2.1. Structure and Optical Transparency of the Ocular Bulb

Light passes through the cornea, aqueous humor, and pupil through the lens and vitreous body, to reach the retina. The pupil regulates light, and the crystalline lens acts as the lens. The retina is a transparent membrane that lines the inside of the eyeball. It mainly consists of a layered structure of four types of cells: retinal ganglion cells, communication/glial cells, photoreceptors, and retinal pigment epithelium (RPE) cells of the vitreous. Light is received by photoreceptors; the pigment cell layer is located outside the photoreceptor layer. Under visible light, the light that passes through the eye differs depending on its wavelength, and UVA (320–400 nm), UVB (280–320 nm), and UVC (100–280 nm) are absorbed by the cornea and lens [184,185], although UV can reach the retina in an aphakic eye (i.e., an eye with no crystalline lens). In the human eye, light with a wavelength of 400–760 nm reaches the retina with little absorption from the cornea to the vitreous body. Since mid- and far-infrared rays are absorbed by water, they are absorbed by water in the cornea, lens, and aqueous humor and do not reach the fundus of the eye. Since near-infrared light has low water absorption, it penetrates the choroid [186].

5.2.2. Photooxidative Stress

Visual cells, which are photoreceptors, contain visual substances of chromoproteins that receive light. They are composed of two types of cone cells that sense colors and rod cells that sense the intensity (brightness and darkness) of light. Cone cells have photopsins, and rod cells have rhodopsins. $^1O_2$ is produced by the oxygen present in photoreceptors [187]. In addition, photoreceptor cells and retinal pigment epithelial cells accumulate waste products, called lipofuscin, with aging [188], and these are also activated by light absorption to produce radicals [189,190].

Lipofuscin is considered a waste product in the lysosome, consisting primarily of 30–58% protein, 19–51% lipid-like substances (oxidation products of PUFA), carbohydrates, and trace metals (2%), including iron, copper, aluminum, zinc, zinc calcium, and manganese [191]. Lipofuscin cannot be degraded by lysosomal hydrolytic enzymes because of the polymerization and cross-linking of peptides with aldehydes, resulting in a plastic-like structure that is not biologically degradable and accumulates in neurons and cardiomyocytes with age. Retinal lipofuscin has been shown to exhibit strong photosensitizing properties when photoexcited in the presence of oxygen [189].

Recently, fluorescent bis-retinoid N-retinyl-N-retinylidene ethanolamine (A2E) (Figure 10A,B) has been implicated in photooxidation in the retina [192,193]. A2E is the main component of lipofuscin that accumulates in RPE cells during aging. Rhodopsin, a pigment protein present in photoreceptor cells, is composed of a protein moiety called opsin and 11-*cis*-retinal (Figure 10A,B). The retina contains vitamin A (retinol). Rhodopsin is broken down into opsin and all-*trans*-retinal when it absorbs light. The RPE regenerates all *trans* retinal to 11 *cis* retinal (a process called the visual cycle) (Figure 10B). During this regeneration process, all-*trans*-retinal reacts with phospholipids and another all-*trans*-retinal generates A2E (Figure 10A,B). A2E, a major component of lipofuscin, generates $^1O_2$ upon light stimulation, especially high-energy blue light [194].

ROS and radicals generated by light act on the cell membrane lipids of photoreceptors to form lipid radicals [195,196]. Recently, photo-oxidative stress has been found to lead to retinal aging and is a factor in the onset of age-related macular degeneration [195,196].

5.2.3. Compounds Protecting against $^1O_2$ in the Macular Pigment and Lens

The macula is susceptible to oxidative damage owing to the lack of a retinal inner layer, which is affected by intense light. Therefore, the retina has a macular pigment composed of lutein and zeaxanthin, which are carotenoids, and meso-zeaxanthin [197], which is converted from lutein by the RPE-65 enzyme [198,199].

The lens contains lutein and vitamin C, which have a protective mechanism against blue light, with an absorption peak at 400–500 nm. In addition, these macular pigments and carotenoids have an antioxidant ability to eliminate $^1O_2$ and are thought to maintain retinal homeostasis [198].

5.2.4. Blue Light Hazard

Visible blue light (400–500 nm) from light-emitting diode (LED), mobile phones, and industrial equipment is increasingly being used, and we are susceptible to this in our daily lives. The accumulation of photo-oxidative stress caused by blue light, called blue light hazard, has been reported to cause chronic retinal changes [200,201].

Marie et al. exposed a 10 nm wide light band, range 390–520 nm, to primary retinal pigment epithelial cells treated with A2E and measured the precise action spectrum that produced the highest amount of ROS in the cells [194]. Of the spectrum of sunlight reaching the retina, blue light at 415–455 nm produced the highest amount of ROS and induced mitochondrial dysfunction. Lipofuscin and A2E accumulate in the RPE with aging, and blue light induce a photooxidative reaction that promotes cell death and angiogenesis. The need for the elderly to filter these wavelengths of light is emphasized.

Exposure to blue light (415–455 nm) for 15 h decreased activities of SOD and catalase and increased oxidized glutathione (GSSG) and ROS in A2E-treated RPE cells [194], suggesting that the balance between the oxidant and antioxidants is readily disrupted by massive blue light exposure [194].

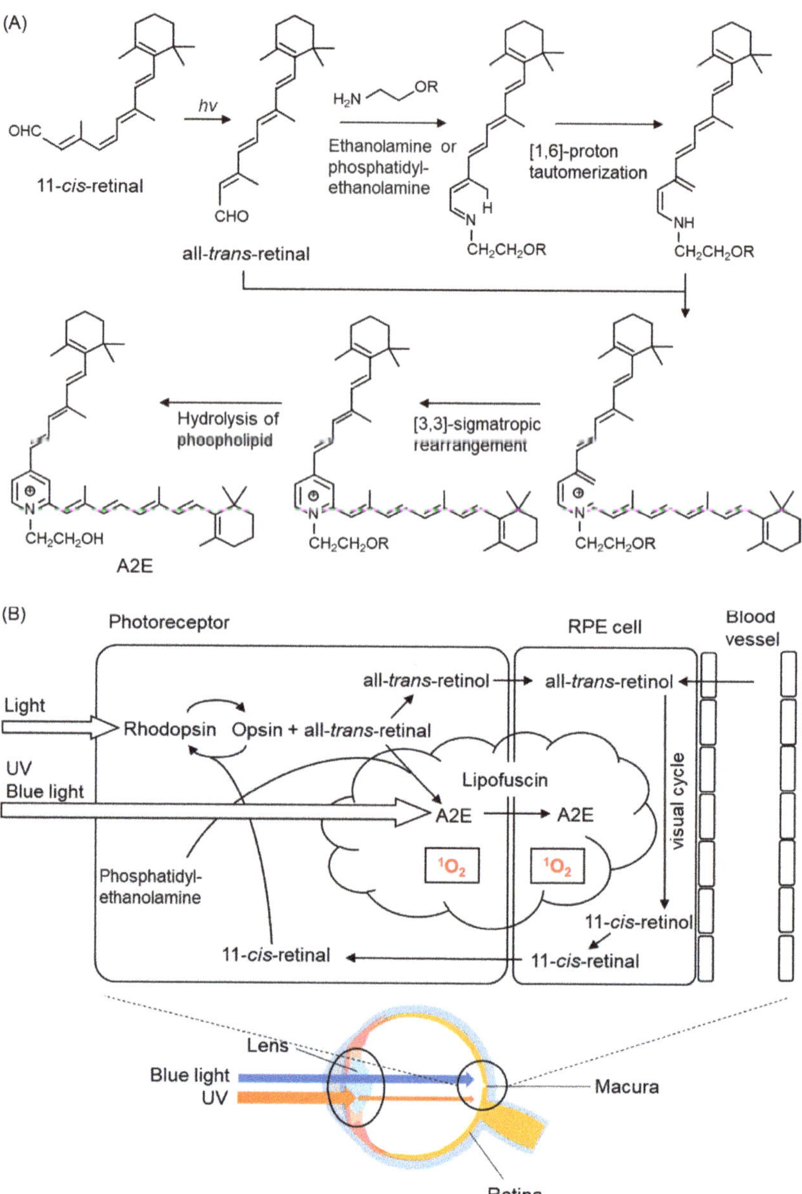

Figure 10. The structures of biosynthesis of A2E. (**A**) After Schiff base formation between all-*trans*-retinal and ethanolamine or phosphatidylethanolamine, a [1,6]-proton tautomerization to enamine follows. After further Schiff base formation with a second molecule of all-*trans*-retinal, a [3,3]-sigmatropic rearrangement is followed by the hydrolysis of the linked phosphatidylethanolamine adduct, resulting in the formation of N-retinyl-N-retinylidene ethanolamine (A2E). (**B**) $^1O_2$ production process mediated by A2E in the retina. Light irradiation of A2E accumulated in lipofuscin generates $^1O_2$.

5.2.5. Age-Related Macular Degeneration

Age-related macular degeneration (AMD) is a disease in which the photoreceptor cells in the macula of the central fundus of the eye are damaged. There are two types: the exudative type with neovascularization generated from the choroid that feeds the retina, and the atrophic type without it. AMD is a multifactorial disease characterized by high oxygen consumption, massive radiation exposure, and high PUFA content in the retina. Photooxidative stress is presumed to be one of the causes of retinal disorders such as AMD [200]. PUFAs in the membranes are targets for ROS, which drastically increases the susceptibility of the retina to photochemical damage. In addition, as described above, A2E-containing lipofuscin, which accumulates in RPE cells with aging, is activated by light, and $^1O_2$ is generated from abundant oxygen in the retina. In the early stages of the disease, the body's defense mechanism acts. However, the lesion begins to progress as the body's defense mechanisms decline with age.

5.2.6. Glaucoma

Glaucoma is a progressive glaucomatous optic neuropathy that causes visual field loss and irreversible blindness [202–204]. The death and axon loss of retinal ganglion cells (RGCs) cause glaucomatous optic neuropathy [202]. Elevated intraocular pressure (IOP) is a primary risk factor for open-angle glaucoma (OAG) including primary OAG (POAG) [202]. In OAGs, the elevation of IOP is explained by a reduction in aqueous humor (AH) outflow at the trabecular meshwork (TM) due to qualitative and quantitative changes in the extracellular matrix in the TM [205]. Numerous reports have shown that various oxidative stressors induce RGC damage [206,207]. For example, antioxidant thioredoxins prevent glaucomatous tissue injury, specifically glutamate- and IOP-induced RGC death [208,209], indicating that oxidative stress is thought to be involved in IOP elevation and then RGC loss in POAG.

We previously demonstrated the involvement of oxidative stress in the pathogenesis of glaucoma by measuring AH and serum HODE levels derived from free radical-mediated oxidation [94,210,211], suggesting that systemic oxidation is at least partially involved in the disease. Furthermore, the levels of HODEs/LA (oxidized/parent molecules ratio, LA; linoleic acid) in AH correlated with those in serum, suggesting that ocular oxidative injury proceeds simultaneously with systemic oxidative stress [211]. The serum levels of 10- and 12-(Z,E)-HODEs/LA formed via $^1O_2$ specific oxidation were correlated with IOP [94], which are indices of glaucoma severity derived from TM cell dysfunction. One possible process by which $^1O_2$ is produced in the eye is type II photooxidation via a sensitizer present in the vicinity of the reaction milieu, similar to a cataract [212]. The specific region of oxidative injury has not been identified because sunlight does not reach TM cells. In addition, the pathways of the excretion and circulation of HODEs formed in the eyes remain undefined. However, our findings suggested that cerebrospinal HODE levels were well reflected in plasma levels [213,214]. Other studies have simultaneously analyzed the systemic and local redox status, suggesting that alterations in systemic oxidant and antioxidant levels reflect local redox status [215–217].

5.2.7. Cataract

Epidemiological approaches have indicated that sun exposure is a risk factor in age-related cataracts [218,219]. UVB is mostly filtered out by the cornea and aqueous humor [220]. For this reason, UVA, which occupies nearly 95% of the UV in sunlight, has been associated with cataracts [221]. It has been proposed that chromophores with UVA-visible light absorption properties, which accumulate in the lens with age, influence the development of cataracts through the ROS generation by photosensitizing reactions [222].

The major protein components of the lens are crystallins, of which there are three main types: α-, β-, and γ-crystallin. The most abundant of these is α-crystallin, which functions to maintain lens transparency and acts as a protective chaperone for the lens [223]. As α-

crystallin is not turned over, damage to the amino acids/proteins constituting α-crystallin accumulates. This leads to changes in refraction and lens opacity (cataract formation) [224].

Tryptophan metabolites (kynurenine (Figure 8G), 3-hydroxykynurenine (Figure 8G), 3-hydroxykynurenine o-β-D-glucoside, and 4-(2-amino-3-hydroxyphenyl)-4-oxobutanoic acid o-β-D glucoside) are present in significant concentrations in the lens and act as endogenous chromophores and UV-absorbing filters [225]. These endogenous chromophores absorb UV and become excited but regenerate to their original ground state without producing radicals or $^1O_2$. These chromophores undergo deamination under physiological conditions and are covalently bound to proteins (crystallin) via Michael addition reactions [226]. As covalent modification progresses gradually with age, the lens turns yellow. Paradoxically, tryptophan derivatives, which act as UV filters, have also been shown to generate $^1O_2$ by acting as photosensitizers when bound to proteins [226]. However, the association between cataract development and $^1O_2$ remains undefined and has not been confirmed in our studies [94,210].

5.2.8. $^1O_2$-Derived Peroxidation Products in the Diagnosis and Evaluation of Ophthalmological Diseases

The measurement of $^1O_2$-derived lipid peroxidation products in tissues removed during cataract and glaucoma surgery may provide assistance in elucidating the pathophysiology and in selecting the use of singlet oxygen scavengers. As our results indicate, measuring lipid oxides in circulation may also be useful for diagnosis and pathophysiological evaluation.

5.3. Diabetes Mellitus
5.3.1. Biomarkers for Diabetes and Diabetic Complications

Diabetes is characterized by a deficiency of the secretion or action of insulin, resulting in microvasculature damage to the retina, renal glomerulus, and heart, as well as peripheral neuropathy. Diabetes can be fatal if its complications include nephritis and atherosclerosis. Type 2 diabetes mellitus (T2D) indicates elevated blood glucose levels caused by the impairment of insulin secretion and insulin resistance. Prediabetic states, including impaired fasting glucose, impaired glucose tolerance (IGT), or slightly elevated blood glucose levels, may precede T2D for the year [227,228]. Progression from prediabetes to T2D can be prevented or delayed by improving diet and increasing physical activity.

Diabetic vascular complications include nephropathy, myocardial infarction, and glaucoma. The early detection of diabetes is important to prevent complications. Several studies have been conducted on the early detection of these complications. Biomarkers of diabetic nephropathy include urinary heme oxygenase-1 (HO-1) [229] and 8-hydroxydeoxyguanosine [230,231], and circulating microRNA 130b [232]. S-glutathionylation [233] and oxidized dityrosine-containing protein [234] are considered candidate markers of vasculopathy in diabetes. Angiotensin-II, protein kinase C (PKC), and advanced glycation end products (AGEs) activate NADPH oxidase and upregulate the production of ROS, leading to cardiac dysfunction [235,236]. The metallic elements selenium [237], copper, and zinc [238] are associated with the detection of diabetes risk in pregnant women. Recently, band 3 anion transport protein, also known as anion exchanger 1 or band 3, or solute carrier family 4 member 1, was proposed for the early detection of the glycation of hemoglobin leading to AGEs [239–241]. To completely detect diabetes and its complications, a combination of several biomarkers is necessary.

5.3.2. Diabetic Biomarkers of Lipid Peroxidation

Several studies have examined the association of oxidation products with diabetes pathology [242–246]. Griesser et al. revealed the interplay between lipid and protein modifications using animal models. A large cohort study showed that imbalances in the redox system contribute to the development of T2D [245]. Leinish et al. clarified that the structural and functional alterations in RNase A result from oxidation by $^1O_2$ and ROO$^\bullet$, not solely from histidine and tyrosine cross-linking [246].

We recently found that 10- and 12-(Z,E)-HODE, $^1O_2$-specific products derived from linoleic acid, significantly correlated with a risk index for impaired glucose tolerance and insulin resistance in oral glucose tolerance tests performed on healthy volunteers [92,247,248]. Free radical-specific lipid oxidation products, such as 9- and 13-(E,E)-HODE and 7β-hydroxycholesterol, and hydroxyeicosatetraenoic acids (HETEs), which are oxidation products derived from arachidonic acid, have not been detected. The process by which these $^1O_2$-specific lipid oxidation products are produced in diabetic pathology is speculated to involve the reaction between H_2O_2 and ClO^- derived from MPO (Figure 2B). This $^1O_2$ generation mechanism is mediated by MPO from activated phagocytes [249,250] or eosinophils peroxidase [251,252]. Several reports have shown that neutrophil-derived MPO plays an important role in diabetic vascular injury (Figure 11) [253–255]. The source of H_2O_2 in diabetic pathology is thought to be NADPH oxidase in the vascular endothelial cells [253]. MPO released from activated neutrophils is known to bind to the vessel wall for several days [256,257]. The H_2O_2 derived from hyperglycemia-activated NADPH oxidase may be used by MPO bound to the vascular endothelium to produce HOCl, resulting in the production of $^1O_2$ and vascular injury. Compounds that inhibit the activity of MPO, such as hydroxamic acids, hydrazides, and azides, have potentially harmful side effects. In contrast, quercetin was recently reported to inhibit MPO-dependent HOCl production and prevent vascular endothelial injury [258]. Onyango et al. reviewed the contribution of $^1O_2$ to insulin resistance and reported that $^1O_2$ generates bioactive aldehydes and induces mitochondrial DNA modification and endoplasmic reticulum stress, which lead to insulin resistance [259].

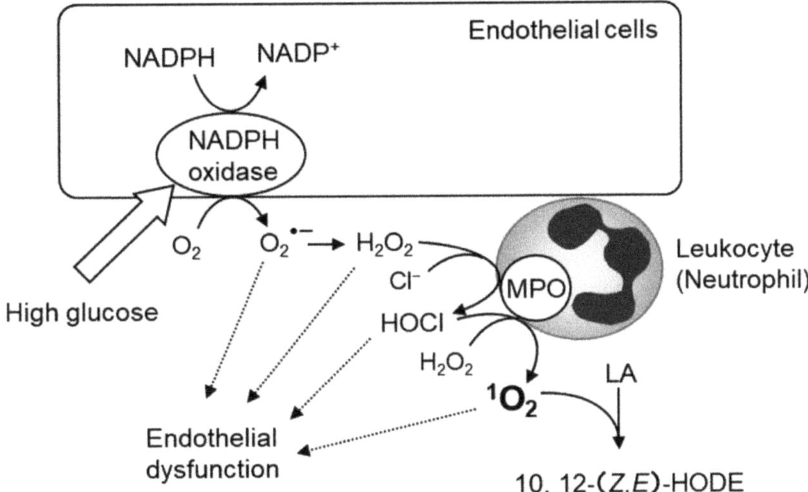

Figure 11. Assumed scheme of $^1O_2$ production via NADPH oxidase and myeloperoxidase (MPO) in vascular injury in diabetes mellitus. High glucose produces hydrogen peroxide (H_2O_2) by NADPH oxidase in vascular endothelial cells. $^1O_2$ is generated via MPO in activated neutrophils bound to the vessel wall.

Clinicians may be able to manage and/or advise patients regarding their food and exercise habits before the onset of diabetes, by evaluating the levels of $^1O_2$-induced lipid peroxidation products in the near future. In any event, more information and studies are needed to determine the pathological significance of $^1O_2$ and 10,12-(Z,E)-HODE.

5.4. Bronchial Asthma

Bronchial asthma is characterized by chronic airway inflammation, reversible airflow obstruction, airway hyperresponsiveness, and structural changes in the airways, and its clinical manifestations include variable airway narrowing, cough, dyspnea, and wheezing [260]. Asthma is caused by multiple interactions between epigenetic regulation and environmental factors. Although this heterogeneity has made it difficult to define biologically distinct subgroups based on the differences in clinical phenotypes, recent advances in genetics and molecular biology have led to the pathogenetic classification of endotypes [261]. These characteristics of asthma may be helpful for understanding the pathophysiology and precision of medicine.

More than 300 million people worldwide suffer from asthma, of which approximately 10% have severe asthma. Symptom control in severe refractory asthma is difficult to achieve despite optimal treatment with high-dose inhaled corticosteroids (ICS) and long-acting adrenergic receptor β2 agonists [262]. Airway inflammation in asthma is typically involved in the submucosal infiltration of activated Th2 lymphocytes, neutrophils, eosinophils, mast cells, and macrophages [263,264]. Although Th2 cytokines and eosinophilic inflammation are typical factors in the development of mild to moderate asthma in response to ICS, a subset of asthmatics with Th2 high-eosinophil-predominant have a refractory course despite optimal treatment [265]. In more severe cases of steroid-resistance, the cellular environment is characterized by airway inflammation induced by Th1/Th2 cytokine expression and neutrophil, with poorly reversible airflow obstruction [265–267]. In adults, elevated neutrophil counts in sputum are related to disease severity [268] and the persistence of symptoms [269]. These reports suggest an important role for neutrophils in asthma and their association with a more steroid-refractory subtype.

Neutrophils are recruited to the site of pathogen invasion for immune defense [270,271]. Several proinflammatory mediators, including cytokines, chemokines, and complements, contribute to the regulation of lung neutrophilic recruitment. Accumulated neutrophils release chemotactic factors that attract monocytes and/or macrophages to the infection site and, subsequently, exacerbate airway inflammation [272]. In particular, Th17/IL-17 and IL-8 are involved in the pathogenesis of steroid-resistant asthma [273,274]. Neutrophil extracellular traps (NETs) comprising neutrophil DNA, whose formation is facilitated by proinflammatory cytokines, are also thought to be implicated in the pathological condition of neutrophilic asthma [274–277]. Activated neutrophils release not only inflammatory cytokines, but also ROS and MPO, which damage airway endothelial cells and exacerbate allergic inflammation [278–281]. Plasma MPO levels in asthmatic patients have been proposed as biomarkers for the evaluation of asthma severity in adults [277,282]. As described in Sections 2 and 5.3.2, MPO contributes to the production of $^1O_2$ mediated by the progression of the reaction with H_2O_2 and Cl^- producing HClO. These suggest that $^1O_2$ may be associated with pathogenesis phenotypes in refractory asthma. We have demonstrated a positive correlation between the levels of $^1O_2$-mediated oxidation products 10- and 12-(Z,E)-HODEs, and the levels of MPO activity and IL-17-derived nerve growth factor (NGF) in the bronchoalveolar lavage fluid (BALF) of an asthmatic mouse model with mixed inflammation [97]. As increased NGF in the BALF of asthmatic patients induces smooth muscle hyperplasia in the airway [283,284] and anti-NGF antibodies improve airway hyperresponsiveness [285], NGF is also considered a therapeutic target for lung disease [286]. Interestingly, the generation of $^1O_2$ with an endoperoxide [(3-(1, 4-epidioxy-4-methyl-1, 4-dihydro-1- naphthyl) propionic acid] increased NGF and IL-8 levels in human bronchial epithelial cells, suggesting that $^1O_2$ generation may be an upstream event of increase in NGF.

Recent studies have described neutrophil-derived MPO as a potential biomarker for neutrophilic asthma [277,282]. Therefore, we would expect that 10, 12-(Z,E)-HODEs, which are elevated according to neutrophil recruitment, could also be a prominent biomarker with more stability than MPO for neutrophilic asthma. Glucocorticoids, which are used as the first choice for asthma therapy, fail to attenuate neutrophilic inflammation and may even promote neutrophil survival [268,287]. It seems reasonable to choose macrolide antibiotics for the treatment of asthma rather than steroids when the levels of 10- and 12-(Z,E)-HODEs are increased. However, the use of macrolides is not appropriate for the long-term treatment of neutrophilic asthma due to the risk of bacterial resistance. Although future studies are needed to determine how $^1O_2$ is related to the pathological condition of asthma patients, 10- and 12-(Z,E)-HODEs may be useful in stratifying patients with refractory asthma and in selecting treatment options.

6. $^1O_2$ Scavengers

The living body contains superoxide dismutase and catalase, which can scavenge ROS such as $O_2^{\bullet-}$ and H_2O_2, but it does not have enzymes that can scavenge $^1O_2$. Substances that can effectively scavenge $^1O_2$ depend on exogenous compounds such as dietary carotenoids.

6.1. Carotenoids

Carotenoids are plant pigments that are pro-vitamin A and potent scavengers of $^1O_2$. The $^1O_2$ scavenging of carotenoids is mainly based on physical scavenging. The excitation energy of $^1O_2$ is transferred to carotenoids (singlet carotenoids; ^{1}Car) to produce triplet oxygen ($^3O_2$) and triplet carotenoids (^{3}Car). ^{3}Car, which receives excitation energy, returns to ground state carotenoids by releasing heat as energy. Thus, carotenoids can repeatedly scavenge $^1O_2$ [288].

$$^1Car + {}^1O_2 \rightarrow {}^3Car + {}^3O_2$$

Carotenoids are the most potent $^1O_2$ scavengers found in nature, and carotenoids with 11 carbon atoms involved in the π-conjugation length, such as lycopene (Figure 12A), β-carotene (Figure 12B), and astaxanthin (Figure 11C), are the most efficient $^1O_2$ scavengers [288]. Among carotenoids, studies have mainly been conducted on the products associated with the $^1O_2$ scavenging of β-carotene, and β-carotene 5,8-endoperoxide (Figure 12B) is a specific product mediated by $^1O_2$ oxidation [102,289,290]. In contrast, β-carotene 5,6-epoxide is formed by a free radical-mediated reaction. β-Carotene-5,8-endoperoxide has been detected in vitro [291,292] and in vivo [289]. Among carotenoids, lycopene has the strongest $^1O_2$ scavenging capacity [293]. The second-order rate constants of physical quenching (kq) in ethanol/chloroform were $31{,}000 \times 10^6$ $M^{-1}s^{-1}$ for lycopene, $14{,}000 \times 10^6$ $M^{-1}s^{-1}$ for β-carotene, 8000×10^6 $M^{-1}s^{-1}$ for lutein, and $10{,}000 \times 10^6$ $M^{-1}s^{-1}$ for zeaxanthin, indicating that lycopene was the most efficient $^1O_2$ quencher [294]. Lycopene and β-carotene also exhibited the fastest $^1O_2$ scavenging rate constants under the conditions of the model membrane system using liposomes [295]. Unlike other carotenoids, lycopene does not possess a ring structure. Since 2-methyl-2-hepten-6-one (Figure 12A) and apo-6′-lycopenal (Figure 12A) were detected under 500 W light irradiation in the presence of methylene blue, a photosensitizer, lycopene is assumed to react with $^1O_2$ via a dioxetane intermediate [296,297].

Carotenoids in plants exist in close proximity to chlorophyll in chloroplasts and scavenge $^1O_2$ generated by the photosensitizing effect of chlorophyll during light absorption. Thus, carotenoids have a protective property against light stress in plants. In addition, carotenoids, such as xanthophylls (lutein (Figure 12D) and zeaxanthin (Figure 12E), specifically accumulate in the macula of the retina, which is an organ exposed to light. In sunlight, the energy of blue light (400 nm), in particular, is 100-fold greater than that of red light (590 nm) and causes severe damage to cells. Since lutein and zeaxanthin have absorption maxima for light at approximately 440 nm, these carotenoids can efficiently absorb blue light. Therefore, the accumulation of lutein and zeaxanthin in the retinal macula can prevent the oxidative degeneration of macular tissue and reduce the risk of AMD and other diseases. Lutein and zeaxanthin are also present in the lens, where they prevent the development of cataract [298,299]. The consumption of tomatoes rich in β-carotene and lycopene can reduce the formation of erythema in human skin due to UV irradiation [300,301].

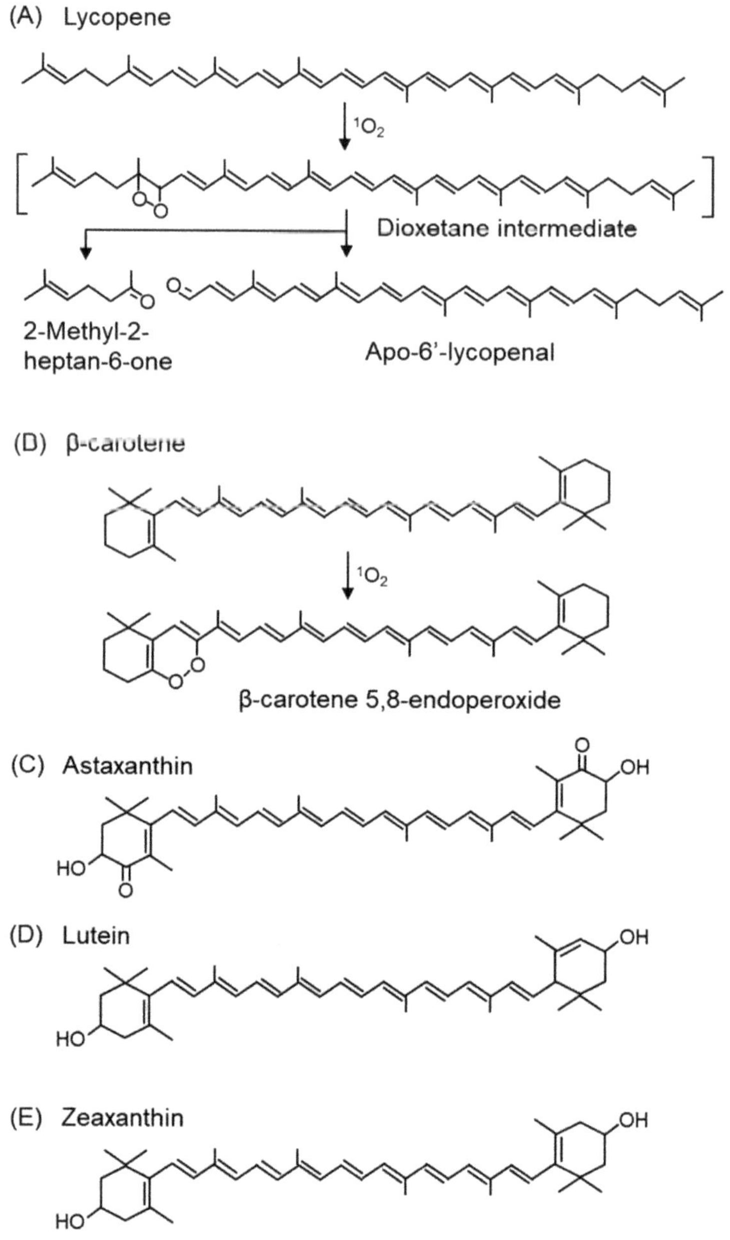

Figure 12. The structures of carotenoid and peroxidation products from lycopene and β-carotene. (**A**) Lycopene, (**B**) β-carotene, (**C**) astaxanthin, (**D**) lutein, (**E**) zeaxanthin.

6.2. Vitamin E (Tocopherol)

α-Tocopherol, a primary fat-soluble antioxidant in vivo, also has $^1O_2$ scavenging capacity that is approximately 30–100 times weaker than that of carotenoids [302]. However, the $^1O_2$ scavenging capacity of carotenoids is also reported to be reduced in liposomes, which are biomembrane models, compared to solution [303].Considering the level of tocopherols in vivo, the in vivo $^1O_2$ scavenging activity of α-tocopherol is considered to be functional. Among tocopherols, α->β->γ->δ-tocopherol have $^1O_2$ capacities in this order [304,305]. Tocotrienols, which are homologues of tocopherols, also have $^1O_2$ scavenging ability in the order of α->β->γ->δ-tocotrienol [304].

The oxidation product of α-tocopherol produced by $^1O_2$ is α-tocopherylquinone, which can be measured in biological samples [306–308]. α-Tocopherylquinone is not an $^1O_2$-specific marker because it is also produced by ROS other than $^1O_2$.

6.3. Other Compounds

As previously mentioned, PUFA is susceptible to oxidation by $^1O_2$ and has $^1O_2$ scavenging capacity, although it is very weak compared to carotenoids [309].

Bakuchiol (Figure 13A), a terpeno-phenolic compound, is a functional analog of retinol that has attracted attention in skincare for its ability to induce retinol gene expression and stimulate collagen production [310]. Bakuchiol protects retinol from degradation by $^1O_2$ produced by H_2O_2 and lithium molybdenum. It was shown to inhibit the $^1O_2$-induced peroxidation of squalene, prevent pore clogging, and reduce the onset of acne by 42% with topical bakuchiol treatment in 54 volunteers [311]. In addition, the 12-week application of a cream containing bakuchiol (0.5%) improved wrinkles and pigmentation [312].

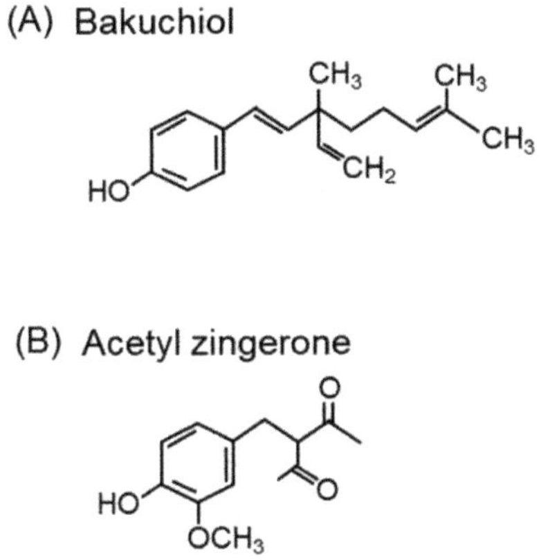

Figure 13. Compounds with $^1O_2$ scavenging activity other than carotenoids. The structures of (**A**) bakuchiol and (**B**) acetyl zingerone.

Zingerone (4-[4-hydroxy-3-methoxyphenyl]-2-butanone) from *Zingiber officinale Roscoe* (ginger) scavenges $^1O_2$ [313]. The structure of zingerone is similar to that of turmeric-derived curcumin (1,7-bis[4-hydroxy-3-methoxyphenyl]-1,6-heptane-3,5-dione). Zingerone and curcumin can both prevent skin aging. Acetyl zingerone (3-(4-hydroxy-3-methoxybenzyl) pentane-2,4-dione) (Figure 13B), a multifunctional skincare ingredient, was synthesized using the molecular structures of zingerone and curcumin [314]. Curcumin has been shown to have $^1O_2$ scavenging activity in an in vitro experimental system by ESR spectroscopy using TEMP as a spin-trap [315]. Acetyl zingerone, which has a higher $^1O_2$ scavenging activity than α-tocopherol [316], decreases the expression of genes associated with extracellular matrix degradation (MMP-3 and cathepsin V) and inhibits the activity of MMP-1, -3, and -12 [314]. Clinical reports have demonstrated that treatment with a cream containing 1% acetyl zingerone applied twice daily for 8 weeks improves wrinkles, hyperpigmentation, and erythema [317].

7. Summary and Outlook

As previously described, diseases and dysfunctions mediated by or, at least, related to $^1O_2$ can be categorized into two groups: those caused by the UV irradiation of photosensitizers accumulated in the living body, and those resulting from the activation of MPO in leukocytes. The former includes cutaneous photosensitivity and retinal diseases, which have been clinically investigated. For the latter, we have previously reported findings in diabetic patients, glaucoma patients, and in an asthma mouse model. MPO is expressed in leukocytes and is implicated in many inflammatory diseases [318]. MPO is associated with cardiovascular disease, athcrosclerosis, glomerulonephritis, arthritis, and Alzheimer's disease. Practical biomarkers are required for the early diagnosis, assessment, and progression of diseases, and for the evaluation of treatment efficacy. However, as summarized in Tables 2 and 3, there are few reports on the analysis of lipid peroxidation products formed by $^1O_2$ in clinical samples, except for SqOOHs. SqOOHs can be measured noninvasively as they are found in sebum; therefore, an increasing number of reports are available (Table 3), especially in the cosmetics and beauty industries. SqOOHs are lipid peroxidation products, but not specific products of $^1O_2$. Since lipid peroxidation is a major oxidative injury in vivo, lipid peroxidation products, especially hydroxides, may be useful biomarkers. It should be noted, however, that lipid peroxidation products are generally not diagnostic indicators of specific diseases. Their usefulness should be enhanced by combination with other biomarkers. Further prospective studies and analysis of the correlation between lipid peroxidation products, specifically from $^1O_2$, and the severity of disease progression should be conducted in the future. When disease risk can be assessed by a prominent biomarker, advice on diet, including carotenoids, and exercise habits can be provided.

Author Contributions: Writing—original draft preparation, K.M., A.U., M.S., M.T. and Y.Y.; writing—review and editing, K.M., A.U., M.S., M.T. and Y.Y.; visualization, K.M. and M.S. All authors have read and agreed to the published version of the manuscript.

Funding: This study was partially supported by a Grant-in-Aid for JSPS KAKENHI Grant Number JP18K17951.

Institutional Review Board Statement: Not applicable.

Informed Consent Statement: Not applicable.

Data Availability Statement: Not applicable.

Conflicts of Interest: Yasukazu Yoshida is an employee of LG Japan Lab, Inc. This paper has not received any form of financial support from LG Japan Lab, Inc. The findings of this paper do not result in any commercial or economic interest to LG Japan Lab, Inc. or any of the authors.

References

1. Chen, Z.H.; Yoshida, Y.; Saito, Y.; Noguchi, N.; Niki, E. Adaptive response induced by lipid peroxidation products in cell cultures. *FEBS Lett.* **2006**, *580*, 479–483. [CrossRef]
2. Haugaard, N. Cellular mechanisms of oxygen toxicity. *Physiol. Rev.* **1968**, *48*, 311–373. [CrossRef]
3. Gottlieb, S.F. Effect of hyperbaric oxygen on microorganisms. *Annu. Rev. Microbiol.* **1971**, *25*, 111–152. [CrossRef]
4. Gilbert, D.L. Symposium on oxygen. Introduction: Oxygen and life. *Anesthesiology* **1972**, *37*, 100–111. [CrossRef]
5. Chance, B.; Sies, H.; Boveris, A. Hydroperoxide metabolism in mammalian organs. *Physiol. Rev.* **1979**, *59*, 527–605. [CrossRef] [PubMed]
6. Hackbarth, S.; Islam, W.; Fang, J.; Šubr, V.; Röder, B.; Etrych, T.; Maeda, H. Singlet oxygen phosphorescence detection in vivo identifies PDT-induced anoxia in solid tumors. *Photochem. Photobiol. Sci.* **2019**, *18*, 1304–1314. [CrossRef] [PubMed]
7. Collin, F. Chemical Basis of Reactive Oxygen Species Reactivity and Involvement in Neurodegenerative Diseases. *Int. J. Mol. Sci.* **2019**, *20*, 2407. [CrossRef] [PubMed]
8. Wang, X.; Lv, H.; Sun, Y.; Zu, G.; Zhang, X.; Song, Y.; Zhao, F.; Wang, J. New porphyrin photosensitizers-Synthesis, singlet oxygen yield, photophysical properties and application in PDT. *Spectrochim. Acta. A Mol. Biomol. Spectrosc.* **2022**, *279*, 121447. [CrossRef]
9. You, Z.Q.; Hsu, C.P.; Fleming, G.R. Triplet-triplet energy-transfer coupling: Theory and calculation. *J. Chem. Phys.* **2006**, *124*, 044506. [CrossRef]
10. Dzebo, D.; Moth-Poulsen, K.; Albinsson, B. Robust triplet-triplet annihilation photon upconversion by efficient oxygen scavenging. *Photochem. Photobiol. Sci. Off. J. Eur. Photochem. Assoc. Eur. Soc. Photobiol.* **2017**, *16*, 1327–1334. [CrossRef] [PubMed]

11. Turro, N.J. Singlet oxygen and chemiluminescent organic reactions. In *Modern Molecular Photochemistry*; University Science Books: Melville, NY, USA, 1991; pp. 583–593.

12. Hatz, S.; Poulsen, L.; Ogilby, P.R. Time-resolved singlet oxygen phosphorescence measurements from photosensitized experiments in single cells: Effects of oxygen diffusion and oxygen concentration. *Photochem. Photobiol.* **2008**, *84*, 1284–1290. [CrossRef] [PubMed]

13. Pedersen, B.W.; Sinks, L.E.; Breitenbach, T.; Schack, N.B.; Vinogradov, S.A.; Ogilby, P.R. Single cell responses to spatially controlled photosensitized production of extracellular singlet oxygen. *Photochem. Photobiol.* **2011**, *87*, 1077–1091. [CrossRef] [PubMed]

14. Liang, P.; Kolodieznyi, D.; Creeger, Y.; Ballou, B.; Bruchez, M.P. Subcellular Singlet Oxygen and Cell Death: Location Matters. *Front. Chem.* **2020**, *8*, 592941. [CrossRef] [PubMed]

15. Denys, G.A.; Devoe, N.C.; Gudis, P.; May, M.; Allen, R.C.; Stephens, J.T., Jr. Mechanism of Microbicidal Action of E-101 Solution, a Myeloperoxidase-Mediated Antimicrobial, and Its Oxidative Products. *Infect. Immun.* **2019**, *87*, e00261-19. [CrossRef]

16. Neale, T.J.; Ullrich, R.; Ojha, P.; Poczewski, H.; Verhoeven, A.J.; Kerjaschki, D. Reactive oxygen species and neutrophil respiratory burst cytochrome b558 are produced by kidney glomerular cells in passive Heymann nephritis. *Proc. Natl. Acad. Sci. USA* **1993**, *90*, 3645–3649. [CrossRef] [PubMed]

17. Russell, G.A. Deuterium-isotope Effects in the Autoxidation of Aralkyl Hydrocarbons. Mechanism of the Interaction of PEroxy Radicals1. *J. Am. Chem. Soc.* **1957**, *79*, 3871–3877. [CrossRef]

18. Gonçalves, L.C.P.; Massari, J.; Licciardi, S.; Prado, F.M.; Linares, E.; Klassen, A.; Tavares, M.F.M.; Augusto, O.; Di Mascio, P.; Bechara, E.J.H. Singlet oxygen generation by the reaction of acrolein with peroxynitrite via a 2-hydroxyvinyl radical intermediate. *Free Radic. Biol. Med.* **2020**, *152*, 83–90. [CrossRef]

19. Sun, S.; Bao, Z.; Ma, H.; Zhang, D.; Zheng, X. Singlet oxygen generation from the decomposition of alpha-linolenic acid hydroperoxide by cytochrome c and lactoperoxidase. *Biochemistry* **2007**, *46*, 6668–6673. [CrossRef]

20. Chen, T.; Cohen, D.; Itkin, M.; Malitsky, S.; Fluhr, R. Lipoxygenase functions in $^1O_2$ production during root responses to osmotic stress. *Plant Physiol.* **2021**, *185*, 1638–1651. [CrossRef]

21. Khan, A.U.; Kasha, M. Singlet molecular oxygen in the Haber-Weiss reaction. *Proc. Natl. Acad. Sci. USA* **1994**, *91*, 12365–12367. [CrossRef]

22. Miyamoto, S.; Ronsein, G.E.; Corrêa, T.C.; Martinez, G.R.; Medeiros, M.H.; Di Mascio, P. Direct evidence of singlet molecular oxygen generation from peroxynitrate, a decomposition product of peroxynitrite. *Dalton Trans.* **2009**, *29*, 5720–5729. [CrossRef] [PubMed]

23. Bauer, G. The synergistic effect between hydrogen peroxide and nitrite, two long-lived molecular species from cold atmospheric plasma, triggers tumor cells to induce their own cell death. *Redox Biol.* **2019**, *26*, 101291. [CrossRef] [PubMed]

24. Kanofsky, J.R.; Sima, P. Singlet oxygen production from the reactions of ozone with biological molecules. *J. Biol. Chem.* **1991**, *266*, 9039–9042. [CrossRef]

25. Cerkovnik, J.; Erzen, E.; Koller, J.; Plesnicar, B. Evidence for HOOO radicals in the formation of alkyl hydrotrioxides (ROOOH) and hydrogen trioxide (HOOOH) in the ozonation of C-H bonds in hydrocarbons. *J. Am. Chem. Soc.* **2002**, *124*, 404–409. [CrossRef]

26. Pershin, A.A.; Torbin, A.P.; Zagidullin, M.V.; Mebel, A.M.; Mikheyev, P.A.; Azyazov, V.N. Rate constants for collision-induced emission of O(2)(a(1)Δ(g)) with He, Ne, Ar, Kr, N(2), CO(2) and SF(6) as collisional partners. *Phys. Chem. Chem. Phys.* **2018**, *20*, 29677–29683. [CrossRef]

27. Tao, F.; Gonzalez-Flecha, B.; Kobzik, L. Reactive oxygen species in pulmonary inflammation by ambient particulates. *Free Radic. Biol. Med.* **2003**, *35*, 327–340. [CrossRef] [PubMed]

28. Gurgueira, S.A.; Lawrence, J.; Coull, B.; Murthy, G.G.; González-Flecha, B. Rapid increases in the steady-state concentration of reactive oxygen species in the lungs and heart after particulate air pollution inhalation. *Environ. Health Perspect.* **2002**, *110*, 749–755. [CrossRef]

29. Jensen, R.L.; Arnbjerg, J.; Ogilby, P.R. Reaction of singlet oxygen with tryptophan in proteins: A pronounced effect of the local environment on the reaction rate. *J. Am. Chem. Soc.* **2012**, *134*, 9820–9826. [CrossRef]

30. Nascimento, R.O.; Prado, F.M.; Massafera, M.P.; Di Mascio, P.; Ronsein, G.E. Dehydromethionine is a common product of methionine oxidation by singlet molecular oxygen and hypohalous acids. *Free Radic. Biol. Med.* **2022**, *187*, 17–28. [CrossRef]

31. Fouquerel, E.; Barnes, R.P.; Uttam, S.; Watkins, S.C.; Bruchez, M.P.; Opresko, P.L. Targeted and Persistent 8-Oxoguanine Base Damage at Telomeres Promotes Telomere Loss and Crisis. *Mol. Cell* **2019**, *75*, 117–130.e116. [CrossRef] [PubMed]

32. Lu, W.; Sun, Y.; Tsai, M.; Zhou, W.; Liu, J. Singlet O(2) Oxidation of a Deprotonated Guanine-Cytosine Base Pair and Its Entangling with Intra-Base-Pair Proton Transfer. *Chemphyschem* **2018**, *19*, 2645–2654. [CrossRef]

33. Kato, S.; Shimizu, N.; Otoki, Y.; Ito, J.; Sakaino, M.; Sano, T.; Takeuchi, S.; Imagi, J.; Nakagawa, K. Determination of acrolein generation pathways from linoleic acid and linolenic acid: Increment by photo irradiation. *NPJ Sci. Food* **2022**, *6*, 21. [CrossRef] [PubMed]

34. Ishikawa, A.; Ito, J.; Shimizu, N.; Kato, S.; Kobayashi, E.; Ohnari, H.; Sakata, O.; Naru, E.; Nakagawa, K. Linoleic acid and squalene are oxidized by discrete oxidation mechanisms in human sebum. *Ann. N. Y. Acad. Sci.* **2021**, *1500*, 112–121. [CrossRef] [PubMed]

35. Carroll, L.; Pattison, D.I.; Davies, J.B.; Anderson, R.F.; Lopez-Alarcon, C.; Davies, M.J. Superoxide radicals react with peptide-derived tryptophan radicals with very high rate constants to give hydroperoxides as major products. *Free Radic. Biol. Med.* **2018**, *118*, 126–136. [CrossRef]

36. Moe, M.M.; Tsai, M.; Liu, J. Singlet Oxygen Oxidation of the Radical Cations of 8-Oxo-2′-deoxyguanosine and Its 9-Methyl Analogue: Dynamics, Potential Energy Surface, and Products Mediated by C5-O(2) -Addition. *Chempluschem* **2021**, *86*, 1243–1254. [CrossRef] [PubMed]
37. Monzo-Beltran, L.; Vazquez-Tarragón, A.; Cerdà, C.; Garcia-Perez, P.; Iradi, A.; Sánchez, C.; Climent, B.; Tormos, C.; Vázquez-Prado, A.; Girbés, J.; et al. One-year follow-up of clinical, metabolic and oxidative stress profile of morbid obese patients after laparoscopic sleeve gastrectomy. 8-oxo-dG as a clinical marker. *Redox Biol.* **2017**, *12*, 389–402. [CrossRef] [PubMed]
38. Ponce, M.A.; Ramirez, J.A.; Galagovsky, L.R.; Gros, E.G.; Erra-Balsells, R. Singlet-oxygen ene reaction with 3β-substituted stigmastanes. An alternative pathway for the classical Schenck rearrangement. *J. Chem. Soc. Perkin Trans.* **2000**, *2*, 2351–2358. [CrossRef]
39. Samuel, D.; Norrell, K.; Hilmey, D.G. Novel ring chemistry of vitamin B6 with singlet oxygen and an activated ene: Isolated products and identified intermediates suggesting an operable [3 + 2] cycloaddition. *Org. Biomol. Chem.* **2012**, *10*, 7278–7281. [CrossRef]
40. Petrou, A.L.; Petrou, P.L.; Ntanos, T.; Liapis, A. A Possible Role for Singlet Oxygen in the Degradation of Various Antioxidants. A Meta-Analysis and Review of Literature Data. *Antioxidants* **2018**, *7*, 35. [CrossRef]
41. Zeinali, N.; Oluwoye, I.; Altarawneh, M.K.; Almatarneh, M.H.; Dlugogorski, B.Z. Probing the Reactivity of Singlet Oxygen with Cyclic Monoterpenes. *ACS Omega* **2019**, *4*, 14040–14048. [CrossRef]
42. Morales, J.; Günther, G.; Zanocco, A.L.; Lemp, E. Singlet oxygen reactions with flavonoids. A theoretical-experimental study. *PLoS ONE* **2012**, *7*, e40548. [CrossRef] [PubMed]
43. Chen, L.H.; Tsai, H.C.; Yu, P.L.; Chung, K.R. A Major Facilitator Superfamily Transporter-Mediated Resistance to Oxidative Stress and Fungicides Requires Yap1, Skn7, and MAP Kinases in the Citrus Fungal Pathogen Alternaria alternata. *PLoS ONE* **2017**, *12*, e0169103. [CrossRef] [PubMed]
44. Di Mascio, P.; Martinez, G.R.; Miyamoto, S.; Ronsein, G.E.; Medeiros, M.H.G.; Cadet, J. Singlet Molecular Oxygen Reactions with Nucleic Acids, Lipids, and Proteins. *Chem. Rev.* **2019**, *119*, 2043–2086. [CrossRef] [PubMed]
45. Kanofsky, J.R. Singlet oxygen production by biological systems. *Chem. Biol. Interact.* **1989**, *70*, 1–28. [CrossRef]
46. Dong, S.; Xu, J.; Jia, T.; Xu, M.; Zhong, C.; Yang, G.; Li, J.; Yang, D.; He, F.; Gai, S.; et al. Upconversion-mediated $ZnFe_2O_4$ nanoplatform for NIR-enhanced chemodynamic and photodynamic therapy. *Chem. Sci.* **2019**, *10*, 4259–4271. [CrossRef]
47. Choi, J.; Kim, S.Y. Synthesis of near-infrared-responsive hexagonal-phase upconversion nanoparticles with controllable shape and luminescence efficiency for theranostic applications. *J. Biomater. Appl.* **2022**, *37*, 646–658. [CrossRef]
48. Takajo, T.; Kurihara, Y.; Iwase, K.; Miyake, D.; Tsuchida, K.; Anzai, K. Basic Investigations of Singlet Oxygen Detection Systems with ESR for the Measurement of Singlet Oxygen Quenching Activities. *Chem. Pharm. Bull.* **2020**, *68*, 150–154. [CrossRef]
49. Yu, Y.Y.; Quan, W.Z.; Cao, Y.; Niu, Q.; Lu, Y.; Xiao, X.; Cheng, L. Boosting the singlet oxygen production from H2O2 activation with highly dispersed Co-N-graphene for pollutant removal. *RSC Adv* **2022**, *12*, 17864–17872. [CrossRef]
50. Ruiz-González, R.; Bresolí-Obach, R.; Gulías, Ò.; Agut, M.; Savoie, H.; Boyle, R.W.; Nonell, S.; Giuntini, F. NanoSOSG: A Nanostructured Fluorescent Probe for the Detection of Intracellular Singlet Oxygen. *Angew Chem. Int. Ed. Engl.* **2017**, *56*, 2885–2888. [CrossRef]
51. Liu, H.; Carter, P.J.H.; Laan, A.C.; Eelkema, R.; Denkova, A.G. Singlet Oxygen Sensor Green is not a Suitable Probe for $(^{1})O_{(2)}$ in the Presence of Ionizing Radiation. *Sci. Rep.* **2019**, *9*, 8393. [CrossRef]
52. Girotti, A.W.; Korytowski, W. Cholesterol as a singlet oxygen detector in biological systems. *Methods Enzymol.* **2000**, *319*, 85–100. [PubMed]
53. Miyamoto, S.; Di Mascio, P. Lipid hydroperoxides as a source of singlet molecular oxygen. *Subcell Biochem.* **2014**, *77*, 3–20. [PubMed]
54. Torinuki, W.; Miura, T. Singlet oxygen emission at 1270 nm in protoporphyrin solution excited by argon laser. *Tohoku J. Exp. Med.* **1983**, *140*, 297–299. [CrossRef] [PubMed]
55. Baker, A.; Kanofsky, J.R. Direct observation of singlet oxygen phosphorescence at 1270 nm from L1210 leukemia cells exposed to polyporphyrin and light. *Arch. Biochem. Biophys.* **1991**, *286*, 70–75. [CrossRef]
56. Baker, A.; Kanofsky, J.R. Time-resolved studies of singlet-oxygen emission from L1210 leukemia cells labeled with 5-(N-hexadecanoyl)amino eosin. A comparison with a one-dimensional model of singlet-oxygen diffusion and quenching. *Photochem. Photobiol.* **1993**, *57*, 720–727. [CrossRef]
57. Oelckers, S.; Ziegler, T.; Michler, I.; Roder, B. Time-resolved detection of singlet oxygen luminescence in red-cell ghost suspensions: Concerning a signal component that can be attributed to $^{1}O_2$ luminescence from the inside of a native membrane. *J. Photochem. Photobiol. B* **1999**, *53*, 121–127. [CrossRef] [PubMed]
58. Schweitzer, C.; Schmidt, R. Physical mechanisms of generation and deactivation of singlet oxygen. *Chem. Rev.* **2003**, *103*, 1685–1757. [CrossRef] [PubMed]
59. Niedre, M.; Patterson, M.S.; Wilson, B.C. Direct near-infrared luminescence detection of singlet oxygen generated by photodynamic therapy in cells in vitro and tissues in vivo. *Photochem. Photobiol.* **2002**, *75*, 382–391. [CrossRef]
60. Nosaka, Y.; Nosaka, A.Y. Generation and Detection of Reactive Oxygen Species in Photocatalysis. *Chem. Rev.* **2017**, *117*, 11302–11336. [CrossRef]
61. Lion, Y.; Delmelle, M.; van de Vorst, A. New method of detecting singlet oxygen production. *Nature* **1976**, *263*, 442–443. [CrossRef]
62. Moan, J.; Wold, E. Detection of singlet oxygen production by ESR. *Nature* **1979**, *279*, 450–451. [CrossRef]

63. Dzwigaj, S.; Pezerat, H. Singlet oxygen-trapping reaction as a method of (1)O2 detection: Role of some reducing agents. *Free Radic. Res.* **1995**, *23*, 103–115. [CrossRef] [PubMed]

64. Victória, H.F.V.; Ferreira, D.C.; Filho, J.B.G.; Martins, D.C.S.; Pinheiro, M.V.B.; Sáfar, G.A.M.; Krambrock, K. Detection of singlet oxygen by EPR: The instability of the nitroxyl radicals. *Free Radic. Biol. Med.* **2022**, *180*, 143–152. [CrossRef] [PubMed]

65. Jung, M.Y.; Choi, D.S. Electron spin resonance and luminescence spectroscopic observation and kinetic study of chemical and physical singlet oxygen quenching by resveratrol in methanol. *J. Agric. Food Chem.* **2010**, *58*, 11888–11895. [CrossRef] [PubMed]

66. Kobayashi, K.; Maehata, Y.; Kawamura, Y.; Kusubata, M.; Hattori, S.; Tanaka, K.; Miyamoto, C.; Yoshino, F.; Yoshida, A.; Wada-Takahashi, S.; et al. Direct assessments of the antioxidant effects of the novel collagen peptide on reactive oxygen species using electron spin resonance spectroscopy. *J. Pharmacol. Sci.* **2011**, *116*, 97–106. [CrossRef] [PubMed]

67. Soh, N.; Katayama, Y.; Maeda, M. A fluorescent probe for monitoring nitric oxide production using a novel detection concept. *Analyst* **2001**, *126*, 564–566. [CrossRef] [PubMed]

68. Tanaka, K.; Miura, T.; Umezawa, N.; Urano, Y.; Kikuchi, K.; Higuchi, T.; Nagano, T. Rational design of fluorescein-based fluorescence probes. Mechanism-based design of a maximum fluorescence probe for singlet oxygen. *J. Am. Chem. Soc.* **2001**, *123*, 2530–2536. [CrossRef] [PubMed]

69. Gomes, A.; Fernandes, E.; Lima, J.L. Fluorescence probes used for detection of reactive oxygen species. *J. Biochem. Biophys. Methods* **2005**, *65*, 45–80. [CrossRef]

70. Corey, E.J.; Taylor, W.C. Study of the peroxidation of organic compounds by externally generated singlet oxygen molecules. *J. Am. Chem. Soc.* **1964**, *86*, 3881–3882. [CrossRef]

71. Francis Wilkinson, W.P.H.; Ross, A.B. Rate Constants for the Decay and Reactions of the Lowest Electronically Excited Singlet State of Molecular Oxygen in Solution. *J. Phys. Chem. Ref.* **1995**, *24*, 663–1021.

72. Umezawa, N.; Tanaka, K.; Urano, Y.; Kikuchi, K.; Higuchi, T.; Nagano, T. Novel Fluorescent Probes for Singlet Oxygen. *Angew. Chem. Int. Ed. Engl.* **1999**, *38*, 2899–2901. [CrossRef]

73. Flors, C.; Fryer, M.J.; Waring, J.; Reeder, B.; Bechtold, U.; Mullineaux, P.M.; Nonell, S.; Wilson, M.T.; Baker, N.R. Imaging the production of singlet oxygen in vivo using a new fluorescent sensor, Singlet Oxygen Sensor Green. *J. Exp. Bot.* **2006**, *57*, 1725–1734. [CrossRef]

74. Price, M.; Reiners, J.J.; Santiago, A.M.; Kessel, D. Monitoring singlet oxygen and hydroxyl radical formation with fluorescent probes during photodynamic therapy. *Photochem. Photobiol.* **2009**, *85*, 1177–1181. [CrossRef]

75. Ragas, X.; Jimenez-Banzo, A.; Sanchez-Garcia, D.; Batllori, X.; Nonell, S. Singlet oxygen photosensitisation by the fluorescent probe Singlet Oxygen Sensor Green. *Chem. Commun.* **2009**, *28*, 2920–2922. [CrossRef]

76. Kim, S.; Tachikawa, T.; Fujitsuka, M.; Majima, T. Far-red fluorescence probe for monitoring singlet oxygen during photodynamic therapy. *J. Am. Chem. Soc.* **2014**, *136*, 11707–11715. [CrossRef]

77. Murotomi, K.; Umeno, A.; Sugino, S.; Yoshida, Y. Quantitative kinetics of intracellular singlet oxygen generation using a fluorescence probe. *Sci. Rep.* **2020**, *10*, 10616. [CrossRef]

78. Chen, B.; Yang, Y.; Wang, Y.; Yan, Y.; Wang, Z.; Yin, Q.; Zhang, Q.; Wang, Y. Precise Monitoring of Singlet Oxygen in Specific Endocytic Organelles by Super-pH-Resolved Nanosensors. *ACS Appl. Mater. Interfaces* **2021**, *13*, 18533–18544. [CrossRef]

79. Nath, P.; Hamadna, S.S.; Karamchand, L.; Foster, J.; Kopelman, R.; Amar, J.G.; Ray, A. Intracellular detection of singlet oxygen using fluorescent nanosensors. *Analyst* **2021**, *146*, 3933–3941. [CrossRef]

80. Niki, E. Lipid peroxidation products as oxidative stress biomarkers. *Biofactors* **2008**, *34*, 171–180. [CrossRef]

81. Niki, E. Assessment of antioxidant capacity in vitro and in vivo. *Free Radic. Biol. Med.* **2010**, *49*, 503–515. [CrossRef]

82. Akazawa-Ogawa, Y.; Shichiri, M.; Nishio, K.; Yoshida, Y.; Niki, E.; Hagihara, Y. Singlet-oxygen-derived products from linoleate activate Nrf2 signaling in skin cells. *Free Radic. Biol. Med.* **2015**, *79*, 164–175. [CrossRef]

83. Murotomi, K.; Umeno, A.; Yasunaga, M.; Shichiri, M.; Ishida, N.; Abe, H.; Yoshida, Y.; Nakajima, Y. Switching from singlet-oxygen-mediated oxidation to free-radical-mediated oxidation in the pathogenesis of type 2 diabetes in model mouse. *Free Radic. Res.* **2015**, *49*, 133–138. [CrossRef] [PubMed]

84. Luchetti, F.; Canonico, B.; Cesarini, E.; Betti, M.; Galluzzi, L.; Galli, L.; Tippins, J.; Zerbinati, C.; Papa, S.; Iuliano, L. 7-Ketocholesterol and 5,6-secosterol induce human endothelial cell dysfunction by differential mechanisms. *Steroids* **2015**, *99*, 204–211. [CrossRef]

85. Dantas, L.S.; Chaves-Filho, A.B.; Coelho, F.R.; Genaro-Mattos, T.C.; Tallman, K.A.; Porter, N.A.; Augusto, O.; Miyamoto, S. Cholesterol secosterol aldehyde adduction and aggregation of Cu,Zn-superoxide dismutase: Potential implications in ALS. *Redox Biol.* **2018**, *19*, 105–115. [CrossRef]

86. Niki, E.; Yoshida, Y. Biomarkers for oxidative stress: Measurement, validation, and application. *J. Med. Invest.* **2005**, *52*, 228–230. [CrossRef]

87. Yoshida, Y.; Niki, E. Bio-markers of lipid peroxidation in vivo: Hydroxyoctadecadienoic acid and hydroxycholesterol. *Biofactors* **2006**, *27*, 195–202. [CrossRef]

88. Yoshida, Y.; Hayakawa, M.; Habuchi, Y.; Itoh, N.; Niki, E. Evaluation of lipophilic antioxidant efficacy in vivo by the biomarkers hydroxyoctadecadienoic acid and isoprostane. *Lipids* **2007**, *42*, 463–472. [CrossRef]

89. Yoshida, Y.; Kodai, S.; Takemura, S.; Minamiyama, Y.; Niki, E. Simultaneous measurement of F2-isoprostane, hydroxyoctadecadienoic acid, hydroxyeicosatetraenoic acid, and hydroxycholesterols from physiological samples. *Anal. Biochem.* **2008**, *379*, 105–115. [CrossRef]

90. Yoshida, Y.; Umeno, A.; Shichiri, M. Lipid peroxidation biomarkers for evaluating oxidative stress and assessing antioxidant capacity in vivo. *J. Clin. Biochem. Nutr.* **2013**, *52*, 9–16. [CrossRef]

91. Yoshida, Y.; Umeno, A.; Akazawa, Y.; Shichiri, M.; Murotomi, K.; Horie, M. Chemistry of lipid peroxidation products and their use as biomarkers in early detection of diseases. *J. Oleo Sci.* **2015**, *64*, 347–356. [CrossRef]

92. Umeno, A.; Shichiri, M.; Ishida, N.; Hashimoto, Y.; Abe, K.; Kataoka, M.; Yoshino, K.; Hagihara, Y.; Aki, N.; Funaki, M.; et al. Singlet oxygen induced products of linoleates, 10- and 12-(Z,E)-hydroxyoctadecadienoic acids (HODE), can be potential biomarkers for early detection of type 2 diabetes. *PLoS ONE* **2013**, *8*, e63542. [CrossRef] [PubMed]

93. Umeno, A.; Yoshino, K.; Hashimoto, Y.; Shichiri, M.; Kataoka, M.; Yoshida, Y. Multi-Biomarkers for Early Detection of Type 2 Diabetes, Including 10- and 12-(Z,E)-Hydroxyoctadecadienoic Acids, Insulin, Leptin, and Adiponectin. *PLoS ONE* **2015**, *10*, e0130971. [CrossRef] [PubMed]

94. Umeno, A.; Tanito, M.; Kaidzu, S.; Takai, Y.; Horie, M.; Yoshida, Y. Comprehensive measurements of hydroxylinoleate and hydroxyarachidonate isomers in blood samples from primary open-angle glaucoma patients and controls. *Sci. Rep.* **2019**, *9*, 2171. [CrossRef] [PubMed]

95. Kiessling, U.; Spiteller, G. The course of enzymatically induced lipid peroxidation in homogenized porcine kidney tissue. *Z. Naturforsch C J. Biosci.* **1998**, *53*, 431–437. [CrossRef]

96. Horie, M.; Miura, T.; Hirakata, S.; Hosoyama, A.; Sugino, S.; Umeno, A.; Murotomi, K.; Yoshida, Y.; Koike, T. Comparative analysis of the intestinal flora in type 2 diabetes and nondiabetic mice. *Exp. Anim.* **2017**, *66*, 405–416. [CrossRef]

97. Ogawa, H.; Azuma, M.; Umeno, A.; Shimizu, M.; Murotomi, K.; Yoshida, Y.; Nishioka, Y.; Tsuneyama, K. Singlet oxygen -derived nerve growth factor exacerbates airway hyperresponsiveness in a mouse model of asthma with mixed inflammation. *Allergol. Int.* **2022**, *71*, 395–404. [CrossRef]

98. Adachi, J.; Asano, M.; Naito, T.; Ueno, Y.; Imamichi, H.; Tatsuno, Y. Cholesterol hydroperoxides in erythrocyte membranes of alcoholic patients. *Alcohol. Clin. Exp. Res.* **1999**, *23*, 96s–100s. [CrossRef]

99. Yamazaki, S.; Ozawa, N.; Hiratsuka, A.; Watanabe, T. Quantitative determination of cholesterol 5alpha-, 7alpha-, and 7beta-hydroperoxides in rat skin. *Free Radic. Biol. Med.* **1999**, *27*, 110–118. [CrossRef]

100. Minami, Y.; Yokoi, S.; Setoyama, M.; Bando, N.; Takeda, S.; Kawai, Y.; Terao, J. Combination of TLC blotting and gas chromatography-mass spectrometry for analysis of peroxidized cholesterol. *Lipids* **2007**, *42*, 1055–1063. [CrossRef]

101. Minami, Y.; Kawabata, K.; Kubo, Y.; Arase, S.; Hirasaka, K.; Nikawa, T.; Bando, N.; Kawai, Y.; Terao, J. Peroxidized cholesterol-induced matrix metalloproteinase-9 activation and its suppression by dietary beta-carotene in photoaging of hairless mouse skin. *J. Nutr. Biochem.* **2009**, *20*, 389–398. [CrossRef]

102. Terao, J.; Minami, Y.; Bando, N. Singlet molecular oxygen-quenching activity of carotenoids: Relevance to protection of the skin from photoaging. *J. Clin. Biochem. Nutr.* **2011**, *48*, 57–62. [CrossRef] [PubMed]

103. Wentworth, P., Jr.; Nieva, J.; Takeuchi, C.; Galve, R.; Wentworth, A.D.; Dilley, R.B.; DeLaria, G.A.; Saven, A.; Babior, B.M.; Janda, K.D.; et al. Evidence for ozone formation in human atherosclerotic arteries. *Science* **2003**, *302*, 1053–1056. [CrossRef] [PubMed]

104. Zhang, Q.; Powers, E.T.; Nieva, J.; Huff, M.E.; Dendle, M.A.; Bieschke, J.; Glabe, C.G.; Eschenmoser, A.; Wentworth, P., Jr.; Lerner, R.A.; et al. Metabolite-initiated protein misfolding may trigger Alzheimer's disease. *Proc. Natl. Acad. Sci. USA* **2004**, *101*, 4752–4757. [CrossRef] [PubMed]

105. Elshafei, M.E.; Minamiyama, Y.; Ichikawa, H. Singlet oxygen from endoperoxide initiates an intracellular reactive oxygen species release in HaCaT keratinocytes. *J. Clin. Biochem. Nutr.* **2022**, *71*, 198–205. [CrossRef]

106. Umar, S.A.; Shahid, N.H.; Nazir, L.A.; Tanveer, M.A.; Divya, G.; Archoo, S.; Raghu, S.R.; Tasduq, S.A. Pharmacological Activation of Autophagy Restores Cellular Homeostasis in Ultraviolet-(B)-Induced Skin Photodamage. *Front. Oncol.* **2021**, *11*, 726066. [CrossRef]

107. Mokrzyński, K.; Krzysztyńska-Kuleta, O.; Zawrotniak, M.; Sarna, M.; Sarna, T. Fine Particulate Matter-Induced Oxidative Stress Mediated by UVA-Visible Light Leads to Keratinocyte Damage. *Int. J. Mol. Sci.* **2021**, *22*, 10645. [CrossRef] [PubMed]

108. Requena, M.B.; Russignoli, P.E.; Vollet-Filho, J.D.; Salvio, A.G.; Fortunato, T.C.; Pratavieira, S.; Bagnato, V.S. Use of dermograph for improvement of PpIX precursor's delivery in photodynamic therapy: Experimental and clinical pilot studies. *Photodiagnosis Photodyn. Ther.* **2020**, *29*, 101599. [CrossRef]

109. Pinto, M.A.F.; Ferreira, C.B.R.; de Lima, B.E.S.; Molon, Â.C.; Ibarra, A.M.C.; Cecatto, R.B.; Dos Santos Franco, A.L.; Rodrigues, M. Effects of 5-ALA mediated photodynamic therapy in oral cancer stem cells. *J. Photochem. Photobiol. B* **2022**, *235*, 112552. [CrossRef]

110. Arakane, K.; Ryu, A.; Hayashi, C.; Masunaga, T.; Shinmoto, K.; Mashiko, S.; Nagano, T.; Hirobe, M. Singlet oxygen (1 delta g) generation from coproporphyrin in Propionibacterium acnes on irradiation. *Biochem. Biophys. Res. Commun.* **1996**, *223*, 578–582. [CrossRef]

111. Patwardhan, S.V.; Richter, C.; Vogt, A.; Blume-Peytavi, U.; Canfield, D.; Kottner, J. Measuring acne using Coproporphyrin III, Protoporphyrin IX, and lesion-specific inflammation: An exploratory study. *Arch. Dermatol. Res.* **2017**, *309*, 159–167. [CrossRef]

112. Zhang, S.; Sun, X.; Wang, Z.; Sun, J.; He, Z.; Sun, B.; Luo, C. Molecularly Self-Engineered Nanoamplifier for Boosting Photodynamic Therapy via Cascade Oxygen Elevation and Lipid ROS Accumulation. *ACS Appl. Mater Interfaces* **2022**, *14*, 38497–38505. [CrossRef] [PubMed]

113. Uliana, M.P.; da Cruz Rodrigues, A.; Ono, B.A.; Pratavieira, S.; de Oliveira, K.T.; Kurachi, C. Photodynamic Inactivation of Microorganisms Using Semisynthetic Chlorophyll a Derivatives as Photosensitizers. *Molecules* **2022**, *27*, 5769. [CrossRef] [PubMed]

114. Sugiura, M.; Azami, C.; Koyama, K.; Rutherford, A.W.; Rappaport, F.; Boussac, A. Modification of the pheophytin redox potential in Thermosynechococcus elongatus Photosystem II with PsbA3 as D1. *Biochim. Biophys. Acta* **2014**, *1837*, 139–148. [CrossRef] [PubMed]

115. Wang, E.; Braun, M.S.; Wink, M. Chlorophyll and Chlorophyll Derivatives Interfere with Multi-Drug Resistant Cancer Cells and Bacteria. *Molecules* **2019**, *24*, 2968. [CrossRef] [PubMed]

116. Zhao, Z.; Poojary, M.M.; Skibsted, L.H.; Lund, M.N. Cleavage of Disulfide Bonds in Cystine by UV-B Illumination Mediated by Tryptophan or Tyrosine as Photosensitizers. *J. Agric. Food Chem.* **2020**, *68*, 6900–6909. [CrossRef]

117. Ionita, G.; Matei, I. Application of Riboflavin Photochemical Properties in Hydrogel Synthesis. In *Biophysical Chemistry Advance Applications*; InTech Open: London, UK, 2020.

118. Wangsuwan, S.; Meephansan, J. Comparative Study Of Photodynamic Therapy With Riboflavin-Tryptophan Gel And 13% 5-Aminolevulinic Acid In The Treatment Of Mild To Moderate Acne Vulgaris. *Clin. Cosmet. Investig. Dermatol.* **2019**, *12*, 805–814. [CrossRef]

119. Chignell, C.F.; Kukielczak, B.M.; Sik, R.H.; Bilski, P.J.; He, Y.Y. Ultraviolet A sensitivity in Smith-Lemli-Opitz syndrome: Possible involvement of cholesta-5,7,9(11)-trien-3 beta-ol. *Free Radic. Biol. Med.* **2006**, *41*, 339–346. [CrossRef]

120. Bode, C.W.; Hänsel, W. 5-(3-Phenylpropoxy)psoralen and 5-(4-phenylbutoxy)psoralen: Mechanistic studies on phototoxicity. *Pharmazie* **2005**, *60*, 225–228.

121. Lee, K.P.; Girijala, R.L.; Chon, S.Y. Phytophotodermatitis due to a Citrus-Based Hand Sanitizer: A Case Report. *Korean J. Fam. Med.* **2022**, *43*, 271–273. [CrossRef]

122. Pappas, A. Epidermal surface lipids. *Dermatoendocrinology* **2009**, *1*, 72–76. [CrossRef]

123. Chapkin, R.S.; Ziboh, V.A.; Marcelo, C.L.; Voorhees, J.J. Metabolism of essential fatty acids by human epidermal enzyme preparations: Evidence of chain elongation. *J. Lipid Res.* **1986**, *27*, 945–954. [CrossRef]

124. Ni Raghallaigh, S.; Bender, K.; Lacey, N.; Brennan, L.; Powell, F.C. The fatty acid profile of the skin surface lipid layer in papulopustular rosacea. *Br. J. Dermatol.* **2012**, *166*, 279–287.

125. Shimizu, N.; Ito, J.; Kato, S.; Otoki, Y.; Goto, M.; Eitsuka, T.; Miyazawa, T.; Nakagawa, K. Oxidation of squalene by singlet oxygen and free radicals results in different compositions of squalene monohydroperoxide isomers. *Sci. Rep.* **2018**, *8*, 9116. [CrossRef] [PubMed]

126. Chiba, K.; Kawakami, K.; Sone, T.; Onoue, M. Characteristics of skin wrinkling and dermal changes induced by repeated application of squalene monohydroperoxide to hairless mouse skin. *Skin Pharmacol. Appl. Skin Physiol.* **2003**, *16*, 242–251. [CrossRef]

127. Chiba, K.; Yoshizawa, K.; Makino, I.; Kawakami, K.; Onoue, M. Comedogenicity of squalene monohydroperoxide in the skin after topical application. *J. Toxicol. Sci.* **2000**, *25*, 77–83. [CrossRef]

128. Khalifa, S.; Enomoto, M.; Kato, S.; Nakagawa, K. Novel Photoinduced Squalene Cyclic Peroxide Identified, Detected, and Quantified in Human Skin Surface Lipids. *Antioxidants* **2021**, *10*, 1760. [CrossRef]

129. Dalrymple, A.; McEwan, M.; Brandt, M.; Bielfeldt, S.; Bean, E.J.; Moga, A.; Coburn, S.; Hardie, G. A novel clinical method to measure skin staining reveals activation of skin damage pathways by cigarette smoke. *Skin Res. Technol.* **2022**, *28*, 162–170. [CrossRef] [PubMed]

130. Minami, Y.; Yokoyama, K.; Bando, N.; Kawai, Y.; Terao, J. Occurrence of singlet oxygen oxygenation of oleic acid and linoleic acid in the skin of live mice. *Free Radic. Res.* **2008**, *42*, 197–204. [CrossRef]

131. Egawa, M.; Kohno, Y.; Kumano, Y. Oxidative effects of cigarette smoke on the human skin. *Int. J. Cosmet. Sci.* **1999**, *21*, 83–98. [CrossRef] [PubMed]

132. Nakagawa, K.; Ibusuki, D.; Suzuki, Y.; Yamashita, S.; Higuchi, O.; Oikawa, S.; Miyazawa, T. Ion-trap tandem mass spectrometric analysis of squalene monohydroperoxide isomers in sunlight-exposed human skin. *J. Lipid Res.* **2007**, *48*, 2779–2787. [CrossRef] [PubMed]

133. Ekanayake-Mudiyanselage, S.; Tavakkol, A.; Polefka, T.G.; Nabi, Z.; Elsner, P.; Thiele, J.J. Vitamin E delivery to human skin by a rinse-off product: Penetration of alpha-tocopherol versus wash-out effects of skin surface lipids. *Skin Pharmacol. Physiol.* **2005**, *18*, 20–26. [CrossRef] [PubMed]

134. Ekanayake Mudiyanselage, S.; Hamburger, M.; Elsner, P.; Thiele, J.J. Ultraviolet a induces generation of squalene monohydroperoxide isomers in human sebum and skin surface lipids in vitro and in vivo. *J. Invest. Dermatol.* **2003**, *120*, 915–922. [CrossRef] [PubMed]

135. Auffray, B. Protection against singlet oxygen, the main actor of sebum squalene peroxidation during sun exposure, using Commiphora myrrha essential oil. *Int. J. Cosmet. Sci.* **2007**, *29*, 23–29. [CrossRef]

136. Lens, M.; Podesta Marty, M.H. Assessment of the kinetics of the antioxidative capacity of topical antioxidants. *J. Drugs Dermatol.* **2011**, *10*, 262–267. [PubMed]

137. Shimizu, N.; Ito, J.; Kato, S.; Eitsuka, T.; Saito, T.; Nishida, H.; Miyazawa, T.; Nakagawa, K. Evaluation of squalene oxidation mechanisms in human skin surface lipids and shark liver oil supplements. *Ann. N. Y. Acad. Sci.* **2019**, *1457*, 158–165. [CrossRef]

138. Jourdain, R.; Moga, A.; Vingler, P.; El Rawadi, C.; Pouradier, F.; Souverain, L.; Bastien, P.; Amalric, N.; Breton, L. Exploration of scalp surface lipids reveals squalene peroxide as a potential actor in dandruff condition. *Arch. Dermatol. Res.* **2016**, *308*, 153–163. [CrossRef] [PubMed]

139. Mountfort, K.A.; Bronstein, H.; Archer, N.; Jickells, S.M. Identification of oxidation products of squalene in solution and in latent fingerprints by ESI-MS and LC/APCI-MS. *Anal. Chem.* **2007**, *79*, 2650–2657. [CrossRef]

140. Curpen, S.; Francois-Newton, V.; Moga, A.; Hosenally, M.; Petkar, G.; Soobramaney, V.; Ruchaia, B.; Lutchmanen Kolanthan, V.; Roheemun, N.; Sokeechand, B.N.; et al. A novel method for evaluating the effect of pollution on the human skin under controlled conditions. *Skin Res. Technol.* **2020**, *26*, 50–60. [CrossRef]

141. Bielfeldt, S.; Jung, K.; Laing, S.; Moga, A.; Wilhelm, K.P. Anti-pollution effects of two antioxidants and a chelator-Ex vivo electron spin resonance and in vivo cigarette smoke model assessments in human skin. *Skin Res. Technol.* **2021**, *27*, 1092–1099. [CrossRef]

142. Someya, K.; Totsuka, Y.; Murakoshi, M.; Kitano, H.; Miyazawa, T. The antioxidant effect of palm fruit carotene on skin lipid peroxidation in guinea pigs as estimated by chemiluminescence-HPLC method. *J. Nutr. Sci. Vitaminol.* **1994**, *40*, 315–324. [CrossRef] [PubMed]

143. Pena, A.M.; Baldeweck, T.; Decencière, E.; Koudoro, S.; Victorin, S.; Raynaud, E.; Ngo, B.; Bastien, P.; Brizion, S.; Tancrède-Bohin, E. In vivo multiphoton multiparametric 3D quantification of human skin aging on forearm and face. *Sci. Rep.* **2022**, *12*, 14863. [CrossRef] [PubMed]

144. Scharffetter, K.; Wlaschek, M.; Hogg, A.; Bolsen, K.; Schothorst, A.; Goerz, G.; Krieg, T.; Plewig, G. UVA irradiation induces collagenase in human dermal fibroblasts in vitro and in vivo. *Arch. Dermatol. Res.* **1991**, *283*, 506–511. [CrossRef]

145. Nakamura, T.; Noma, A.; Shimada, S.; Ishii, N.; Bando, N.; Kawai, Y.; Terao, J. Non-selective distribution of isomeric cholesterol hydroperoxides to microdomains in cell membranes and activation of matrix metalloproteinase activity in a model of dermal cells. *Chem. Phys. Lipids* **2013**, *174*, 17–23. [CrossRef] [PubMed]

146. Cadet, J.; Douki, T.; Ravanat, J.L. Oxidatively generated damage to the guanine moiety of DNA: Mechanistic aspects and formation in cells. *Acc. Chem. Res.* **2008**, *41*, 1075–1083. [CrossRef] [PubMed]

147. Cadet, J.; Douki, T.; Ravanat, J.L. Oxidatively generated damage to cellular DNA by UVB and UVA radiation. *Photochem. Photobiol.* **2015**, *91*, 140–155. [CrossRef] [PubMed]

148. Ravanat, J.L.; Di Mascio, P.; Martinez, G.R.; Medeiros, M.H.; Cadet, J. Singlet oxygen induces oxidation of cellular DNA. *J. Biol. Chem.* **2000**, *275*, 40601–40604. [CrossRef] [PubMed]

149. Yagura, T.; Schuch, A.P.; Garcia, C.C.M.; Rocha, C.R.R.; Moreno, N.C.; Angeli, J.P.F.; Mendes, D.; Severino, D.; Bianchini Sanchez, A.; Di Mascio, P.; et al. Direct participation of DNA in the formation of singlet oxygen and base damage under UVA irradiation. *Free Radic. Biol. Med.* **2017**, *108*, 86–93. [CrossRef]

150. Cadet, J.; Sage, E.; Douki, T. Ultraviolet radiation-mediated damage to cellular DNA. *Mutat. Res.* **2005**, *571*, 3–17. [CrossRef]

151. Mouret, S.; Baudouin, C.; Charveron, M.; Favier, A.; Cadet, J.; Douki, T. Cyclobutane pyrimidine dimers are predominant DNA lesions in whole human skin exposed to UVA radiation. *Proc. Natl. Acad. Sci. USA* **2006**, *103*, 13765–13770. [CrossRef]

152. Prat, F.; Hou, C.-C.; Foote, C.S. Determination of the Quenching Rate Constants of Singlet Oxygen by Derivatized Nucleosides in Nonaqueous Solution. *J. Am. Chem. Soc.* **1997**, *119*, 5051–5052. [CrossRef]

153. David, S.S.; O'Shea, V.L.; Kundu, S. Base-excision repair of oxidative DNA damage. *Nature* **2007**, *447*, 941–950. [CrossRef] [PubMed]

154. Hsu, G.W.; Ober, M.; Carell, T.; Beese, L.S. Error-prone replication of oxidatively damaged DNA by a high-fidelity DNA polymerase. *Nature* **2004**, *431*, 217–221. [CrossRef] [PubMed]

155. Greenman, C.; Stephens, P.; Smith, R.; Dalgliesh, G.L.; Hunter, C.; Bignell, G.; Davies, H.; Teague, J.; Butler, A.; Stevens, C.; et al. Patterns of somatic mutation in human cancer genomes. *Nature* **2007**, *446*, 153–158. [CrossRef] [PubMed]

156. Lindahl, T.; Barnes, D.E. Repair of endogenous DNA damage. *Cold Spring Harb. Symp. Quant. Biol.* **2000**, *65*, 127–133. [CrossRef]

157. Greinert, R.; Volkmer, B.; Henning, S.; Breitbart, E.W.; Greulich, K.O.; Cardoso, M.C.; Rapp, A. UVA-induced DNA double-strand breaks result from the repair of clustered oxidative DNA damages. *Nucleic Acids Res.* **2012**, *40*, 10263–10273. [CrossRef] [PubMed]

158. Agar, N.S.; Halliday, G.M.; Barnetson, R.S.; Ananthaswamy, H.N.; Wheeler, M.; Jones, A.M. The basal layer in human squamous tumors harbors more UVA than UVB fingerprint mutations: A role for UVA in human skin carcinogenesis. *Proc. Natl. Acad. Sci. USA* **2004**, *101*, 4954–4959. [CrossRef]

159. Kunisada, M.; Sakumi, K.; Tominaga, Y.; Budiyanto, A.; Ueda, M.; Ichihashi, M.; Nakabeppu, Y.; Nishigori, C. 8-Oxoguanine formation induced by chronic UVB exposure makes Ogg1 knockout mice susceptible to skin carcinogenesis. *Cancer Res.* **2005**, *65*, 6006–6010. [CrossRef]

160. Kunisada, M.; Kumimoto, H.; Ishizaki, K.; Sakumi, K.; Nakabeppu, Y.; Nishigori, C. Narrow-band UVB induces more carcinogenic skin tumors than broad-band UVB through the formation of cyclobutane pyrimidine dimer. *J. Invest. Dermatol.* **2007**, *127*, 2865–2871. [CrossRef]

161. Zhang, R.B.; Eriksson, L.A. A triplet mechanism for the formation of cyclobutane pyrimidine dimers in UV-irradiated DNA. *J. Phys. Chem. B* **2006**, *110*, 7556–7562. [CrossRef]

162. You, Y.; Zhu, F.; Li, Z.; Zhang, L.; Xie, Y.; Chinnathambi, A.; Alahmadi, T.A.; Lu, B. Phyllanthin prevents diethylnitrosamine (DEN) induced liver carcinogenesis in rats and induces apoptotic cell death in HepG2 cells. *Biomed. Pharmacother.* **2021**, *137*, 111335. [CrossRef]

163. Yuan, H.; Guo, L.; Su, Q.; Su, X.; Wen, Y.; Wang, T.; Yang, P.; Xu, M.; Li, F. Afterglow Amplification for Fast and Sensitive Detection of Porphyria in Whole Blood. *ACS Appl. Mater. Interfaces* **2021**, *13*, 27991–27998. [CrossRef]

164. Ventura, P.; Brancaleoni, V.; Di Pierro, E.; Graziadei, G.; Macrì, A.; Carmine Guida, C.; Nicolli, A.; Rossi, M.T.; Granata, F.; Fiorentino, V.; et al. Clinical and molecular epidemiology of erythropoietic protoporphyria in Italy. *Eur. J. Dermatol.* **2020**, *30*, 532–540. [CrossRef] [PubMed]

165. Balwani, M. Erythropoietic Protoporphyria and X-Linked Protoporphyria: Pathophysiology, genetics, clinical manifestations, and management. *Mol. Genet. Metab.* **2019**, *128*, 298–303. [CrossRef] [PubMed]

166. Mathews-Roth, M.M.; Pathak, M.A.; Fitzpatrick, T.B.; Harber, L.C.; Kass, E.H. Beta-carotene as a photoprotective agent in erythropoietic protoporphyria. *N. Engl. J. Med.* **1970**, *282*, 1231–1234. [CrossRef] [PubMed]

167. Black, H.S.; Boehm, F.; Edge, R.; Truscott, T.G. The Benefits and Risks of Certain Dietary Carotenoids that Exhibit both Anti- and Pro-Oxidative Mechanisms-A Comprehensive Review. *Antioxidants* **2020**, *9*, 264. [CrossRef]

168. Mathews-Roth, M.M.; Pathak, M.A.; Fitzpatrick, T.B.; Harber, L.H.; Kass, E.H. Beta carotene therapy for erythropoietic protoporphyria and other photosensitivity diseases. *Arch. Dermatol.* **1977**, *113*, 1229–1232. [CrossRef]

169. Maitra, D.; Bragazzi Cunha, J.; Elenbaas, J.S.; Bonkovsky, H.L.; Shavit, J.A.; Omary, M.B. Porphyrin-Induced Protein Oxidation and Aggregation as a Mechanism of Porphyria-Associated Cell Injury. *Cell Mol Gastroenterol. Hepatol.* **2019**, *8*, 535–548. [CrossRef]

170. Maitra, D.; Carter, E.L.; Richardson, R.; Rittié, L.; Basrur, V.; Zhang, H.; Nesvizhskii, A.I.; Osawa, Y.; Wolf, M.W.; Ragsdale, S.W.; et al. Oxygen and Conformation Dependent Protein Oxidation and Aggregation by Porphyrins in Hepatocytes and Light-Exposed Cells. *Cell Mol. Gastroenterol. Hepatol.* **2019**, *8*, 659–682.e651. [CrossRef]

171. Maitra, D.; Elenbaas, J.S.; Whitesall, S.E.; Basrur, V.; D'Alecy, L.G.; Omary, M.B. Ambient Light Promotes Selective Subcellular Proteotoxicity after Endogenous and Exogenous Porphyrinogenic Stress. *J. Biol. Chem.* **2015**, *290*, 23711–23724. [CrossRef]

172. Tomita, H.; Hines, K.M.; Herron, J.M.; Li, A.; Baggett, D.W.; Xu, L. 7-Dehydrocholesterol-derived oxysterols cause neurogenic defects in Smith-Lemli-Opitz syndrome. *Elife* **2022**, *11*, e67141. [CrossRef]

173. Luo, Y.; Zhang, C.; Ma, L.; Zhang, Y.; Liu, Z.; Chen, L.; Wang, R.; Luan, Y.; Rao, Y. Measurement of 7-dehydrocholesterol and cholesterol in hair can be used in the diagnosis of Smith-Lemli-Opitz syndrome. *J. Lipid Res.* **2022**, *63*, 100228. [CrossRef]

174. Koczok, K.; Horváth, L.; Korade, Z.; Mezei, Z.A.; Szabó, G.P.; Porter, N.A.; Kovács, E.; Mirnics, K.; Balogh, I. Biochemical and Clinical Effects of Vitamin E Supplementation in Hungarian Smith-Lemli-Opitz Syndrome Patients. *Biomolecules* **2021**, *11*, 1228. [CrossRef] [PubMed]

175. Quiec, D.; Mazière, C.; Auclair, M.; Santus, R.; Gardette, J.; Redziniak, G.; Franchi, J.; Dubertret, L.; Mazière, J.C. Lovastatin enhances the photocytotoxicity of UVA radiation towards cultured N.C.T.C. 2544 human keratinocytes: Prevention by cholesterol supplementation and by a cathepsin inhibitor. *Biochem. J.* **1995**, *310 Pt 1*, 305–309. [CrossRef] [PubMed]

176. Alrashidi, A.; Rhodes, L.E.; Sharif, J.C.H.; Kreeshan, F.C.; Farrar, M.D.; Ahad, T. Systemic drug photosensitivity-Culprits, impact and investigation in 122 patients. *Photodermatol. Photoimmunol. Photomed.* **2020**, *36*, 441–451. [CrossRef]

177. Chen, H.L.; Chang, H.M.; Wu, H.J.; Lin, Y.C.; Chang, Y.H.; Chang, Y.C.; Lee, W.H.; Chang, C.T. Effect of hydrophilic and lipophilic statins on early onset cataract: A nationwide case-control study. *Regul. Toxicol. Pharmacol.* **2021**, *124*, 104970. [CrossRef] [PubMed]

178. Yang, J.; Zhang, B.W.; Lin, L.N.; Zan, X.L.; Zhang, G.C.; Chen, G.S.; Ji, J.Y.; Ma, W.H. Key factors affecting photoactivated fungicidal activity of sodium pheophorbide a against Pestalotiopsis neglecta. *Pestic. Biochem. Physiol.* **2020**, *167*, 104584. [CrossRef]

179. Kobayashi, T.; Sugaya, K.; Onose, J.I.; Abe, N. Peppermint (*Mentha piperita* L.) extract effectively inhibits cytochrome P450 3A4 (CYP3A4) mRNA induction in rifampicin-treated HepG2 cells. *Biosci. Biotechnol. Biochem.* **2019**, *83*, 1181–1192. [CrossRef]

180. Puschner, B.; Chen, X.; Read, D.; Affolter, V.K. Alfalfa hay induced primary photosensitization in horses. *Vet. J.* **2016**, *211*, 32–38. [CrossRef]

181. Bayoumy, A.B.; Crouwel, F.; Chanda, N.; Florin, T.H.J.; Buiter, H.J.C.; Mulder, C.J.J.; de Boer, N.K.H. Advances in Thiopurine Drug Delivery: The Current State-of-the-Art. *Eur. J. Drug Metab. Pharmacokinet.* **2021**, *46*, 743–758. [CrossRef]

182. Euvrard, S.; Kanitakis, J.; Claudy, A. Skin cancers after organ transplantation. *N. Engl. J. Med.* **2003**, *348*, 1681–1691. [CrossRef]

183. Bignon, E.; Marazzi, M.; Besancenot, V.; Gattuso, H.; Drouot, G.; Morell, C.; Eriksson, L.A.; Grandemange, S.; Dumont, E.; Monari, A. Ibuprofen and ketoprofen potentiate UVA-induced cell death by a photosensitization process. *Sci. Rep.* **2017**, *7*, 8885. [CrossRef] [PubMed]

184. Zawadzka, M.; Ràcz, B.; Ambrosini, D.; Görbitz, C.H.; Morth, J.P.; Wilkins, A.L.; Østeby, A.; Elgstøen, K.B.P.; Lundanes, E.; Rise, F.; et al. Searching for a UV-filter in the eyes of high-flying birds. *Sci. Rep.* **2021**, *11*, 273. [CrossRef] [PubMed]

185. Guan, L.L.; Lim, H.W.; Mohammad, T.F. Sunscreens and Photoaging: A Review of Current Literature. *Am. J. Clin. Dermatol.* **2021**, *22*, 819–828. [CrossRef]

186. Behar-Cohen, F.; Martinsons, C.; Viénot, F.; Zissis, G.; Barlier-Salsi, A.; Cesarini, J.P.; Enouf, O.; Garcia, M.; Picaud, S.; Attia, D. Light-emitting diodes (LED) for domestic lighting: Any risks for the eye? *Prog. Retin. Eye Res.* **2011**, *30*, 239–257. [CrossRef] [PubMed]

187. Tao, J.X.; Zhou, W.C.; Zhu, X.G. Mitochondria as Potential Targets and Initiators of the Blue Light Hazard to the Retina. *Oxid. Med. Cell. Longev.* **2019**, *2019*, 6435364. [CrossRef] [PubMed]

188. Kaarniranta, K.; Blasiak, J.; Liton, P.; Boulton, M.; Klionsky, D.J.; Sinha, D. Autophagy in age-related macular degeneration. *Autophagy* **2023**, *19*, 388–400. [CrossRef] [PubMed]

189. Różanowska, M.B.; Pawlak, A.; Różanowski, B. Products of Docosahexaenoate Oxidation as Contributors to Photosensitising Properties of Retinal Lipofuscin. *Int. J. Mol. Sci.* **2021**, *22*, 3525. [CrossRef]

190. Höhn, A.; Grune, T. Lipofuscin: Formation, effects and role of macroautophagy. *Redox Biol.* **2013**, *1*, 140–144. [CrossRef]

191. Singh Kushwaha, S.; Patro, N.; Kumar Patro, I. A Sequential Study of Age-Related Lipofuscin Accumulation in Hippocampus and Striate Cortex of Rats. *Ann. Neurosci.* **2018**, *25*, 223–233. [CrossRef]
192. Gao, Z.; Liao, Y.; Chen, C.; Liao, C.; He, D.; Chen, J.; Ma, J.; Liu, Z.; Wu, Y. Conversion of all-trans-retinal into all-trans-retinal dimer reflects an alternative metabolic/antidotal pathway of all-trans-retinal in the retina. *J. Biol. Chem.* **2018**, *293*, 14507–14519. [CrossRef]
193. Parish, C.A.; Hashimoto, M.; Nakanishi, K.; Dillon, J.; Sparrow, J. Isolation and one-step preparation of A2E and iso-A2E, fluorophores from human retinal pigment epithelium. *Proc. Natl. Acad. Sci. USA* **1998**, *95*, 14609–14613. [CrossRef] [PubMed]
194. Marie, M.; Bigot, K.; Angebault, C.; Barrau, C.; Gondouin, P.; Pagan, D.; Fouquet, S.; Villette, T.; Sahel, J.A.; Lenaers, G.; et al. Light action spectrum on oxidative stress and mitochondrial damage in A2E-loaded retinal pigment epithelium cells. *Cell Death Dis.* **2018**, *9*, 287. [CrossRef] [PubMed]
195. Bernstein, P.S.; Li, B.; Vachali, P.P.; Gorusupudi, A.; Shyam, R.; Henriksen, B.S.; Nolan, J.M. Lutein, zeaxanthin, and meso-zeaxanthin: The basic and clinical science underlying carotenoid-based nutritional interventions against ocular disease. *Prog. Retin. Eye Res.* **2016**, *50*, 34–66. [CrossRef] [PubMed]
196. Lem, D.W.; Davey, P.G.; Gierhart, D.L.; Rosen, R.B. A Systematic Review of Carotenoids in the Management of Age-Related Macular Degeneration. *Antioxidants* **2021**, *10*, 1255. [CrossRef] [PubMed]
197. Shyam, R.; Gorusupudi, A.; Nelson, K.; Horvath, M.P.; Bernstein, P.S. RPE65 has an additional function as the lutein to meso-zeaxanthin isomerase in the vertebrate eye. *Proc. Natl. Acad. Sci. USA* **2017**, *114*, 10882–10887. [CrossRef] [PubMed]
198. Johra, F.T.; Bepari, A.K.; Bristy, A.T.; Reza, H.M. A Mechanistic Review of β-Carotene, Lutein, and Zeaxanthin in Eye Health and Disease. *Antioxidants* **2020**, *9*, 1046. [CrossRef] [PubMed]
199. Mizdrak, J.; Hains, P.G.; Truscott, R.J.; Jamie, J.F.; Davies, M.J. Tryptophan-derived ultraviolet filter compounds covalently bound to lens proteins are photosensitizers of oxidative damage. *Free Radic. Biol. Med.* **2008**, *44*, 1108–1119. [CrossRef]
200. Ouyang, X.; Yang, J.; Hong, Z.; Wu, Y.; Xie, Y.; Wang, G. Mechanisms of blue light-induced eye hazard and protective measures: A review. *Biomed. Pharmacother.* **2020**, *130*, 110577. [CrossRef]
201. Boulton, M.; Rozanowska, M.; Rozanowski, B.; Wess, T. The photoreactivity of ocular lipofuscin. *Photochem. Photobiol. Sci. Off. J. Eur. Photochem. Assoc. Eur. Soc. Photobiol.* **2004**, *3*, 759–764. [CrossRef]
202. Weinreb, R.N.; Khaw, P.T. Primary open-angle glaucoma. *Lancet* **2004**, *363*, 1711–1720. [CrossRef]
203. Foster, A.; Resnikoff, S. The impact of Vision 2020 on global blindness. *Eye* **2005**, *19*, 1133–1135. [CrossRef] [PubMed]
204. Tanito, M.; Kaidzu, S.; Takai, Y.; Ohira, A. Status of systemic oxidative stresses in patients with primary open-angle glaucoma and pseudoexfoliation syndrome. *PLoS ONE* **2012**, *7*, e49680. [CrossRef] [PubMed]
205. Alvarado, J.A.; Murphy, C.G. Outflow obstruction in pigmentary and primary open angle glaucoma. *Arch. Ophthalmol.* **1992**, *110*, 1769–1778. [CrossRef] [PubMed]
206. Himori, N.; Yamamoto, K.; Maruyama, K.; Ryu, M.; Taguchi, K.; Yamamoto, M.; Nakazawa, T. Critical role of Nrf2 in oxidative stress-induced retinal ganglion cell death. *J. Neurochem.* **2013**, *127*, 669–680. [CrossRef] [PubMed]
207. Yokoyama, Y.; Maruyama, K.; Yamamoto, K.; Omodaka, K.; Yasuda, M.; Himori, N.; Ryu, M.; Nishiguchi, K.M.; Nakazawa, T. The role of calpain in an in vivo model of oxidative stress-induced retinal ganglion cell damage. *Biochem. Biophys. Res. Commun.* **2014**, *451*, 510–515. [CrossRef]
208. Inomata, Y.; Nakamura, H.; Tanito, M.; Teratani, A.; Kawaji, T.; Kondo, N.; Yodoi, J.; Tanihara, H. Thioredoxin inhibits NMDA-induced neurotoxicity in the rat retina. *J. Neurochem.* **2006**, *98*, 372–385. [CrossRef]
209. Munemasa, Y.; Ahn, J.H.; Kwong, J.M.; Caprioli, J.; Piri, N. Redox proteins thioredoxin 1 and thioredoxin 2 support retinal ganglion cell survival in experimental glaucoma. *Gene Ther.* **2009**, *16*, 17–25. [CrossRef]
210. Umeno, A.; Tanito, M.; Kaidzu, S.; Takai, Y.; Yoshida, Y. Involvement of free radical-mediated oxidation in the pathogenesis of pseudoexfoliation syndrome detected based on specific hydroxylinoleate isomers. *Free Radic. Biol. Med.* **2020**, *147*, 61–68. [CrossRef] [PubMed]
211. Umeno, A.; Yoshida, Y.; Kaidzu, S.; Tanito, M. Positive Association between Aqueous Humor Hydroxylinoleate Levels and Intraocular Pressure. *Molecules* **2022**, *27*, 2215. [CrossRef]
212. Schafheimer, N.; King, J. Tryptophan cluster protects human γD-crystallin from ultraviolet radiation-induced photoaggregation in vitro. *Photochem. Photobiol.* **2013**, *89*, 1106–1115. [CrossRef] [PubMed]
213. Yoshida, Y.; Hayakawa, M.; Habuchi, Y.; Niki, E. Evaluation of the dietary effects of coenzyme Q in vivo by the oxidative stress marker, hydroxyoctadecadienoic acid and its stereoisomer ratio. *Biochim. Biophys. Acta* **2006**, *1760*, 1558–1568. [CrossRef] [PubMed]
214. Hirashima, Y.; Doshi, M.; Hayashi, N.; Endo, S.; Akazawa, Y.; Shichiri, M.; Yoshida, Y. Plasma platelet-activating factor-acetyl hydrolase activity and the levels of free forms of biomarker of lipid peroxidation in cerebrospinal fluid of patients with aneurysmal subarachnoid hemorrhage. *Neurosurgery* **2012**, *70*, 602–609. [CrossRef]
215. Fan Gaskin, J.C.; Shah, M.H.; Chan, E.C. Oxidative Stress and the Role of NADPH Oxidase in Glaucoma. *Antioxidants* **2021**, *10*, 238. [CrossRef] [PubMed]
216. Garcia-Medina, J.J.; Rubio-Velazquez, E.; Lopez-Bernal, M.D.; Cobo-Martinez, A.; Zanon-Moreno, V.; Pinazo-Duran, M.D.; Del-Rio-Vellosillo, M. Glaucoma and Antioxidants: Review and Update. *Antioxidants* **2020**, *9*, 1031. [CrossRef] [PubMed]
217. Tang, B.; Li, S.; Cao, W.; Sun, X. The Association of Oxidative Stress Status with Open-Angle Glaucoma and Exfoliation Glaucoma: A Systematic Review and Meta-Analysis. *J. Ophthalmol.* **2019**, *2019*, 1803619. [CrossRef]

218. Neale, R.E.; Purdie, J.L.; Hirst, L.W.; Green, A.C. Sun exposure as a risk factor for nuclear cataract. *Epidemiology* **2003**, *14*, 707–712. [CrossRef]
219. Garźon-Chavez, D.R.; Quentin, E.; Harrison, S.L.; Parisi, A.V.; Butler, H.J.; Downs, N.J. The geospatial relationship of pterygium and senile cataract with ambient solar ultraviolet in tropical Ecuador. *Photochem. Photobiol. Sci. Off. J. Eur. Photochem. Assoc. Eur. Soc. Photobiol.* **2018**, *17*, 1075–1083. [CrossRef]
220. Dillon, J.; Atherton, S.J. Time resolved spectroscopic studies on the intact human lens. *Photochem. Photobiol.* **1990**, *51*, 465–468. [CrossRef]
221. Kessel, L.; Eskildsen, L.; Lundeman, J.H.; Jensen, O.B.; Larsen, M. Optical effects of exposing intact human lenses to ultraviolet radiation and visible light. *BMC Ophthalmol.* **2011**, *11*, 41. [CrossRef]
222. Davies, M.J.; Truscott, R.J. Photo-oxidation of proteins and its role in cataractogenesis. *J. Photochem. Photobiol. B* **2001**, *63*, 114–125. [CrossRef]
223. Ehrenshaft, M.; Zhao, B.; Andley, U.P.; Mason, R.P.; Roberts, J.E. Immunological detection of N-formylkynurenine in porphyrin-mediated photooxided lens α-crystallin. *Photochem. Photobiol.* **2011**, *87*, 1321–1329. [CrossRef] [PubMed]
224. Benedek, G.B. Theory of transparency of the eye. *Appl. Opt.* **1971**, *10*, 459–473. [CrossRef] [PubMed]
225. Ávila, F.; Ravello, N.; Zanocco, A.L.; Gamon, L.F.; Davies, M.J.; Silva, E. 3-Hydroxykynurenine bound to eye lens proteins induces oxidative modifications in crystalline proteins through a type I photosensitizing mechanism. *Free Radic. Biol. Med.* **2019**, *141*, 103–114. [CrossRef]
226. Parker, N.R.; Jamie, J.F.; Davies, M.J.; Truscott, R.J. Protein-bound kynurenine is a photosensitizer of oxidative damage. *Free Radic. Biol. Med.* **2004**, *37*, 1479–1489. [CrossRef]
227. Ratter-Rieck, J.M.; Maalmi, H.; Trenkamp, S.; Zaharia, O.P.; Rathmann, W.; Schloot, N.C.; Straßburger, K.; Szendroedi, J.; Herder, C.; Roden, M. Leukocyte Counts and T-Cell Frequencies Differ Between Novel Subgroups of Diabetes and Are Associated With Metabolic Parameters and Biomarkers of Inflammation. *Diabetes* **2021**, *70*, 2652–2662. [CrossRef] [PubMed]
228. Penno, G.; Solini, A.; Orsi, E.; Bonora, E.; Fondelli, C.; Trevisan, R.; Vedovato, M.; Cavalot, F.; Zerbini, G.; Lamacchia, O.; et al. Insulin resistance, diabetic kidney disease, and all-cause mortality in individuals with type 2 diabetes: A prospective cohort study. *BMC Med.* **2021**, *19*, 66. [CrossRef] [PubMed]
229. Li, Z.; Xu, Y.; Liu, X.; Nie, Y.; Zhao, Z. Urinary heme oxygenase-1 as a potential biomarker for early diabetic nephropathy. *Nephrology* **2017**, *22*, 58–64. [CrossRef]
230. Xu, G.W.; Yao, Q.H.; Weng, Q.F.; Su, B.L.; Zhang, X.; Xiong, J.H. Study of urinary 8-hydroxydeoxyguanosine as a biomarker of oxidative DNA damage in diabetic nephropathy patients. *J. Pharm. Biomed. Anal.* **2004**, *36*, 101–104. [CrossRef]
231. Maschirow, L.; Khalaf, K.; Al-Aubaidy, H.A.; Jelinek, H.F. Inflammation, coagulation, endothelial dysfunction and oxidative stress in prediabetes–Biomarkers as a possible tool for early disease detection for rural screening. *Clin. Biochem.* **2015**, *48*, 581–585. [CrossRef] [PubMed]
232. Lv, C.; Zhou, Y.H.; Wu, C.; Shao, Y.; Lu, C.L.; Wang, Q.Y. The changes in miR-130b levels in human serum and the correlation with the severity of diabetic nephropathy. *Diabetes Metab. Res. Rev.* **2015**, *31*, 717–724. [CrossRef]
233. Sánchez-Gómez, F.J.; Espinosa-Díez, C.; Dubey, M.; Dikshit, M.; Lamas, S. S-glutathionylation: Relevance in diabetes and potential role as a biomarker. *Biol. Chem.* **2013**, *394*, 1263–1280. [CrossRef] [PubMed]
234. Liang, M.; Wang, J.; Xie, C.; Yang, Y.; Tian, J.W.; Xue, Y.M.; Hou, F.F. Increased plasma advanced oxidation protein products is an early marker of endothelial dysfunction in type 2 diabetes patients without albuminuria 2. *J. Diabetes* **2014**, *6*, 417–426. [CrossRef]
235. Ceriello, A.; Testa, R.; Genovese, S. Clinical implications of oxidative stress and potential role of natural antioxidants in diabetic vascular complications. *Nutr. Metab. Cardiovasc. Dis.* **2016**, *26*, 285–292. [CrossRef] [PubMed]
236. Paneni, F.; Costantino, S.; Cosentino, F. Molecular mechanisms of vascular dysfunction and cardiovascular biomarkers in type 2 diabetes. *Cardiovasc. Diagn. Ther.* **2014**, *4*, 324–332. [PubMed]
237. Molnar, J.; Garamvolgyi, Z.; Herold, M.; Adanyi, N.; Somogyi, A.; Rigo, J., Jr. Serum selenium concentrations correlate significantly with inflammatory biomarker high-sensitive CRP levels in Hungarian gestational diabetic and healthy pregnant women at mid-pregnancy. *Biol. Trace Elem. Res.* **2008**, *121*, 16–22. [CrossRef] [PubMed]
238. Basu, A.; Yu, J.Y.; Jenkins, A.J.; Nankervis, A.J.; Hanssen, K.F.; Henriksen, T.; Lorentzen, B.; Garg, S.K.; Menard, M.K.; Hammad, S.M.; et al. Trace elements as predictors of preeclampsia in type 1 diabetic pregnancy. *Nutr. Res.* **2015**, *35*, 421–430. [CrossRef]
239. Morabito, R.; Remigante, A.; Spinelli, S.; Vitale, G.; Trichilo, V.; Loddo, S.; Marino, A. High Glucose Concentrations Affect Band 3 Protein in Human Erythrocytes. *Antioxidants* **2020**, *9*, 365. [CrossRef]
240. Chai, J.F.; Kao, S.L.; Wang, C.; Lim, V.J.; Khor, I.W.; Dou, J.; Podgornaia, A.I.; Chothani, S.; Cheng, C.Y.; Sabanayagam, C.; et al. Genome-Wide Association for HbA1c in Malay Identified Deletion on SLC4A1 that Influences HbA1c Independent of Glycemia. *J. Clin. Endocrinol. Metab.* **2020**, *105*, dgaa658. [CrossRef]
241. Zhekova, H.R.; Jiang, J.; Wang, W.; Tsirulnikov, K.; Kayık, G.; Khan, H.M.; Azimov, R.; Abuladze, N.; Kao, L.; Newman, D.; et al. CryoEM structures of anion exchanger 1 capture multiple states of inward- and outward-facing conformations. *Commun. Biol.* **2022**, *5*, 1372. [CrossRef]
242. Kostopoulou, E.; Kalaitzopoulou, E.; Papadea, P.; Skipitari, M.; Rojas Gil, A.P.; Spiliotis, B.E.; Georgiou, C.D. Oxidized lipid-associated protein damage in children and adolescents with type 1 diabetes mellitus: New diagnostic/prognostic clinical markers. *Pediatr. Diabetes* **2021**, *22*, 1135–1142. [CrossRef]

243. Veitch, S.; Njock, M.S.; Chandy, M.; Siraj, M.A.; Chi, L.; Mak, H.; Yu, K.; Rathnakumar, K.; Perez-Romero, C.A.; Chen, Z.; et al. MiR-30 promotes fatty acid beta-oxidation and endothelial cell dysfunction and is a circulating biomarker of coronary microvascular dysfunction in pre-clinical models of diabetes. *Cardiovasc. Diabetol.* **2022**, *21*, 31. [CrossRef] [PubMed]

244. Griesser, E.; Vemula, V.; Raulien, N.; Wagner, U.; Reeg, S.; Grune, T.; Fedorova, M. Cross-talk between lipid and protein carbonylation in a dynamic cardiomyocyte model of mild nitroxidative stress. *Redox Biol* **2017**, *11*, 438–455. [CrossRef]

245. Schöttker, B.; Xuan, Y.; Gào, X.; Anusruti, A.; Brenner, H. Oxidatively Damaged DNA/RNA and 8-Isoprostane Levels Are Associated with the Development of Type 2 Diabetes at Older Age: Results From a Large Cohort Study. *Diabetes Care* **2020**, *43*, 130–136. [CrossRef]

246. Leinisch, F.; Mariotti, M.; Hägglund, P.; Davies, M.J. Structural and functional changes in RNAse A originating from tyrosine and histidine cross-linking and oxidation induced by singlet oxygen and peroxyl radicals. *Free Radic. Biol. Med.* **2018**, *126*, 73–86. [CrossRef] [PubMed]

247. Miyamoto, S.; Martinez, G.R.; Medeiros, M.H.; Di Mascio, P. Singlet molecular oxygen generated by biological hydroperoxides. *J. Photochem. Photobiol. B* **2014**, *139*, 24–33. [CrossRef]

248. Umeno, A.; Fukui, T.; Hashimoto, Y.; Kataoka, M.; Hagihara, Y.; Nagai, H.; Horie, M.; Shichiri, M.; Yoshino, K.; Yoshida, Y. Early diagnosis of type 2 diabetes based on multiple biomarkers and non-invasive indices. *J. Clin. Biochem. Nutr.* **2018**, *62*, 187–194. [CrossRef]

249. Hampton, M.B.; Kettle, A.J.; Winterbourn, C.C. Inside the neutrophil phagosome: Oxidants, myeloperoxidase, and bacterial killing. *Blood* **1998**, *92*, 3007–3017. [CrossRef]

250. Badwey, J.A.; Karnovsky, M.L. Active oxygen species and the functions of phagocytic leukocytes. *Annu. Rev. Biochem.* **1980**, *49*, 695–726. [CrossRef] [PubMed]

251. Kanofsky, J.R. Singlet oxygen production from the peroxidase-catalyzed oxidation of indole-3-acetic acid. *J. Biol. Chem.* **1988**, *263*, 14171–14175. [CrossRef] [PubMed]

252. Miyamoto, S.; Martinez, G.R.; Rettori, D.; Augusto, O.; Medeiros, M.H.; Di Mascio, P. Linoleic acid hydroperoxide reacts with hypochlorous acid, generating peroxyl radical intermediates and singlet molecular oxygen. *Proc. Natl. Acad. Sci. USA* **2006**, *103*, 293–298. [CrossRef]

253. Tian, R.; Ding, Y.; Peng, Y.Y.; Lu, N. Myeloperoxidase amplified high glucose-induced endothelial dysfunction in vasculature: Role of NADPH oxidase and hypochlorous acid. *Biochem. Biophys. Res. Commun.* **2017**, *484*, 572–578. [CrossRef] [PubMed]

254. Zhang, C.; Yang, J.; Jennings, L.K. Leukocyte-derived myeloperoxidase amplifies high-glucose–induced endothelial dysfunction through interaction with high-glucose—Stimulated, vascular non—Leukocyte-derived reactive oxygen species. *Diabetes* **2004**, *53*, 2950–2959. [CrossRef] [PubMed]

255. Zhang, C.; Yang, J.; Jacobs, J.D.; Jennings, L.K. Interaction of myeloperoxidase with vascular NAD(P)H oxidase-derived reactive oxygen species in vasculature: Implications for vascular diseases. *Am. J. Physiol. Heart Circ. Physiol.* **2003**, *285*, H2563–H2572. [CrossRef] [PubMed]

256. Eiserich, J.P.; Baldus, S.; Brennan, M.L.; Ma, W.; Zhang, C.; Tousson, A.; Castro, L.; Lusis, A.J.; Nauseef, W.M.; White, C.R.; et al. Myeloperoxidase, a leukocyte-derived vascular NO oxidase. *Science* **2002**, *296*, 2391–2394. [CrossRef] [PubMed]

257. Nussbaum, C.; Klinke, A.; Adam, M.; Baldus, S.; Sperandio, M. Myeloperoxidase: A leukocyte-derived protagonist of inflammation and cardiovascular disease. *Antioxid. Redox Signal* **2013**, *18*, 692–713. [CrossRef]

258. Tian, R.; Jin, Z.; Zhou, L.; Zeng, X.P.; Lu, N. Quercetin Attenuated Myeloperoxidase-Dependent HOCl Generation and Endothelial Dysfunction in Diabetic Vasculature. *J. Agric. Food Chem.* **2021**, *69*, 404–413. [CrossRef]

259. Onyango, A.N. The Contribution of Singlet Oxygen to Insulin Resistance. *Oxid. Med. Cell. Longev.* **2017**, *2017*, 8765972. [CrossRef]

260. Nakamura, Y.; Tamaoki, J.; Nagase, H.; Yamaguchi, M.; Horiguchi, T.; Hozawa, S.; Ichinose, M.; Iwanaga, T.; Kondo, R.; Nagata, M.; et al. Japanese guidelines for adult asthma 2020. *Allergol. Int.* **2020**, *69*, 519–548. [CrossRef]

261. Kaur, R.; Chupp, G. Phenotypes and endotypes of adult asthma: Moving toward precision medicine. *J. Allergy Clin. Immunol.* **2019**, *144*, 1–12. [CrossRef]

262. Hekking, P.W.; Wener, R.R.; Amelink, M.; Zwinderman, A.H.; Bouvy, M.L.; Bel, E.H. The prevalence of severe refractory asthma. *J. Allergy Clin. Immunol.* **2015**, *135*, 896–902. [CrossRef]

263. Hammad, H.; Lambrecht, B.N. The basic immunology of asthma. *Cell* **2021**, *184*, 1469–1485. [CrossRef]

264. Lambrecht, B.N.; Hammad, H. The immunology of asthma. *Nat. Immunol.* **2015**, *16*, 45–56. [CrossRef] [PubMed]

265. Trevor, J.L.; Deshane, J.S. Refractory asthma: Mechanisms, targets, and therapy. *Allergy* **2014**, *69*, 817–827. [CrossRef] [PubMed]

266. Woodruff, P.G.; Khashayar, R.; Lazarus, S.C.; Janson, S.; Avila, P.; Boushey, H.A.; Segal, M.; Fahy, J.V. Relationship between airway inflammation, hyperresponsiveness, and obstruction in asthma. *J. Allergy Clin. Immunol.* **2001**, *108*, 753–758. [CrossRef] [PubMed]

267. Shaw, D.E.; Berry, M.A.; Hargadon, B.; McKenna, S.; Shelley, M.J.; Green, R.H.; Brightling, C.E.; Wardlaw, A.J.; Pavord, I.D. Association between neutrophilic airway inflammation and airflow limitation in adults with asthma. *Chest* **2007**, *132*, 1871–1875. [CrossRef]

268. Moore, W.C.; Hastie, A.T.; Li, X.; Li, H.; Busse, W.W.; Jarjour, N.N.; Wenzel, S.E.; Peters, S.P.; Meyers, D.A.; Bleecker, E.R.; et al. Sputum neutrophil counts are associated with more severe asthma phenotypes using cluster analysis. *J. Allergy Clin. Immunol.* **2014**, *133*, 1557–1563. [CrossRef]

269. Jatakanon, A.; Uasuf, C.; Maziak, W.; Lim, S.; Chung, K.F.; Barnes, P.J. Neutrophilic inflammation in severe persistent asthma. *Am. J. Respir. Crit. Care Med.* **1999**, *160*, 1532–1539. [CrossRef]

270. Rosales, C. Neutrophil: A Cell with Many Roles in Inflammation or Several Cell Types? *Front. Physiol.* **2018**, *9*, 113. [CrossRef]

271. Li, Y.; Wang, W.; Yang, F.; Xu, Y.; Feng, C.; Zhao, Y. The regulatory roles of neutrophils in adaptive immunity. *Cell Commun. Signal* **2019**, *17*, 147. [CrossRef]

272. Giacalone, V.D.; Margaroli, C.; Mall, M.A.; Tirouvanziam, R. Neutrophil Adaptations upon Recruitment to the Lung: New Concepts and Implications for Homeostasis and Disease. *Int. J. Mol. Sci.* **2020**, *21*, 851. [CrossRef]

273. Wills-Karp, M. Neutrophil ghosts worsen asthma. *Sci. Immunol.* **2018**, *3*, eaau0112. [CrossRef] [PubMed]

274. Ray, A.; Kolls, J.K. Neutrophilic Inflammation in Asthma and Association with Disease Severity. *Trends Immunol.* **2017**, *38*, 942–954. [CrossRef] [PubMed]

275. Chen, F.; Yu, M.; Zhong, Y.; Wang, L.; Huang, H. Characteristics and Role of Neutrophil Extracellular Traps in Asthma. *Inflammation.* **2022**, *45*, 6–13. [CrossRef]

276. Chen, X.; Li, Y.; Qin, L.; He, R.; Hu, C. Neutrophil Extracellular Trapping Network Promotes the Pathogenesis of Neutrophil-associated Asthma through Macrophages. *Immunol. Invest.* **2021**, *50*, 544–561. [CrossRef]

277. Varricchi, G.; Modestino, L.; Poto, R.; Cristinziano, L.; Gentile, L.; Postiglione, L.; Spadaro, G.; Galdiero, M.R. Neutrophil extracellular traps and neutrophil-derived mediators as possible biomarkers in bronchial asthma. *Clin. Exp. Med.* **2022**, *22*, 285–300. [CrossRef] [PubMed]

278. Lu, Y.; Huang, Y.; Li, J.; Huang, J.; Zhang, L.; Feng, J.; Li, J.; Xia, Q.; Zhao, Q.; Huang, L.; et al. Eosinophil extracellular traps drive asthma progression through neuro-immune signals. *Nat. Cell Biol.* **2021**, *23*, 1060–1072. [CrossRef]

279. Michaeloudes, C.; Abubakar-Waziri, H.; Lakhdar, R.; Raby, K.; Dixey, P.; Adcock, I.M.; Mumby, S.; Bhavsar, P.K.; Chung, K.F. Molecular mechanisms of oxidative stress in asthma. *Mol. Aspects Med.* **2022**, *85*, 101026. [CrossRef]

280. Crisford, H.; Sapey, E.; Rogers, G.B.; Taylor, S.; Nagakumar, P.; Lokwani, R.; Simpson, J.L. Neutrophils in asthma: The good, the bad and the bacteria. *Thorax* **2021**, *76*, 835–844. [CrossRef]

281. Andreadis, A.A.; Hazen, S.L.; Comhair, S.A.; Erzurum, S.C. Oxidative and nitrosative events in asthma. *Free Radic Biol. Med.* **2003**, *35*, 213–225. [CrossRef]

282. Obaid Abdullah, S.; Ramadan, G.M.; Makki Al-Hindy, H.A.; Mousa, M.J.; Al-Mumin, A.; Jihad, S.; Hafidh, S.; Kadhum, Y. Serum Myeloperoxidase as a Biomarker of Asthma Severity Among Adults: A Case Control Study. *Rep. Biochem. Mol. Biol.* **2022**, *11*, 182–189. [CrossRef]

283. Olgart Hoglund, C.; de Blay, F.; Oster, J.P.; Duvernelle, C.; Kassel, O.; Pauli, G.; Frossard, N. Nerve growth factor levels and localisation in human asthmatic bronchi. *Eur. Respir. J.* **2002**, *20*, 1110–1116. [CrossRef] [PubMed]

284. Ogawa, H.; Azuma, M.; Tsunematsu, T.; Morimoto, Y.; Kondo, M.; Tezuka, T.; Nishioka, Y.; Tsuneyama, K. Neutrophils induce smooth muscle hyperplasia via neutrophil elastase-induced FGF-2 in a mouse model of asthma with mixed inflammation. *Clin. Exp. Allergy* **2018**, *48*, 1715–1725. [CrossRef] [PubMed]

285. Chen, J.; Kou, L.; Kong, L. Anti-nerve growth factor antibody improves airway hyperresponsiveness by down-regulating RhoA. *J. Asthma* **2018**, *55*, 1079–1085. [CrossRef] [PubMed]

286. Liu, P.; Li, S.; Tang, L. Nerve Growth Factor: A Potential Therapeutic Target for Lung Diseases. *Int. J. Mol. Sci.* **2021**, *22*, 9112. [CrossRef] [PubMed]

287. Saffar, A.S.; Ashdown, H.; Gounni, A.S. The molecular mechanisms of glucocorticoids-mediated neutrophil survival. *Curr. Drug Targets* **2011**, *12*, 556–562. [CrossRef]

288. Tamura, H.; Ishikita, H. Quenching of Singlet Oxygen by Carotenoids via Ultrafast Superexchange Dynamics. *J. Phys. Chem. A* **2020**, *124*, 5081–5088. [CrossRef]

289. Stratton, S.P.; Schaefer, W.H.; Liebler, D.C. Isolation and identification of singlet oxygen oxidation products of beta-carotene. *Chem. Res. Toxicol.* **1993**, *6*, 542–547. [CrossRef]

290. Zbyradowski, M.; Duda, M.; Wisniewska-Becker, A.; Heriyanto; Rajwa, W.; Fiedor, J.; Cvetkovic, D.; Pilch, M.; Fiedor, L. Triplet-driven chemical reactivity of β-carotene and its biological implications. *Nat. Commun.* **2022**, *13*, 2474. [CrossRef]

291. Montenegro, M.A.; Nazareno, M.A.; Durantini, E.N.; Borsarelli, C.D. Singlet molecular oxygen quenching ability of carotenoids in a reverse-micelle membrane mimetic system. *Photochem. Photobiol.* **2002**, *75*, 353–361. [CrossRef]

292. Ramel, F.; Birtic, S.; Cuiné, S.; Triantaphylidès, C.; Ravanat, J.L.; Havaux, M. Chemical quenching of singlet oxygen by carotenoids in plants. *Plant Physiol.* **2012**, *158*, 1267–1278. [CrossRef]

293. Kruk, J.; Szymańska, R. Singlet oxygen oxidation products of carotenoids, fatty acids and phenolic prenyllipids. *J. Photochem. Photobiol. B* **2021**, *216*, 112148. [CrossRef]

294. Di Mascio, P.; Devasagayam, T.P.; Kaiser, S.; Sies, H. Carotenoids, tocopherols and thiols as biological singlet molecular oxygen quenchers. *Biochem. Soc. Trans.* **1990**, *18*, 1054–1056. [CrossRef] [PubMed]

295. Cantrell, A.; McGarvey, D.J.; Truscott, T.G.; Rancan, F.; Böhm, F. Singlet oxygen quenching by dietary carotenoids in a model membrane environment. *Arch. Biochem. Biophys.* **2003**, *412*, 47–54. [CrossRef]

296. Ukai, N.; Lu, Y.; Etoh, H.; Yagi, A.; Ina, K.; Oshima, S.; Ojima, F.; Sakamoto, H.; Ishiguro, Y. Photosensitized Oxygenation of Lycopene. *Biosci. Biotechnol. Biochem.* **1994**, *58*, 1718–1719. [CrossRef]

297. Naik, P.; Kumar, M. Receptor interactions of constituents of Zingiber officinalis and Solanum lycopersicum on COX. *J. Pharmacogn. Phytochem.* **2019**, *8*, 4637–4641.

298. Hayashi, R.; Hayashi, S.; Machida, S. Changes in Aqueous Humor Lutein Levels of Patients with Cataracts after a 6-Week Course of Lutein-Containing Antioxidant Supplementation. *Curr. Eye Res.* **2022**, *47*, 1016–1023. [CrossRef]

299. Sideri, O.; Tsaousis, K.T.; Li, H.J.; Viskadouraki, M.; Tsinopoulos, I.T. The potential role of nutrition on lens pathology: A systematic review and meta-analysis. *Surv. Ophthalmol.* **2019**, *64*, 668–678. [CrossRef]

300. Groten, K.; Marini, A.; Grether-Beck, S.; Jaenicke, T.; Ibbotson, S.H.; Moseley, H.; Ferguson, J.; Krutmann, J. Tomato Phytonutrients Balance UV Response: Results from a Double-Blind, Randomized, Placebo-Controlled Study. *Skin Pharmacol. Physiol.* **2019**, *32*, 101–108. [CrossRef]

301. Fam, V.W.; Holt, R.R.; Keen, C.L.; Sivamani, R.K.; Hackman, R.M. Prospective Evaluation of Mango Fruit Intake on Facial Wrinkles and Erythema in Postmenopausal Women: A Randomized Clinical Pilot Study. *Nutrients* **2020**, *12*, 3381. [CrossRef]

302. Ouchi, A.; Aizawa, K.; Iwasaki, Y.; Inakuma, T.; Terao, J.; Nagaoka, S.; Mukai, K. Kinetic study of the quenching reaction of singlet oxygen by carotenoids and food extracts in solution. Development of a singlet oxygen absorption capacity (SOAC) assay method. *J. Agric. Food Chem.* **2010**, *58*, 9967–9978. [CrossRef]

303. Fukuzawa, K.; Inokami, Y.; Tokumura, A.; Terao, J.; Suzuki, A. Rate constants for quenching singlet oxygen and activities for inhibiting lipid peroxidation of carotenoids and alpha-tocopherol in liposomes. *Lipids* **1998**, *33*, 751–756. [CrossRef] [PubMed]

304. Foote, C.S.; Ching, T.Y.; Geller, G.G. Chemistry of singlet oxygen. XVIII. Rates of reaction and quenching of alpha-tocopherol and singlet oxygen. *Photochem. Photobiol.* **1974**, *20*, 511–513. [PubMed]

305. Gruszka, J.; Pawlak, A.; Kruk, J. Tocochromanols, plastoquinol, and other biological prenyllipids as singlet oxygen quenchers-determination of singlet oxygen quenching rate constants and oxidation products. *Free Radic. Biol. Med.* **2008**, *45*, 920–928. [CrossRef] [PubMed]

306. Shichiri, M.; Yoshida, Y.; Ishida, N.; Hagihara, Y.; Iwahashi, H.; Tamai, H.; Niki, E. α Tocopherol suppresses lipid peroxidation and behavioral and cognitive impairments in the Ts65Dn mouse model of Down syndrome. *Free Radic. Biol. Med.* **2011**, *50*, 1801–1811. [CrossRef] [PubMed]

307. Shichiri, M.; Harada, N.; Ishida, N.; Komaba, L.K.; Iwaki, S.; Hagihara, Y.; Niki, E.; Yoshida, Y. Oxidative stress is involved in fatigue induced by overnight deskwork as assessed by increase in plasma tocopherylhydroqinone and hydroxycholesterol. *Biol. Psychol.* **2013**, *94*, 527–533. [CrossRef]

308. Shichiri, M.; Ishida, N.; Hagihara, Y.; Yoshida, Y.; Kume, A.; Suzuki, H. Probucol induces the generation of lipid peroxidation products in erythrocytes and plasma of male cynomolgus macaques. *J. Clin. Biochem. Nutr.* **2019**, *64*, 129–142. [CrossRef]

309. Krasnovsky, A.A.; Kagan, V.E.; Minin, A.A. Quenching of singlet oxygen luminescence by fatty acids and lipids: Contribution of physical and chemical mechanisms. *FEBS Lett.* **1983**, *155*, 233–236. [CrossRef]

310. Chaudhuri, R.K.; Bojanowski, K. Bakuchiol: A retinol-like functional compound revealed by gene expression profiling and clinically proven to have anti-aging effects. *Int. J. Cosmet. Sci.* **2014**, *36*, 221–230. [CrossRef]

311. Chaudhuri, R.K.; Marchio, F. Bakuchiol in the Management of Acne-affected Skin. *Cosmet. Toilet.* **2011**, *126*, 502–510.

312. Dhaliwal, S.; Rybak, I.; Ellis, S.R.; Notay, M.; Trivedi, M.; Burney, W.; Vaughn, A.R.; Nguyen, M.; Reiter, P.; Bosanac, S.; et al. Prospective, randomized, double-blind assessment of topical bakuchiol and retinol for facial photoageing. *Br. J. Dermatol.* **2019**, *180*, 289–296. [CrossRef]

313. Krishnakantha, T.P.; Lokesh, B.R. Scavenging of superoxide anions by spice principles. *Indian J. Biochem. Biophys.* **1993**, *30*, 133–134.

314. Swindell, W.R.; Bojanowski, K.; Chaudhuri, R.K. A Zingerone Analog, Acetyl Zingerone, Bolsters Matrisome Synthesis, Inhibits Matrix Metallopeptidases, and Represses IL-17A Target Gene Expression. *J. Invest. Dermatol.* **2020**, *140*, 602–614.e615. [CrossRef] [PubMed]

315. Das, K.C.; Das, C.K. Curcumin (diferuloylmethane), a singlet oxygen ((^{1})O$_{(2)}$) quencher. *Biochem. Biophys. Res. Commun.* **2002**, *295*, 62–66. [CrossRef] [PubMed]

316. Chaudhuri, R.K.; Meyer, T.; Premi, S.; Brash, D. Acetyl zingerone: An efficacious multifunctional ingredient for continued protection against ongoing DNA damage in melanocytes after sun exposure ends. *Int. J. Cosmet. Sci.* **2020**, *42*, 36–45. [CrossRef]

317. Dhaliwal, S.; Rybak, I.; Pourang, A.; Burney, W.; Haas, K.; Sandhu, S.; Crawford, R.; Sivamani, R.K. Randomized double-blind vehicle controlled study of the effects of topical acetyl zingerone on photoaging. *J. Cosmet. Dermatol.* **2021**, *20*, 166–173. [CrossRef] [PubMed]

318. Aratani, Y. Myeloperoxidase: Its role for host defense, inflammation, and neutrophil function. *Arch. Biochem. Biophys.* **2018**, *640*, 47–52. [CrossRef] [PubMed]

International Journal of
Molecular Sciences

Article

Evaluation of Oxidative Stress and Metabolic Profile in a Preclinical Kidney Transplantation Model According to Different Preservation Modalities

Simona Mrakic-Sposta [1], Alessandra Vezzoli [1,*], Emanuela Cova [2], Elena Ticcozzelli [3], Michela Montorsi [4], Fulvia Greco [5], Vincenzo Sepe [2], Ilaria Benzoni [3], Federica Meloni [6], Eloisa Arbustini [7], Massimo Abelli [3] and Maristella Gussoni [5,*]

[1] Institute of Clinical Physiology, National Research Council (IFC-CNR), 20159 Milano, Italy; simona.mrakicsposta@cnr.it
[2] Department of Molecular Medicine, IRCCS Foundation Policlinico San Matteo, 27100 Pavia, Italy; e.cova@smatteo.pv.it (E.C.); v.sepe@smatteo.pv.it (V.S.)
[3] Department of Surgery, IRCCS Foundation Policlinico San Matteo, 27100 Pavia, Italy; e.ticozzelli@smatteo.pv.it (E.T.); ilaria.benzoni@asst-cremona.it (I.B.); m.abelli@smatteo.pv.it (M.A.)
[4] Department of Human Sciences and Promotion of the Quality of Life, San Raffaele Roma Open University, 00166 Roma, Italy; michela.montorsi@uniroma5.it
[5] Institute of Chemical Sciences and Technologies "G. Natta", National Research Council (SCITEC-CNR), 20133 Milan, Italy; fulvia.greco@cnr.it
[6] Section of Pneumology, Department of Internal Medicine, University of Pavia, 27100 Pavia, Italy; f.meloni@smatteo.pv.it
[7] Centre for Inherited Cardiovascular Diseases, IRCCS Foundation Policlinico San Matteo, 27100 Pavia, Italy; e.arbustini@smatteo.pv.it
*   Correspondence: alessandra.vezzoli@cnr.it (A.V.); maristella.gussoni@unimi.it (M.G.)

Citation: Mrakic-Sposta, S.; Vezzoli, A.; Cova, E.; Ticcozzelli, E.; Montorsi, M.; Greco, F.; Sepe, V.; Benzoni, I.; Meloni, F.; Arbustini, E.; et al. Evaluation of Oxidative Stress and Metabolic Profile in a Preclinical Kidney Transplantation Model According to Different Preservation Modalities. *Int. J. Mol. Sci.* **2023**, 24, 1029. https://doi.org/10.3390/ijms24021029

Academic Editors: Rossana Morabito and Alessia Remigante

Received: 30 November 2022
Revised: 2 January 2023
Accepted: 3 January 2023
Published: 5 January 2023

Abstract: This study addresses a joint nuclear magnetic resonance (NMR) and electron paramagnetic resonance (EPR) spectroscopy approach to provide a platform for dynamic assessment of kidney viability and metabolism. On porcine kidney models, ROS production, oxidative damage kinetics, and metabolic changes occurring both during the period between organ retrieval and implantation and after kidney graft were examined. The ^{1}H-NMR metabolic profile—valine, alanine, acetate, trimetylamine-N-oxide, glutathione, lactate, and the EPR oxidative stress—resulting from ischemia/reperfusion injury after preservation (8 h) by static cold storage (SCS) and ex vivo machine perfusion (HMP) methods were monitored. The functional recovery after transplantation (14 days) was evaluated by serum creatinine (SCr), oxidative stress (ROS), and damage (thiobarbituric-acid-reactive substances and protein carbonyl enzymatic) assessments. At 8 h of preservation storage, a significantly ($p < 0.0001$) higher ROS production was measured in the SCS vs. HMP group. Significantly higher concentration data ($p < 0.05$–0.0001) in HMP vs. SCS for all the monitored metabolites were found as well. The HMP group showed a better function recovery. The comparison of the areas under the SCr curves (AUC) returned a significantly smaller (-12.5 %) AUC in the HMP vs. SCS. EPR-ROS concentration ($\mu mol \cdot g^{-1}$) from bioptic kidney tissue samples were significantly lower in HMP vs. SCS. The same result was found for the NMR monitored metabolites: lactate: -59.76%, alanine: -43.17%; valine: -58.56%; and TMAO: -77.96%. No changes were observed in either group under light microscopy. In conclusion, a better and more rapid normalization of oxidative stress and functional recovery after transplantation were observed by HMP utilization.

Keywords: kidney transplant; hypothermic machine perfusion; organ preservation; ROS; metabolomic; EPR; ^{1}H-NMR; oxidative damage

1. Introduction

Kidney transplantation can be considered the best treatment for end-stage renal failure, with longer life expectancy and superior quality of life [1–3].

However, the lack of suitable organ donors is a major constraint to transplantation.

Thus, to decrease the disparity in supply and demand, to the Standard Criteria Donors (SCD), the Donation after Cardiac Death (DCD) [4] and the Expanded Criteria Donor (ECD) [5–8] groups have been added [9]. This implies that organs are most often in marginal conditions because the donors present co-morbidity factors (diabetes, obesity, cardiovascular disease, and/or hypertension). These factors weaken the organ, making it more susceptible to developing lesions correlated to the graft outcome. To overcome these problems, innovative means have been established to obtain kidneys suitable for transplantation and, at the same time, to be able to determine the kidney quality. From this new perspective, the effort is devoted to assessing the best method of organ preservation while at the same time, developing methodologic criteria leading to the determination of the reimplant suitability or discharging.

Successful transplantation depends on a sequence of events related to donor selection, graft preservation, surgical implantation, and recipient treatment during which the kidney is exposed to a series of non-physiologic insults. Particularly, at the time of organ retrieval from the donor, blood supply is interrupted, leading to ischemic insults to renal function. Thus, kidneys from all donor types are exposed to ischemia. However, the cause of ischemia differs according to the donor source: in the after-cardiac-arrest donor group, the ischemic period is divided into warm and cold ischemic times. In the first period, removal of the kidney from the circulatory system leads to the absence of oxygen and anaerobic metabolism [10,11].

Nonetheless, even under hypothermic conditions—the metabolic rate at levels of approximately 10%—the need for oxygen persists, and hypoxia remains the main source of lesions induced in the context of preservation. This is termed cold ischemic injury together with a strong catalyst for the generation of oxygen free radicals [12]. In synthesis, warm ischemia leads to hypoxic lesions, while cold ischemia leads to both hypoxic and hypothermic lesions that strongly correlate with immediate and long-term kidney function [13–16].

In all cases, ischemia is followed by reperfusion, which occurs when the graft is connected to the recipient vascular system, and this is well known to exacerbate the cellular injuries initiated by ischemia. Ischemia–reperfusion is a complex pathophysiological process involving hypoxia and/or reoxygenation, ionic-imbalance-induced edema and acidosis, oxidative stress, mitochondrial uncoupling, coagulation, and endothelium activation that is associated with a proinflammatory immune response [17,18].

Particularly, there is strong evidence that reactive oxygen species (ROS) are important mediators of ischemia reperfusion injury in organ transplantation [19,20]. Indeed, the observed oxidative burst triggers inflammation and tubular cell injury. ROS–molecule interactions, including (a) oxidation of amino acids, resulting in the loss of important functional properties; (b) lipid peroxidation of cell membranes [20], resulting in decreased membrane viability; and (c) cleavage and crosslinking of renal DNA, resulting in harmful mutations, promote renal injury through damage to molecular kidney components. Lipid oxidation gives rise to malondialdehyde (MDA), which is a good indicator of nephrotoxicity and tissue damage and is a reliable diagnostic biomarker of initial graft injury and dysfunction earlier than serum creatinine. Moreover, levels of MDA in plasma within the first week after reperfusion of the graft predict long-term graft outcome [21].

Static cold storage (SCS) is undoubtedly the simplest and most widely adopted method of hypothermic preservation [22,23].

However, with the increasing use of marginal kidneys, there has been a renewed interest into the use of hypothermic machine perfusion (HMP) [24–26]. The evidence suggests that HMP may be more beneficial in reducing delayed graft function (DGF) rates in marginal kidneys [27]. A recent study by Nath et al. [28], comparing porcine kidneys' metabolic profiles, suggested that beneficial effects exerted by HMP might be related to the removal of ROS.

As is well known, electron paramagnetic resonance (EPR) is the only technique capable of providing direct quantitative detection of the 'instantaneous' presence of ROS [29,30] in a variety of biological samples [31–35], resulting in its being the "gold standard".

At the same time, nuclear magnetic resonance is the technique of choice to obtain metabolic profiles in the field of medicine, biology, and food science [36–40].

By a joint EPR and ^{1}HNMR analysis, the present study aims to examine the ROS production, the oxidative damage kinetics, and the metabolic changes occurring both during the period between organ retrieval and implantation and after kidney graft, according to the mode of preservation of the transplanted organ.

All experiments were carried out on a porcine model that, for its intrinsic characteristics, is well adapted to the modeling of kidney transplantation [15,41–43]. ROS production, oxidative damage kinetics, and metabolic changes occurring both during the period between organ retrieval and implantation and after kidney graft were examined. A better and more rapid normalization of oxidative stress and functional recovery after transplantation were observed by HMP utilization.

2. Results

2.1. NMR and EPR Procedure

A joint use of the NMR and EPR techniques constitutes one of the highlights of the present study. A sketch of the procedures and typical EPR and ^{1}H NMR spectra acquired from perfusate and tissues showing the main assigned resonances are displayed in Figure 1.

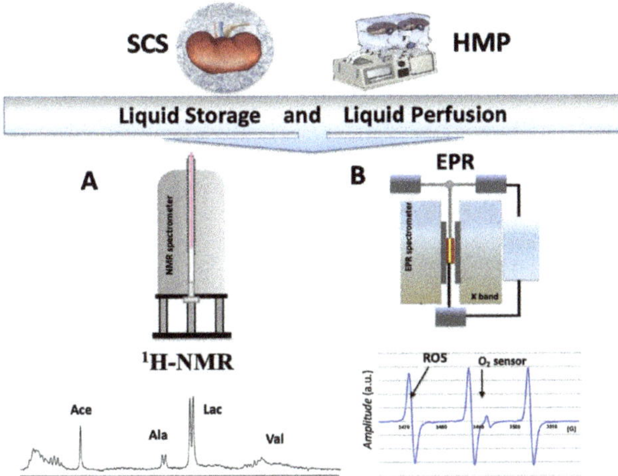

Figure 1. Static cold storage (SCS, the kidney stored in solution surrounded by crushed ice to keep the temperature between 4 and 0 °C is shown in the photo) and hypothermic machine perfusion (HMP, the kidneys in the machine are shown in the photo) preservation conditions assessed by NMR (**A**) and EPR (**B**) instruments. Under the instruments: (left) a typical high field ^{1}HNMR spectrum obtained from the perfusion solution during a SCS or HMP experiment. The main assigned resonances are indicated by arrows: Valine: proximal tubules disfunction; Lactate: global ischemia; Acetate: cortical lesion; and TMAO: medullar lesion. Right: a typical EPR spectrum recorded from the perfusion solution at 37 °C. The triplet signal comes from the reaction of 1-hydroxy-3-carboxymethyl-2,2,5,5-tetramethyl-pyrrolidine spin probe (CMH, EPR silent) to 3-carboxymethyl-2,2,5,5-tetramethyl-pyrrolidinyloxy radical (CM, EPR active). When the signal is sequentially acquired, the ROS production rate can be calculated. Using a stable radical compound, such as 3-Carboxy-2,2,5,5-tetramethyl-1-pyrrolidinyloxy (CP), as a reference, the absolute concentration levels are obtained. The singlet is the signal from oxygen label (O$_2$ sensor).

2.2. Analysis during the Preservation Procedure

The intra-renal resistance (IRR) resulted in the physiological range throughout the duration of time of the HMP preservation [44]. The recorded data are reported in Figure 2A. Starting from 4 mmHg/mL·min, the IRR values gradually but significantly (range $p = 0.05$–0.001) decreased, following a sigmoidal trend ($R^2 = 0.998$), reaching a value of 0.75 ± 0.14 mmHg/mL·min at 4 h, then slowly decreasing up to the final level of 0.51 ± 0.12 mmHg/mL·min at the end of the perfusion.

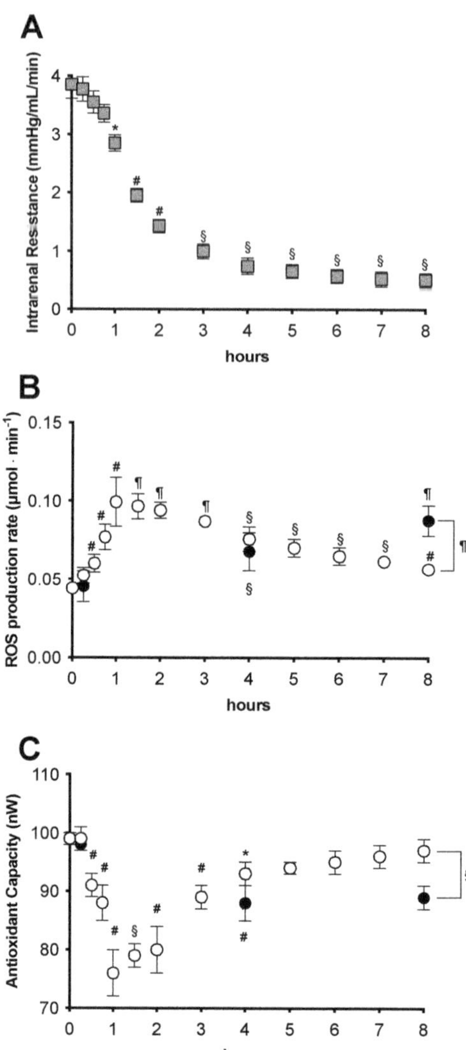

Figure 2. Intra-operative monitoring data. (**A**) Intra-renal resistance during hypothermic kidney machine perfusion (HMP, mmHg/mL·min); (**B**) EPR-measured reactive oxygen species production rate (ROS, µmol·min^{-1}); and (**C**) Antioxidant capacity (TAC, nW) during static cold storage (SCS, full symbols) and hypothermic machine perfusion (HMP, empty symbols). The results are expressed as mean $\pm$ SD. Significant differences: * $p < 0.05$; # $p < 0.01$, § $p < 0.001$, and ¶ $p < 0.0001$.

The changes in the ROS production rate recorded in perfusion solution during HMP perfusion and in liquid storage solution during SCS are reported in Figure 2B. As shown in the figure, during the HMP perfusion, the ROS production rate data, after 15 min from the beginning, began to linearly increase at a rate of 0.054 ± 0.007 µmol·min^{-1} ($R^2 = 0.95$), reaching the peak value of 0.099 ± 0.015 µmol·min^{-1} at 1 h, then slowly returning toward the initial level of 0.044 ± 0.003, following a single exponential trend ($K = 0.1072$, $R^2 = 0.99$). All data resulted in being significantly (range $p < 0.05$–0.0001) higher than the initial level. Significant ($p < 0.0001$) increases of the ROS production rate with respect to the initial level were likewise measured at 4 (0.067 ± 0.012 µmol·min^{-1}) and 8 h (0.087 ± 0.009 µmol·min^{-1}) of SCS storage. As is shown in the figure, at 15 min and 4 h, the SCS data did not significantly differ from those measured in HMP. However, at 8 h of storage, a significantly ($p < 0.0001$) higher ROS production was measured in the SCS compared with the HMP group (0.056 ± 0.003 vs. 0.087 ± 0.009 µmol·min^{-1}).

Antioxidant capacity data recorded during HMP perfusion and in SCS storage are reported in Figure 2B. Showing a specular trend with respect to the ROS production data reported in 1B, HMP values started to significantly and linearly decrease at a rate of 28.8 nW/h ($R^2 = 0.95$) after 30 min of perfusion, that is, a little later than ROS (initial level 91.0 ± 2.0 nW), reaching a minimum at 1 h (76.0 ± 4.0 nW), then slowly returning toward the initial levels, in turn, following a single exponential trend ($K = 0.4023$, $R^2 = 0.98$). All data recorded between 30 min and 4 h resulted in being significantly (range $p < 0.05$–0.001) lower than the initial value. A significant ($p < 0.01$) decrease of the TAC was recorded at 4 h in SCS (88.0 ± 3.0 nW) as well. In a similar fashion, as seen in Figure 2C, at 8 h of storage, the TAC measured from the SCS group resulted in being significantly ($p < 0.001$) lower than that of the HMP group (89.0 ± 2.0 vs. 97.0 ± 2.0 nW).

From a comparison of the ROS production rate increase with respect to the contemporary TAC decrease, during the first hour of HMP perfusion, the ROS level was found to have increased by approximately 60%, while the TAC decreased by approximately 23%. At the same time, at the end of the storage, the difference in the ROS levels measured from the SCS vs. HMP groups was approximately 35%, while the TAC level resulted in being approximately 8.2% lower.

^{1}H-NMR Analysis from Preservation Solutions

The absolute concentration (mM) data of the most important metabolites calculated by the ^{1}H-NMR spectra recorded from the preservation solution during 8 h of storage at 4 °C for both the HMP and SCS groups are reported in Figure 3: lactate (A), trimethylamine *N*-oxide (TMAO, B), valine (C), alanine (D), and acetate (E).

All these metabolites present de novo (therefore, likely produced by the kidney) in the solution showed an overall increase over time (T0 vs. T8): lactate: 0.0168 ± 0.0023 vs. 0.8440 ± 0.0300; TMAO: 0.0018 ± 0.0004 vs. 0.0500 ± 0.0081; valine: 0.0011 ± 0.0001 vs. 0.0176 ± 0.0002; alanine: 0.0014 ± 0.0001 vs. 0.0470 ± 0.0037; and acetate: 0.0185 ± 0.0028 vs. 0.0211 ± 0.0135. Moreover, as is shown in the figure, a significantly higher concentration data ($p < 0.05$–0.0001) was calculated for the HMP with respect to the SCS group during the entire experimental time and for all the monitored metabolites. In a similar fashion, the amount of all metabolites in the HMP group suddenly increased during essentially the first hour of storage, following a single exponential trend, even though at a different rate (see Figure 3). The rate of concentration change was greater in HMP kidneys than those of the SCS group, where a significant increase ($p < 0.05$–0.01) of the metabolite concentration (mM) was observed during the 8 h of storage (T0 vs. T8): lactate (0.017 ± 0.002 vs. 0.220 ± 0.051), TMAO (0.0019 ± 0.0019 vs. 0.0120 ± 0.0065), valine (0.0018 ± 0.0005 vs. 0.066 ± 0.0012), alanine (0.0016 ± 0.0001 vs. 0.0107 ± 0.0059), and acetate (0.0189 ± 0.0094 vs. 0.0384 ± 0.0194).

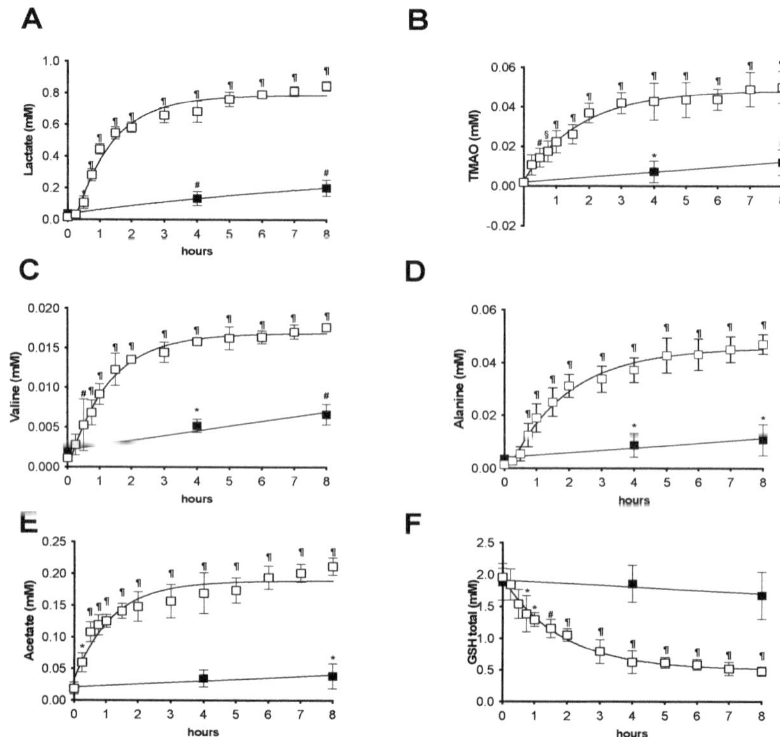

Figure 3. ^{1}H-NMR metabolite concentration (mM) during hypothermic kidney perfusion (HMP, empty symbols) and static cold storage (SCS, full symbols). The rate (mM/h) of the fitted trend resulted as follows: (**A**) lactate (HMP: K = 0.83, R^2 = 0.98; SCS: K = 0.02, R^2 = 0.97), (**B**) trimethylamine N-oxide (TMAO; HMP: K = 0.57, R^2 = 0.98; SCS: K = 0.001, R^2 = 0.99), (**C**) valine (HMP: K = 0.78, R^2 = 0.99; SCS: K = 0.0005, R^2 = 0.95), (**D**) alanine (HMP: K = 0.55, R^2 = 0.98; SCS: K = 0.001, R^2 = 0.90), (**E**) acetate (HMP: K = 0.85, R^2 = 0.93; SCS: K = 0.002, R^2 = 0.88), and (**F**) total glutathione (GSH; HMP: K = 0.55, R^2 = 0.99; SCS: K = −0.026, R^2 = 0.85). The results are expressed as mean ± SD. Significant differences: * $p < 0.05$; # $p < 0.01$, § $p < 0.001$, and ¶ $p < 0.0001$.

Finally, glutathione compound was one of the most important constituents of both HMP and SCS preservation solutions. The total glutathione (GSH) concentration levels calculated during the HMP and SCS storage are reported in Figure 3F. As is shown in the figure, a significant ($p < 0.05$–0.0001) single exponential GSH consumption was found during the HMP (T0 1.95 ± 0.14 vs. T8 0.47 ± 0.07 mM), while throughout the 8 h of SCS storage (T0 1.88 ± 0.28 vs. T8 1.67 ± 0.37 mM), the level remained almost unchanged.

Finally, positive correlations were found between ROS production by EPR and NMR metabolites: lactate (r = 0.97, p = 0.0008), TMAO (r = 0.94, p = 0.004), valine (r = 0.95, p = 0.003), alanine (r = 0.96, p = 0.001), and acetate (r = 0.88, p = 0.02), while inverse correlation was found with oxy-GSH (r = −0.94, p = 0.004).

2.3. Functional Analysis during the Survival Period

As an inclusion criterion, in the present study, it was established that the animal had to survive without any complications until the 14th day after kidney transplantation. Therefore, all animals that prematurely died or showed any type of complication during this period were excluded from the study. Indeed, all animals belonging to the HMP group reached the end of the follow up, while only five animals of the SCS group survived. One animal death was attributable to a renal failure possibly related to primary non-function

(PNF), and the other animal died after a massive haemorrhage due to the traumatic removal of the jugular central venous line produced by the animal itself.

2.3.1. Early Functional Recovery

The functional recovery after transplantation was evaluated by serum creatinine concentration assessment (SCr). Figure 4A shows the SCr values obtained from the HMP (dashed line) and SCS (continuous line) groups during the follow-up period. No SCr differences were found between HMP and SCS at Day 1. In the following days, until the sacrifice, compared with the animals of the SCS group, those belonging to the HMP group showed a better function recovery, even though the levels of the SCr peak (approximately 12.7 mg/dL) were similar and were reached at approximately the same time (5th day). The comparison of the areas under the curves (AUC) returned a significantly smaller (-12.5%) AUC in the HMP compared with the SCS group.

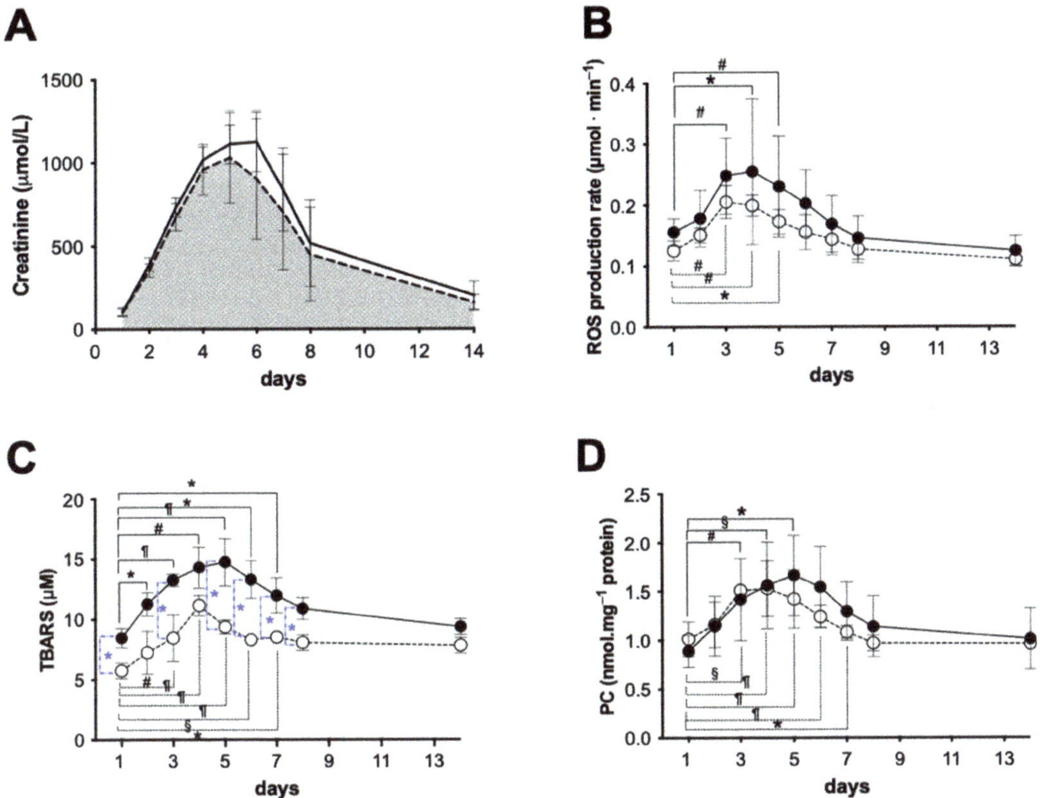

Figure 4. Renal functionality and oxidative stress after kidney re-implantation: (**A**) Serum creatinine concentration (mg/dL) measured from the HMP (dashed line) and SCS (continuous line) groups and areas under the curves (AUC); (**B**) Reactive oxygen species production rate (ROS, $\mu mol \cdot min^{-1}$) detected by EPR technique; (**C**) Thiobarbituric-acid-reactive substances (TBARS, μM); and (**D**) Protein carbonyl (PC, $nmol \cdot mg^{-1}$ protein) from 1st day to 14th day after kidney transplantation in static cold storage (SCS, full symbols) and hypothermic machine perfusion (HMP, empty symbols) groups. Continuous brackets indicate the significance of the intra-group data with respect to the first day. Blue stars (**C**) indicate the significance between HMP and SCS data at the same day. The data are expressed as mean $\pm$ SD. Significant differences: * $p < 0.05$; # $p < 0.01$, § $p < 0.001$, and ¶ $p < 0.0001$.

2.3.2. Oxidative Stress Evaluation

The plasmatic concentration levels of ROS, TBARS, and PC, measured during the post transplantation, are, respectively, shown in Figure 4B–D for the HMP (empty symbols) and SCS (full symbols) groups.

As can be observed in Figure 4B, the ROS production rate (μmol·min^{-1}) started from a higher level and remained significantly higher (range $p < 0.05$–0.0001) during the entire period in the SCS in comparison with the HMP group. Starting from a concentration of 0.14 ± 0.04 and 0.12 ± 0.01 μmol·min^{-1} for the SCS and HMP, respectively, the differences became greater in the following days, reaching a peak at the 4th day, then the values slowly returned, measuring below the initial levels at the 14th day: SCS 0.13 ± 0.04 and HMP 0.11 ± 0.01 μmol·min^{-1}.

Moreover, with respect to the 1st day, significant increases were found within each group as follows: SCS: at 3rd day 0.23 ± 0.09, 4th day 0.24 ± 0.06, and 5th day 0.22 ± 0.06; HMP: at 3rd day 0.21 ± 0.03 and 4th day 0.20 ± 0.06.

As is shown in Figure 4C, an even greater difference of the values of TBARS concentration (μM), a biomarker of lipids peroxidation, recorded in the two groups was found. With respect to Day 1, SCS: 8.45 ± 0.78 and HMP: 5.75 ± 0.63 were reported during the monitored period; the data calculated for the HMP group resulted in being significantly (range $p < 0.01$–0.0001) lower compared with the other group.

At the same time, with respect to the 1st day, significant increases (range $p < 0.05$–0.0001) were found within each group, reaching a peak at the 4th day, as follows: SCS: at 2nd day 11.26 ± 0.53, 3rd day 13.24 ± 0.53, 4th day 14.30 ± 1.71, 5th day 14.76 ± 1.99, 6th day 13.29 ± 1.55, and 7th day 11.96 ± 1.45, thereafter slowly returning to the initial level at the 14th day 9.37 ± 0.64. HMP: 3rd day 8.45 ± 0.53, 4th day 11.15 ± 0.84, 5th day 9.38 ± 0.58, 6th day 8.30 ± 1.55, and 7th day 8.52 ± 1.45, thereafter slowly returning toward the initial level at 14th day 7.83 ± 0.64 as well.

Regarding PC (nmol·mg^{-1} protein), a biomarker of proteins oxidation, an almost similar trend can be observed for the two groups within the first 4 days (Figure 4D). Then, the difference became greater, reaching a peak at the 5th day. Thereafter, the values slowly returned towards the initial level at the 14th day: SCS: 1.01 ± 0.31 and HMP: 0.96 ± 0.07, even though the data recorded for the HMP treated kidneys resulted in being lower than for the SCS group. Significant differences in concentration in the post-surgery days were found. Again, with respect to the 1st day level: SCS: 0.89 ± 0.16 and HMP: 1.02 ± 0.18, significant increases were found within each group as follows: SCS: at 3rd day 1.42 ± 0.42, 4th day 1.56 ± 0.44, 5th day 1.67 ± 0.41, and 6th day 1.54 ± 0.42; HMP: at 3rd day 1.51 ± 0.06, 4th day 1.53 ± 0.29, and 5th day 1.42 ± 0.30.

Finally, the comparison between the areas calculated under the curves (AUC) of the two groups resulted in being lower (ROS $= -19\%$, TBARS $= -28\%$, and PC $= -7.6\%$) for the HMP than the SCS group.

Finally, a positive correlation between the mean values of the creatinine levels and all the determined oxidative stress biomarkers in plasma could be established as follows: ROS ($r = 0.73$, $p < 0.03$), TBARS ($r = 0.81$, $p = 0.001$), and PC ($r = 0.77$, $p = 0.03$), see Figure 5.

2.4. ^{1}HNMR and EPR Tissue Analysis

ROS concentration (μmol·g^{-1}) calculated by EPR spectra collected from bioptic kidney tissue samples (0.0031 ± 0.0016 g, $n = 14$) at different experimental times—T0 (anesthesia), T1(end of warm ischemia -WI-, 75 min), and T2 (end of preservation, 8 h)—in SCS (full symbols) and HMP (empty symbols) are displayed in the histogram bars of Figure 6A.

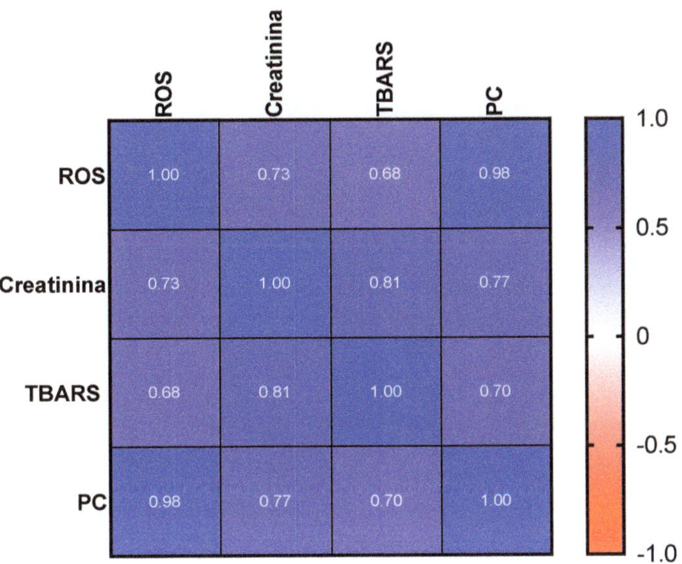

Figure 5. Heat map chart of creatinine levels and oxidative stress biomarkers in plasma: ROS, TBARS, and PC. The correlation coefficient (r) is reported in the squares.

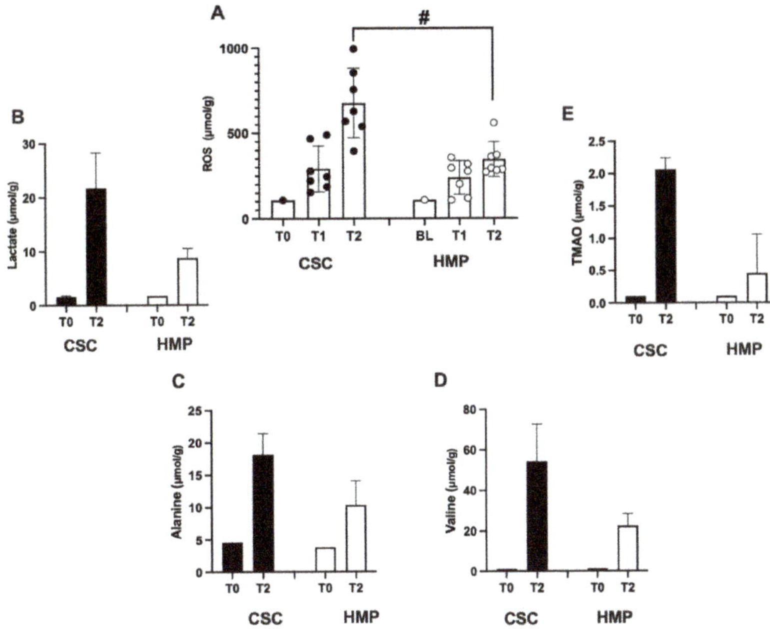

Figure 6. Histogram bars of: (**A**) ROS concentration (μmol/g) in kidney tissue at different experimental times: T0 (anesthesia), T1 (end of WI, 75min), and T2 (end of perfusion, 8 h) in SCS (full symbols) and HMP (empty symbols) groups. (**B–E**) metabolite concentration (μmol/g) calculated from ^{1}HNMR spectra collected on kidney tissue samples in SCS (full bars) and HMP (empty bars) groups at T0 and T2 experimental times. Significant difference # $p < 0.01$.

Histogram bars showing the concentration (μmol/g) of the metabolites of greatest interest calculated from ^{1}HNMR spectra collected on kidney tissue samples in SCS (full bars) and HMP (empty bars) groups are shown as well: lactate (Figure 6B), alanine (Figure 6C), valine (Figure 6D), and TMAO (Figure 6E).

With respect to T0, a great increase in the ROS levels (range 340–162%) was found in both groups at the end of ischemia (T1), and a great increase of ROS production was found in the two groups: SCS +525% and HMP 219+% at the end of perfusion (T2) as well. The difference between the two investigated groups at T2 was significant ($p < 0.01$).

NMR experiments were carried out on a smaller number of samples ($n = 6$) due to the greater amount of tissue needed for NMR with respect EPR acquisitions. Therefore, only two experimental times (T0 and T2) were analyzed. For all the analyzed metabolites, a great increase of the concentration could be observed at T2 vs. T0. However, as is shown in the figure, the increase was lower for the tissues belonging to the HMP group as follows: lactate: −59.76%, alanine: −43.17%; valine: −58.56%; and TMAO: −77.96%.

2.5. Histological Evaluation

No changes were observed in either experimental animal tissue sections groups under light microscopy. Representative images of tissue samples obtained from kidneys belonging to SCS and HMP groups at T1 and T2 are shown in Figure 7. As is shown in the figure, light microscopy did not show any significant differences between the groups of kidneys preserved by SCS or HMP. In detail, the following were observed in both groups: at T2 (end of preservation), well-preserved glomeruli, patent, and dilated tubular lumina, and at Tend (sacrifice), focal interstitial mononuclear infiltration.

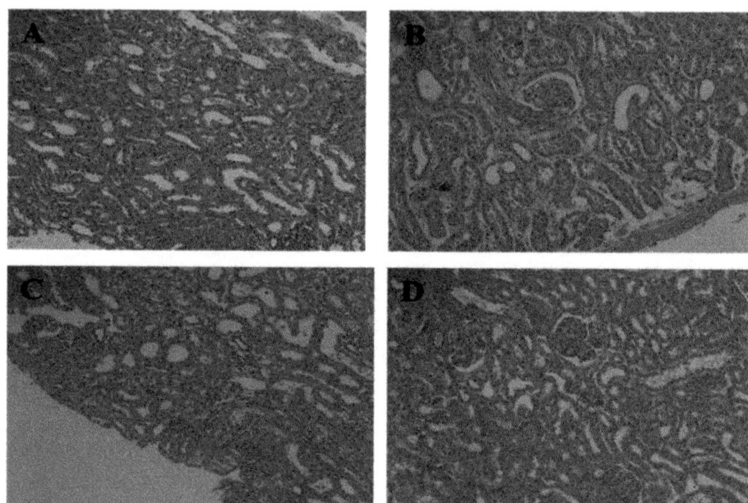

Figure 7. Histological examination. Representative light microscopy images of kidney tissue samples from SCS at T2 (**A**) and Tend (**B**) and HMP groups at T2 (**C**) and Tend (**D**). Magnification $\times 100$.

3. Discussion

As always occurs in medical science, early diagnosis and timely intervention will also improve outcomes in organ transplantation [43]. The association between DGF and worse outcomes has led to increased efforts to better understand the mechanisms of ischemia–reperfusion injury, the major cause of dysfunction and/or non-function upon reperfusion in the recipient, and to develop interventions to reduce its occurrence and impact [45,46].

Build-up of ROS and lactate secondary to anaerobic cellular metabolism, occurring during donor organ procurement and static cold storage, is the mediator of the ischemia–

reperfusion injury [47]. Cellular swelling and acidosis are the consequences of the rapid depletion of ATP under anaerobic metabolism. This leads to adenosine degradation, causing the accumulation of toxic substances, such as hypoxanthine and xanthine oxidase generating ROS [48].

Nevertheless, oxygen free radicals are generated both during preservation and during reperfusion [49], stimulating oxidative damage and promoting apoptosis and necrosis. The severity of tissue injury, caused by hypothermic preservation, influences the level of I/R injury and the subsequent functionality of the kidney. Decreasing ROS, as observed by HMP, might increase successful kidney transplantations, resulting in breaking the vicious cycle of cell damage, inflammation, and distal organ impairment [50,51].

Metabolic support may be another important factor in the observed benefit of HMP vs. SCS for preserving kidneys prior to transplantation. Experimental studies detail the metabolic activity of kidneys under various storage conditions [28,29,42].

In the present study, the combined oxidative and metabolic approach achieved by the simultaneous use of EPR and NMR techniques confirmed previous findings demonstrating the benefits of HMP vs. SCS on chronic kidney graft outcome.

It is believed that HMP protects against cold ischemic injury by providing a better flush of the kidney, clearing of red blood cells, and prevention of the accumulation of waste products. Compared with static conditions, it was demonstrated to support higher levels of metabolism [52,53].

These figures appear to be confirmed by the results of this study, as shown in Figures 2–4. In fact, as can be observed in Figure 2, starting from almost the same level, during the HMP preservation period, ROS concentration linearly increased in the first two hours, then followed a low exponential decrease. TAC, in turn, showed an expected specular behavior, even though it was not of the same amount: ROS increased by 60%, while TAC decreased by 23%. By contrast, in the SCS group, during the entire preservation period, both ROS and TAC remained at a slightly lower level, reaching a significant difference ($p < 0.001$) at the end of the period. In accordance with that reported above, these findings suggest to us that the flush realized by the HMP promotes, over time, a gradual decrease of the ROS production, suddenly increasing in the first hours while, at the same time, allowing for a restoration of the TAC towards the initial level. One of the major factors in the pathogenesis of complications is the imbalance between the ROS formation and clearance by the antioxidative system. This disparity can cause endothelial dysfunction and impairment of the regulatory functions of endothelium for vasculation [54].

In this respect, the advantage of HMP vs. SCS is to eliminate from the organ the ROS amount produced during preservation. Indeed, ROS produced as a by-product of cellular respiration are important in many cell-signalling cascades, but their increased production leads to peroxidation of membrane lipids, proteins, and DNA damage, up to cellular dysfunction and death [55,56].

The advantage of the HMP vs. SCS found even more confirmation when the metabolic kidney response was considered, as is shown in Figure 3, where the absolute quantitative concentration trend of the most important metabolites is displayed. Alanine and valine, the most important metabolites indicators of proximal tubules disfunctions; acetate, whose level mainly increases in case of cortical lesions; trimethylamine-N-oxide (TMAO), a common metabolite in animals and humans, oxidation product of trimethylamine, and derived from Choline, indicator of medullar lesions; and finally, lactate, the most important indicator of an anaerobic cellular metabolic status and global ischemia [57] can be observed in the figure.

A great difference between the two preservation methods is evident: the HMP machine allowed for a great amount of the metabolites, produced by the kidney itself, to flow out, thus washing waste products from the organ. An exponential increase of all self-produced metabolites was found, even though with a different rate constant. For each of them, an almost constant level can be observed for the SCS group.

At the same time, in the HMP group, total GSH followed an exponential decay, while it remained almost at the same level in SCS.

The antioxidant capacity of GSH, can be consumed when using the HMP machine, thus promoting a redox status restoration [58].

All the achieved results lead us to conclude that, unlike SCS, HMP restores, at least partially, the fundamental processes of glomerular filtration, which eliminate toxic solutes, including oxidative stress.

On the other hand, delayed graft function is the most common complication in the immediate post-transplantation period, mainly in deceased renal allografts, almost invariably in the non-heart beating transplant [59]. The functional results of this study showed that, in the SCS group, there was a severe level of renal dysfunction with lower levels of creatinine clearance and an accentuated unbalance of redox homeostasis with higher levels of oxidative damage (see Figure 4).

The observed significantly increased levels of lipid peroxidation (TBARS, Figure 4C) and protein carbonyls (PC, Figure 4D) were associated with significantly increased ROS production (Figure 4B). During the two weeks of observation, the transplanted animals presented a peak of ROS production and of damage biomarkers concentration at Days 3–4, remaining almost at the plateau level until Days 5–6, and then it decreased from Day 7 onwards. The time of the peak did not differ in the two monitored groups; however, the peak levels were significantly lower (range $p < 0.01$–0.0001), and the restoring kinetics were more rapid in the HMP group.

Lipid peroxidation is a product of the free-radical-mediated oxidation of arachidonic acid, which is associated with membrane damage and is known to instigate proinflammatory mediators and stimulate the release of cytokines and chemokines, thus causing tissue injury [60–63] and endoplasmic alteration and deterioration of tissue integrity after reperfusion [64].

At the same time, higher levels of protein carbonyl were found in the static storage groups compared with the other storage conditions. Other studies supported these findings, with increased oxidative damage [65,66].

Even more encouraging results were obtained by the EPR and NMR analysis carried out on bioptic kidney samples. Needless to say, the most important issue was to maximally preserve the integrity of the kidney, and, as is well known, the main limitation of NMR technique, compared with others, as for example EPR, is the sample amount needed to obtain the spectra. Therefore, for each kidney, one tissue sample was excised to be analyzed by EPR at T1 and T2, while only three samples for each group at T2 were biopted for NMR analysis.

Nonetheless, the histogram bars of Figure 6 show an ROS increase at the different experimental times, but at T2, the ROS level resulted in being significantly lower ($p < 0.01$) in the kidneys perfused by the HMP machine compared with SCS. The same was observed for all the NMR analyzed metabolites, which showed an increase at T2 with respect to T0, but the increase was lower in the tissues perfused by the HMP machine. Interestingly, the trend was the opposite when compared with the perfusion liquid NMR analysis, as shown in Figure 3. This figure confirmed that when ROS and metabolites, inevitably increasing during the preservation time, are washed out by the HMP machine flow, their concentration logically increases in the perfusion liquid, but at the same time, lowers in the tissue with respect to the cold storage maintenance.

Finally, it is relevant to note, in the present study, that despite the lower ROS and TBARS productions in HMP vs. SCS and all the other measured parameters that indicated a better performance of HMP machine, the observed light microscopy images did not suggest a difference between the tissue samples excised from the two groups nor relevant impairment of the renal or tubule cell function during the 14-day post re-implantation phase (see Figure 7). These findings underline the relevance of the measurements performed by advanced techniques, such as EPR and NMR, to monitor transplanted organs during the entire period and even beforehand to be able to determine whether the organ will be reimplanted or discharged. On the other hand, the transplanted kidney would have been monitored for a longer period.

In addition, in order to further enhance the effectiveness of perfusates, some additives to the solution could be used. Many substances are reported in the literature to counteract the ischemic injury by improving both the metabolic response to anaerobiosis and the oxidative stress. Oxygen and/or energy substrates can be supplemented to the perfusion solution. Particularly, oxygen carriers have been applied experimentally in kidney preservation. At the same time, the addition of free radical scavengers, such as superoxide dismutase (SOD), to the preservation solution has been found to be beneficial in preventing the generation of oxygen free radicals in this highly oxygenated environment [15,67].

4. Materials and Methods

4.1. Animal Model and Surgical Procedure

Fourteen healthy 30-kg female domestic pigs were sedated by an intramuscular injection of Zolazepam/Tiletamine 6 mg/kg. A general anaesthesia was induced (Induction: Propofol 2 mg/kg and Atracurium 1 mg/kg. Maintenance: Propofol 5–10 mg/kg/h and Atracurium 1 mg/kg bolus every 30 min. Analgesia: Diflunixin Meglumine 100 mg) after the placement of a peripheral catheter in a vein of the ear. Later, the surgical placement of a jugular central venous line took place, and a xyphopubic laparotomy was performed. The animals were intubated and mechanically ventilated with a 60% O_2–air mix. The left renal artery and vein were isolated and clamped to mimic the condition of a DCD donor [68]. The clamping was maintained for 75 min (warm ischemia, WI); this timing was chosen to reach the maximum tissue damage, adopting an ischemic period much longer than the maximum warm ischemia time (WIT) allowed by the clinical protocols (40 min) [69]. The organ was then removed and immediately cold-flushed with 500–600 mL of preservation solution (Belzer Machine Perfusion Solution (MPS) or IGL-1) [70] at the constant pressure of 75 mmHg. Kidneys were preserved for 8 h at 4 °C and subdivided in two randomized groups (see Figure 7) as follows: Group 1: SCS ($n = 7$) by static conservation in IGL-1; Group 2: HMP ($n = 7$) using the RM3 Waters Medical Systems pulsatile machine in MPS. Pressure, flow, and resistance parameters were monitored by the perfusion machine. The systolic pressure was settled by the operator, then gradually increased during the first hour and stabilized until the end of the perfusion. Specifically, the starting pressure was set at 20 mmHg, then 2 mmHg was added every 8 min, reaching the maximal level of 35 mmHg at the end of the first hour. The flow would have ideally been 0.5 mL·g^{-1} tissue; it is insufficient for values less than 60 mL·min^{-1}. Resistances had to be set in the range of 0.10–0.40 mmHg·mL^{-1}·min^{-1}. Renal blood flow (RBF) and systolic pressure (MAP) were recorded continuously. Intra-renal resistance (IRR) could be calculated as MAP/RBF. These cut-off values were established on the basis of both the experiments on animal models and the clinical experience of international transplant centres [44].

After 8 h of storage/perfusion, a renal orthotopic auto transplantation was performed by an end-to-side arterial anastomosis between the renal artery and the aorta and an end-to-side venous anastomosis between the renal vein and the inferior vena cava. Finally, the ureterovesical anastomosis (Lich–Gregoire technique) was performed. The contralateral right nephrectomy was carried out before the closure of the abdominal wall. All animals received humane care and all studies were carried out in accordance with policies and guidelines of the Italian Ministry of Health for the use and care of laboratory animals. All procedures were carried out under the animal use protocols approved by the ethical committee (IRCCS Policlinico San Matteo, Pavia, 27100 Italy, "Comitato Etico per la Sperimentazione Animale" 18/03/2014, N. 4/2014).

4.2. Post-Surgical Evaluation Procedure

The transplanted animals were kept under observation for 14 days and then sacrificed by an anaesthetic lethal injection procedure. During the first 7 post-operative days, analgesic (Ketorolac 15 mg), diuretic (Furosemide 10 mg), antibiotic (Cefotaxime 1 g), and gastroprotective (Omeprazole 20 mg) therapies were administered every 12 h.

4.3. Sampling Methods

4.3.1. Perfusate Sampling

For each kidney, 2 mL of perfusate was sampled, following the timing reported in Figure 7. The perfusate was transferred to a cryogenic vial and stored at $-80\ °C$ until NMR analysis. Samples were thawed only once for the analyses, which was performed within two weeks of collection. ROS production and total antioxidant capacity assessment in perfusate were performed immediately after collection.

4.3.2. Blood Sampling

Approximately 15 mL of blood was drawn from the jugular central venous line and collected in heparinized and serum vacutainer tubes (Becton Dickinson and Company, Oxford, UK). Plasma was separated by centrifuge (5702R, Eppendorf, Hamburg, Germany) at $3000\times g$ for 5 min at $4\ °C$ and immediately stored in multiple aliquots at $-80\ °C$ until assayed. Samples were thawed only once for the analyses, which was performed within two weeks of collection. SCr concentration assessment was performed immediately after collection.

4.3.3. Tissue Sampling

The renal tissue was collected for histological, EPR, and NMR experiments after each of the following: anaesthesia (T0), warm ischemia (75 min) and flushing (T1), at the end of preservation (8 h, T2), and at the sacrifice (14th day: Tend).

4.4. Measurements

4.4.1. EPR Measurements

An X-band EPR instrument (E-Scan—Bruker BioSpin, GmbH, Billerica, MA, USA) was utilized for ROS measurements. The instrument allows us to handle very low concentrations of paramagnetic species in small (50 µL) samples. Among the spin probe molecules suitable for biological utilization, the CMH (1-hydroxy-3-methoxycarbonyl-2,2,5,5-tetramethylpyrrolidine) probe was adopted.

ROS production rate was determined in perfusate and plasma samples by means of a well-established EPR method [31,32].

Acquisition EPR parameters were microwave frequency: 9.652 GHz; modulation frequency: 86 kHz; modulation amplitude: 2.28 G; center field: 3456.8 G; sweep width: 60 G; microwave power: 21.90 mW; number of scans: 10; and receiver gain: 3.17×10^1. Sample temperature was firstly stabilized and then maintained at $37\ °C$ by a temperature and gas controller "Bio III" unit, interfaced to the spectrometer. From the EPR spectra, relative quantitative determination of ROS production was obtained then converted into absolute concentration by using the stable radical CP* (3-Carboxy-2,2,5,5-tetramethyl-1-pyrrolidinyloxy) as external reference [71–73].

A small amount of tissue (0.0031 ± 0.0016 g, $n = 14$) was excised from each kidney of both groups at T0, T1, and T2 for EPR experiments. For ROS detection in kidney tissue, methods were previously described [34,35,74]. Briefly, samples were biopted, weighed, and immediately incubated at $37\ °C$ in Krebs-HEPES buffer (KHB) with 25 µM deferoxamine methane–sulfonate salt (DF) chelating agent and 5 µM sodium diethyldithio-carbamate trihydrate (DETC) at pH 7.4 with 1 mM of spin probe CMH. After 30 min, the isolated tissues were placed in the center of a 1 mL plastic syringe according to Dikalov et al. [74], snap-frozen, and stored at $-80\ °C$. Then, the frozen block was removed by gentle pushing from the warmed-up syringe and analyzed in the quartz Dewar with liquid N_2. Spectra were recorded at 77 K; the acquisition parameters were modulation amplitude: 5 G; centered field: 2.0023 g; sweep time: 10 s; field sweep: 60 G; microwave power: 1 mW; number of scans: 10; and receiver gain: 1×10^3. Data were, in turn, converted into absolute concentration levels (micromoles per gram) by adopting CP• stable radical as external reference.

Spectra were recorded and analyzed by using the Win EPR software (2.11 version) standardly supplied by Bruker. An example of the recorded EPR signal from perfusate and tissue showing the triplet coming from the interaction of the ^{14}N–OH group of CMH with the ROS oxygen unpaired electron (NOH + O•$_2$ → NO• + H$_2$O$_2$) is displayed in Figure 8A.

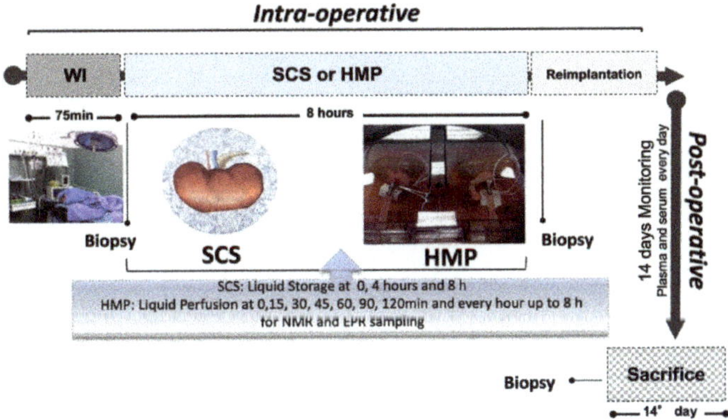

Figure 8. Sketch of the experimental protocol adopted to measure ROS production rate by EPR, metabolite concentration changes by ^{1}H-NMR, and oxidative damage during the intra-operative (organ preservation) and post-surgery sessions. The storage solution was sampled at 0 (T1), 4, and 8 h (T2) in SCS and at 0, 15, 30, 45, 60, 90, 120 min, and every hour up to 8 h in HMP. Blood samples were collected every day until the 7th day after re-implantation and at the sacrifice (14th day: Tend). Kidney tissues were biopted after each of the following: anesthesia (T0), 75 min of WI (T1), after 8 h of preservation (T2), and at the 14th day (Tend: sacrifice). WI: warm ischemia; SCS: static cold storage; and HMP: hypothermic machine perfusion.

4.4.2. Antioxidant Capacity

The amount of 10 µL of perfusate was used to assess the reducing capacity by means of a commercial EDEL potentiostat electrochemical analyser (Edel Therapeutics, Lausanne, Switzerland). The instrument was equipped with a redox sensor in a three-electrode arrangement able to respond to all water-soluble compounds present in biological fluids and oxidized within a defined potential range [72,75,76]. Data were expressed in nW.

4.4.3. Enzymatic Assays

Thiobarbituric-Acid-Reactive Substances (TBARS)

TBARS determination method was utilized to detect lipid peroxidation in plasma samples. By means of the TBARS assay kit (Cayman Chemical, Ann Arbor, MI, USA), a rapid photometric detection of the thiobarbituric acid malondialdehyde (TBAMDA) adduct at 532 nm was performed according to the manufacturer's instruction. A calibration curve was obtained from pure malondialdehyde-containing reactions.

Protein Carbonyls (PC)

The accumulation of oxidized proteins in plasma samples was measured by their reactive carbonyls content. A Protein Carbonyl assay kit (Cayman Chemical, Ann Arbor, MI, USA) was used to evaluate colorimetrically oxidized proteins as described in detail in the kit's user manual. All tests were performed in duplicate, and all data were obtained by a microplate reader spectrophotometer (Infinite M200, Tecan, 8708 Männedorf, Switzerland).

4.4.4. ^{1}H-NMR Analysis

All ^{1}H-NMR experiments were carried out at 400 MHz using a Bruker Avance (1H/BB BBI) vertical bore instrument with a z-field gradient accessory capable of delivering gradients up to 500 mT/m.

A selective Watergate sequence was used with sine-shaped gradients of 1 ms duration. The composite pulse was formed by a 3–9–19 binomial selective pulse (interpulse delay = 70 us). The p/2 hard and soft pulses were calibrated against the water signal. All perfusate solution (500 µL) samples collected were placed into a 5 mm glass NMR tube. Acquisition parameters were size: 32 K; spectral width: 20 ppm; acquisition time: 2.04 s; and relaxation delay: 1 s. Chemical shifts are reported from Sodium Trimethylsilyl Propionate (TSP, 0.2 mM) added to the solution as internal standard. All experiments were carried out at the controlled temperature of 28 °C.

All NMR data are referred to the TSP signal, added in a known amount (10 mM) in 100% D_2O solution into a capillary coaxially placed into the NMR tube for the acquisitions performed on the perfusion solution or added to the solution for tissue samples. Moreover, to monitor the kinetic trend of the metabolites during the 8 h of storage, the spectra were acquired every hour (HMP group) or every four hours (SCS group). The kinetical data could be obtained by scaling each spectrum to the TSP area. A specific NMR spectrometer automated calibration routine was used for this aim. This procedure was not applied to the metabolite signals acquired from tissues, where the data of each spectrum were calibrated against the TSP signal area.

Tissue samples (0.15 ± 0.03 g; $n = 8$) were excised from kidneys at T0 and T2, frozen, then thawed and gently placed into a 5 mm NMR tube. Phosphate Ringer solution in 100% D2O and TSP (10 mM) was added to completely fill the sensitive region of the coils. The same sequence and acquisition parameters (NS = 2048) were used.

Resonances assignation was based on a combined extraction of polar and lipophilic metabolites obtained from cortical and medullar kidney samples (512 ± 36.77 mg) according to the Protocol by Beckonert et al. [77]: metabolic profiling, metabolomic, and metabonomic procedures for NMR spectroscopy of urine, plasma, serum, and tissue extract [77].

The extract samples were resuspended in 500 µL Phosphate buffer Ringer (100% D_2O, TSP 1 mM), placed into a 5 mm NMR tube, and acquired by using the parameters used for the perfusion solutions (NS = 128).

All the examined metabolite signals were quantified, at each perfusion time, by absolute integration of the NMR signals reported to the reference signal [38].

All spectra were processed with the standard TopSpin 2.1 Bruker software. All the experimental data were fitted with the Maquardt–Levenberg algorithm implemented in SigmaPlot 9.0 software (Systat Software Inc., San Jose, CA, USA).

Typical high-field ^{1}HNNMR spectra collected on perfusate at the end of the perfusion time by MPH machine and from kidney tissue at the same experimental time are displayed in Figure 9B(a,b). The most relevant metabolites are indicated by arrows.

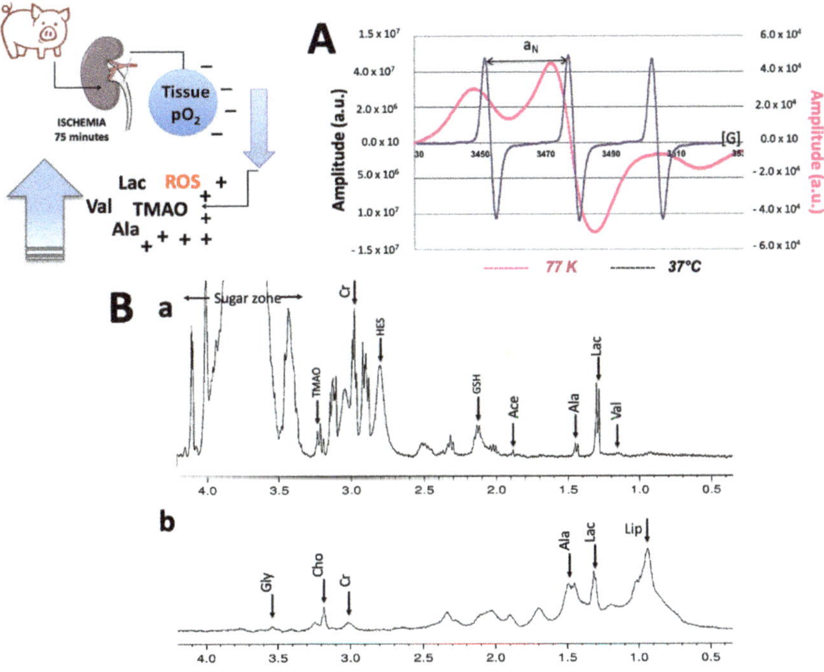

Figure 9. (**A**) EPR spectra recorded from both frozen kidney tissue (77 K) and from perfusion solution at 37 °C. The signals come from the reaction of 1-hydroxy-3-carboxymethyl-2,2,5,5-tetramethyl-pyrrolidine spin probe (CMH, EPR silent) to 3-carboxymethyl-2,2,5,5-tetramethyl-pyrrolidinyloxy radical (CM–, EPR active). (**B**) High-field ^{1}HNMR spectra of (**a**) perfusate from a machine perfused in Belzer group at the end of the perfusion time (8 h, T2); (**b**) same expansion from a kidney biopsy at the same time. The correspondent metabolites are indicated by arrows.

4.5. Histological Evaluation

At the T2 and Tend experimental times, corticomedullary kidney tissue was excised and fixed in 10% formaldehyde, dehydrated in an ascending grade of ethanol, cleared in xylene, and embedded in paraffin according to standard methods. The paraffin-embedded samples were cut into 3 µm sections for histopathological examination. Tissue sections were de-paraffinized in xylene and stained with hematoxylin and eosin (HE) (Sigma-Aldrich S.r.l., Milan, Italy, hematoxylin, MHS1; eosin, 230251), periodic acid-Schiff (PAS) (Sigma-Aldrich S.r.l., Milan, Italy, 395B), and methenamine silver–periodic acid-Schiff stain (Silver) (Sigma-Aldrich S.r.l., Milan, Italy PROTSIL1 Silver Stain Kit). Histopathological examinations of the kidneys were performed by an independent pathologist and one of the investigators.

All tissue sections were examined by light microscopy (Nikon Eclipse E200). Glomerular sclerosis, mesangial and capillary wall changes, tubular, and interstitial and vascular pathologic lesions were investigated.

4.6. Renal Function Recovery Evaluation

To monitor the function recovery, blood samples were collected from the pig ear vein before the left kidney removal every day during the first week and at the 14th day after auto-transplantation. Serum was used to determine the creatinine levels using an Advia Chemistry XPT analyzer (Siemens Healthcare GmbH, Erlangen, Germany).

4.7. Statistical Analysis

Statistical analysis was performed using SPSS Statistics 17.0 (Software Inc., Chicago, IL, USA) and the GraphPad Prism version 9.3.1 for Mac OS X (Software Inc., San Diego, CA, USA). The Shapiro–Wilks W test was used to test variables for normal distribution.

Data were compared by one-way ANOVA for repeated measures, followed by Bonferroni's multiple comparison test to further check the among-group significance.

The two-way analysis of variance (ANOVA) was run with "condition" (static cold storage: SCS; hypothermic machine perfusion: HMP) and "time" (for example, intra-operative from 0 to 8 h or post-surgery from 1st to 14th days). The Sidak's multiple comparisons test was used to test the significance of the differences. Areas under the curve (AUC) by creatinine measurements were calculated for the entire monitoring period. Pearson product moment correlation coefficient (r) with 90% confidence intervals (CI) was used to examine the relationships among selected parameters (creatinine levels vs. oxidative stress biomarkers and/or the metabolite concentrations assessed by NMR versus the ROS production by EPR).

Quantitative data are presented as mean $+$ SD. A p-value < 0.05 was assumed as statistically significant. Change $\Delta\%$ estimation $(((\text{postvalue} - \text{prevalue})/\text{prevalue}) \times 100)$ is also reported in the text.

5. Conclusions

All these findings lead us to conclude that since the surgical procedure of transplantation and ischemic injury to the organ during the procurement and transplantation procedures cause increased oxidative stress, it seems that successful kidney transplant might result in better and more rapid normalization of the antioxidant status and lipid metabolism by eliminating free radicals and decreasing oxidative stress. These latter, joined with the possibility of flowing out metabolites and discharge products de novo produced by the kidney, might be valid reasons suggesting the perfusion by pulsatile machine, that has been shown to reduce the incidence of delayed graft function after transplantation of deceased donor kidneys from brainstem death (DBD) and circulatory death (DCD) donors [68,78], as the best preservation mode.

The promising results of the present study reached by the joint use of EPR and NMR techniques and enzymatic assays determinations encourage us to continue in this research direction.

Author Contributions: Conceptualization: S.M.-S., A.V., E.T., E.C., M.A. and M.G.; investigation: S.M.-S., A.V., E.C., E.T., M.M., F.G., V.S., I.B., F.M., E.A., M.A. and M.G.; writing: S.M.-S., A.V., V.S. and M.G.; funding: M.A., E.T., S.M.-S., A.V. and M.G.; data analysis: S.M.-S., A.V., M.M., F.G., E.A. and M.G. All authors have read and agreed to the published version of the manuscript.

Funding: This work was supported in part by grants from the Fondazione IRCCS Policlinico San Matteo di Pavia (Italy), Progetto Ricerca Corrente 080-63008/63015-2013.

Institutional Review Board Statement: Ethical Committee approval from IRCCS Policlinico San Matteo, Pavia, 27100 Italy, the Ethical Committee for Research on Animals (Comitato Etico per la Sperimentazione Animale 18/03/2014, N. 4/2014).

Informed Consent Statement: Not applicable.

Data Availability Statement: Data are available at request from the authors.

Acknowledgments: Authors are grateful to the students of the school in medicine and surgery of the University of Pavia who participated in the experimentation. Thanks to the veterinarian who followed the animals, and a special thanks to the little pigs, who unknowingly sacrificed themselves. The authors thank the Fondazione Antonio De Marco for the support.

Conflicts of Interest: The authors declare no conflict of interest.

Abbreviations

ATP	Adenosine Triphosphate
^{1}H NMR	Proton Nuclear Magnetic Resonance
CMH	1-hydroxy-3-methoxycarbonyl-2,2,5,5-tetramethylpyrrolidine
CP	3-Carboxy-2,2,5,5-tetramethyl-1-pyrrolidinyloxy
DCD	Donation after Cardiac Death
DETC	Sodium Diethyldithio-Carbamate Trihydrate
DF	Deferoxamine Methane-Sulfonate Salt
DGF	Delayed Graft Function
EPR	Electron Paramagnetic Resonance
ECD	Expanded Criteria Donor
HE	Hematoxylin and Eosin
MDA	Malondialdehyde
PC	Protein Carbonyls
HMP	Hypothermic Machine Perfusion
ROS	Reactive Oxygen Species
SCD	Standard Criteria Donors
SCS	Static Cold Storage
TAC	Total Antioxidant Capacity
TBARS	Thiobarbituric-Acid-Reactive Substances
WIT	Warm Ischemia Time

References

1. Wolfe, R.A.; Ashby, V.B.; Milford, E.L.; Ojo, A.O.; Ettenger, R.E.; Agodoa, L.Y.; Held, P.J.; Port, F.K. Comparison of mortality in all patients on dialysis, patients on dialysis awaiting transplantation, and recipients of a first cadaveric transplant. *N. Engl. J. Med.* **1999**, *41*, 1725–1730. [CrossRef]
2. Tonelli, M.; Wiebe, N.; Knoll, G.; Bello, A.; Browne, S.; Jadhav, D.; Klarenbach, S.; Gill, J. Systematic review: Kidney trans‚Äê plantation compared with dialysis in clinically relevant outcomes. *Am. J. Transplant.* **2011**, *11*, 2093. [CrossRef] [PubMed]
3. Rana, A.; Gruessner, A.; Agopian, V.G.; Khalpey, Z.; Riaz, I.B.; Kaplan, B.; Halazun, K.J.; Busuttil, R.W.; SGruessner, R.W.G. Survival benefit of solid-organ transplant in the United States. *JAMA Surg.* **2015**, *150*, 252–259. [CrossRef]
4. Morrissey, P.E.; Monaco, A.P. Donation after circulatory death: Cur- rent practices, ongoing challenges, and potential improve-ments. *Transplantation* **2014**, *97*, 258–264. [CrossRef] [PubMed]
5. Rao, P.S.; Ojo, A. The alphabet soup of kidney transplantation: SCD, DCD, ECD–fundamentals for the practicing nephrologist. *Clin. J. Am. Soc. Nephrol.* **2009**, *4*, 1827–1831. [CrossRef] [PubMed]
6. Darius, T.; Gianello, P.; Vergauwen, M.; Mourad, N.; Buemi, A.; De Meyer, M.; Mourad, M. The effect on early renal function of various dynamic preservation strategies in a preclinical pig ischemia-reperfusion autotransplant model. *Am. J. Transplant.* **2019**, *19*, 752–762. [CrossRef]
7. Alshahrani, M.; Alotaibi, M.; Bhutto, B. Assessing the Outcome of Adult Kidney Transplantation from a Deceased Expanded Criteria Donor: A Descriptive Study. *Cureus* **2020**, *12*, e11199. [CrossRef]
8. Barreda, M.; Redondo-Pachón, D.; Miñambres García, E.; Rodrigo Calabia, E. Kidney transplant outcome of expanded criteria donors after circulatory death. *Nefrologia* **2022**, *42*, 135–144.
9. Akoh, J.A. Kidney donation after cardiac death. *World J. Nephrol.* **2012**, *1*, 79–91. [CrossRef] [PubMed]
10. Salahudeen, A.K. Consequences of cold ischemic injury of kidneys in clinical transplantation. *J. Investig. Med.* **2004**, *52*, 296–298. [CrossRef]
11. Salahudeen, A.K. Cold ischemic injury of transplanted organs: Some new strategies against an old problem. *Am. J. Transplant.* **2004**, *4*, 1. [CrossRef]
12. Norma, A.; Palomeque, J.; Carbonell, T. Oxidative stress and antioxidant activity in hypothermia and rewarming: Can RONS modulate the beneficial effects of therapeutic hypothermia? *Oxid. Med. Cell. Longev.* **2013**, *2013*, 957054.
13. Koetting, M.; Frotscher, C.; Minor, T. Hypothermic reconditioning after cold storage improves postischemic graft function in isolated porcine kidneys. *Transplant. Int.* **2010**, *23*, 538–542. [CrossRef] [PubMed]
14. Hoyer, D.P.; Gallinat, A.; Swoboda, S.; Wohlschlaeger, J.; Rauen, U.; Paul, A.; Minor, T. Influence of oxygen concentration during hypothermic machine perfusion on porcine kidneys from donation after circulatory death. *Transplantation* **2014**, *98*, 944–950. [CrossRef]
15. Thuillier, R.; Allain, G.; Celhay, O.; Hebrard, W.; Barrou, B.; Badet, L.; Leuvenink, H.; Hauet, T. Benefits of active oxygenation during hypothermic machine perfusion of kidneys in a preclinical model of deceased after cardiac death donors. *J. Surg. Res.* **2013**, *184*, 1174–1181. [CrossRef]

16. Darius, T.; Vergauwen, M.; Smith, T.; Gerin, I.; Joris, V.; Mueller, M.; Aydin, S.; Muller, X.; Schlegel, A.; Nath, J.; et al. Brief O uploading during continuous hypothermic machine perfusion is simple yet effective oxygenation method to improve initial kidney function in a porcine autotransplant model. *Am. J. Transplant.* **2020**, *20*, 030–2043. [CrossRef] [PubMed]

17. Perico, N.; Cattaneo, D.; Sayegh, M.H.; Remuzzi, G. Delayed graft function in kidney transplantation. *Lancet* **2004**, *364*, 1814–1827. [CrossRef] [PubMed]

18. Hofmann, J.; Otarashvili, G.; Meszaros, A.; Ebner, S.; Weissenbachwer, A.; Cardini, B.; Oberhuber, R.; Resch, T.; Ofner, D.; Schneeberger, S.; et al. Restoring Mitochondrial Function While Avoiding Redox Stress: The Key to Preventing Ischemia/Reperfusion Injury in Machine Perfused Liver Grafts? *Int. J. Mol. Sci.* **2020**, *21*, 3132. [CrossRef]

19. Theodore, K.; Baines, C.P.; Krenz, M.; Korthuis, R.J. Ischemia/Reperfusion. *Compr. Physiol.* **2016**, *7*, 113–170.

20. Micò-Carnero, M.; Zaouali, M.A.; Rojano-Alfonso, C.; Maroto-Serrat, C.; Abdennebi, H.B.; Peralta, C. A Potential Route to Reduce Ischemia/Reperfusion Injury in Organ Preservation. *Cells* **2022**, *11*, 2763. [CrossRef]

21. Fonseca, I.; Reguengo, H.; Almeida, M.; Dias, L.; Martins, L.S.; Pedroso, S.; Santos, J.; Lobato, L.; Henriques, A.C.; Mendonça, D. Oxidative stress in kidney transplantation: Malondialdehyde is an early predictive marker of graft dysfunction. *Transplantation* **2014**, *97*, 1058–1065. [CrossRef]

22. O'Callaghan, J.M.; Knight, S.R.; Morgan, R.D.; Morris, P.J. Preservation solutions for static cold storage of kidney allografts: A systematic review and meta-analysis. *Am. J. Transplant.* **2012**, *12*, 896–906. [CrossRef]

23. Tingle, S.J.; Figueiredo, R.S.; Moir, J.A.; Goodfellow, M.; Talbot, D.; Wilson, C.H. Machine perfusion preservation versus static cold storage for deceased donor kidney transplantation. *Cochrane Database Syst. Rev.* **2019**, *3*, CD011671. [CrossRef]

24. Belzer, F.O.; Southard, J.H. Principles of solid-organ preservation by cold storage. *Transplantation* **1988**, *45*, 673–676. [CrossRef]

25. Peng, P.; Ding, Z.; He, Y.; Zhang, J.; Wang, X.; Yang, Z. Hypothermic machine perfusion versus static cold storage in deceased donor kidney transplantation: A systematic review and meta-analysis of randomized controlled trials. *Artif. Organs* **2019**, *43*, 478–489. [CrossRef] [PubMed]

26. Jiang, X.; Feng, L.; Pan, M.; Gao, Y. Optimizing livers for transplantation using machine perfusion versus cold storage in large animal studies and human studies: A systematic review and meta-analysis. *Biomed. Res. Int.* **2018**, *2018*, 9180757. [CrossRef] [PubMed]

27. Weberskirch, S.; Katou, S.; Reuter, S.; Kneifel, F.; Morgul, M.H.; Becker, F.; Houben, P.; Pascher, A.; Vogel, T.; Radunz, S. Dynamic Parameters of Hypothermic Machine Perfusion—An Image of Initial Graft Function in Adult Kidney Transplantation? *J. Clin. Med.* **2022**, *11*, 5698. [CrossRef] [PubMed]

28. Nath, J.; Smith, T.B.; Patel, K.; Ebbs, S.R.; Hollis, A.; Tennant, D.A.; Ludwig, C.; Ready, A.R. Metabolic differences between cold stored and machine perfused porcine kidneys: A 1H NMR based study. *Cryobiology* **2017**, *74*, 115–120. [CrossRef]

29. Dikalov, S.; Griendling, K.K.; Harrison, D.G. Measurement of reactive oxygen species in cardiovascular studies. *Hypertension* **2007**, *49*, 717–727. [CrossRef]

30. Suzen, S.; Gurer-Orhan, H.; Saso, L. Detection of Reactive Oxygen and Nitrogen Species by Electron Paramagnetic Resonance (EPR) Technique. *Molecules* **2017**, *22*, 181. [CrossRef]

31. Mrakic-Sposta, S.; Gussoni, M.; Montorsi, M.; Porcelli, S.; Vezzoli, A. A quantitative method to monitor reactive oxygen species production by electron paramagnetic resonance in physiological and pathological conditions. *Oxid. Med. Cell. Longev.* **2014**, *2014*, 306179. [CrossRef]

32. Mrakic-Sposta, S.; Vezzoli, A.; Maderna, L.; Gregorini, F.; Montorsi, M.; Moretti, S.; Greco, F.; Cova, M.; Gussoni, M. R(+)-Thioctic Acid Effects on Oxidative Stress and Peripheral Neuropathy in Type II Diabetic Patients: Preliminary Results by Electron Paramagnetic Resonance and Electroneurography. *Oxid. Med. Cell. Longev.* **2018**, *10*, 2018, 1767265. [CrossRef]

33. Cova, E.; Inghilleri, S.; Pandolfi, L.; Morosini, M.; Magni, S.; Colombo, M.; Piloni, D.; Finetti, C.; Ceccarelli, G.; Benedetti, L.; et al. Bioengineered gold nanoparticles targeted to mesenchymal cells from patients with bronchiolitis obliterans syndrome does not rise the inflammatory response and can be safely inhaled by rodents. *Nanotoxicology* **2017**, *11*, 534–545. [CrossRef] [PubMed]

34. Rivellini, C.; Porrello, E.; Dina, G.; Mrakic-Sposta, S.; Vezzoli, A.; Bacigaluppi, M.; Gullotta, G.S.; Chaabane, l.; Leocani, L.; Marenna, S.; et al. JAB1 deletion in oligodendrocytes causes senescence-induced inflammation and neurodegeneration in mice. *J. Clin. Investig.* **2022**, *132*, e145071. [CrossRef]

35. Malacrida, S.; De Lazzari, F.; Mrakic-Sposta, S.; Vezzoli, A.; Zordan, M.A.; Bisaglia, M.; Menti, G.M.; Meda, N.; Frighetto, G.; Bosco, G.; et al. Lifespan and ROS levels in different Drosophila melanogaster strains after 24 h hypoxia exposure. *Biol. Open* **2022**, *11*, bio059386. [CrossRef]

36. Vezzoli, A.; Gussoni, M.; Greco, F.; Zetta, L. Effects of temperature and extracellular pH on metabolites: Kinetics of anaerobic metabolism in resting muscle by 31P- and ^{1}H-NMR spectroscopy. *J. Exp. Biol.* **2003**, *206 Pt 17*, 3043–3052. [CrossRef]

37. Vezzoli, A.; Gussoni, M.; Greco, F.; Zetta, L.; Cerretelli, P. Temperature and pH dependence of energy balance by (31)P- and (1)H-MRS in anaerobic frog muscle. *Biochim. Biophys. Acta* **2004**, *1608*, 163–170. [CrossRef]

38. Gussoni, M.; Cremonini, M.A.; Vezzoli, A.; Greco, F.; Zetta, L. A quantitative method to assess muscle tissue oxygenation in vivo by monitoring 1H nuclear magnetic resonance myoglobin resonances. *Anal. Biochem.* **2010**, *400*, 33–45. [CrossRef] [PubMed]

39. Scorciapino, M.A.; Spiga, E.; Vezzoli, A.; Mrakic-Sposta, S.; Russo, R.; Fink, B.; Casu, M.; Gussoni, M.; Ceccarelli, M. Structure-function paradigm in human myoglobin: How a single-residue substitution affects NO reactivity at low pO2. *Am. Chem. Soc.* **2013**, *135*, 7534–7544. [CrossRef]

40. Pasini, G.; Greco, F.; Cremonini, M.A.; Brandolini, A.; Consonni, R.; Gussoni, M. Structural and Nutritional Properties of Pasta from Triticum monococcum and Triticum durum Species. A Combined ^{1}H NMR, MRI, and Digestibility Study. *J. Agric. Food Chem.* **2015**, *63*, 5072–5082. [CrossRef] [PubMed]

41. Rossard, L.; Favreau, F.; Demars, J.; Robert, R.; Nadeau, C.; Cau, J.; Thuillier, R.; Hauet, T. Evaluation of early regenerative processes in a preclinical pig model of acute kidney injury. *Curr. Mol. Med.* **2012**, *12*, 502–505. [PubMed]

42. Codas, R.; Thuillier, R.; Hauet, T.; Badet, L. Renoprotective effect of pulsatile perfusion machine RM3: Pathophysiological and kidney injury biomarker characterization in a preclinical model of autotransplanted pig. *BJU Int.* **2012**, *109*, 141–147. [CrossRef] [PubMed]

43. Nasr, M.; Sigdel, T.; Sarwal, M. Advances in diagnostics for transplant rejection. *Expert Rev. Mol. Diagn.* **2016**, *16*, 1121–1132. [CrossRef] [PubMed]

44. Available online: http://www.accessdata.fda.gov/cdrh_docs/pdf5/K053169.pdf (accessed on 1 January 2022).

45. Situmorang, G.R.; Sheerin, N.S. Ischaemia reperfusion injury: Mechanisms of progression to chronic graft dysfunction. *Pediatr. Nephrol.* **2019**, *34*, 951–963. [CrossRef]

46. Zhao, H.; Alam, A.; Pac Soo, A.; George, A.J.T.; Ma, D. Ischemia-Reperfusion Injury Reduces Long Term Renal Graft Survival: Mechanism and Beyond. *EBioMedicine* **2018**, *28*, 31–42. [CrossRef]

47. Patrono, D.; Roggio, D.; Mazzeo, A.T.; Catalano, G.; Mazza, E.; Rizza, G.; Gambella, A.; Rigo, F.; Leone, N.; Elia, V.; et al. Clinical assessment of liver metabolism during hypothermic oxygenated machine perfusion using microdialysis. *Artif. Organs* **2022**, *46*, 281–295. [CrossRef]

48. Forrester, S.J.; Kikuchi, D.S.; Hernandes, M.S.; Xu, Q.; Griendling, K.K. Reactive Oxygen Species in Metabolic and Inflammatory Signaling. *Circ. Res.* **2018**, *122*, 877–902. [CrossRef]

49. Granger, N.I.; Kvietys, P.R. Reperfusion injury and reactive oxygen species: The evolution of a concept. *Redox. Biol.* **2015**, *6*, 524–551. [CrossRef]

50. Tejchman, K.; Sierocka, A.; Kotfis, K.; Kotowski, M.; Dolegowska, B.; Ostrowski, M.; Sienko, J. Assessment of Oxidative Stress Markers in Hypothermic Preservation of Transplanted Kidneys. *Antioxidants* **2021**, *10*, 1263. [CrossRef] [PubMed]

51. Carcy, R.; Cougnon, M.; Poet, M.; Durandy, M.; Sicard, A.; Counillon, L.; Blondeau, N.; Hauet, T.; Tauc, M.; Pisani, D.F. Targeting oxidative stress, a crucial challenge in renal transplantation outcome. *Radic. Biol. Med.* **2021**, *69*, 258–270. [CrossRef]

52. Darius, T.; Vergauwen, M.; Smith, T.B.; Patel, K.; Craps, J.; Joris, V.; Aydin, S.; Ury, B.; Buemi, A.; De Meyer, M.; et al. Influence of Different Partial Pressures of Oxygen During Continuous Hypothermic Machine Perfusion in a Pig Kidney Ischemia-reperfusion Autotransplant Model. *Transplantation* **2020**, *104*, 731–774. [CrossRef] [PubMed]

53. Hosgood, S.A.; Yang, B.; Bagul, A.; Mohamed, I.H.; Nicholson, M.L. A comparison of hypothermic machine perfusion versus static cold storage in an experimental model of renal ischemia reperfusion injury. *Transplantation* **2010**, *89*, 830–837. [CrossRef] [PubMed]

54. Endemann, D.H.; Schiffrin, E.L. Endothelial Dysfunction. *JASN* **2004**, *15*, 1983–1992. [CrossRef] [PubMed]

55. Dröge, W. Free radicals in the physiological control of cell function. *Physiol. Rev.* **2002**, *82*, 47–95. [CrossRef] [PubMed]

56. Kohen, R.; Nyska, A. Oxidation of biological systems: Oxidative stress phenomena, antioxidants, redox reactions, and methods for their quantification. *Toxicol. Pathol.* **2002**, *30*, 620–650. [CrossRef]

57. Wishart, D.S. Metabolomics in monitoring kidney transplants. *Curr. Opin. Nephrol. Hyperten.* **2006**, *15*, 637–642. [CrossRef] [PubMed]

58. Aquilano, K.; Baldelli, S.; Ciriolo. M.R. Glutathione: New roles in redox signaling for an old antioxidant. *Front. Pharmacol.* **2014**, *5*, 196. [CrossRef]

59. Shokes, D.A.; Halloran, P.F. Delayed Graft Function in Renal Transplantation: Etiology, Management and Long-term Significance. *J. Urol.* **1996**, *155*, 1831–1840. [CrossRef] [PubMed]

60. Gutteridge, J.M. Lipid peroxidation and antioxidants as biomarkers of tissue damage. *Clin. Chem.* **1995**, *41 Pt 2*, 1819–1828. [CrossRef]

61. Ayala, A.; Muñoz, M.F.; Argüelles, S. Lipid Peroxidation: Production, Metabolism, and Signaling Mechanisms of Malondialdehyde and 4-Hydroxy-2-Nonenal. *Oxid. Med. Cell. Longev.* **2014**, *2014*, 360438. [CrossRef]

62. Mylonas, C.; Kouretas, D. Lipid peroxidation and tissue damage. *In Vivo* **1999**, *13*, 295–309.

63. Gianazza, E.; Brioschi, M.; Fernandez, A.M.; Banfi, C. Lipoxidation in cardiovascular diseases. *Redox. Biol.* **2019**, *23*, 101119. [CrossRef] [PubMed]

64. Silver, E.H.; Szabo, S. Role of lipid peroxidation in tissue injury after hepatic ischemia. *Exp. Mol. Pathol.* **1983**, *38*, 69–76. [CrossRef]

65. Semmelmann, A.; Neeff, H.; Sommer, O.; Thomusch, O.; Hopt, U.T.; Von Dobschuetz, E. Evaluation of preservation solutions by ESR-spectroscopy: Superior effects of University of Wisconsin over Histidine-Tryptophan-Ketoglutarate in reducing renal reactive oxygen species. *Kidney Int.* **2007**, *71*, 875–881. [CrossRef]

66. Chen, Y.; Shi, J.; Xia, T.C.; Xu, R.; He, X.; Xia, Y. Preservation Solution for kidney transplantation: History, advances and mechanisms. *Cell Transplantation.* **2019**, *28*, 1472–1489. [CrossRef]

67. Ostroóżka-Cieslik, A. The Effect of Antioxidant Added to Preservation Solution on the Protection of Kidneys before Transplantation. *Int. J. Mol. Sci.* **2022**, *23*, 3141. [CrossRef] [PubMed]

68. Rampino, T.; Abelli, M.; Ticozzelli, E.; Gregorini, M.; Bosio, F.; Piotti, G.; Bedino, G.; Esposito, P.; Balenzano, C.T.; Geraci, P.; et al. Non-heart-beating-donor transplant: The first experience in Italy. *G Ital. Nefrol.* **2010**, *27*, 56–68.

69. Treska, V.; Kobr, J.; Hasman, D.; Racek, J.; Trefil, L.; Reischig, T.; Hes, O.; Kuntscher, V.; Molacek, J.; Treska, I. Ischemia-reperfusion injury in kidney transplantation from non-heart-beating donor–do antioxidants or antiinflammatory drugs play any role? *Bratisl. Lek. Listy* **2009**, *110*, 133–136. [PubMed]
70. Catena, F.; Coccolini, F.; Montori, G.; Vallicelli, C.; Amaduzzi, A.; Ercolani, G.; Ravaioli, M.; Del Gaudio, M.; Schiavina, R.; Brunocilla, E.; et al. Kidney Preservation: Review of Present and Future Perspective. *Transplant. Proc.* **2013**, *45*, 3170–3177. [CrossRef]
71. Mrakic-Sposta, S.; Gussoni, M.; Montorsi, M.; Porcelli, S.; Vezzoli, A. Assessment of a standardized ROS production profile in humans by electron paramagnetic resonance. *Oxid. Med. Cell. Longev.* **2012**, *2012*, 973927. [CrossRef]
72. Mrakic-Sposta, S.; Gussoni, M.; Porcelli, S.; Pugliese, L.; Pavei, G.; Bellistri, G.; Montorsi, M.; Tacchini, P.; Vezzoli, A. Training effects on ROS production determined by electron paramagnetic resonance in master swimmers. *Oxid. Med. Cell. Longev.* **2015**, *2015*, 804794. [CrossRef]
73. Mrakic Sposta, S.; Montorsi, M.; Porcelli, S.; Marzorati, M.; Healey, B.; Dellanoce, C.; Vezzoli, A. Effects of Prolonged Exposure to Hypobaric Hypoxia on Oxidative Stress: Overwintering in Antarctic Concordia Station. *Oxid. Med. Cell. Longev.* **2022**, *2022*, 4430032. [CrossRef] [PubMed]
74. Dikalov, S.I.; Polienko, Y.F.; Kirilyuk, I. Electron paramagnetic resonance measurements of reactive oxygen species by cyclic hydroxylamine spin probes. *Antioxid. Redox. Signal.* **2018**, *28*, 1433–1443. [CrossRef] [PubMed]
75. Liu, J.; Roussel, C.; Lagger, G.; Tacchini, P.; Girault, H.H. Antioxidant sensors based on DNA-modified electrodes. *Anal. Chem.* **2005**, *77*, 7687–7694. [CrossRef] [PubMed]
76. Liu, J.; Su, B.; Lagger, G.; Tacchini, P.; Girault, H.H. Antioxidant redox sensors based on DNA modified carbon screen-printed electrodes. *Anal. Chem.* **2006**, *78*, 6879–6884. [CrossRef]
77. Beckonert, O.; Keun, H.C.; Ebbels, T.M.D.; Bundy, J.; Holmes, E.; Lindon, J.C.; Nicholson, J.K. Metabolic profiling, metabolomic and metabonomic procedures for NMR spectroscopy of urine, plasma, serum and tissue extracts. *Nat. Protoc.* **2007**, *2*, 2692–2703. [CrossRef]
78. Bathini, V.; McGregor, T.; McAlister, V.C.; Luke, P.P.; Sener, A. Renal perfusion pump vs cold storage for donation after cardiac death kidneys: A systematic review. *J. Urol.* **2013**, *189*, 2214–2220. [CrossRef]

International Journal of
Molecular Sciences

Article

Microplastics Exacerbate Cadmium-Induced Kidney Injury by Enhancing Oxidative Stress, Autophagy, Apoptosis, and Fibrosis

Hui Zou [†], Yan Chen [†], Huayi Qu, Jian Sun, Tao Wang, Yonggang Ma, Yan Yuan, Jianchun Bian [iD] and Zongping Liu *[iD]

Jiangsu Co-Innovation Center for Prevention and Control of Important Animal Infectious Diseases and Zoonoses, College of Veterinary Medicine, Yangzhou University, Yangzhou 225009, China
* Correspondence: liuzongping@yzu.edu.cn
† These authors contributed equally to the work.

Abstract: Cadmium (Cd) is a potential pathogenic factor in the urinary system that is associated with various kidney diseases. Microplastics (MPs), comprising of plastic particles less than 5 mm in diameter, are a major carrier of contaminants. We applied 10 mg/L particle 5 μm MPs and 50 mg/L CdCl2 in water for three months in vivo assay to assess the damaging effects of MPs and Cd exposure on the kidney. In vivo tests showed that MPs exacerbated Cd-induced kidney injury. In addition, the involvement of oxidative stress, autophagy, apoptosis, and fibrosis in the damaging effects of MPs and Cd on mouse kidneys were investigated. The results showed that MPs aggravated Cd-induced kidney injury by enhancing oxidative stress, autophagy, apoptosis, and fibrosis. These findings provide new insights into the toxic effects of MPs on the mouse kidney.

Keywords: microplastic; cadmium; kidney; oxidative stress; autophagy

check for **updates**

Citation: Zou, H.; Chen, Y.; Qu, H.; Sun, J.; Wang, T.; Ma, Y.; Yuan, Y.; Bian, J.; Liu, Z. Microplastics Exacerbate Cadmium-Induced Kidney Injury by Enhancing Oxidative Stress, Autophagy, Apoptosis, and Fibrosis. *Int. J. Mol. Sci.* **2022**, 23, 14411. https://doi.org/10.3390/ijms232214411

Academic Editors: Rossana Morabito and Alessia Remigante

Received: 6 October 2022
Accepted: 17 November 2022
Published: 19 November 2022

Publisher's Note: MDPI stays neutral with regard to jurisdictional claims in published maps and institutional affiliations.

1. Introduction

Cadmium (Cd) is a highly toxic heavy metal in the environment. Its harm to human health mainly comes from environmental pollution caused by industrial and agricultural production. Cadmium has a long biological half-life and low excretion rate in the human body [1]. The kidney is one of the most important accumulation sites and target organs of Cd. Chronic Cd exposure causes renal damage, and this process is generally considered irreversible [2,3]. There is no effective treatment available, and the number of deaths from kidney damage caused by Cd exposure is reported to be increasing every year [4,5]. Therefore, it is necessary to clarify the relationship between Cd in the environment and kidney damage.

Microplastics (MPs) are plastic particles with a diameter of less than 5 mm [6] and are major pollutants. Due to plastic's debris durability, these particles are ubiquitous in the terrestrial and marine environment [7]. NPs and MPs were detected in global surface water, continental waters, soil, atmosphere, and even polar regions [8]. Microplastics might cause unpredictable adverse effects on human health through food chain transport [8,9]. Microplastics have been reported to accumulate in the intestine, liver, and kidneys of mice [10,11]. The kidney can excrete ultrafine particles, while the deposition and accumulation of MPs in the kidney will cause damage, including physical damage and an immune response [12–14]. A study showed that microplastics (4 μm in diameter) could be found in the intercellular space and renal tubule using transmission electron microscopy, and the number of 4 μm particles entering the tissue was lower than that of 600 nm particles [15]. In that study, the organ indices and serum biochemical parameters of the kidney varied more significantly than those of other organs, such as the liver and heart [15]. Therefore, the kidney is an important target organ to study the effects of MPs on mammals (unlike aquatic animals). However, the toxic effects of MPs on mammalian kidneys remain unclear and require further study.

Furthermore, MPs in the environment can serve as vectors for organic and metallic contaminants owing to their remarkable binding capacity [16,17]. According to a report on the soil contamination status in China issued in 2014, Cd ranks first among the metallic contaminants in the farmland exceeding Chinese soil environmental quality standards [18]. Recent studies have confirmed the occurrence of various MPs in farmlands [19], as well as in freshwater [20]. Microplastics and the heavy metals adhering to their surfaces can be taken up by organisms, leading to heavy metal deposition. Microplastic ingestion provides a pathway for the transfer of biological contaminants such as heavy metals into organisms through ingestion [21]. Experiments have shown that exposure of fish to MPs and Cd, either singly or in combination, for 30 days resulted in an increase in MP concentrations that reduced Cd accumulation in the fish. Co-exposure led to severe oxidative stress, which could stimulate innate immunity [22]. Although there have been many studies on the effects of MPs and heavy metals on aquatic animals, there have been few studies on the combined effects of MPs and Cd on mammals.

Studies have shown that co-treatment with MPs with Cd enhances early growth, oxidative stress, and apoptosis in zebrafish, resulting in tissue damage [23]. In addition, MP-induced myocardial dysplasia in birds is mainly attributed to the endoplasmic reticulum stress-mediated autophagic pathway [24]. These results suggest that the effects of exposure to plastic particles and Cd contamination are complex, with toxic effects varying among organisms. Furthermore, studies on the exposure of MPs and Cd in mammals and their potential toxicity mechanisms are still limited. Therefore, further study is needed to determine the toxic effects of co-exposure of MPs and Cd on mammals.

2. Results

2.1. Microplastics Aggravate Cadmium-Induced Kidney Damage

Mice were treated with Cd (50 mg/L) and/or 5 µm MPs (10 mg/L) for 90 days. During the whole study, no abnormal behavior or symptoms were observed in any group of mice. The food intake and body weight of the mice were monitored daily, and the organ coefficients decreased more significantly in the mice in the Cd and MP co-treatment group than in either single treatment group (Figure 1A,B). To investigate the role of Cd in kidney damage with MPs, we assessed the histological changes in the kidneys using H&E staining (Figure 1C). Compared with that in the control group, Cd resulted in kidney tissue with more visible dilatation of the renal capsule lumen (black arrow), necrosis and detachment of the tubular epithelium (blue arrow), cytoplasmic sparing and light staining, and increased erythrocyte (red arrow) and monocyte infiltration (green arrow) around a few vessels. These findings confirmed that the model of Cd-induced nephrotoxicity had been successfully established. In addition, co-treatment with MPs and Cd significantly increased the Cd-induced tubular injury, with occasional dilatation of the renal peritoneal lumen (yellow arrow). In addition, we examined the renal ultrastructure using transmission electron microscopy. As shown in Figure 1D,E, in the control group, the nucleus was intact and the mitochondrial structures and mitochondrial cristae were clearly visible. However, the Cd group showed significant ultrastructural changes in terms of nuclear depression and deformation (red arrow). In addition, Cd also induced mitochondrial damage, as evidenced by significant vacuolization (blue arrow) and cristae disruption (black arrow). The nuclear and mitochondrial damage were increased in the co-treatment group. In addition, more lipid droplets (green arrow) were observed in the co-treatment group, and it is suspected that lipid accumulation occurred, which could be further investigated. In addition, we assessed the accumulation and localization of fluorescent polystyrene microspheres in kidney tissue. The results showed that the microspheres accumulated in the mouse renal tubular epithelium, suggesting that microplastics with a 5 µm particle size can enter the kidney (Figure 1F). Taken together, these results confirmed that MPs exacerbated Cd-induced kidney injury in vivo.

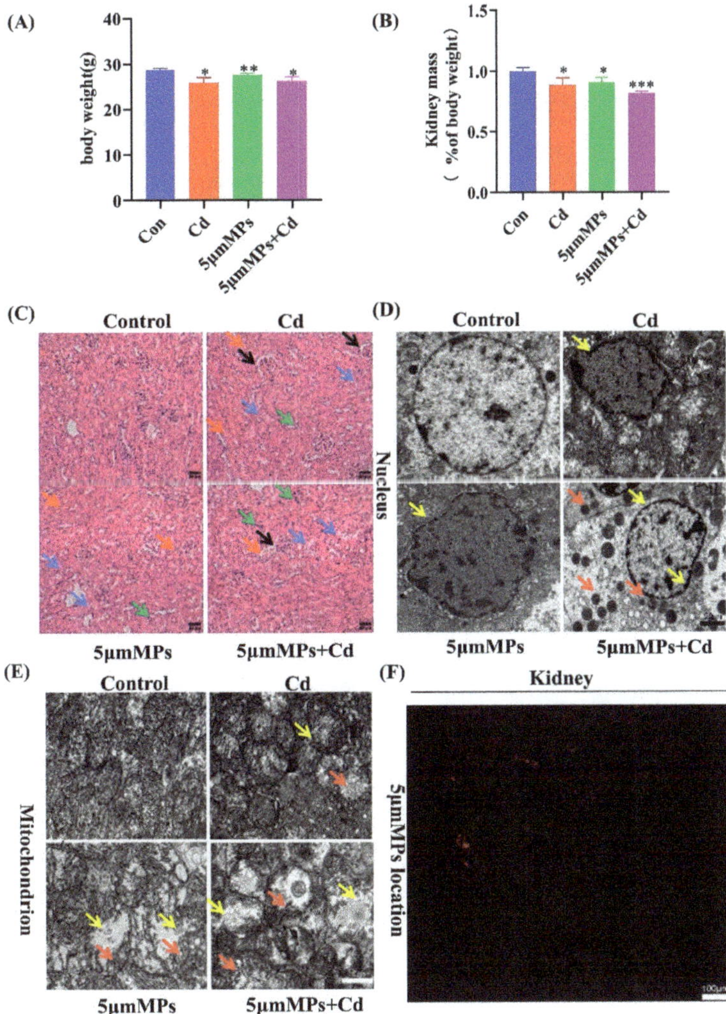

Figure 1. Microplastics (MPs) aggravate cadmium (Cd)-induced kidney injury. Mice were treated with Cd (50 mg/L) and/or 5 μm MPs (10 mg/L) for 90 days. (**A**) Body weight (g), weight of the animal before dissection. (**B**) Kidney weight (% body weight) (**C**) Hematoxylin and eosin (H&E) staining of kidney samples from the mice in each group. a: Glomerulus, b: Renal tubule. A period of 90 days of Cd exposure induced dilatation of renal capsule (black arrow), necrosis and shedding of the renal tubular epithelium (blue arrow), and increased red blood cell (red arrow) and monocyte infiltration (green arrow). Treatment with MPs aggravated Cd-induced kidney injury in mice. Scale bars: 50 μm. Representative electron microscope images showing the nuclei (**D**) and mitochondria (**E**) in the kidneys of mice in each group. Damage to nuclear, as evidenced by nuclear depression and deformation (yellow arrow) and lipid droplets (red arrow). Damage to mitochondria, as evidenced by significant vacuolization (yellow arrow) and cristae disruption (red arrow). Scale bars: 20 μm and 10 μm. (**F**) Bioaccumulation of MPs in mouse kidneys was determined using laser scanning confocal microscopy (5 μm). Red fluorescence indicates the location of MPs. Scale bars: 100 μm. Data are presented as means ± SD from three independent experiments. * $p < 0.05$, ** $p < 0.01$, *** $p < 0.001$ compared to the control group.

2.2. Microplastics Exacerbate Cd-Induced Oxidative Stress in Mouse Serum and Kidneys

Mice were treated with Cd (50 mg/L) and/or 5 µm MPs (10 mg/L) for 90 days. To explore whether MPs can enhance Cd-induced systemic and renal oxidative stress in vivo, we assessed the oxidative stress parameters in the serum and kidney in the mouse model. The results showed that Cd exposure significantly elevated the serum MDA levels and decreased CAT, GSH, and SOD levels compared with those in the control group (Figure 2A–D). However, the co-treatment group had significantly higher MDA levels and no significant changes in CAT, GSH, and SOD levels compared with those in the Cd group. In addition, we examined the levels of oxidative stress-related proteins. Western blotting showed that Cd exposure significantly increased the levels of SOD2, HO-1, and Sirt3 in the kidney compared with those in the control group. Compared with those in the Cd group, SOD2 and Sirt3 levels were significantly elevated in the co-treatment group, while the HO-1 levels did not change significantly (Figure 2E–H). These results confirmed that co-treatment with MPs and Cd increased the levels of systemic and renal oxidative stress in mice in vivo.

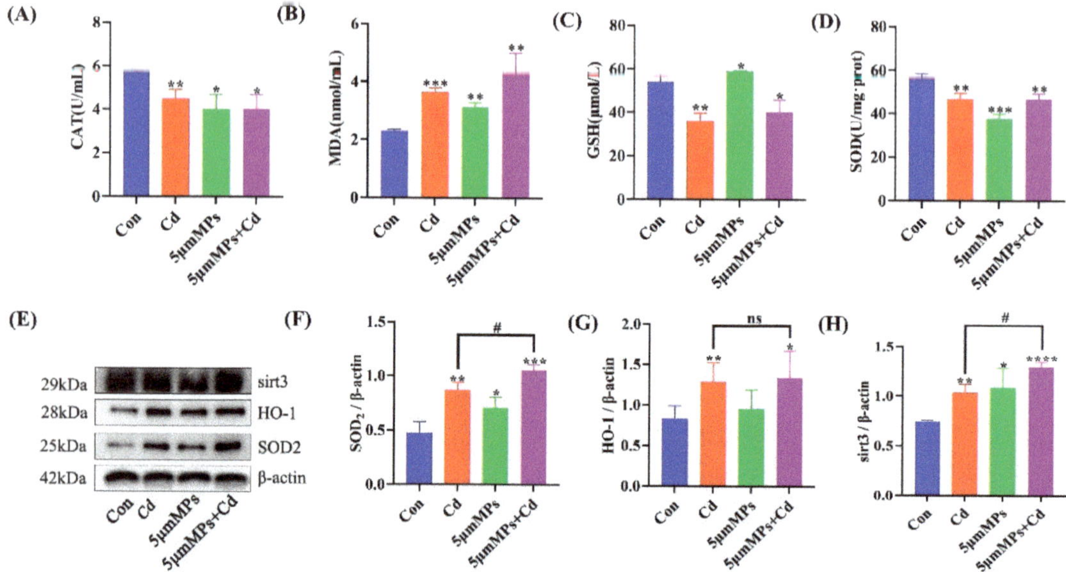

Figure 2. Effect of MPs and Cd on oxidative stress in mouse kidneys. Mice were treated with Cd (50 mg/L) and/or 5 µm MPs (10 mg/L) for 90 days. (**A–D**): The levels of catalase (CAT) activity, glutathione (GSH), and malondialdehyde (MDA were evaluated in the mouse serum. (**E–H**): the levels of sirtuin 3 (Sirt3), heme oxygenase 1 (HO-1), and superoxide dismutase 2 (SOD2) were detected using western blotting. (**E**): Representative western blot image; (**F–H**): Quantitative analysis. Data are presented as means ± SD from three independent experiments. * $p < 0.05$, ** $p < 0.01$, *** $p < 0.001$, **** $p < 0.0001$ compared to the control group; # $p < 0.05$ compared to the Cd group.

2.3. Effect of Co-Treatment with MPs and Cd on Autophagy in Mouse Kidneys

Mice were treated with Cd (50 mg/L) and/or 5 µm MPs (10 mg/L) for 90 days. To investigate the effect of MPs and Cd on the level of autophagy in the kidney, we detected autophagy-related proteins by western blotting and observed the number of autophagosomes using transmission electron microscope. Western blotting showed that, compared with those in the Cd group, the levels of the autophagy marker LC3 and the early autophagy proteins ATG5, Beclin-1, and ATG7 were significantly higher in the co-treatment group (Figure 3A–E). In addition, the number of autophagosomes in the co-treatment group was higher than that in the Cd group (Figure 3F). These results suggest that the co-treatment with MPs with Cd increased the level of autophagy in mouse kidneys in vivo.

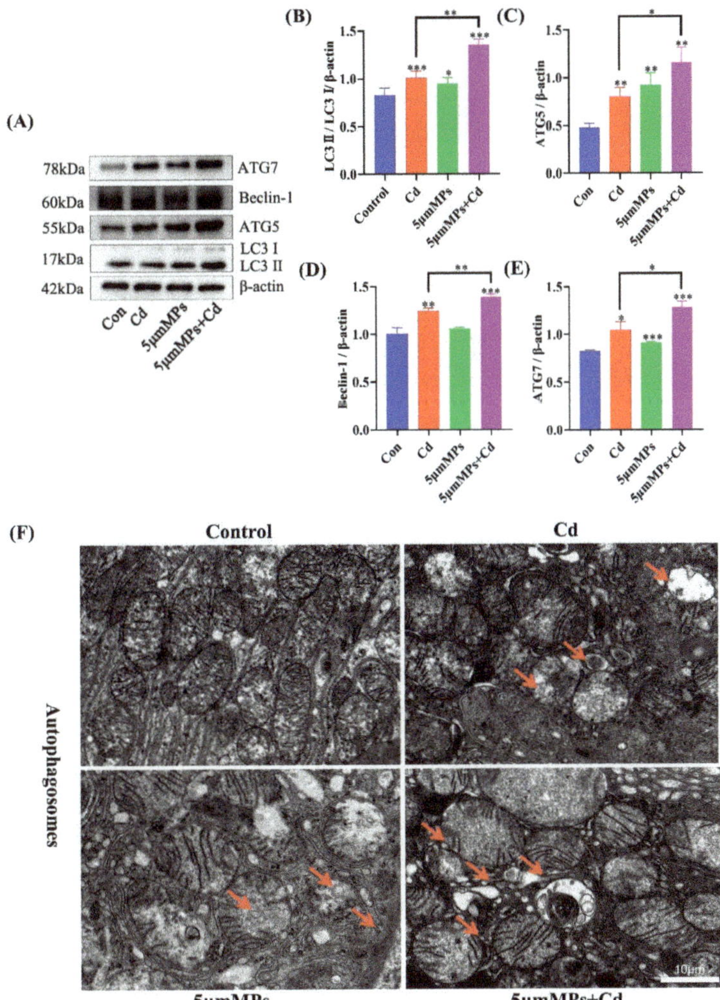

Figure 3. Effect of MPs and Cd on autophagy in mouse kidneys. Mice were treated with Cd (50 mg/L) and/or 5 μm MPs (10 mg/L) for 90 days. (**A–E**): The levels of LC3, ATG5, Beclin-1, and ATG7 were detected by Western blot. (**A**): representative Western blot image; (**B–E**): quantitative analysis. (**F**): Representative electron microscope images show the autophagosomes (red arrow) in the kidneys of mice in each group. Scale bars: 1 μm. Data are presented as means ± SD from three independent experiments. * $p < 0.05$, ** $p < 0.01$, *** $p < 0.001$ compared to the control group.

2.4. Effect of Co-Treatment with MPs and Cd on Apoptosis in Mouse Kidneys

Mice were treated with Cd (50 mg/L) and/or 5 μm MPs (10 mg/L) for 90 days. To investigate the effect of MPs and Cd on the level of apoptosis in the kidney, we detected apoptosis-related proteins by IHC and western blotting. The IHC results showed that compared with those in the Cd group, the areas that were positive for Bax, Cytc, and Caspase-3 were significantly increased in the co-treatment group (Figure 4A–D). Western blotting showed that the Bax/Bcl-2 ratio was increased in the co-treatment group compared with that in the Cd group, and the expression levels of their downstream regulatory proteins, Caspase-3 and Cleaved Caspase-3, were both significantly increased, whereas the levels of Cleaved Caspase-9 did not change significantly (Figure 4E–I), which was consistent with the IHC results. These results confirmed that co-treatment with MPs and Cd increased the level of apoptosis in mouse kidneys in vivo.

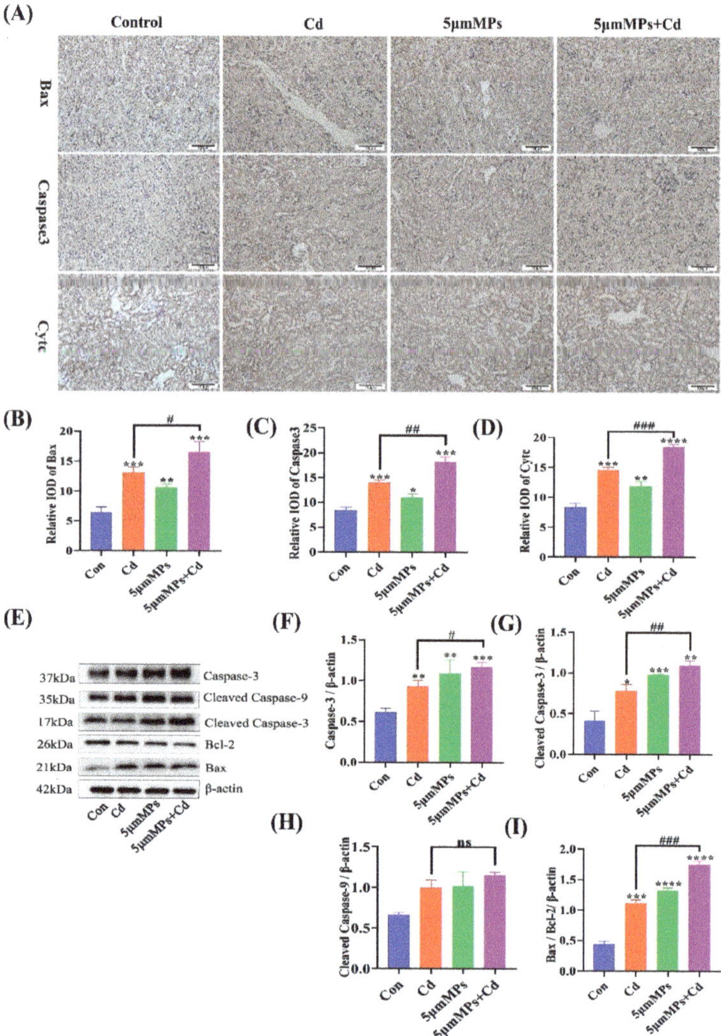

Figure 4. Effect of MPs and Cd on apoptosis in mouse kidneys. Mice were treated with Cd (50 mg/L) and/or 5 μm MPs (10 mg/L) for 90 days. (**A–D**): BCL2 associated X protein (Bax), Caspase-3, and cytochrome C (Cytc) levels in different groups assessed using an immunohistochemical (IHC) assay. Scale bars: 50 μm. (**C,D**): Quantitative analysis. (**E–I**): The levels of Bax, B-cell CLL/lymphoma 2 (Bcl-2), Caspase-3, Cleaved Caspase-3, and Cleaved Caspase-9 were detected by western blotting. (**E**): Representative western blot image; (**F–I**): Quantitative analysis. Data are presented as means ± SD from three independent experiments. * $p < 0.05$, ** $p < 0.01$, *** $p < 0.001$, **** $p < 0.0001$ compared to the control group; # $p < 0.05$, ## $p < 0.01$, ### $p < 0.001$ compared to the Cd group.

2.5. Effect of Co-Treatment with MPs and Cd on Kidney Fibrosis in Mice

Mice were treated with Cd (50 mg/L) and/or 5 μm MPs (10 mg/L) for 90 days. To investigate the effect of MPs and Cd on the extent and severity of renal fibrosis, we assessed the level of renal fibrosis using Sirius Red and Masson staining. Both staining results showed a significant increase in collagen fibers in the kidney tissue of mice in the MPs and Cd co-treatment group compared with that in the Cd group (Figure 5A–C). In addition, we examined the levels of the fibrosis marker proteins α-SMA, TGF-β1, and COL4A1 (Figure 5D–G). Western blotting showed that the levels of α-SMA and COL4A1 were significantly higher in the co-treatment group than in the Cd group, while no significant differences were observed in the levels of TGF-β1. The above results indicated that MPs exacerbated Cd-induced renal fibrosis in vivo.

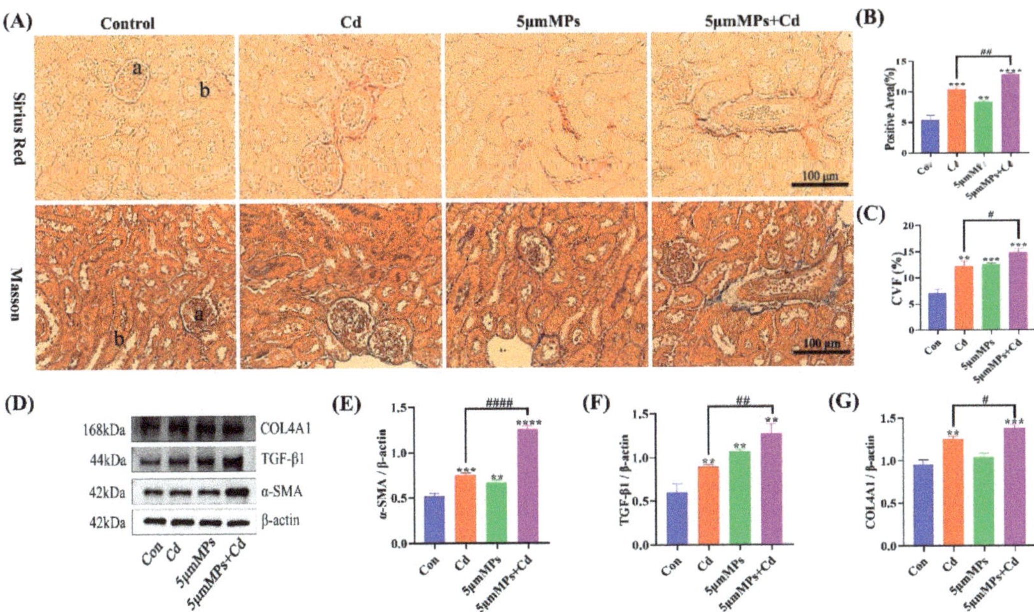

Figure 5. Effect of MPs and Cd on fibrosis in mouse kidneys. (**A–C**): Representative images of Sirius Red staining and Masson staining of the kidney after MP and/or Cd exposure. a: Glomerulus, b: Renal tubule. (**B,C**): Quantitative analysis. (**D–G**): The levels of collagen type IV alpha 1 chain (COL4A1), transforming growth factor beta 1 (TGF-β1), and alpha smooth muscle actin (α-SMA) were detected using western blotting. (**D**): Representative western blot image; (**E–G**): Quantitative analysis. Data are presented as means ± SD from three independent experiments. ** $p < 0.01$, *** $p < 0.001$, **** $p < 0.0001$ compared to the control group; # $p < 0.05$, ## $p < 0.01$, #### $p < 0.0001$ compared to the Cd group.

3. Discussion

Cd is a non-degradable heavy metal in the environment that is relevant to human health and industrial production. There is growing evidence that renal damage is a major feature of Cd-induced toxicity [25]. Cd causes nephrotoxicity through different mechanisms, such as oxidative stress [26], autophagy [27], apoptosis [28], and fibrosis [29]. Microplastics can attract other polluting particles from the surrounding environment and the combined effect can increase their contamination, posing a potential hazard to the ecological environment [30]. The toxicity of microplastic sorption depends on the nature of the sorbent, the particle size and the composition of the microplastic [31]. Considering the intrinsic link between MPs and heavy metal toxicity, this study aimed to evaluate nephrotoxic effects of MPs on Cd-induced nephrotoxicity.

The changes in body weight and organ coefficients of experimental animals can reflect the toxic effects of toxins on animal bodies and organs. H&E staining and transmission electron microscopy can reveal kidney damage at the ultrastructural level. fluorescent polystyrene microspheres allow observation of the accumulation and localization of MPs in kidney tissue. The results for the co-treatment group showed a highly significant decrease in kidney coefficients compared with those in the control group. H&E staining in the Cd group showed dilatation of the renal capsule lumen, necrosis and detachment of the tubular epithelium, and infiltration of mononuclear cells. Transmission electron microscopy showed nuclei degeneration, mitochondrial vacuolation, and cristae disruption. It was determined that 5 μm MPs accumulated in the mouse renal tubular epithelium and entered the renal tissue, which is consistent with previous findings [32]. These results suggested that the Cd-mediated kidney injury model was successfully established and that MPs co-treatment enhances its deleterious effects.

One of the most important causes of kidney injury is oxidative stress caused by excessive production of reactive oxygen species (ROS) and inhibition of the antioxidant system [33,34]. In a laboratory experiment, early juveniles of Symphysodon aequifasciatus (discus fish) were exposed to single and combined effect of polystyrene microplastic (0, 50, or 500 mg/L) and cadmium (0 or 50 mg/L). Co-exposure resulted in severe oxidative stress and could stimulate innate immunity; however, there was no effect on growth and survival rate of the juveniles [31]. Earlier studies showed that Cd promotes ROS production in renal cells through a variety of indirect mechanisms, such as disruption of electron flow in the mitochondrial respiratory chain, transition metal release, and inactivation of endogenous redox defense molecules through binding to sulfhydryl groups [35,36]. Elevated MDA levels have also been reported as an indicator of lipid peroxidation [37]. As protection against oxidant stress, the antioxidant enzymes SOD, CAT, and GSH are widely considered as powerful defense mechanisms against oxidative kidney damage [38]. In this study, Cd or MPs caused significant oxidative stress, as evidenced by increased MDA levels and decreased antioxidant enzyme activities (SOD and CAT) and GSH levels in serum. However, apart from increasing MDA levels, co-treatment with MPs and Cd did not further decrease SOD, CAT, and GSH activities, which might be a response to the excess ROS stimulated by them in cells [6]. Research has shown that SOD2 is the downstream mediator of Sirt3, protecting the nucleus and mitochondrial DNA, as well as other cellular macromolecules, from ROS-related damage through its attenuation of ROS levels [39,40]. The results of the present study showed that MPs significantly upregulated Sirt3 and SOD2 levels following Cd exposure. The different effects on Sirt3 levels might be related to different stages of the body's tolerance to oxidative stress. In conclusion, these data suggest that MPs play an important role in Cd-induced oxidative stress in mouse kidneys.

Autophagy is a dynamic process in which damaged macromolecules and organelles within cells are degraded and recirculated to synthesize new cellular components. Recent studies have begun to reveal the role of autophagy in progressive kidney disease and subsequent fibrosis. In this study, MPs increased the levels of Cd-mediated autophagy marker LC3-II and the autophagy early proteins ATG5, Beclin-1, and ATG7 and increased the levels of the renal fibrosis marker proteins α-SMA, TGF-β1, and COL4A1. This suggests that MPs play an important role in Cd-mediated renal autophagy and fibrosis.

Apoptosis is a controlled form of cell death characterized by cell shrinkage, chromosome condensation, and breakage [41]. The cell death receptor (exogenous) and mitochondrial (endogenous) pathways are the two main signaling pathways that trigger apoptotic cell death. It is reported that in mitochondrial pathway-mediated apoptosis, Caspase-9 can be activated to induce apoptosis by initiating the downstream Caspase-3 pathway. In addition, Bax and Bcl-2 were the first proteins identified to be involved in the positive and negative regulation of apoptosis. Bax induces apoptosis when it forms a homodimer; whereas, when Bax forms a heterodimer with Bcl-2, it activates the function of Bcl-2 to inhibit apoptosis. Both of these pathways contribute to oligomerization of pro-apoptotic proteins in mitochondria and induce the release of mitochondrial cytochrome c [42]. In this study, we found that MPs significantly upregulated Caspase-3, Cleaved Caspase-3, and Bax levels and downregulated Bcl-2 levels following Cd exposure. In addition, the IHC results confirmed that mitochondrial cytochrome c was released into the cytoplasm. In conclusion, these data suggest that MPs play an important role in Cd-induced apoptosis in mouse kidney cells.

Previous studies showed that oxidative stress affects the production and balance of autophagy, apoptosis, and fibrosis, thereby exacerbating tissue damage [43–45]. Therefore, further studies on autophagy, apoptosis, and fibrosis might confirm the combined toxicity of MPs and Cd. Autophagy degrades long-lived cytoplasmic proteins and organelles, provides substrates for energy metabolism, and restores amino acids, fatty acids, and nucleotides to meet the needs of cellular biosynthesis. In the present study, combined exposure to MPs and Cd resulted in elevated levels of autophagy. This might have occurred through excessive accumulation of ROS, which induces sustained autophagy in the mouse kidney [46]. However, further experimental studies are needed to confirm this. Recent studies have also begun to reveal the role of autophagy in progressive nephropathy and subsequent fibrosis. Livingston et al. (2016) first reported the pro-fibrotic role of autophagy in the mouse kidney. Studies showed that overexpression of WNT1-inducible signaling pathway protein 1 (WISP-1) increased LC3-II and Beclin-1 expression and exacerbated renal fibrosis in the unilateral ureteral obstruction (UUO) model and in TGF-β-treated renal tubular epithelial cells [47]. In this study, we hypothesized that there might be an association between sustained autophagy and exacerbation of fibrosis in the mouse kidney; however, further studies are needed. Apoptosis is a tightly regulated intracellular program in which cells are destined to die and is activated to degrade cellular DNA, nuclear proteins, and cytoplasmic proteins [48]. In this study, combined exposure to MPs and Cd induced excessive apoptosis, which might lead to normal tissue cell death, possibly inhibiting tissue repair and thus exacerbating tissue damage [49,50]. In this model, excessive autophagy might lead to apoptosis. The above results suggest that oxidative stress induced by combined exposure to MPs and Cd affects the levels of autophagy, apoptosis, and fibrosis, which leads to further kidney damage in mice. However, further experimental studies are needed to confirm this.

4. Materials and Methods

4.1. Chemicals and Reagents

Cadmium chloride ($CdCl_2$) was purchased from Sigma Aldrich (St. Louis, MO, USA). Polystyrene fluorescent microspheres (5 μm; 7-1-0500) and polystyrene microspheres (5 μm; 6-1-0050) were purchased from BaseLine Chromatographic Technology Development Center (Tianjin, China). All antioxidant enzyme detection kits were obtained from Nanjing Jiancheng Bioengineering Institute (Nanjing, China). The following primary antibodies were used: Anti-BCL2 associated X protein (Bax) (T40051), anti-B-cell CLL/lymphoma 2 (Bcl-2) (T40056), anti-autophagy related 7 (ATG7) (T57051M), anti-autophagy related 5 (ATG5) (T55766M), anti-Beclin-1 (T55092), anti-Caspase-3 (T40044), and anti-Caspase-9 (T40046) were obtained from Abmart (Shanghai, China). Anti-microtubule associated protein 1 light chain 3 beta (LC3B) (#83506), anti-alpha smooth muscle actin (α-SMA) (#19245), anti-transforming growth factor beta 1 (TGF-β1) (#3711), anti-collagen type IV alpha 1 chain (COL4A1) (#50273), and anti-β-actin (#4970) antibodies were obtained from Cell Signaling Technology (Danvers, MA, USA). Anti-sirtuin 3 (Sirt3) (sc-365175), anti-heme oxygenase 1 (HO-1) (sc-136960), and anti-superoxide dismutase 2 (SOD2) (sc-137254) antibodies were obtained from Santa Cruz Biotechnology (Dallas, TX, USA). All secondary antibodies are from Jackson ImmunoResearch (Philadelphia, PA, USA). Other chemicals and reagents were of analytical grade and were purchased locally.

4.2. Animals and Treatments

A total of 32 6-week-old male C57BL/6 mice were collected from the experimental animal center of Jiangsu University (Zhenjiang, China). The mice were housed in a well-controlled temperature environment (23 ± 2 °C) and subjected to a 12-h light-dark cycle, with water and food provided ad libitum. After one week of adaption to these conditions, the 32 mice were randomly divided into four groups ($n = 8$ per group): (1) The control group (given purified water as drinking water); (2) the Cd group (given purified water containing 50 mg/L Cd); (3) the MP group (given purified water containing 10 mg/L MPs of 5 μm particle size); and (4) the MP and Cd co-treatment group (given purified water containing 50 mg/L Cd and 10 mg/L of 5 μm particle size MPs). Dosage was determined according to previous studies [10,51,52]. All groups drank water ad libitum and after 3 months of contamination, all mice were weighed and anesthetized with 2% sodium pentobarbital and then sacrificed by cervical decapitation. Blood samples were taken from the ventral side of the aorta, and serum was obtained by centrifuging the samples at $2000\times g$ for 15 min. The kidneys were immediately removed, and the kidney cortex was isolated. The tissues were fixed in 2.5% glutaraldehyde or 4% paraformaldehyde (PFA) or stored in a -80 °C freezer until further analysis.

4.3. Hematoxylin and Eosin (H&E) Staining and Histological Analysis

The kidney tissues collected from mice were fixed in 4% PFA at 4 °C for 24 h and then cut into 3 mm-thick sections. The samples were dehydrated in graded solutions of ethanol, soaked in xylene, embedded in paraffin, and sectioned to 4 μm thickness. The obtained tissue sections were assembled on slides and stained with hematoxylin and eosin (H&E) for histological analysis. All samples were observed and photographed under a Leica light microscope equipped with a digital camera (DMI3000B, Leica, Wetzlar, Germany).

4.4. Transmission Electron Microscopy

For transmission electron microscopy observation, small (1 mm^3) pieces of kidney tissue from each group were fixed overnight at 4 °C in pre-chilled 2.5% glutaraldehyde. The samples were washed three times in phosphate-buffered saline (PBS) and then fixed using 1% osmium tetroxide. Thereafter, the samples were washed three times in PBS, dehydrated in an ethanol graded solution, and embedded in epoxy resin. Ultrathin sections were obtained using an ultramicrotome (EM UC7, Leica), which were stained with uranyl acetate and lead citrate. The sections were then examined using a transmission electron microscope (Tecnai 12, Philips, Holland).

4.5. Localization of Fluorescent Polystyrene Microspheres in Renal Tissue

To study the accumulation and localization of MPs in the kidney, we fed mice with 1 mg/mL fluorescent polystyrene microspheres for one month (0.2 mL/day via gavage). The mice were sacrificed, fresh kidney tissues were excised, and optimal cutting temperature (OCT)-embedded/fixed tissue was obtained and OCT-embedded after dehydration in a 15–30% gradient sucrose solution. The sections were stored at low temperature after completion. Sections were observed under a fluorescence microscope (TCS SP8 STED, Leica) to assess the aggregation and localization of fluorescent MPs.

4.6. Oxidative Stress Assessment

Fresh mouse kidney tissue samples were collected and stored at -80 °C. Catalase (CAT) activity and glutathione (GSH), malondialdehyde (MDA), and superoxide dismutase (SOD) levels were measured using commercial kits according to the manufacturer's instructions (Nanjing Jiancheng Bioengineering Institute).

4.7. Western Blotting

Kidney tissue was ground and centrifuged to extract the precipitate and the tissue was lysed on ice for 15 min using Radioimmunoprecipitation assay (RIPA) lysis solution containing protease inhibitors. After ultrasonic lysis on ice for 15 min, the supernatant was collected by centrifugation at $12,000\times g$ for 10 min at 4 °C and the protein concentration was quantified using the bicinchoninic acid (BCA) method. Equal amounts of proteins were separated using sodium dodecyl sulfate polyacrylamide gel electrophoresis (SDS-PAGE) and then the proteins were electrotransferred to polyvinylidene difluoride (PVDF) membranes. After blocking in 5% skim milk powder for 2 h at room temperature, the membranes were incubated with primary antibodies overnight at 4 °C. The next day, the membranes were washed three times with Tris-buffered saline-Tween 20 (TBST) on a shaker for 10 min each time and then incubated for 2 h at room temperature with the appropriate horseradish peroxidase-conjugated secondary antibody. After incubation, the membranes were washed three times using TBST for 10 min each time on a shaker and then detected using enhanced chemiluminescence. Immunoreactive protein levels were analyzed using Image Lab software (National Institutes of Health, Bethesda, MD, USA). The density of each band was normalized to its corresponding loading control (β-actin).

4.8. Immunohistochemical Analysis

Kidney tissue was fixed in 4% PFA and embedded in paraffin. The embedded tissue was cut into 2 μm sections. For immunohistochemical (IHC) staining, the sections were incubated with anti-Bax antibodies (mouse, 1:250, Abcam, Cambridge, MA, USA); anti-cytochrome C (Cytc) antibodies (mouse, 1:250, Abcam), and anti-Caspase-3 antibodies (mouse, 1:100, Abcam). Slides were scanned and imaged using an SCN400 slide scanner (Leica). The quantitative analysis of the images was measured by Image-Pro Plus 6.0 software.

4.9. Sirius Red Staining

Kidney tissue was fixed in 4% PFA and embedded in paraffin. The embedded tissue was cut into 2 μm sections and stained with Sirius Red. Sirius Red staining kits were purchased from Abcam. After the sections were dewaxed, stained, dehydrated, and made transparent, they were sealed for observation. Sirius Red staining was used to determine collagen deposition as a measure of the extent of liver fibrosis. The slides were scanned and imaged using a SCN400 slide scanner (Leica). The quantitative analysis of the images was measured by Image-Pro Plus 6.0 software.

4.10. Masson Staining

A 4% PFA-fixed kidney tissue block was cut to 0.5 cm^3 size for paraffin embedding, followed by dewaxing and rinsing in distilled water. Tissues were stained with hematoxylin for 5–10 min and then washed with distilled water. The tissue was then stained with Lichon Red acidic magenta staining solution for 5–10 min, followed by a period of immersion in 2% glacial acetic acid aqueous solution and 1% phosphomolybdic acid aqueous solution for 3–5 min. After drying, the stained tissue was incubated with Aniline Blue solution for 5 min. After further incubation in 0.2% glacial acetic acid for 1 min, the pieces were quickly washed and dehydrated with 95% ethanol and anhydrous ethanol. Finally, the pieces were sectioned, infiltrated with xylene, dried, and sealed with neutral glue. The sections were then viewed under a microscope (Leica). The quantitative analysis of the images was measured by Image-Pro Plus 6.0 software.

4.11. Statistical Analyses

The results are expressed as the mean ± standard deviation (SD) of at least three independent experiments. Significance was calculated by one-way analysis of variance (ANOVA) using SPSS 26.0 software (IBM Corp., Armonk, NY, USA), followed by Tukey's test. The results were considered statistically significant at a threshold of $p < 0.05$.

5. Conclusions

We revealed the localization and toxicological effects of MPs in a Cd-mediated mouse kidney injury model. To investigate the damaging effects of MPs and cadmium on the kidney, we applied 10 mg/L particle 5 μm MPs and 50 mg/L CdCl$_2$ singly and in combination in an in vivo assay. The results showed that polystyrene particles of 5 μm diameter could enter the systemic circulation and accumulate in the kidneys of mice where they induced severe biological responses. The kidneys of mice exposed to MPs and Cd exhibited a state of oxidative stress, autophagy, apoptosis, and fibrosis, leading to kidney damage, alteration of kidney tissue structure, and ultimately, nephrotoxicity. Hence, these findings provide further evidence for the threat of MPs and their adsorbed heavy metals.

Author Contributions: Conceptualization, Z.L., H.Z. and Y.C.; funding acquisition, Z.L., H.Z., Y.Y. and J.B.; resources, Y.M., T.W. and J.S.; investigation and formal analysis, Y.C. and H.Q.; writing—original draft, Y.C.; writing—review and editing, Z.L. All authors have read and agreed to the published version of the manuscript.

Funding: This work was supported by the National Natural Science Foundation of China (grant numbers 31702305, 31872533, 31802260 and 31772808]) and a project Funded by Priority Academic Program Development of Jiangsu Higher Education Institutions (PAPD).

Institutional Review Board Statement: All experiments and procedures using animals in this study were approved by the Animal Care and Use Committee of Yangzhou University (approval ID: SYXK (Su) 2017-0044) and were carried out in strict accordance with the National Research Council's Guide for the Care and Use of Laboratory Animals.

Informed Consent Statement: Not applicable.

Data Availability Statement: The data presented in this study are available on request from the corresponding author Z.L.

Acknowledgments: The authors sincerely acknowledge Yangzhou University, China.

Conflicts of Interest: The authors declare no conflict of interest.

References

1. Jones, M.M.; Cherian, M.G. The search for chelate antagonists for chronic cadmium intoxication. *Toxicology* **1990**, *62*, 1–25. [CrossRef]
2. Luo, T.; Liu, G.; Long, M.; Yang, J.; Song, R.; Wang, Y.; Yuan, Y.; Bian, J.; Liu, X.; Gu, J.; et al. Treatment of cadmium-induced renal oxidative damage in rats by administration of alpha-lipoic acid. *Environ. Sci. Pollut. Res.* **2017**, *24*, 1832–1844. [CrossRef] [PubMed]
3. Momeni, H.R.; Eskandari, N. Curcumin protects the testis against cadmium-induced histopathological damages and oxidative stress in mice. *Hum. Exp. Toxicol.* **2019**, *39*, 653–661. [CrossRef] [PubMed]
4. Nogawa, K.; Kido, T. Biological monitoring of cadmium exposure in itai-itai disease epidemiology. *Int. Arch. Occup. Environ. Health* **1993**, *65*, S43–S46. [CrossRef] [PubMed]
5. Teruhiko, K.; Koji, N. Dose-response relationship between total cadmium intake and β2-microglobulinuria using logistic regression analysis. *Toxicol. Lett.* **1993**, *69*, 113–120. [CrossRef]
6. Lei, L.; Wu, S.; Lu, S.; Liu, M.; Song, Y.; Fu, Z.; Shi, H.; Raley-Susman, K.M.; He, D. Microplastic particles cause intestinal damage and other adverse effects in zebrafish Danio rerio and nematode Caenorhabditis elegans. *Sci. Total Environ.* **2018**, *619–620*, 1–8. [CrossRef]
7. Waring, R.H.; Harris, R.M.; Mitchell, S.C. Plastic contamination of the food chain: A threat to human health? *Maturitas* **2018**, *115*, 64–68. [CrossRef]
8. González-Pleiter, M.; Velázquez, D.; Edo, C.; Carretero, O.; Gago, J.; Barón-Sola, Á.; Hernández, L.E.; Yousef, I.; Quesada, A.; Leganés, F.; et al. Fibers spreading worldwide: Microplastics and other anthropogenic litter in an Arctic freshwater lake. *Sci. Total Environ.* **2020**, *722*, 137904. [CrossRef]
9. Fadare, O.O.; Wan, B.; Guo, L.-H.; Zhao, L. Microplastics from consumer plastic food containers: Are we consuming it? *Chemosphere* **2020**, *253*, 126787. [CrossRef]
10. Deng, Y.; Zhang, Y.; Lemos, B.; Ren, H. Tissue accumulation of microplastics in mice and biomarker responses suggest widespread health risks of exposure. *Sci. Rep.* **2017**, *7*, 46687. [CrossRef]
11. Yang, Y.F.; Chen, C.Y.; Lu, T.H.; Liao, C.M. Toxicity-based toxicokinetic/toxicodynamic assessment for bioaccumulation of polystyrene microplastics in mice. *J. Hazard. Mater.* **2019**, *366*, 703–713. [CrossRef] [PubMed]
12. De Matteis, V. Exposure to Inorganic Nanoparticles: Routes of Entry, Immune Response, Biodistribution and In Vitro/In Vivo Toxicity Evaluation. *Toxics* **2017**, *5*, 29. [CrossRef] [PubMed]
13. Li, Z.; Feng, C.; Wu, Y.; Guo, X. Impacts of nanoplastics on bivalve: Fluorescence tracing of organ accumulation, oxidative stress and damage. *J. Hazard. Mater.* **2020**, *392*, 122418. [CrossRef] [PubMed]

14. Wardoyo, A.Y.P.; Juswono, U.P.; Noor, J.A.E. Varied dose exposures to ultrafine particles in the motorcycle smoke cause kidney cell damages in male mice. *Toxicol. Rep.* **2018**, *5*, 383–389. [CrossRef]

15. Meng, X.; Zhang, J.; Wang, W.; Gonzalez-Gil, G.; Vrouwenvelder, J.S.; Li, Z. Effects of nano- and microplastics on kidney: Physicochemical properties, bioaccumulation, oxidative stress and immunoreaction. *Chemosphere* **2022**, *288*, 132631. [CrossRef]

16. Hodson, M.E.; Duffus-Hodson, C.A.; Clark, A.; Prendergast-Miller, M.T.; Thorpe, K.L. Plastic Bag Derived-Microplastics as a Vector for Metal Exposure in Terrestrial Invertebrates. *Environ. Sci. Technol.* **2017**, *51*, 4714–4721. [CrossRef]

17. Zhang, G.S.; Liu, Y.F. The distribution of microplastics in soil aggregate fractions in southwestern China. *Sci. Total Environ.* **2018**, *642*, 12–20. [CrossRef]

18. Zhao, F.J.; Ma, Y.; Zhu, Y.G.; Tang, Z.; McGrath, S.P. Soil contamination in China: Current status and mitigation strategies. *Environ. Sci. Technol.* **2015**, *49*, 750–759. [CrossRef]

19. Liu, M.; Lu, S.; Song, Y.; Lei, L.; Hu, J.; Lv, W.; Zhou, W.; Cao, C.; Shi, H.; Yang, X.; et al. Microplastic and mesoplastic pollution in farmland soils in suburbs of Shanghai, China. *Environ. Pollut.* **2018**, *242*, 855–862. [CrossRef]

20. Turner, A.; Holmes, L.A. Adsorption of trace metals by microplastic pellets in fresh water. *Environ. Chem.* **2015**, *12*, 600–610. [CrossRef]

21. Wang, J.; Peng, J.; Tan, Z.; Gao, Y.; Zhan, Z.; Chen, Q.; Cai, L. Microplastics in the surface sediments from the Beijiang River littoral zone: Composition, abundance, surface textures and interaction with heavy metals. *Chemosphere* **2017**, *171*, 248–258. [CrossRef] [PubMed]

22. Wen, B.; Jin, S.R.; Chen, Z.Z.; Gao, J.Z.; Liu, Y.N.; Liu, J.H.; Feng, X.S. Single and combined effects of microplastics and cadmium on the cadmium accumulation, antioxidant defence and innate immunity of the discus fish (Symphysodon aequifasciatus). *Environ. Pollut.* **2018**, *243*, 462–471. [CrossRef]

23. Yang, H.; Zhu, Z.; Xie, Y.; Zheng, C.; Zhou, Z.; Zhu, T.; Zhang, Y. Comparison of the combined toxicity of polystyrene microplastics and different concentrations of cadmium in zebrafish. *Aquat. Toxicol.* **2022**, *250*, 106259. [CrossRef] [PubMed]

24. Zhang, Y.; Wang, D.; Yin, K.; Zhao, H.; Lu, H.; Meng, X.; Hou, L.; Li, J.; Xing, M. Endoplasmic reticulum stress-controlled autophagic pathway promotes polystyrene microplastics-induced myocardial dysplasia in birds. *Environ. Pollut.* **2022**, *311*, 119963. [CrossRef]

25. Tokumoto, M.; Lee, J.-Y.; Satoh, M. Transcription Factors and Downstream Genes in Cadmium Toxicity. *Biol. Pharm. Bull.* **2019**, *42*, 1083–1088. [CrossRef]

26. Zhuang, J.; Nie, G.; Yang, F.; Dai, X.; Cao, H.; Xing, C.; Hu, G.; Zhang, C. Cadmium induces cytotoxicity through oxidative stress-mediated apoptosis pathway in duck renal tubular epithelial cells. *Toxicol. In Vitro* **2019**, *61*, 104625. [CrossRef]

27. Liu, F.; Wang, X.Y.; Zhou, X.P.; Liu, Z.P.; Song, X.B.; Wang, Z.Y.; Wang, L. Cadmium disrupts autophagic flux by inhibiting cytosolic Ca(2+)-dependent autophagosome-lysosome fusion in primary rat proximal tubular cells. *Toxicology* **2017**, *383*, 13–23. [CrossRef]

28. Komoike, Y.; Inamura, H.; Matsuoka, M. Effects of salubrinal on cadmium-induced apoptosis in HK-2 human renal proximal tubular cells. *Arch. Toxicol.* **2012**, *86*, 37–44. [CrossRef]

29. Joardar, S.; Dewanjee, S.; Bhowmick, S.; Dua, T.K.; Das, S.; Saha, A.; De Feo, V. Rosmarinic Acid Attenuates Cadmium-Induced Nephrotoxicity via Inhibition of Oxidative Stress, Apoptosis, Inflammation and Fibrosis. *Int. J. Mol. Sci.* **2019**, *20*, 2027. [CrossRef] [PubMed]

30. Zhang, R.; Wang, M.; Chen, X.; Yang, C.; Wu, L. Combined toxicity of microplastics and cadmium on the zebrafish embryos (*Danio rerio*). *Sci. Total Environ.* **2020**, *743*, 140638. [CrossRef]

31. Naqash, N.; Prakash, S.; Kapoor, D.; Singh, R. Interaction of freshwater microplastics with biota and heavy metals: A review. *Environ. Chem. Lett.* **2020**, *18*, 1813–1824. [CrossRef]

32. Liang, B.; Zhong, Y.; Huang, Y.; Lin, X.; Liu, J.; Lin, L.; Hu, M.; Jiang, J.; Dai, M.; Wang, B.; et al. Underestimated health risks: Polystyrene micro- and nanoplastics jointly induce intestinal barrier dysfunction by ROS-mediated epithelial cell apoptosis. *Part. Fibre Toxicol.* **2021**, *18*, 20. [CrossRef] [PubMed]

33. Demir, Y. The behaviour of some antihypertension drugs on human serum paraoxonase-1: An important protector enzyme against atherosclerosis. *J. Pharm. Pharmacol.* **2019**, *71*, 1576–1583. [CrossRef] [PubMed]

34. Gur, C.; Kandemir, O.; Kandemir, F.M. Investigation of the effects of hesperidin administration on abamectin-induced testicular toxicity in rats through oxidative stress, endoplasmic reticulum stress, inflammation, apoptosis, autophagy, and JAK2/STAT3 pathways. *Environ. Toxicol.* **2022**, *37*, 401–412. [CrossRef] [PubMed]

35. Brzóska, M.M.; Borowska, S.; Tomczyk, M. Antioxidants as a Potential Preventive and Therapeutic Strategy for Cadmium. *Curr. Drug Targets* **2016**, *17*, 1350–1384. [CrossRef] [PubMed]

36. Dewanjee, S.; Gangopadhyay, M.; Sahu, R.; Karmakar, S. Cadmium induced pathophysiology: Prophylactic role of edible jute (*Corchorus olitorius*) leaves with special emphasis on oxidative stress and mitochondrial involvement. *Food Chem. Toxicol.* **2013**, *60*, 188–198. [CrossRef] [PubMed]

37. Nielsen, F.; Mikkelsen, B.B.; Nielsen, J.B.; Andersen, H.R.; Grandjean, P. Plasma malondialdehyde as biomarker for oxidative stress: Reference interval and effects of life-style factors. *Clin. Chem.* **1997**, *43*, 1209–1214. [CrossRef]

38. Liao, W.; Ning, Z.; Chen, L.; Wei, Q.; Yuan, E.; Yang, J.; Ren, J. Intracellular Antioxidant Detoxifying Effects of Diosmetin on 2,2-Azobis(2-amidinopropane) Dihydrochloride (AAPH)-Induced Oxidative Stress through Inhibition of Reactive Oxygen Species Generation. *J. Agric. Food Chem.* **2014**, *62*, 8648–8654. [CrossRef]

39. Carafa, V.; Rotili, D.; Forgione, M.; Cuomo, F.; Serretiello, E.; Hailu, G.S.; Jarho, E.; Lahtela-Kakkonen, M.; Mai, A.; Altucci, L. Sirtuin functions and modulation: From chemistry to the clinic. *Clin. Epigenetics* **2016**, *8*, 61. [CrossRef]

40. Pantazi, E.; Zaouali, M.A.; Bejaoui, M.; Folch-Puy, E.; Ben Abdennebi, H.; Roselló-Catafau, J. Role of sirtuins in ischemia-reperfusion injury. *World J. Gastroenterol.* **2013**, *19*, 7594–7602. [CrossRef]

41. Orrenius, S.; Zhivotovsky, B.; Nicotera, P. Regulation of cell death: The calcium–apoptosis link. *Nat. Rev. Mol. Cell Biol.* **2003**, *4*, 552–565. [CrossRef] [PubMed]

42. Yamada, K.; Yoshida, K. Mechanical insights into the regulation of programmed cell death by p53 via mitochondria. *Biochim. Et Biophys. Acta (BBA) Mol. Cell Res.* **2019**, *1866*, 839–848. [CrossRef]

43. Kanno, H.; Ozawa, H.; Sekiguchi, A.; Yamaya, S.; Itoi, E. Induction of Autophagy and Autophagic Cell Death in Damaged Neural Tissue After Acute Spinal Cord Injury in Mice. *Spine* **2011**, *36*, E1427–E1434. [CrossRef] [PubMed]

44. Sekiguchi, A.; Kanno, H.; Ozawa, H.; Yamaya, S.; Itoi, E. Rapamycin Promotes Autophagy and Reduces Neural Tissue Damage and Locomotor Impairment after Spinal Cord Injury in Mice. *J. Neurotrauma* **2011**, *29*, 946–956. [CrossRef]

45. Zheng, D.; Chen, L.; Li, G.; Jin, L.; Wei, Q.; Liu, Z.; Yang, G.; Li, Y.; Xie, X. Fucoxanthin ameliorated myocardial fibrosis in STZ-induced diabetic rats and cell hypertrophy in HG-induced H9c2 cells by alleviating oxidative stress and restoring mitophagy. *Food Funct.* **2022**, *13*, 9559–9575. [CrossRef]

46. Zhang, L.; Zhang, X.; Che, D.; Zeng, L.; Zhang, Y.; Nan, K.; Zhang, X.; Zhang, H.; Guo, Z. 6-Methoxydihydrosanguinarine induces apoptosis and autophagy in breast cancer MCF-7 cells by accumulating ROS to suppress the PI3K/AKT/mTOR signaling pathway. *Phytother. Res. PTR* **2022**, 1–16. [CrossRef] [PubMed]

47. Yang, X.; Wang, H.; Tu, Y.; Li, Y.; Zou, Y.; Li, G.; Wang, L.; Zhong, X. WNT1-inducible signaling protein-1 mediates TGF-β1-induced renal fibrosis in tubular epithelial cells and unilateral ureteral obstruction mouse models via autophagy. *J. Cell. Physiol.* **2020**, *235*, 2009–2022. [CrossRef]

48. Youle, R.J.; Strasser, A. The BCL-2 protein family: Opposing activities that mediate cell death. *Nat. Rev. Mol. Cell Biol.* **2008**, *9*, 47–59. [CrossRef]

49. Cui, Y.; Zhang, Z.; Zhang, B.; Zhao, L.; Hou, C.; Zeng, Q.; Nie, J.; Yu, J.; Zhao, Y.; Gao, T.; et al. Excessive apoptosis and disordered autophagy flux contribute to the neurotoxicity induced by high iodine in Sprague-Dawley rat. *Toxicol. Lett.* **2018**, *297*, 24–33. [CrossRef]

50. Li, P.; Liu, L.; Zhou, G.; Tian, Z.; Luo, C.; Xia, T.; Chen, J.; Niu, Q.; Dong, L.; Zhao, Q.; et al. Perigestational exposure to low doses of PBDE-47 induces excessive ER stress, defective autophagy and the resultant apoptosis contributing to maternal thyroid toxicity. *Sci. Total Environ.* **2018**, *645*, 363–371. [CrossRef] [PubMed]

51. Jin, Y.; Lu, L.; Tu, W.; Luo, T.; Fu, Z. Impacts of polystyrene microplastic on the gut barrier, microbiota and metabolism of mice. *Sci. Total Environ.* **2019**, *649*, 308–317. [CrossRef] [PubMed]

52. Lu, L.; Wan, Z.; Luo, T.; Fu, Z.; Jin, Y. Polystyrene microplastics induce gut microbiota dysbiosis and hepatic lipid metabolism disorder in mice. *Sci. Total Environ.* **2018**, *631–632*, 449–458. [CrossRef] [PubMed]

International Journal of
Molecular Sciences

Article

Protective Effect of Resveratrol in an Experimental Model of Salicylate-Induced Tinnitus

Anji Song [1,2,†], Gwang-Won Cho [1,2,†], Changjong Moon [3], Ilyong Park [4] and Chul Ho Jang [5,*]

1 Department of Biology, College of Natural Science, Chosun University, Gwangju 61452, Republic of Korea
2 BK21 FOUR Education Research Group for Age-Associated Disorder Control Technology, Department of Integrative Biological Science, Chosun University, Gwangju 61452, Republic of Korea
3 Department of Veterinary Anatomy, College of Veterinary Medicine and BK21 FOUR Program, Chonnam Natioanal University, Gwanjgu 61186, Republic of Korea
4 Department of Biomedical Engineering, College of Medicine, Dankook University, Cheonan 31116, Republic of Korea
5 Department of Otolaryngology, Chonnam National University Medical School, Gwangju 61469, Republic of Korea
* Correspondence: chulsavio@hanmail.net; Tel.: +82-622206774
† These authors contributed equally to this work.

Abstract: To date, the effect of resveratrol on tinnitus has not been reported. The attenuative effects of resveratrol (RSV) on a salicylate-induced tinnitus model were evaluated by in vitro and in vivo experiments. The gene expression of the activity-regulated cytoskeleton-associated protein (*ARC*), tumor necrosis factor-alpha (*TNFα*), and NMDA receptor subunit 2B (*NR2B*) in SH-SY5Y cells was examined using qPCR. Phosphorylated cAMP response element-binding protein (p-CREB), apoptosis markers, and reactive oxygen species (ROS) were evaluated by in vitro experiments. The in vivo experiment evaluated the gap-prepulse inhibition of the acoustic startle reflex (GPIAS) and auditory brainstem response (ABR) level. The *NR2B* expression in the auditory cortex (AC) was determined by immunohistochemistry. RSV significantly reduced the salicylate-induced expression of *NR2B*, *ARC*, and *TNFα* in neuronal cells; the GPIAS and ABR thresholds altered by salicylate in rats were recovered close to their normal range. RSV also reduced the salicylate-induced NR2B overexpression of the AC. These results confirmed that resveratrol exerted an attenuative effect on salicylate-induced tinnitus and may have a therapeutic potential.

Keywords: resveratrol; NMDA receptor; salicylate; excitotoxicity; tinnitus

Citation: Song, A.; Cho, G.-W.; Moon, C.; Park, I.; Jang, C.H. Protective Effect of Resveratrol in an Experimental Model of Salicylate-Induced Tinnitus. *Int. J. Mol. Sci.* 2022, 23, 14183. https://doi.org/10.3390/ijms232214183

Academic Editors: Rossana Morabito and Alessia Remigante

Received: 17 September 2022
Accepted: 11 November 2022
Published: 16 November 2022
Corrected: 30 May 2023

Publisher's Note: MDPI stays neutral with regard to jurisdictional claims in published maps and institutional affiliations.

1. Introduction

Tinnitus is the conscious perception of sound without any corresponding external acoustic stimulus, commonly described as a phantom sound [1]. In many cases, tinnitus is caused by damage to the cochlear or auditory nerve or the central auditory pathway [2]. It is accompanied by symptoms such as anxiety, emotional disturbances, sleep disturbances, and work disorders, among others [1,3]. So far, there is no FDA-approved treatment for tinnitus. Although tinnitus can be transiently suppressed by lidocaine, no drugs have been shown to reverse the neural hyperactivity at the root of tinnitus. Existing medications cannot cure tinnitus, but can relieve a few of the serious symptoms of tinnitus.

Salicylate, one of the most widely used drugs, is prescribed for treating pain, headaches, a high fever, and inflammation. Nowadays, salicylate is commonly used in the secondary prevention of chronic coronary syndromes. However, its long-term use can cause tinnitus and eventually hearing loss due to an auditory nerve dysfunction [4,5].

A consequential nerve malfunction is the result of abnormal excitability at the brainstem and cerebral cortex [6], which is induced by the overactivity of the N-methyl-D-aspartate (NMDA) receptor subunit 2B (NR2B) [7–10]. Several studies have suggested that an increase in the expression of NMDA receptors causes ototoxicity in the cochlea [11,12].

A high dose of salicylate inhibits cyclooxygenase (COX) activity and increases the level of arachidonic acid in the lipid bilayers, which increases the open probability of NMDA receptors [10].

Resveratrol (RSV) is a polyphenol substance contained in most berries. It is known to have antioxidant, anti-aging, and anti-inflammatory effects [13]. RSV also mediates neuroprotective effects by maintaining the expression of glutamatergic and cholinergic receptors via influencing the sirtuin1 (SIRT1) expression [14]. The protective effect of RSV against cisplatin-induced ototoxicity using the HEI-OC1 cell line via antioxidants has been reported [15]. Although the neuroprotective role of RSV is well-reported, there are no reports on the attenuative effect of RSV on salicylate- or noise-induced tinnitus. In this study, we evaluated the protective effect of RSV in salicylate-induced in vitro and in vivo models.

2. Results

2.1. RSV Induces a Decrease in the NR2B Expression in Neuronal Cells

SH-SY5Y cells were differentiated into neuron-like cells and treated with RSV or salicylate. The expression of NR2B, ARC, and TNFα increased in the salicylate-treated cells, but decreased in the RSV-pretreated cells, as measured by qPCR (Figure 1A). The NR2B expression was determined by immunocytochemical staining and an immunoblot analysis; consistent results were obtained (Figure 1B–D).

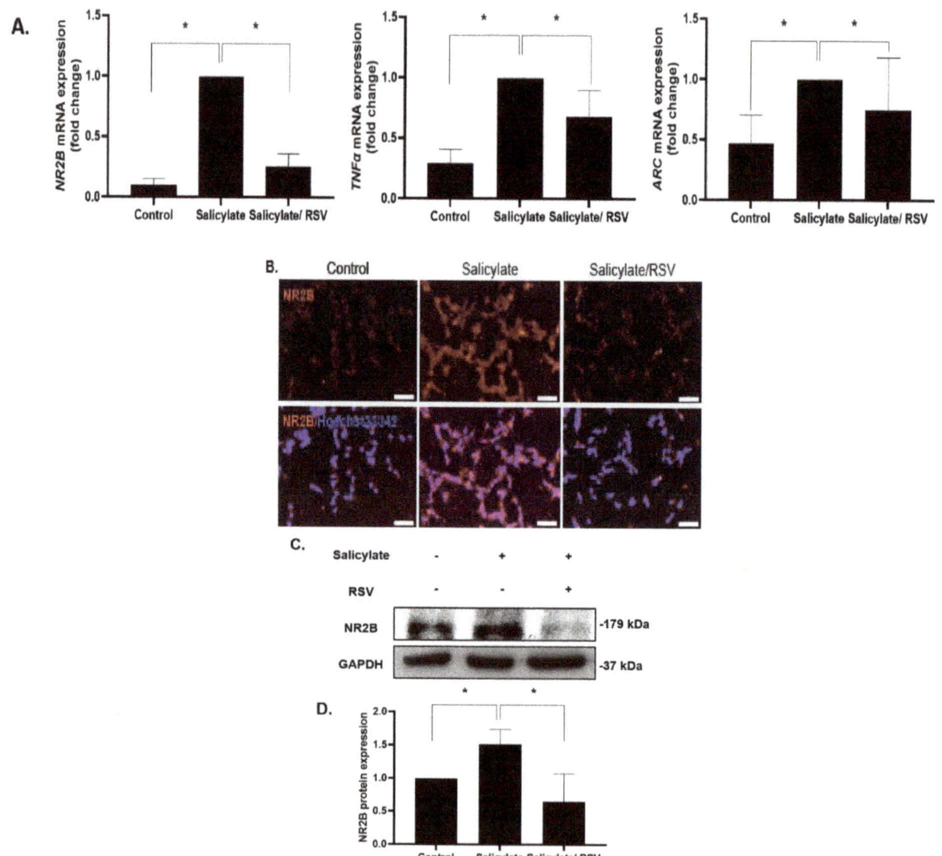

Figure 1. Expression of NR2B, *TNFα*, and the immediate early gene *ARC* in RSV-treated neuronal

cells. Salicylate significantly increased the expression of *NR2B*, *TNFα*, and the immediate early gene *ARC*, whose expression was decreased in resveratrol (RSV) pretreated SH-SY5Y cells, as measured by quantitative polymerase chain reaction (**A**); * $p < 0.05$, mean ± SD. NR2B expression in SH-SY5Y cells increased with salicylate treatment and decreased with RSV pretreatment, as demonstrated by immunocytochemical staining (**B**) and immunoblot analysis (**C**). Western blot experiments were repeated and the bands were quantified using ImageJ software (**D**); * $p < 0.05$, mean ± SD. ARC: activity-regulated cytoskeleton-associated protein; TNFα: tumor necrosis factor-alpha; NR2B: NMDA receptor subunit 2B. Scale bar = 50 um.

2.2. RSV Pretreatment Decreases Phosphorylated CREB in Salicylate-Treated Neuronal Cells

Immunocytochemical staining and an immunoblotting analysis demonstrated that the p-CREB content increased in the salicylate-treated cells. There was no decrease with the RSV pretreatment (Figure 2A,B). Immunoblot assays were repeated three times and the p-CREB expression was quantified (Figure 2C).

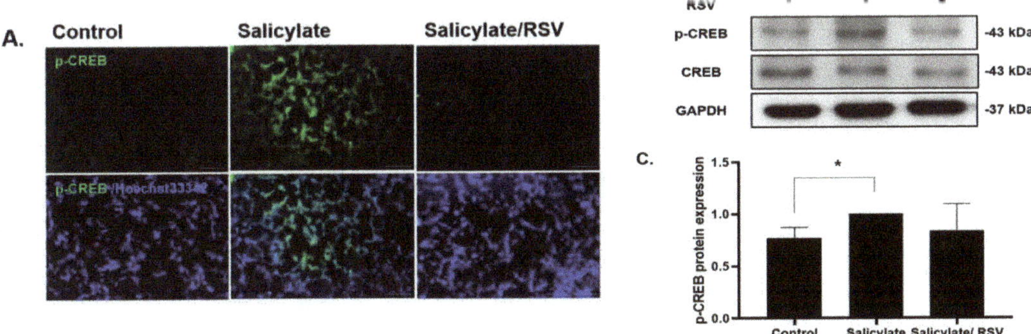

Figure 2. Phosphorylation of cAMP response element-binding protein (CREB) was reduced in resveratrol (RSV)-pretreated neuronal cells. Salicylate treatment induced CREB phosphorylation, which was reduced in RSV-pretreated SH-SY5Y cells as shown by immunocytochemistry staining (**A**) (scale bar = 100 μm) and immunoblot analysis (**B**). Western blot data were quantified and presented as a graph (**C**); * $p < 0.05$, mean ± SD. GAPDH was used as the internal standard. Scale bar = 100 μm.

2.3. RSV Inhibits a ROS Increase in Salicylate-Treated Neuronal Cells

To determine the intracellular ROS levels in salicylate-treated neuronal cells, ROS-detecting fluorescent DCFH-DA was used. The ROS levels increased in the salicylate-treated cells, but decreased with the RSV pretreatment (Figure 3A,B). The immunoblot analysis revealed that the salicylate-induced increase in p53 and cleaved caspase-3 expression was reduced when the neuronal cells were pretreated with RSV (Figure 3C).

2.4. Effect of RSV on Salicylate-Treated Rat Cortical Neurons

We then confirmed the protective effects of RSV in the primary culture cortical neurons. Rat cortical neurons were isolated and cultured in conditioned media and subjected to an immunoblot analysis with antibodies specific for NR2B and the apoptotic protein cleaved caspase-3. The NR2B and cleaved caspase-3 expression was increased in the salicylate-treated cells; it decreased with the RSV pretreatment (Figure 4).

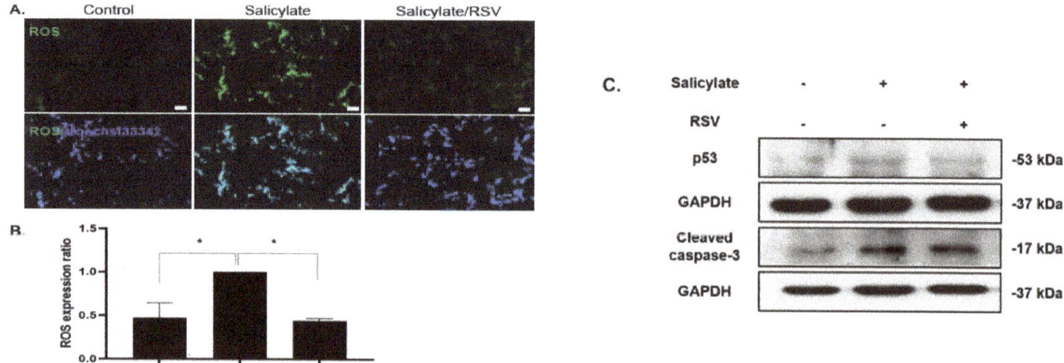

Figure 3. Pretreatment of resveratrol (RSV) reduced intracellular reactive oxygen species (ROS) levels in salicylate-treated neuronal cells. ROS levels increased with salicylate treatment, but decreased in RSV-pretreated neuronal cells as observed by fluorescence microscopy (**A**). The ROS levels were quantified using ImageJ (**B**); * $p < 0.05$, mean ± SD. Salicylate increased the expression of cleaved caspase-3 and p53, which was deceased in RSV-pretreated cells (**C**); scale bar = 100 μm. GAPDH was used as the internal standard. Scale bar = 100 μm.

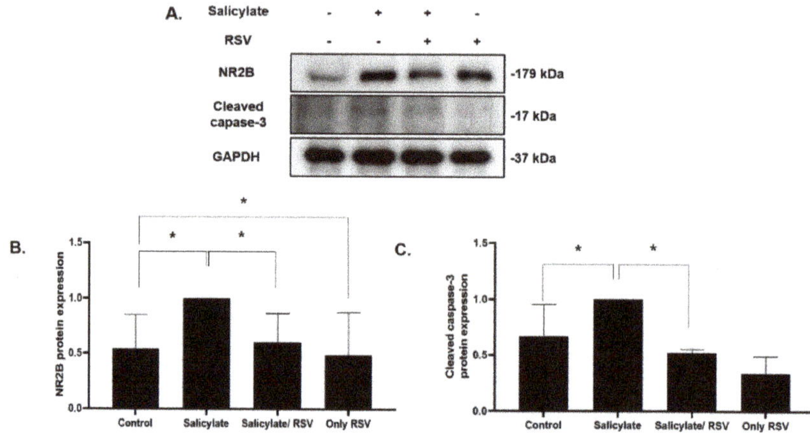

Figure 4. Effect of resveratrol (RSV) on the expression of NR2B and cleaved caspase-3 in rat cortical neuronal cells. Immunoblot analysis of the expression of NR2B and cleaved caspase-3 (**A**). Expression of NR2B (**B**) and cleaved caspase-3 (**C**) quantified using ImageJ (* $p < 0.05$, mean ± SD). GAPDH was used as the internal standard. NR2B: NMDA receptor subunit 2B.

2.5. RSV Attenuates a Salicylate-Induced GPIAS Decrease in Rats

As shown in Figure 5 there was no significant difference in the mean GPIAS value between the control group (52%) and the study group (50%) during the baseline session (day 0). However, there was a significant difference in the mean GPIAS values between the two groups during the drug administration; the mean GPIAS values decreased in the control group, but were attenuated by the RSV treatment in the study group. These results suggested that RSV protected against salicylate-induced tinnitus ($p < 0.05$) (Figure 5).

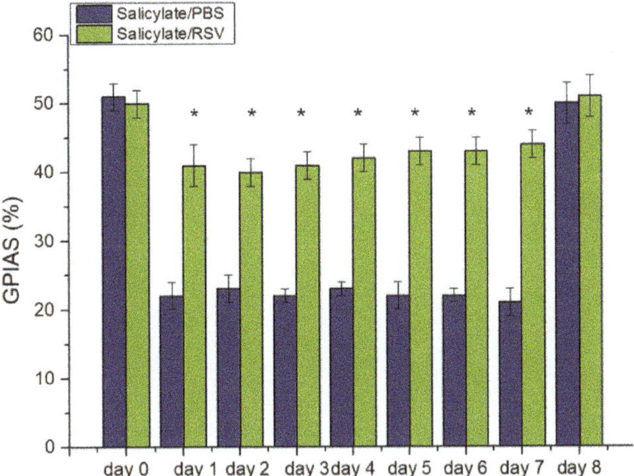

Figure 5. Resveratrol (RSV) attenuated a salicylate-induced decrease in gap-prepulse inhibition of the acoustic startle reflex (GPIAS) ratio. From day 1 to 7, RSV treatment protected against a salicylate-induced reduction of the GPIAS ratio in Sprague Dawley rats. * $p < 0.05$. PBS: phosphate-buffered saline.

2.6. RSV Attenuates a Salicylate-Induced Elevation of the ABR Threshold in Rats

As shown in Figure 6, a temporary elevated ABR threshold at 8, 16, and 32 kHz was confirmed in the control group on day 7 compared with that before the treatment (day 0), but RSV attenuated the elevation of the ABR threshold on day 7 in the study group. The mean ABR threshold in both groups did not significantly differ between before the treatment and one day after the end of the treatment ($p > 0.05$) (Figure 6). These results suggested that RSV could potentially protect against a salicylate-induced threshold shift.

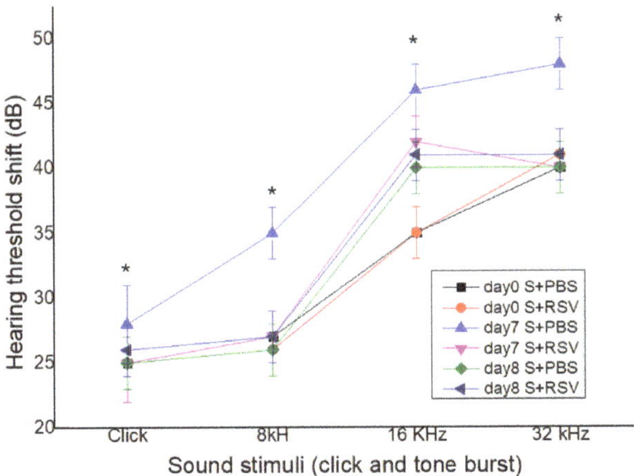

Figure 6. Resveratrol (RSV) attenuated a salicylate-induced elevation of auditory brainstem response (ABR) threshold in rats. RSV attenuated a salicylate-induced elevated ABR threshold shift except in the click group; salicylate injection given for seven consecutive days. * $p < 0.05$.

2.7. Spatial Expression of NR2B in the Auditory Cortex after Salicylate and RSV Treatments

As shown in Figure 7A, the salicylate-treated control group showed an increased NR2B expression in the auditory cortex whereas RSV significantly attenuated the NR2B expression (Figure 7B). In the group treated with salicylate and RSV, the NR2B expression decreased to the level of the vehicle-treated group.

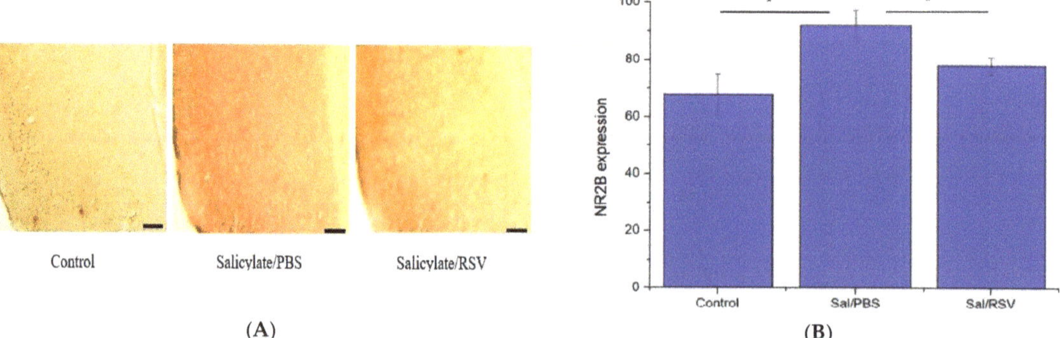

(A)   (B)

Figure 7. Resveratrol (RSV) attenuated NR2B expression in Sprague Dawley rats. NR2B expression in the control, salicylate/phosphate-buffered saline (PBS), and salicylate/RSV groups (**A**); scale bar: 100 μm, 100 × magnification. NR2B expression quantified using ImageJ. The salicylate/PBS group showed increased NR2B expression whereas RSV significantly attenuated NR2B expression (**B**); * $p < 0.05$. NR2B: NMDA receptor subunit 2B; PBS: phosphate-buffered saline.

3. Discussion

Salicylate is a widely prescribed drug due to its beneficial analgesic and anti-inflammatory properties [16]. However, its long-term, high-dose usage may lead to adverse effects such as tinnitus [17,18]. Salicylate abuse can stimulate the NMDA receptor and accelerate the intracellular Ca^{2+} influx, causing the excitotoxicity of the auditory nervous system, eventually leading to cellular damage [19]. Here, we examined the effects of salicylate on neuronal cells and identified that salicylate increased the expression of *NR2B* and its related genes, *TNFα* and *ARC*.

RSV is a well-known natural compound that has several beneficial effects [20,21]. RSV mitigates neurodegenerative disorders that involve Sirtuin 1 activation [22]. In a recently published paper, it was shown that the neuroprotective effect induced by RSV was mediated by a resistance to glutamate-induced excitotoxicity. The authors suggested that RSV could moderate glutamate release at the synaptic terminal by decreasing the protein kinase activity, which in turn relieved the voltage-gating Ca^{2+} channel activity [23]. Thus, RSV could be expected to have a protective effect on salicylate-induced neurotoxicity. In this study, we demonstrated that the increased expression of *NR2B* and its related genes induced by salicylate was reduced with an RSV pretreatment. At the cochlea level, the protective effect of RSV against cisplatin-induced ototoxicity in HEI-OC1 auditory cells has been reported [15]. RSV was decreased in cisplatin-induced ROS associated with resveratrol. Excessive ROS production is one of the most important causes of hair cell damage [24]. Non-steroidal anti-inflammatory drugs produce excessive ROS as a side effect of long-term use and can induce ototoxicity of the peripheral/central auditory neurons [25]. Salicylate is also known to increase ROS levels in patients if used for long periods and can eventually damage the cochlear spiral ganglion neurons [26]. Our in vitro experiments confirmed that salicylate increased the ROS levels, which were decreased by a pretreatment with RSV. CREB controls plasticity, neurogenesis, and survival in neurons [27]. Previous studies have shown that CREB was phosphorylated via the Ca^{2+}/calmodulin-dependent protein kinase II (CaMKII) signaling pathway in the auditory cortex of mice with salicylate-induced tinnitus [28,29]. In accordance with these studies, we observed an increase in CREB phosphorylation in salicylate-treated neuronal cells, which decreased when the cells were pretreated with RSV. Son et al. [30] reported that glutamate at high concentrations induced neuronal cell death and ROS formation using HT22 neuronal cells. They observed that piceatannol reduced glutamate-induced cell death and ROS formation. RSV and piceatannol have been shown to have antioxidant and anti-apoptotic effects [31].

Short-term tinnitus develops shortly after a high-dose salicylate injection. The known mechanism of salicylate-induced tinnitus in animals is NMDA receptor stimulation [32]. The development of a GPIAS measurement has enabled the study of animal research in a tinnitus model.

In this study, we focused on 16 kHz for the GPIAS. According to the review paper of Galazyuk et al. [33], most salicylate studies were performed on rats for a tinnitus induction. It has been shown that 1–2 h after a systemic salicylate injection ranging in dosage from 150 to 400 mg/kg, rats exhibit GPIAS deficits typically around 16 kHz [32] and rarely at a wider range of frequencies [33]. Ralli et al. [34] reported that rats treated with salicylate alone (SAL) showed a significant reduction in GPIAS at 16 kHz, which was consistent with tinnitus-like behavior with a pitch near 16 kHz. These findings were confirmed by our previous paper [14].

High doses of salicylate cause hyperactivation in the hippocampus and the cerebral cortex (including the auditory cortex). The plasticity of these cortical excitatory neurons is involved in salicylate-induced tinnitus [6,34–36]. We observed that NR2B overexpression and excessive ROS production in salicylate-treated cells were reversed when they were pretreated with RSV. These beneficial effects of RSV were confirmed in the primary cortical neurons. Thus, these results suggest that RSV protects against cell excitotoxicity. Based on our results in a tinnitus animal model, RSV therapy could be proposed as an effective treatment for tinnitus.

4. Materials and Methods

4.1. Cell Culture and Treatment

SH-SY5Y cells were cultured in Dulbecco's Modified Eagle's Medium/F12 supplemented with 10% fetal bovine serum, L-glutamine, and 1% antibiotics and incubated in a humidified atmosphere of 5% CO_2 at 37 °C. The cells were differentiated by a retinoic acid (1 μM) treatment in a 0.1% serum-supplemented medium for 2 days [37]. The differentiated cells were treated with 2 μM RSV in a 0.1% serum-supplemented medium for 12 h and then treated with 40 μg/mL salicylate in a 0.1% serum-supplemented medium for 8 h.

The cerebral cortex was dissected from Sprague Dawley rat pups at embryonic day 18 and prepared for the cell culture. The primary rat cortical neuron cells were seeded (0.5×10^6 cells/well) on poly-L-lysine (150 μg/mL; Sigma-Aldrich)-coated 6-well plates. The cells were cultured in a growth medium containing Neurobasal A, 1X B27 supplement (Invitrogen), 100 units/mL penicillin, 0.1 mg/mL streptomycin, and 0.5 mM glutamine (Invitrogen) and incubated in a humidified incubator set at 37 °C supplied with 5% CO_2. These cells were subjected to the same treatment method as that for the SH-SY5Y cells, except for the use of serum media.

4.2. Quantitative Polymerase Chain Reaction (qPCR)

The total RNA was extracted from the treated cells using RNAiso Plus (TAKARA, Tokyo, Japan). The cDNA was synthesized using PrimeScript II 1st Strand cDNA Synthesis kits (TAKARA) with the supplied buffer (0.2 μg random primers and 1 mM dNTPs). The PCR was performed using human and rat gene-specific primers specific for *NR2B*, tumor necrosis factor-alpha (*TNFα*), activity-regulated cytoskeleton-associated protein (*ARC*), and *β-actin* using the Power SYBR Green PCR Master Mix (Applied Biosystems, Carlsbad, CA, USA). The primers were synthesized by Genotech (Daejeon, Korea) and Integrated DNA Technologies Inc. (Coralville, IA, USA). Details about the primers are summarized in Table 1.

Table 1. Primers used for the quantitative polymerase chain reaction.

Gene	Forward Primer (5′→3′)	Reverse Primer (3′→5′)	Gene Accession
Human β-actin	ATCCGCAAAGACCTGTACGC	TCTTCATTGTGCTGGGTGCC	NM_001101
Human NMDA (NR2B)	GGAGAGGTGGTCATGAAGAG	CATTGCTGCGTGACACCATG	NM_000834.4
Human TNFα	GTTGTAGCAAACCCTCAAGCTG	CCAGCTGGTTATCTCTCAGCTC	NM_000594.3
Human ARC	ACAACAGGTCTCAAGGTTCCC	AGCCGACTCCTCTCTGTAGC	NM_015193.4
Rat GAPDH	CTGCCACTCAGAAGACTGTGG	TTCAGCTCTGGGATGACCTTG	NM_017008.4
Rat NMDA (Nr2b)	GGAGATGGAAGAACTGGAAGCTC	GACACCTGCCATATTGTCGATG	NM_012574.1
Rat TNFα	CCACCACGCTCTTCTGTCTAC	GATGATCTGAGTGTGAGGGTCTG	NM_012675.3
Rat ARC	GTCTGCTGCATAGAAGGAACCAG	AGGGTGCCCACCACATACTGA	NM_019361.1

4.3. Immunocytochemistry Staining

SH-SY5Y cells were cultured on poly-D-lysine-coated coverslips (Thermo Fisher Scientific, Waltham, MA, USA) and differentiated into neuronal cells. Following the treatment, the cells were fixed with 4% paraformaldehyde and incubated with primary antibodies specific for NR2B (1:200; Santa-Cruz Biotechnology, CA, USA) and p-CREB (1:200; Santa Cruz Biotechnology) for 90 min at room temperature (RT) followed by incubation with Alexa 555-conjugated donkey anti-mouse IgG (1:500; Molecular Probes Inc., Eugene, OR, USA) and Alexa 488-conjugated donkey anti-rabbit IgG secondary antibodies (1:500; Molecular Probes Inc.), respectively, along with Hoechst 33342 (1:1000; Molecular Probes Inc.) in phosphate-buffered saline (PBS) for 90 min at RT. The cells were then mounted (ProLong Gold anti-fade reagent; Molecular Probes Inc.) and visualized under a Nikon Eclipse Ti2 fluorescence microscope (Nikon, Tokyo, Japan). The images were taken by a DS-Ri2 digital camera (Nikon).

4.4. Immunoblotting

The proteins were isolated from the cells using a radioimmunoprecipitation assay (RIPA) buffer along with a protease inhibitor cocktail, 2 mM phenylmethylsulphonyl fluoride (PMSF), and 1 mM sodium orthovanadate. The cell lysates were centrifuged at $16,000\times g$ for 20 min at 4 °C and the protein was quantified. Equal amounts of proteins were resolved by SDS-PAGE using 6–15% resolving gels and electro-transferred onto a polyvinylidene difluoride (PVDF) membrane. The membrane was blocked with 5% normal horse serum (NHS) in TBS-T and then incubated overnight at 4 °C with primary antibodies specific for NR2B (1:200; Santa Cruz Biotechnology), p-CREB (1:500; Santa Cruz Biotechnology), CREB (1:500; Santa Cruz Biotechnology), cleaved caspase-3 (1:150; Merck Millipore, CA, USA), p53 (1:500; Santa Cruz Biotechnology), and GAPDH (1:1000; Santa Cruz Biotechnology) as well as horseradish peroxidase-conjugated anti-rabbit, anti-mouse, and anti-goat (Santa Cruz Biotechnology). The protein bands were visualized by enhanced chemiluminescence detection (GE Healthcare, Buckinghamshire, UK) and quantified using ImageJ software 1.53t, Bethesda, MD, USA). All the bands were normalized with GAPDH and p-CREB was normalized with CREB.

4.5. ROS Assay

The ROS levels were analyzed using 2′,7′-dichlorofluorescein diacetate (DCFH-DA). Briefly, the differentiated cells (approximately 70% confluence) were incubated with 2 µM RSV in 0.1% FBS for 12 h and then treated with 40 µg/mL salicylate for 8 h at 37 °C. The cells were then incubated with 20 µM DCFH-DA at 37 °C for 20 min. They were mounted and visualized under a Nikon Eclipse Ti2 fluorescence microscope. The ROS levels were measured using ImageJ software.

4.6. In Vivo Experiments

4.6.1. Animals

Twenty adult male Sprague Dawley rats (weighing 250–300 g) with normal eardrums and Preyer's reflex were used in this study. This study was approved by the Institutional Animal Care and Use Committee (C IACUC-2020-S0019). The rats were randomly classified into two groups, a control group (*n* = 10, salicylate treatment group) and a study group (*n* = 10, salicylate and RSV co-treatment group). The ABR and GPIAS were evaluated in both groups. The control group received an intraperitoneal (IP) injection of sodium salicylate (Sigma-Aldrich, 400 mg·kg^{-1}·day^{-1}) combined with PBS (0.2 mL) whereas the study group received an IP injection of the same dose of sodium salicylate combined with RSV (Sigma, 250 mg·kg^{-1}·day^{-1}). The IP injections were administered daily for 7 days (Figure 8).

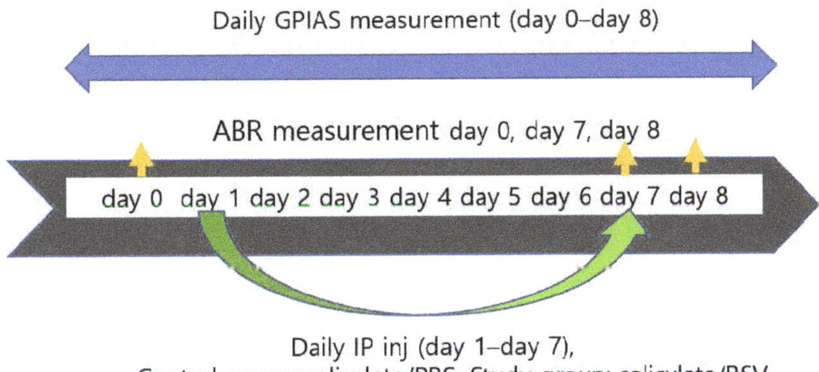

Figure 8. Schematic diagram of in vivo experimental procedure for investigating the tinnitus-attenuating effect of resveratrol in Sprague Dawley (SD) rats. Drugs were administered from day 1 to day 7 in the control and study groups (salicylate, RSV (resveratrol), and PBS (phosphate-buffered saline)). Gap-prepulse inhibition of acoustic startle reflex (GPIAS) was measured from day 1 to 8. Auditory brainstem response (ABR) thresholds were evaluated on day 0 as well as on days 7 and 8.

4.6.2. GPIAS Measurement

The GPIAS measurement method used in this study to measure the startle response was similar to that reported in our previous study [32]. In brief, the GPIAS system consisted of a mesh cage with a vibration sensor, a noise box with an anechoic inner wall, an acoustic stimulator with a full-range loudspeaker (PM-5004 amplifier; Marantz, Kawasaki, Japan), a reference microphone (40 PH; GRAS, Holte, Denmark), a sensor signal acquisition hardware tool (PC and NI PCIe-6321; National Instruments, Austin, TX, USA), and LabVIEW-based custom graphical user interface (GUI) software. Each rat was subjected to a session comprising 15 gap-conditioned stimuli-evoked startle responses and 15 responses with no gap. Additionally, two types of acoustic stimuli (gap and no gap) were given to rats in a random order. The inter-stimulus interval (ISI) of these acoustic stimuli was randomized between 17 and 23 s to eliminate the habituation effect that might arise from a fixed startle-stimulus interval [32]. In addition, the rats were placed in the cage for about 2 min before the first session to acclimatize them to the measurement environment. The two types of acoustic stimuli mentioned above included a continuous narrowband background noise with a 1 kHz bandwidth, 16 kHz center frequency, and 60 dB sound pressure level (SPL) and a short high-level sound as a startle stimulus (broadband noise burst, 50 ms in length, 105 dB SPL) [32,38]. The central frequency of the narrowband background noise used in the stimulation condition followed the typical GPIAS measurement conditions for the salicylic acid-induced SD-rat tinnitus model used in several previous studies [33,39].

The gap-conditioned stimulus included a silence prepulse and a gap starting 100 ms before the startle stimulus and lasting for 50 ms [32]. The GPIAS value was defined as a percentage of the reduction in the gap-conditioned stimulus response relative to the magnitude of the no-gap-conditioned one [32].

4.6.3. ABR Measurement

ABR was evaluated using a TDT system (Tucker-Davis Technologies, Miami, FL, USA) in an electrically sound-shielded sound-proof box. As shown in Figure 1, the measurement of the ABR thresholds was performed on day 0 (before the drug administration), day 7 (end of the administration), and day 8 (one day after the drug administration). The measurement method was similar to that used in our previous study [30]. In brief, subdermal needle electrodes were placed at the vertex (active) below the left pinna (reference) and the electrode was inserted under the right ear (ground). The stimuli consisted of click and tone bursts (8, 16, and 32 kHz) with reducing levels in the range of 10–90 dB of 5 dB intervals to determine the lowest intensity level. Each measurement point was recorded and averaged 1000 times. The stimuli were presented at a rate of 19 s and the acoustic stimuli were calibrated before the measurement of each group.

4.6.4. Preparation of Free-Floating Sections

For the histological examination of their brain tissue, the rats were sacrificed 2 days after the last injection of each drug or vehicle. They were anesthetized and perfused with 4% paraformaldehyde (PFA) in PBS (pH 7.4). Their brains were immediately removed, stored in 4% (w/v) PFA in PBS for 2 days at 4 °C, suspended in 30% (w/v) sucrose for 4 days, and then embedded in an optimum cutting temperature compound (Miles Inc., Elkhart, IN, USA). Using a sliding microtome (SM2010R; Leica Microsystems, Wetzlar, Germany), the brain hemispheres were coronally sectioned for the auditory cortex at approximately 4.30–5.30 mm caudal to the bregma. Free-floating serial sections (30 µm thick) were collected into 10 PBS-filled wells.

4.6.5. Immunohistochemistry

The brain samples were collected and fixed using 4% (w/v) PFA in PBS. The samples were immersed in 30% (w/v) sucrose for at least 4 days, following which they were sectioned into coronal slices (30 µm thick) using a frozen sliding microtome (SM2010R) and stored in PBS at 4 °C. To disable the intrinsic peroxidase activity, the sections were incubated in 0.3% (v/v) hydrogen peroxide in distilled water for 20 min. This was followed by 1 h of blocking with 5% (v/v) normal goat serum (Vector ABC Elite Kit; Vector Laboratories, Burlingame, CA, USA) in 0.3% (v/v) Triton X-100. The sections were then incubated with a rabbit anti-NMDAR2B antibody (1:200; Abcam, Cambridge, UK) diluted with an antibody dilution buffer (Invitrogen) at 4 °C overnight. After washing, the sections were reacted with biotinylated goat anti-rabbit IgG (Vector ABC Elite Kit; Vector Laboratories) for 1 h and washed again. The sections were then incubated for 1 h at RT with an avidin–biotin–peroxidase complex (Vector ABC Elite Kit; Vector Laboratories) according to the manufacturer's instructions. After washing, the diaminobenzidine substrate (DAB kit; Vector Laboratories) was used as the chromogen to visualize the signal.

The immunoreactivity of NMDAR2B in the auditory cortex subregions was analyzed by measuring the intensity of the NMDAR2B-immunopositive reaction with ImageJ software. Coronal sections (30 µm thick) were selected approximately 2.5–3.6 mm posterior to the bregma in each brain and the intensities in the subregion were assessed. The intensity levels were expressed as a mean ± standard error (SE; $n = 4$ rats/group).

4.7. Statistical Analysis

An analysis of variance (ANOVA) with Tukey's HSD post hoc test (three or more groups) was conducted using GraphPad Prism to determine the statistical significance. If the p-value was less than 0.05, it was considered to be statistically significant.

5. Conclusions

Our results provide evidence that RSV may reduce salicylate-induced tinnitus in rats. Further studies should determine the potential therapeutic effects of RSV in clinical settings.

Author Contributions: Conceptualization, G.-W.C. and C.H.J.; methodology, A.S., C.M., I.P. and C.H.J.; data curation, A.S.; writing—original draft preparation, A.S., C.M. and C.H.J.; writing—review and editing, G.-W.C., C.M. and C.H.J.; funding acquisition, C.H.J. All authors have read and agreed to the published version of the manuscript.

Funding: This work was supported by the National Research Foundation of Korea (NRF) grant funded by the Korea government (MSIT) (No. 2021RIF1A1060212).

Institutional Review Board Statement: The study was conducted according to the guidelines of the Declaration of Helsinki and approved by Ethics Committee of Chosun University CIACUC-2020-S0019.

Informed Consent Statement: Not applicable.

Data Availability Statement: The data presented in this study are available on request from the corresponding author.

Conflicts of Interest: The authors declare no conflict of interest.

References

1. Tang, D.; Li, H.; Chen, L. Advances in Understanding, Diagnosis, and Treatment of Tinnitus. *Adv. Exp. Med. Biol.* **2019**, *1130*, 109–128. [CrossRef] [PubMed]
2. Yi, B.; Hu, S.; Zuo, C.; Jiao, F.; Lv, J.; Chen, D.; Ma, Y.; Chen, J.; Mei, L.; Wang, X.; et al. Effects of long-term salicylate administration on synaptic ultrastructure and metabolic activity in the rat CNS. *Sci. Rep.* **2016**, *6*, 24428. [CrossRef] [PubMed]
3. Gonçalves, J.T.; Schafer, S.T.; Gage, F.H. Adult Neurogenesis in the Hippocampus: From Stem Cells to Behavior. *Cell* **2016**, *167*, 897–914. [CrossRef] [PubMed]
4. Alvan, G.; Berninger, E.; Gustafsson, L.L.; Karlsson, K.K.; Paintaud, G.; Wakelkamp, M. Concentration-Response Relationship of Hearing Impairment Caused by Quinine and Salicylate: Pharmacological Similarities but Different Molecular Mechanisms. *Basic. Clin. Pharmacol. Toxicol.* **2017**, *120*, 5–13. [CrossRef] [PubMed]
5. Palmer, B.F.; Clegg, D.J. Salicylate Toxicity. *N. Engl. J. Med.* **2020**, *382*, 2544–2555. [CrossRef]
6. Wu, C.; Wu, X.; Yi, B.; Cui, M.; Wang, X.; Wang, Q.; Wu, H.; Huang, Z. Changes in GABA and glutamate receptors on auditory cortical excitatory neurons in a rat model of salicylate-induced tinnitus. *Am. J. Transl. Res.* **2018**, *10*, 3941–3955.
7. Bing, D.; Lee, S.C.; Campanelli, D.; Xiong, H.; Matsumoto, M.; Panford-Walsh, R.; Wolpert, S.; Praetorius, M.; Zimmermann, U.; Chu, H.; et al. Cochlear NMDA receptors as a therapeutic target of noise-induced tinnitus. *Cell Physiol. Biochem.* **2015**, *35*, 1905–1923. [CrossRef]
8. Noreña, A.J.; Farley, B.J. Tinnitus-related neural activity: Theories of generation, propagation, and centralization. *Hear. Res.* **2013**, *295*, 161–171. [CrossRef]
9. Cui, W.; Wang, H.; Cheng, Y.; Ma, X.; Lei, Y.; Ruan, X.; Shi, L.; Lv, M. Long-term treatment with salicylate enables NMDA receptors and impairs AMPA receptors in C57BL/6J mice inner hair cell ribbon synapse. *Mol. Med. Rep.* **2019**, *19*, 51–58. [CrossRef]
10. Hong, J.; Chen, Y.; Zhang, Y.; Li, J.; Ren, L.; Yang, L.; Shi, L.; Li, A.; Zhang, T.; Li, H.; et al. N-Methyl-D-Aspartate Receptors Involvement in the Gentamicin-Induced Hearing Loss and Pathological Changes of Ribbon Synapse in the Mouse Cochlear Inner Hair Cells. *Neural. Plast.* **2018**, *2018*, 3989201. [CrossRef]
11. Segal, J.A.; Harris, B.D.; Kustova, Y.; Basile, A.; Skolnick, P. Aminoglycoside neurotoxicity involves NMDA receptor activation. *Brain Res.* **1999**, *815*, 270–277. [CrossRef]
12. Xia, N.; Daiber, A.; Förstermann, U.; Li, H. Antioxidant effects of resveratrol in the cardiovascular system. *Br. J. Pharmacol.* **2017**, *174*, 1633–1646. [CrossRef]
13. Karthick, C.; Periyasamy, S.; Jayachandran, K.S.; Anusuyadevi, M. Intrahippocampal Administration of Ibotenic Acid Induced Cholinergic Dysfunction via NR2A/NR2B Expression: Implications of Resveratrol against Alzheimer Disease Pathophysiology. *Front. Mol. Neurosci.* **2016**, *9*, 28. [CrossRef] [PubMed]
14. Wu, Y.; Pang, Y.; Wei, W.; Shao, A.; Deng, C.; Li, X.; Chang, H.; Hu, P.; Liu, X.; Zhang, X. Resveratrol protects retinal ganglion cell axons through regulation of the SIRT1-JNK pathway. *Exp. Eye Res.* **2020**, *200*, 108249. [CrossRef] [PubMed]
15. Lee, S.H.; Kim, H.S.; An, Y.S.; Chang, J.; Choi, J.; Im, G.J. Protective effect of resveratrol against cisplatin-induced ototoxicity in HEI-OC1 auditory cells. *Int. J. Pediatr. Otorhinolaryngol.* **2015**, *79*, 58–62. [CrossRef] [PubMed]
16. Jastreboff, P.J.; Brennan, J.F.; Coleman, J.K.; Sasaki, C.T. Phantom auditory sensation in rats: An animal model for tinnitus. *Behav. Neurosci.* **1988**, *102*, 811–822. [CrossRef] [PubMed]
17. Qin, D.; Liu, P.; Chen, H.; Huang, X.; Ye, W.; Lin, X.; Wei, F.; Su, J. Salicylate-Induced Ototoxicity of Spiral Ganglion Neurons: Ca^{2+}/CaMKII-Mediated Interaction Between NMDA Receptor and GABA(A) Receptor. *Neurotox. Res.* **2019**, *35*, 838–847. [CrossRef]
18. Sheppard, A.; Hayes, S.H.; Chen, G.D.; Ralli, M.; Salvi, R. Review of salicylate-induced hearing loss, neurotoxicity, tinnitus and neuropathophysiology. *Acta Otorhinolaryngol. Ital.* **2014**, *34*, 79–93.
19. Gupta, K.; Hardingham, G.E.; Chandran, S. NMDA receptor-dependent glutamate excitotoxicity in human embryonic stem cell-derived neurons. *Neurosci. Lett.* **2013**, *543*, 95–100. [CrossRef]

20. Nunes, S.; Danesi, F.; Del Rio, D.; Silva, P. Resveratrol and inflammatory bowel disease: The evidence so far. *Nutr. Res. Rev.* **2018**, *31*, 85–97. [CrossRef]

21. Shaito, A.; Posadino, A.M.; Younes, N.; Hasan, H.; Halabi, S.; Alhababi, D.; Al-Mohannadi, A.; Abdel-Rahman, W.M.; Eid, A.H.; Nasrallah, G.K.; et al. Potential Adverse Effects of Resveratrol: A Literature Review. *Int. J. Mol. Sci.* **2020**, *21*, 2084. [CrossRef] [PubMed]

22. Corpas, R.; Griñán-Ferré, C.; Palomera-Ávalos, V.; Porquet, D.; García de Frutos, P.; Franciscato Cozzolino, S.M.; Rodríguez-Farré, E.; Pallàs, M.; Sanfeliu, C.; Cardoso, B.R. Melatonin induces mechanisms of brain resilience against neurodegeneration. *J. Pineal. Res.* **2018**, *65*, e12515. [CrossRef] [PubMed]

23. Chang, Y.; Wang, S.J. Inhibitory effect of glutamate release from rat cerebrocortical nerve terminals by resveratrol. *Neurochem. Int.* **2009**, *54*, 135–141. [CrossRef] [PubMed]

24. Scasso, F.; Sprio, A.E.; Canobbio, L.; Scanarotti, C.; Manini, G.; Berta, G.N.; Bassi, A.M. Dietary supplementation of coenzyme Q10 plus multivitamins to hamper the ROS mediated cisplatin ototoxicity in humans: A pilot study. *Heliyon* **2017**, *3*, e00251. [CrossRef]

25. Tabuchi, K.; Nishimura, B.; Nakamagoe, M.; Hayashi, K.; Nakayama, M.; Hara, A. Ototoxicity: Mechanisms of cochlear impairment and its prevention. *Curr. Med. Chem.* **2011**, *18*, 4866–4871. [CrossRef]

26. Deng, L.; Ding, D.; Su, J.; Manohar, S.; Salvi, R. Salicylate selectively kills cochlear spiral ganglion neurons by paradoxically up-regulating superoxide. *Neurotox. Res.* **2013**, *24*, 307–319. [CrossRef]

27. Pardo, L.; Valor, L.M.; Eraso Pichot, A.; Barco, A.; Golbano, A.; Hardingham, G.E.; Masgrau, R.; Galea, E. CREB Regulates Distinct Adaptive Transcriptional Programs in Astrocytes and Neurons. *Sci. Rep.* **2017**, *7*, 6390. [CrossRef]

28. Zhao, J.; Wang, B.; Wang, X.; Shang, X. Up-regulation of Ca(2+)/CaMKII/CREB signaling in salicylate-induced tinnitus in rats. *Mol. Cell Biochem.* **2018**, *448*, 71–76. [CrossRef]

29. Yi, B.; Wu, C.; Shi, R.; Han, K.; Sheng, H.; Li, B.; Mei, L.; Wang, X.; Huang, Z.; Wu, H. Long-term Administration of Salicylate-induced Changes in BDNF Expression and CREB Phosphorylation in the Auditory Cortex of Rats. *Otol. Neurotol.* **2018**, *39*, e173–e180. [CrossRef]

30. Son, Y.; Byun, S.J.; Pae, H.O. Involvement of heme oxygenase-1 expression in neuroprotection by piceatannol, a natural analog and a metabolite of resveratrol, against glutamate-mediated oxidative injury in HT22 neuronal cells. *Amino Acids* **2013**, *45*, 393–401. [CrossRef]

31. Hosoda, R.; Hamada, H.; Uesugi, D.; Iwahara, N.; Nojima, I.; Horio, Y.; Kuno, A. Different Antioxidative and Antiapoptotic Effects of Piceatannol and Resveratrol. *J. Pharmacol. Exp. Ther.* **2021**, *376*, 385–396. [CrossRef] [PubMed]

32. Jang, C.H.; Lee, S.; Park, I.Y.; Song, A.; Moon, C.; Cho, G.W. Memantine Attenuates Salicylate-induced Tinnitus Possibly by Reducing NR2B Expression in Auditory Cortex of Rat. *Exp. Neurobiol.* **2019**, *28*, 495–503. [CrossRef] [PubMed]

33. Galazyuk, A.; Hébert, S. Gap-Prepulse Inhibition of the Acoustic Startle Reflex (GPIAS) for Tinnitus Assessment: Current Status and Future Directions. *Front. Neurol.* **2015**, *6*, 88. [CrossRef] [PubMed]

34. Chen, G.D.; Stolzberg, D.; Lobarinas, E.; Sun, W.; Ding, D.; Salvi, R. Salicylate-induced cochlear impairments, cortical hyperactivity and re-tuning, and tinnitus. *Hear. Res.* **2013**, *295*, 100–113. [CrossRef] [PubMed]

35. Gong, N.; Zhang, M.; Zhang, X.B.; Chen, L.; Sun, G.C.; Xu, T.L. The aspirin metabolite salicylate enhances neuronal excitation in rat hippocampal CA1 area through reducing GABAergic inhibition. *Neuropharmacology* **2008**, *54*, 454–463. [CrossRef]

36. Liu, J.; Li, X.; Wang, L.; Dong, Y.; Han, H.; Liu, G. Effects of salicylate on serotoninergic activities in rat inferior colliculus and auditory cortex. *Hear. Res.* **2003**, *175*, 45–53. [CrossRef]

37. Kim, H.T.; Ohn, T.; Jeong, S.G.; Song, A.; Jang, C.H.; Cho, G.W. Oxidative stress-induced aberrant G9a activation disturbs RE-1-containing neuron-specific genes expression, leading to degeneration in human SH-SY5Y neuroblastoma cells. *Korean J. Physiol. Pharmacol.* **2021**, *25*, 51–58. [CrossRef]

38. Ralli, M.; Troiani, D.; Podda, M.V.; Paciello, F.; Eramo, S.L.; de Corso, E.; Salvi, R.; Paludetti, G.; Fetoni, A.R. The effect of the NMDA channel blocker memantine on salicylate-induced tinnitus in rats. *Acta Otorhinolaryngol. Ital.* **2014**, *34*, 198–204.

39. Lobarinas, E.; Hayes, S.H.; Allman, B.L. The gap-startle paradigm for tinnitus screening in animal models: Limitations and optimization. *Hear. Res.* **2013**, *295*, 150–160. [CrossRef]

International Journal of
Molecular Sciences

Article

Usefulness of Urinary Biomarkers for Assessing Bladder Condition and Histopathology in Patients with Interstitial Cystitis/Bladder Pain Syndrome

Yuan-Hong Jiang [1], Jia-Fong Jhang [1], Yuan-Hsiang Hsu [2] and Hann-Chorng Kuo [1,*]

[1] Department of Urology, Hualien Tzu Chi Hospital, Buddhist Tzu Chi Medical Foundation, Tzu Chi University, Hualien 970, Taiwan
[2] Department of Pathology, Hualien Tzu Chi Hospital, Buddhist Tzu Chi Medical Foundation, Tzu Chi University, Hualien 970, Taiwan
* Correspondence: hck@tzuchi.com.tw; Tel.: +886-3-8561825 (ext. 2117); Fax: +886-3-8560794

Abstract: This study investigated the usefulness of urinary biomarkers for assessing bladder condition and histopathology in patients with interstitial cystitis/bladder pain syndrome (IC/BPS). We retrospectively enrolled 315 patients (267 women and 48 men) diagnosed with IC/BPS and 30 controls. Data on clinical and urodynamic characteristics (visual analog scale (VAS) score and bladder capacity) and cystoscopic hydrodistention findings (Hunner's lesion, glomerulation grade, and maximal bladder capacity (MBC)) were recorded. Urine samples were utilized to assay inflammatory, neurogenic, and oxidative stress biomarkers, including interleukin (IL)-8, C-X-C motif chemokine ligand 10 (CXCL10), monocyte chemoattractant protein-1 (MCP-1), brain-derived neurotrophic factor (BDNF), eotaxin, IL-6, macrophage inflammatory protein 1 beta (MIP-1β), regulated on activation, normal T cell expressed and secreted (RANTES), tumor necrosis factor-alpha (TNF-α), prostaglandin E2 (PGE2), 8-hydroxy-2′-deoxyguanosine (8-OHdG), and 8-isoproatane, and total antioxidant capacity. Further, specific histopathological findings were identified via bladder biopsy. The associations between urinary biomarker levels and bladder conditions and histopathological findings were evaluated. The results reveal that patients with IC/BPS had significantly higher urinary MCP-1, eotaxin, TNF-α, PGE2, 8-OHdG, and 8-isoprostane levels than controls. Patients with Hunner's IC (HIC) had significantly higher IL-8, CXCL10, BDNF, eotaxin, IL-6, MIP-1β, and RANTES levels than those with non-Hunner's IC (NHIC). Patients with NHIC who had an MBC of $\leq$760 mL had significantly high urinary CXCL10, MCP-1, eotaxin, IL-6, MIP-1β, RANTES, PGE2, and 8-isoprostane levels and total antioxidant capacity. Patients with NHIC who had a higher glomerulation grade had significantly high urinary MCP-1, IL-6, RANTES, 8-OHdG, and 8-isoprostane levels. A significant association was observed between urinary biomarkers and glomerulation grade, MBC, VAS score, and bladder sensation. However, bladder-specific histopathological findings were not well correlated with urinary biomarker levels. The urinary biomarker levels can be useful for identifying HIC and different NHIC subtypes. Higher urinary inflammatory and oxidative stress biomarker levels are associated with IC/BPS. Most urinary biomarkers are not correlated with specific bladder histopathological findings; nevertheless, they are more important in the assessment of bladder condition than bladder histopathology.

Keywords: interstitial cystitis; bladder pain syndrome; biomarker; cytokine

check for **updates**

Citation: Jiang, Y.-H.; Jhang, J.-F.; Hsu, Y.-H.; Kuo, H.-C. Usefulness of Urinary Biomarkers for Assessing Bladder Condition and Histopathology in Patients with Interstitial Cystitis/Bladder Pain Syndrome. *Int. J. Mol. Sci.* **2022**, 23, 12044. https://doi.org/10.3390/ijms231912044

Academic Editors: Rossana Morabito and Alessia Remigante

Received: 28 August 2022
Accepted: 7 October 2022
Published: 10 October 2022

Publisher's Note: MDPI stays neutral with regard to jurisdictional claims in published maps and institutional affiliations.

1. Introduction

Interstitial cystitis/bladder pain syndrome (IC/BPS) is a condition characterized by urinary urgency and frequency, nocturia, and commonly pelvic pain but without bacterial infection or identifiable lower urinary tract pathology [1]. IC/BPS has different subtypes (ulcerative and non-ulcerative), which may have distinct pathophysiological and clinical presentations [2,3]. The prominent pathological findings of IC/BPS include urothelial denudation and bladder wall inflammation. The incidence of bladder inflammation is higher in Hunner's IC (HIC) than in non-Hunner's (NHIC) [4,5]. Previous research has shown that IC/BPS involves an aberrant differentiation of bladder urothelium leading to the decreased production of cell surface proteoglycans, adhesion and tight junction proteins, and bacterial defense molecules [6]. Moreover, based on our previous study, patients with IC/BPS have increased production of apoptotic signaling molecules, including Bad, Bax, and cleaved caspase-3 [7]. In addition to bladder pathophysiology that may contribute to the clinical presentations of bladder-centered IC/BPS, several somatic and functional disorders coexist with IC/BPS. A recent report revealed that systemic inflammatory diseases might play an important role in the pathogenesis of IC/BPS [8].

In IC/BPS, chronic pain symptomatology could be attributed to persistent urinary bladder abnormalities, which activate the afferent sensory system and central nervous system (CNS) sensitization [9]. A recent study has proposed several pathophysiological mechanisms with underlying pain in IC/BPS. These include epithelial dysfunction, mast cell activation, neurogenic inflammation, autoimmunity, and occult infection. Further, IC/BPS is considered a heterogeneous syndrome. The diagnosis of IC/BPS might not be solely based on clinical and cystoscopic hydrodistention and the exclusion of other bladder disorders [10]. The use of urinary or serum biomarkers that can be used along with clinical symptoms could improve the early and accurate diagnoses of IC/BPS subtypes [11].

Based on our recent studies, the use of urinary biomarkers for diagnosing IC/BPS and for differentiating HIC from NHIC has progressed. Patients with IC/BPS have significantly higher levels of urinary cytokines and chemokines such as monocyte chemoattractant protein 1 (MCP-1), eotaxin, tumor necrosis factor-alpha (TNF-α), and prostaglandin E2 (PGE2) [11]. Most urinary biomarkers were significantly associated with maximal bladder capacity (MBC), glomerulation grade, and treatment outcome evaluated using the global response assessment [12]. Patients with HIC had significantly higher levels of urinary interleukin-8 (IL-8), C-X-C motif chemokine ligand 10 (CXCL10), brain-derived neurotrophic factor (BDNF), and eotaxin, and regulated, on activation, normal T cell expression and secretion (RANTES) than those with NHIC and controls. Among all urinary biomarkers, TNF-α has the best sensitivity, specificity, and positive and negative predictive values [12]. In addition to inflammatory cytokines, urinary oxidative stress biomarkers can be novel biomarkers in patients with IC/BPS. 8-hydroxy-2'-deoxyguanosine (8-OHdG) and 8-isoprostane have high diagnostic values (area under the curve (AUC): >0.7) in distinguishing patients with European Society for The Study of Interstitial Cystitis (ESSIC) type 2 IC/BPS from controls. These urinary biomarkers were positively correlated with glomerulation grade and were negatively correlated with MBC [13].

In a previous study on the histopathology of different IC/BPS subtypes, IC/BPS had different characteristic pathologies in bladder specimens, which include bladder inflammation, urothelium denudation, eosinophil and plasma cell infiltration, lamina propria hemorrhage, and granulation of different severity [4]. Bladder histopathological findings were associated with clinical parameters, and there were differences in patient-reported treatment outcomes. Whether urinary cytokine, chemokine, and oxidative stress biomarker levels are associated with clinical and histopathological characteristics, and whether these biomarkers can be used to identify different IC/BPS phenotypes must be confirmed [14]. The current study aimed to retrospectively analyze the associations between urinary biomarker levels and bladder conditions, clinical presentations, and histopathological findings in patients with IC/BPS.

2. Results

In total, 315 patients clinically diagnosed with IC/BPS (291 with NHIC and 24 with HIC) and 30 controls were included in the analysis. The mean age was 53.3 $\pm$ 13.3 and 57.7 $\pm$ 10.1 years old in the IC/BPS and control groups, respectively (p = 0.076). Among the IC/BPS patients, the mean baseline ICSI was 11.0 $\pm$ 4.6, ICPI was 10.8 $\pm$ 3.8, VAS was 4.46 $\pm$ 2.87, MBC was 723.9 $\pm$ 189.5, and glomerulation grade was 1.49 $\pm$ 0.9. All patients received cystoscopic hydrodistention under intravenous general anesthesia. Then, bladder biopsies and urine sample collection were performed. The development of petechia, glomerulations, splotch hemorrhage, mucosal fissures, or ulceration in the bladder was cautiously examined [15]. The glomerulation grade was classified as follows: 0, none; 1, less than half of the bladder wall; 2, more than half of the bladder wall; or 3, severe waterfall bleeding [15]. Hunner's lesions, with or without glomerulation, were classified as HIC. Cystoscopy and bladder hydrodistention results were obtained from the surgical report in the patient's chart. Table 1 shows the urinary biomarker levels of all patients with IC/BPS and controls. Patients with IC/BPS had significantly higher urinary MCP-1, eotaxin, TNF-α, PGE2, 8-OHdG, and 8-isoprostane levels than controls.

Table 1. The levels of urine biomarkers in patients with interstitial cystitis and normal controls.

Urine Cytokines	IC/BPS (N = 315)	Control (N = 30)	p-Value	AUC	Cut-Off Value	Odd Ratio	95% CI
IL-8	17.6 ± 26.6	12.5 ± 21.0	0.328	0.587	≥2.1	1.010	0.990–1.031
CXCL 10	11.5 ± 20	13.8 ± 18.4	0.583	0.590	≤1.60	0.995	0.978–1.012
MCP-1	295 ± 300	147 ± 110	<0.001	0.639	≥283	1.429	1.088–1.875
BDNF	0.58 ± 0.16	0.55 ± 0.12	0.310	0.551	≥0.54	1.161	0.870–1.550
Eotaxin	7.56 ± 7.51	4.98 ± 3.7	0.002	0.587	≥1.92	1.077	0.992–1.168
IL-6	3.43 ± 8.28	1.29 ± 1.35	0.160	0.534	≤0.52	1.125	0.945–1.338
MIP-1β	1.23 ± 1.72	2.52 ± 1.82	<0.001	0.774	≤0.81	0.753	0.642–0.883
RANTES	5.63 ± 8.12	6.04 ± 5.15	0.820	0.636	≤1.50	0.994	0.951–1.039
TNF-α	1.66 ± 0.38	0.82 ± 0.33	<0.001	0.920	≥1.05	2.368	1.783–3.146
PGE2	290 ± 239	161 ± 105	<0.001	0.679	≥175	1.669	1.166–2.389
8-OHDG	32 ± 21.8	18 ± 13.73	<0.001	0.688	≥25.0	1.468	1.163–1.853
8-isoprostane	54.1 ± 62.7	16.8 ± 11.8	<0.001	0.721	≥22.3	1.526	1.199–1.943
TAC	1105 ± 937	1078 ± 925	0.861	0.533	≥745	1.003	0.963–1.045

Abbreviations: IC/BPS: interstitial cystitis/bladder pain syndrome, AUC: area under curve, CI: confidence interval, IL-8: interleukin-8, CXCL10: C-X-C motif chemokine ligand 10, MCP-1: monocyte chemoattractant protein-1, BDNF: brain-derived neurotrophic factor, MIP-1β: macrophage inflammatory proteins, RANTES: regulated on activation, normal T-cell expressed and secreted, TNF-α: tumor necrosis factor –alpha, PGE2: prostaglandin E2, 8-OHdG: 8-hydroxydeoxyguanosine. TAC: total antioxidant capacity.

When we divide IC/BPS patients into NHIC with MBC > 760 mL, NHIC with MBC ≤ 760 mL, and HIC subgroups, patients with HIC were found to have significantly higher urinary IL-8, CXCL10, BDNF, eotaxin, IL-6, MIP-1β, and RANTES levels than patients with NHIC. Patients with IC/BPS who had an MBC of ≤760 mL had significantly high urinary CXCL-10, MCP-1, eotaxin, IL-6, MIP-1β, RANTES, PGE2, 8-isoprostane levels, and TAC (Table 2).

After classifying NHIC into different subtypes according to glomerulation grade under cystoscopic hydrodistention and HIC, patients with HIC were found to have a significantly high urinary IL-8, CXCL-10, BDNF, eotaxin, IL-6, and RANTES levels. Patients with NHIC who had a higher glomerulation grade had significantly elevated urinary MCP-1, IL-6, RANTES, 8-OHdG, and 8-isoprostane levels than those with lower glomerulation grade (Table 3).

We performed a correlation analysis between urinary biomarker levels and bladder characteristics, such as glomerulation grade and MBC under cystoscopic hydrodistention, VAS score, FSF, FS, and CBC on the urodynamic study, in patients with IC/BPS. Table 4 shows the correlation coefficients. Glomerulation grade was significantly associated with CXCL10, MCP-1, IL-6, RANTES, PGE2, and 8-OHdG levels. MBC was correlated with all urinary biomarkers except IL-8, BDNF, and TAC. Further, there was an association between VAS score and BDNF, IL-6, PGE2, and 8-OHdG levels. Bladder FS and CBC were correlated with MCP-1, IL-6, and 8-isoprostane levels.

In total, 187 patients with NHIC had available data on bladder histopathological findings and urinary biomarker levels. Table 5 shows the associations between urinary biomarker levels and the histopathological classification of IC/BPS, including inflammatory cell infiltration, urothelial denudation grade, and the presence of eosinophil infiltration, plasma cell infiltration, suburothelial granulation, and lamina propria hemorrhage. (Figure 1) There was no significant difference in all urinary biomarkers except CXCL10 and eotaxin among the different histopathological subgroups. Patients with a higher inflammatory cell infiltration grade had elevated CXCL10 levels, and patients with eosinophil cell infiltration had high eotaxin levels. Further, patients without suburothelial granuloma had high IL-8 levels. After analyzing bladder histopathological findings according to the ESSIC classification [16], only elevated CXCL10 level was noted in ESSIC type C histopathological subgroup. The area under curve (AUC) and cut-off value (COV) for each biomarker in differentiate specific histopathological finding are: ESSIC classification type C: CXCL10 (AUC 0.652, COV $\geq$ 9.12); Gr 1–3 inflammatory cell infiltration: CXCL10 (AUC 0.627, COV $\geq$ 5.035); presence of eosinophil infiltration: eotaxin (AUC 0.681, COV $\geq$ 3.655); presence of suburothelial granulation: IL-8 (AUC 0.509, COV $\leq$ 14.29); presence of lamina propria hemorrhage: eotaxin (AUC 0.778, COV $\leq$ 3.8).

Table 2. Comparison of urine levels of biomarkers among IC/BPS subtypes with different maximal bladder capacities.

Urine Cytokines	IC/BPS			Control (N = 30)	p-Value [#]	p-Value [@]
	NHIC MBC > 760 (N = 130)	NHIC MBC $\leq$ 760 (N = 161)	HIC (N = 24)			
IL-8	15.1 ± 25.8	17.3 ± 24.3	34.4 ± 39.7 *	12.5 ± 21.0	0.047	0.448
CXCL 10	6.25 ± 11.7	12.9 ± 20.2	35.1 ± 38.2 *	13.8 ± 18.4	0.001	0.001
MCP-1	224 ± 220 *	354 ± 349 *	289 ± 238	147 ± 110	<0.001	<0.001
BDNF	0.58 ± 0.13	0.56 ± 0.14	0.71 ± 0.3	0.55 ± 0.12	0.020	0.305
Eotaxin	5.81 ± 5.81	8.32 ± 7.67 *	12 ± 11.47	4.98 ± 3.7	0.007	0.002
IL-6	1.94 ± 5.95	3.63 ± 7.51 *	10.8 ± 17.4	1.29 ± 1.35	0.020	0.033
MIP-1β	0.86 ± 1.22 *	1.42 ± 1.8 *	1.96 ± 2.8	2.52 ± 1.82	0.034	0.002
RANTES	4.05 ± 8.27	6.24 ± 7.35	10.2 ± 10.1	6.04 ± 5.15	0.006	0.019
TNF-α	1.62 ± 0.35 *	1.66 ± 0.35 *	1.85 ± 0.64 *	0.82 ± 0.33	0.149	0.380
PGE2	253 ± 211 *	320 ± 242 *	302 ± 335	161 ± 105	0.139	0.013
8-OHDG	30.2 ± 21.9 *	34.6 ± 20.5 *	23.9 ± 27.4	18 ± 13.73	0.039	0.079
8-isoprostane	43.2 ± 50.8 *	66.2 ± 72.5 *	34.3 ± 27.5 *	16.8 ± 11.8	<0.001	0.002
TAC	918 ± 782	1290 ± 1055	941 ± 651	1078 ± 925	<0.001	0.001

Abbreviations: same as in Table 1; NHIC: non-Hunner's interstitial cystitis, HIC: Hunner's interstitial cystitis, MBC: maximal bladder capacity, * p values < 0.05 when compared with controls, [#] p values between NHIC subtypes of HIC patients, [@] p values between different NHIC subgroup patients.

Table 3. Comparison of urine levels of biomarkers among NHIC subtypes with different grades of glomerulation and HIC.

Urine Cytokines	(A) GR $\leq$ 1 (N = 159)	(B) GR > 1 (N = 132)	(C) HIC (N = 24)	Total (N = 315)	Control (N = 30)	p-Value [#]	p-Value [$]	Poshoc [$]
IL-8	18.7 ± 29.1	13.54 ± 18.38	34.4 ± 39.7 *	17.6 ± 26.6	12.5 ± 21.0	0.071	0.024	C v AB
CXCL 10	8.5 ± 16	11.61 ± 18.56	35.1 ± 38.2 *	11.5 ± 20	13.8 ± 18.4	0.128	0.003	C v AB
MCP-1	234 ± 226 *	367 ± 365 *	289 ± 239	295 ± 300	147.1 ± 110	<0.001	<0.001	A v B
BDNF	0.57 ± 0.14	0.56 ± 0.14	0.71 ± 0.3	0.58 ± 0.16	0.55 ± 0.12	0.431	0.021	
Eotaxin	6.76 ± 6.73	7.73 ± 7.32 *	12 ± 11.47	7.56 ± 7.51	4.98 ± 3.7	0.245	0.051	
IL-6	2.35 ± 5.64	3.51 ± 8.14 *	10.8 ± 17.4	3.43 ± 8.28	1.29 ± 1.35	0.155	0.027	
MIP-1β	1.14 ± 1.65 *	1.21 ± 1.52 *	1.96 ± 2.8	1.23 ± 1.72	2.52 ± 1.82	0.686	0.262	
RANTES	4.55 ± 8.24	6.1 ± 7.27	10.2 ± 10.1	5.63 ± 8.12	6.04 ± 5.15	0.094	0.014	A v C
TNF-α	1.64 ± 0.32 *	1.65 ± 0.38 *	1.85 ± 0.64 *	1.66 ± 0.38	0.82 ± 0.33	0.854	0.177	
PGE2	263 ± 224 *	323 ± 235 *	302 ± 335	290 ± 239	161 ± 105	0.029	0.213	
8-OHDG	28.2 ± 19.7 *	37.9 ± 21.8 *	23.9 ± 27.4	32 ± 21.8	18.0 ± 13.7	<0.001	<0.001	B v AC
8-isoprostane	49 ± 62.5 *	63.9 ± 66.2 *	34.3 ± 27.5 *	54.1 ± 62.7	16.8 ± 11.8	0.052	0.036	B v AC
TAC	1014 ± 853	1251 ± 1061	941 ± 651	1105 ± 937	1078 ± 925	0.038	0.073	

Abbreviations: same as in Table 2, GR: grade of glomerulation, * p values < 0.05 when compared with controls, [$] p values between NHIC subtypes of HIC patients, [#] p values between different NHIC patients.

Table 4. Correlation analysis of the association between urine biomarker level and bladder characteristics of IC/BPS patients.

		IL-8	CXCL10	MCP-1	BDNF	Eotaxin	IL-6	MIP-1β	RANTES	TNF-α	PGE2	8-OHdG	8-isoprostane	TAC
Glomerulation	Pearson	0.039	0.158	0.141	0.044	0.108	0.126	0.021	0.137	0.086	0.113	0.198	0.067	0.062
	p-value	0.492	0.005	0.013	0.441	0.058	0.026	0.710	0.016	0.126	0.047	0.000	0.242	0.283
MBC	Pearson	−0.093	−0.244	−0.261	−0.042	−0.255	−0.232	−0.185	−0.242	−0.134	−0.161	−0.130	−0.162	−0.095
	p-value	0.103	0.000	0.000	0.456	0.000	0.000	0.001	0.000	0.018	0.005	0.022	0.004	0.099
VAS	Pearson	−0.036	0.096	−0.056	0.161	0.057	0.209	0.008	0.036	0.019	−0.163	−0.148	−0.024	0.027
	p-value	0.601	0.166	0.421	0.019	0.408	0.002	0.903	0.539	0.777	0.018	0.031	0.721	0.02
VUDS-FSF	Pearson	−0.054	−0.032	−0.175	0.086	−0.088	−0.105	−0.073	−0.098	−0.053	−0.061	−0.057	−0.132	−0.011
	p-value	0.358	0.588	0.003	0.141	0.131	0.073	0.215	0.093	0.363	0.301	0.331	0.024	0.855
VUDS-FS	Pearson	−0.110	−0.052	−0.149	0.051	−0.023	−0.146	−0.064	−0.093	−0.055	0.026	−0.030	−0.078	0.038
	p-value	0.061	0.383	0.010	0.379	0.698	0.012	0.273	0.112	0.348	0.663	0.612	0.184	0.518
VUDS-CBC	Pearson	−0.075	−0.077	−0.122	−0.048	−0.077	−0.089	−0.100	−0.076	−0.076	−0.033	−0.014	−0.022	−0.039
	p-value	0.187	0.178	0.031	0.395	0.174	0.118	0.078	0.179	0.182	0.558	0.804	0.700	0.493

Abbreviations: same as in Table 3, MBC: maximal bladder capacity, VAS: visual analog score of pain, VUDS: videourodynamic study, FSF: fist sensation of filling, FS: full sensation, CBC: cystometric bladder capacity.

Table 5. Association between urinary biomarker levels and ESSIC classification and specific bladder histopathological findings in patients with non-Hunner's interstitial cystitis.

Histopathology	Subtype	N	IL-8	CXCL10	MCP-1	BDNF	Eotaxin	IL-6	MIP-1β	RANTES	TNF-α	PGE2	8-OHDG	8-Isoprostane	TAC
Control		30	12.5 ± 21.0	13.8 ± 18.4	147 ± 110	0.55 ± 0.12	4.98 ± 3.7	1.29 ± 1.35	2.52 ± 1.82	6.04 ± 5.15	0.82 ± 0.33	161 ± 105	18 ± 13.73	16.8 ± 11.8	1078 ± 925
ESSIC Classification	Type A	35	18.7 ± 28.0	3.42 ± 4.56	264 ± 287	0.57 ± 0.17	5.99 ± 5.58	3.05 ± 9.80	0.94 ± 1.02	3.46 ± 3.74	1.63 ± 0.36	253 ± 199	36.6 ± 22.2	72.5 ± 78.3	1242 ± 1079
	Type C	152	18.4 ± 25.5	10.9 ± 14.9	328 ± 32	0.57 ± 0.14	7.25 ± 6.68	3.11 ± 7.32	1.26 ± 1.81	5.66 ± 8.90	1.67 ± 0.35	302 ± 237	31.3 ± 19.7	55.7 ± 68.3	1143 ± 928
Inflammatory cell infiltration	Gr 0	56	18.8 ± 28.8	6.27 ± 12.4	275 ± 306	0.57 ± 0.16	6.14 ± 5.48	3.22 ± 9.17	1.06 ± 1.47	5.12 ± 11.8	1.64 ± 0.33	305 ± 275	34.3 ± 21.3	65.9 ± 76.9	1330 ± 1093
	Gr 1-3	131	18.4 ± 24.6	10.9 ± 14.3	333 ± 332	0.57 ± 0.14	7.4 ± 6.87	3.05 ± 7.19	1.26 ± 1.78	5.29 ± 6.1	1.67 ± 0.37	287 ± 209	31.4 ± 19.8	55.9 ± 67.5	1089 ± 885
Urothelial denudation	Gr 0	108	18.3 ± 25.1	9.02 ± 12.2	324 ± 352	0.58 ± 0.15	7.36 ± 6.97	3.23 ± 9.11	1.1 ± 1.45	4.52 ± 6.09	1.67 ± 0.4	276 ± 230	33.7 ± 20.0	65.3 ± 78.9	1215 ± 1030
	Gr 1-3	79	18.7 ± 27.1	10.2 ± 15.9	303 ± 282	0.56 ± 0.14	6.53 ± 5.77	2.93 ± 5.66	1.33 ± 1.99	5.68 ± 10.5	1.66 ± 0.28	314 ± 230	30.2 ± 20.5	49.9 ± 55.2	1091 ± 847
Eosinophil cell infiltration	Presence	16	15.27 ± 15.3	15.1 ± 18.6	412 ± 442	0.54 ± 0.14	10.3 ± 8.99	3.31 ± 6.05	1.59 ± 1.51	7.58 ± 6.36	1.59 ± 0.31	313 ± 187	30.2 ± 22.8	51.1 ± 35.5	837 ± 416
	no	171	18.8 ± 26.7	8.97 ± 13.3	307 ± 311	0.57 ± 0.15	6.72 ± 6.17	3.08 ± 7.98	1.16 ± 1.71	5.02 ± 8.35	1.67 ± 0.36	291 ± 234	32.5 ± 20.0	59.7 ± 72.9	1193 ± 988
Plasma cell infiltration	Presence	22	18.7 ± 17.3	13.7 ± 18.4	377 ± 366	0.53 ± 0.14	8.77 ± 8.45	2.69 ± 4.77	1.78 ± 1.95	6.58 ± 6.28	1.62 ± 0.27	278 ± 170	30.9 ± 24.4	38.1 ± 30.7	983 ± 570
	no	165	18.5 ± 26.9	8.93 ± 13.1	307 ± 318	0.57 ± 0.15	6.79 ± 6.19	3.16 ± 8.15	1.12 ± 1.65	5.06 ± 8.44	1.67 ± 0.37	294 ± 237	32.5 ± 19.7	61.8 ± 73.8	1187 ± 996
Suburothelial granulation	Presence	9	13.3 ± 15.9	16.8 ± 17.5	345 ± 271	0.55 ± 0.18	11.7 ± 10.1	2.22 ± 3.15	2.04 ± 1.89	7.17 ± 6.37	1.54 ± 0.17	299 ± 173	33.6 ± 25.8	42.3 ± 50.1	1237 ± 1195
	no	178	18.7 ± 26.3	9.12 ± 13.6	314 ± 327	0.57 ± 0.15	6.80 ± 6.24	3.15 ± 7.98	1.15 ± 1.68	5.14 ± 8.3	1.67 ± 0.36	292 ± 233	32.2 ± 20.0	59.8 ± 71.3	1158 ± 947
Lamina propria hemorrhage	Presence	5	12.5 ± 13.8	2.24 ± 1.85	221 ± 156	0.50 ± 0.04	2.93 ± 0.65	0.47 ± 0.31	0.23 ± 0.07	0.52 ± 0.65	1.62 ± 0.11	340 ± 278	28.6 ± 20.7	58.0 ± 68.4	671 ± 334
	no	182	18.7 ± 26.2	9.7 ± 14.0	318 ± 328	0.57 ± 0.15	7.13 ± 6.54	3.18 ± 7.91	1.22 ± 1.71	5.36 ± 8.3	1.67 ± 0.36	291 ± 230	32.4 ± 20.2	59.0 ± 70.6	1176 ± 965

Abbreviations: same as in Table 3.

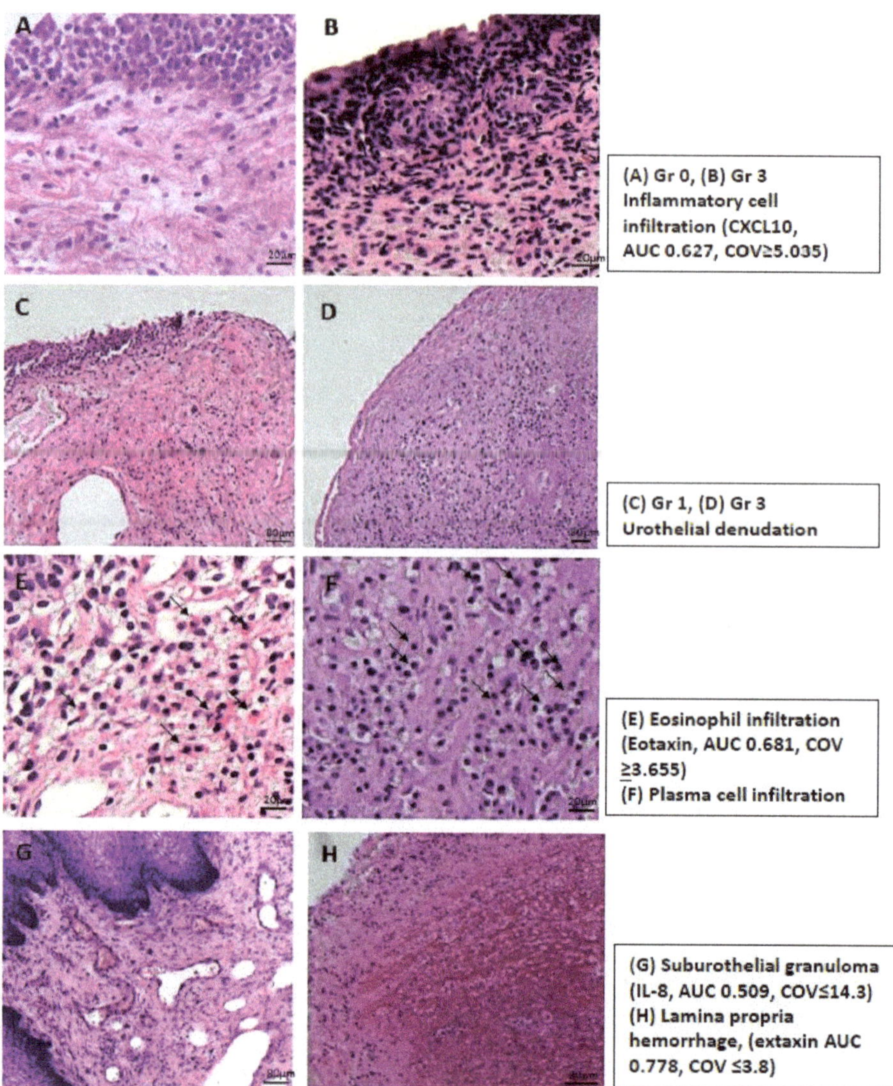

Figure 1. The histopathological findings of interstitial cystitis/bladder pain syndrome and the urine biomarker to differentiate the histopathological finding: (**A,B**) inflammatory cell infiltration, (**C,D**) urothelial denudation grade, (**E**) eosinophil infiltration, (**F**) plasma cell infiltration, (**G**) suburothelial granulation, and (**H**) lamina propria hemorrhage. Bar scale: (**A,B,E,F**) 20 μm, (**C,D,G,H**) 80 μm.

Table 6 shows the urinary levels of oxidative stress biomarkers in controls, patients with HIC, and different NHIC subgroups. Patients with all NHIC subtypes regardless of glomerulation grade and MBC had higher 8-isoprostane levels than controls. NHIC patients with a higher grade of glomerulation and low MBC had the highest 8-isoprostane level. Patients with NHIC who had a low glomerulation grade and high MBC had significantly higher 8-isoprostane levels but not 8-OHdG levels than controls.

Table 6. Urinary oxidative stress biomarker levels of the different IC/BPS subgroups.

Oxidative Stress Biomarker	IC/BPS					(F) Control N = 30	*p*-Value [#]	*p*-Value [$]	Post-Hoc [$]
	(A) GR $\leq$ 1 MBC > 760 N = 85	(B) GR $\leq$ 1 MBC $\leq$ 760 N = 70	(C) GR > 1 MBC > 760 N = 41	(D) GR > 1 MBC $\leq$ 760 N = 89	(E) Hunner's Ulcer N = 24				
8-OHdG	27.0 ± 20.1	29.7 ± 18.8 *	37.3 ± 24.1 *	38.4 ± 20.9 *	23.9 ± 27.4	18 ± 13.7	<0.001	0.001	D vs. ABE, C vs. AE
8-isoprostane	39.1 ± 47.3 *	61.9 ± 77.4 *	52.3 ± 58.8 *	70.2 ± 69.4 *	34.3 ± 27.5 *	16.8 ± 11.8	<0.001	0.003	D vs. AE
TAC	899 ± 713	1165 ± 999	948 ± 941	1413 ± 1101	941 ± 651	1078 ± 925	0.007	0.003	A vs. D

Abbreviations: same as in Table 3, * *p* < 0.05 compared with controls; [#] *p* values between patients with IC/BPS and controls; [$] *p* values between patients with IC/BP.

3. Discussion

IC/BPS is a functional bladder disorder without a characteristic pathology of the bladder or its associated nerves. Previous studies have assessed the mechanisms underlying IC/BPS symptoms, including bladder-centric manifestation, complex processes, and psychological and physical stress [17]. The bladder histopathological features include non-specific chronic inflammation of the urothelium and mast cell infiltration. However, growing evidence has revealed a histological distinction between HIC and NHIC. Pathological evaluation plays an important role in the classification of IC/BPS subtype and clinical management [18].

In the current study, patients with IC/BPS had significantly higher urinary MCP-1, eotaxin, TNF-α, PGE2, 8-OHdG, and 8-isoprostane levels than controls. Patients with HIC had significantly higher urinary IL-8, CXCL10, BDNF, eotaxin, IL-6, MIP-1β, and RANTES levels than those with NHIC. Therefore, there are common pathways associated with bladder dysfunction in HIC and NHIC. However, the overall incidence of bladder inflammatory conditions is higher in HIC than in NHIC. In addition, patients with NHIC who had a low MBC had a higher incidence of bladder inflammation than those with NHIC who had a high MBC. However, these urinary biomarkers are not significant between patients with NHIC with different glomerulation grades after hydrodistention. Therefore, clusters of urinary biomarkers might be more useful to identify HIC and NHIC subtypes; however, currently, we have not found a satisfactory cluster yet.

The pathological findings of IC/BPS include chronic inflammation and urothelial denudation [1,2]. The severity of bladder-wall inflammation is significantly associated with the severity of IC symptoms, glomerulation grade, and MBC under cystoscopic hydrodistention in patients with bladder-centered IC/BPS. Hence, pathological lesions are located inside the urinary bladder in IC/BPS [4]. Increased microvascular endothelial cell apoptosis causes glomerulations in IC, and impaired urothelial homeostasis is associated with chronic bladder inflammation [3,4]. Chronic bladder pain is likely caused by CNS sensitization and the activation of sensory afferent nerves in patients with IC/BPS [5]. Based on the current study, each urinary biomarker was associated with specific clinical features of IC/BPS, such as MBC, glomerulation grade, bladder pain VAS score, bladder filling sensation, and CBC. Patients with NHIC who had higher urinary biomarker levels presented with a lower MBC and CBC, higher glomerulation grade, and more hypersensitive bladder. Therefore, the abnormal clinical features in IC/BPS could be represented by alterations in one or more urinary biomarkers. Using disease-specific urine biomarkers, bladder-centered NHIC could be distinguished from non-bladder-centered NHIC.

Previous studies have shown that cytokines and chemokines play important roles in the pathogenesis of several chronic inflammatory diseases. These cytokines may have a cross-talk with the nervous system, thereby resulting in the hypersensitization of pain receptors and causing pain via neurogenic inflammation. Hence, urine biomarkers that could represent the pathophysiological mechanism of IC/BPS should be identified [19]. However, findings including increased serum or urinary cytokines and chemokines vary widely in patients with IC/BPS. A recent study showed that patients with IC/BPS had significantly high levels of serum proinflammatory cytokines (IL-1β, IL-6, and TNF-α) and chemokines (IL-8) [10].

Regarding cytokines with high diagnostic values for ESSIC type 2 IC/BPS, RANTES, MIP-1β, and IL-8 had a high sensitivity, and MCP-1, CXCL10, and eotaxin had a high specificity. MCP-1, CXCL10, eotaxin, and RANTES were positively correlated with glomerulation grade and negatively associated with MBC [11]. In addition to bladder-centered IC/BPS, some IC/BPS symptoms might be associated with non-bladder-centered psychosomatic syndrome. Using different urinary biomarkers including IL-10, RANTES, eotaxin, CXCL10, IL-12p70, NGF, IL-6, IL-17A, MCP-1, and IL-1RA, IC/BPS could be distinguished from overactive bladder [20]. The current study found that patients with NHIC who have a lower MBC and higher glomerulation grade had elevated levels of urinary inflammatory cytokines and oxidative stress biomarkers. Hence, more higher urinary biomarker levels are likely to indicate worse bladder conditions.

The most common pathological findings of IC/BPS are urothelial denudation and bladder inflammation. However, their histopathological findings are inconsistent [6,7]. Recent studies on electron microscopy and immunohistochemistry have revealed increased umbrella cell pleomorphism and decreased microplicae of the cell membrane in IC/BPS. Patients with moderate to severe defects in umbrella cell integrity and cell membrane microplicae had more severe bladder pain and lower bladder capacity [21]. Our recent study showed that the incidence of inflammatory cell infiltration (69.6%) was significantly higher than that of urothelium denudation (44.6%, $p < 0.001$). There was a significant correlation between inflammatory cell infiltration grade and urothelium denudation. Therefore, patients with chronic inflammation might likely present with urinary frequency and urgency but not bladder pain [4].

Patients with HIC have distinct pathological findings compared with those with NHIC. Patients with HIC have an elevated expression of macrophage migration inhibitory factors and urinary proinflammatory genes [22]. Patients with HIC had significantly high-inflammatory and endoplasmic reticulum stress protein levels [23]. Moreover, they presented with significant overexpression of HIF1α and upregulation of its related biological pathways. These findings indicated that combined ischemia and inflammation might play a pathophysiological role in HIC [24]. Increased urinary CXCL10 levels can be used to differentiate HIC from NHIC with modest sensitivity and high specificity [25]. In this study, patients with HIC had significantly higher urinary CXCL10 levels than those with NHIC, and a high urinary CXCL10 level was associated with a greater inflammatory cell infiltration and ESSIC type C IC/BPS subgroup. NHIC is characterized by severe fibrosis and increased mast cell infiltration and HIC by severe inflammation and urothelial denudation in the whole bladder [26]. In this study, patients with HIC had significantly higher urinary inflammatory and neurogenic protein levels than those with NHIC. However, the oxidative stress biomarker levels were similar between patients with HIC and those with NHIC. Thus, some common pathways might exist between bladder-centered IC/BPS.

In our previous paper, we did not measure urine levels of oxidative stress biomarkers. We also did not compare these urine biomarkers with bladder conditions and histopathological findings [12]. The results of this study help us understand the pathophysiology of IC/BPS. In the current research, patients with IC/BPS had significantly higher urinary MCP-1, eotaxin, TNF-α, PGE2, 8-OHdG, and 8-isoprostane levels than controls. However, there were no significant differences in terms of specific bladder histopathological findings. Patients with IC/BPS who had a high inflammatory cell infiltration grade had elevated urine CXCL10 levels, and urinary eotaxin levels were associated with eosinophil cell infiltration. Nevertheless, other histopathological findings were not correlated well with urinary biomarker levels. Hence, random biopsies of samples collected from patients with IC/BPS might not represent the whole bladder pathology. Based on this result, we hypothesize that the use of a cluster of urinary biomarker levels might be a better tool for identifying IC/BPS, differentiating HIC from NHIC, understanding underlying pathophysiologies, decision-making about treatment strategy, and possibly predicting treatment outcomes in patients with IC/BPS. Although bladder histopathological examination could obtain specific findings, the diagnostic value of these results is limited and cannot provide information on bladder conditions in IC/BPS.

A recent study showed that some inflammatory mediators, such as those in systemic inflammatory diseases, might play important roles in the pathogenesis of IC/BPS [10]. Patients with IC/BPS who have a low MBC more commonly present with acute and chronic inflammation based on histological examination. Therefore, a low bladder capacity might be associated with a distinct bladder-centric IC/BPS phenotype [11]. According to these clinical and proteomic results, IC/BPS might be caused by not only conditions confined to the bladder but also mental factors such as internal conflict and stress disorders [12]. Bladder symptoms might be, in part, attributed to the effects of systemic medical comorbidities. Using a cluster of urinary biomarker levels, we might identify patients with IC/BPS symptoms without bladder-centered pathophysiology. Hence, non-bladder targeting treatments, such as psychiatric consultation and physical therapy, which can help relieve urinary frequency and urgency and pelvic pain symptoms, can be considered.

4. Materials and Methods

The current study enrolled 315 patients (267 women and 48 men) diagnosed with IC/BPS from February 2010 to December 2021. The diagnostic criteria for IC/BPS were based on the ESSIC guidelines and the exclusion of similar diseases [3,16]. This study was approved by the institutional review board and ethics committee of the hospital (approval no.: 105-25-B, 105-31-A, 107-175-A). All patients participated in different clinical trials for the treatment of IC/BPS. Further, they were informed about the purpose of the study, and informed consent were obtained. However, the need for informed consent was waived if urine samples were collected in previous clinical trials.

All patients with IC/BPS were assessed using the O'Leary–Saint symptom score (OSS), which comprised the IC symptom index (ICSI), IC problem index (ICPI), and visual analog scale (VAS) for bladder pain. Patients were admitted for cystoscopic hydrodistention under general anesthesia. Hydrodistention was performed under an intravesical pressure of 80 cm H_2O for 10 min. Next, cystoscopic findings such as petechia, glomerulations, splotch hemorrhage, mucosal fissures, and Hunner's lesions were recorded. Glomerulation grade was classified in accordance with the Asian IC guidelines [15]. In total, 30 women with genuine stress urinary incontinence but without other storage or voiding dysfunctions were included in the control group. The detailed inclusion and exclusion criteria were similar to those in our previous study [11]. Based on glomerulation grade and MBC, patients with NHIC were further classified into different clinical subgroups for comparison [27].

4.1. Evaluation of Bladder Histopathology

A bladder biopsy was performed after cystoscopic hydrodistention. All patients underwent endoscopic cold-cup biopsies of the bladder wall, including the mucosa and submucosa. Biopsy and histopathological evaluations were performed, as described in our previous report [4]. We performed hematoxylin and eosin staining of biopsy specimens collected from patients with IC/BPS. Thereafter, a single pathologist blinded to the clinical results reviewed all bladder histopathological findings. Only bladder specimens large enough for analysis were included. Inflammatory cell infiltration and urothelium denudation were graded using a four-point scale (0: none, 1: mild, 2: moderate, and 3: severe). Eosinophil infiltration, plasma cell infiltration, lamina propria hemorrhage, suburothelial granulation, and nerve hyperplasia in the specimens were classified according to the presence or absence of this finding. The inflammation grade was in accordance with that in our previous report [4].

4.2. Assessment of Urinary Biomarkers

The urine biomarker assessments were in accordance with our previous study [20]. In brief, 50 mL of urine samples were collected from the patients and controls before cystoscopic hydrodistention. The urine samples were obtained via self-voiding upon a full bladder. Urine samples collected from patients with confirmed urinary tract infections were excluded. The samples were placed immediately on ice before being transported to the laboratory. Next, they were centrifuged at 1800 rpm for 10 min at 4 °C. The supernatant was preserved in a freezer at −80 °C. The frozen urine samples were centrifuged at 12,000 rpm for 15 min at 4 °C before further analyses were performed, and were used for subsequent measurements.

4.3. Cytokine and Chemokine Assay

Commercial microspheres were used to assay inflammation-associated urinary cytokines and chemokines with the Milliplex® Human cytokine/chemokine magnetic bead-based panel kit (Millipore, Darmstadt, Germany). Urinary cytokines and chemokines were considered important in the diagnosis of IC/BPS. Thus, 10 targeted analytes, namely, IL-8, CXCL10, MCP-1, eotaxin, IL-6, macrophage inflammatory protein 1 beta (MIP-1β), RANTES, TNF-α (catalog number HCYTA-60K), BDNF (catalog number HNDG3MAG-36K), and PGE2 (Cayman Chemical Co., Ann Arbor, MI, USA, No. 514010), were selected [8,11,12,20]. Then, these analytes were measured using the multiplex kit (catalog number: HCYTMAG-60K-PX30). The procedures used to measure urinary cytokine and chemokine levels were performed based on the manufacturer's instructions and the method utilized in previous studies [12,20]. A total of 25 μL assay buffer, 25 μL urine sample, and 25 μL beads were added sequentially into 96-well plates (panel kits), and the plates were incubated overnight in the dark at 4 °C. After the removal of well contents, the plates were washed twice with 200 μL wash buffer. We added 25 μL detection antibody into each well, and the plates were incubated in the dark on a shaker plate for 1 h at room temperature. Next, 25 μL streptavidin-phycoerythrin solution was added into each well (to form a capture sandwich immunoassay) followed by incubation in the dark for 30 min at room temperature. Repeatedly, the well contents were removed, and the plates were washed twice with 200 μL wash buffer. Finally, 150 μL of sheath fluid was added, and the plates were run on the MAGPIX® instrument with xPONENT® 4.3 software. The median fluorescence intensity of each cytokine/chemokine target was recorded and analyzed to calculate the individual corresponding cytokine/chemokine concentration in urinary samples.

4.4. Urinary Oxidative Stress Assay

The quantifications of 8-OHdG, 8-isoprostane, and total antioxidant capacity in urine samples were performed in accordance with the manufacturer's instructions (8-OHdG ELISA kit, Biovision, Waltham, MA, USA, K4160-100; 8-isoprostane ELIZA kit, Enzo, Farmingdale, NY, USA, DI-900-010; and Total Antioxidant Capacity Assay Kit, Abcam, Cambridge, MA, USA, ab52635). The procedures used in the urine biomarker assay were in accordance with our previous report [13].

The quantification of 8-OHdG in urine was performed; briefly, 50 μL biotin-detection antibody working solution and 50 μL of the sample, were sequentially added to 96-well plates (panel kits), and the plates were incubated for 45 min at 37 °C. The contents of the wells were removed and the plates were washed 3 times with 350 μL wash buffer. The 100 μL of the HRP-streptavidin conjugate working solution was added to each well, and the plates were incubated for 30 min at 37 °C. The solution was discarded the and 350 μL wash buffer were used to wash 5 times. The 90 μL of TMB substrate was added into each well; incubation was then performed in the dark for 30 min at 37 °C. Finally, 50 μL stop solution was added, and the plates were evaluated on the microplate reader at 450 nm.

The quantification of 8-isoprostane in samples was performed in accordance with the manufacturer's instructions. Briefly, 50 μL 8-iso-PGF2α conjugation solution, 50 μL 8-iso-PGF2α antibody solution and 100 μL of the sample were sequentially added to 96-well plates (panel kits), and the plates were incubated for 2 h at room temperature on a plate shaker at 500 rpm. The contents of the wells were removed and the plates were washed 3 times with 400 µL wash buffer. A total of 200 µL of the pNpp substrate solution were added to each well, and the plates were incubated for 45 min at room temperature without shaking. Finally, 50 µL of stop solution was added, and the plates should be read immediately at 405 nm. The measurements of urine 8-isoprostane levels were standardized based on urinary creatinine levels measured using a commercial kit (Enzo Life Sciences Inc., Farmingdale, NY, USA, ADI-907-030A). The median fluorescence intensities of the targets were analyzed to calculate the corresponding concentrations of urinary biomarkers in the samples.

The quantification of TAC in samples was performed; briefly, 100 µL copper (2+) containing working solution and 100 µL sample were sequentially added to 96-well plates (panel kits), and the plates were incubated for 90 min at room temperature on a shaker protected from light. Finally, the plates were evaluated on the microplate reader at 570 nm. The median fluorescence intensities of the target were analyzed to calculate the corresponding TAC concentrations in the samples.

4.5. Statistical Analysis

Continuous variables were expressed as means $\pm$ standard deviations and categorical data as numbers and percentages. Correlation analysis between each urinary biomarker and VAS score, cystoscopic hydrodistention findings (MBC and glomerulation grade), and urodynamic parameters, such as the first sensation of filling (FSF), fullness sensation (FS), and cystometric bladder capacity (CBC), were performed. The urinary biomarker levels between patients with IC/BPS and controls and among the HIC and different NHIC subgroups according to clinical characteristics, urodynamic parameters, and specific histopathological findings were analyzed using analysis of variance. Urinary biomarkers with a mean value below the minimum detectable concentrations were not included in the final analysis. The Statistical Package for the Social Sciences software for Windows version 20.0 (IBM Corp., Armonk, NY, USA) was used for statistical analysis, and p values of < 0.05 were considered statistically significant.

5. Conclusions

Urinary inflammatory, neurogenic, and oxidative stress biomarker levels can be useful for identifying HIC and the different subtypes of NHIC. Higher urinary inflammatory and oxidative stress biomarker levels are associated with poor bladder conditions in patients with IC/BPS. However, most urinary biomarker levels are not correlated with specific bladder histopathological findings. The diagnostic value of urinary biomarkers might be higher than that of bladder histopathology in bladder conditions.

Author Contributions: Conceptualization, Y.-H.J. and H.-C.K.; Data curation, Y.-H.H. and H.-C.K.; Investigation, J.-F.J., Y.-H.J., Y.-H.H. and H.-C.K.; Methodology, J.-F.J., Y.-H.J. and Y.-H.H.; Writing—original draft, J.-F.J. and Y.-H.J.; Writing—review & editing, H.-C.K. All authors have read and agreed to the published version of the manuscript.

Funding: This research was founded by Buddhist Tzu Chi Medical Foundation, grant TCMF-SP-108-01, and grant TCMF-MP-110-03-01.

Institutional Review Board Statement: This study was approved by the institutional review board and ethics committee of the hospital (approval no.: 105-25-B, 105-31-A, 107-175-A).

Informed Consent Statement: The informed consent form was waived because the urine samples were obtained from previous clinical trials.

Data Availability Statement: Data are available by contact with the corresponding author.

Conflicts of Interest: The authors declare no conflict of interest.

References

1. Chancellor, M.B.; Yoshimura, N. Treatment of interstitial cystitis. *Urology.* **2004**, *63* (Suppl. S1), 85–92. [CrossRef] [PubMed]
2. Bouchelouche, K.; Nordling, J. Recent developments in the management of interstitial cystitis. *Curr. Opin. Urol.* **2003**, *13*, 309–313. [CrossRef]
3. Hanno, P.M.; Sant, G.R. Clinical highlights of the National Institute of Diabetes and Digestive and Kidney Diseases/Interstitial Cystitis Association scientific conference on interstitial cystitis. *Urology.* **2001**, *57* (Suppl. S6A), 2–6. [CrossRef]
4. Jhang, J.F.; Hsu, Y.H.; Jiang, Y.H.; Ho, H.C.; Kuo, H.C. Clinical relevance of bladder histopathological findings and their impact on treatment outcomes among patients with interstitial cystitis/bladder pain syndrome: An investigation of the European Society for the study of interstitial cystitis histopathological classification. *J. Urol.* **2021**, *205*, 226–235.
5. Kim, H.J. Update on the pathology and diagnosis of interstitial cystitis/bladder pain syndrome: A review. *Int. Neurourol. J.* **2016**, *20*, 13–17. [CrossRef]
6. Hurst, R.E.; Moldwin, R.M.; Mulholland, S.G. Bladder defense molecules, urothelial differentiation, urinary biomarkers, and interstitial cystitis. *Urology* **2007**, *69* (Suppl. S4), 17–23. [CrossRef] [PubMed]
7. Shie, J.H.; Liu, H.T.; Kuo, H.C. Increased cell apoptosis of the urothelium is mediated by Inflammation in interstitial cystitis/painful bladder syndrome. *Urology* **2012**, *79*, 484.e7–484.e13. [CrossRef] [PubMed]
8. Jiang, Y.H.; Peng, C.H.; Liu, H.T.; Kuo, H.C. Increased pro-inflammatory cytokines, C-reactive protein and nerve growth factor expressions in serum of patients with interstitial cystitis/bladder pain syndrome. *PLoS ONE* **2013**, *8*, e76779. [CrossRef]
9. Dupont, M.C., Spitsbergen, J.M., Kim, K.D., Tuttle, J.D., Steers, W.D. Histological and neurotrophic changes triggered by varying models of bladder inflammation. *J. Urol.* **2001**, *166*, 1111–1118. [CrossRef]
10. Akiyama, Y. Biomarkers in Interstitial cystitis/bladder pain syndrome with and without hunner lesion: A review and future perspectives. *Diagnostics* **2021**, *11*, 2238. [CrossRef]
11. Jiang, Y.H.; Jhang, J.F.; Hsu, Y.H.; Ho, H.C.; Wu, Y.H.; Kuo, H.C. Urine cytokines as biomarkers for diagnosing interstitial cystitis/bladder pain syndrome and mapping its clinical characteristics. *Am. J. Physiol. Renal. Physiol.* **2020**, *318*, F1391–F1399. [CrossRef] [PubMed]
12. Yu, W.R.; Jiang, Y.H.; Jhang, J.F.; Kuo, H.C. Use of urinary cytokine and chemokine levels for identifying bladder conditions and Predicting treatment outcomes in patients with interstitial cystitis/bladder pain syndrome. *Biomedicines* **2022**, *10*, 1149. [CrossRef] [PubMed]
13. Jiang, Y.H.; Jhang, J.F.; Ho, H.C.; Chiou, D.Y.; Kuo, H.C. Urine Oxidative stress biomarkers as novel biomarkers in interstitial cystitis/bladder pain syndrome. *Biomedicines* **2022**, *10*, 1701. [CrossRef] [PubMed]
14. Jhang, J.F.; Kuo, H.C. Pathomechanism of interstitial cystitis/bladder pain syndrome and mapping the heterogeneity of disease. *Int. Neurourol. J.* **2016**, *20* (Suppl. S2), S95–S104. [CrossRef]
15. Homma, Y.; Akiyama, Y.; Tomoe, H.; Furuta, A.; Ueda, T.; Maeda, D.; Lin, A.T.; Kuo, H.C.; Lee, M.H.; Oh, S.J.; et al. Clinical guidelines for interstitial cystitis/bladder pain syndrome. *Int. J. Urol.* **2020**, *27*, 578–589. [CrossRef] [PubMed]
16. van de Merwe, J.P.; Nordling, J.; Bouchelouche, P.; Bouchelouche, K.; Cervigni, M.; Daha, L.K.; Elneil, S.; Fall, M.; Hohlbrugger, G.; Irwin, P.; et al. Diagnostic criteria, classification, and nomenclature for painful bladder syndrome/interstitial cystitis: An ESSIC proposal. *Eur. Urol.* **2008**, *53*, 60–67. [CrossRef]
17. Birder, L.A. Pathophysiology of interstitial cystitis. *Int. J. Urol* **2019**, *26* (Suppl. S1), 12–15. [CrossRef]
18. Akiyama, Y.; Homma, Y.; Maeda, D. Pathology and terminology of interstitial cystitis/bladder pain syndrome: A review. *Histol. Histopathol.* **2019**, *34*, 25–32.
19. Duh, K.; Funaro, M.G.; DeGouveia, W.; Bahlani, S.; Pappas, D.; Najjar, S.; Tabansky, I.; Moldwin, R.; Stern, J.N.H. Crosstalk between the immune system and neural pathways in interstitial cystitis/bladder pain syndrome. *Discov. Med.* **2018**, *25*, 243–250.
20. Jiang, Y.H.; Jhang, J.F.; Hsu, Y.H.; Ho, H.C.; Wu, Y.H.; Kuo, H.C. Urine biomarkers in ESSIC type 2 interstitial cystitis/bladder pain syndrome and overactive bladder with developing a novel diagnostic algorithm. *Sci. Rep.* **2021**, *11*, 914. [CrossRef]
21. Jhang, J.F.; Ho, H.C.; Jiang, Y.H.; Lee, C.L.; Hsu, Y.H.; Kuo, H.C. Electron microscopic characteristics of interstitial cystitis/bladder pain syndrome and their association with clinical condition. *PLoS ONE* **2018**, *13*, e0198816. [CrossRef]
22. Whitmore, K.E.; Fall, M.; Sengiku, A.; Tomoe, H.; Logadottir, Y.; Kim, Y.H. Hunner lesion versus non-Hunner lesion interstitial cystitis/bladder pain syndrome. *Int. J. Urol.* **2019**, *26* (Suppl. S1), 26–34. [CrossRef] [PubMed]
23. Ward, E.P.; Bartolone, S.N.; Chancellor, M.B.; Peters, K.M.; Lamb, L.E. Proteomic analysis of bladder biopsies from interstitial cystitis/bladder pain syndrome patients with and without Hunner's lesions reveals differences in expression of inflammatory and structural proteins. *BMC Urol.* **2020**, *20*, 180. [CrossRef] [PubMed]
24. Akiyama, Y.; Miyakawa, J.; O'Donnell, M.A.; Kreder, K.J.; Luo, Y.; Maeda, D.; Ushiku, T.; Kume, H.; Homma, Y. Overexpression of HIF1α in Hunner Lesions of Interstitial Cystitis: Pathophysiological Implications. *J. Urol.* **2022**, *207*, 635–646. [CrossRef] [PubMed]
25. Niimi, A.; Igawa, Y.; Aizawa, N.; Honma, T.; Nomiya, A.; Akiyama, Y.; Kamei, J.; Fujimura, T.; Fukuhara, H.; Homma, Y. Diagnostic value of urinary CXCL10 as a biomarker for predicting Hunner type interstitial cystitis. *Neurourol. Urodyn.* **2018**, *37*, 1113–1119. [CrossRef]

26. Kim, A.; Han, J.Y.; Ryu, C.M.; Yu, H.Y.; Lee, S.; Kim, Y.; Jeong, S.U.; Cho, Y.M.; Shin, D.M.; Choo, M.S. Histopathological characteristics of interstitial cystitis/bladder pain syndrome without Hunner lesion. *Histopathology* **2017**, *71*, 415–424. [CrossRef]
27. Yu, W.R.; Jhang, J.F.; Ho, H.C.; Jiang, Y.H.; Lee, C.L.; Hsu, Y.H.; Kuo, H.C. Cystoscopic hydrodistention characteristics provide clinical and long-term prognostic features of interstitial cystitis after treatment. *Sci. Rep.* **2021**, *11*, 455. [CrossRef]

International Journal of
Molecular Sciences

Article

Lead Exposure Causes Spinal Curvature during Embryonic Development in Zebrafish

Xueting Li [1,†], Ce Chen [1,†], Mingyue He [1], Lidong Yu [2], Renhao Liu [1], Chunmeng Ma [3,4], Yu Zhang [3,4], Jianbo Jia [5], Bingsheng Li [6] and Li Li [1,*]

1 School of Life Science and Technology, Harbin Institute of Technology, Harbin 150080, China
2 School of Physics, Harbin Institute of Technology, Harbin 150080, China
3 State Key Laboratory of Environmental Aquatic Chemistry, Research Center for Eco-Environmental Sciences, Chinese Academy of Sciences, Beijing 100085, China
4 University of Chinese Academy of Sciences, Beijing 100049, China
5 School of Biotechnology and Health Sciences, Wuyi University, Jiangmen 529020, China
6 Key Laboratory of UV Light Emitting Materials and Technology of Ministry of Education, Northeast Normal University, Changchun 130024, China
* Correspondence: lilili@hit.edu.cn
† These authors contributed equally to this work.

Abstract: Lead (Pb) is an important raw material for modern industrial production, they enter the aquatic environment in several ways and cause serious harm to aquatic ecosystems. Lead ions (Pb^{2+}) are highly toxic and can accumulate continuously in organisms. In addition to causing biological deaths, it can also cause neurological damage in vertebrates. Our experiment found that Pb^{2+} caused decreased survival, delayed hatching, decreased frequency of voluntary movements at 24 hpf, increased heart rate at 48 hpf and increased malformation rate in zebrafish embryos. Among them, the morphology of spinal malformations varied, with 0.4 mg/L Pb^{2+} causing a dorsal bending of the spine of 72 hpf zebrafish and a ventral bending in 120 hpf zebrafish. It was detected that spinal malformations were mainly caused by Pb^{2+}-induced endoplasmic reticulum stress and apoptosis. The genetic changes in somatic segment development which disrupted developmental polarity as well as osteogenesis, resulting in uneven myotomal development. In contrast, calcium ions can rescue the series of responses induced by lead exposure and reduce the occurrence of spinal curvature. This article proposes new findings of lead pollution toxicity in zebrafish.

Keywords: Pb^{2+}; toxicology; endoplasmic reticulum stress; apoptosis; somatic segment development

Citation: Li, X.; Chen, C.; He, M.; Yu, L.; Liu, R.; Ma, C.; Zhang, Y.; Jia, J.; Li, B.; Li, L. Lead Exposure Causes Spinal Curvature during Embryonic Development in Zebrafish. *Int. J. Mol. Sci.* **2022**, *23*, 9571. https://doi.org/10.3390/ijms23179571

Academic Editors: Rossana Morabito and Alessia Remigante

Received: 4 August 2022
Accepted: 23 August 2022
Published: 24 August 2022

Publisher's Note: MDPI stays neutral with regard to jurisdictional claims in published maps and institutional affiliations.

1. Introduction

The heavy metal Pb is widely used in manufacturing as well as metallurgy, and with the variety of lead-containing wastes, it has become one of the most prevalent and toxic environmental pollutants. Although lead emissions have been reduced to a great extent globally, the impact of trace amounts of lead cannot be underestimated [1]. The hazard of trace amounts of lead to living organisms has received continuous attention. Studies have shown that Pb can cause damage to the human nervous system, hematopoietic system, digestive system, cardiovascular system, urinary system, reproductive system and immune system [2,3]. Heavy metals have been shown to compromise antioxidant activity. Lead forms a complex with glutathione, inhibits glutathione peroxidase, and interferes with oxidative phosphorylation. Saturation or compromise of these antioxidant processes results in accumulation of ROS and damage to macromolecules (proteins, lipids, and nucleotides), failure in cell function and death [4]. In some cases, very low doses of Pb can cause irreversible damage to the human body. Spinal malformations have a high environmental sensitivity [5] and are one of the common diseases caused by the long-term accumulation of heavy metals in living organisms. Witeska [6] concluded that heavy metals in living organisms first accumulate in bone tissue, leading to a decrease in Ca and P content causing changes in bone mechanical properties, including body and skeletal deformation.

Zebrafish have high reproductive rates, distinct developmental features, transparent embryos for easy observation, and a high degree of homology with advanced vertebrates in genetics and embryonic development [7]. Therefore, zebrafish has been widely used for toxicological assessment of heavy metals, drugs and organic pollutants [8]. Through a previous study, we found that lead-treated zebrafish produced severe spinal curvature, which is consistent with previous findings [9]. However, there is no convincing explanation for the mechanism by which Pb^{2+} cause spinal curvature. Here, we further explain the cause of spinal curvature in developing zebrafish due to Pb^{2+} by studying different time points, different lead ion concentrations, and the addition of rescue drugs. This study provides new ideas on the causes of spinal curvature in developing zebrafish due to lead exposure, and provides a scientific basis for evaluating the hazard of Pb^{2+} to aquatic organisms and guiding the monitoring of lead pollution in water bodies.

2. Results

2.1. Toxicity of Pb^{2+} to Zebrafish Embryos

Exercise frequency is an important index of zebrafish nervous system development. The frequency of exercise can reflect the stimulation or damage to the nervous system of zebrafish. As shown in Figure 1A, Pb^{2+} at concentrations higher than 0.2 mg/L significantly reduced the level of exercise frequency in zebrafish embryos, indicating some effects on the nervous system of zebrafish. The heart of zebrafish was basically developed by 48 hpf, and the blood circulation system was gradually established [10]. As shown in Figure 1B, Pb^{2+} (0.1–0.8 mg/L) caused heart rate acceleration. Low concentrations of Pb^{2+} (0.1 mg/L and 0.2 mg/L) significantly increased the heart rate and the stimulation to the heart was the most obvious, indicating some effect on the early development of the heart and the circulatory system in zebrafish. As shown in Figure 1C, the hatching rate decreased significantly with increasing Pb^{2+} concentration in a concentration-dependent manner. Delayed hatching occurred in the high concentration group. It indicates that Pb^{2+} delays the developmental period of zebrafish embryos.

The mortality rate reflects the acute toxicity of lead ions to zebrafish embryos. The cumulative survival rate of each group within 120 hpf was recorded. As shown in Figure 1D, the cumulative survival rates of zebrafish embryos were reduced under the exposure of Pb^{2+}. The survival rate of low concentration of Pb^{2+} was higher. The survival rate of low concentration of Pb^{2+} was higher. The highest survival rate was 80% in the 0.4 mg/L group, while the highest concentration of 1.6 mg/L group survived only 20%, indicating that high concentrations of lead ions directly affected the survival of zebrafish. Subsequently, we observed the surviving zebrafish malformations and recorded five common malformations at 120 hpf, as shown in Figure 1E. The overall malformation rate was found to be 100% in each Pb^{2+} experimental group. The main malformations were curvature of the spine and absence of the swim bladder, and there was a co-occurrence of both. Some other malformations were observed in the Pb^{2+} group except for the 0.1 mg/L Pb^{2+} group. The incidence of pericardial cavity edema increased with increasing concentration in a concentration-dependent trend. In addition, a large number of brain hemorrhages were observed in the 1.6 mg/L group, and a unique pigment deficiency was observed in the 0.4 mg/L group. The above results indicate that both low and high concentrations of lead ions are toxic to zebrafish embryos, affecting cardiovascular and neurological development, leading to malformations and even death.

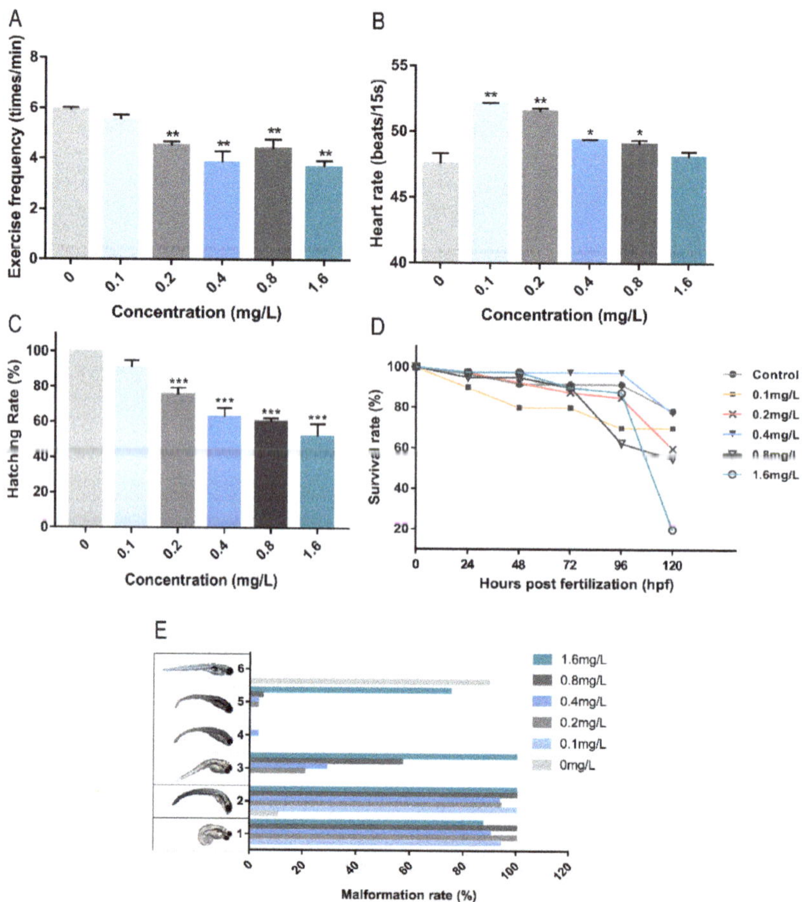

Figure 1. The effects of different concentrations of Pb^{2+} on zebrafish embryos. (**A**) 24 hpf exercise frequency; (**B**) 48 hpf heart rate; (**C**) 57 hpf hatching rate; (**D**) 120 hpf survival rate; (**E**) 120 hpf Malformation rate (1, spinal curvature; 2, absence of swim bladder; 3, pericardial cavity edema; 4, pigment deficiency; 5, cerebral hemorrhage; 6, normal juvenile zebrafish). (* $p < 0.05$, ** $p < 0.01$, *** $p < 0.001$, $n = 15$, repeat three times).

Based on the results of mortality and malformation rate, we found that the survival rate of 0.4 mg/L Pb^{2+} solution was high, and the malformation rate of spinal curvature was high at 120 hpf. Therefore, we chose 0.4 mg/L Pb^{2+} to study the reason of causing spinal curvature. The results showed that the spinal curvature caused by Pb^{2+} started at 72 hpf, and the type of malformation was upward curvature, with 52% of juvenile zebrafish showing upward curvature (Figure 2A). At 96 hpf, the main malformation changed to S-shaped curvature, with 85% of juvenile zebrafish showing S-shaped curvature (Figure 2B). At 120 hpf, the malformation changed to mainly downward curvature, with 100% of zebrafish showing downward curvature (Figure 2C). The specific statistical results are shown in Table 1.

It has been reported that calcium phosphate salts can be used as rescue drugs for Pb exposure [11]. Zebrafish embryos co-treated with 2.0 mmol/L calcium ions (Ca^{2+}) and 0.4 mg/L Pb^{2+} showed almost the same malformation rate as the control group, indicating the elimination of the effect of Pb^{2+} at 120 hpf (Figure 2D).

In this study, the zebrafish spinal curvature model is established using Pb^{2+}. Based on the statistics of malformation and lethality, it is clear that Pb^{2+} caused deformities such as curvature of the spine, and high concentrations have lethal effects, which is consistent with previous studies. From the statistics of malformation rate, it is found that the deformation of swim bladder and spinal curvature are more likely and concurrent. It is found that Pb^{2+} slows down the bone development of zebrafish embryos and causes irregularities in the musculature. Pb^{2+} can cause the loss of calcium ions in the body. The toxicity caused by lead ions can be mitigated by calcium ion rescue experiments.

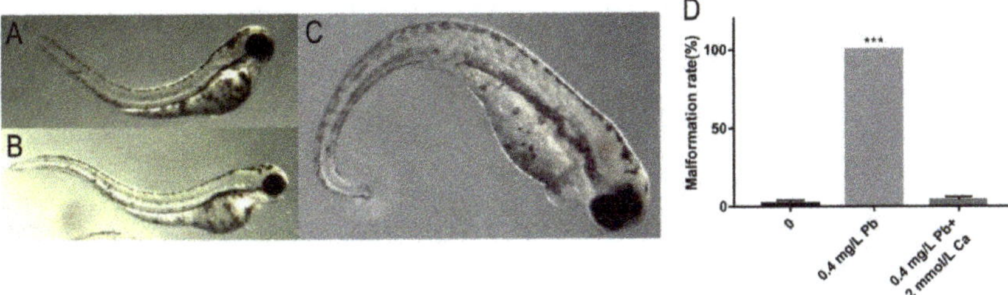

Figure 2. Spinal malformation in zebrafish embryos at different periods of 0.4 mg/L Pb^{2+} exposure. (**A**) Bend upwards at 72 hpf; (**B**) S-shaped curved at 96 hpf; (**C**) Bend downward at 120 hpf; (**D**) Malformation rate at 120 hpf. (*** $p < 0.001$, $n = 50$, Repeat three times).

Table 1. Statistics of Pb^{2+}(0.4 mg/L) on spinal curvature of zebrafish in different periods.

Malformation	72 hpf	96 hpf	120 hpf
upward curvature	$52.0 \pm 1.5\%$	0%	0%
S-shaped curvature	0%	$85.0 \pm 2.5\%$	0%
downward curvature	0%	0%	100%
Total	52%	85%	100%

$n = 50$, repeat three times.

2.2. The Mechanism of Spinal Curvature

Calcein binds Ca^{2+} in tissues and will appear yellow-green under fluorescence. It does not have a large effect on zebrafish and can be used as a good dye for in vivo staining of skeletal development to detect the degree of skeletal calcification in juvenile zebrafish. After staining, hard bones will appear yellow-green, and tissues containing calcium such as muscles can be seen clearly under the fluorescence microscope. This is a very convenient use for observing bones as well as muscle morphology.

Calcein staining was performed on zebrafish at 120 hpf to observe the skeletal ossification and somites development (Figure 3A). It was found that the ossification area of the Pb^{2+} group (0.4 mg/L) was smaller than that of the control group and the calcium rescue group (Figure 3B), and the number of somites was also smaller than that of the other two groups (Figure 3C). In addition, the Pb^{2+} group (0.4 mg/L) showed an uneven distribution of sarcomeres at the site of spinal curvature, showing shortened sarcomeres at one end and longer sarcomeres at the other, which differed significantly from the regular herringbone sarcomeres of the control group (Figure 3A). This suggests that Pb^{2+} contributes to spinal curvature in zebrafish by affecting sarcomeres development and bone formation. Subsequently, we look for the reason at the gene level.

lect1 (chondromodulin) is expressed in cartilage and skull structures [12]. *her12* (hairy-related 12) is involved in the Notch signaling pathway and regulates nervous system development [12]. *rankal* is expressed in head and scale, regulating osteoblast differentiation [12]. Bone morphogenetic protein 2b (*bmp2b*) [13] and RUNX family transcription factor 2b (*runx2b*) [14] together promote osteoblast maturation and differentiation. These genes play an important role in the development of the zebrafish skeletal system and represent the degree of development of the skeletal system. The expression of *lect1* and *her12* was significantly down-regulated after Pb^{2+} treatment, and rankal also showed a down-regulation trend (Figure 3D–F). This demonstrates that Pb^{2+} has serious effects on zebrafish somites and skeletal development. During osteoblast formation, *bmp2b* and *runx2b* were also significantly down-regulated (Figure 3G,H). The decreased expression levels of *lect1*, *her12*, *bmp2b* and *runx2b* together affect the skeletal and somatic segment development in zebrafish. This may be one of the important reasons for the spinal curvature.

In addition, we found that zebrafish with spinal curvature caused by Pb^{2+} were often accompanied by swim bladder deficiency. There is a linkage relationship between these two deformities. The Hedgehog signaling pathway is an important pathway that controls swim bladder development [15]. As shown in Figure 3I, the expression of Indian hedgehog (*ihh*) gene was not significantly different from the control until 96 hpf. However, this gene was significantly upregulated at 120 hpf. Meanwhile, the Sonic hedgehog (*shh*) gene was significantly decreased at 120 hpf (Figure 3J), indicating that Pb^{2+} affected swim bladder development through the Hedgehog pathway. Changes in the expression levels of *ihh* and *shh* may affect developmental polarity leading to the appearance of spinal curvature [16]. The above results suggest that Pb^{2+} regulates the development of zebrafish somites mainly through *lept1* and *her12*, inhibits skeletal development in zebrafish through *bmp2b* and *runx2b*, and affects the process of skeletal and swim bladder development through *ihh* and *shh*.

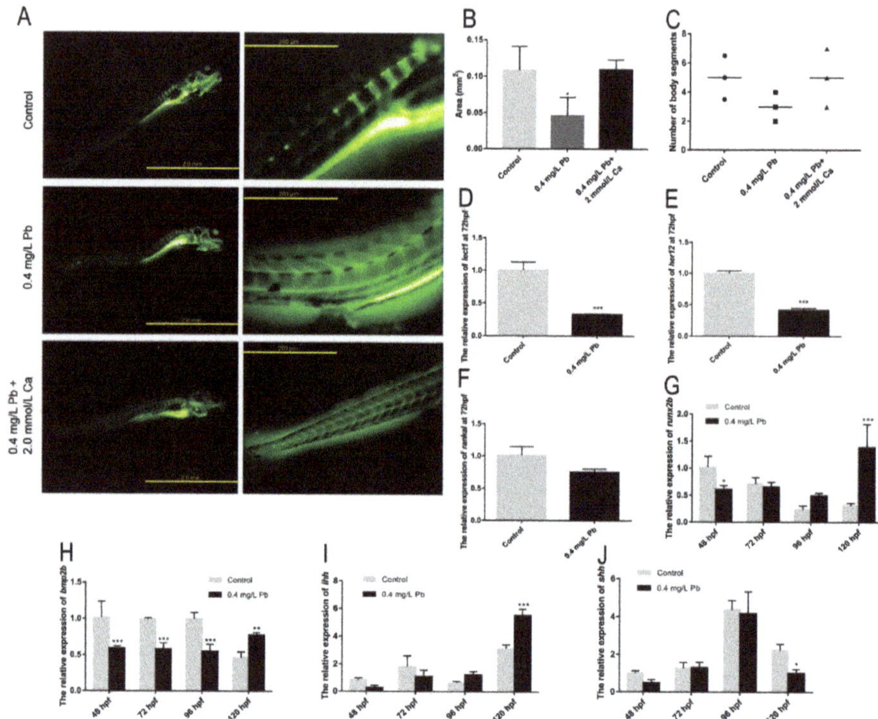

Figure 3. Skeletal and somites development. (**A**) Calcein staining; (**B**) Area of ossification in juvenile zebrafish; (**C**) Body segment statistics. The expression of the somites development-related genes is shown for (**D**) *lect1*, (**E**) *her12*, (**F**) *rankal* (**G**) *runx2b*, (**H**) *bmp2b*. The expression of Hedgehog signaling pathway gene of zebrafish embryos is shown for (**I**) *ihh* and (**J**) *shh*, with *ef1α* as reference. (* $p < 0.05$, ** $p < 0.01$, *** $p < 0.001$, $n = 50$).

2.3. Pb²⁺ Induce Apoptosis by Activating Endoplasmic Reticulum (ER) Stress

Current studies have shown that oxidative damage is one of the important causes of organismal damage caused by Pb^{2+}. It is also reported that the ER stress pathway is also involved in the regulation of *runx2b*, a gene related to bone development. Among them, *perk/chop* are important genes involved in ER stress, and they can regulate *runx2b* with osteoblast differentiation and maturation. It was found that both *perk* and *chop* were significantly upregulated in the Pb^{2+}-treated group compared with the control group at 48 hpf, as shown in Figure 4A,B. Therefore, it is demonstrated that Pb^{2+} induce ER stress in zebrafish embryos. Activation of ER stress chop signaling pathway triggers apoptosis.

In apoptotic cells, apoptotic bodies are formed due to chromatin consolidation or breakage into pieces of varying sizes. Acridine orange stained them with yellow-green fluorescence to detect apoptotic cells, as shown in Figure 4C. It was found that a large amount of green fluorescence appeared in the zebrafish muscular segment part in the Pb^{2+} experimental group. This indicates that Pb^{2+} promotes the occurrence of apoptosis in zebrafish muscle cells. Combined with the previous deformation and uneven distribution of muscle nodes, it suggests that abnormal muscle cell development and apoptosis may be closely associated with their spinal curvature.

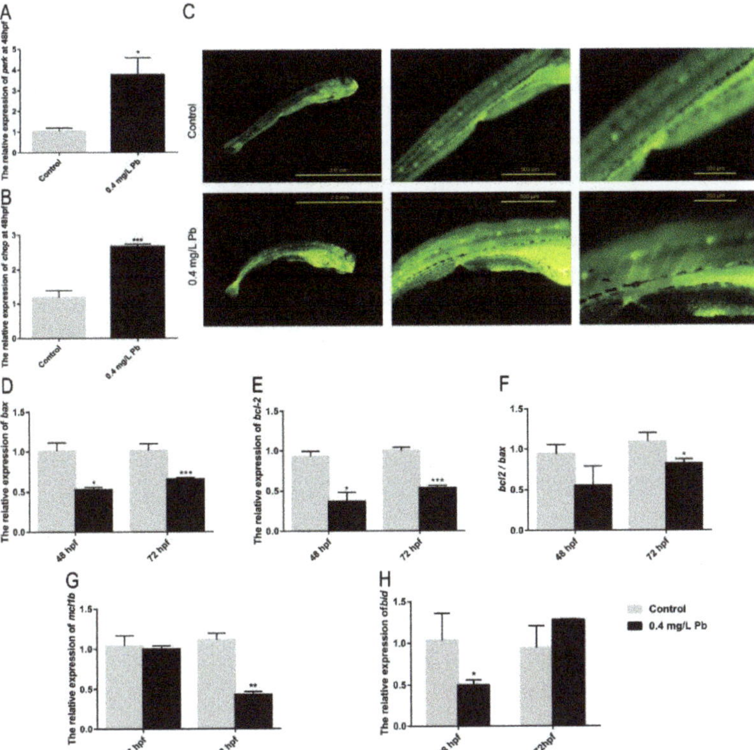

Figure 4. mRNA expression of ER stress and apoptosis-related genes in zebrafish. mRNA expression of ER stress-related genes in zebrafish embryos at 48 hpf using ef1α as reference: (**A**) *perk*, (**B**) *chop*. (**C**) AO staining; mRNA expression of apoptosis-related genes in zebrafish embryos at 48 and 72 hpf; (**D**) *bax*; (**E**) *bcl-2*; (**F**) *bcl-2/bax* ratio; (**G**) *mcl1b*; (**H**) *bid*. (* $p < 0.05$, ** $p < 0.01$, *** $p < 0.001$, $n = 50$).

One of the key indicators of apoptosis is whether the ratio of anti-apoptotic gene *bcl-2* to the pro-apoptotic gene *bax* is down-regulated. The expression of *bcl-2* and *bax* genes was detected by RT-PCR (Figure 4D,E). Both genes showed significant down-regulation at

48 hpf as well as 72 hpf relative to the control group. Additionally, at 72 hpf, the *bcl-2* and *bax* ratios appeared significantly down-regulated (Figure 4F). This indicates that apoptosis occurred at 72 hpf. Meanwhile the anti-apoptotic gene *mcl1b* was significantly decreased at 72 hpf, and the pro-apoptotic gene *bid* was increased (Figure 4G,H). It further demonstrated that Pb^{2+} promoted apoptosis in myocytes.

3. Discussion

Studies have shown that Pb^{2+} increases oxidative stress levels, causing endoplasmic reticulum stress early in development [3]. Pb^{2+} trigger ER stress, which is involved in protein folding and calcium homeostasis. ER stress leads to abnormal endoplasmic reticulum function, triggering apoptosis and inflammation [17]. The molecular chaperone of PERK, BIP, is one of the most abundant proteins in the ER. PERK is considered a major sensor of the unfolded protein response, and BIP separates from PERK when the ER stress response is received, activating PERK to cause apoptosis [18]. CHOP is considered a marker protein to promote apoptosis [19]. The increase in PERK and CHOP in this study indicates that the ER stress response activates the onset of apoptosis.

On the one hand, the accumulation of Pb^{2+} in vivo triggers apoptosis, which deforms the musculature and leads to spinal curvature. On the other hand, Pb^{2+} stimulates the Hedgehog signaling pathway, causing the downward curvature of the spine. During endochondral ossification, secreted signal, Ihh has been shown to regulate the onset of hypertrophic differentiation of chondrocytes. BMPs, family of secreted factors regulating bone formation, have been implicated as potential interactors of the IHH [20]. Functional integration or Synergistic regulation between Hedgehog signaling pathway and BMPs signaling pathway promotes ALP expression and osteogenic differentiation [21]. In addition, Pb^{2+} affects the formation of bones and muscle nodes by suppressing the notochord-associated *shh*. It is because lead ions can lead to the loss of bone calcium [22]. We suppose that the cause of Pb^{2+}-induced spinal curvature is the loss of calcium in vivo, which is caused by the competition between Pb^{2+} and Ca^{2+}. The reduced activity of calcium-binding calmodulin protein causes downregulation of the *shh*, which affects *bmp*, causing changes in *ihh*, *runx2b*, and leading to spinal curvature.

This study provides two possible mechanisms of Pb^{2+}-induced spinal curvature, namely, endoplasmic reticulum stress–apoptosis pathway and bone development-related gene regulation pathway. Both exhibit obvious changes, but which one is the former needs further research. It provides a scientific basis for evaluating the hazard of Pb^{2+} to aquatic organisms and guiding the monitoring of lead pollution in water bodies.

4. Materials and Methods

4.1. Materials and Reagents

Lead nitrate, calcium phosphate, Tricaine and paraformaldehyde used in the experiments were purchased from Biopped, Paterson, NJ, USA. Alizarin red, calcein and acridine orange were purchased from Solarbio Co., Ltd. (Beijing, China). Trizol was purchased from TaKaRa, Kyoto, Japan.

4.2. Breeding and Spawning of Zebrafish

The AB strain zebrafish used in this experiment was obtained from the Heilongjiang Fisheries Research Institute under the Chinese Academy of Fisheries Science, and was raised in our laboratory for more than two months. The water temperature was 28 °C, and the light/dark cycle was 14 h/10 h. The food was fed twice a day, at 9:00 and 15:00, with live *Artemia* cysts (Beijing, China).

4.3. Zebrafish Pb²⁺ Exposure Experiment

Deionized water (DW) was used to wash fertilized eggs and then transferred to 24-well cell culture plates for Pb^{2+} exposure experiments. The maximum allowed discharge concentration of Pb^{2+} is 1.0 mg/L according to the National Comprehensive Discharge

Standard GB 8978-1996 of the People's Republic of China. Therefore, this experiment used lead nitrate at concentrations of 0.1 mg/L, 0.2 mg/L, 0.4 mg/L, 0.8 mg/L and 1.6 mg/L according to the national standard concentration (1.0 mg/L), with deionized water as the control. Three parallel replicates were set up with 45 fertilized eggs per group using DW as control. The culture plates were placed in a biochemical constant-temperature incubator, at a constant 28 °C. The culture medium was changed every 24 h during the experiment, and dead individuals were removed.

The frequency of exercise was observed at 24 hpf, the heart rate was measured for 15 s at 48 hpf, and the hatching rate was counted at 54 hpf. The number of surviving individuals was recorded every 24 h during the entire cycle of the experiment. Hatched larvae were anesthetized with Tricaine and fixed in 4% paraformaldehyde. Photographs and statistics of deformities were taken with stereomicroscope.

4.4. Alizarin Red Staining

Zebrafish embryos were fixed with 4% PFA for 2 h, rinsed twice with PBS, bleached with 1.5% H_2O_2/1% KOH for 5 min, rinsed twice with 25% glycerol/0.1% KOH for 20 min, stained with 0.05% alizarin red for 30 min avoiding light, rinsed twice with 50% glycerol/0.1% KOH, and stored at 4 °C with 50% glycerol/0.1% KOH.

4.5. Calcein Staining

Zebrafish embryos were added with 0.2% calcein solution, kept in the dark for 1 h, washed with E3 solution and left for 2 h to release the excess dye in the embryos, and washed twice with PBS. After zebrafish embryos were anesthetized with 0.05% Tricaine for 5 min, the development of fish skeleton was observed and photographed by fluorescence microscope.

4.6. Acridine Orange Staining

The zebrafish embryos were stained with 2 mg/L AO dye, and stained at room temperature for 20 min in the dark, and then washed twice with PBS solution. Zebrafish embryos were anesthetized with 0.05% Tricaine for 5 min. Apoptosis was observed in fluorescence microscopy and photographed.

4.7. Semi-Quantitative Polymerase Chain Reaction (PCR)

In this experiment, RNA was extracted by Trizol method. The purity and concentration of RNA samples were detected by gel electrophoresis and UV spectrophotometer and stored at −80 °C. PrimeScriptTM RT reagent Kit with gDNA Eraser was purchased from TaKaRa, Inc. The Premix rTag kit (TaKaRa) was used, using reverse-transcribed cDNA as the template, and the housekeeping gene used was ef1α. The primer sequences are shown in Table S1.

Supplementary Materials: The following supporting information can be downloaded at: https://www.mdpi.com/article/10.3390/ijms23179571/s1.

Author Contributions: All authors contributed to the study conception and design. Conceptualization: L.L., methodology: R.L. and C.M., formal analysis and investigation: L.Y., writing—original draft preparation: X.L., writing—review and editing: C.C. and M.H., resources: J.J., B.L. and Y.Z., supervision: L.L. and J.J. All authors have read and agreed to the published version of the manuscript.

Funding: This work was supported by the National Key Technologies R&D Program of China under Grant Number: 2019YFA0705202, the National Natural Science Foundation of China (No. 31701296, 11474076 and 21974097), the Education Department of Guangdong Province (No. 2020KSYS004), the National Key Research and Development Program of China (No. 2021YFC3200804), and the Open Project of State Key Laboratory of Urban Water Resources and Water Environment, Harbin Institute of Technology (No. ES202116).

Institutional Review Board Statement: Ethical review and approval were waived for this study due to the zebrafish used in this experiment is under the 5 dpf (European Animal Research Association 2010/63/EU).

Data Availability Statement: Data are available on request due to restrictions, e.g., privacy or ethical. The data presented in this study are available on request from the corresponding author. The data are not publicly available as part of the data will be used for another unpublished article.

Conflicts of Interest: The authors declare that they have no known competing financial interest or personal relationship that could have appeared to influence the work reported in this paper.

References

1. Resongles, E.; Dietze, V.; Green, D.C.; Harrison, R.M.; Ochoa-Gonzalez, R.; Tremper, A.H.; Weiss, D.J. Strong evidence for the continued contribution of lead deposited during the 20th century to the atmospheric environment in London of today. *Proc. Natl. Acad. Sci. USA* **2021**, *118*, e2102791118. [CrossRef] [PubMed]
2. Rossi, E. Low level environmental lead exposure—A continuing challenge. *Clin. Biochem. Rev.* **2008**, *29*, 63. [PubMed]
3. Roy, N.M.; De Wolf, S.; Carneiro, B. Evaluation of the developmental toxicity of lead in the *Danio rerio* body. *Aquat. Toxicol.* **2015**, *158*, 138–148. [CrossRef]
4. Dunham, J. Encyclopedia of Environmental Health. *Libr. J.* **2011**, *136*, 104.
5. Kolstad, K.; Thorland, I.; Refstie, T.; Gjerde, B. Genetic variation and genotype by location interaction in body weight, spinal deformity and sexual maturity in Atlantic cod (*Gadus morhua*) reared at different locations off Norway. *Aquaculture* **2006**, *259*, 66–73. [CrossRef]
6. Witeska, M.; Jezierska, B. The effects of environmental factors on metal toxicity to fish (review). *Fresen. Environ. Bull* **2003**, *12*, 824–829.
7. Hill, A.J.; Teraoka, H.; Heideman, W.; Peterson, R.E. Zebrafish as a model vertebrate for investigating chemical toxicity. *Toxicol. Sci.* **2005**, *86*, 6–19. [CrossRef]
8. Dai, Y.J.; Jia, Y.F.; Chen, N.; Bian, W.P.; Li, Q.K.; Ma, Y.B.; Chen, Y.L.; Pei, D.S. Zebrafish as a Model System to Study Toxicology. *Env. Toxicol. Chem.* **2014**, *33*, 11–17. [CrossRef]
9. Dou, C.M.; Zhang, J. Effects of lead on neurogenesis during zebrafish embryonic brain development. *J. Hazard Mater.* **2011**, *194*, 277–282. [CrossRef]
10. Kimmel, C.B.; Ballard, W.W.; Kimmel, S.R.; Ullmann, B.; Schilling, T.F. Stages of Embryonic-Development of the Zebrafish. *Dev. Dynam.* **1995**, *203*, 253–310. [CrossRef]
11. Cao, X.D.; Ma, L.Q.; Chen, M.; Singh, S.P.; Harris, W.G. Impacts of phosphate amendments on lead biogeochemistry at a contaminated site. *Environ. Sci. Technol.* **2002**, *36*, 5296–5304. [CrossRef] [PubMed]
12. Mundlos, S.; Olsen, B.R. Heritable diseases of the skeleton. Part I: Molecular insights into skeletal development-transcription factors and signaling pathways. *FASEB J.* **1997**, *11*, 125–132. [CrossRef] [PubMed]
13. Gan, S.; Huang, Z.; Liu, N.; Su, R.; Xie, G.; Zhong, B.; Zhang, K.; Wang, S.; Hu, X.; Zhang, J.; et al. MicroRNA-140-5p impairs zebrafish embryonic bone development via targeting BMP-2. *FEBS Lett.* **2016**, *590*, 1438–1446. [CrossRef] [PubMed]
14. van der Meulen, T.; Kranenbarg, S.; Schipper, H.; Samallo, J.; van Leeuwen, J.L.; Franssen, H. Identification and characterisation of two runx2 homologues in zebrafish with different expression patterns. *Biochim. Biophys. Acta* **2005**, *1729*, 105–117. [CrossRef] [PubMed]
15. Winata, C.L.; Korzh, S.; Kondrychyn, I.; Zheng, W.L.; Korzh, V.; Gong, Z.Y. Development of zebrafish swimbladder: The requirement of Hedgehog signaling in specification and organization of the three tissue layers. *Dev. Biol.* **2009**, *331*, 222–236. [CrossRef] [PubMed]
16. Braunstein, J.A.; Robbins, A.E.; Stewart, S.; Stankunas, K. Basal epidermis collective migration and local Sonic hedgehog signaling promote skeletal branching morphogenesis in zebrafish fins. *Dev. Biol.* **2021**, *477*, 177–190. [CrossRef] [PubMed]
17. Musatov, A.; Robinson, N.C. Susceptibility of mitochondrial electron-transport complexes to oxidative damage. Focus on cytochrome c oxidase. *Free Radic. Res.* **2012**, *46*, 1313–1326. [CrossRef]
18. Huang, R.; Hui, Z.; Wei, S.; Li, D.; Li, W.; Daping, W.; Alahdal, M. IRE1 signaling regulates chondrocyte apoptosis and death fate in the osteoarthritis. *J. Cell Physiol.* **2022**, *237*, 118–127. [CrossRef]
19. Schonthal, A.H. Endoplasmic reticulum stress: Its role in disease and novel prospects for therapy. *Science* **2012**, *2012*, 857516. [CrossRef]
20. Minina, E.; Wenzel, H.; Kreschel, C.; Karp, S.; Gaffield, W.; McMahon, A.; Vortkamp, A. BMP and Ihh/PTHrP signaling interact to coordinate chondrocyte proliferation and differentiation. *Development* **2001**, *128*, 4523–4534. [CrossRef]
21. Reichert, J.; Schmalzl, J.; Prager, P.; Gilbert, F.; Quent, V.; Steinert, A.; Rudert, M.; Nöth, U. Synergistic effect of Indian hedgehog and bone morphogenetic protein-2 gene transfer to increase the osteogenic potential of human mesenchymal stem cells. *Stem Cell Res. Ther.* **2013**, *4*, 105. [CrossRef] [PubMed]
22. Polak-Juszczak, L. Impact of strontium on skeletal deformities in Baltic cod (*Gadus morhua callaris* L.). *Chemosphere* **2011**, *83*, 486–491. [CrossRef] [PubMed]

International Journal of
Molecular Sciences

Review

Recent Developments in the Understanding of Immunity, Pathogenesis and Management of COVID-19

Aram Yegiazaryan [1], Arbi Abnousian [1], Logan J. Alexander [1], Ali Badaoui [1], Brandon Flaig [1], Nisar Sheren [1], Armin Aghazarian [1], Dijla Alsaigh [1], Arman Amin [1], Akaash Mundra [1], Anthony Nazaryan [1], Frederick T. Guilford [2,*] and Vishwanath Venketaraman [1,*]

1   College of Osteopathic Medicine of the Pacific, Western University of Health Sciences, Pomona, CA 91766, USA
2   Your Energy System, LLC 555 Bryant St. #305, Palo Alto, CA 94301, USA
*   Correspondence: drg@readisorb.com (F.T.G.); vvenketaraman@westernu.edu (V.V.); Tel.: +1-909-706-3736 (V.V.); Fax: +1-909-469-5698 (V.V.)

Abstract: Coronaviruses represent a diverse family of enveloped positive-sense single stranded RNA viruses. COVID-19, caused by Severe Acute Respiratory Syndrome Coronavirus-2, is a highly contagious respiratory disease transmissible mainly via close contact and respiratory droplets which can result in severe, life-threatening respiratory pathologies. It is understood that glutathione, a naturally occurring antioxidant known for its role in immune response and cellular detoxification, is the target of various proinflammatory cytokines and transcription factors resulting in the infection, replication, and production of reactive oxygen species. This leads to more severe symptoms of COVID-19 and increased susceptibility to other illnesses such as tuberculosis. The emergence of vaccines against COVID-19, usage of monoclonal antibodies as treatments for infection, and implementation of pharmaceutical drugs have been effective methods for preventing and treating symptoms. However, with the mutating nature of the virus, other treatment modalities have been in research. With its role in antiviral defense and immune response, glutathione has been heavily explored in regard to COVID-19. Glutathione has demonstrated protective effects on inflammation and downregulation of reactive oxygen species, thereby resulting in less severe symptoms of COVID-19 infection and warranting the discussion of glutathione as a treatment mechanism.

Keywords: glutathione; coronavirus; SARS-CoV-2; COVID-19; vaccines; reactive oxygen species

check for
updates

Citation: Yegiazaryan, A.; Abnousian, A.; Alexander, L.J.; Badaoui, A.; Flaig, B.; Sheren, N.; Aghazarian, A.; Alsaigh, D.; Amin, A.; Mundra, A.; et al. Recent Developments in the Understanding of Immunity, Pathogenesis and Management of COVID-19. *Int. J. Mol. Sci.* **2022**, *23*, 9297. https://doi.org/10.3390/ijms23169297

Academic Editors: Rossana Morabito and Alessia Remigante

Received: 5 August 2022
Accepted: 16 August 2022
Published: 18 August 2022

Publisher's Note: MDPI stays neutral with regard to jurisdictional claims in published maps and institutional affiliations.

1. Introduction

In December of 2019, a rise in pneumonia cases of unknown origin were identified and linked to a wholesale market for seafood and wet animals in Wuhan, Hubei Province, China. Through the isolation and sequencing of human airway epithelial cells, the Chinese Center for Disease Control and Prevention (Chinese CDC) discovered a previously unknown betacoronavirus, coined 2019 novel CoronaVirus (2019-nCoV) [1]. The 2019-nCoV pathogen, now referred to as Severe Acute Respiratory Syndrome Coronavirus-2 (SARS-CoV-2), is responsible for causing the coronavirus disease infamously known around the world as COVID-19 [2]. Declared a global pandemic by the World Health Organization (WHO), SARS-CoV-2 is a highly contagious respiratory disease transmissible mainly via close contact and respiratory droplets, and less commonly through airborne, bloodborne, and animal-to-human modes of transmission. Clinical features vary from asymptomatic to flu-like symptoms (fever and cough) to severe pneumonia resulting in acute respiratory distress, multiorgan failure, and potentially death [3]. As of June 2022, the WHO reports over 528 million confirmed cases of SARS-CoV-2 and over 6 million SARS-CoV-2 related deaths globally [4]. In addition, the Centers for Disease Control and Prevention (CDC) reports a total of over 84 million confirmed cases of SARS-CoV-2 and over 1 million SARS-CoV-2 related deaths in the United States [5]. The information reviewed in this publication is intended to highlight a support for a host pathway, exploited by viruses for their replication. Liposomal glutathione offers a host directed adjunct for therapy that can be used against respiratory viruses.

2. Results

2.1. SARS-CoV-2

Coronaviruses (CoVs) are a diverse family of enveloped positive-sense single stranded RNA viruses and consist of four genera: alpha, beta, gamma, and delta [2]. Alpha and beta CoVs exclusively infect mammalian species, whereas gamma and delta CoVs are additionally capable of infecting avian species. Currently, there are seven CoV species that are known to cause human disease, referred to as human coronaviruses (HCoVs). Four of these HCoVs—HCoV-229E, HCoV-OC43, HCoV-NL63, and HCoV-HKU1—result in seasonal and mild respiratory tract infections similar to symptoms of the common cold. On the other hand, three HCoVs, i.e., Severe Acute Respiratory Syndrome Coronavirus (SARS-CoV) and Middle East Respiratory Syndrome Coronavirus (MERS-CoV), and most recently SARS-CoV-2, can result in severe, life-threatening respiratory pathologies via infection of bronchial epithelial cells, pneumocytes and upper respiratory tract cells [1,6].

The large genome of CoVs is packed within a helical capsid formed from the nucleocapsid protein (N) and surrounded further by an envelope. The envelope is further associated with three additional structural proteins. The membrane (M) protein and the envelope (E) protein are responsible for virus assembly. The club-shaped spike (S) protein projections extend from the CoVs spherical surface and are responsible for entrance of the virus into host cells [7]. The S protein consists of three segments: an ectodomain, a single-pass transmembrane anchor, and a short intracellular tail. The ectodomain further consists of a subunit responsible for receptor-binding (S1) and a subunit responsible for membrane-fusion (S2). Following attachment of the receptor binding domain (RBD) of the S1 subunit to its angiotensin-converting enzyme 2 (ACE2) receptor on the host cell surface, the S2 subunit fuses the host and viral membranes resulting in entrance of the viral genome into human cells [2,7]. Upon entrance to the cytoplasm of the host cell, the products of replication, transcription, and translation of the viral structural proteins are assembled and released via exocytosis, allowing for the virus to continue to spread within an infected organism and to other humans via horizontal transmission [2].

2.2. Evolving into Multiple Variants

Due to its ability to adapt to environments through mutations and recombination, and therefore altering host range and tissue tropism, SARS-CoV-2 has evolved into numerous variants referred to as variants of concern (VOCs). Evolution of CoVs is made possible due to their large genome, high mutation rate, and high recombination frequency. Mutations that result in greater fitness are selected for, resulting in VOCs. VOCs are recognized by their increased transmissibility, immuno-escape from neutralizing antibodies or T-cell immunity, and pathogenicity [7–9]. According to the WHO, to be classified as a SARS-CoV-2 variant, the virus of concern must demonstrate one or more of the following: increase in transmissibility or detrimental change in SARS-CoV-2 epidemiology, increase in virulence or change in clinical disease presentation, and/or decrease in effectiveness of public health and social measures, available diagnostics, vaccines, or therapeutics. The WHO recognizes the following five variants of SARS-CoV-2: alpha (PANGO lineage B.1.1.7), beta (PANGO lineage B1.1.351), gamma (PANGO lineage P.1), delta (PANGO lineage B.1.617.2), and most recently omicron (PANGO lineage B.1.1.529) [10].

The alpha variant, which emerged out of the United Kingdom in September 2020, is understood to have functional mutations to the S protein resulting in increased transmission and increased resistance to antibody-mediated neutralization [9,11]. First appearing in South Africa in October of 2020 is the beta variant, which has functional mutations to the S protein, similarly resulting in increased transmission [9,12]. The gamma variant emerged out of Brazil in July of 2020 and is believed to have mutations in the RBD of the S protein resulting in increased transmission. However, less is understood about this variant [9,13]. The delta variant, emerging out of India in October of 2020, contains mutations to the S protein resulting in increased electrostatic interactions between the RBD and ACE2 receptor (RBD-ACE2), resulting in an increase in RBD-ACE2 stability and therefore increased viral infectivity and virus replication. In addition, the delta variant is noted to have increased resistance to monoclonal antibodies, convalescent sera, and vaccinated sera, all of which have been vital treatments for CoVs infections [9]. The Omicron variant, emerging out of multiple countries in November of 2021, is characterized by mutations to the S protein, similar to the Delta variant, leading to increased binding affinity between the RBD and ACE2 receptor and ultimately increased spread of the virus [8,10]. The omicron variant has undergone further changes, leading to descendent lineages with different genetic makeup. This has led to reduction in neutralizing titers post immunization or prior infection, with currently available vaccines. However, data shows significant reduction in hospitalization rate with Omicron infection compared to the Delta variant [10].

2.3. Increased Susceptibility to COVID-19 among Risk Groups

Although SARS-CoV-2 virus has been infectious to all age groups, regardless of their health status, age and pre-morbidities have been shown to increase the risk of adverse consequences when infected [14]. Comorbidities, such as diabetes, chronic obstructive pulmonary disease (COPD), cardiovascular diseases (CVD), hypertension, malignancies, human immunodeficiency virus (HIV), and obesity, among others, have also been correlated with increased risk of severe symptom presentation when infected with SARS-CoV-2 [14,15]. Individuals with certain comorbidities highly express surface ACE2 receptors on cells [15], which bind the S1 protein found on SARS-CoV-2 virus [14]. Once bound, the S2 protein on SARS-CoV-2 facilitates the fusion of viral membrane with host cell membrane [14]. The high release of proprotein convertase in individuals with comorbidities facilitates the endocytosis of the virus into the host cells [15]. The enhanced facilitation of viral entry into host cells in individuals with comorbidities leads to significantly higher risk of morbidity and mortality [15]. People suffering from diabetes mellitus (DM) and/or obesity have impaired innate and adaptive immune systems which produce more pro-inflammatory cytokines and fewer anti-inflammatory cytokines than healthy individuals [16]. This suboptimal level of immunity can be further exacerbated when infected with SARS-CoV-2, leading to higher mortality and morbidity among these individuals [16].

2.4. Pathogenesis of the Disease Induced by Cytokine Storm, Increased IL-6, Decreased Glutathione

It appears that all viruses deplete reduced glutathione (GSH), an intracellular antioxidant, as part of their strategy for replication [17]. In a study designed to analyze the different pathways exploited by viruses for replication, the influenza viruses increased the production of reactive oxygen species (ROS) and decreased the production of GSH. This led to a state of oxidative stress [18–20]. In response to an infection, the body releases Interleukin-6 (IL-6) and other proinflammatory cytokines. As shown in Figure 1, IL-6 blocks the enzymes Iodothyronine deiodinases type I (D_1) and II (D_2) responsible for conversion of prohormone thyroxine T4 to active T3 [21]. In astrocytes, thyroid hormones upregulate glutamate cysteine ligase (GCL), which enhances GSH synthesis [22]. Therefore, blocking thyroid synthesis due to a viral infection can suppress GSH synthesis and lead to increased oxidative stress in cells.

Nicotinamide adenine dinucleotide phosphate (NADPH) oxidase (NOX) family consists of seven members: NOX1 to NOX5 and the two dual oxidases, Duox1 and Duox2, which are expressed in most cell types [23]. The NOX family mediates ROS production in cells which plays an important role in the survival of these cells. During a viral infection, inflammatory cells release NOX2 which enhances the pathogenicity of viruses, similar to influenza A viruses [24]. It has been shown that inhibition of NOX2 oxidase activity lessens influenza A virus-induced lung inflammation in mice [25]. GSH acts as the main intracellular antioxidant against ROS through oxidation of the thiol group of its cysteine to a disulfide bond (GSSG), hence reducing the oxidized species and maintaining the redox state of the cell. GSSG is then converted back to GSH through the action of glutathione reductase. GSH also neutralizes the potentially harmful metals and xenobiotics in the body [26]. In addition, GSH can act as a signaling molecule to the innate immune system [26–28]. The decrease in GSH levels in infected individuals allowed viral glycoprotein haemagglutinin folding and maturation and hence viral replication [29].

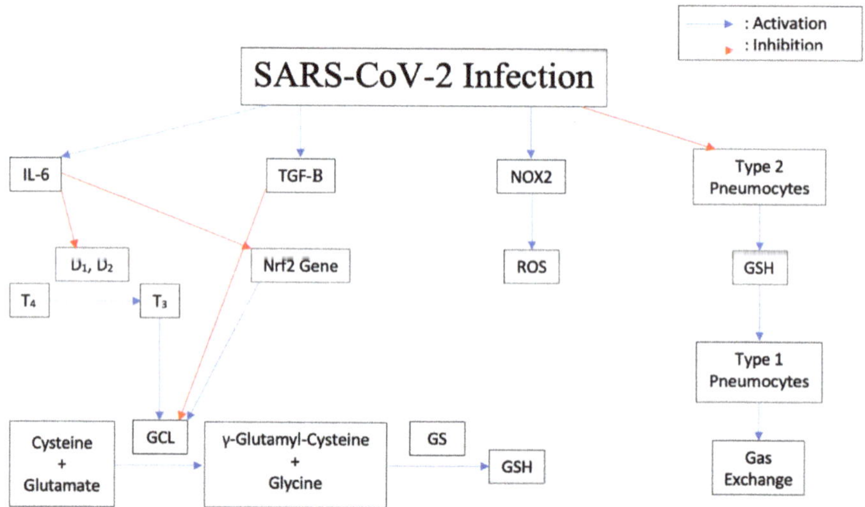

Figure 1. Downstream effects of SARS-CoV-2 Infection on cytokine release, thyroid conversion, ROS production, GSH inhibition, and pneumocyte damage in humans.

Elevated serum concentrations of IL-6 and TGF-β were noted in patients with severe SARS-CoV-2 symptoms compared to those with less severe symptoms [30]. It has been shown that IL-6 induced a dose-dependent decrease in intracellular GSH levels in a number of human cell lines, including lung cells [21]. Additionally, an animal study has shown that IL-6 can down regulate the expression of Nrf2 target genes resulting in decreased expression of the enzyme GCL, which is needed to form GSH in cells, leading to a decreased availability of GSH [31]. Research showed that transforming growth factor β (TGF-β) is present in SARS-CoV-2 patients and can be classified as a strong predictor of SARS-CoV-2 disease severity [32]. Notably, high levels of monocyte chemoattractant protein-1 (MCP-1) and TGF-β1 were identified in an autopsy for SARS-CoV-2 infected lung cells [33]. It has been shown that TGF-β suppresses GCL gene expression and induces oxidative stress in a lung fibrosis model [34]. GCL controls GSH formation intracellularly [35]. The elevation of TGF-β adds additional suppression of GSH formation during SARS-CoV-2 infection. High levels of TGF-β1 may trigger fibrotic changes and may account for the typical early CT scan features of SARS-CoV-2 pneumonia such as the "ground-glass opacity" appearance [36,37]. Fibrosis is a major complication related to SARS-CoV-2 infection [37,38] and is related to TGF-β elevation [34]. The high levels of MCP-1 explain the findings of a mononuclear inflammatory infiltrate with lymphocytes and CD68+ macrophages in the lungs of SARS-CoV-2 patients [38–41].

Both respiratory syncytial virus (RSV) and SARS-CoV-2 infections deplete GSH and decrease the formation of the enzymes needed to form GSH in the cells they invade [17], which occurs in immune cells as well as other cells such as lung Type II pneumocytes. Type II pneumocytes are responsible for the production of GSH, which is found at a very high level in the epithelial lining fluid (ELF) of the lung [42]. When SARS-CoV-2 reaches the lower respiratory tract, it will bind to ACE2 on Type II pneumocytes, leading to viral infection and decreased function. When the Type II pneumocyte function is compromised by viral infection, the level of GSH in the ELF can become diminished, which is associated with compromise of lung function. This can lead to acute respiratory distress syndrome (ARDS) [43], which is associated with SARS-CoV-2 infection [44]. SARS-CoV-2 ARDS appears to have worse outcomes [44]. Type I pneumocytes are supported by Type II pneumocytes, and damage leads to decreased function of the Type I pneumocyte pool. Damage to Type I pneumocytes leads to impaired gas exchange (1). This compromise of gas exchange with a buildup of carbon dioxide tension (Pa_{CO2}) blunts the brain's response to hypoxia and leads to the presentation of individuals with "silent hypoxia." Thus, patients with SARS-CoV-2 are described as exhibiting oxygen levels incompatible with life without dyspnea [45].

2.5. Increased Susceptibility to TB Due to Decrease in Glutathione

Research has shown that the survival of intracellular *Mycobacterium tuberculosis* (Mtb) is enhanced due to the increased uncleared free radicals and inflammatory cytokines as a result of GSH depletion [46]. Therefore, immunocompromised individuals, such as those with Human immunodeficiency virus (HIV) and Type 2 Diabetes Mellitus (T2DM), are at higher risk of contracting Mtb infection [46]. HIV infection is associated with an accumulation of ROS [47] mediated by the envelope protein gp120 [48] and Tat proteins [49]. Furthermore, NOX2 and NOX4-induced ROS overproduction has been reported in HIV gp120 treated astrocytes [50]. An excess of ROS persists in HIV+ individuals even after successful HAART therapy and results in a depletion of GSH [35]. It appears that the decrease in GSH is a result of the production of cytokines including IL-1, TNF-α and IL-17, and TGF-β [35]. TGF-β interferes with the biosynthesis of GSH [51]. Elevation of both IL-6 and TGF-β has also been shown to accompany the loss of GSH in HIV+ individuals and to be decreased by the administration of liposomal GSH [52,53].

According to research, immunological abnormality plays a significant role in the increased susceptibility to Mtb infection in T2DM individuals [54]. Through innate immunity, the elevated release of cytokines in diabetic individuals leads to a decrease in GSH production, and hence an increase in oxidative stress in cells [35]. In addition, compared to healthy individuals, people with T2DM have lower levels of GCL, the rate-limiting enzyme in GSH synthesis [46]. This leads to a significantly lower plasma GSH:GSSG ratio in diabetic patients when compared to controls [55]. Consequently, ROS accumulate in cells, and have been associated with further complications of T2DM [55]. Through adaptive immunity, research revealed that a decrease in Th1:Th2 cytokines can increase the susceptibility to Mtb infection in T2DM individuals [54]. However, maintenance of normal GSH levels promotes Th-1 differentiation via IL-12 and IL-27 cytokines that were otherwise downregulated in immunocompromised patients [46].

2.6. Modulation of Expression of Antioxidant Genes

Nuclear factor erythroid 2p45-related factor 2 (Nrf2) is a transcription factor, which has evolved as an oxidant-sensitive molecule that is activated and will transcriptionally stimulate a series of genes responsible for cytoprotection and detoxification [56]. Nrf2 is one of the best-characterized antioxidative transcription factors with an oxidant/electrophile-sensor function [57]. Under normal conditions, it forms a complex with Kelch-like ECH-associated protein 1 (KEAP1), a well-known negative regulator of Nrf2 [58]. Since KEAP1 serves as an adaptor protein for cullin-3-based E3 ubiquitin ligase, this dimeric Nrf2/KEAP1 complex subjects Nrf2 to constant ubiquitination and proteasomal degradation [59]. In regard to their oxidant-sensing mechanisms, redox-sensitive twenty-five cysteine residues of KEAP1 were shown to play a key role in the regulation of the E3 ubiquitin ligase activity [60]. Essentially, these cysteine residues are very susceptible to conjugation of a variety of ROS-inducing agents. Once conjugated, the KEAP1-mediated ubiquitination of Nrf2 is severely diminished [61]. This leads to liberation of Nrf2 from the KEAP1-mediated restraint. Once stabilized, Nrf2 is able to get inside the nucleus and form complexes with one of small musculoaponeurotic fibrosarcoma proteins and other coactivators. As shown in Figure 2, binding of this trimeric complex to the antioxidant response elements (AREs) in the promoter regions facilitates the transcription of a series of cytoprotective and detoxifying genes, such as heme oxygenase-1 (HO-1), NAD(P)H quinone oxidoreductase-1 (NQO-1), GCL catalytic subunit (GCLC) and GCL regulatory subunit (GCLM), glutathione S-transferase (GST), uridine diphosphate glucuronosyltransferase (UDPGT), superoxide dismutase isoform 1 (SOD1), catalase (CAT), glucose 6 phosphate dehydrogenase (G6PD), and glutathione peroxidase-1 (GPx) [62–65]. Importantly, GCLC and GCLM are critical components of the production of GSH [35].

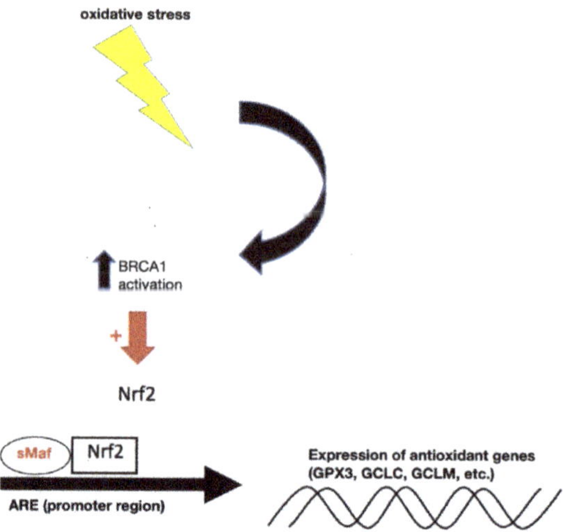

Figure 2. Under normal conditions, BRCA1 located in the nucleus facilitates the expression of Nrf2, leading to upregulation of the ARE to produce antioxidant genes important for GSH synthesis. Abbreviations: BRCA1, breast Cancer gene 1; Nrf2, nuclear factor erythroid 2–related factor 2; ARE, antioxidant responsive element; GPX3, glutathione peroxidase 3; GCLC, glutamate-cysteine ligase catalytic subunit; GCLM, glutamate-cysteine ligase regulatory subunit.

Viruses possess a variety of adaptive mechanisms to deplete GSH in host cells. The HIV virus can decrease the expression of GCLC and GCLM in HIV+ macrophages to about half and GSH is known to be deficient in individuals with HIV [35,66]. Respiratory Syncytial Virus (RSV) was shown to use NOX2 as an essential regulator of RSV-induced NF-κB activation. RSV virus infection led to continuous activation of NF-κB, which likely caused excessive NF-κB-mediated inflammatory gene expression [67]. A later study found that RSV infection down-regulates Nrf2 expression in airway epithelial cells and that a decrease in the expression of airway antioxidant enzymes led to additional oxidative stress [68]. Nrf2 mRNA levels were decreased following RSV infection and the nuclear localization of the protein was decreased in infected cells compared to uninfected ones, resulting in increased oxidation stress and a significant decrease in the GSH:GSSG ratio [69].

In COVID-19 infection, depletion of GSH begins with the binding of SARS-CoV-2 S protein to ACE2, which results in inhibition and decrease of ACE2 expression in infected cells and leads to toxic overaccumulation of ANGII [70]. The increased ANGII, through binding to AT1R, activates NADPH oxidases (NOX) that transfer an electron from NADPH to O_2, generating several radical species which are scavenged by GSH and deplete GSH [70]. As GSH is depleted, ROS-mediated oxidation increases. Although SARS-CoV-2 proteins are synthesized in the cytosol, some viral proteins are also detectable in the nucleus, including Nsp1, Nsp5, Nsp9, Nsp13, Nsp14, and Nsp16 [71]. This becomes important as recent research has shown that the S protein also localizes in the nucleus and inhibits DNA damage repair by impeding key DNA repair protein BRCA1, and recruitment of 53BP1 to the damage site [72].

BRCA1 is known as a tumor suppressor as mutations in this gene confer an increased risk for breast, ovarian, and prostate cancers [73]. Lesser well known is BRCA1's function in stimulating the antioxidant response element (ARE), driving transcriptional activity and its enhancement of the antioxidant response transcription factor Nrf2 [73]. Thus, protecting against oxidative stress is important in BRCA1's role as a caretaker gene [73]. BRCA1 increased the expression of some of these Nrf2-regulated genes (NQO1, MGST1/2, Gsta2, G6PD, and ME2), and BRCA1 also induced (and BRCA1 mutation inhibited) expression of glutathione peroxidase (GPX3) [73]. In addition to endogenous reactive oxygen species, which contribute to carcinogenesis, many DNA-damaging agents and xenobiotics cause oxidative stress, resulting in DNA damage, protein oxidation, and lipid peroxidation [74]. Some of these lesions are detoxified by BRCA1-regulated genes (e.g., GSTs, GPXs, oxidoreductases, and paraoxonases) [73]. BRCA1 stimulates GSH production under oxidizing conditions, possibly by increasing the levels of glucose-6-phosphate dehydrogenase (G6PD). G6PD can then stimulate NADPH production, which plays a role in the regeneration of GSH [73].

These findings suggest the S protein may decrease both the intracellular and intranuclear antioxidant protection of GSH. If the S protein of SARS-CoV-2 localizes to the nucleus and decreases BRCA1 recruitment and thus the benefit of Nrf2 function, evidence of oxidation of nuclear material might be evident. A decrease in available BRCA1 may contribute to the finding of increased oxidized guanine species that have been identified at higher levels in the serum of non-surviving individuals than in those surviving SARS-CoV-2 [75]. Additionally, it was shown that IL-6 induced a dose-dependent decrease in intracellular GSH levels in a number of human cell lines, including lung cells [21]. IL-6 was shown to increase the levels of superoxide radicals, leading to the decrease of intracellular GSH [21]. Increased expression of IL-6 is seen in severe SARS-CoV-2 patients, and its levels are even used to gauge the severity of the disease [76]. An animal study showed that IL-6 can down regulate the expression of Nrf2 target genes resulting in decreased expression of the enzyme glutamyl cysteine ligase (GCL), leading to decreased availability of GSH [31]. These findings highlight the importance of IL-6 and the S protein in maintaining a pro-inflammatory state and reducing the antioxidant protection mechanisms needed to deal with ROS.

2.7. COVID-19 Can Increase the Risk for Development of TB

As previously mentioned, the S protein of SARS-CoV-2 plays a significant role in the viral pathogenicity and its downstream consequences. When BRCA1 is inhibited, ARE production is downregulated, resulting in decreases in antioxidant molecules such as GSH [77]. Due to the excessive oxidative stress caused by SARS-CoV-2 infection, cells have a diminished capacity to undergo de novo GSH synthesis. GSH depletion has been associated with increased risk of opportunistic infections such as Mtb [77]. Large increases in ROS, in addition to GSH depletion, place individuals at increased risk of Mtb. The ability to ward off active Mtb is dependent on CD4+ T-cells. These cells play an integral role in adaptive immunity as well as providing "memory", to fight off repeat infections more effectively. Opportunistic infections like HIV and Mtb infections have been shown to rely heavily on CD4+ T-cells, which serve as predictors for risk of infections [78]. Both Mtb and SARS-CoV-2 infections invade lung pneumocytes. Patients infected with SARS-CoV-2 have an increased susceptibility to Mtb. SARS-CoV-2 inhibits several mediators which can help fight off Mtb infection. Under normal circumstances, Mtb enters respiratory airways as either droplets or aerosols and ultimately ends up in the alveoli. At this point the bacteria is phagocytosed by alveolar macrophages (AMs). The pathogenicity of Mtb has been well described in the literature and includes the interruption of key signaling pathways including cyclic GMP-AMP synthase (cGAS), stimulator of interferon genes (STING), which involve interferons type I and II. Type II interferon, also called interferon-gamma (IFN-γ), is the major cytokine involved in protection against Mtb infections [79].

Management of Mtb infection involves both innate and adaptive immune cells. The first line of defense includes phagocytic cells which are found in the lung epithelium. Integral adaptive immunity responses include CD4+ T-cells that produce IFN-γ, which serves to activate AMs [80]. These AMs within the lung epithelium serve as an important entry point for SARS-CoV-2 viruses as they possess the ACE2 receptors. In the presence of SARS-CoV-2 infection, these AMs become damaged and are less capable of producing normal amounts of IFN-γ [81]. A specific type of SARS-CoV-2 infection, called ORF9b, involves a unique accessory protein (ORF9b) that antagonizes the cGAS-STING pathway and leads to a decreased ability to combat many inflammatory processes, including but not limited to infections [82]. As shown in Figure 3, SARS-CoV-2 ORF9b inhibits phosphorylation of TANK binding kinase 1 (TBK1, a serine/threonine protein kinase) leading to the inhibition of IRF3 and its nuclear translocation, and hence type I interferon transcription. The antagonism of the cGAS-STING pathway ultimately leads to a down regulation of de-novo interferon I synthesis, as well as IFN-γ which is produced by activated lymphocytes, including AMs [83]. With these decreased levels of both IFN-γ and cGAS-STING pathway activity, Mtb is better able to proliferate in the presence of SARS-CoV-2 infection. Additionally, Mtb is able to exit phagosomes via ESX-1, a type VII secretion system utilized by the bacteria [84]. Once in the cytoplasm, the bacterial DNA can be recognized by the cGAS-STING pathway and upregulate type I interferons and IFN-γ. One study in mice found that cGAS-STING pathway is greatly increased during Mtb infections in vitro, which is critical for the pulmonary expression of IFN-β (a type I interferon) [85].

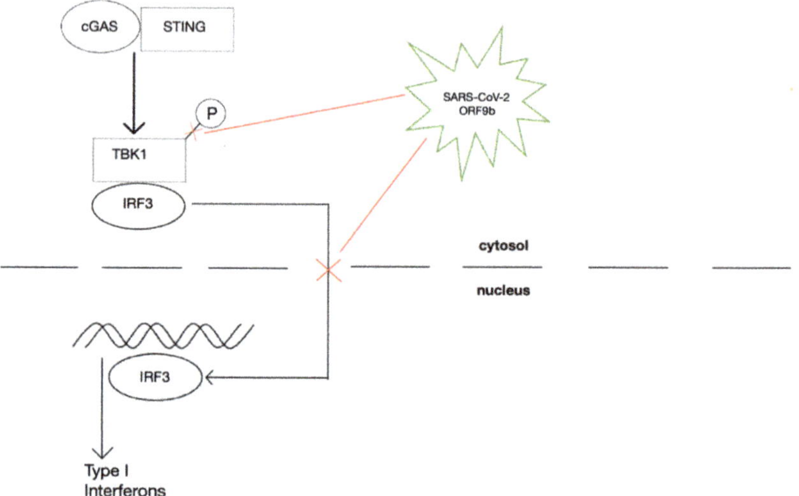

Figure 3. cGAS-STING pathway inhibition by SARS-CoV-2 ORF9b which consequently reduces normal production of type I interferons. Abbreviations: cGAS, cyclic GMP-AMP synthase; STING, stimulator of interferon genes; TBK1, TANK-binding Kinase 1; IRF3, Interferon Regulatory Factor 3.

Overall, there are several key factors that are negatively affected by SARS-CoV-2 infection, including GSH synthesis, IFN-γ and cGAS-STING pathways. The interference with these pathways by SARS-CoV-2 infection places patients at a greater risk for the development of Mtb. It is important to note the concurrent infections of SARS-CoV-2 and Mtb, as they are both respiratory illnesses, can rapidly spread and have large socioeconomic impacts on a global scale. The mortality rate of Mtb and SARS-CoV-2 coinfections is 12.3%, which is much higher than SARS-CoV-2 infection alone [86]. Mtb is the second leading cause of deaths caused by respiratory illness, after SARS-CoV-2 [87]. With a large number of Mtb infections going unnoticed, it would be beneficial to improve screening and testing for latent Mtb to avoid more severe cases of people developing concurrent

Mtb and SARS-CoV-2 infections. As previously mentioned, SARS-CoV-2 primarily infects the respiratory epithelium, subsequently invading and activating alveolar macrophages (AMs) [88]. AM activation occurs via signals such as IFN-γ and causes the release of a host of pro-inflammatory cytokines such as IL-6, which has been shown to be elevated in SARS-CoV-2 infected individuals.

2.8. Thrombus Formation during SARS-CoV-2 Infection

One potential major complication of more severe stages of SARS-CoV-2 infection is the increased risk of thrombus formation. GSH serves an integral role in protecting against ROS, helping to lower ACE activity, and has demonstrated the ability to downregulate the production of Nuclear Factor Kappa B (NF-kB) [71]. NF-kB serves as an important mediator of inflammation and coagulation. Vascular endothelium, monocytes, and even neutrophils all respond to NF-kB in a pro-coagulatory manner and contribute to thrombus formation [89]. NF-kB signaling leads to numerous physiological changes, including vascular endothelial changes promoting adhesion molecules that allow leukocyte migration into tissues. Additionally, neutrophils can be signaled to release their DNA from the cell and become neutrophil extracellular traps (NETs). It is thought that NETs' main function is to trap and kill microbes, but this has also been associated with clot formation [89]. These NETs interact with the complement system and interestingly also provide a framework for thrombus formation to begin [90]. In the pulmonary vasculature, the hyperinflammatory response seen with SARS-CoV-2 infection results in the accumulation and localization of NETs along with platelets, which contribute to thrombus formation. One suggested treatment to reduce these NET/platelet complexes and even reduce the activation of endothelial cells is by the administration of N-acetyl cysteine (NAC), a precursor molecule for GSH [89].

Matrix metalloproteinases (MMPs) are important enzymes involved in processes including degradation of the extracellular matrix (ECM) and can be activated by NF-kB [91,92]. These MMPs are widespread throughout the body and are important in epithelial tissues, such as the gastrointestinal tract, respiratory system, and vascular endothelium. As they relate to vascular endothelium, MMPs degrade the ECM to allow for vessel remodeling under both normal and pathological conditions. With MMPs activated, epithelial cells become more permeable, allowing for healthy tissue turnover, but also lends the opportunity for pathogenesis, including cardiovascular diseases and infiltration by host defenses such as NET/platelet complexes. In one study looking at patients with systemic lupus erythematosus (SLE), MMP-2 inhibition resulted in lower levels of NETs and improved endothelial function [93]. We can see that when GSH levels are within normal ranges, less NF-kB will be produced, ultimately leading to lower levels of MMPs and NETs, which can prevent thrombus formation.

Furthermore, during high oxidative stress situations, such as during SARS-CoV-2 infection, RBCs can adhere to the endothelial lining of blood vessels and contribute to the clot formation [94]. When GSH levels are low, or there is a low GSH:GSSG ratio, platelet aggregation is potentiated [95]. There is evidence that additional factors involved in the clotting cascade become elevated during SARS-CoV-2 infection, such as von Willebrand factor, Factor III (tissue factor), and free thrombin [96]. Romagnoli et al. (2012) found that correction of the GSH:GSSG ratio led to decreased production and activity of MMP-2 in intestinal subepithelial myofibrobalsts (ISEMFs) [97]. This suggests that restoration of normal levels of GSH (or GSH:GSSG ratio) can serve as protection against exaggerated increases in MMP-2 levels and other elements that relate to the clotting cascade.

Lower GSH levels are associated with elevation in many clotting elements and pro-coagulatory components, such as NETs, MMPs, NF-kB, and vessel endothelial changes, placing patients at a greater risk for microvascular thrombosis and platelet aggregation. Thus, it is important to maintain adequate levels of GSH as it can protect against thrombosis in more severe forms of SARS-CoV-2 infections.

2.9. Vaccines, Vaccine Evasion, and Protection against COVID-19

Development of the major vaccines used for SARS-CoV-2 were based on the experience with severe acute respiratory syndrome coronavirus (SARS-CoV) and Middle East respiratory syndrome coronavirus (MERS-CoV). Currently, there are multiple vaccines that have emergency-use approval in a large number of countries, which differ in their type and technique used to elicit an immune response. There are mRNA vaccines, such as BNT162b2 (Pfizer) and mRNA-1273 (Moderna), and adenoviral-vectored vaccines, including Ad26.COV2.S (Janssen) and ChAdOx1nCoV-19 (University of Oxford/AstraZeneca), all of which focus on providing antibodies against the S protein of SARS-CoV-2 [98–100]. As previously mentioned, the S protein allows the virus to bind to the ACE2 receptor and enter its host cell, making the S protein an important target in vaccine development [101]. Once antibodies have been created against the S protein, the body can produce a robust immune response to fight off future infection.

The immune response to SARS-CoV-2 in humans involves both humoral and cell-mediated immunity [102,103]. Antibodies directed at the S protein increase the chances of survival, and neutralizing antibodies (NAbs) targeted at the S protein are present in most individuals [104]. Studies have shown that the amount of NAbs directed at the S protein is a strong indicator for survival, and it is imperative for vaccines to elicit NAb responses [105,106]. Indeed, both mRNA vaccines, the BNT162b2 and mRNA-1273, have been shown to increase levels of RBG-binding IgG antibodies and neutralizing antibody titers when compared to those without the vaccine [107,108]. These antibodies play an important role in neutralizing and preventing viral infection and have been correlated with increased survival [109]. In addition, BNT162b2 can stimulate the production of antiviral cytokines, such as IFN-γ, which has been shown to inhibit the replication of SARS-CoV-2 [110,111]. Both mRNA vaccines elicit a Th1 immune response, with levels of antigen specific CD8+ memory cells and CD4+ T-cells greatly increased [112]. Clinical trials with these mRNA vaccines showed two doses of BNT162b2 and mRNA-1273 were 95% and 94.1% effective at preventing the original SARS-CoV-2 virus, respectively [113,114].

The Ad26.COV2.S and ChAdOx1nCoV-19 vaccines are adenoviral-vectored that target the full-length S protein of SARS-CoV-2. Overall, they provide a very similar immune response as the mRNA vaccines, with both adenovirus vaccines increasing levels of spike antibodies and RBG specific IgG, IgA, and IgM antibodies [115,116]. Likewise, they also induce a largely Th1 cellular immune response, increasing the levels of CD8+ and CD4+ T-cells to fight off current infection and confer immune memory for future infection [115,116]. The Ad26.COV2.S vaccine has a single dose and showed a 69.7% efficacy against SARS-CoV-2, while ChAdOx1nCoV-19 is two doses giving a 74% efficacy against the virus [117,118].

Even with the development of multiple vaccines, research has shown their efficacy decreases five months after initial vaccination [119]. This has led to the creation of additional booster shots that are given at this time to provide more immune protection. They have been shown to be effective at increasing levels of antibodies against SARS-CoV-2, with the BNT162b2 and mRNA-1272 booster shots giving higher protection than their original vaccines [119]. Furthermore, despite these recent advancements in vaccine technology, there have been mutations in the SARS-CoV-2 virus that have helped it evade vaccines. These recombination events surrounding the emergence of vaccine evading variants suggests that these changes could also lead to the virus evading pharmacologic therapies.

Viral escape mutations on the S protein receptor-binding domain (RBD) provide us with insights into infectivity and antibody resistance caused by new variants of SARS-CoV-2 [101]. RBD site facilitates binding between the S protein and the host ACE2. One example of a variant with mutations on the S protein is the highly transmissible B.1.1.529 Omicron (Omicron) variant of SARS-CoV-2. The Omicron variant has been shown to exhibit as many as 15 mutations at the RBD site and 32 on the S protein, allowing for stronger binding to the ACE2 receptor and the ability to evade current vaccines [101].

A recent study tested the ability of monoclonal antibodies (mAbs) to neutralize an infection Omicron isolate. Several mAbs lost their neutralizing activity, and several others had reduced capacity to neutralize the isolate [120]. This study shows that some antibodies in clinical use could lose efficacy against emerging variants. This highlights the importance of effective responses against a wide range of epitopes on the S protein and surveillance of vaccine effectiveness in response to emerging variants.

2.10. Vaccine Hesitancy

In addition to antibody resistance and escape mutations that emerging variants thrive upon, vaccine hesitancy is also another burden towards achieving herd immunity, in which a large enough proportion of a community, area, or location are immune to a specific cause of a disease, such as a virus or bacteria. According to The SAGE Working Group, vaccine hesitancy is the delay in acceptance or refusal of vaccination regardless of the availability or accessibility of vaccination services [121]. Hesitancy can manifest itself in a wide spectrum of actions, including refusal of some vaccines but agreeing to others, delaying vaccination, or accepting vaccination but not sure in doing so. Thus, the term hesitancy is considered to be a continuum between those individuals that accept all vaccines without doubts, to individuals who completely refuse vaccines with no doubts, with hesitant individuals falling in between these two extremes [122].

Vaccine hesitancy can increase the risk of vaccine-preventable disease outbreaks and epidemics and also delay achieving herd immunity even when there are no accessibility or availability issues [122]. Even though vaccination is considered as one of the most successful public health measures, it is still perceived as unsafe and unnecessary by some individuals. There are socio-cultural aspects to vaccine hesitancy, which includes conversations regarding controversies and vaccination scares, media, and other aspects of daily living that encompass some individuals.

In today's world, technology, particularly media, plays a powerful role in spreading misinformation and causing fear. One example of an extensive and large-scale vaccination scare that technology allowed to become widespread was the correlation between vaccines and autism [122]. The misconception surrounding this correlation led to difficulties for the scientific community to educate the public effectively and adequately about precautionary health measures. These types of doubts towards the scientific community regarding vaccination persist throughout the world and pose challenges towards reaching herd immunity early on and preventing epidemics as efficiently as they are possible.

Furthermore, developing countries possess additional challenges in addition to vaccine hesitancy. Many developing countries rely on vaccines developed in other nations due to economic burdens and lack of technological advancements. Thus, one of the best options is buying from other developed nations, but most global vaccine availability is also taken up by wealthier nations in the beginning. Thus, this pushes developing nations to the back of the list and prolongs wait time for accessing vaccines [123]. Some of the vaccines also require extra safety measures for viability. Some require cold temperatures for storage and transfer, and thus require more advanced technology and equipment for viable administration and vaccine services [124]. It is important to implement the lessons learned here in the future, to prevent epidemics and expedite global cooperation and information sharing.

2.11. Newly Approved Pills for SARS-CoV-3 and Their Adverse Effects

Since the onset of the SARS-CoV-2 pandemic, there has been extensive research and discoveries made in the search to find therapies for the prevention of, and progression to, severe SARS-CoV-2 infections. Different classes of drugs have implemented potential targets of the virus, and of these, nirmatrelvir, remdesivir, molnupiravir, baricitinib, and sotrovimab have shown promising results. They have gained Food and Drug Administration (FDA) approval for use in SARS-CoV-2 affected patients through the Emergency Use Authorization (EUA) and/or eventually filing for a New Drug Application to be continuously used [125].

Pfizer has recently been granted an emergency use authorization (EUA) to make a novel oral antiviral drug candidate under the name of PAXLOVID, consisting of nirmatrelvir (PF-07321332) and ritonavir tablets, available for the treatment of SARS-CoV-2. It is used to treat mild-to-moderate SARS-CoV-2 infections in adults and children (12 or older and weighing at least 88 pounds) with a positive test for SARS-CoV-2 and who are at risk for progression to severe SARS-CoV-2 [126]. Nirmatrelvir primarily functions as a SARS-CoV-2 protease inhibitor that inhibits viral replication at the stage of proteolysis by targeting the SARS-CoV-2 3-chymotrypsin-like cysteine protease enzyme (Mpro) right before DNA replication [127,128]. It is designed to be administered orally at the first sign of exposure. Nirmatrelvir co-administered with a low dose of ritonavir has shown to slow the metabolism of the drug in the human body, allowing it to be more effective in combating the virus efficiently. Nirmatrelvir is mainly metabolized by CYP3A4, and therefore co-administration with a CYP3A4 inhibitor such as ritonavir enhances its pharmacokinetics [129].

Results from phase 2–3 trial in unvaccinated individuals given nirmatrelvir (300 mg) with ritonavir (100 mg) every 12 h for five days showed an 89.1% relative risk reduction in SARS-CoV-2 related hospitalization deaths from any cause, and an additional reduction in SARS-CoV-2 viral load at day 5 by a factor of 10 compared to placebo [130]. The most frequent adverse effects reported were dysgeusia, diarrhea, myalgia, and hypertension [131]. Treatment of symptomatic SARS-CoV-2 patients with nirmatrelvir plus ritonavir lowers the risk of progression to severe SARS-CoV-2 without any evident safety concerns and is a good candidate drug for first line treatment of SARS-CoV-2 in symptomatic individuals. Although PAXLOVID is currently authorized for emergency use, Pfizer announced on 30 June 2022, the submission of a New Drug Application to the U.S. Food and Drug Administration (FDA) for the approval of PAXLOVID for patients who are at high risk for progression to severe illness from SARS-CoV-2 after recording available safety data that are consistent in more than 3500 PAXLOVID-treated participants across the EPIC clinical development program [132].

Veklury (remdesivir) was approved by the FDA on 25 April 2022, for the treatment of SARS-CoV-2 in adults and pediatric patients over the age of 28 days who are either hospitalized with or have mild-to-moderate SARS-CoV-2 and are considered to be high risk for severe progression of SARS-CoV-2 [133]. Veklury is a prodrug of an analog to adenosine nucleoside triphosphate that targets the viral RNA dependent RNA polymerase [134,135]. In human airway epithelial cell assays, remdesivir has been shown to inhibit replication of coronaviruses. In mouse infection models, it was effective therapeutically against SARS-CoV-2 [136].

For adults and pediatrics weighing $\geq$40 kg, remdesivir should be initiated as early as possible following diagnosis of symptomatic SARS-CoV-2 and within seven days of symptom onset. Treatment should begin with 200 mg on Day 1 followed by a once-daily maintenance dose of 100 mg from day 2 onwards to be administered intravenously. Treatment duration varies based on SARS-CoV-2 severity status ranging from three to 10 days. The most common adverse reactions in $\geq$5% were nausea, headache, cough and increases in the liver enzymes, alanine transaminase (ALT) and aspartate transaminase (AST). The most serious side effect of this drug is renal impairment and is not recommended in persons with an eGFR < 30 mL/min [133,137].

For non-hospitalized patients with SARS-CoV-2, a three-day course of remdesivir was found to decrease the risk of SARS-CoV-2 related hospitalization and/or death by 87% [137]. The Adaptive SARS-CoV-2 Treatment Trial (ACTT) was implemented to study remdesivir for SARS-CoV-2 treatment. According to the ACTT, hospitalized patients who received a 10-day course of remdesivir were found to have a shorter median time to recovery (10 days) than those in a placebo group (15 days) as well as lowering the risk of progression to mechanical ventilation [138]. In the pathway towards finding ways to manage SARS-CoV-2, remdesivir adds another tool to help combat the progression of high-risk SARS-CoV-2.

Molnupiravir is an RdRp (RNA-Dependent RNA-Polymerase) inhibitor. In December 2021, the drug was granted FDA emergency use authorization for the treatment of mild to moderate SARS-CoV-2 patients with at least one risk factor for progression, and for whom other FDA approved treatments are not available. This authorization was granted based on the results of the MOVe-OUT study which showed that treatment with Molnupiravir, initiated within 5 days after the onset of symptoms, reduced the risk of hospitalization or death from SARS-CoV-2 [139]. Molnupiravir exerts its effect through its active form, EIDD-2061, being utilized by RdRp as a substrate in place of cytidine/uridine triphosphate, resulting in the generation of mutated RNA copies that leave the virus unable to replicate [140]. Molnupiravir is administered orally and well absorbed with minimal interference from food, making it useful in an outpatient setting. Results from Phase 1 studies, such as that of Painter et al. (2021), demonstrated that Molnupiravir is well tolerated, with mild adverse events mainly consisting of diarrhea and headache [141]. No accumulation was seen in multiple-ascending-doses and little of the drug or its metabolites were detected in urine. Hiremath 2021 further affirmed that Molnupiravir has a strong safety profile [142]. Although the mutagenic antiviral activity of Molnupiravir is speculated to be a risk for being incorporated into human DNA, several preclinical studies concluded that treatment with Molnupiravir cannot induce mutations in host cell DNA [136]. Of note, the FDA authorization letter stated that Molnupiravir should not be used in pregnant women due to the fetal toxicity reported in preclinical studies. The letter also reports that Molnupiravir may cause adverse effects for infants via breastfeeding. No drug-drug interactions have been identified [143].

Janus kinases (JAK) are non-receptor tyrosine kinase intracellular molecules that are involved in signaling pathways for cytokine receptors, such as IL-6 receptor, and growth factors. They are important in maintaining immune function [144]. Selective JAK inhibitors, such as baricitinib, inhibit the production of cytokines elevated in SARS-CoV-2, such as IL-2, IL-6, IL-10, and IFN-γ [145]. This drug has been shown to improve lymphocyte count and inhibit the entry of SARS-CoV-2 into cells [145]. A double blind randomized controlled study in 2021, showed that baricitinib combined with remdesivir was more effective in improving recovery times in SARS-CoV-2 infected patients who were receiving oxygenation [145]. The Food and Drug administration (FDA) has authorized the use of baricitinib in SARS-CoV-2 patients and pediatric patients >2 years old who are on supplemental oxygen. However, adverse effects of this medication should be noted. Most serious adverse effects are found in Rheumatoid arthritis patients who are on concomitant immunosuppressants. Adverse effects include cardiovascular disease, thrombosis, and malignancies such as lymphoma [146]. Additionally, adverse effects in SARS-CoV-2 patients include thrombocytosis, neutropenia, deep vein thrombosis, and pulmonary embolism. Baricitinib use is contraindicated with upadacitinib, as immunosuppressive risks are enhanced. Its use is not recommended with other potent immunosuppressive agents, JAK inhibitors, or disease modifying anti-rheumatic drugs (DMARD) [147].

Sotrovimab (VIR-7831) is a human neutralizing monoclonal antibody targeting the receptor binding domain (RBD) of the spike protein of SARS-CoV-2. The variable region of Sotrovimab binds to a highly conserved epitope of the SARS-CoV-2 spike protein in a region outside the highly mutagenic ACE2 receptor binding motif (Kd = 0.21 nM) and does not compete with human ACE2 receptor binding [148]. The highly conserved epitope targeted by Sotrovimab has been found on 99.8% of SARS-CoV-2 viruses which allows its use against mutagenic variants, including Omicron [149]. In May 2021, Sotrovimab received an Emergency Use Authorization in the United States for the treatment of mild to moderate SARS-CoV-2 in higher risk patients to prevent progression to severe symptoms [148]. Sotrovimab dosing is recommended within days of experienced symptoms. It inhibits viral fusion which ultimately reduces viral load and subsequently progression to severe symptoms [150]. The efficacy of Sotrovimab was studied by COMET-ICE investigators in a randomized, double-blind, placebo-controlled trial study. The enrolled symptomatic patients tested positive for SARS-CoV-2 and were at a higher susceptibility to progression of severe symptoms. A total of 291 patients in the treatment group received Sotrovimab 500 mg IV given once while 292 patients served as a control. The placebo group demonstrated 21 patients hospitalized for SARS-CoV-2 and one mortality. The Sotrovimab group only experienced three hospitalizations, with no cases of mortality, a relative risk reduction of 85%, and a statistically significant difference (p value: 0.002) [151]. Additionally, there were no significant differences in adverse effects between the two groups. Zheng et al. (2022), conducted a cohort study to compare the effectiveness of sotrovimab to molnupiravir in preventing severe SARS-CoV-2 outcomes in high-risk SARS-CoV-2 patients [152]. The study included 3288 patients receiving sotrovimab and 2663 receiving molnupiravir. While both groups showed a reduction in SARS-CoV-2 hospitalizations and deaths, patients with sotrovimab treatment were at a lower risk of severe SARS-CoV-2 symptoms compared to molnupiravir. There were 53 hospitalizations in the molnupiravir group and 31 in the sotrovimab group.

3. Conclusions

The role of GSH as an antiviral molecule has been explored in studies regarding SARS-CoV-2 and Influenza. A study performed by De Flora et al. showed that patients who received doses of N-acetylcysteine, a precursor of GSH, for six months had a significant decrease in clinically apparent disease [153]. A smaller study demonstrated that SARS-CoV-2 patients with a lower baseline GSH level had more severe symptoms while SARS-CoV-2 patients with a higher baseline GSH level had milder symptoms [154]. GSH is well known for its anti-inflammatory functions while SARS-CoV-2 is known to exacerbate inflammatory processes within the body. The potential use and functions of GSH to enhance the immune response against SARS-CoV-2 represent a topic that warrants further discussion.

One of the major protective effects of GSH on inflammation is driven by the modulation and balance of ACE/ACE2 activity in cells infected by SARS-CoV-2. The upregulation of ACE activity and concurrent downregulation of ACE2 activity leads to the production of ROS within the cell [71]. The production of ROS and subsequent change in cellular redox status increases activity of NF-kB [155]. In a study using SARS infected mice, it was shown that drugs which inhibit NF-kB signaling reduce inflammation and increase survival in mice after infection with SARS-CoV [156]. GSH has inhibitory effects on the activity of ACE and has the ability to decrease the production of ROS via inhibition of ACE, leading to decreased NF-kB signaling, providing an avenue for decreased inflammation in SARS-CoV infected cells [157].

Author Contributions: A.Y., A.A. (Arbi Abnousian), L.J.A., A.B., B.F., N.S., A.A. (Armin Aghazarian), D.A., A.A. (Arman Amin), A.M. and A.N. contributed equally to drafting this review. F.T.G. and V.V. conceived the framework, provided guidance and assistance, and made edits to the draft. All authors have read and agreed to the published version of the manuscript.

Funding: We appreciate the funding support from NIH (HL143545-01A1) and Your Energy Systems.

Institutional Review Board Statement: Not applicable.

Informed Consent Statement: Not applicable.

Conflicts of Interest: The authors declare no conflict of interest.

References

1. Zhu, N.; Zhang, D.; Wang, W.; Li, X.; Yang, B.; Song, J.; Zhao, X.; Huang, B.; Shi, W.; Lu, R.; et al. A Novel Coronavirus from Patients with Pneumonia in China, 2019. *N. Engl. J. Med.* **2020**, *382*, 727–733. [CrossRef]
2. Ochani, R.; Asad, A.; Yasmin, F.; Shaikh, S.; Khalid, H.; Batra, S.; Sohail, M.R.; Mahmood, S.F.; Ochani, R.; Arshad, M.H.; et al. COVID-19 pandemic: From origins to outcomes. A comprehensive review of viral pathogenesis, clinical manifestations, diagnostic evaluation, and management. *Infez. Med.* **2021**, *29*, 20–36. Available online: https://pubmed.ncbi.nlm.nih.gov/33664170/ (accessed on 14 April 2022).
3. Acar, D.; Kıcali, Ü.Ö. An Integrated Approach to COVID-19 Preventive Behaviour Intentions: Protection Motivation Theory, Information Acquisition, and Trust. *Soc. Work. Public Health* **2022**, *37*, 419–434. [CrossRef] [PubMed]

4. World Health Organization. WHO Coronavirus (COVID-19) Dashboard. Available online: https://covid19.who.int/info (accessed on 14 April 2022).

5. Centers for Disease Control and Prevention. CDC COVID Data Tracker. Available online: https://covid.cdc.gov/covid-data-tracker/#datatracker-home (accessed on 14 April 2022).

6. V'Kovski, P.; Kratzel, A.; Steiner, S.; Stalder, H.; Thiel, V. Coronavirus biology and replication: Implications for SARS-CoV-2. *Nat. Rev. Microbiol.* **2020**, *19*, 155–170. [CrossRef] [PubMed]

7. Li, F. Structure, Function, and Evolution of Coronavirus Spike Proteins. *Annu. Rev. Virol.* **2020**, *3*, 237–261. [CrossRef] [PubMed]

8. Pascarella, S.; Ciccozzi, M.; Bianchi, M.; Benvenuto, D.; Cauda, R.; Cassone, A. The electrostatic potential of the Omicron variant spike is higher than in Delta and Delta-plus variants: A hint to higher transmissibility? *J. Med. Virol.* **2021**, *94*, 1277–1280. [CrossRef] [PubMed]

9. Singh, J.; Pandit, P.; McArthur, A.G.; Banerjee, A.; Mossman, K. Evolutionary trajectory of SARS-CoV-2 and emerging variants. *Virol. J.* **2021**, *18*, 166. [CrossRef] [PubMed]

10. World Health Organization. Tracking SARS-CoV-2 Variants. 2022. Available online: https://www.who.int/activities/tracking-SARS-CoV-2-variants (accessed on 10 June 2022).

11. Kemp, S.; Collier, D.; Datir, R.; Ferreira, I.; Gayed, S.; Jahun, A.; Hosmillo, M.; Rees-Spear, C.; Mlcochova, P.; Lumb, I.U.; et al. Neutralizing antibodies in Spike mediated SARS-CoV-2 adaptation. *medRxiv* **2020**. [CrossRef]

12. Tegally, H.; Wilkinson, E.; Giovanetti, M.; Iranzadeh, A.; Fonseca, V.; Giandhari, J.; Doolabh, D.; Pillay, S.; San, E.J.; Msomi, N.; et al. Emergence and rapid spread of a new severe acute respiratory syndrome-related coronavirus 2 (SARS-CoV-2) lineage with multiple spike mutations in South Africa. *medRxiv* **2020**. [CrossRef]

13. Centers for Disease Control and Prevention. Emerging SARS-CoV-2 Variants. COVID-19. 2021. Available online: https://www.cdc.gov/coronavirus/2019-ncov/more/science-and-research/scientific-brief-emerging-variants.html (accessed on 10 June 2022).

14. Tellez, D.; Dayal, S.; Phan, P.; Mawley, A.; Shah, K.; Consunji, G.; Tellez, C.; Ruiz, K.; Sabnis, R.; Dayal, S.; et al. Analysis of COVID-19 on Diagnosis, Vaccine, Treatment, and Pathogenesis with Clinical Scenarios. *Clin. Pract.* **2021**, *11*, 309–321. [CrossRef]

15. Ejaz, H.; Alsrhani, A.; Zafar, A.; Javed, H.; Junaid, K.; Abdalla, A.E.; Abosalif, K.O.A.; Ahmed, Z.; Younas, S. COVID-19 and comorbidities: Deleterious impact on infected patients. *J. Infect. Public Health* **2020**, *13*, 1833–1839. [CrossRef]

16. Abu-Farha, M.; Al-Mulla, F.; Thanaraj, T.A.; Kavalakatt, S.; Ali, H.; Abdul Ghani, M.; Abubaker, J. Impact of Diabetes in Patients Diagnosed with COVID-19. *Front. Immunol.* **2020**, *11*, 576818. [CrossRef]

17. Checconi, P.; De Angelis, M.; Marcocci, M.E.; Fraternale, A.; Magnani, M.; Palamara, A.T.; Nencioni, L. Redox-Modulating Agents in the Treatment of Viral Infections. *Int. J. Mol. Sci.* **2020**, *21*, 4084. [CrossRef]

18. Nencioni, L.; Iuvara, A.; Aquilano, K.; Ciriolo, M.R.; Cozzolino, F.; Rotilio, G.; Garaci, E.; Palamara, A.T. Influenza A virus replication is dependent on an antioxidant pathway that involves GSH and Bcl-2. *FASEB J.* **2003**, *17*, 758–760. [CrossRef]

19. Amatore, D.; Sgarbanti, R.; Aquilano, K.; Baldelli, S.; Limongi, D.; Civitelli, L.; Nencioni, L.; Garaci, E.; Ciriolo, M.R.; Palamara, A.T. Influenza virus replication in lung epithelial cells depends on redox-sensitive pathways activated by NOX4-derived ROS. *Cell. Microbiol.* **2014**, *17*, 131–145. [CrossRef]

20. Celestino, I.; Checconi, P.; Amatore, D.; De Angelis, M.; Coluccio, P.; Dattilo, R.; Alunni Fegatelli, D.; Clemente, A.M.; Matarrese, P.; Torcia, M.G.; et al. Differential Redox State Contributes to Sex Disparities in the Response to Influenza Virus Infection in Male and Female Mice. *Front. Immunol.* **2018**, *9*, 1747. [CrossRef]

21. Wajner, S.M.; Goemann, I.M.; Bueno, A.L.; Larsen, P.R.; Maia, A.L. IL-6 promotes nonthyroidal illness syndrome by blocking thyroxine activation while promoting thyroid hormone inactivation in human cells. *J. Clin. Investig.* **2011**, *121*, 1834–1845. [CrossRef]

22. Dasgupta, A.; Das, S.; Sarkar, P.K. Thyroid hormone promotes glutathione synthesis in astrocytes by up regulation of glutamate cysteine ligase through differential stimulation of its catalytic and modulator subunit mRNAs. *Free Radic. Biol. Med.* **2007**, *42*, 617–626. [CrossRef]

23. Bedard, K.; Krause, K.-H. The NOX Family of ROS-Generating NADPH Oxidases: Physiology and Pathophysiology. *Physiol. Rev.* **2007**, *87*, 245–313. [CrossRef]

24. To, E.E.; Vlahos, R.; Luong, R.; Halls, M.L.; Reading, P.C.; King, P.T.; Chan, C.; Drummond, G.R.; Sobey, C.G.; Broughton, B.R.S.; et al. Endosomal NOX2 oxidase exacerbates virus pathogenicity and is a target for antiviral therapy. *Nat. Commun.* **2017**, *8*, 69. [CrossRef]

25. Vlahos, R.; Stambas, J.; Bozinovski, S.; Broughton, B.R.S.; Drummond, G.R.; Selemidis, S. Inhibition of Nox2 Oxidase Activity Ameliorates Influenza A Virus-Induced Lung Inflammation. *PLoS Pathog.* **2011**, *7*, e1001271. [CrossRef] [PubMed]

26. Forman, H.J. Glutathione—From antioxidant to post-translational modifier. *Arch. Biochem. Biophys.* **2016**, *595*, 64–67. [CrossRef] [PubMed]

27. Diotallevi, M.; Checconi, P.; Palamara, A.T.; Celestino, I.; Coppo, L.; Holmgren, A.; Abbas, K.; Peyrot, F.; Mengozzi, M.; Ghezzi, P. Glutathione Fine-Tunes the Innate Immune Response toward Antiviral Pathways in a Macrophage Cell Line Independently of Its Antioxidant Properties. *Front. Immunol.* **2017**, *8*, 1239. [CrossRef] [PubMed]

28. Checconi, P.; Limongi, D.; Baldelli, S.; Ciriolo, M.R.; Nencioni, L.; Palamara, A.T. Role of Glutathionylation in Infection and Inflammation. *Nutrients* **2019**, *11*, 1952. [CrossRef]

29. Sgarbanti, R.; Nencioni, L.; Amatore, D.; Coluccio, P.; Fraternale, A.; Sale, P.; Mammola, C.L.; Carpino, G.; Gaudio, E.; Magnani, M.; et al. Redox regulation of the influenza hemagglutinin maturation process: A new cell-mediated strategy for anti-influenza therapy. *Antioxid. Redox Signal.* **2011**, *15*, 593–606. [CrossRef]

30. Boni, M.F.; Lemey, P.; Jiang, X.; Lam, T.T.Y.; Perry, B.W.; Castoe, T.A.; Rambaut, A.; Robertson, D.L. Evolutionary origins of the SARS-CoV-2 sarbecovirus lineage responsible for the COVID-19 pandemic. *Nat. Microbiol.* **2020**, *5*, 1408–1417. [CrossRef]

31. Forcina, L.; Miano, C.; Scicchitano, B.M.; Rizzuto, E.; Berardinelli, M.G.; De Benedetti, F.; Pelosi, L.; Musarò, A. Increased Circulating Levels of Interleukin-6 Affect the Redox Balance in Skeletal Muscle. *Oxidative Med. Cell. Longev.* **2019**, *2019*, 3018584. [CrossRef]

32. Ghazavi, A.; Ganji, A.; Keshavarzian, N.; Rabiemajd, S.; Mosayebi, G. Cytokine profile and disease severity in patients with COVID-19. *Cytokine* **2021**, *137*, 155323. [CrossRef]

33. He, L.; Ding, Y.; Zhang, Q.; Che, X.; He, Y.; Shen, H.; Wang, H.; Li, Z.; Zhao, L.; Geng, J.; et al. Expression of elevated levels of pro-inflammatory cytokines in SARS-CoV-infected ACE2+cells in SARS patients: Relation to the acute lung injury and pathogenesis of SARS. *J. Pathol.* **2006**, *210*, 288–297. [CrossRef]

34. Liu, R.-M.; Vayalil, P.K.; Ballinger, C.; Dickinson, D.A.; Huang, W.-T.; Wang, S.; Kavanagh, T.J.; Matthews, Q.L.; Postlethwait, E.M. Transforming growth factor β suppresses glutamate-cysteine ligase gene expression and induces oxidative stress in a lung fibrosis model. *Free Radic. Biol. Med.* **2012**, *53*, 554–563. [CrossRef]

35. Morris, D.; Guerra, C.; Donohue, C.; Oh, H.; Khurasany, M.; Venketaraman, V. Unveiling the Mechanisms for Decreased Glutathione in Individuals with HIV Infection. *Clin. Dev. Immunol.* **2012**, *2012*, 734125. [CrossRef] [PubMed]

36. Hu, Q.; Guan, H.; Sun, Z.; Huang, L.; Chen, C.; Ai, T.; Pan, Y.; Xia, L. Early CT features and temporal lung changes in COVID-19 pneumonia in Wuhan, China. *Eur. J. Radiol.* **2020**, *128*, 109017. [CrossRef] [PubMed]

37. Lechowicz, K., Drożdżal, S.; Machaj, F.; Rosik, J.; Szostak, B.; Zegan-Barańska, M.; Biernawska, J.; Dabrowski, W.; Rotter, I.; Kotfis, K. COVID-19: The Potential Treatment of Pulmonary Fibrosis Associated with SARS-CoV-2 Infection. *J. Clin. Med.* **2020**, *9*, 1917. [CrossRef] [PubMed]

38. Xu, Z.; Shi, L.; Wang, Y.; Zhang, J.; Huang, L.; Zhang, C.; Liu, S.; Zhao, P.; Liu, H.; Zhu, L.; et al. Pathological findings of COVID-19 associated with acute respiratory distress syndrome. *Lancet Respir. Med.* **2020**, *8*, 420–422. [CrossRef]

39. Fox, S.E.; Akmatbekov, A.; Harbert, J.L.; Li, G.; Quincy Brown, J.; Vander Heide, R.S. Pulmonary and cardiac pathology in African American patients with COVID-19: An autopsy series from New Orleans. *Lancet Respir. Med.* **2020**, *8*, 681–686. [CrossRef]

40. Carsana, L.; Sonzogni, A.; Nasr, A.; Rossi, R.S.; Pellegrinelli, A.; Zerbi, P.; Rech, R.; Colombo, R.; Antinori, S.; Corbellino, M.; et al. Pulmonary post-mortem findings in a series of COVID-19 cases from northern Italy: A two-centre descriptive study. *Lancet Infect. Dis.* **2020**, *20*, 1135–1140. [CrossRef]

41. Agrati, C.; Sacchi, A.; Bordoni, V.; Cimini, E.; Notari, S.; Grassi, G.; Casetti, R.; Tartaglia, E.; Lalle, E.; D'Abramo, A.; et al. Expansion of myeloid-derived suppressor cells in patients with severe coronavirus disease (COVID-19). *Cell Death Differ.* **2020**, *27*, 3196–3207. [CrossRef]

42. van Klaveren, R.J.; Demedts, M.; Nemery, B. Cellular glutathione turnover in vitro, with emphasis on type II pneumocytes. *Eur. Respir. J.* **1997**, *10*, 1392–1400. [CrossRef]

43. Rahman, I.; MacNee, W. Oxidative stress and regulation of glutathione in lung inflammation. *Eur. Respir. J.* **2000**, *16*, 534. [CrossRef]

44. Gibson, P.G.; Qin, L.; Puah, S.H. COVID-19 acute respiratory distress syndrome (ARDS): Clinical features and differences from typical pre-COVID-19 ARDS. *Med. J. Aust.* **2020**, *213*, 54–56.e1. [CrossRef]

45. Tobin, M.J.; Laghi, F.; Jubran, A. Why COVID-19 Silent Hypoxemia Is Baffling to Physicians. *Am. J. Respir. Crit. Care Med.* **2020**, *202*, 356–360. [CrossRef]

46. Singh, M.; Vaughn, C.; Sasaninia, K.; Yeh, C.; Mehta, D.; Khieran, I.; Venketaraman, V. Understanding the Relationship between Glutathione, TGF-β, and Vitamin D in Combating *Mycobacterium tuberculosis* Infections. *J. Clin. Med.* **2020**, *9*, 2757. [CrossRef]

47. Ivanov, A.V.; Valuev-Elliston, V.T.; Ivanova, O.N.; Kochetkov, S.N.; Starodubova, E.S.; Bartosch, B.; Isaguliants, M.G. Oxidative Stress during HIV Infection: Mechanisms and Consequences. *Oxidative Med. Cell. Longev.* **2016**, *2016*, 8910396. [CrossRef]

48. Shatrov, V.A.; Ratter, F.; Gruber, A.; Dröoge, W.; Lehmann, V. HIV Type 1 Glycoprotein 120 Amplifies Tumor Necrosis Factor-Induced NF-kB Activation in Jurkat Cells. *AIDS Res. Hum. Retrovir.* **1996**, *12*, 1209–1216. [CrossRef]

49. Gu, Y.; Wu, R.F.; Xu, Y.C.; Flores, S.C.; Terada, L.S. HIV Tat Activates c-Jun Amino-terminal Kinase through an Oxidant-Dependent Mechanism. *Virology* **2001**, *286*, 62–71. [CrossRef]

50. Shah, A.; Kumar, S.; Simon, S.D.; Singh, D.P.; Kumar, A. HIV gp120- and methamphetamine-mediated oxidative stress induces astrocyte apoptosis via cytochrome P450 2E1. *Cell Death Dis.* **2013**, *4*, e850. [CrossRef]

51. Liu, R.M.; Gaston Pravia, K.A. Oxidative stress and glutathione in TGF-β-mediated fibrogenesis. *Free Radic. Biol. Med.* **2010**, *48*, 1–15. [CrossRef]

52. Ly, J.; Lagman, M.; Saing, T.; Singh, M.K.; Tudela, E.V.; Morris, D.; Anderson, J.; Daliva, J.; Ochoa, C.; Patel, N.; et al. Liposomal Glutathione Supplementation Restores TH1 Cytokine Response to *Mycobacterium tuberculosis* Infection in HIV-Infected Individuals. *J. Interferon Cytokine Res.* **2015**, *35*, 875–887. [CrossRef]

53. Valdivia, A.; Ly, J.; Gonzalez, L.; Hussain, P.; Saing, T.; Islamoglu, H.; Pearce, D.; Ochoa, C.; Venketaraman, V. Restoring Cytokine Balance in HIV-Positive Individuals with Low CD4 T Cell Counts. *AIDS Res. Hum. Retrovir.* **2017**, *33*, 905–918. [CrossRef]

54. Chumburidze-Areshidze, N.; Kezeli, T.; Avaliani, Z.; Mirziashvili, M.; Avaliani, T.; Gongadze, N. The Relationship between Type-2 Diabetes and Tuberculosis. *Georgian Med. News* **2020**, *300*, 69–74.

55. Calabrese, V.; Cornelius, C.; Leso, V.; Trovato-Salinaro, A.; Ventimiglia, B.; Cavallaro, M.; Scuto, M.; Rizza, S.; Zanoli, L.; Neri, S.; et al. Oxidative stress, glutathione status, sirtuin and cellular stress response in type 2 diabetes. *Biochim. Biophys. Acta—Mol. Basis Dis.* **2012**, *1822*, 729–736. [CrossRef] [PubMed]

56. Lee, C. Therapeutic Modulation of Virus-Induced Oxidative Stress via the Nrf2-Dependent Antioxidative Pathway. *Oxidative Med. Cell. Longev.* **2018**, *2018*, 6208067. [CrossRef] [PubMed]

57. Moi, P.; Chan, K.; Asunis, I.; Cao, A.; Kan, Y.W. Isolation of NF-E2-related factor 2 (Nrf2), a NF-E2-like basic leucine zipper transcriptional activator that binds to the tandem NF-E2/AP1 repeat of the beta-globin locus control region. *Proc. Natl. Acad. Sci. USA* **1994**, *91*, 9926–9930. [CrossRef] [PubMed]

58. Kobayashi, M.; Li, L.; Iwamoto, N.; Nakajima-Takagi, Y.; Kaneko, H.; Nakayama, Y.; Eguchi, M.; Wada, Y.; Kumagai, Y.; Yamamoto, M. The Antioxidant Defense System Keap1-Nrf2 Comprises a Multiple Sensing Mechanism for Responding to a Wide Range of Chemical Compounds. *Mol. Cell. Biol.* **2009**, *29*, 493–502. [CrossRef]

59. Itoh, K.; Wakabayashi, N.; Katoh, Y.; Ishii, T.; Igarashi, K.; Engel, J.D.; Yamamoto, M. Keap1 represses nuclear activation of antioxidant responsive elements by Nrf2 through binding to the amino-terminal Neh2 domain. *Genes Dev.* **1999**, *13*, 76–86. [CrossRef]

60. Kobayashi, A.; Kang, M.I.; Okawa, H.; Ohtsuji, M.; Zenke, Y.; Chiba, T.; Igarashi, K.; Yamamoto, M. Oxidative Stress Sensor Keap1 Functions as an Adaptor for Cul3-Based E3 Ligase to Regulate Proteasomal Degradation of Nrf2. *Mol. Cell. Biol.* **2004**, *24*, 7130–7139. [CrossRef]

61. Kobayashi, A.; Kang, M.-I.; Watai, Y.; Tong, K.I.; Shibata, T.; Uchida, K.; Yamamoto, M. Oxidative and Electrophilic Stresses Activate Nrf2 through Inhibition of Ubiquitination Activity of Keap1. *Mol. Cell. Biol.* **2006**, *26*, 221–229. [CrossRef]

62. Solis, W.A.; Dalton, T.P.; Dieter, M.Z.; Freshwater, S.; Harrer, J.M.; He, L.; Shertzer, H.G.; Nebert, D.W. Glutamate–cysteine ligase modifier subunit: Mouse Gclm gene structure and regulation by agents that cause oxidative stress. *Biochem. Pharmacol.* **2002**, *63*, 1739–1754. [CrossRef]

63. Cho, H.Y.; Reddy, S.P.; DeBiase, A.; Yamamoto, M.; Kleeberger, S.R. Gene expression profiling of NRF2-mediated protection against oxidative injury. *Free Radic. Biol. Med.* **2005**, *38*, 325–343. [CrossRef]

64. Das, B.N.; Kim, Y.W.; Keum, Y.S. Mechanisms of Nrf2/Keap1-Dependent Phase II Cytoprotective and Detoxifying Gene Expression and Potential Cellular Targets of Chemopreventive Isothiocyanates. *Oxidative Med. Cell. Longev.* **2013**, *2013*, 839409. [CrossRef]

65. Jin, C.H.; So, Y.K.; Han, S.N.; Kim, J.B. Isoegomaketone Upregulates Heme Oxygenase-1 in RAW264.7 Cells via ROS/p38 MAPK/Nrf2 Pathway. *Biomol. Ther.* **2016**, *24*, 510–516. [CrossRef]

66. Herzenberg, L.A.; De Rosa, S.C.; Dubs, J.G.; Roederer, M.; Anderson, M.T.; Ela, S.W.; Deresinski, S.C.; Herzenberg, L.A. Glutathione deficiency is associated with impaired survival in HIV disease. *Proc. Natl. Acad. Sci. USA* **1997**, *94*, 1967–1972. [CrossRef]

67. Fink, K.; Duval, A.; Martel, A.; Soucy-Faulkner, A.; Grandvaux, N. Dual Role of NOX2 in Respiratory Syncytial Virus- and Sendai Virus-Induced Activation of NF-κB in Airway Epithelial Cells. *J. Immunol.* **2008**, *180*, 6911–6922. [CrossRef]

68. Komaravelli, N.; Ansar, M.; Garofalo, R.P.; Casola, A. Respiratory syncytial virus induces NRF2 degradation through a promyelocytic leukemia protein - ring finger protein 4 dependent pathway. *Free Radic. Biol. Med.* **2017**, *113*, 494–504. [CrossRef]

69. Hosakote, Y.M.; Liu, T.; Castro, S.M.; Garofalo, R.P.; Casola, A. Respiratory Syncytial Virus Induces Oxidative Stress by Modulating Antioxidant Enzymes. *Am. J. Respir. Cell Mol. Biol.* **2009**, *41*, 348–357. [CrossRef]

70. Silvagno, F.; Vernone, A.; Pescarmona, G.P. The Role of Glutathione in Protecting against the Severe Inflammatory Response Triggered by COVID-19. *Antioxidants* **2020**, *9*, 624. [CrossRef]

71. Zhang, J.; Cruz-cosme, R.; Zhuang, M.W.; Liu, D.; Liu, Y.; Teng, S.; Wang, P.H.; Tang, Q. Correction: A systemic and molecular study of subcellular localization of SARS-CoV-2 proteins. *Signal Transduct. Target. Ther.* **2021**, *6*, 192. [CrossRef]

72. De Michele, M.; d'Amati, G.; Leopizzi, M.; Iacobucci, M.; Berto, I.; Lorenzano, S.; Mazzuti, L.; Turriziani, O.; Schiavo, O.G.; Toni, D. Evidence of SARS-CoV-2 spike protein on retrieved thrombi from COVID-19 patients. *J. Hematol. Oncol.* **2022**, *16*, 108. [CrossRef]

73. Bae, I.; Fan, S.; Meng, Q.; Rih, J.K.; Kim, H.J.; Kang, H.J.; Xu, J.; Goldberg, I.D.; Jaiswal, A.K.; Rosen, E.M. BRCA1 Induces Antioxidant Gene Expression and Resistance to Oxidative Stress. *Cancer Res.* **2004**, *64*, 7893–7909. [CrossRef]

74. Kovacic, P.; Jacintho, J. Mechanisms of Carcinogenesis Focus on Oxidative Stress and Electron Transfer. *Curr. Med. Chem.* **2001**, *8*, 773–796. [CrossRef]

75. Lorente, L.; Martín, M.M.; González-Rivero, A.F.; Pérez-Cejas, A.; Cáceres, J.J.; Perez, A.; Ramos-Gómez, L.; Solé-Violán, J.; Ramos, J.A.M.Y.; Ojeda, N.; et al. DNA and RNA Oxidative Damage and Mortality of Patients with COVID-19. *Am. J. Med. Sci.* **2021**, *361*, 585–590. [CrossRef]

76. Grifoni, E.; Valoriani, A.; Cei, F.; Lamanna, R.; Gelli, A.M.G.; Ciambotti, B.; Vannucchi, V.; Moroni, F.; Pelagatti, L.; Tarquini, R.; et al. Interleukin-6 as prognosticator in patients with COVID-19. *J. Infect.* **2020**, *81*, 452–482. [CrossRef]

77. Venketaraman, V.; Rodgers, T.; Linares, R.; Reilly, N.; Swaminathan, S.; Hom, D.; Millman, A.C.; Wallis, R.; Connell, N.D. Glutathione and growth inhibition of *Mycobacterium tuberculosis* in healthy and HIV infected subjects. *AIDS Res. Ther.* **2006**, *3*, 5. [CrossRef]

78. Geremew, D.; Melku, M.; Endalamaw, A.; Woldu, B.; Fasil, A.; Negash, M.; Baynes, H.W.; Geremew, H.; Teklu, T.; Deressa, T.; et al. Tuberculosis and its association with CD4+ T cell count among adult HIV positive patients in Ethiopian settings: A systematic review and meta-analysis. *BMC Infect. Dis.* **2020**, *20*, 325. [CrossRef]

79. Cavalcanti, Y.V.N.; Brelaz, M.C.A.; de Andrade Lemoine Neves, J.K.; Ferraz, J.C.; Pereira, V.R.A. Role of TNF-Alpha, IFN-Gamma, and IL-10 in the Development of Pulmonary Tuberculosis. *Pulm. Med.* **2012**, *2012*, 745483. [CrossRef]

80. Sia, J.K.; Rengarajan, J. Immunology of *Mycobacterium tuberculosis* Infections. *Gram-Posit. Pathog. Third Ed.* **2019**, *7*, 1056–1086. [CrossRef] [PubMed]

81. Abassi, Z.; Knaney, Y.; Karram, T.; Heyman, S.N. The Lung Macrophage in SARS-CoV-2 Infection: A Friend or a Foe? *Front. Immunol.* **2020**, *11*, 1312. [CrossRef] [PubMed]

82. Han, L.; Zhuang, M.; Deng, J.; Zheng, Y.; Zhang, J.; Nan, M.; Zhang, X.; Gao, C.; Wang, P. SARS-CoV-2 ORF9b antagonizes type I and III interferons by targeting multiple components of the RIG-I/MDA-5–MAVS, TLR3–TRIF, and cGAS–STING signaling pathways. *J. Med. Virol.* **2021**, *93*, 5376–5389. [CrossRef] [PubMed]

83. Fenton, M.J.; Vermeulen, M.W.; Kim, S.; Burdick, M.; Strieter, R.M.; Kornfeld, H. Induction of gamma interferon production in human alveolar macrophages by *Mycobacterium tuberculosis. Infect. Immun.* **1997**, *65*, 5149–5156. [CrossRef]

84. Conrad, W.H.; Osman, M.M.; Shanahan, J.K.; Chu, F.; Takaki, K.K.; Cameron, J.; Hopkinson-Woolley, D.; Brosch, R.; Ramakrishnan, L. Mycobacterial ESX-1 secretion system mediates host cell lysis through bacterium contact-dependent gross membrane disruptions. *Proc. Natl. Acad. Sci. USA* **2017**, *114*, 1371–1376. [CrossRef]

85. Marinho, F.V.; Benmerzoug, S.; Rose, S.; Campos, P.C.; Marques, J.T.; Báfica, A.; Barber, G.; Ryffel, B.; Oliveira, S.C.; Quesniaux, V.F. The cGAS/STING Pathway Is Important for Dendritic Cell Activation but Is Not Essential to Induce Protective Immunity against *Mycobacterium tuberculosis* Infection. *J. Innate Immun.* **2018**, *10*, 239–252. [CrossRef]

86. Khurana, A.K.; Aggarwal, D. The (in)significance of TB and COVID-19 co-infection. *Eur. Respir. J.* **2020**, *56*, 2002105. [CrossRef]

87. World Health Organization. Tuberculosis Deaths Rise for the First Time in More than a Decade Due to the COVID-19 Pandemic. 2021. Available online: https://www.who.int/news/item/14-10-2021-tuberculosis-deaths-rise-for-the-first-time-in-more-than-a-decade-due-to-the-covid-19-pandemic (accessed on 20 June 2022).

88. Morris, G.; Bortolasci, C.C.; Puri, B.K.; Olive, L.; Marx, W.; O'Neil, A.; Athan, E.; Carvalho, A.; Maes, M.; Walder, K.; et al. Preventing the development of severe COVID-19 by modifying immunothrombosis. *Life Sci.* **2021**, *264*, 118617. [CrossRef]

89. Mussbacher, M.; Salzmann, M.; Brostjan, C.; Hoesel, B.; Schoergenhofer, C.; Datler, H.; Hohensinner, P.; Basílio, J.; Petzelbauer, P.; Assinger, A.; et al. Cell Type-Specific Roles of NF-κB Linking Inflammation and Thrombosis. *Front. Immunol.* **2019**, *10*, 85. [CrossRef]

90. de Bont, C.M.; Boelens, W.C.; Pruijn, G.J.M. NETosis, complement, and coagulation: A triangular relationship. *Cell. Mol. Immunol.* **2018**, *16*, 19–27. [CrossRef]

91. Bogani, P.; Canavesi, M.; Hagen, T.M.; Visioli, F.; Bellosta, S. Thiol supplementation inhibits metalloproteinase activity independent of glutathione status. *Biochem. Biophys. Res. Commun.* **2007**, *363*, 651–655. [CrossRef]

92. Chen, Q.; Jin, M.; Yang, F.; Zhu, J.; Xiao, Q.; Zhang, L. Matrix Metalloproteinases: Inflammatory Regulators of Cell Behaviors in Vascular Formation and Remodeling. *Mediat. Inflamm.* **2013**, *2013*, 928315. [CrossRef]

93. Carmona-Rivera, C.; Zhao, W.; Yalavarthi, S.; Kaplan, M.J. Neutrophil extracellular traps induce endothelial dysfunction in systemic lupus erythematosus through the activation of matrix metalloproteinase-2. *Ann. Rheum. Dis.* **2014**, *74*, 1417–1424. [CrossRef]

94. Wang, Q.; Zennadi, R. Oxidative Stress and Thrombosis during Aging: The Roles of Oxidative Stress in RBCs in Venous Thrombosis. *Int. J. Mol. Sci.* **2020**, *21*, 4259. [CrossRef]

95. Essex, D.W.; Li, M.; Feinman, R.D.; Miller, A. Platelet surface glutathione reductase-like activity. *Blood* **2004**, *104*, 1383–1385. [CrossRef]

96. Tang, N.; Li, D.; Wang, X.; Sun, Z. Abnormal coagulation parameters are associated with poor prognosis in patients with novel coronavirus pneumonia. *J. Thromb. Haemost.* **2020**, *18*, 844–847. [CrossRef]

97. Romagnoli, C.; Marcucci, T.; Picariello, L.; Tonelli, F.; Vincenzini, M.T.; Iantomasi, T. Role of N-acetylcysteine and GSH redox system on total and active MMP-2 in intestinal myofibroblasts of Crohn's disease patients. *Int. J. Colorectal Dis.* **2012**, *28*, 915–924. [CrossRef]

98. Sadarangani, M.; Marchant, A.; Kollmann, T.R. Immunological mechanisms of vaccine-induced protection against COVID-19 in humans. *Nat. Rev. Immunol.* **2021**, *21*, 475–484. [CrossRef]

99. Smith, T.R.F.; Patel, A.; Ramos, S.; Elwood, D.; Zhu, X.; Yan, J.; Gary, E.N.; Walker, S.N.; Schultheis, K.; Purwar, M.; et al. Immunogenicity of a DNA vaccine candidate for COVID-19. *Nat. Commun.* **2020**, *11*, 2601. [CrossRef] [PubMed]

100. Sadoff, J.; Gray, G.; Vandebosch, A.; Cárdenas, V.; Shukarev, G.; Grinsztejn, B.; Goepfert, P.A.; Truyers, C.; Fennema, H.; Spiessens, B.; et al. Safety and Efficacy of Single-Dose Ad26.COV2.S Vaccine against COVID-19. *N. Engl. J. Med.* **2021**, *384*, 2187–2201. [CrossRef] [PubMed]

101. Chen, J.; Wang, R.; Gilby, N.B.; Wei, G.-W. Omicron Variant (B.1.1.529): Infectivity, Vaccine Breakthrough, and Antibody Resistance. *J. Chem. Inf. Modeling* **2022**, *62*, 412–422. [CrossRef] [PubMed]

102. Gudbjartsson, D.F.; Norddahl, G.L.; Melsted, P.; Gunnarsdottir, K.; Holm, H.; Eythorsson, E.; Arnthorsson, A.O.; Helgason, D.; Bjarnadottir, K.; Ingvarsson, R.F.; et al. Humoral Immune Response to SARS-CoV-2 in Iceland. *N. Engl. J. Med.* **2020**, *383*, 1724–1734. [CrossRef]

103. Del Valle, D.M.; Kim-Schulze, S.; Huang, H.-H.; Beckmann, N.D.; Nirenberg, S.; Wang, B.; Lavin, Y.; Swartz, T.H.; Madduri, D.; Stock, A.; et al. An inflammatory cytokine signature predicts COVID-19 severity and survival. *Nat. Med.* **2020**, *26*, 1636–1643. [CrossRef]

104. Atyeo, C.; Fischinger, S.; Zohar, T.; Slein, M.D.; Burke, J.; Loos, C.; McCulloch, D.J.; Newman, K.L.; Wolf, C.; Yu, J.; et al. Distinct Early Serological Signatures Track with SARS-CoV-2 Survival. *Immunity* **2020**, *53*, 524–532.e4. [CrossRef]

105. Yu, J.; Tostanoski, L.H.; Peter, L.; Mercado, N.B.; McMahan, K.; Mahrokhian, S.H.; Nkolola, J.P.; Liu, J.; Li, Z.; Chandrashekar, A.; et al. DNA vaccine protection against SARS-CoV-2 in rhesus macaques. *Science* **2020**, *369*, 806–811. [CrossRef]

106. Chandrashekar, A.; Liu, J.; Martinot, A.J.; McMahan, K.; Mercado, N.B.; Peter, L.; Tostanoski, L.H.; Yu, J.; Maliga, Z.; Nekorchuk, M.; et al. SARS-CoV-2 infection protects against rechallenge in rhesus macaques. *Science* **2020**, *369*, 812–817. [CrossRef]

107. Levin, E.G.; Lustig, Y.; Cohen, C.; Fluss, R.; Indenbaum, V.; Amit, S.; Doolman, R.; Asraf, K.; Mendelson, E.; Ziv, A.; et al. Waning Immune Humoral Response to BNT162b2 COVID-19 Vaccine over 6 Months. *N. Engl. J. Med.* **2021**, *385*, e84. [CrossRef]

108. Gilbert, P.B.; Montefiori, D.C.; McDermott, A.B.; Fong, Y.; Benkeser, D.; Deng, W.; Zhou, H.; Houchens, C.R.; Martins, K.; Jayashankar, L.; et al. Immune correlates analysis of the mRNA-1273 COVID-19 vaccine efficacy clinical trial. *Science* **2022**, *375*, 43–50. [CrossRef]

109. Dispinseri, S.; Secchi, M.; Pirillo, M.F.; Tolazzi, M.; Borghi, M.; Brigatti, C.; De Angelis, M.L.; Baratella, M.; Bazzigaluppi, E.; Venturi, G.; et al. Neutralizing antibody responses to SARS-CoV-2 in symptomatic COVID-19 is persistent and critical for survival. *Nat. Commun.* **2021**, *12*, 2670. [CrossRef]

110. Sahin, U.; Muik, A.; Derhovanessian, E.; Vogler, I.; Kranz, L.M.; Vormehr, M.; Baum, A.; Pascal, K.; Quandt, J.; Maurus, D.; et al. COVID-19 vaccine BNT162b1 elicits human antibody and TH1 T-cell responses. *Nature* **2020**, *586*, 594–599. [CrossRef]

111. Sainz, B.; Mossel, E.C.; Peters, C.J.; Garry, R.F. Interferon-beta and Interferon-gamma synergistically inhibit the replication of severe acute respiratory syndrome-associated coronavirus (SARS-CoV). *Virology* **2004**, *329*, 11–17. [CrossRef]

112. Polack, F.P.; Thomas, S.J.; Kitchin, N.; Absalon, J.; Gurtman, A.; Lockhart, S.; Perez, J.L.; Pérez Marc, G.; Moreira, E.D.; Zerbini, C.; et al. Safety and Efficacy of the BNT162b2 mRNA COVID-19 Vaccine. *N. Engl. J. Med.* **2020**, *383*, 2603–2615. [CrossRef]

113. Baden, L.R.; El Sahly, H.M.; Essink, B. Efficacy and Safety of the mRNA-1273 SARS-CoV-2 Vaccine. *N. Engl. J. Med.* **2020**, *384*, 403–416. [CrossRef]

114. Stephenson, K.E.; Le Gars, M.; Sadoff, J.; de Groot, A.M.; Heerwegh, D.; Truyers, C.; Atyeo, C.; Loos, C.; Chandrashekar, A.; McMahan, K.; et al. Immunogenicity of the Ad26.COV2.S Vaccine for COVID-19. *JAMA* **2021**, *325*, 1535–1544. [CrossRef]

115. Barrett, J.R.; Belij-Rammerstorfer, S.; Dold, C.; Ewer, K.J.; Folegatti, P.M.; Gilbride, C.; Halkerston, R.; Hill, J.; Jenkin, D.; Stockdale, L.; et al. Phase 1/2 trial of SARS-CoV-2 vaccine ChAdOx1 nCoV-19 with a booster dose induces multifunctional antibody responses. *Nat. Med.* **2021**, *27*, 279–288. [CrossRef]

116. Sadoff, J.; Gray, G.; Vandebosch, A.; Cárdenas, V.; Shukarev, G.; Grinsztejn, B.; Goepfert, P.A.; Truyers, C.; van Dromme, I.; Spiessens, B.; et al. Final Analysis of Efficacy and Safety of Single-Dose Ad26.COV2.S. *N. Engl. J. Med.* **2022**, *386*, 847–860. [CrossRef]

117. Falsey, A.R.; Sobieszczyk, M.E.; Hirsch, I.; Sproule, S.; Robb, M.L.; Corey, L.; Neuzil, K.M.; Hahn, W.; Hunt, J.; Mulligan, M.J.; et al. Phase 3 Safety and Efficacy of AZD1222 (ChAdOx1 nCoV-19) COVID-19 Vaccine. *N. Engl. J. Med.* **2021**, *385*, 2348–2360. [CrossRef]

118. Menni, C.; May, A.; Polidori, L.; Louca, P.; Wolf, J.; Capdevila, J.; Hu, C.; Ourselin, S.; Steves, C.J.; Valdes, A.M.; et al. COVID-19 vaccine waning and effectiveness and side-effects of boosters: A prospective community study from the ZOE COVID Study. *Lancet Infect. Dis.* **2022**, *22*, 1002–1010. [CrossRef]

119. VanBlargan, L.A.; Errico, J.M.; Halfmann, P.J.; Zost, S.J.; Crowe, J.E.; Purcell, L.A.; Kawaoka, Y.; Corti, D.; Fremont, D.H.; Diamond, M.S. An infectious SARS-CoV-2 B.1.1.529 Omicron virus escapes neutralization by therapeutic monoclonal antibodies. *Nat. Med.* **2022**, *28*, 490–495. [CrossRef]

120. MacDonald, N.E. SAGE Working Group on Vaccine Hesitancy. Vaccine hesitancy: Definition, scope and determinants. *Vaccine* **2015**, *33*, 4161–4164. [CrossRef]

121. Dubé, E.; Laberge, C.; Guay, M.; Bramadat, P.; Roy, R.; Bettinger, J.A. Vaccine hesitancy. *Hum. Vaccines Immunother.* **2013**, *9*, 1763–1773. [CrossRef]

122. Mnookin, S. The panic virus: A true story of medicine, science, and fear. *J. Clin. Investig.* **2011**, *121*, 2533.

123. Sheikh, A.B.; Pal, S.; Javed, N.; Shekhar, R. COVID-19 Vaccination in Developing Nations: Challenges and Opportunities for Innovation. *Infect. Dis. Rep.* **2021**, *13*, 429–436. [CrossRef]

124. Crommelin, D.J.A.; Anchordoquy, T.J.; Volkin, D.B.; Jiskoot, W.; Mastrobattista, E. Addressing the Cold Reality of mRNA Vaccine Stability. *J. Pharm. Sci.* **2020**, *110*, 997–1001. [CrossRef]

125. Research, Center for Drug Evaluation and "Coronavirus (COVID-19) | Drugs". Available online: https://www.fda.gov/drugs/emergency-preparedness-drugs/coronavirus-covid-19-drugs (accessed on 10 May 2022).

126. Emergency Use Authorization (EUA) of Paxlovid for Coronavirus Disease 2019 (COVID-19). Available online: https://labeling.pfizer.com/ShowLabeling.aspx?id=16473 (accessed on 10 May 2022).

127. Pfizer's Novel COVID-19 Oral Antiviral Treatment Candidate Reduced Risk of Hospitalization or Death by 89% in Interim Analysis of Phase 2/3 EPIC-HR Study. Available online: https://www.pfizer.com/news/press-release/press-release-detail/pfizers-novel-covid-19-oral-antiviral-treatment-candidate (accessed on 10 May 2022).

128. Hilgenfeld, R. From SARS to MERS: Crystallographic studies on coronaviral proteases enable antiviral drug design. *FEBS J.* **2014**, *281*, 4085–4096. [CrossRef]

129. Owen, D.R.; Allerton, C.M.N.; Anderson, A.S.; Aschenbrenner, L.; Avery, M.; Berritt, S.; Boras, B.; Cardin, R.D.; Carlo, A.; Coffman, K.J.; et al. An oral SARS-CoV-2 M pro inhibitor clinical candidate for the treatment of COVID-19. *Science* **2021**, *374*, 1586–1593. [CrossRef] [PubMed]

130. Hammond, J.; Leister-Tebbe, H.; Gardner, A.; Abreu, P.; Bao, W.; Wisemandle, W.; Baniecki, M.; Hendrick, V.M.; Damle, B.; Simón-Campos, A.; et al. Oral Nirmatrelvir for High-Risk, Nonhospitalized Adults with Covid-19. *N. Engl. J. Med.* **2022**, *386*, 1397–1408. [CrossRef] [PubMed]

131. Emergency Use Authorization for Paxlovid. Available online: https://labeling.pfizer.com/ShowLabeling.aspx?id=16474 (accessed on 2 July 2022).

132. Pfizer Announces Submission of New Drug Application to the U.S FDA for PAXLOVID. Available online: https://www.pfizer.com/news/press-release/press-release-detail/pfizer-announces-submission-new-drug-application-us-fda (accessed on 2 July 2022).

133. Veklury (Remdesivir) Use for Pediatric Patients. Available online: https://www.gilead.com/remdesivir (accessed on 3 July 2022).

134. Brown, A.J.; Won, J.J.; Graham, R.L.; Dinnon, K.H.; Sims, A.C.; Feng, J.Y.; Cihlar, T.; Denison, M.R.; Baric, R.S.; Sheahan, T.P. Broad spectrum antiviral remdesivir inhibits human endemic and zoonotic deltacoronaviruses with a highly divergent RNA dependent RNA polymerase. *Antivir. Res.* **2019**, *169*, 104541. [CrossRef] [PubMed]

135. Saber-Ayad, M.; Saleh, M.A.; Abu Gharbieh, E. The Rationale for Potential Pharmacotherapy of COVID-19. *Pharmaceuticals* **2020**, *13*, 96. [CrossRef]

136. Sheahan, T.P.; Sims, A.C.; Graham, R.L.; Menachery, V.D.; Gralinski, L.E.; Case, J.B.; Leist, S.R.; Pyrc, K.; Feng, J.Y.; Trantcheva, I.; et al. Broad-spectrum antiviral GS-5734 inhibits both epidemic and zoonotic coronaviruses. *Sci. Transl. Med.* **2017**, *9*, eaal3653. [CrossRef]

137. Gottlieb, R.L.; Vaca, C.E.; Paredes, R.; Mera, J.; Webb, B.J.; Perez, G.; Oguchi, G.; Ryan, P.; Nielsen, B.U.; Brown, M.; et al. Early Remdesivir to Prevent Progression to Severe COVID-19 in Outpatients. *N. Engl. J. Med.* **2022**, *386*, 305–315. [CrossRef]

138. Beigel, J.H.; Tomashek, K.M.; Dodd, L.E.; Mehta, A.K.; Zingman, B.S.; Kalil, A.C.; Hohmann, E.; Chu, H.Y.; Luetkemeyer, A.; Kline, S.; et al. Remdesivir for the Treatment of COVID-19—Final Report. *N. Engl. J. Med.* **2020**, *383*, 1813–1826. [CrossRef]

139. Jayk Bernal, A.; Gomes Da Silva, M.M.; Musungaie, D.B.; Kovalchuk, E.; Gonzalez, A.; Delos Reyes, V.; Martín-Quirós, A.; Caraco, Y.; Williams-Diaz, A.; Brown, M.L.; et al. Molnupiravir for Oral Treatment of COVID-19 in Nonhospitalized Patients. *N. Engl. J. Med.* **2022**, *386*, 509–520. [CrossRef]

140. Imran, M.; Kumar Arora, M.; Asdaq, S.M.B.; Khan, S.A.; Alaqel, S.I.; Alshammari, M.K.; Alshehri, M.M.; Alshrari, A.S.; Mateq Ali, A.; Al-shammeri, A.M.; et al. Discovery, Development, and Patent Trends on Molnupiravir: A Prospective Oral Treatment for COVID-19. *Molecules* **2021**, *26*, 5795. [CrossRef]

141. Painter, W.P.; Holman, W.; Bush, J.A.; Almazedi, F.; Malik, H.; Eraut, N.C.J.E.; Morin, M.J.; Szewczyk, L.J.; Painter, G.R. Human Safety, Tolerability, and Pharmacokinetics of Molnupiravir, a Novel Broad-Spectrum Oral Antiviral Agent with Activity against SARS-CoV-2. *Antimicrob. Agents Chemother.* **2021**, *65*, e02428-20. [CrossRef]

142. Hiremath, C.N. Abbreviated Profile of Drugs (APOD): Modeling drug safety profiles to prioritize investigational COVID-19 treatments. *Heliyon* **2021**, *7*, e07666. [CrossRef]

143. Kamal, L.; Ramadan, A.; Farraj, S.; Bahig, L.; Ezzat, S. The pill of recovery; Molnupiravir for treatment of COVID-19 patients; a systematic review. *Saudi Pharm. J.* **2022**, *30*, 508–518. [CrossRef]

144. Choy, E.H. Clinical significance of Janus Kinase inhibitor selectivity. *Rheumatology* **2018**, *58*, 953–962. [CrossRef]

145. Kalil, A.C.; Patterson, T.F.; Mehta, A.K.; Tomashek, K.M.; Wolfe, C.R.; Ghazaryan, V.; Marconi, V.C.; Ruiz-Palacios, G.M.; Hsieh, L.; Kline, S.; et al. Baricitinib plus Remdesivir for Hospitalized Adults with COVID-19. *N. Engl. J. Med.* **2021**, *384*, 795–807. [CrossRef]

146. Baricitinib: Drug Information. Available online: https://www.uptodate.com/contents/baricitinib-drug-information?topicRef=5666&source=see_link (accessed on 5 July 2022).

147. Baricitinib (Rx). Available online: https://reference.medscape.com/drug/olumiant-baricitinib-1000107#3 (accessed on 5 July 2022).

148. Heo, Y.A. Sotrovimab: First Approval. *Drugs* **2022**, *82*, 477–484. [CrossRef]

149. Passariello, M.; Ferrucci, V.; Sasso, E.; Manna, L.; Lembo, R.R.; Pascarella, S.; Fusco, G.; Zambrano, N.; Zollo, M.; de Lorenzo, C. A Novel Human Neutralizing mAb Recognizes Delta, Gamma and Omicron Variants of SARS-CoV-2 and Can Be Used in Combination with Sotrovimab. *Int. J. Mol. Sci.* **2022**, *23*, 5556. [CrossRef]

150. Emergency Use Authorization (EUA) of Sotrovimab. Available online: https://www.fda.gov/media/149534/download (accessed on 5 July 2022).

151. Gupta, A.; Gonzalez-Rojas, Y.; Juarez, E.; Crespo Casal, M.; Moya, J.; Falci, D.R.; Sarkis, E.; Solis, J.; Zheng, H.; Scott, N.; et al. Early Treatment for COVID-19 with SARS-CoV-2 Neutralizing Antibody Sotrovimab. *N. Engl. J. Med.* **2021**, *385*, 1941–1950. [CrossRef]

152. Zheng, B.; Green, A.C.; Tazare, J.; Curtis, H.J.; Fisher, L.; Nab, L.; Schultze, A.; Mahalingasivam, V.; Parker, E.P.; Hulme, W.J.; et al. Comparative effectiveness of sotrovimab and molnupiravir for prevention of severe COVID-19 outcomes in non-hospitalised patients: An observational cohort study using the OpenSAFELY platform. *medRxiv* **2022**. [CrossRef]

153. de Flora, S.; Grassi, C.; Carati, L. Attenuation of influenza-like symptomatology and improvement of cell-mediated immunity with long-term N-acetylcysteine treatment. *Eur. Respir. J.* **1997**, *10*, 1535–1541. [CrossRef]

154. Polonikov, A. Endogenous Deficiency of Glutathione as the Most Likely Cause of Serious Manifestations and Death in COVID-19 Patients. *ACS Infect. Dis.* **2020**, *6*, 1558–1562. [CrossRef]
155. Baeuerle, P.A.; Henkel, T. Function and Activation of NF-kappaB in the Immune System. *Annu. Rev. Immunol.* **1994**, *12*, 141–179. [CrossRef]
156. DeDiego, M.L.; Nieto-Torres, J.L.; Regla-Nava, J.A.; Jimenez-Guardeño, J.M.; Fernandez-Delgado, R.; Fett, C.; Castaño-Rodriguez, C.; Perlman, S.; Enjuanes, L. Inhibition of NF-κB-Mediated Inflammation in Severe Acute Respiratory Syndrome Coronavirus-Infected Mice Increases Survival. *J. Virol.* **2014**, *88*, 913–924. [CrossRef]
157. Basi, Z.; Turkoglu, V. In vitro effect of oxidized and reduced glutathione peptides on angiotensin converting enzyme purified from human plasma. *J. Chromatogr. B* **2019**, *1104*, 190–195. [CrossRef]

MDPI AG
St. Alban-Anlage 66
4052 Basel
Switzerland
www.mdpi.com

International Journal of Molecular Sciences Editorial Office
E-mail: ijms@mdpi.com
www.mdpi.com/journal/ijms